Algebraic and Differential Topology of Robust Stability

Algebraic and Differential Topology of Robust Stability

Edmond A. Jonckheere

University of Southern California
Department of Electrical Engineering—Systems
and Center for Applied Mathematical Sciences
Los Angeles, California

with 79 pictures,
computer generated by
Chih-Yung Cheng, Chung-Kuang Chu, and Murilo G. Coutinho

New York Oxford • Oxford University Press 1997

Oxford University Press

Oxford New York
Athens Auckland Bangkok Bogota Bombay Buenos Aires
Calcutta Cape Town Dar es Salaam Delhi Florence Hong Kong
Istanbul Karachi Kuala Lumpur Madras Madrid Melbourne
Mexico City Nairobi Paris Singapore Taipei Tokyo Toronto

and associated companies in
Berlin Ibadan

Published by Oxford University Press, Inc.,
198 Madison Avenue, New York, New York 10016

Oxford is a registered trademark of Oxford University Press

Library of Congress Cataloging-in-Publication Data
Jonckheere, Edmond A., 1954–
Algebraic and differential topology of robust stability / Edmond A. Jonckheere :
with over 50 illustrations by Chih-Yung Cheng, Chung-Kuang Chu, Murilo G. Coutinho.
p. cm.
Includes bibliographical references and index.
ISBN 0-19-509301-1
1. Control theory. 2. Algebraic topology. 3. Differential topology. I. Title.
QA402.3.J66 1996
629.8'312'015142—dc20 95-49073

1 3 5 7 9 8 6 4 2

Printed in the United States of America
on acid-free paper

In conclusion I would like to suggest that
algebraic topology has now reached a sufficient
degree of maturity so that it should be regarded
as a tool available for use in appropriate
branches of analysis. At least I hope it will
be the natural thing for a mathematician to ask:
if I vary the coefficients or parameters of my problem,
what sort of a topological space do I get –
is it for example contractible and if not,
what is the significance of its topological invariants?

M.F. ATIYAH

PREFACE

In this book, two seemingly unrelated fields of intellectual endeavor—algebraic/differential topology and robust control—are brought together. The terminology "seemingly unrelated" might be somewhat inappropriate, because this link already appears at the most fundamental level of multivariable theory. It is indeed well-known, at least to control engineers, that the stable multivariable loop transmission $L(s)$ remains stable after closing the loop if and only if $\det(I + L(j\omega)) \neq 0, \forall\omega$ together with the condition that the plot of $\det(I + L(j\omega))$ does not circle around $0 + j0$. What is less well-known, at least to control engineers, is that the latter is (generically) nothing other than a necessary and sufficient condition for invertibility of the Toeplitz operator $I + T_L$. Invertibility of a Toeplitz operator relies crucially on the concept of *index* of a *Fredholm* operator. The problem of finding an index for a *family* of Fredholm operators—in other words, an "uncertain" Fredholm operator—historically led to the so-called (topological) *K-theory*. It is fair to say that K-theory has been one of the most recent and important developments in algebraic topology. The fact that index of a family of Fredholm operators was a milestone in the historical development of K-theory was very convincingly argued by Atiyah in a 1969 monograph and reiterated in a 1977 survey paper. There, in an effort to provide a *most natural motivation for K-theory*, Atiyah introduced what in the control jargon would be called the "uncertainty" by defining a *family* $\Delta \mapsto I + T_{L\Delta}$ of operators, indexed by Δ running in some topological space D, the "uncertainty space." To make the problem of mathematical interest, the family is restricted to be *Fredholm*, which in essence means that whenever $I + T_{L\Delta}$ is invertible for *one single* Δ, then it is invertible for *all* Δ's in the same connected component. In other words, Atiyah restricted the system to be robustly stable. He then showed that the fundamental K-theory concept of *Grothendieck group of vector bundles over the uncertainty space*, $K^0(D)$, comes out most naturally as the isomorph of the set of homotopy classes of maps from the uncertainty space into the space of Fredholm operators. It is therefore fair to say that the robust stability problem, disguised as a "Fredholm family of Toeplitz operators" problem, has been one of the driving motivations of the whole K-theory (besides algebraic geometry). We hasten to say that, in this book, we will enter the algebraic topology arena from a much more applications-oriented path approach, although we will return to the Fredholm family problem.

More recently, about three decades after the emergence of K-theory, the robust stability problem—that is, the problem of checking stability for a great many uncertain parameter values—has received several ad hoc solutions, gravitating around the idea of breaking the set of uncertainties D into pieces and then localizing the stability boundary within some pieces. What does not seem to have been perceived is the fact that breaking a domain of uncertainty into pieces is a *basic* technique of algebraic topology—usually referred to as *triangulation*—and has roots tracing back to Poincaré and Lefschetz. In such a "structured" problem as the multivariable phase margin, the uncertainty space becomes a nontrivial manifold, the n-torus or the unitary group, depending on the formulation. Triangulating a manifold and then looking at the combinatorial properties of the resulting assembly of simplexes is the *basic* technique for computing such *global* properties of manifolds as number of connected components, loops, holes, twists, and so on. The same triangulation technique is also standard when one has to ascertain how loops, holes, twists, and so on, of a manifold are mapped by a continuous function into loops, holes, twists, and so on, in the image manifold.

In the early robust stability techniques, it never came out quite clearly what is being accomplished by breaking the domain of uncertainty into pieces. These robust stability techniques followed in the footsteps of the "Horowitz template" approach. The domain of uncertainty D is mapped into the complex plane by means of the Nyquist map to yield the template $N = f(D)$. In algebraic topology, decomposing the domain of definition of such a continuous map as the Nyquist map f is a process aimed at deriving an approximate map $\tilde{f}$ that has the remarkable property that, restricted to each simplex σ of the triangulation, it "commutes" with the boundary, in the sense that $\partial\tilde{f}(\sigma) \subseteq \tilde{f}(\partial\sigma)$. The latter has the computationally interesting property that, if we need to check the boundary of the image, $\partial\tilde{f}(\sigma)$, which is typical of a homotopy procedure, it suffices to check the boundary of the domain, $\tilde{f}(\partial\sigma)$. This decomposition technique culminates in the celebrated *simplicial approximation theorem* that asserts that $\tilde{f}$ can be chosen arbitrarily close and homotopic to f.

Horowitz should be credited as being probably the first who perceived this boundary issue. He indeed pointed out that if the perimeter of a rectangular domain of uncertain parameters is mapped into the perimeter of a polygonal Horowitz template—$\partial f\sigma = f\partial\sigma$ in our notation—then the robust stability check can be substantially simplified.

To proceed more formally, assume that, after an application of the simplicial approximation theorem, the approximate Nyquist map $\tilde{f}$ commutes with the boundary over each simplex σ of the domain of uncertainty. Since the map commutes with the boundary, it suffices to check stability for all parameter values on the *boundary* to ensure stability for *all* parameters.

Unfortunately, if we follow the path of a *direct* application of the simplicial approximation theorem, because of the triangulation it entails, we are likely to end up with what has usually been referred to as *combinatorial explosion*. Indeed, the drawback of the simplicial approximation theorem is that the higher the required accuracy of the approximation, the more finely the manifold of uncertainty should be triangulated. Clearly, requiring a high accuracy simplicial approximation could lead to prohibitively complicated gridding, if a uniform gridding procedure is adopted.

Along parallel lines of investigation, a substantial amount of effort was devoted to the avoidance of the combinatorial explosion, until it was shown in [Coxson and DeMarco 1991] that such a representative problem as the computation of the real μ-function is NP-hard.

A procedure to alleviate the odds of combinatorial explosion stems from the observation that it is not necessary to further triangulate a big chunk of the uncertainty manifold, if we can make reasonably sure that no instability will develop in that region. This leads us to the idea of implementing a *variable grid refinement restart algorithm*: Start with a coarse triangulation, check for stability in each piece using a boundary test, and refine, in a nested fashion, only those simplexes where an instability could develop. Operations researchers have been very successful at implementing these kinds of schemes, following in the footsteps of the celebrated *Sperner lemma*, and leading to the so-called *simplicial algorithms*. Many other computationally intensive mathematical problems, such as the Brouwer fixed-point theorem, have received solutions along this line of ideas. Our approach to the structured stability problem makes no exception to this rule.

Of course, since we are dealing with NP-hard problems, some sort of combinatorial nastiness must appear *somewhere*. It turns out that the bottleneck of the variable grid refinement restart procedure is to guarantee Sperner properness. Nevertheless, with a bit of heuristics, the procedure can be made to work very well.

In the course of writing this book, it became evident that we could not possibly develop the algorithms all the way to the details of their implementation and prove anything about their complexity status. This would require two other books and a few more years of research. In this book, we are limiting ourselves to show the *topological roots* of the simplicial algorithms and their combinatorial nastiness, and we are leaving the details of the implementation and the complexity analyses to our students. The [Cheng 1994] dissertation contains the details of the simplicial algorithms on V-triangulation and the so-called vector labeling algorithm. The stochastic complexity of simplicial algorithms along with the root lattice properties of regular pattern triangulations have been developed by [Chu 1996].

The above computational implementations were developed as Matlab m-files, relying on matrix data bases and primitive instructions. However, since the fundamental problem is *geometric*, it appeared more natural to develop an implementation based on *geometric* data bases and primitive instructions, as it is done in the emerging science of *computational geometry.* This led our student, Mr. Coutinho, to develop the SimplicialVIEW project. SimplicialView is a $C/C++$ language code, the core of which is a computational implementation of the simplicial approximation theorem using specialized data bases and primitive instructions to perform such geometrical constructions as the Delaunay triangulation of the Horowitz template; from the simplicial approximation of the Nyquist map, the code constructs the crossover hypersurface as an assembly of simplexes and provides "visualization" for it using state-of-the-art graphical display. The original Nyquist map can be generated either by a built in script language or via a Matlab interface. For further information about the SimplicialVIEW project, the reader is invited to contact Dr. Coutinho at `coutinho@eudoxus.usc.edu`. The reader can also get the relevant information about the SimplicialVIEW project from `http://www.isi.edu/people/coutinho`.

Trying to catch ∂N by restricting the search to ∂P is an *algorithmic* way to get to the heart of the problem. It certainly answers the question as to whether or not the system is robustly stable. However, to get the "picture" in the space of uncertainty (for example, to get the stability/instability regions), we have to trace the Horowitz template, including its boundary, back into the parameter space. We therefore reach the *more conceptual* issue as to *where* the inverse image of the boundary, $f^{-1}(\partial N)$, lay. Is it included in the boundary of P? In other words, do we have $\tilde{f}^{-1}\partial \subseteq \partial \tilde{f}^{-1}$? It turns out that a simplicial approximation, in general, falls short of achieving $\tilde{f}^{-1}\partial \subseteq \partial \tilde{f}^{-1}$. What is true, however, is that the commutativity relation $\tilde{f}^{-1}\partial\tau \subseteq \partial \tilde{f}^{-1}\tau$ holds provided that $\tilde{f}$ is interpreted as the restriction of the simplicial map to a maximum dimension simplex σ in the inverse image of a 2-simplex τ of the Nyquist template. Thus, a local version of the relation is guaranteed and is useful for algorithmic purposes. More formally, in order to be able to deal with the *global inverse image* properties of the Nyquist map, we have to resort to *fiber bundles, semisimplicial bundles, fibrations, semisimplicial fibration*, and spectral sequences. What in general prevents the commutativity relation is the "twisting of the fiber," typical of an open-loop unstable system, and any "hole" in the fiber space.

In the simple Kharitonov case, the inverse Nyquist mapping commutes with the boundary, while in the overwhelming majority of other cases, it does not. This explains the discrepancy between the amazingly simple Kharitonov test for robust stability and the amazingly complicated solution to the other structured stability problems.

From all of the above discussion, the way we would like to present this book is from the perspective that it establishes the trilogy

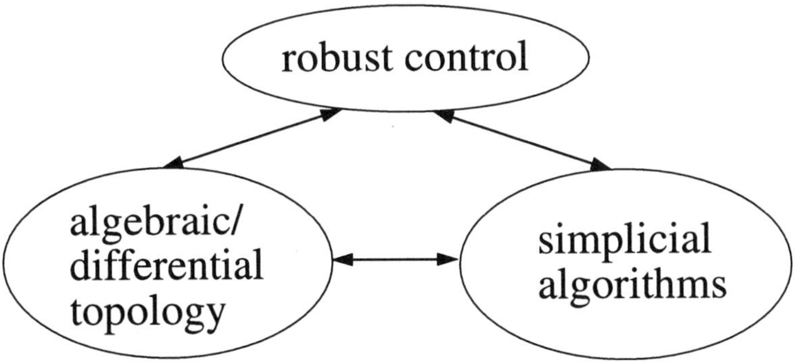

The easily motivated boundary behavior problem is just one among many avenues of approach to algebraic/differential topology. Actually, the boundary behavior problem would only lead to that part of algebraic topology referred to as *homology theory*, which can be defined as the formal treatment of the boundary operator. In fact, in addition to the simplicial approximation theorem, there are two other avenues of approach: obstruction theory and Morse theory.

Obstruction theory is a mathematical conceptualization of the intuitive idea that, as we attempt to extend the Nyquist map from the vertex of "nominal" parameters, we are likely to encounter an "obstruction" on our way to the full parameter space. The "obstruction" is a topological hurdle that manifests itself by the impossibility to extend the map to higher dimension without the image covering $0 + j0$. From this point of view, computing the stability boundary appears equivalent to locating the obstruction. A valuable insight provided by the obstruction approach is that the most critical step of a robust stability test is to ensure that stability remains preserved while going from the *edges* to the *faces* of the polyhedron of uncertainties (meant to include the frequency).

Obstruction theory is just one avenue of approach to the more general issue of homotopy of robust stability. Loosely speaking, homotopy of robust stability deals with the question as to *what* partial Nyquist maps can be defined from a subset of uncertainty to the complex plane devoid of $0+j0$. If the partial map avoids $0 + j0$, then either it is homotopic to a constant map or it "wraps around" $0 + j0$, the "wrapping" being defined formally by the Brouwer degree. Nyquist maps into the space of return difference *matrices* can also be defined, and the homotopy question is *what* subset of uncertainty maps to the general linear group.

A closely related homotopy question is how a partial Nyquist map maps into the space of Fredholm operators (induced by the return difference). This is the path along which we will introduce K-theory.

The *Morse theory* comes into the picture as a differential topology technique that tells us *something* about $f^{-1}\partial N$. More precisely, $f^{-1}\partial N$ is characterized as a subset of *critical points* of the Nyquist map. The basic boundary preserving paradigm therefore can be rephrased as the issue as to *where* are the critical points in the space of uncertainty. Using the language of *stratified spaces* as applied to P, the Morse theory essentially says that a robust stability problem is "easy" whenever the so-called *critical points* of the Nyquist map are in the lower-dimensional strata of the space of uncertainties.

A byproduct of the Morse theory is that it reveals the *singularity structure* of the problem—that is, plots of critical points on the uncertainty manifold together with plots of critical values on the Nyquist template. It turns out, and this is a general pattern, that robust stability problems, "simple" by their external appearance, could have a very complicated singularity structure. From there, we jump to yet another issue: structural stability. Roughly speaking, structural stability is the issue as to what could happen if some parameters, which have been quite subjectively declared "certain," begin to drift. It turns out that lack of structural stability manifests itself as lack of continuity of, say, the real μ-function relative to rounding errors, as it typically happens when the crossover intersects the singularity set.

The Morse theory also comes as a follow-up on the inverse image properties of the semisimplicial approximation of the Nyquist map. Indeed, the reader will agree that the results derived by resorting to semisimplicial bundles lack aesthetic appeal. The reason is that the inverse image properties are closely related to the singularity structure, and a simplicial decomposition cannot be expected to be able to reproduce the complex singularity phenomena typical of the Nyquist map. Getting out of the confines of rectilinear subdivisions, and allowing for more general decompositions, one can get a *cell decomposition* of the Nyquist map, articulated on the critical values plots, such that, for any cell N^i, we have the aesthetically appealing property that $f^{-1}\partial N^i = \partial f^{-1} N^i$.

Therefore, if we amplify the "algebraic/differential topology" vertex of the above diagram we find yet another trilogy:

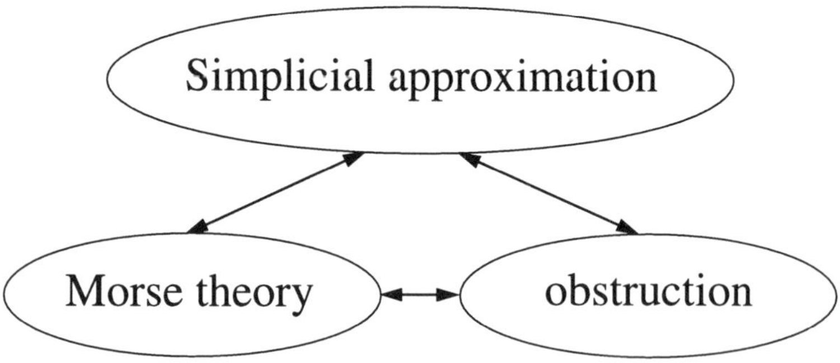

Finally, the last part of the book deals with the crossover—that is, the set of solutions to $\det(I + L(j\omega)\Delta(q)) = 0$, viewed as an *real algebraic set*. Complexity bounds on the crossover are derived, and these bounds corroborate NP-hardness of the problem of catching the crossover. Projecting the crossover parallel to Ω is viewed as a Tarski–Seidenberg elimination of the *real variable* ω and yields the separating boundary in the space of uncertain q-parameters.

It might be argued that by covering such diverse and apparently irreconcilable approaches as *combinatorial* topology and *differential* topology, we have spread ourselves too thin. To this, we would rebut that some puzzling robust geometric computation issues find their natural explanations in the differential topology of the Nyquist map. To be specific, many combinatorial/geometric problems occur in the vicinity of the critical curves drawn on the manifold of uncertainty. The images of the sample vertices tend to cluster around the critical value curves drawn on the Horowitz template.

SIMPLICIAL VERSUS SINGULAR HOMOLOGY

A question that mathematicians are prompt to ask is whether any use is made of *singular* homology. Here, we have all the way attempted to remain within the nonsingular, simplicial homology theory. Nonsingular, simplicial homology can be loosely defined as the tool of decomposing topological spaces into *rectilinear simplexes*. The limitation of simplicial theory can be easily understood, for indeed *only polyhedra* can be decomposed into simplexes. A way to go around this difficulty, which we shall exploit, is to approximate a compact topological space with a polyhedron and then decompose the polyhedron into simplexes. The singular theory, on the other hand, allows for more flexibility in the allowable "building blocks" into which a space is decomposed. In singular theory, the space is decomposed into *singular simplexes or cubes*, a singular simplex or cube being defined as the continuous image (not necessarily the homeomorph!) of a standard simplex or cube. Singular theory is by far more powerful a tool than the nonsingular theory, allowing us to deal with topological spaces that could not be tackled using the nonsingular theory. Despite the fact that this book is application-oriented and is henceforth dealing with reasonably simple topological spaces, we have to admit that we have not been entirely successful at avoiding the singular theory.

The Cartesian product of spaces is an issue dealt with by the celebrated Eilenberg–Zilber and Künneth theorems. It is a typical pattern that, precisely at that level, difficulties with the simplicial theory crop up, for indeed the Cartesian product of two simplexes is not a simplex. We have avoided a full-blown singular approach by resorting to an ad hoc recipe called *prismatic triangulation*. The prismatic triangulation, and its further conceptualization within the so-called *semisimplicial* theory, has roots tracing back to the early days of algebraic topology. However, there are signs that the semisimplicial theory has survived the test of time, as exemplified by the recent re-edition [May 1992] of the original monograph [May 1967], the objective of which is to provide a "grand unified" semisimplicial approach to algebraic topology.

Probably the place where the need for singular homology is most badly felt is the justification of the semisimplicial bundle analysis of the Nyquist map. A semisimplicial bundle is a combinatorial map, hence amenable to *finitary* analysis, yet it reveals such deep topological features as the links between the domain of uncertainty, the template, and the crossover. The problem is that a mere simplicial approximation to the Nyquist map won't, in general, be a semisimplicial bundle. To obtain a semisimplicial bundle, one has to construct a topological map, called Nyquist fibration, homotopically equivalent to the Nyquist map, go to the singular complexes of the uncertainty and the template, and obtain the singular chain map, which is a semisimplicial bundle.

The fact that we needed a *cellular*, rather than *simplicial*, decomposition of the Nyquist template to get a neat commutativity relation between the inverse image and the boundary is yet another manifestation of the limitation of the simplicial theory.

CUBES OR SIMPLEXES?

This is another question that has fueled the debate in both the mathematical and the operations research communities. Historically, algebraic topology emerged as a "simplicial" theory with the work of Poincaré. The work of Brouwer and Lefschetz, as well as the axiomatic foundation of [Eilenberg and Steenrod 1952], remained very much in the spirit of the simplicial theory of Poincaré. However, two milestones in the history of topology indicate that new concepts are sometimes more easily formulated using cubes rather than simplexes: The ground-breaking work of Kan on homotopy theory emerged in the semicubical setting and was later reformulated by [Gugenheim and Moore 1957] in the semisimplicial setting. The Serre spectral sequence of a fibration was first developed in the singular cubical set up and later reformulated in the semisimplicial setting by [May 1967]. [Massey 1980] developed an "all-cubical" approach to algebraic topology that has some pedagogical advantages. Some control problems are rather of the "cubical" type, the best example being Kharitonov's theorem characterized by its "cube" of uncertainties. In this book, we have opted for the *simplicial* theory, not because of the simplicial approximation theorem (Massey has a "cubical" approximation theorem), but for the most compelling reason that a great many combinatorial–geometrical manipulations with simplexes can be implemented using modern techniques of *computa-*

tional geometry.

STYLE

In preparing this book, we faced the problem of how the three different subjects of robust control, algebraic/differential topology, and simplicial algorithms should be presented in such a matter that they result in a symbiotic approach. While the basic ideas of simplicial algorithms should be relatively easy to grasp for anyone with average control engineering background, unfortunately algebraic topology has traditionally been perceived as a not easily accessible domain of mathematics that has the notorious reputation of requiring a long string of definitions, constructions, and theorems before significant results are achieved. For that reason, we have renounced developing algebraic topology as a separate subject and then showing its relevance to control, because we are sure the reader would already have given up before he had a chance to reach the robust stability applications.

We are instead proposing a pedagogical innovation: We develop *both* subjects *in parallel*. We proceed from the robust stability problem, conceptualizing what we believe is the essential issue of robust control—the boundary problem—and then show how this *naturally* leads us right into the heart of algebraic and differential topology. Our objective is to guide the reader into the maze of definitions and proofs, introducing every single concept precisely when there is engineering motivation for it.

Developing both robust stability and algebraic/differential topology in parallel has the drawback that algebraic/differential topological concepts appear confined within the robust stability arena with the difficulty of providing definitions and proofs of sufficient generality. We have, at least partially, solved this problem by presenting in a series of appendices a more *austere* exposition of some of the algebraic/differential topological concepts. Also relegated to the appendices are some highly technical concepts that would be too overwhelming to develop in the main body and that would unduly interrupt the control-theoretic train of thoughts.

Another criterion we have used to draw the line between the main body and the appendices is that *purely algebraic* results should be in the appendices. The main body of the book develops the geometric intuition, while the appendices provide the purely algebraic approach, deliberately stripped of geometric content. An exception to this rule are spectral sequences which are left in the main body, because the geometry and the algebra are so inextricably intertwined.

All along this book, we pay great attention to motivate the concepts, to proceed from engineering intuition to its precise mathematical formulation, while at the same time securing the correct sequencing of the mathematical concepts. Reformulating robust stability in topological terms leads to a host of mathematically standard results shedding new light on robust stability. Occasionally, some new mathematical results crop up. Of course, all new results are proved. Regarding the standard mathematical results, we feel that it is not completely satisfactory to ask the reader to believe the standard results without proofs, even though we provide precise references to the places where the proofs can be found. We feel it is important for the reader to develop the "feeling" for the proofs. The problem is that the proofs of several instrumental results, like the natural stratification of the space of differentiable functions, involve arguments of such an "incredible complexity"—even by Mathematicians' standards (see [Goresky and MacPherson 1988, page 22])—that we could not possibly prove everything. We have attempted to achieve a delicate balance; formal proofs of standard results are given provided that they do not take us too much off tracks. When it is impossible to provide a formal proof, then we supply plenty of references and give a sketch, or at least the *idea*, of the proof. At some other instances, we leave the proof as an exercise and provide some hints.

ORGANIZATION

The overall organization of the book is meant to reflect the gradual transition from elementary to more advanced mathematical concepts. The book is therefore divided into five parts:

- Part I. Simplicial approximation and algorithms
- Part II. Homology of robust stability
- Part III. Homotopy of robust stability
- Part IV. Differential topology of robust stability
- Part V. Algebraic geometry of crossover

Part I is for the application-oriented reader whose primary objective is to understand the boundary issue and how it applies to simplicial algorithms. This part is organized around the simplicial approximation theorem and its "computational geometry" and "simplicial algorithms" implementations. Such spinoffs as triangulation, chain complexes, chain maps, and semisimplicial maps are already introduced and serve as the forerunners to more advanced algebraic topological concepts.

Part II is a genuine introduction to the algebraic topology of the Nyquist map. The concept of homology groups and Betti numbers of the uncertainty, the template, the crossover, and other spaces are introduced. Throughout this entire Part, the (co)homology groups are used as algebraic description of the various relevant spaces. Fiber bundle, fibration, and related techniques are introduced as tools to investigate the inverse image properties of the Nyquist map. This Part culminates with the spectral sequence of the Nyquist map that establishes the connection between the topology—that is, the homology groups—of the Horowitz template, the uncertainty space, and the crossover $f^{-1}(0 + j0)$.

Part III is motivated by the obstruction approach to recursive robust stability tests over higher-and-higher-dimensional subsets of uncertainties. The fundamental concept of *homotopy groups* are introduced and motivated as value groups for the obstructions in various dimensions. The basic obstruction problem, and such spinoffs as the Brouwer degree of the Nyquist map and the Hopf classification of Nyquist curves, are introduced in the simple context of a map from the uncertainty to $\mathbb{C} \setminus \{0 + j0\}$. The latter topics serve as an introduction to the more advanced homotopy concepts cropping up with a map from the uncertainty to the GL-valued return difference matrix and eventually the map into the space of Fredholm operators, at which point this Part culminates in K-theory.

In addition to being of interest in its own right, Part III reveals yet another way to get to the simplicial algorithms. Indeed, it turns out that the simplicial algorithms of Part I can be viewed as algorithms to determine where the obstruction to extending the Nyquist map is. More broadly speaking, this leads to the homotopy approach to algorithms, usually referred to as *continuation or embedding methods*, that are introduced as applications of the Brouwer degree.

In Part IV, the book takes a definite differential flavor. We rephrase the basic boundary preserving paradigm in differential topological terms: The boundary of the Horowitz template is a subset of the critical values of the Nyquist map. Should the Horowitz template exhibit such pathology as a "kink," this means that the Nyquist map has higher-codimensional singularities. Next, we address the issue as to whether the singularity sets are structurally stable under variation of "certain" parameters. This is the issue of stability of Nyquist map. A closely related issue is structural stability of the crossover $f^{-1}(0+j0)$ under variation of certain parameters.

Finally, Part V deals with some algebraic geometry concepts applied to the crossover. Assuming that the Nyquist map is rational, we view the crossover $f^{-1}(0 + j0)$ as an algebraic set.

HOW TO READ THIS BOOK

The nominal way to read the book is just to follow the orderly sequence of parts. Parts I, II, and III are tightly interrelated and should be read in the proper sequence. The reader with some background in differential geometry might want to first browse through Part IV. Some very specialized topics in Part IV require background in homology and homotopy, covered in Parts II and III, respectively; however, that should not preclude the differential-geometry-oriented reader to first try to get the overall picture of Part IV and then come back to Parts I, II, and III. The same remark applies to Part IV; some background in homology is required but can be easily dispensed of at a first reading.

PREREQUISITES

It is assumed that the reader has an elementary working knowledge of modern multivariable robust control and is aware of Kharitonov's theorem. A certain level of mathematical maturity—that is, what one would expect from a graduate student in either Mathematics or the "Systems" side of Electrical Engineering—is essential. The elementary concepts of point set topology—in particular the notion of *topological space*—are taken for granted. Also taken for granted are such elementary algebraic concepts as groups, quotient groups, group homomorphisms, fields, and so on. Some background in polynomial matrices, which is pretty standard background for the average control scientist, would be beneficial. Differential calculus, the concepts of C^k and C^∞ spaces, the implicit function theorem, complex variables, and analytic functions are also taken for granted. A little bit of background in operator theory, which is pretty standard in control these days, would also be beneficial. Beyond that, we have attempted to define all relevant concepts in algebraic topology, differential topology, and algebraic geometry.

BIBLIOGRAPHICAL GUIDE

Some readers might want to get some guidance as to what would be some good companion texts in their venture through this book.

For the reader who needs to "start from scratch" we recommend the excellent text by [Jänich 1984], especially for its exposition of homotopy theory, the fundamental group, and the covering space. For the reader interested in a quick reading of Part I, we recommend the monograph by [Alexandroff 1961]. In about 50 pages of vivid-style, highly motivated, easy-to-read text, the author exposes the basic algebraic topological facts from the intuitive point of view of polyhedral theory, stressing the central role played by the simplicial approximation theorem. Another excellent, introductory text to combinatorial topology is [Henle 1979].

The basic algebraic facts needed to read this book are outlined in [Hilton and Wu 1974].

There are several "classics" that are about at the same level as this book. The classic by [Hilton and Wylie 1965] is still one of the best, despite the fact that it is somewhat outdated. It is very intuitive in nature and the reader could perceive the historical development of algebraic topology, from elementary simplicial theory to advanced algebraic concepts. Another valuable text, which has the advantage of being up to date, is the text by [Munkres 1984]. It proceeds from the intuitive simplicial homology and proceed all the way to the modern exposition of singular homology and categories. We have attempted to follow as closely as possible the common notation and terminology of [Hilton and Wylie 1965] and [Munkres 1984], and we have made ample reference to these two texts, probably at the expense of being redundant, but leaving to the reader the choice of his reference text.

Another valuable text, to which we sometimes refer, is the one by [Massey 1980]. This text has the major advantage of providing a relatively concise, yet powerful, approach to homology theory. This is achieved by systematic utilization, from the beginning on, of singular cubes as the basic building block. ([Hilton and Wylie 1965] and [Munkres 1984] proceed from intuitive simplicial decomposition and then proceed to more abstract decompositions.) Of course the question is whether decomposition in simplexes and decomposition in singular cubes lead to the same result. These are the *topological invariance results*, and they are guaranteed for polyhedra and some other relatively simple topological spaces. Therefore, when we refer to [Massey 1980] we should always keep in mind that some topological invariance results are invoked. Another "classic," to which we sometimes refer, is [Spanier 1989]. This latter text is very detailed, probably beyond what we need here; however, for some tedious results, it is the only readily available reference. Finally, we could not be complete in our bibliographical survey without mentioning the monumental work of [Whitehead 1978] on homotopy theory; despite its impressive size (700 pages) it offers a fairly intuitive exposition of homotopy theory, to which we often refer.

Another valuable addition to the reader's library is [McCleary 1985], which has the advantage of putting under the same cover some advanced concepts of both homological algebra and topology, and, by the same token, unraveling the mysteries of spectral sequences.

Regarding differential topology, all we really need is the classic by [Hirsch 1976]. The mathematically more sophisticated reader might want to rather go by [Golubitzky and Guillemin 1973]. Regarding the closely related field of singularity theory, an intuitive yet fairly complete text is [Arnold, Gusein-Zade, and Varchenko 1985]; the more algebraically oriented reader might instead want to use [Castrigiano and Hayes 1993].

Our general style reflects our belief that it always helps to go back to the historical source of results that otherwise do not have the foggiest motivation. For that reason, at the risk of being redundant, we have also made ample reference to the historical survey of algebraic and differential topology by [Dieudonné 1989].

Most chapters conclude with a *Bibliographical and Historical Notes* section. These by no means claim to be exhaustive. Their only objective is to provide the reader with some reference points.

ACKNOWLEDGMENTS

I am very grateful to Sir Michael Atiyah, Master of Trinity College, Cambridge, who, despite his busy schedule, welcomed me in Trinity College to discuss that part of the book dealing with K-theory. Not only did I benefit from his expertise in K-theory, but I also benefited from his very broad view on topology in general and how it should be applied to real-life problems.

I am also grateful to Professor Max Karoubi, Department of Mathematics, University of Paris VII, for his deep insight into some of the subtle technicalities of real K-theory.

Many thanks to Professor John P. D'Angelo, Department of Mathematics, University of Illinois at Urbana–Champaign, for several discussions on holomorphic functions of several complex variables and the Whitney root system.

I would like to take this opportunity to thank my colleague and friend, Professor Irving S. Reed, for many discussions on the role of Algebra in Engineering. I have come to adopt his point of view that sometimes it pays off to think *purely algebraically*, and I therefore dedicate the appendices to him.

I also benefited a lot from informal discussions with several colleagues of the Department of Mathematics of the University of Southern California: Professor Francis Bonahon, who answered numerous questions over a span of several years; Professor Eugene Gutkin with whom I had numerous beneficial "lunch-time discussions" on invariant differential forms on Lie groups; and Professor Paul Yang, who was always available to answer questions, including late in the evening!

Many years of informal discussions with many friends and colleagues have helped shape parts of this book: Dr. Gary Hewer, Naval Weapons Center, China Lake, California; Dr. Doup, Koninklijke/Shell Laboratory, Amsterdam, the Netherlands; Bill Helton, UC San Diego; Keith Glover, Cambridge University; R. Ober; and many others.

Several Electrical Engineering Ph.D. students of the University of Southern California became involved in one way or another in this project—Jonathan Bar-on, whose dissertation ([Bar-on 1990]) revealed the early intrusion of the classical groups in the multivariable phase margin problem; Richard Colgren, a senior Lockheed–Martin Skunk Works engineer, who developed a nonlinear $\mathcal{H}^\infty$ design procedure using describing functions which led to several topologically interesting examples of stability domains; and, last but not least, Chih-Yung Cheng, Chung-Kuang (Peter) Chu, and Murilo G. Coutinho who volunteered their time to draw the pictures.

During spring 1994, a graduate course based on an early draft of this book was offered at the University of Southern California, and the active participation of several students—in particular, Mrs. Nanaz Fathpour, Mr. Alan Meyer, and Mr. Reid Reynolds—helped improve some parts of the book.

This research was financially supported by National Science Foundation under grants ECS-91-13088, ECS-93-00016, and ECS-95-10656.

This research was also supported, in part, by the National Aeronautics and Space Administration under grant NASA-NCC-2-4002 of NASA Ames Research Center.

Part of this book was written during summer 1992 and summer 1993, when the author was holding a visiting professorship at the Department of Mathematics of the University of Namur, Belgium. I take this opportunity to thank Professor F. Callier for arranging this visit and to acknowledge the financial support of the Belgian Programme on Inter-university Poles of Attraction, initiated by the Belgian State, Prime Minister's Office for Science, Technology and Culture. I also thank Professor Callier and the Namur control team for many helpful discussions and seminars.

In August 1994, a series of lectures based on a draft of this book were presented at the Workshop on Robust Control in Taiwan, Republic of China, and helped shape up part of this book. I thank Professor Bing-Fei Wu, National Chiao-Tung University, Hinschu, Taiwan and Professor Jyh-Ching Juang, National Cheng-Kung University, Tainan for organizing this Workshop and acknowledge the financial assistance of the Government of Taiwan, R.O.C.

In July 1995, another series of lectures based on a draft of the book were organized by Huibert Kwakernaak, Jan van Schuppen, and Malo Hautus, under the sponsorship of the Dutch Institute of Systems and Control (DISC).

The very last parts of this book were drafted in November 1995, when the author was a Departmental Visitor to the Department of Systems Engineering, Australian National University, Canberra, Australia.

Last but not least, many thanks to my wife, Barbara, for her editorial assistance and her patience during many years of labor.

E.A.J.

Los Angeles, California
March 1996

CONTENTS

III HOMOTOPY OF ROBUST STABILITY

FIGURES

LIST OF SYMBOLS

$a^0, a^1, \ldots$	vertices (of P, in general)
$a^0 a^1 \ldots a^n$	algebraic n-simplex, which can be viewed as a skew-commutative product (no repetition of vertices allowed)
$[a^0 a^1 \ldots a^n]$	total simplex (repetition allowed)
$[z_0, z_1, \ldots, z_n]$	homogeneous coordinates of a point of $\mathbb{CP}^n$
$\underline{a}_i$	lower bound on the parameter a_i
$\overline{a}_i$	upper bound on parameter a_i
adj	the adjoint of a matrix
$b^0, b^1, \ldots$	vertices (of N, in general)
b_k	kth Betti number
B	base space of a bundle
$\mathbb{B}$	algebra of bounded operators in Hilbert space
BU	classifying space of unitary group
B_n	group of n-boundaries
B^n	group of n-coboundaries
$B_x(r)$	the open ball with center at x and radius r
$B(A)$	the open unit ball of the normed space A
$\mathbb{C}$	the complex field
$\mathbb{CP}^n$	the complex projective space
c_n	an n-chain—that is, a formal linear combination with integral coefficients of n-simplexes
c^n	an n-cochain
C_n	n-chain group
C^n	n-cochain group
$c_i(E, \pi, B)$	Chern classes of the bundle (E, π, B)
$c(E, \pi, B)$	total Chern class of the bundle (E, π, B)
$ch(E, \pi, B)$	Chern character of the bundle (E, π, B)
$\mathcal{C}^k(A, B)$	space of functions $A \to B$, k times continuously differentiable
C_x	a germ of smooth functions around $x = 0$
codim	the codimension of a function
coker	the cokernel of a mapping or a morphism

$\text{conv}(b^0, b^1, ...)$	the convex hull of the set $\{b^i\}$ of vertices
d	exterior differentiation
$d_p f$	differential of f at p
$d(\cdot)$	denotes, in general, a diffeomorphism
$\text{Diff}(M)$	group of diffeomorphisms of the manifold M
$\text{Diff}_+(M)$	group of orientation-preserving diffeomorphisms of the ori manifold M
$\dim_{\mathbb{F}} V$	dimension of V as a vector space over the field $\mathbb{F}$
D	the space of uncertainties, excluding frequency
D^i	the ith degeneracy operator of semisimplicial theory
$\mathbb{D}$	the open unit disk of the complex plane
$\mathbf{D}$	denotes, in general a disk, other than $\mathbb{D}$
Δ	structured uncertainty
δ_{ij}	Kronecker symbol
∂	topological boundary
$\partial_\bullet$	distinguished boundary of a polydisk
∂_n, ∂_*	algebraic boundary
∂_*	the connecting morphism
$\frac{\partial f_\theta}{\partial p}$	$= \begin{pmatrix} \frac{\partial f_\theta}{\partial p_1} & \frac{\partial f_\theta}{\partial p_2} & \cdots \end{pmatrix}$
δ^n, δ^*	n-coboundary operator
$DT(\{b^i\})$	Delaunay triangulation of vertex set $\{b^i\}$
e	$= (1, 1, 1, ...)^T$
e^i	Euclidean basis vector $e^i = (0, 0, ..., 0, 1, 0, ..., 0)^T$, where 1 is i ith position
E	total space of a bundle
E_p	the fiber above p in a bundle
$E^r_{u,v}$	the module indexed by (u, v) at the rth step of a (homology) tral sequence
$E_r^{u,v}$	the (u, v)-module at the rth step of a (cohomology) sp sequence
f	Nyquist mapping
$\tilde{f}$	simplicial approximation to Nyquist mapping
f_{PL}	piecewise-linear extension of vertex transformation $a^i \mapsto b^i$
$f_\theta(p)$	$= \cos\theta \Re f(p) + \sin\theta \Im f(p)$
f_ω	fixed-frequency Nyquist mapping
φ	chain map
$\mathcal{F}$	space of Fredholm operators

F	fiber of a bundle or a fibration
F^i	ith face operator in semisimplicial theory
G	generic notation for a group
gcd	greatest common divisor
$G_{n,m}$	(real or complex) Grassmannian manifold of all m-D subspaces in n-D space
$GL(n_l, \mathbb{C})$	general linear group of $n_l \times n_l$ matrices with complex entries
$h(\cdot)$	in general, homeomorphism
$H_n(P)$	nth homology group of P
$H^n(P)$	nth cohomology group of P
$\mathcal{H}$	a Hilbert space
$\mathcal{H}^2, \mathcal{H}^\infty$	the classical Hardy spaces
$\mathbb{HR}^n$	the n-dimensional Euclidean half-space—that is, $\{(x_1, x_2, ..., x_n) : x_i \in \mathbb{R}, i = 1, ..., n; x_n \geq 0\}$
hom(A, B)	the set of homomorphisms $A \to B$
H_x	ring of germs of analytic functions around $x = 0$
$\Im$	imaginary part of a complex entity
im	image of a homomorphism or mapping
int$(\cdot)$	the interior of a set
$J_f(p)$	the Jacobian of the map f at the point p
$J(R)$	the Jacobson ideal or Jacobson radical of the ring R
$j^k f(x)$	the k-jet of f around x
$j^k f$	the jet section $x \mapsto j^k f(x)$ of the mapping f
$J^k(X, Y)$	the space of k-jets from X to Y
K	generic notation for a simplicial complex
$\|K\|$	the polyhedron of the simplicial complex K
K^n	the n-skeleton of a simplicial complex—that is, the set of simplexes of dimension $\leq n$
$K^0(D)$	topological K-group of D
$K^{-n}(D)$	higher K-groups of space D
$KO(\cdot)$	real K-groups
$KR(\cdot)$	K-groups of space with involution—for example, complex conjugation
$\mathbb{K}$	ideal of compact operators on Hilbert space
ker	kernel of a homomorphism
ℓ	labeling function
$L(s)$	the loop matrix
$\mathcal{L}^2, \mathcal{L}^\infty$	the classical Lebesgue spaces

ℓ^2	Hilbert space of square summable sequences
λ	Morse index of a nondegenerate critical point of a differen function
$(\lambda_0, ..., \lambda_n)$	barycentric coordinates of a point in the simplex σ_n
M_n	generic notation for an n-manifold
$\mathbb{M}(m, n; \mathbb{C})$	set of $m \times n$ matrices with complex entries
$\mathcal{M}, \mathcal{M}_x, \mathcal{M}[[x]]$	(unique) maximal ideals of the (local) rings R, C_x, respectively
μ	slack variable of linear programming
N	Nyquist supertemplate
N_ω	Horowitz template—that is, $f(D, \omega)$
n_l	number of feedback loops
n_p	dimension of polyhedron P of uncertainty (including frequer
n_q	dimension of space D of uncertainty (excluding frequency)
$\mathbb{N}$	the set of natural numbers
O, O_i	open sets
$OLHP$	the open left half-plane of $\mathbb{C}$
$ORHP$	the open right half-plane of $\mathbb{C}$
$OUHP$	the open upper half-plane of $\mathbb{C}$
P	polyhedron of uncertainty, $D \times \Omega$
PDH	cone of positive definite Hermitian matrices
$\mathbb{P}$	projective space
p	a point in the polyhedron of uncertainties
$\pi_n(A, *)$	nth homotopy group of space A with base point $*$
π	generic notation for a projection
π_i	projection on the ith factor
RHP	the (closed) right half-plane of $\mathbb{C}$
q	a point in D
$\mathbb{Q}$	the field of rational numbers
s	Laplace symbol
$\mathbb{R}$	the real line
$\mathbb{RP}^n$	the real projective space
$\mathbb{R}[x]$	ring of polynomials in x, with real coefficients
$\mathbb{R}[[x]]$	ring of formal power series in x, with real coefficients
$\Re$	the real part of ...
S^n	the n-dimensional sphere
$S_r(f)$	Thom–Boardman singularity set of f
$SL(n, \mathbb{C})$	special linear group—that is, subgroup of matrices with det

$SO(n, \mathbb{R})$	special orthogonal group—that is, group of $n \times n$ real, orthogonal matrices with det $= 1$
$\mathrm{star}(\sigma)$	the star of the simplex σ—that is, the set of simplexes having σ as a face
$SU(n, \mathbb{C})$	special unitary group—that is, group of $n \times n$ complex, unitary matrices with det $= 1$
σ_n	an n-simplex, of $D \times \Omega$, in general
Σ_n	the standard n-simplex
$\sigma_{\max}$	the maximum singular value of a matrix
t	in general, homotopy parameter
τ	a simplex of N
T_pX	tangent space to manifold X at $p \in X$
T_f	Toeplitz operator induced by f
$\mathrm{Tor}(A, B)$	torsion product of A, B
Trace	the trace of a matrix
$\mathbb{T}$	unit circle of complex plane
$\mathbb{T}^n$	the n-torus
$\mathcal{T}(A)$	a topology for the set A
$\chi(A)$	the Euler characteristic of the space A
$U(n_l)$	group of $n_l \times n_l$ unitary matrices
$V_{n,k}$	Stiefel manifold of k-frames in $\mathbb{C}^n$
$VD(\{b^i\})$	Voronoi diagram of set of points $\{b^i\}$
$\mathrm{Vect}(X)$	equivalent classes of vector bundles over the space X
X	generic notation for domain of definition of a map
$\mathcal{X}$	a smooth manifold encompassing a stratified space (domain of definition of a map)
X^+	space obtained by adjoining a point to X, $X^+ = X \cup \{\text{point}\}, \text{point} \notin X$; if point $= \infty$, then X^+ is the 1-point compactification
Y	range space of a map
$\mathcal{Y}$	a smooth manifold encompassing a stratified range space of a map
Y^X	the space of all continuous maps $X \to Y$
Z_n	group of n-cycles
Z^n	groups of n-cocyles
$\mathbb{Z}$	the Abelian group of integers
ω	frequency
ϖ	an exterior differential form
$\varpi^n(A)$	a bi-invariant differential form around $A \in GL$, of homogeneous degree n

Ω	frequency spectrum
ΩA	space of loops in the space A
$\mho$	a domain—that is, an open, connected subset of $\mathbb{C}^n$
$B \subseteq A$	B is a subset of A ($A = B$ allowed)
$B \subset A$	B is a proper subset of A ($A \neq B$)
$A \setminus B$	set of elements of A not in B
$\overline{A}$	the closure of the space A
$\overline{z}$	the complex conjugate of $z \in \mathbb{C}$
$\sqcup$	disjoint union
$\sim$	equivalence relation
$\equiv$	is identical to ...
$\cong$	is isomorphic or homeomorphic to ...
$\simeq$	is homotopic to ...
$\approx$	is approximately equal to ...
$\preceq$	facing relation of an abstract simplicial complex or stratificatio
#	in general, denotes the number of critical points of a certain ty of a smooth function, or the number of n-simplexes of a compl and so on
$\overline{z}$	the complex conjugate of $z \in \mathbb{C}$
p^*	denotes a critical point in the space of p-parameters
$X \to Y$	mapping or morphism from X to Y
$X, A \to Y, B$	a mapping or morphism $X \to Y$ that maps the subset $A \subseteq X$ i the subset $B \subseteq Y$
$\xrightarrow{\equiv}$	identity mapping or morphism
$\xrightarrow{\cong}$	homeomorphic mapping or isomorphic morphism
$\xrightarrow{?}$	a mapping, the existence of which is in question
$\searrow$	collapse
$[A, B]$	the set of homotopy classes of maps $A \to B$
$[A, B]_*$	the set of homotopy classes of base point preserving maps A, a_* B, b_*
1_A	identity mapping or morphism on A
$\wedge$	wedge product of exterior differential forms
$\oplus$	direct sum of groups, vector spaces, or Hilbert spaces
$\otimes$	tensor product of Abelian groups
$\otimes_R$	tensor product of R-modules
$\otimes_\Lambda$	tensor product of modules, defined over the differential graded gebra Λ
$\perp$	orthogonal complement

$\blacksquare$	end of proof
$*$	base point of a space
$\{a\}$	the equivalence class of a
$f\|A$	restriction of $f : X \to Y$ to $A \subseteq X$
A/B	quotient space or quotient group
$\circ$	composition of maps
$(\cdot, \cdot)$	inner product or bilinear form
$\|\| \cdot \|\|$	Euclidean norm
$\smile$	cup product
$A * B$	torsion product of A, B
$\times_D$	fiber product
∇f_θ	gradient of f_θ, $\begin{pmatrix} \frac{\partial f_\theta}{\partial p_1} \\ \frac{\partial f_\theta}{\partial p_2} \\ \vdots \end{pmatrix}$

REMARKS ABOUT NOTATION AND TERMINOLOGY

The notation S^1 denotes the one-dimensional unit sphere—that is, the unit circle; S^1 also denotes anything topologically equivalent (homeomorphic) to the unit circle; on the other hand, the notation $\mathbb{T}$ more specifically means the unit circle of the *complex plane*.

An *onto* or *surjective* group homomorphism will sometimes be called an **epimorphism**. An *injective* group homomorphism will sometimes be called a **monomorphism**.

Algebraic and Differential Topology of Robust Stability

1

PROLOGUE

In conclusion I would like to suggest that algebraic topology has now reached a sufficient degree of maturity so that it should be regarded as a tool available for use in appropriate branches of analysis. At least I hope it will be the natural thing for a mathematician to ask: if I vary the coefficients or parameters of my problem, what sort of a topological space do I get—is it for example contractible and if not, what is the significance of its topological invariants?

M.F. Atiyah,
"The role of algebraic topology in mathematics,"
J. London Mathematical Society,
vol. 41, pp. 63–69, 1966.

Consider a string of three masses, m_1, m_2, and m_3. The mass m_i is connected to the mass m_j by a spring of stiffness k_{ij}. The left and right ends of the string are attached to "walls" ($m_0 = \infty$, $m_4 = \infty$) via springs of stiffnesses k_{01} and k_{34}, respectively. Let x_i denote the position of the mass m_i relative to its equilibrium position and let F_i denote the external force applied to m_i. The open-loop dynamics is

$$\underbrace{\begin{pmatrix} m_1 & 0 & 0 \\ 0 & m_2 & 0 \\ 0 & 0 & m_3 \end{pmatrix}}_{M} \begin{pmatrix} \ddot{x}_1 \\ \ddot{x}_2 \\ \ddot{x}_3 \end{pmatrix}$$

$$= \underbrace{\begin{pmatrix} -k_{01} - k_{12} & k_{12} & 0 \\ k_{12} & -k_{12} - k_{23} & k_{23} \\ 0 & k_{23} & -k_{23} - k_{34} \end{pmatrix}}_{-K} \begin{pmatrix} x_1 \\ x_2 \\ x_3 \end{pmatrix} + \begin{pmatrix} F_1 \\ F_2 \\ F_3 \end{pmatrix}$$

In this experiment, we feed the *rate* sensor outputs back to the input, and this yields the open-loop transfer matrix $L(s)$,

$$\dot{x}(s) = L(s)F(s)$$

Such a system is said to have *colocated rate sensors and force actuators* in the sense that the control effort $F_i(s)$ is applied at the precise point where the rate $\dot{x}_i(s)$ is measured. The specific property of this kind of systems is that

$$L(j\omega) + L^T(-j\omega) \geq 0, \ \forall \omega \in \mathbb{R}$$

Such a system is said to be *positive real.* This is a corollary of the colocation property together with the fact that the mechanical system could only dissipate energy. Many distributed parameters systems, provided they be dissipative with their rate sensors colocated with the force actuators, satisfy that property, so that through the simple spring-mass-dashpot example a large class of systems is invoked.

This system is closed by the feedback control

$$F = -\Delta \dot{x}$$

A fundamental result of hyperstability theory asserts that the feedback remains stable for all Δ positive definite Hermitian. In other words, the system has *infinite gain margin.* In particular, taking Δ, a diagonal matrix of positive gains, one obtains a *decentralized* (stabilizing) control from the colocated rate sensors to the force actuators.

In this experiment, we take $\Delta = I$ to be the nominal, stabilizing control.

However robust the feedback is under positive definite Hermitian perturbation, it is unfortunately very sensitive to such phase errors as

$$\Delta = \begin{pmatrix} \exp(-j\theta_1) & 0 & 0 \\ 0 & \exp(-j\theta_2) & 0 \\ 0 & 0 & \exp(-j\theta_3) \end{pmatrix}, \quad \theta_i \in [0, 2\pi]$$

This feedback configuration is shown in Figure 1.1. This phase error is meant to model the mismatch between the point of application of the control and the point of measurement of the rate in distributed parameters systems. Allowing θ_i to run in $[0, 2\pi]$ might be thought to be too much of a global analysis; however, in a distributed parameters system, a very small colocation error could easily result in a 180-degree phase shift in $L(j\omega)$ in the frequency range of the higher vibration modes.

The collection of all diagonal phase perturbations can be identified with $[0, 2\pi]^3$. However, the mapping $\theta \mapsto \Delta$ is many-to-one. Clearly, θ and $(\theta_1 + 2\pi k, \theta_2 + 2\pi l, \theta_3 + 2\pi m)$, $k, l, m \in \mathbb{Z}$, are mapped to the same Δ. To remove this singularity, the easiest technical fix is to identify the opposite faces of the cube $[0, 2\pi]^3$ of parameters. Identification of a pair of opposite faces yields a solid, square "donut." Identification of two pairs of opposite faces of the cube results in a hollow sphere, still embedded in the Euclidean space $\mathbb{R}^3$. The identification of the remaining pair of opposite faces of the cube is equivalent to identification of the inner boundary and the outer boundary of the hollow sphere. Clearly, this cannot be done in three-dimensional space. Therefore, the most intuitive low-dimensional model of the uncertainty space in case of three uncertain phase angles is a hollow sphere where the inner boundary points and the outer boundary points

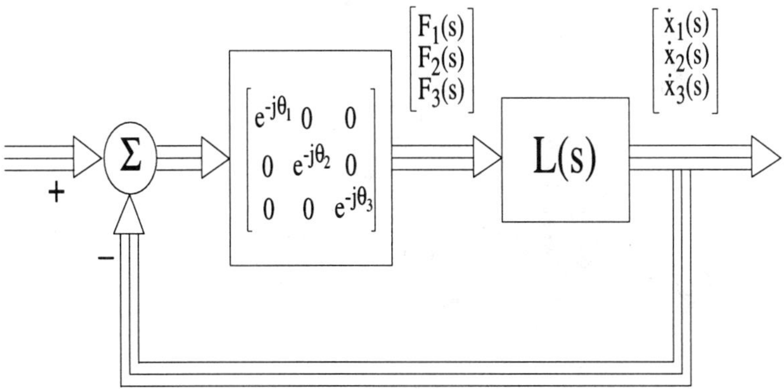

Fig. 1.1. Positive real feedback perturbed by diagonal phase uncertainty.

on the same radial are identified. The resulting geometric structure is the *3-torus* $\mathbb{T}^3$.

Another procedure for removing the singularity of the mapping $\theta \mapsto \Delta$ is to make $(\theta_1, \theta_2, \theta_3)$ local Riemann coordinates by covering $\{\Delta\}$ with open sets O_i such that the restriction of $\Delta \mapsto (\theta_1, \theta_2, \theta_3)$ to each open set O_i is a homeomorphism h_i. Such a homeomorphic mapping of an open subset of $\{\Delta\}$ to an open subset of $\mathbb{R}^3$ is called a *chart*. For two overlapping open sets O_i, O_j of $\{\Delta\}$, we define the (bijective) *gluing map* $h_j \circ h_i^{-1} : h_i(O_i \cap O_j) \to h_j(O_i \cap O_j)$ between open subsets of $\mathbb{R}^3$. With the collection of all such charts, and provided that the gluing maps are continuous, $\mathbb{T}^3$ becomes a *manifold*. Actually, the gluing maps can be made analytic; therefore, $\mathbb{T}^3$ becomes an *analytic manifold*.

From an algebraic point of view, the collection $\mathbb{T}^3$ of all phase perturbations is easily seen to be a subgroup of the unitary group

$$U(3) = \{\Delta \in \mathbb{M}(3, \mathbb{C}) : \Delta^* \Delta = I\}$$

Hence $\mathbb{T}^3$ is at the same time a manifold and a group. Furthermore, the two structures are compatible in the sense that the multiplication map

$$\left(\begin{pmatrix} \exp -j\theta_1 & 0 & 0 \\ 0 & \exp -j\theta_2 & 0 \\ 0 & 0 & \exp -j\theta_3 \end{pmatrix}, \begin{pmatrix} \exp -j\theta_1' & 0 & 0 \\ 0 & \exp -j\theta_2' & 0 \\ 0 & 0 & \exp -j\theta_3' \end{pmatrix} \right)$$
$$\mapsto \begin{pmatrix} \exp -j(\theta_1 - \theta_1') & 0 & 0 \\ 0 & \exp -j(\theta_2 - \theta_2') & 0 \\ 0 & 0 & \exp -j(\theta_3 - \theta_3') \end{pmatrix}$$

is continuous. Clearly, the multiplication map is analytic.

To summarize, $\mathbb{T}^3$ is an analytic manifold with analytic multiplication and inverse; this is the *analytic structure* that makes $\mathbb{T}^3$ a *Lie group*.

In the cube $[0, 2\pi]^3$, with its opposite faces properly identified, there are stability and instability regions, separated by "crossover" surfaces. Nu-

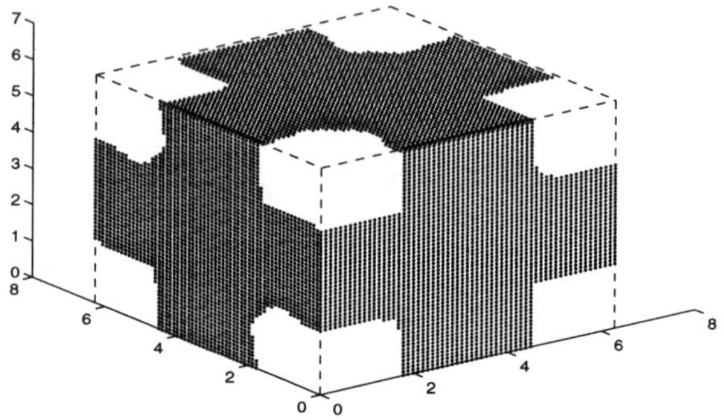

Fig. 1.2. The unstable (shaded) areas in the cubical model of the 3-torus.

merical exploration yields the picture of Figure 1.2. The shaded areas are regions of instability while the white areas are regions of stability. Clearly, these stability/instability regions and their separating boundaries can get complicated.

The aim of this book is to formulate these robust stability problems in precise topological terms, to develop fast algorithms for computing the "crossover," and finally to develop algebraic procedures for computing the "topography" of the problem without the need for numerical exploration.

Part I

SIMPLICIAL APPROXIMATION AND ALGORITHMS

2

ROBUST MULTIVARIABLE NYQUIST CRITERION

SUMMARY

In this chapter, we briefly review multivariable closed-loop stability and the related robustness issue. The material is fairly standard, except possibly for the tedious issue of a varying number of open-loop right half-plane poles. We already provide a bit of the topological flavor to come by emphasizing the fundamental problem of mapping the *topological space* of uncertainty to the complex plane, resulting in a *template*, that could exhibit strange singularity phenomena, and that should stay away from $0 + j0$.

2.1 Multivariable Nyquist Criterion

We first look at the *multivariable* Nyquist criterion *without uncertainty*. In other words, we freeze the uncertainty to some nominal value, say Δ_0. The multivariable Nyquist stability criterion involves several subtleties related to pole/zero cancellation. We approach this problem from the Youla–Desoer rational coprime factorization:

$$L(s)\Delta_0 = D^{-1}(s)N(s)$$

In the above, $D(s)$ and $N(s)$ are square, rational, $\mathcal{H}^\infty$ matrices, and in addition they are coprime. With this factorization, the closed-loop transfer matrix of the feedback system of Figure 2.1 can be written

$$\begin{aligned}(I + L\Delta_0)^{-1}L\Delta_0 &= (D(s) + N(s))^{-1}N(s) \\ &= \frac{\operatorname{adj}(D(s) + N(s))}{\det(D(s) + N(s))}N(s)\end{aligned}$$

To get a clear picture as to what the poles of $(D(s) + N(s))^{-1}$ are, we use the Smith–McMillan form of $D(s) + N(s)$,

$$D + N \cong \begin{pmatrix} \frac{a_1(s)}{b_1(s)b_2(s)\ldots b_{n_l}(s)} & & & \\ & \frac{a_1(s)a_2(s)}{b_1(s)b_2(s)\ldots b_{n_l-1}(s)} & & \\ & & \ddots & \\ & & & \frac{a_1(s)a_2(s)\ldots a_{n_l}(s)}{b_1(s)} \end{pmatrix}$$

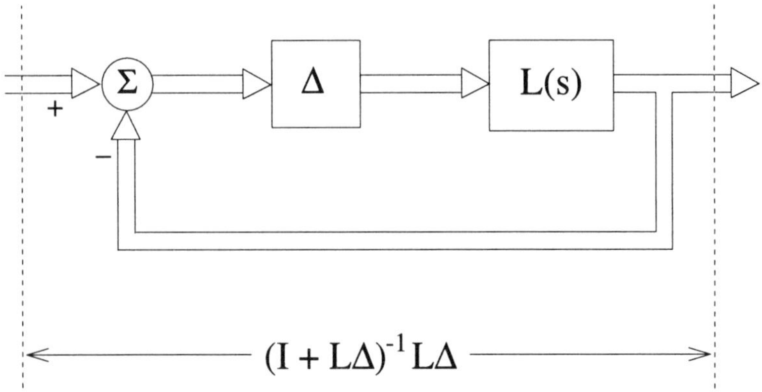

Fig. 2.1. The multivariable, uncertain feedback.

In this Smith–McMillan form, the $a_i(s)$'s and the $b_i(s)$'s are polynomials, and there are no poles/zeros cancellations in any of the rational fractions

$$\frac{a_1(s)a_2(s)...a_{n_l-i+1}(s)}{b_1(s)b_2(s)...b_i(s)}, \quad i = 1, ..., n_l$$

appearing on the diagonal. Furthermore, the Youla–Desoer construction guarantees that $N(s)$ and $D(s)$ are stable, so that the $b_i(s)$'s are Hurwitz. Clearly, the poles of $(D(s) + N(s))^{-1}$ are the zeros of the numerator polynomials of the Smith–McMillan form of $D+N$. The next problem is to determine how the poles of $(D(s)+N(s))^{-1}$ and the zeros of $\det(D(s)+N(s))$ are related. Since the Smith–McMillan form is obtained from $D + N$ by unimodular transformations that have unit determinant, it follows that

$$\det(D + N) = \frac{a_1^{n_l} a_2^{n_l-1} ...a_{n_l}}{b_1^{n_l} b_2^{n_l-1} ...b_{n_l}}$$

The crucial feature is to observe that there could be pole/zero cancellations in the above, even though there are no poles/zero cancellation in the Smith–McMillan form. For example, there could be a cancellation between $b_2(s)$ and $a_{n_l}(s)$. More generally, observe that the following fractions,

$$\frac{a_{n_l-i+2}(s)}{b_i(s)}, \quad i = 2, ..., n_l$$

do not appear in the diagonal entries of the Smith–McMillan form, but do appear in $\det(D(s) + N(s))$ where some cancellation might occur. It follows that there could be a discrepancy between the zeros of $\det(D + N)$ and the poles of $(D + N)^{-1}$; to be more specific,

$$\{\text{poles of } (D + N)^{-1}\} =$$

$$\{\text{zeros of } \det(D + N)\} \cup \{\text{poles/zeros canceled in } \Pi_{i=2}^{n_l} \frac{a_{n_l-i+2}(s)}{b_i(s)}\}$$

However, if any cancellation occurs, this could only be a LHP pole/zero cancellation because the $b_i(s)$'s are Hurwitz. Therefore,

$$\{\text{RHP poles of } (D+N)^{-1}\} = \{\text{RHP zeros of } \det(D+N)\}$$

At this stage we have reached the following result: The multivariable closed-loop matrix $(I + L\Delta_0)^{-1}L\Delta_0$ is stable iff $\det(D(s) + N(s))$ has no *right half-plane* zeros.

The next step is to reduce the RHP zero test on $\det(D+N)$ to a test involving the usual Nyquist plot of $\det(I + L\Delta_0)$. By the principle of the arguments, $\det(D(s) + N(s))$ has no RHP zeros iff

$$\frac{1}{2\pi j}\oint d\log\det(D(s)+N(s)) = \frac{1}{2\pi j}\oint \frac{(\det(D+N))'}{\det(D+N)}ds = 0$$

where the integral with respect to s is taken from $-j\infty$ to $+j\infty$ followed by a large semicircle enclosing the right half-plane. The above integral is usually referred to as *winding number* or *topological index* of $\det(D(s) + N(s))$ relative to $0 + j0$. To convert the above into a criterion involving the Nyquist plot of $\det(I + L(j\omega)\Delta_0)$, observe that

$$\det(D+N) = \det(I + L\Delta_0)\det(D)$$

Therefore, the winding number of the plot of $\det(D+N)$ around $0+j0$ is

$$\frac{1}{2\pi j}\oint d\log\det(D+N) = \frac{1}{2\pi j}\oint d\log\det(I+L\Delta_0) + \frac{1}{2\pi j}\oint d\log\det D$$

At this stage, let us agree to call *number of open-loop right half-plane poles* the number, multiplicity counted, of right half-plane zeros of $\det(D(s))$. Clearly, $\frac{1}{2\pi j}\oint d\log\det D$ is equal to minus the number of open-loop right half-plane poles. Therefore, for closed-loop robust stability, we must have

$$\frac{1}{2\pi j}\oint d\log\det(I + L\Delta_0) = \#(\text{RHP poles})$$

In other words, the Nyquist plots of $\det(I+L\Delta_0)$ must circle #(RHP poles) times counterclockwise around $0 + j0$.

Theorem 2.1. (Multivariable Nyquist Stability Criterion) *The closed-loop transfer matrix $(I+L\Delta_0)^{-1}L\Delta_0$ is stable iff $\det(I+L(j\omega)\Delta_0) \neq 0, \forall\omega$ and furthermore the plot circles #(open-loop RHP poles) times around $0 + j0$, where #(open-loop RHP poles) is defined as the number of RHP poles of the determinant of the denominator matrix of the rational coprime factorization of the loop matrix.*

2.2 Robust Multivariable Nyquist Criterion

The crucial issue in the *robust* multivariable Nyquist criterion is the effect of the uncertainty on the factorization. If the uncertainty Δ enters as a square, nonsingular matrix that pre- or postmultiplies the loop matrix, the dependency of the factorization on Δ is straightforward. Just absorb Δ in either $D(s)$ or $N(s)$.

However, in the general case, the analytical dependency of the factorization on the uncertainty, $L(s,\Delta) = D^{-1}(s,\Delta)N(s,\Delta)$, is very difficult to assess. The problem is even worse when it comes to the dependency of the Smith–McMillan form on Δ. We therefore attack the problem from a more tractable path of approach. The idea is to go through the argument of the preceding section for every $\Delta \in D$. Clearly, for such an argument to go through, it is necessary that the Nyquist plot of $\det(I + L\Delta)$ be continuously deformed when Δ sweeps the space D. To guarantee this, it is necessary that the Nyquist map be continuous *and that the space D be connected.* If, under such a deformation, proceeding from a stabilizing Δ_0, the plot doesn't cross $0 + j0$, *and if in addition the number of open-loop RHP poles does not change*, then closed-loop stability is preserved. Here we are at a *crucial point*: *Under variation of Δ, the number of open-loop RHP poles could change, without the winding number of the Nyquist plot changing.* This can easily be seen from the following counterexample:

$$L\Delta = \begin{pmatrix} \frac{\Delta}{s+10} & \frac{1}{s+0.5-\Delta} \\ 0 & \frac{1}{s+0.1} \end{pmatrix}, \quad \Delta \in [0,1]$$

It is easily seen that as Δ crosses 0.5 the closed-loop system becomes unstable and the Nyquist plot doesn't cross $0+j0$. Of course, in this example, it could be argued that if we look at the open-loop *matrix* $I + L\Delta$ it shows an open-loop pole crossing the imaginary axis for $\Delta = 0.5$. Unfortunately, it is *even not enough* to watch for poles of the open-loop matrix crossing the imaginary axis, as shown by the following counterexample:

$$L\Delta = \begin{pmatrix} \frac{1}{s+10} & \frac{\exp(-\frac{1}{\Delta^2})}{s-1} \\ 0 & \frac{1}{s+0.1} \end{pmatrix}$$

The closed-loop system is stable for $\Delta = 0$. However, it is unstable for $\forall\Delta \neq 0$, even though nothing wrong can be perceived from the Nyquist plot of $\det(I + L\Delta)$. To make things worse, in this case, the open-loop matrix does not even show a pole crossing the $j\omega$-axis around $\Delta = 0$.

Clearly, in the *robust* multivariable Nyquist criterion, it is *not enough* to check $\det(I + L\Delta) \neq 0, \forall\omega$ together with the encirclement condition. In addition, we have to keep track of the number of open-loop RHP poles.

The issue of robust stabilization of a family of plants with a varying number of RHP poles was considered by [Verma, Helton, and Jonckheere 1986]. The essential point is the following: Let $L(s,\Delta) = G(s,\Delta)K(s)$, where $G(s,\Delta)$ is the plant and $K(s)$ the compensator. Assume the effect of the uncertainty is constrained in a disk centered at $G(j\omega,\Delta_0)$,

$\Delta \in D \Leftrightarrow$

$$\sigma_{\max}(G(j\omega, \Delta) - G(j\omega, \Delta_0)) \leq |r(j\omega)|, \quad \forall \omega$$
$$\#(\text{RHP poles of } G(s, \Delta)) \leq \#(\text{RHP poles of } G(s, \Delta_0))$$

Then for the family of plants to be robustly stabilizable by a single compensator it is necessary that the family $\{G(s, \Delta) : \Delta \in D\}$ has a constant number of RHP poles. [Verma, Helton, and Jonckheere 1986]'s approach to this issue is functional-analytic—it relies on the Ball–Helton theory. However, it might already be clear to the reader that there is some deep homotopy interpretation to this result, on which we shall come back later.

From the [Verma, Helton, and Jonckheere 1986] result, it is reasonable to assume that the number of RHP open-loop poles remains constant under $\mathcal{L}^\infty$-bounded uncertainty. (To formulate robustness results for the case of a varying number of open-loop RHP poles, we would have to use the gap metric instead of the $\mathcal{L}^\infty$ norm.) These considerations motivate the following:

Theorem 2.2. (Robust Multivariable Nyquist Stability Criterion) *Assume the closed-loop system is stable for some nominal $\Delta_0 \in D$, where D is a connected set of uncertainties. Assume the* **Nyquist mapping** *$(\omega, \Delta) \mapsto \det(I + L\Delta)$ is continuous. Assume the number of open-loop RHP poles of $L(s, \Delta)$ remains constant as Δ runs in D. Then the closed-loop system is robustly stable—that is, is stable $\forall \Delta \in D$—iff $\det(I + L\Delta) \neq 0, \forall \Delta \in D, \forall \omega$.*

We will formulate more sophisticated Nyquist-based stability criteria in later chapters. In particular, in Chapter 19, motivated by the first counterexample, we map the uncertainty to the return difference *matrix*; as an illustration, observe that the return difference matrix of the first counterexample is not continuous around $(\Delta = 0, \omega = 0)$. In Chapter 20, we take the point of view that the return difference matrix is just a representation of the return difference *operator*; to illustrate this, observe that the return difference operator of the second counterexample lacks continuity in Δ around $\Delta = 0$.

2.3 Uncertainty Space

The concept of uncertainty space D is very general. In the simplest case of *real* parameters, each lying in a compact interval, $q_i \in [\underline{q}_i, \overline{q}_i]$, D is the topological product $\Pi_i[\underline{q}_i, \overline{q}_i]$—that is, a *compact* cube or parallelepiped.

However simple the closed cube is, it leads to some difficulties because it is *not* a manifold: A point on the boundary of the cube *does not* have a neighborhood homeomorphic to the Euclidean space or its open unit ball. As we shall see later, it is sometimes necessary to disassemble the *closed* cube into the disjoint union of the *open* cube, its open faces, open edges, and so on, in order to do, for example, a Morse singularity analysis of the map.

The SISO phase margin problem provides a first clue that parameterization of the uncertainty might not be the easiest possible problem. Indeed, we could take the phase angle θ as parameter (running in $\mathbb{R}$ or $[0, 2\pi]$) and absorb $\exp(-j\theta)$ in the Nyquist map. The problem is that the mapping $\theta \mapsto \exp(-j\theta)$ is *many-to-one* (it is actually a *covering map* if θ runs in $\mathbb{R}$). Therefore, absorbing $\exp(-j\theta)$ in the Nyquist map makes the domain of definition more trivial, but at the expense of making the Nyquist map more "singular"; that is, more points of the domain are mapped to the same point in $\mathbb{C}$. Nevertheless, if the latter is not our concern, it is clearly more economical to use $[0, 2\pi]$ rather than $\mathbb{R}$ as domain of definition. The problem with such a domain of definition as $[0, 2\pi]$ is that it is *not* a manifold. As already argued, a domain of definition with manifold structure is essential for singularity analysis of the Nyquist map, itself related to the extremely important problem of continuity of robustness margin relative to "certain" parameters. Clearly, we could use the manifold $(0, 2\pi)$ as domain, but we would lose the information as to how the map behaves around $\theta = 0, 2\pi$. We could think of using $[0, 2\pi)$ or $(0, 2\pi]$, but again these fail to be manifolds. Therefore, the only way to make the domain of definition $[0, 2\pi]$ a *manifold* is to identify the end points $0, 2\pi$—this is not harmful since $0, 2\pi$ are mapped to the same point of $\mathbb{C}$—resulting in a topological structure equivalent to the circle S^1 which *is* a manifold.

Equivalently, one could argue that the phase margin uncertainty is $\exp(-j\theta)$ and that the collection of all such uncertainties is the unit circle of the complex plane, $\mathbb{T}$, which we could adopt as domain of definition. This choice makes the domain of definition of the Nyquist map a *manifold* and also removes the unnecessary many-to-one mappings typical of parameterization by phase angle.

Defining the uncertainty space of a phase margin problem as the *full* unit circle might be deemed an "overkill." However, as already argued in Chapter 1, small sensor-actuator colocation errors in distributed parameters flexible space structures result in drastic phase variations in the frequency range of the higher structural eigenfrequencies. Besides this, a more compelling reason to consider the full unit circle is that such relevant *local* properties as the *analytical degree* of the Nyquist map are actually *global* invariants (see Section 18.3); conversely, the global properties of the map, defined over $\mathbb{T}$, sometimes shed light on relevant local properties.

In a multichannel feedback system, each channel being corrupted by a phase margin term, the uncertainty is the product of n_l copies of the unit circle—that is, the n_l torus, $\mathbb{T}^{n_l}$.

We leave it to the reader as an exercise to capture the difference among Nyquist maps defined over $\mathbb{R}^{n_q}$, $[0, 2\pi]^{n_q}$, and $\mathbb{T}^{n_q}$ for the multichannel phase margin problem.

We can certainly combine uncertainties of different topological structure. For example, combining a real parameter q together with a phase margin term yields the topological product,

$$[\underline{q}, \overline{q}] \times \mathbb{T}$$

that is, a *cylinder* of uncertainty.

In the SISO gain margin problem, the uncertainty is the half-interval $[0, \infty)$, which is *not compact.* Leaving the domain noncompact raises some questions as to what is "happening" at infinity. To clarify the behavior of the map at infinity, it is convenient to *compactify* the domain. We defer the compactification issue to the next section, where we will do it explicitly on the imaginary axis $(-j\infty, +j\infty)$.

One way to define the multivariable gain margin is to introduce a positive definite Hermitian multiplicative uncertainty matrix somewhere along the feedback path. The collection of all positive definite Hermitian matrices PDH is, topologically, a cone. Remember, a **cone** is a subset of the Euclidean space such that if q is in the cone, then λq is also in the cone $\forall \lambda \geq 0$. This approach to the multivariable gain margin problem is developed in [Bar-on and Jonckheere 1991].

Besides the multichannel phase margin problem, we could also define the multivariable phase margin problem by putting a *unitary* matrix along the open-loop transmission. The collection of all unitary $n_l \times n_l$ matrices is the **unitary group** $U(n_l)$. A fundamental result by [Bar-on and Jonckheere 1990] is that, at every frequency, the smallest phase angle destabilizing perturbation is of the form

$$\begin{pmatrix} SU(2,\mathbb{C}) & 0 \\ 0 & I_{(n_l-2)\times(n_l-2)} \end{pmatrix}$$

In the above, $SU(2,\mathbb{C})$ denotes the **special unitary (Lie) group**—that is, the collection of all 2×2 unitary matrices with det $= 1$. Observe that there exists a surjective mapping

$$\begin{aligned} \pi : \mathbb{T}^3 &\rightarrow SU(2,\mathbb{C}) \\ (\exp(j\theta), \exp(j\phi_1), \exp(j\phi_2)) &\mapsto \begin{pmatrix} \cos\theta \exp(j\phi_1) & \sin\theta \exp(j\phi_2) \\ -\sin\theta \exp(-j\phi_2) & \cos\theta \exp(-j\phi_1) \end{pmatrix} \end{aligned}$$

The above can be viewed as a "parameterization" of $SU(2,\mathbb{C})$ by three phase angles. Observe, however, that the mapping is many-to-one. Indeed, $(\theta, \phi_1, 0)$ and $(2\pi - \theta, \phi_1, \pi)$ are mapped to the same special unitary matrix; furthermore, the points $(\theta, 0, \phi_2)$ and $(\pi - \theta, \pi, \phi_2)$ of the 3-torus are also mapped to the same special unitary matrix. Again, removing these singularities in order to make the mapping one-to-one and continuous (i.e., homeomorphic) requires identification of all points of T^3 mapped to the same point of $SU(2, C)$.

In case of a two-channel feedback with an $SU(2,\mathbb{C})$ perturbation, the nominal fixed-frequency Nyquist map is $f_\omega : SU(2,\mathbb{C}) \rightarrow \mathbb{C}$. Writing it *explicitly* requires choosing charts for the manifold $SU(2,\mathbb{C})$. The problem can be simplified by defining the Nyquist map as $f_\omega \circ \pi : \mathbb{T}^3 \rightarrow \mathbb{C}$. The problem is that the latter is a Nyquist map that is more "many-to-one" than it really has to be.

To further motivate $D = U(n_l)$, let us proceed from the premise that the fundamental multivariable robustness problem is whether a system can sustain multiplicative perturbation of its loop matrix by a nonsingular matrix Δ. The collection of all such matrices is the **general linear group** $GL(n_l, \mathbb{C})$. The nonsingular perturbation matrix has a **polar decomposition**, $\Delta = HU$, where H is positive definite Hermitian and U is unitary. Therefore, in a certain sense, it can be asserted that

$$D = U(n_l) \times \text{PDH}$$

The relevant feature is that the cone of all positive definite Hermitian matrices is *contractible to the identity matrix.* This means that there exists a homotopy $h_t : \text{PDH} \times [0,1] \to \text{PHD}$ such that h_0 is the identity mapping $\text{PDH} \to \text{PDH}$ and h_1 is the constant mapping $\text{PDH} \to I$. Although we shall expand on this point later, the reader might already perceive that the Nyquist map has "homotopy properties" related to, for example, the existence of "holes" in the template. It turns out that contracting the PDH part of the uncertainty to a point does not affect some of the homotopy properties of the map. In other words, it is possible to study the Nyquist map defined over GL, *up to homotopy equivalence*, by restricting its domain to $U(n_l)$. Therefore, the multivariable phase margin problem—that is, plugging a $U(n_l)$ uncertainty in the feedback path—is a way to capture some of the homotopy properties of the Nyquist map under GL-uncertainty.

In general, we will assume that D is *connected.* The opposite case where D can be broken down into many connected components can be interpreted as the potential for catastrophic failures.

2.4 "Punctured" Uncertainty Spaces

Besides phase-margin-related problems, there are other processes through which topologically nontrivial uncertainty spaces naturally crop up. Consider, for example, the counterexample of Section 23.16, showing that the real structured singular value is not always a continuous function of the problem data. The uncertain, real parameter (q_1, q_2) is running in $\mathbb{R}^2$. However, it is easily seen that, for

$$(\hat{q}_1, \hat{q}_2, \hat{\omega}) = (-0.4, 1, \sqrt{a}), \;\; a = 3 + \sqrt{2}$$

the loop transmission is *undefined* in the sense that

$$L\Delta(\hat{p}) = \frac{0}{0}$$

Furthermore, a homotopy argument shows that, whatever value we assign to $L\Delta(\hat{q}_1, \hat{q}_2, \hat{\omega})$, the resulting loop transmission $L\Delta(p)$ will be discontinuous at $\hat{p}$. Under those circumstances the only way to proceed with some sort of mathematical analysis is to remove the point $\hat{q}$ from D, so that the resulting domain of uncertainty is the punctured plane,

$$D = \mathbb{R}^2 \setminus \{\hat{q}\}$$

The problem is that the punctured plane is homotopically quite different from the plane. Consequently, a map defined on the punctured plane could exhibit specific phenomena that could not happen on the full plane.

2.5 Compactification of Imaginary Axis

Having recognized that the set of uncertainties could have a subtle topology, there appears the question as to what the set Ω of all real frequencies really is. If we have an a priori knowledge as to where the crossover is, Ω could be chosen a compact interval around the frequency band of interest. However, if no such a priori knowledge is available, or, more importantly, if an analysis *in the large* is needed, then the natural choice is $\Omega = (-j\infty, +j\infty)$, a *noncompact set*. Under mild roll-off condition, $\lim_{\omega \to \pm\infty} f(\Delta, \omega) = 1$, so that, insofar as $(-j\infty, +j\infty)$ is the domain of the Nyquist map, we could think of identifying $+j\infty$ and $-j\infty$, which means considering Ω to be the circle. Mathematically, this process is called **one-point compactification**:

Theorem 2.3. (Alexandroff) *Any locally compact space X—for example, $(-j\infty, +j\infty)$—can be embedded in a compact space X^+, such that $X^+ \setminus X$ is one single point, $\{\infty\}$, the "point at infinity." This space is defined by $X^+ = X \cup \{\infty\}$, for some $\{\infty\} \notin X$; and the open sets of X^+ are the open sets of X and the sets of the form $X^+ \setminus K$, where K is compact in X. Moreover, X^+ is unique up to homeomorphism.*

Proof. See [Dugundji 1970, Theorem 8.4] or [Bredon 1993, Theorem 11.3]. ■

The one-point compactification of $(-j\infty, +j\infty)$ is well-known to be the unit circle S^1, see [Dugundji 1970],

$$(-j\infty, +j\infty)^+ = (-j\infty, +j\infty) \cup \{\infty\} \cong S^1$$

This unit circle can the taken to be the imaginary section through the Riemann sphere since the Riemann sphere is the one-point compactification of the complex plane,

$$\mathbb{C} \cup \{\infty\} \cong S^2$$

Since the one-point compactification is unique, we can use any recipe to map the line to $S^1 \setminus \{\text{point}\}$. One such recipe is the bilinear transformation:

$$j\omega = \frac{1 - \exp(j\zeta)}{1 + \exp(j\zeta)}$$

One could use the Riemann stereographic projection as well (see Figure 2.3).

Now, we would like to figure out the extent to which the Nyquist map can be extended to $(-j\infty, +j\infty)^+$.

Theorem 2.4. *The continuous map $f : X \to Y$ between locally compact Hausdorff spaces can be extended to a map $f^+ : X^+ \to Y^+$ with $f(\infty_X) = \infty_Y$ iff the map f is* **proper**—*that is, iff $f^{-1}(K)$ is compact (in X) whenever K is compact (in Y).*

Proof. See [Bredon 1993, Theorem 11.4]. ■

Clearly, because of the roll-off condition, the extension of the Nyquist map exists; that is, the following diagram commutes:

$$\begin{array}{ccc} (-j\infty,+j\infty) & \stackrel{f(\Delta,\cdot)}{\rightarrow} & \mathbb{C} \\ i\downarrow & \stackrel{f^+}{\nearrow} & \\ S^1 & & \end{array}$$

Most of the time, we shall not make a distinction between f and f^+.

Unfortunately, after this compactification procedure, Ω inherits a nontrivial topology due to the "closing of the loop." Taking the Cartesian product $D \times \Omega$ of two objects with nontrivial topology makes the resulting topology even more subtle (we are alluding to the Eilenberg–Zilber theorem [Munkres 1984, 59.2] and the Künneth theorem [Munkres 1984, 59.3]).

2.6 Horowitz Supertemplate Approach

2.6.1 No Imaginary-Axis Open-Loop Poles

To check stability for all perturbations with a given structure, we draw the **Horowitz template**,

$$N_\omega = \{\det(I + L(j\omega)\Delta) : \Delta \in D\}$$

Then we draw the Nyquist plot of Horowitz templates, which we call the **supertemplate**,

$$N = \cup_{\omega \in R} N_\omega$$

The problem is to make sure that at no frequency does the template cross over the origin,

$$N_\omega \not\ni 0 + j0, \ \forall \omega,$$

and that the plots of templates circles #(open-loop RHP poles) counterclockwise around $0 + j0$ (assuming #(open-loop RHP poles) does not depend on Δ).

An equivalent way to go about the Horowitz template and the supertemplate is to define the **fixed-frequency Nyquist mapping**

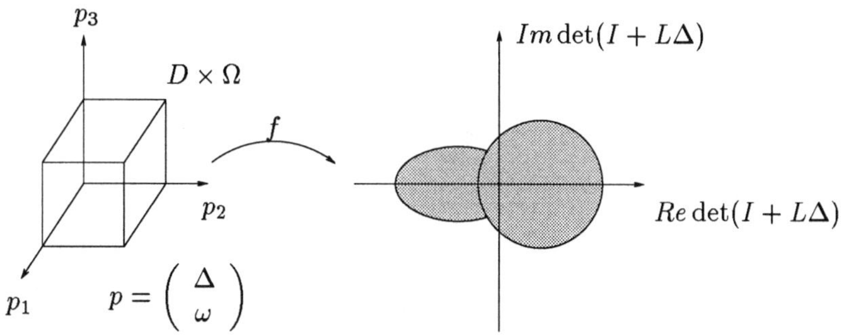

Fig. 2.2. Space of uncertainty, Nyquist mapping, and supertemplate.

$$\begin{array}{rcl} f_\omega : D & \to & \mathbb{C} \\ \Delta & \mapsto & \det(I + L(j\omega)\Delta) \end{array}$$

so that the Horowitz template becomes the image of the map:

$$N_\omega = f_\omega(D)$$

Likewise, if we define the **Nyquist mapping**

$$\begin{array}{rcl} f : D \times \Omega & \to & \mathbb{C} \\ (\Delta, \omega) & \mapsto & \det(I + L(j\omega)\Delta) \end{array}$$

we have

$$N = f(D \times \Omega)$$

and the basic problem is to check whether

$$N \not\ni 0 + j0$$

together with the encirclement condition. In this approach, illustrated in Figure 2.2, the frequency is treated as an extra parameter. This approach is particularly useful for open-loop stable systems, since the encirclement issue is not critical.

2.6.2 Imaginary-Axis Open-Loop Poles

In the important case of an integrator in the open-loop system to achieve zero tracking error, the Nyquist map, as defined in the previous subsection, does not exists. To go about this difficulty, we add the point $\{\infty\}$ to the complex plane and as such we obtain the *one-point compactification of the complex plane*, $\mathbb{C} \cup \{\infty\}$. The Riemann stereographic projection (see Figure 2.3) allows for a representation of $\mathbb{C} \cup \{\infty\}$ as *Riemann sphere* S^2. We define the Nyquist map, as taking value in S^2,

$$\pi_s \circ f : D \times \Omega \to S^2$$

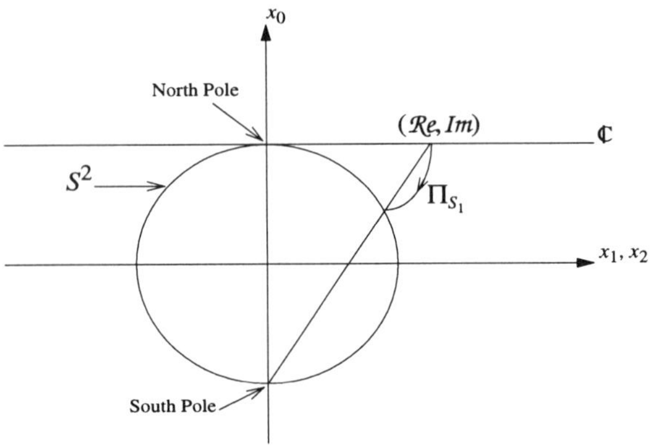

Fig. 2.3. The Riemann stereographic projection of the complex plane, tangent to the sphere at its north pole, onto the sphere devoid of its south pole.

where f is the usual map into the complex plane and π_s is the stereographic projection of the compactified complex plane onto the sphere. In case of an open-loop integrator, $\omega = 0$ is mapped into ∞ on the Riemann sphere and the map is continuous around $\omega = 0$. The Horowitz templates N_ω and the supertemplate N are meant to be drawn on the Riemann sphere.

To do complex analysis on the Riemann sphere, we need to endow it with a **complex-analytic structure**; that is, S^2 must be covered by open sets O_i, each O_i is homeomorphic to $\mathbb{C}$, and the gluing maps are holomorphic with nonvanishing derivatives. To be more specific, let $S^2 = \{x \in \mathbb{R}^3 : x_0^2 + x_1^2 + x_2^2 = 1\}$. Call "north pole" the $x_0 = 1$ point on the sphere and "south pole" the point $x_0 = -1$. It is easily seen that an open covering of the sphere is

$$\begin{aligned} O_1 &= S^2 \setminus \{\text{north pole}\} \\ O_2 &= S^2 \setminus \{\text{south pole}\} \end{aligned}$$

The Riemann stereographic projections are used as homeomorphisms between the punctured spheres and the complex plane:

$$\begin{aligned} \pi_{si} : \mathbb{C} &\to O_i \\ (\Re, \Im) &\mapsto \begin{pmatrix} x_0 &=& \frac{1-\frac{1}{4}(\Re^2+\Im^2)}{1+\frac{1}{4}(\Re^2+\Im^2)}(-1)^i \\ x_1 &=& \frac{\Re}{1+\frac{1}{4}(\Re^2+\Im^2)} \\ x_2 &=& \frac{\Im}{1+\frac{1}{4}(\Re^2+\Im^2)} \end{pmatrix} \end{aligned}$$

Now, let us check the gluing map. Observe that both π_{s1} and π_{s2} induce a homeomorphism $\mathbb{C} \setminus \{0 + j0\} \to O_1 \cap O_2$. Therefore, the gluing map is easily found to be

$$\pi_{s2}^{-1}\pi_{s1} : \mathbb{C} \setminus \{0 + j0\} \;\to\; \mathbb{C} \setminus \{0 + j0\}$$

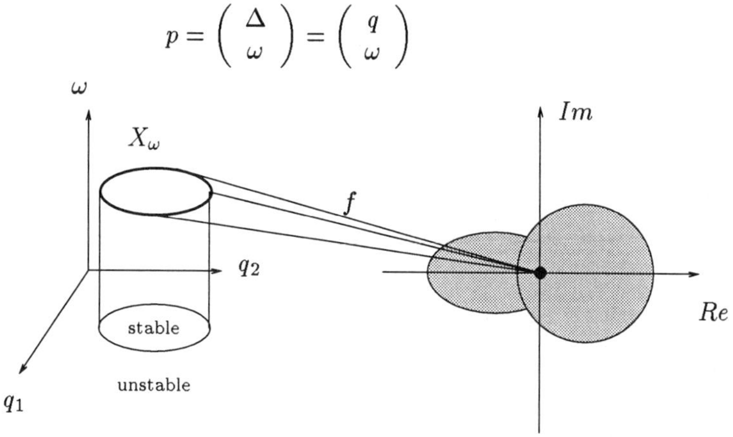

Fig. 2.4. Crossover or neutral stability region.

$$z \quad \mapsto \quad \frac{1}{z}$$

Clearly this map is holomorphic with nonvanishing derivative; hence S^2 has a complex-analytic structure.

2.7 Crossover

Besides the problem of "checking stability for all parameters," there is another relevant problem. Assume D is "too big" and that it cannot be reasonably expected that the system remains stable $\forall \Delta \in D$. In this case, the template covers the origin, namely, $N \ni 0 + j0$. We call $f^{-1}(0 + j0) \subseteq D \times \Omega$ the **crossover** or **neutral stability region**, and we often write it as X_ω. Projecting $f^{-1}(0 + j0)$ parallel to Ω into D decomposes D into several regions, each region having a constant number of closed-loop RHP poles. Clearly the projection of $f^{-1}(0 + j0)$ encompasses the **stability boundary**, X—that is, the subset of X_ω separating the stabilizing region from the destabilizing region. This is illustrated in Figure 2.4.

Closely related to the concept of stability boundary, there is the problem of quantifying *how much* variation of the nominal, stabilizing parameter vector q^0 can be afforded before we reach the stability boundary. Here we need to split the problem in two cases: The first case is when the parameter vector is running in an Euclidean structure; the second case is when q is running on a manifold.

If the parameter space is Euclidean, we define the **robustness margin** to be the infinity distance between q^0 and X,

$$r_{\max} \quad = \quad \inf_{q \in X} \max_i |q_i - q_i^0|$$

$$= \inf_{q \in X} ||q - q^0||_\infty$$

$r_{\max}$ is clearly 1/2 of the length of the edge of the largest cube that can be fitted within the stabilizing region. Clearly, with this concept, $\forall q$ such that $|q_i^0 - q_i| < r, \forall i$, the system remains stable.

In case D is a manifold, we could recover the above case by invoking Whitney's theorem and embedding D in a Euclidean space of dimension $2n_q + 1$.

Another possibility in case of a manifold is to define the margin as $r = \inf_\gamma \int_\gamma \sqrt{ds^2}$, where γ is a geodesic arc joining the nominal parameter to X and ds^2 is the Riemann metric. With this concept, for any $q \in D$ such that $\int_q^{q^0} \sqrt{ds^2} < r$, the integral being evaluated along a geodesic arc, the system remains stable. For an application of this concept to the phase margin, see [Bar-on and Jonckheere 1992]. The problem with this concept is that it does not mesh very well with the usual multivariable definition of the margin that is meant to be the maximum allowable variation on *each component* of q.

2.8 Mapping into Other Spaces

The above is the *basic* approach that we will take in this book, since it is the most popular in control engineering. However, we shall occasionally depart from this strict line, because there are robust stability criteria other than the Nyquist criterion. Their common feature is that all of them involve mapping the uncertainty into a "performance evaluation" space which could be more complicated than the complex plane or the Riemann sphere, and checking whether the image is in the correct subset of the performance evaluation space. Therefore, we will have to deal with "templates" in the general linear group, "templates" in the space of Fredholm operators, and so on.

BIBLIOGRAPHICAL AND HISTORICAL NOTES

The multivariable Nyquist stability criterion is due to [Callier and Desoer 1982]. The *template* approach, also referred to as *Quantitative Feedback Theory (QFT)*, has roots tracing back to the classic of [Horowitz 1963]; for a more recent exposition, see [Horowitz 1982]. The topological formulation of the multivariable phase margin as a unitary perturbation problem is due to [Bar-on and Jonckheere 1990, 1992]. For the multivariable gain margin, see [Bar-on and Jonckheere 1991]. The general topological formulation of robust stability is in [Jonckheere and Bar-on 1991].

The concept of punctured parameter space has roots tracing back to the early work on parameterization of linear systems [Brockett 1976]; see also [Clark 1976]. The space of SISO transfer functions with a fixed degree is obtained after removal of those parameter values yielding pole/zero cancellation. Subsequently, this line of investigation lead to [Segal 1979], [Lomadze 1990], [Mann and Milgram 1991], [Cohen, Cohen, Mann, and Milgram 1991], [Helmke 1993], [Ravi and Rosenthal 1994], and so on, culminating in the homology groups of the space of transfer matrices with a fixed McMillan degree and its compactification using singular systems.

A more concrete approach to parameterization of linear systems based on balanced coordinates was initiated, in the SISO case, by [Jonckheere and Silverman 1983] and [Jonckheere 1984], and eventually culminated, in the multivariable case, in [Furhmann and Ober 1993].

3

A BASIC TOPOLOGICAL PROBLEM

Brouwer's theorem on invariance of domain is a powerful theorem and it deserves to be better known.

William S. Massey,
Singular Homology Theory,
Springer-Verlag, New York, 1980,
Chapter III, Section 6, page 67.

SUMMARY

This chapter provides just one of the many possible ways to motivate the use of algebraic topological tools in robust control. This avenue of approach is the most intuitive one. It deals with the behavior of the boundary under the Nyquist map. This problem is easily motivated by practical robust stability issues. In this chapter, we present a blending of examples, borrowed from both robust control and traditional mathematics, from which the "boundary problem" emerges. Interestingly enough, while the boundary problem, as perceived by Horowitz in the 1960s, was motivated by very practical computational robust stability issues, the same problem was deemed to have enough mathematical depth to attract the attention of such Giants of Mathematics as Poincaré, Brouwer, and Carathéodory, to name but a few!

3.1 The Boundary Problem

It is argued that an important issue is the interplay between the Nyquist mapping f and the boundary. To illustrate this in a simple setting, assume that D is a two-dimensional space, say a rectangle, from which it follows that $D \times \Omega$ is a 3-D space. On the other hand, the Nyquist template N is a 2-D space. Assume that the closed-loop system is stable for all Δ's and that *the problem* is to compute the stability margin, understood for illustrative purposes to be the distance between $0 + j0$ and the template,

$$d(0 + j0, N) = \inf\{|s| : s \in N\} = \inf\{|f(p)| : p \in D \times \Omega\}$$

It is easily proved that the distance is achieved for s on the "perimeter" of the Nyquist template. The above computation would be much simplified if the inverse image of the "perimeter" of N would be included in the "walls" of $D \times \Omega$. Indeed, in this case, for stability margin calculation, it would suffice to restrict the search to the "walls" of $D \times \Omega$,

$$d(0 + j0, N) = \inf\{|f(p)| : p \in \text{ "walls" of } D \times \Omega\}$$

As another example, consider the problem of checking whether, for an open-loop stable system, $0 + j0 \notin N$. Clearly, if we know the perimeter of N, it is possible to determine whether $0 + j0 \notin N$. Now the question is whether it is possible to generate the perimeter of N by sweeping the walls of $D \times \Omega$. Unless the problem is **equidimensional**—that is, $\dim(D \times \Omega) = \dim(\mathbb{C}) = 2$—the best that one can hope for is that the image of the walls of $D \times \Omega$ *covers* the perimeter of N. From a subset known to cover the perimeter of N, it is possible, under mild conditions, to reconstruct N and hence infer whether $0 + j0 \notin N$.

In mathematical language, the "wall" of $D \times \Omega$ or the "perimeter" of N is called the *boundary*. The concept of the boundary can be approached by appealing to our intuition: It is clear what the boundary of the unit disk is. It is clear what the boundary of the Kharitonov cube of uncertainty is. However, the formal definition of the boundary is topological. The formal definition of the boundary of such a set as the template N requires a topology for the bigger space $\mathbb{C}$ in which N is embedded. Also, the continuity of the Nyquist mapping requires a topology for both $D \times \Omega$ and $\mathbb{C}$. Clearly there are several point set topological issues to be settled before we can proceed further.

3.2 Topology for Boundary and Continuity

Given a set A, a **topology** for A, $\mathcal{T}(A)$, is a collection of subsets of A closed under union and finite intersection. A subset in the collection $\mathcal{T}(A)$ is said to be **open**. A subset is **closed** if it is the complement of an open set. $\emptyset$ and A are the only subsets that are simultaneously open and closed. A set A together with a topology $\mathcal{T}(A)$ is said to be a **topological space**.

Consider the Nyquist template N. It is embedded in the complex plane, equivalent for this matter to the Euclidean plane $\mathbb{R}^2$, equipped with the usual metric topology $\mathcal{T}(\mathbb{R}^2)$ generated by the open disks. As a subset of $\mathbb{R}^2$, N can be given a **(topological) boundary** defined as

$$\partial N = \overline{N} \setminus \text{int}(N)$$

The closure of N, $\overline{N}$, is the intersection of all closed sets of $\mathbb{R}^2$ covering N, while the interior, $\text{int}(N)$, is the union of all open sets contained in N. This formal definition coincides, for reasonably well-behaved templates, with the intuitive notion that one has of the "boundary."

It is important to observe that, to get the correct boundary ∂N, we *cannot* utilize some topology $\mathcal{T}(N)$ restricted to the set N. To make our point, let us topologize N using, for example, the **intersection** or **relative topology**; that is, an open set of $\mathcal{T}(N)$ is defined as the intersection of N with an open set of $\mathcal{T}(\mathbb{R}^2)$. For the $\mathcal{T}(N)$ topology, N is open and closed and as such one has $\partial N = \overline{N} \setminus \text{int}(N) = N \setminus N = \emptyset$. This clearly does not correspond to what we want. This is to say that we need to embed N into a *bigger* space and use the topology of the bigger space to define ∂N.

The same remark applies to the space D of uncertain parameters

$$q_1, \ldots, q_{n_q}$$

Assume D is a subset of $\mathbb{R}^{n_q}$, typically a polyhedron. For the formula $\partial D = \overline{D} \setminus \text{int}(D)$ to provide the expected result, we need to treat D as a subset of $\mathbb{R}^{n_q}$ and use the topology $\mathcal{T}(\mathbb{R}^{n_q})$.

Regarding the topological product $D \times \Omega$, it is a general result (see [Dugundji 1970, page 91]) that

$$\partial(D \times \Omega) = ((\partial D) \times \overline{\Omega}) \cup (\overline{D} \times (\partial \Omega))$$

Some extra difficulties crop up while attempting to define $\partial(D \times \Omega)$ when $D \times \Omega$ is a manifold. We will come back to this issue in Part IV, Section 21.1 and Section 22.1, so that here we will only attempt to get an idea of the difficulty involved. Assume, for example, that $D \times \Omega = (\underline{q}, \overline{q}) \times S^1$. If we use the neighborhood topology, $\mathcal{T}(D \times \Omega)$, of the manifold we would get $\partial(D \times \Omega) = \emptyset$. Again, this is not what we want. If we invoke Whitney's celebrated theorem saying that $D \times \Omega$ can be embedded in the Euclidean space $\mathbb{R}^{2n_p+1}$, and if this Euclidean space comes with the usual metric topology, $\mathcal{T}(\mathbb{R}^{2n_p+1})$, the latter topology would yield $\partial((\underline{q}, \overline{q}) \times S^1) = [\underline{q}, \overline{q}] \times S^1$. Again, this is not what we want. A relatively easy way to define the boundary of such a manifold as an open cylinder is to close the cylinder and make it a *manifold-with-boundary* ([Spivak 1965]). By definition, every point p of an n_p-dimensional manifold-with-boundary has a neighborhood O_p homeomorphic to either the Euclidean space $\mathbb{R}^{n_p}$ or the Euclidean half-space

$$\mathbb{HR}^{n_p} = \{(x_1, x_2, \ldots, x_{n_p}) \in \mathbb{R}^{n_p} : x_{n_p} \geq 0\}$$

Let $h_p : O_p \rightarrow \mathbb{HR}^{n_p}$ be any such local homeomorphism. We then define

$$\partial(D \times \Omega) = \cup_{h_p} h_p^{-1}(\partial \mathbb{HR}^{n_p})$$

where the boundary of the Euclidean half-space is defined for the topology $\mathcal{T}(\mathbb{R}^{n_p})$.

Recall that a map, typically $f : D \times \Omega \rightarrow N$ is **continuous** if for any open set O of $\mathcal{T}(N)$, $f^{-1}(O)$ is open for $\mathcal{T}(D \times \Omega)$.

An issue intertwined with the boundary problem is choosing topologies for N (or $\mathbb{R}^2$) and $D\times\Omega$ such that a reasonably well-behaved Nyquist map $f : D \times \Omega \to N$ (or $\mathbb{R}^2$) is continuous for the chosen topologies. For the template, we could use either the intersection topology $\mathcal{T}(N)$ or the usual metric topology of $\mathcal{T}(\mathbb{R}^2)$; as far as the preimage is concerned, the two options are equivalent; indeed, take $O \in \mathcal{T}(\mathbb{R}^2)$; it follows that $f^{-1}(O) = f^{-1}(O \cap N)$ and clearly $O \cap N \in \mathcal{T}(N)$. Regarding the domain, we are much more restricted; indeed, unless there is some (in general nontrivial!) way to extend f beyond $D\times\Omega$, our only choice is $\mathcal{T}(D\times\Omega)$. Unfortunately, as we already know, the $\mathcal{T}(D \times \Omega)$ topology fails to reveal the boundary. The topology of a space encompassing $D\times\Omega$ is needed to get the boundary; this is one of the features of the concept of Whitney stratified space; see Section 22.1.

The lesson that we have learned is that some caution should be exercised with the *point set* topological concept of the boundary and its behavior under continuous map.

3.3 Mathematical Formulation of Boundary Problem

With the topological apparatus developed in the previous section, the desirable property can be written

$$f^{-1}\partial N \subseteq \partial(D \times \Omega)$$

Equivalently, the inverse image f^{-1} and the boundary commute:

$$f^{-1}\partial N \subseteq \partial f^{-1} N$$

The above is a *conceptualization* of the basic paradigm of commutativity between the inverse image and the boundary. It is a *conceptualization* because the inverse image operation prohibits this relation, as it is, to be of any computational significance. We, therefore, have to seek a reformulation of the basic paradigm that is more amenable to computation. The problem can be tackled as follows: For obvious reasons, it is important to be able to efficiently compute ∂fP. If ∂fP can be gotten by checking the boundary of P, in other words, if

$$f\partial P \supseteq \partial fP$$

then a substantial amount of computation is saved. The problem is that the *conceptual* relationship $f^{-1}\partial N \subseteq \partial f^{-1}(N)$ and the *algorithmic* relation $f\partial P \supseteq \partial fP$ are not equivalent. We, however, have the following result.

Theorem 3.1. *If $f^{-1}\partial N \subseteq \partial f^{-1}N$, then $f\partial P \supseteq \partial fP$.*

Proof. Since $f^{-1}N = P$ we immediately get $f^{-1}\partial N \subseteq \partial P$. It further follows that $ff^{-1}\partial N \subseteq f\partial P$. It is easily seen, and it is actually a general fact, that $ff^{-1}\partial N = \partial N$. Therefore, $\partial fP \subseteq f\partial P$, as claimed. ■

Does the map f occurring in a structured stability margin problem enjoy the desirable "boundary preserving" property? To grasp the pattern of this issue, we look at several examples.

3.4 Example (Continuous Fraction Criterion)

Consider the problem of checking Hurwitzness of the polynomial

$$p(s) = a_n s^n + a_{n-1}s^{n-1} + ... + a_1 s + a_0, \quad (a_n \neq 0)$$

The so-called *continued fraction criterion* relies on the continued fraction expansion of

$$\frac{a_n s^n + a_{n-2}s^{n-2} + ...}{a_{n-1}s^{n-1} + a_{n-3}s^{n-3} + ...}$$

into

$$C(s) = h_1 s + \frac{1}{h_2 s + \frac{1}{h_3 s + \frac{1}{...}}}$$

The criterion for Hurwitzness is given by the following well-known theorem:

Theorem 3.2. *The polynomial is Hurwitz iff* $h_i > 0$.

After normalization $a_n = 1$, the continued fraction criterion reveals another mapping relevant to robust control,

$$(a_{n-1}, a_{n-2}, ..., a_1, a_0) \mapsto (h_1, h_2, ..., h_n)$$

This one has the advantage of being **equidimensional**; that is, the domain and the image have the same dimension. If, in particular, the coefficients run in a cube,

$$(a_{n-1}, ..., a_0) \in \Pi_{i=0}^{n-1}[\underline{a}_i \overline{a}_i]$$

the problem is to make sure that the cube is mapped into the positive Euclidean space $(\mathbb{HR})^n$.

In particular, in the case of a polynomial of degree 3, the mapping is

$$(a_2, a_1, a_0) \mapsto (\frac{1}{a_2}, \frac{a_2}{a_1 - \frac{a_0}{a_2}}, \frac{(a_1 - \frac{a_0}{a_2})}{a_0})$$

Clearly, the continued fraction criterion is equivalent to checking that the mapping

$$(a_2, a_1, a_0) \mapsto (a_2, a_1 a_2 - a_0, a_0)$$

is into the positive Euclidean space. For this latter mapping, it is easily seen that

$$f(\partial \text{cube}) = \partial f(\text{cube}).$$

Therefore to check robust stability it suffices to check it on the boundary of the cube. Furthermore, since the map is multilinear, it actually suffices to check the vertices.

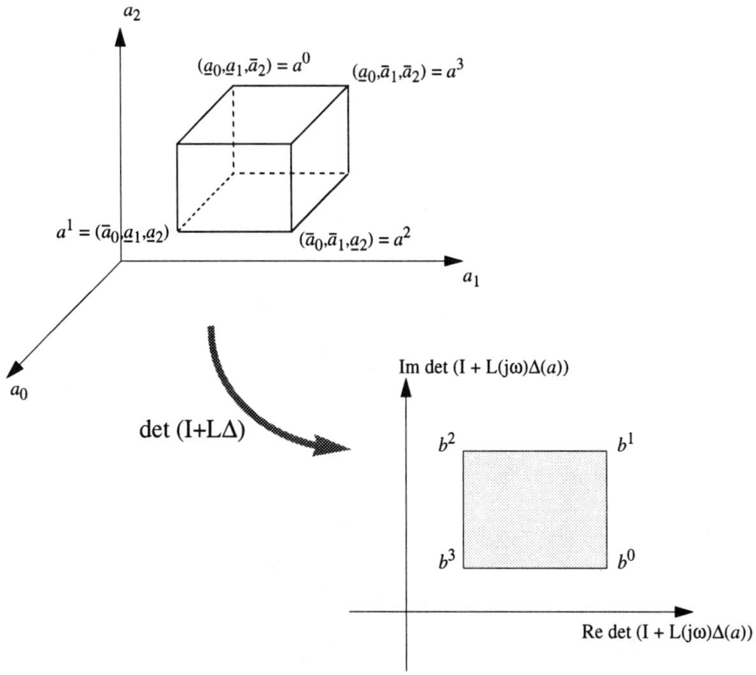

Fig. 3.1. Nyquist mapping formulation of Kharitonov's theorem.

3.5 Example (Kharitonov)

Consider a Kharitonov polynomial

$$s^3 + a_2 s^2 + a_1 s + a_0, \quad a_i \in [\underline{a}_i, \overline{a}_i]$$

of degree three. The extension to higher degree is trivial. To reformulate Kharitonov's theorem in the strict Nyquist context, define

$$L(s) = \frac{1}{s^3} \begin{pmatrix} 1 & s & s^2 \end{pmatrix}$$

$$\Delta = \begin{pmatrix} a_0 \\ a_1 \\ a_2 \end{pmatrix}$$

The uncertainty is constrained to lie in a cube

$$\Delta \in D = \text{Cube} = \prod_i [\underline{a}_i, \overline{a}_i]$$

It follows that

$$(I + L\Delta)^{-1} = \frac{s^3}{s^3 + a_2 s^2 + a_1 s + a_0}$$

Let us freeze the frequency. The fixed-frequency mapping is illustrated in Figure 3.1. The Horowitz template becomes a dilated, +90-degree tilted version of the usual Kharitonov rectangle

$$\det(I + L\Delta)(D) = \\ \{(j\omega)^3 + a_2(j\omega)^2 + a_1(j\omega) + a_0 \colon (a_0, a_1, a_2) \in \text{Cube}\} \times (1/j\omega)^3$$

Define the vertices of the dilated, tilted rectangle:

$$\begin{aligned} b^0 &= ((j\omega)^3 + \overline{a}_2(j\omega)^2 + \underline{a}_1(j\omega) + \underline{a}_0) \times (1/j\omega)^3 \\ b^1 &= ((j\omega)^3 + \underline{a}_2(j\omega)^2 + \underline{a}_1(j\omega) + \overline{a}_0) \times (1/j\omega)^3 \\ b^2 &= ((j\omega)^3 + \underline{a}_2(j\omega)^2 + \overline{a}_1(j\omega) + \overline{a}_0) \times (1/j\omega)^3 \\ b^3 &= ((j\omega)^3 + \overline{a}_2(j\omega)^2 + \overline{a}_1(j\omega) + \underline{a}_0) \times (1/j\omega)^3 \end{aligned}$$

The following inverse images are easily computed:

$$\begin{aligned} (\det(I + L\Delta))^{-1}(\overline{b^0 b^1}) &= \text{Face } (a_1 = \underline{a}_1) \text{ of Cube} \\ (\det(I + L\Delta))^{-1}(\overline{b^0 b^3}) &= \text{Edge } (a_0 = \underline{a}_0, a_2 = \overline{a}_2) \text{ of Cube} \\ (\det(I + L\Delta))^{-1}(\overline{b^3 b^2}) &= \text{Face } (a_1 = \overline{a}_1) \text{ of Cube} \\ (\det(I + L\Delta))^{-1}(\overline{b^2 b^1}) &= \text{Edge } (a_0 = \overline{a}_0, a_2 = \underline{a}_2) \text{ of Cube} \end{aligned}$$

($\overline{b^i b^j}$ denotes the *closed* line segment.) Now, observe that

$$\overline{b^0 b^1} \cup \overline{b^1 b^2} \cup \overline{b^2 b^3} \cup \overline{b^3 b^0} = \partial \text{ Rectangle}$$

$$\text{Faces, Edges of Cube } \in \partial \text{ Cube}$$

$$\text{Cube} = (\det(I + L\Delta))^{-1}(\text{Rectangle})$$

Therefore,

$$(\det(I + L\Delta))^{-1}(\partial \text{ Rectangle}) \subseteq \partial\ ((\det(I + L\Delta))^{-1} \text{ Rectangle})$$

In other words, in the Kharitonov case, the inverse image and the boundary commute. We can *somehow* improve the accuracy of this formula by computing the inverse images of the *vertices*. (Clearly, the vertices are part of the boundary.) The following is easily proved:

$$(\det(I + L\Delta))^{-1}\{\text{vertices of Rectangle}\} \subseteq \{\text{vertices of Cube}\}$$

The deeper interpretation of the above is not completely straightforward and is postponed to Section 13.1.5 and Theorem 22.8. Note, however, that the inverse images of the four vertices of the rectangle are nothing other than the four critical vertices at which stability of the polynomial is to be checked to guarantee stability all over the cube.

With this prerequisite we can easily prove the following theorem.

Theorem 3.3. (Kharitonov) *Assume the four polynomials*

$$\underline{a_0}, \underline{a_1}, \overline{a_3}, \ldots$$

$$\underline{a_0}, \overline{a_1}, \overline{a_3}, \ldots$$

$$\overline{a_0}, \underline{a_1}, \underline{a_3}, \ldots$$

$$\overline{a_0}, \overline{a_1}, \underline{a_3}, \ldots$$

are Hurwitz. Then the whole family is Hurwitz.

Proof. Consider the rectangle $b^0b^1b^2b^3$ as we sweep the frequency. For a frequency large enough, this rectangle is almost shrunk to $1 + j0$ and does not cover $0 + j0$. It remains to show that, as we sweep the frequency, the rectangle never absorbs $0 + j0$. Assume by contradiction that it does. The rectangle could not hit $0 + j0$ at one of its vertices because indeed this would contradict Hurwitzness of one of the four polynomials. Assume now that, as we sweep the frequency, the rectangle absorbs $0 + j0$ through an edge, say b^3b^0. At that frequency, we have $\Im(b^3) = \Im(b^0) = 0$, which yields

$$\omega^2 = \frac{\underline{a_0}}{\overline{a_2}}$$

In this situation we would also have $\Re(b^0) > 0$, and this yields

$$1 + \frac{\underline{a_1}}{-\omega^2} > 0$$

Substituting for the frequency yields

$$\underline{a_1}\,\overline{a_2} - \underline{a_0} < 0$$

This contradicts the continued fraction criterion for Hurwitzness of the polynomial $\underline{a_0}, \underline{a_1}, \overline{a_2}$. In the other conceivable cases of absorption of $0 + j0$ by the rectangle through one of the three remaining edges, we reach a similar kind of contradiction. ∎

3.6 Example (Real Structured Singular Value)

The real structured singular value problem of [deGaston and Safonov 1988] is characterized by

$$\Delta = \begin{pmatrix} q_1 & 0 & 0 \\ 0 & q_2 & 0 \\ 0 & 0 & q_3 \end{pmatrix}, \quad q_i \in [-1, +1]$$

In this case, the domain of uncertainty D is a *cube*. It is further assumed that the loop transfer matrix $L(s)$ is stable. The problem is to compute the largest gain k that can be tolerated, namely,

$$k_{\max} = \inf\{k \in [0, \infty) : \det(I + kL(j\omega)\Delta) = 0, \text{ for some } \Delta, \omega\}$$

With this $k_{\max}$, one can indeed guarantee that the feedback remains stable $\forall k < k_{\max}, \forall \Delta \in D$.

This problem formulation typically arises when one considers a SISO loop function subject to *real* parameter variation—for example,

$$\frac{(s+Z)}{(s+P)} \cdot \frac{K_{10}(1+q_1)}{s(s+P_{20}(1+q_2))(s+P_{30}(1+q_3))}$$

where q_1, q_2, and q_3 represent the relative variations of the gain K_1 and the real poles P_2 and P_3, respectively; K_{10}, P_{20}, and P_{30} denote the nominal values; $\frac{(s+Z)}{(s+P}$ denotes a compensator with fixed pole/zero pair. By a well-known loop shifting procedure, the uncertainty can be pulled out and put in the above diagonal perturbation form.

One way to organize the computation of $k_{\max}$ is to do it selectively at every frequency and then do a frequency sweep:

1. Select ω.
2. Select k.
3. Compute fixed-frequency template,
$$N_\omega = \det(I + kL(j\omega)\Delta(D))$$
4. If $0 + j0 \in N_\omega$, go back to 2 and reduce k; if $0 + j0 \notin N_\omega$, go back to 2 and increase k.
5. After cycling through 2,3,4, get
$$k_{\max}(\omega) = \sup\{k \in [0, \infty) : \det(I + kL(j\omega)\Delta(D)) \not\ni 0 + j0, \forall\Delta\}$$
6. Go to 1 with the next frequency sample.
7. When frequency sweep is terminated, compute
$$k_{\max} = \inf_\omega k_{\max}(\omega)$$

This whole algorithm entails a substantial amount of computation, so that it is imperative to look for any possible ways to save on computation. In particular, the construction of the fixed-frequency template, and up to a certain extent the cycling step 4, can be simplified. The salient fact is the multilinear property of the fixed-frequency mapping:

$$\begin{array}{rcl} f_\omega : D & \to & \mathbb{C} \\ (q_1, q_2, q_3) & \mapsto & \det(I + kL(j\omega)\begin{pmatrix} q_1 & 0 & 0 \\ 0 & q_2 & 0 \\ 0 & 0 & q_3 \end{pmatrix}) \end{array}$$

Indeed, by using the well-known result that the determinant of the sum of two matrices is equal to the sum of the determinants of all matrices obtained from some columns of the first matrix together with the complementary columns of the other, we derive

$$\begin{aligned} &\det(I + kL(j\omega)\Delta) \\ &= 1 + kq_1L_{11} + kq_2L_{22} + kq_3L_{33} \\ &\quad + k^2q_2q_3 \det\begin{pmatrix} L_{22} & L_{23} \\ L_{32} & L_{33} \end{pmatrix} + k^2q_1q_3 \det\begin{pmatrix} L_{11} & L_{13} \\ L_{31} & L_{33} \end{pmatrix} \end{aligned}$$

$$+k^2 q_1 q_2 \det \begin{pmatrix} L_{11} & L_{12} \\ L_{21} & L_{22} \end{pmatrix}$$
$$+q_1 q_2 q_3 k^3 \det(L)$$

As a trivial corollary, it follows that *any line segment drawn in the cube, parallel to a coordinate axis, is mapped into a line segment in the template.* From this simple fact, we deduce the following:

Theorem 3.4. (Multilinear Mapping Theorem) *Let a^i be the vertices of the cube D. Let $b^i = f_\omega(a^i)$ be the images. Then*

$$N_\omega \subseteq conv\ (\{b^i\})$$

Proof. This theorem is proved by recursion on the dimension of the cube. First of all, the theorem is obvious if D has dimension 1. For the inductive step, assume the theorem is proved in dimension $(n-1)$. Let us prove that it holds in dimension n. Take a point $q \in D$. If q is in a lower-dimensional face, the theorem is proved by inductive hypothesis. If q is within the interior of the cube, draw a line, parallel to a coordinate axis. This line intersects two opposite faces at points, say A and B. By the fundamental multilinear property, it suffices to show that A and B are in the convex hull of the b^i's. But A and B are in $(n-1)$-dimensional cubes and by the inductive hypothesis $f_\omega(A)$ and $f_\omega(B)$ are in the convex hull of the images of the vertices of the $(n-1)$-dimensional cube. Hence the proof. ■

The preceding result can be strengthened. Indeed, it could happen that part of the boundary of the template coincides with the boundary of the convex hull.

Theorem 3.5. *If an edge $a^i a^j$ of the cube has its vertices mapped to two consecutive vertices $b^i = f_\omega(a^i), b^j = f_\omega(a^j)$ of the convex hull $conv\ (\{b^k\})$, then $f_\omega(a^i a^j) = b^i b^j$. Consequently, $f_\omega(a^i a^j) \subseteq \partial N_\omega$.*

Therefore, if b^i, b^j denote the images of two vertices of the cube adjacent in the template, that part of ∂N_ω that connects b^i, b^j is either a curve going inside the convex hull or the straight edge $b^i b^j$. In case it is a curve, its preimage could be either within a face of the cube or could be in the interior of the cube, depending on circumstances. If the boundary is $b^i b^j$, its preimage is an edge of the cube.

The idea in [deGaston and Safonov 1988] is to get the convex hull of $\{f_\omega(a^i)\}$, at fixed-frequency ω, for a sequence of k's converging to $k_{\max}(\omega)$. A first estimate of $k_{\max}$ is obtained by dilating the template, by increasing k, until the boundary of the convex hull hits $0 + j0$, say $b^i b^j \ni 0 + j0$. As observed by [de Gaston and Safonov 1988], the issue is whether $b^i b^j \in \partial N_\omega$. If $b^i b^j \in \partial N_\omega$, the smallest destabilizing perturbation, that is in $f_\omega^{-1}(0+j0)$, is within an edge of the cube, the $k_{\max}$ obtained is exact, and the problem is "easy." If, however, $b^i b^j \notin \partial N_\omega$, the estimate of $k_{\max}$ is conservative, and k has to be further increased until the boundary arc joining b^i to b^j hits $0 + j0$. In this case, the problem is more difficult. For indeed, the smallest destabilizing perturbation is, in the simplest case, within a face of the cube or, in the more difficult case, in the interior of the cube. In the latter case, it is necessary to go inside the cube to locate the smallest destabilizing perturbation, and this can be computationally intensive.

Observe that the procedure of "dissecting the cube" of [deGaston and Safonov 1988] is central in cubical homology theory [Massey 1980, II.7], where it is referred to as "subdivision of singular cubes."

3.7 Example (Brouwer Domain Invariance)

Consider a SISO control system where the uncertainty is a complex pole, say P, laying in a bounded domain D. The typical example is a large space structure with a pole having an uncertain eigenfrequency together with an uncertain damping. The nominal value of the pole is written $P_0(\in D)$. We write the uncertainty as a multiplicative perturbation of the nominal loop transmission,

$$\Delta = \frac{(j\omega - P_0)(j\omega - \overline{P}_0)}{(j\omega - P)(j\omega - \overline{P})}$$

Here, it turns out that, at fixed frequency, the mapping $f_\omega : P \mapsto L(j\omega)\Delta(P)$ is *not* conformal; it is not even holomorphic. However, observe that it can be viewed as a mapping from a subset of the 2-D Euclidean space $\mathbb{R}^2$ into the 2-D Euclidean space. The celebrated *Brouwer theorem on the invariance of the domain* tells us when the boundary is preserved.

Theorem 3.6. (Brouwer Domain Invariance) *Let D be an open subset of $\mathbb{R}^n$ and let $f : D \to f(D) \subseteq \mathbb{R}^n$ be a homeomorphism. Then $f(D)$ is open in $\mathbb{R}^n$.*

Proof. The original proofs of Brouwer, including the algebraic proof based on degree theory, are outlined in [Dieudonné 1989, page 183]. A breakthrough came with [Sperner 1928] where, by clever utilization of what became known as *Sperner's lemma*, the proof is reduced to pure combinatorics. The modern topological proofs rely on the *Jordan–Brouwer separation theorem*, which is more fundamental in the sense that it can be disposed of *algebraically* thereby bypassing the intricacies of point set topology. Four our purposes, we state the separation theorem as follows: The homeomorphic image of S^{n-1} in $\mathbb{R}^n$ breaks $\mathbb{R}^n \setminus h(S^{n-1})$ into the disjoint union of exactly two disjoint, connected, open components and $h(S^{n-1})$ is the common boundary of the two components.

We now show that the separation theorem implies the domain invariance. Given $q \in D$, we have to show that $q_2 = f(q)$ has an open neighborhood in $D_2 = f(D)$. Let $B_q \subset D$ be an open ball with its center at q. Observe the following:

$$\begin{aligned}\mathbb{R}^n \setminus f(\partial B_q) &= (\mathbb{R}^n \setminus f(\overline{B_q})) \cup (f(\overline{B_q}) \setminus f(\partial B_q)) \\ &= (\mathbb{R}^n \setminus f(\overline{B_q})) \cup f(\overline{B_q} \setminus \partial B_q)\end{aligned}$$

The above reveals that $\mathbb{R}^n \setminus f(\partial B_q)$, with $\partial B_q \cong S^{n-1}$, is the disjoint union of nonempty, connected subsets of $\mathbb{R}^n$. $f(\partial B_q)$ is said to *separate* $\mathbb{R}^n$. By the Jordan–Brouwer theorem, both sets in the disjoint union are open; in particular, $f(\overline{B_q} \setminus \partial B) = f(B_q)$ is open and provides the required open neighborhood of $f(q)$. For more details, see [Massey 1980, Chapter III, Corollary 6.7], [Dugundji 1970, Chapter XVII, Corollary 3.2], or [Spanier 1989, pages 198–199]. ■

The next issue is whether boundary points are mapped to boundary points. Since the boundary is not part of the domain, there are some questions as to how the image of a boundary point should be defined.

Theorem 3.7. *Let $f : D \to D_2$ be a homeomorphic mapping between bounded, connected, open subsets D, D_2 of $\mathbb{R}^n$, where $D_2 = f(D)$. If a sequence $\{q^k\}$ converges to ∂D, the image sequence $\{f(q^k)\}$ converges to ∂D_2.*

Proof. Take a compact set $K \subset D_2$. By continuity of f and boundedness of D, $f^{-1}(K)$ is compact. Since $q^k \to \partial D$, it follows that the sequence $\{q^k\}$ has a tail outside $f^{-1}(K)$. Since the map is homeomorphic, it follows that $\{f(q^k)\}$ has a tail outside K. Repeating this argument for an arbitrary compact set $K \subset D_2$, it follows that $f(q^k) \to \partial D_2$. ■

The above theorem can be rewritten

$$f^+(\partial D) \subseteq \partial f(D)$$

where $f^+(q)$, $q \in \partial D$, is defined as the set of points of ∂D_2 to which converging subsequences of $\{f(q^k)\}$, $q^k \to q$, converge. Since f is homeomorphic, the same argument can be repeated for f^{-1} and it follows that

$$f^{-1}(\partial D_2) \subseteq \partial f^{-1}(D_2)$$

where the superscript $+$ has been omitted for notational convenience.

Observe that the above theorem falls short of asserting that the homeomorphism f between open sets can be extended to a homeomorphism between the closures of the sets! What a mere homeomorphism is missing is the kind of metrical invariance typically enjoyed by conformal maps. However, under some extra conditions, we can dispose of the extension problem without resorting to conformal mapping theory.

Theorem 3.8. *Under the same conditions as the preceding theorem, if $\{f(q^k)\}$ converges to a point of ∂D_2 whenever $\{q^k\}$ converges to a point of ∂D, then the map can be extended to a continuous map $\overline{D} \to \overline{D_2}$. If, in addition, $\{f^{-1}(q_2^k)\}$ converges to a point of ∂D whenever $\{q_2^k\}$ converges* *to a point of ∂D_2, then the map can be extended to a homeomorphism $\overline{D} \to \overline{D_2}$.*

Proof. See [Siegel 1969, page 129]. See also [Pommerenke 1992, Corollary 2.10] for a related result. ■

To get the feeling for the subtleties involved, we invite the reader to look at the map $(0, 2\pi) \to \mathbb{T} \setminus \{1 + j0\}$, $\theta \mapsto \exp(j\theta)$. This map does not extend to a homeomorphism $[0, 2\pi] \to \mathbb{T}$. The reason is that a sequence converging to $1 + j0$, with alternating positive and negative arguments, does not have a preimage sequence converging to a point of $\partial[0, 2\pi]$.

Finally, if the homeomorphism is defined between *closed* sets, then the extension problem does not exists and the situation is much easier.

Theorem 3.9. *Let D be closed in $\mathbb{R}^n$ and let $f : D \to f(D) \subseteq \mathbb{R}^n$ be a homeomorphism. Then interior points of D are mapped to interior points of $f(D)$ and boundary points of D are mapped to boundary points of $f(D)$. Furthermore, the same statements hold for the inverse mapping.*

Proof. By the Brouwer domain invariance, interior points are mapped, homeomorphically, to interior points, and this can be rewritten as

$$f(\text{int}(D)) = \text{int}(f(D))$$

Regarding the boundary behavior of the map, observe that

$$\begin{aligned} f(\partial D) &= f(D \setminus \text{int}(D)) \\ &= f(D) \setminus f(\text{int}(D)) \\ &= f(D) \setminus \text{int}(f(D)) \\ &= \partial f(D) \end{aligned}$$

The same argument holds for f^{-1}. ■

Coming back to the robust stability problem, let the domain D of variation of the complex pole be compact. For the boundary to be preserved in case of an uncertain pole, it suffices to verify that the map $f_\omega : P \mapsto 1 + L(j\omega)\Delta(P)$ is a homeomorphism. It is easily seen that this is the case if $\omega \neq 0$ and $D \subseteq$ (OUHP $\cap$ OLHP). (OUHP and OLHP denote the open upper half-plane and the open left half-plane, respectively.) In this case,

$$\begin{aligned} \partial f_\omega(D) &= f_\omega(\partial D) \\ \partial f_\omega^{-1}(N_\omega) &= f_\omega^{-1}(\partial N_\omega) \end{aligned}$$

3.8 Example (Covering Map)

Fundamental in algebraic topology is the concept of *covering map*:

Definition 3.10. *Let E, B be path connected topological spaces. A map $\pi : E \to B$ is said to be a* **covering map** *iff*

- *π is surjective;*
- *$\forall b \in B$, $\pi^{-1}(b)$ is a discrete (possibly countably infinite) set $F = \{f^1, f^2, ...\}$;*
- *$\forall b \in B$ there exists an open neighborhood O_b of b such that $\pi^{-1}(O_b)$ is homeomorphic to $O_b \times F$; that is, $\pi^{-1}(O_b)$ the disjoint union of at most countably infinitely many homeomorphic copies of O_b.*

E is called **covering space** *or* **total space**, *B is called* **base space**, *π is called* **covering projection**, *and F is called* **fiber**.

The motivation for this definition is not as yet clear, but it is easy to tie it up with the general line of ideas of this chapter:

Theorem 3.11. *Under the conditions and notation of the previous definition,*

$$\pi^{-1}\partial O_b \subseteq \partial\pi^{-1}(O_b)$$

Proof. Clearly, the covering map behaves locally like a homeomorphism. To be specific, let e^i be a preimage of b, and let $O_{e^i} \subseteq \pi^{-1}(O_b)$; clearly, $\pi|O_{e^i} : O_{e^i} \to O_b$ is a homeomorphism. From the Brouwer invariance of the domain, it follows that

$$(\pi|O_{e^i})^{-1}(\partial O_b) \subseteq \partial(\pi|O_{e^i})^{-1}(O_b)$$

Taking the countable union over all preimages of b yields the result. ∎

Given a space B, there are in general many covering spaces E that can be placed "on top" of B and linked to B via a covering projection. For the covering space to be of any interest, it must enjoy some additional properties. In particular, given a space B, there exists a so-called **universal covering space**—that is, a covering space that has the additional property of being simply connected.

The most typical example of a covering map is $\mathbb{R} \to S^1$, $x \mapsto \exp 2\pi jx$. Since $\mathbb{R}$ is simply connected, it is actually a *universal* cover of S^1. Covering maps occur quite naturally in complex analysis: For example, the Riemann surface of $\log z$ is a covering space (it is actually the *universal* covering space) of $\mathbb{C} \setminus \{0 + j0\}$. Besides these simple illustrative examples, covering maps play a crucial role in topology. Roughly speaking, because the universal covering space is simply connected, it is sometimes easier to work in the covering space than in the base space. This is the generalization of the philosophy behind the Riemann surface: To sort out multivalued function issues, it is sometimes convenient to consider a function as being defined on a Riemann surface rather than on the complex plane after removal of the branch points.

Two far-reaching generalizations of the concept of covering map—fiber bundle and fibration (see Chapter 13)—will play a crucial role in the analysis of the Nyquist map.

3.9 Example (Holomorphic Mapping)

Consider a SISO problem with combined gain and phase uncertainty. In other words, Δ is running in a template $D \subset \mathbb{C}$ that has the shape of a conic sector with bounded magnitude, namely,

$$D = \{r \exp(j\zeta) : \underline{\zeta} \leq \zeta \leq \overline{\zeta}, \underline{r} \leq r \leq \overline{r}\}$$

In robust control, it is customary to define the uncertainty space as *closed*, as it is done above. However, for mathematical reasons, we will tacitly assume that D is embedded in a **domain** $\mho \subseteq \mathbb{C}$ —that is, an open and connected subset of $\mathbb{C}$ —and that the relevant map can be extended to $\mho$, which we shall write as

$$D \subset \mho \rightarrow \mathbb{C}$$

For *fixed* ω, the mapping $\Delta \mapsto 1 + L(j\omega)\Delta$ is *complex-analytic* or *holomorphic*. We can slightly generalize and consider a complex-analytic mapping $q \mapsto \Delta(q)$, where q lies in some domain of the complex plane, say $\mho$, in which case the mapping $q \mapsto 1 + L(j\omega)\Delta(q)$ is still holomorphic. Recall the following definition:

Definition 3.12. *The continuously differentiable function $f : \mho \rightarrow \mathbb{C}$ of one complex variable $q = x + jy$ defined over the domain $\mho$ is said to be* **holomorphic** *or* **complex-analytic** *iff it satisfies the Cauchy–Riemann conditions:*

$$\frac{\partial}{\partial \bar{q}} f = \frac{1}{2}\left(\frac{\partial}{\partial x} + j\frac{\partial}{\partial y}\right) f = 0, \quad \forall q \in \mho$$

An often overlooked result is the following:

Theorem 3.13. *A nonconstant holomorphic mapping $f : \mho \rightarrow \mathbb{C}$ defined over a domain $\mho$ is* **open**—*that is, for any open subset $O \subseteq \mho$, $f(O)$ is open.*

Proof. We must show that for any open subset $O \subseteq \mho$, $f(O)$ is open. Take $z_0 \in f(O)$ and let $q_0 \in O$ be a preimage. We must show that z_0 has an open neighborhood in $f(O)$. In other words, we must show that any z close enough to z_0 has a preimage in O. We break the proof into two cases.

Assume first that $f'(q_0) \neq 0$. In this case, it follows from the complex implicit function theorem that the map f is locally homeomorphic and the result follows at once from the Brouwer invariance of the domain.

If $f'(q_0) = 0$, we differentiate the function f until a nonvanishing higher-order derivative is found:

$$f'(q_0) = 0, \;\; f''(q_0) = 0, \ldots, \;\; f^{(k-2)}(q_0) = 0, \;\; f^{(k-1)}(q_0) \neq 0$$

(This is possible, for otherwise the function would be constant.) It follows from the Weierstrass preparation theorem (see Appendix D) that the equation $f(z) = q$ is locally equivalent to

$$(q-q_0)^k + (q-q_0)^{k-1} r_{k-1}(z-z_0) + \ldots + (q-q_0) r_1(z-z_0) + r_0(z-z_0) = 0$$

for some functions r_i, $i = 0, \ldots, k-1$, holomorphic in a neighborhood of 0, and vanishing at 0. By the fundamental theorem of algebra, all z's in a neighborhood of z_0 have preimages in a neighborhood of q_0.

(See [Rudin 1987, Theorem 10.32] for a more elementary proof that does not appeal to the rather involved Weierstrass preparation theorem.) ■

The relevance of the above result to robust stability is the following:

Theorem 3.14. *Let D be a closed template embedded in a domain $\mho$. If the mapping*

$$\begin{aligned} f : \mho &\to \mathbb{C} \\ q &\mapsto 1 + L(j\omega)\Delta(q) \end{aligned}$$

is holomorphic and nonconstant, then the inverse images of boundary points of $N = 1 + L(j\omega)\Delta(D)$ are included within the boundary of D; that is,

$$f^{-1}(\partial N) \subseteq \partial D$$

Proof. Take $q \in \partial N$ and assume by contradiction that, for some $z \in f^{-1}(q)$, $z \notin \partial D$. It follows that $z \in \text{int}(D)$. By the open mapping theorem, $f(\text{int}(D))$ is open. Furthermore, $f(\text{int}(D)) \subset N$, and by definition of the boundary $q \notin f(\text{int}(D))$. Hence we reach the obvious contradiction that $z \in \text{int}(D)$ and $q = f(z) \notin f(\text{int}(D))$. ■

To conclude, in case the uncertainty maps *holomorphically* to the template, the inverse image of the boundary of the Horowitz template is contained within the boundary of the uncertainty. Observe that it cannot be claimed, in general, that the image of the boundary is within the boundary. A simple counterexample is given by the holomorphic map $z \mapsto z^2$ defined over the open disk with center at $-1 + j0$ and with a radius of 2. The boundary point $1 + j0$ of the domain of definition is not mapped to a boundary point of the image.

3.10 Example (Proper Mapping)

Let $\mho_1, \mho_2$ be **domains**—that is, open and connected subsets of $\mathbb{C}^n$. A map $f : \mho_1 \to \mho_2$ is said to be **proper** if for any compact subset K of $\mho_2$, $f^{-1}(K)$ is compact. This kind of map was already encountered in Section 2.5, where it was characterized by existence of a continuous extension to the 1-point compactification of the domain and the target. The difficulty with such a map is that, technically speaking, it is defined over an *open* set $\mho_1$, so that the boundary is not part of the domain, and therefore there are problems at conceptualizing such a thing as the "image of the boundary." Nevertheless, the following theorem states a precise "boundary behavior" result for proper maps:

Theorem 3.15. *The map* $f : \mho_1 \to \mho_2$ *is proper iff for any sequence* $\{z^k\}$ *in* $\mho_1$ *converging to* $\partial\mho_1$ *the image sequence* $\{f(z^k)\}$ *converges to* $\partial\mho_2$.

Proof. The proof follows the same line of ideas as that of Theorem 3.7. For details, see [D'Angelo 1993, page 13, Lemma 1]. ∎

A very deep issue in the theory of several complex variables is whether a proper holomorphic map (or a proper biholomorphic map) extends smoothly (or diffeomorphically) to the boundary (see [D'Angelo 1993, Chapter 7] and Section 3.16).

If the map extends to the boundary, then the extension f^+ maps the boundary to the boundary, namely,

$$f^+(\partial\mho_1) \subseteq \partial\mho_2 = \partial f^+(\mho_1)$$

3.11 Example (Conformal Mapping)

A holomorphic map $f : \mho_1 \to \mathbb{C}$ defined over a domain $\mho_1$ is **conformal** iff it is injective. In other words, a holomorphic map is conformal iff it is univalent. (Recall that a map f is **univalent** iff $f(z) = f(z') \Rightarrow z = z'$.)

By the Weierstrass preparation theorem (see Appendix D), it follows that, if $f'(z_{10}) = 0$ for some $z_{10} \in \mho_1$, the equation $f(z_1) = z_2$ is locally (around z_{10}) equivalent to a polynomial equation in z_1 of degree ≥ 2, so that by the fundamental theorem of algebra the map f could not possibly be one-to-one. It follows that a conformal map $f : \mho_1 \to \mathbb{C}$ has the property that $f'(z_1) \neq 0$, $\forall z_1 \in \mho_1$. This latter property is easily seen to be equivalent to the map preserving the angles. As such, we have recovered the original definition of a conformal map to be an *angle preserving* map.

It should be stressed, however, that a holomorphic map that has the mere property that its derivative is everywhere nonvanishing *need not* be univalent. Actually, it is a traditional problem of the geometric theory of holomorphic functions to determine what extra conditions, if any, are sufficient to guarantee univalence. For example, one such result reads as follows: If $\Re f'(z_1) > 0$ all over a *convex* domain, the map f is univalent over the full convex domain; see [Pommerenke 1991, Proposition 1.10].

From the definition of a conformal map, it is not hard to show that the inverse $g = f^{-1}$ of a conformal map exists and is conformal.

In this section, we will be concerned with a conformal map $f : \mho_1 \to \mathbb{C}$ defined over a simply connected domain $\mho_1$. It follows from the Brouwer domain invariance that $\mho_2 = f(\mho_1)$ is open—that is, $\mho_2$ is a domain, which is easily seen to be simply connected. Therefore, in this section, we will be concerned with conformal maps between two *simply connected* domains, $f : \mho_1 \to \mho_2$, where $\mho_2 = f(\mho_1)$.

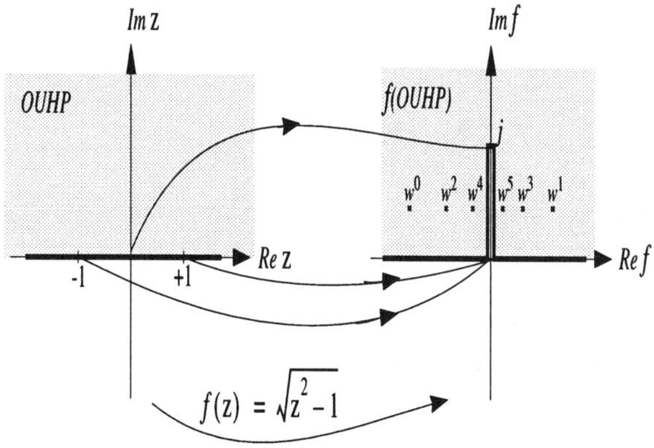

Fig. 3.2. Example of a Schwarz–Christoffel conformal mapping exhibiting some pathological boundary behavior.

As we have already alluded to in Section 3.7, a homeomorphic map between open sets does not always extend, even continuously, to the boundary.

The so-called *boundary behavior* of a conformal mapping was initiated by [Carathéodory 1913]; see also [Ahlfors 1973, Chapter 6, Section 1.2], [Cohn 1967, Chapter 11], and the book by [Pommerenke 1992] specially dedicated to the subject of boundary behavior of conformal maps.

The difficulty with the *general* boundary behavior problem is that, even for such simple domains as $\mathbb{D}$ or OUHP, the boundary of the image, $\partial f(\mathbb{D})$ or $\partial f(\text{OUHP})$, might get quite complicated. The well-known Schwarz–Christoffel conformal mapping provides plenty of examples of pathological boundary behavior. For example, Figure 3.2 shows a situation where $f(\text{OUHP})$ is $\text{OUHP} \setminus [0, j]$. We will come back to the nature of this pathology later.

We will therefore distinguish two cases. The "easy" case is when $\partial f(\mathbb{D})$ is *locally connected.* Such a space as $\partial f(\mathbb{D})$ is said to be **locally connected** whenever for any $\epsilon > 0$ there exists a $\delta > 0$ such that for any two points $z, z' \in \partial f(\mathbb{D})$ such that $|z - z'| < \delta$ there exists a compact connected subset of $\partial f(\mathbb{D})$ of diameter $< \epsilon$ joining z and z'. Recall that the Hahn–Mazurkiewicz theorem asserts that a space is locally connected iff it is a curve. (A **curve** is, by definition, the continuous image of $[0, 1]$. A (closed) **Jordan arc** is, by definition, the homeomorphic image of $[0, 1]$. A **Jordan curve** is, by definition, the homeomorphic image of $\mathbb{T}$. A **Jordan region** is, by definition, the bounded domain bounded by a Jordan curve.) For the details, see [Hocking and Young 1961, Section 3.5]. The "difficult" case of boundary behavior is when $\partial f(\mathbb{D})$ fails to be locally connected. For example, by compounding Schwarz-Christoffel transformations, or by invoking the Riemann mapping theorem, it is possible to find a conformal map:

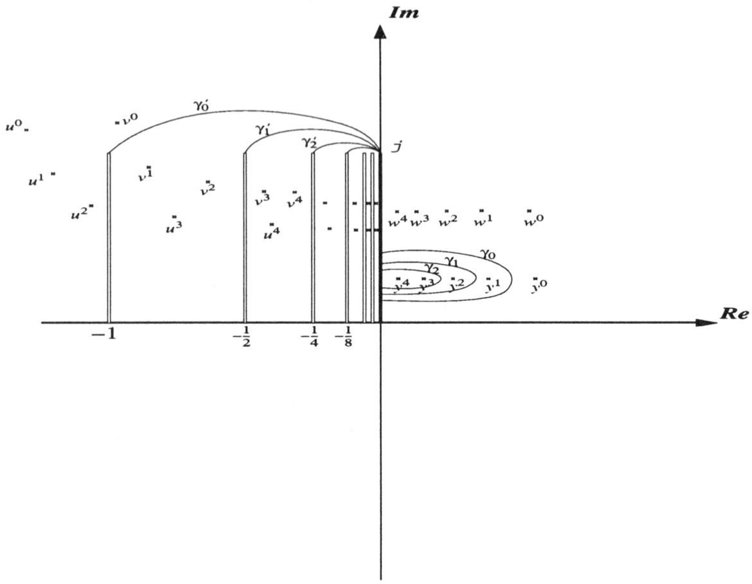

Fig. 3.3. A domain with a boundary that is not locally connected.

$$\text{OUHP} \to \text{OUHP} \setminus (\cup_{n=0}^{\infty}[-\frac{1}{2^n}, -\frac{1}{2^n} + j]) \setminus [0, j] \tag{3.1}$$

the image of which has a boundary that is *not* locally connected. This situation is shown in Figure 3.3. To show that the boundary is not locally connected, take the points $-\frac{1}{2^n} + j$ and $-\frac{1}{2^{n+1}} + j$, both of which are on the boundary; as $n \to \infty$, the distance between the two points decreases to zero; yet, however large n is, any subset of the boundary connecting the two points has a diameter of at least 1. This "difficult" case is dealt with by resorting to the concepts of *fundamental sequences* and *prime ends* introduced by Carathéodory to develop a metric–geometric characterization of $\partial f(\mathbb{D})$.

3.11.1 Locally Connected Boundary

The fundamental result for this "easy" case is the following:

Theorem 3.16. (Carathéodory) *Let $f : \mathbb{D} \to \mathbb{C}$ be a conformal map and let $f(\mathbb{D})$ be bounded. Then,*

- *f extends continuously to $\overline{\mathbb{D}}$ iff $\partial f(\mathbb{D})$ is locally connected.*
- *f extends continuously and injectively to $\overline{\mathbb{D}}$ iff $\partial f(\mathbb{D})$ is a Jordan curve.*

Proof. These results are metric–geometric. Their proofs rely on length distortion under conformal mapping. For the details, see [Pommerenke 1992, Theorems 2.1 and 2.6]. ∎

Back to the robust control problem, if the fixed-frequency mapping

$$\begin{array}{rcl} f_\omega : \mathbb{D} & \to & N_\omega \\ q & \mapsto & \det(I + L(j\omega)\Delta(q)) \end{array}$$

maps the uncertainty $\mathbb{D}$ onto the Jordan region Horowitz template $N_\omega = f_\omega(\mathbb{D})$, the fixed-frequency mapping can be extended to a homeomorphism:

$$\overline{\mathbb{D}}, \partial\mathbb{D} \to \overline{N_\omega}, \partial N_\omega$$

3.11.2 General Boundary

To deal with the "difficult" case where $\partial f(\mho)$ is not locally connected, we need a reformulation of $\partial f(\mho)$ in terms the metric–geometric concepts typical of conformal mapping. In this reformulation, we deal with the concept of *prime ends* rather than that of boundary. A *prime end* is an equivalence class of *fundamental sequences.* A *fundamental sequence* in, say $f(\mho)$, is an infinite sequence of points of $f(\mho)$, finitely many elements of which can be separated from the tail by crosscuts of arbitrarily small length. We now proceed to make these statements precise.

A **crosscut** γ through the domain $\mho$ is an open Jordan arc $(0,1) \to \mho$ such that $\lim_{t\downarrow 0}\gamma(t)$ and $\lim_{t\uparrow 1}\gamma(t)$ both exist and are in $\partial\mho$. It follows that $\mho \setminus \gamma$ can be written as the disjoint union $\mho' \cup \mho''$ of open, connected sets. Two points $z', z'' \in \mho$ are said to be **separated** by γ if they lie in different connected components of $\mho' \cup \mho''$. Finitely many elements of a sequence $\{z^k\}$ are said to be **separated** by γ iff the finite subset and its complement are in different connected components of $\mho' \cup \mho''$.

If the map $f : \mathbb{D} \to f(\mathbb{D})$ is conformal between simply connected domains, it can be proved that the preimage of a crosscut in $f(\mathbb{D})$ is a crosscut in $\mathbb{D}$; however, the image of a crosscut in $\mathbb{D}$ is not always a crosscut in $f(\mathbb{D})$; see [Pommerenke 1992, page 29].

We now proceed to the concept of a fundamental sequence. To make this concept a conformal invariant, we need a metric–geometric formulation of a fundamental sequence as a sequence such that finitely many of its elements can be separated by crosscuts γ, of arbitrarily small "length," through the domain $\mho$. The difficulty is to define what the "length" is, and formulate its conformal invariance. Here, we follow [Ahlfors 1973].

Consider in the domain $\mho$ the Riemann metric

$$ds^2 = \rho^2(x,y)(dx^2 + dy^2), \;\; \rho \geq 0, \;\; (z = x + jy)$$

subject to the condition that

$$\int\int_{\mho} \rho^2(x,y)dxdy < \infty$$

Under the conformal mapping $z = g(z_2)$ the metric ρ is mapped to $\rho_2(z_2) = \rho(g(z_2))|g'(z_2)|$ so that under conformal mapping metrics are mapped to metrics and vice versa. Define a **cluster of crosscuts**, $\cup_i \gamma_i$, to be a finite union of crosscuts forming a connected set. Next, consider a sequence $\{z^k\}$ and consider the family Γ_z of all clusters of crosscuts of $\mho$, each cluster separating finitely many elements of the sequence from the tail. The **length** of the family Γ_z relative to the metric ρ is defined as

$$L(\Gamma_z, \rho) = \inf_{\gamma \in \Gamma_z} \int_\gamma \rho ds$$

(Observe that this length can be defined for *any* family of Jordan arcs in $\mho$.) Next the **area** of the domain is defined as

$$A(\mho, \rho) = \int\int_\mho \rho^2 dx dy$$

The problem is that there are many square integrable metrics $\rho \geq 0$ that can be defined over the domain $\mho$. To remove the dependency on ρ, we define the **extremal length** of the family as

$$\lambda_\mho(\Gamma_z) = \sup_\rho \frac{\inf_{\gamma \in \Gamma}(L(\Gamma_z, \rho))^2}{A(\mho, \rho)}$$

The quantity $\frac{1}{\lambda_\mho(\cdot)}$, called **module**, can be shown to be a subadditive measure on the set of Jordan arcs defined in $\mho$; see [Pommerenke 1992, Section 9.2]. The conformal invariance result is the following:

Theorem 3.17. *$\lambda_\cdot(\cdot)$ is a conformal invariant. To be more specific, under a conformal mapping $\mho \to \mho_2$, an arbitrary family Γ of Jordan arcs in $\mho$ is mapped to a family Γ_2 of Jordan arcs in $\mho_2$ and furthermore,*

$$\lambda_\mho(\Gamma) = \lambda_{\mho_2}(\Gamma_2)$$

Proof. This result is fairly easy to prove from the definition; for details, see [Ahlfors 1973] or [Pommerenke 1992, Proposition 9.1]. ■

Now, we can provide the precise metric–geometric definition of a fundamental sequence:

Definition 3.18. *A* **fundamental sequence** *$\{z^k\}$ is a sequence such that $\lambda_\mho(\Gamma_z) = 0$.*

Corollary 3.1. (Carathéodory) *Fundamental sequences are conformally invariant.*

Proof. It relies on the conformal invariance of the extremal length. ■

The metric–geometric concept of fundamental sequence can be shown to be equivalent to a more intuitive concept:

Theorem 3.19. *A sequence in a bounded domain $\mho$ is fundamental iff any point $z \in \mho$ can be separated from almost all points of the sequence by crosscuts of arbitrarily small diameters.*

Proof. The proof hinges on two steps. First, we prove that whenever the tail of the fundamental sequence can be separated from its complement by a *cluster* of crosscuts, the same can be done with *one single* crosscut. Second, from the definition of the extremal length, it follows that, for a Jordan arc γ in $\mho$,

$$\left(\int_\gamma \sqrt{dx^2 + dy^2}\right)^2 \leq \int\int_\mho dxdy \cdot \lambda_\mho(\gamma)$$

so that, if the extremal length is vanishingly small, so is the Euclidean length. (In case of an unbounded domain, use the spherical length—that is, the length as measured on the Riemann sphere after stereographic projection of $\mathbb{C}$ on the Riemann sphere; the Euclidean and spherical lengths are equivalent in case of a finite domain.) For the details, see [Ahlfors 1973]. ∎

The *metric–geometric* concept of fundamental sequence and the *topological* concept of the boundary can be identified for reasonably well-defined domains:

Corollary 3.2. *When $\mho$ is the open unit disk or any other Jordan region, a sequence is fundamental iff it converges to a point of $\partial\mho$.*

Proof. See [Ahlfors 1973, Section 4-6]. ∎

Intuitively, a fundamental sequence converges to a boundary point; however, two fundamental sequences might converge to the same boundary point. This ambiguity is removed by introducing an equivalence relation. Two fundamental sequences z_k, w_k are **equivalent**, $z \sim w$, iff $\lambda_\mho(\Gamma_{z \cup w}) = 0$. It is not hard to show that this is an equivalence relation; see [Ahlfors 1973]. This leads to the concept of *prime ends*:

Definition 3.20. *A* **prime end** *is an equivalence class of fundamental sequences for the relation $z \sim w \Leftrightarrow \lambda_\mho(\Gamma_{z \cup w}) = 0$.*

Clearly, for $\mathbb{D}$ or any other Jordan region, prime ends can be identified with boundary points. Therefore we have,

Corollary 3.3. *For a conformal map $f : \mho_1 \to \mho_2$ between Jordan regions, the map can be extended to a homeomorphism:*

$$\overline{\mho_1}, \partial\mho_1 \to \overline{\mho_2}, \partial\mho_2$$

Back to the control problem, if the fixed-frequency mapping

$$\begin{aligned} f_\omega : D &\to N_\omega \\ q &\mapsto \det(I + L(j\omega)\Delta(q)) \end{aligned}$$

maps the Jordan region uncertainty D onto the Jordan region Horowitz template N_ω, the map can be extended to a homeomorphism:

$$\overline{D}, \partial D \to \overline{N_\omega}, \partial N_\omega$$

For domains that are not Jordan regions, it should be stressed that the concepts of "prime end" and "boundary point" do not coincide. One such discrepancy is the case where *several* prime ends are associated with *one* given boundary point. For example, there are two prime ends associated with the boundary point $\frac{j}{2}$ of OUHP $\setminus [0, j]$ shown in Figure 3.2; one prime end is represented by the fundamental sequence $\{w^{2k}\}$ in OLHP while the other prime end is represented by the fundamental sequence $\{w^{2k+1}\}$ in ORHP. A much more serious discrepancy occurs with the image of the conformal mapping 3.1. *All* fundamental sequences in OLHP converging to a point of $[0, j]$ are equivalent, so that the whole OLHP side of $[0, j]$ is *one single* prime end. For example, the fundamental sequences $\{u^k\}$, $\{v^k\}$ define a *unique* prime end because both their tails can be separated by crosscuts γ_i' of arbitrary small length. On the ORHP side, however, two fundamental sequences in ORHP converging to points of $[0, j]$ are equivalent iff they converge to the same point, so that every point on the ORHP side of $[0, j]$ is one single prime end. For example, the sequence $\{y^k\}$ that has its tail separated by γ_i defines a *unique* prime end.

For domains that are not Jordan regions, the celebrated prime-end theorem asserts existence of a bijective mapping between *prime ends*, and not between boundary points! We formulate the prime end theorem for the case where the domain is $\mathbb{D}$, because we can be a little more specific as to what the prime end correspondence is.

Theorem 3.21. (Carathéodory) *Let $f : \mathbb{D} \to f(\mathbb{D})$ be a conformal map. Then there exists a bijective mapping*

$$\partial f : \{\text{prime ends of } \mathbb{D}\} = \mathbb{T} \to \{\text{prime ends of } f(\mathbb{D})\}$$

such that, given a sequence $\{\gamma_n\}$ of crosscuts representing the prime end $\partial f(\exp(j\theta))$, in other words,

- $\lambda_{f(\mathbb{D})}(\gamma_n) \to 0$ *as* $n \to \infty$,
- *for a fundamental sequence $\{z^k\}$ representing the prime end $\partial f(\exp(j\theta))$, γ_n separates finitely many elements of the sequence from its tail,*

the sequence $\{f^{-1}(\gamma_n)\}$ is a sequence of crosscuts representing the prime end $\exp(j\theta)$.

Proof. This a reformulation of the statement of [Pommerenke 1992, Theorem 2.15] in terms of the extremal length concept of [Ahlfors 1973]. ■

Observe that the boundary behavior of a conformal map f : OUHP $\to$ f(OUHP) is easily formulated by trivial modification of the above.

Back to the control problem, if the fixed-frequency mapping

$$f_\omega : \mathbb{D} \to N_\omega$$

$$q \quad \mapsto \quad \det(I + L(j\omega)\Delta(q))$$

maps the uncertainty $\mathbb{D}$ *conformally* onto the simply connected Horowitz template N_ω, then the fixed-frequency mapping f_ω induces a bijective correspondence

$$\{\text{prime ends of } \mathbb{D}\} \leftrightarrow \{\text{prime ends of } N_\omega\}$$

It is important to observe that the above theorem falls short of claiming that the function can be continuously extended to a map between the closed domains. Whether the extension exists depends on the nature of the prime end. Let $\{\gamma_n\}$ be defined as in the above theorem and let C_n be the connected component of $f(\mathbb{D}) \setminus \gamma_n$ containing the tail of the fundamental sequence representing the prime end. The prime end $\partial f(\exp(j\theta))$ is **degenerate** if $\cap_n \overline{C_n}$ is a single point. It can be shown that f can be extended to be continuous at $\exp(j\theta)$ iff the prime end $\partial f(\exp(j\theta))$ is degenerate; see [Pommerenke 1992, page 35].

If we look at Figure 3.2, the prime end $\frac{j}{2}$ represented by the fundamental sequence $\{w^{2k}\}$ is degenerate and the map f can be extended continuously to include $\frac{\sqrt{3}}{2} \mapsto \frac{j}{2}$. The same applies to the prime end represented by the fundamental sequence $\{w^{2k+1}\}$.

On the other hand, if we look at the inverse (conformal) mapping $g = f^{-1}$ of Figure 3.2, it is not hard to see that g does not extend continuously to such boundary points as $\frac{j}{2}$. The technicality involved is that $\frac{j}{2}$ is not a simple boundary point of OUHP $\setminus [0, j]$. This means that given a sequence $\{w^k\}$ in OUHP $\setminus [0, j]$, there does not exists a Jordan arc in OUHP $\setminus [0, j]$ that passes through all of the points of the sequence; see [Rudin 1987, Theorem 14.18]. This technical condition that the boundary points be simple is satisfied for such well-behaved domains as the unit disk or any Jordan region; under those circumstances, we are back to the "easy" case of the previous subsection.

3.12 Examples (Horowitz)

Horowitz [Horowitz 1963, Section 4.12] considers a series of structured stability margin problems, proceeding from simple examples where the boundary is preserved to more difficult cases where the map, allegedly, does not preserve the boundary.

3.12.1 Uncertain Gain/Real Pole

The first example [Horowitz 1963, Figure 4.12-2] is characterized by an uncertain real pole P together with an uncertain real gain K. Writing the uncertainty as a multiplicative perturbation of the loop transmission yields

$$\Delta = \frac{K(j\omega - P_0)}{K_0(j\omega - P)}$$

where K_0 and P_0 are the nominal parameter values. To establish the link between our notation and Horowitz's, we split our map as

$$(K, P) \overset{\text{Horowitz } h}{\mapsto} \frac{1}{\Delta} \overset{\text{conformal}}{\mapsto} 1 + L(\frac{1}{\Delta})^{-1}$$

In the above, h is the inverse Nyquist array map of the uncertainty analyzed by Horowitz. In this simple case, Horowitz correctly observes that the boundary of the rectangle D of allowable values of (K, P) is mapped bijectively onto the boundary of $h(D)$, that appears to be a polygon. Since the mapping $\frac{1}{\Delta} \mapsto \Delta$ is conformal ($\Delta \neq 0$), and since both the domain and image boundaries are Jordan curves, it follows that the mapping $(K, P) \mapsto 1 + L(j\omega)\Delta$ preserves the boundary, to be more specific,

$$\begin{aligned} f_\omega \partial D &= \partial f_\omega D \\ f_\omega^{-1} \partial N_\omega &= \partial f_\omega^{-1} N_\omega \end{aligned}$$

As correctly observed by Horowitz, checking stability for all parameters is an easy task.

3.12.2 Uncertain Pole/Zero Pair

Another example [Horowitz 1963, Figure 4.12-6] is the case of an uncertain real pole/zero pair (P, Z) laying in a rectangle D; that is, in our notation,

$$\Delta = \frac{(j\omega - Z)(j\omega - P_0)}{(j\omega - Z_0)(j\omega - P)}$$

[Horowitz 1963, Figure 4.12-6] seems to indicate that part of $h\partial D$ lies in the *interior* of N. Actually, closer examination * reveals that $h\partial D = \partial h D$. However, the stronger condition $h^{-1}\partial N \subseteq \partial h^{-1} N$ may fail. This can be seen as follows: If the rectangle D intersects the bisecting line $Z = P$ in the line segment a^0a^1, it follows that $h(a^0a^1)$ is one single point, say b, and the Horowitz template is easily seen to be "pinched" at that point. Clearly, $b \in \partial N$ and obviously $h^{-1}(b) = a^0a^1 \not\subseteq \partial D$.

Since the mapping $\frac{1}{\Delta} \mapsto \Delta$ is conformal, and since both the domain and image boundaries are Jordan curves, the situation for f remains the same as that for h. To sum up, we do have

$$f_\omega \partial D = \partial f_\omega D$$

However, the stronger condition $f_\omega^{-1} \partial N_\omega \subseteq \partial f_\omega^{-1} N_\omega$ may or may not hold.

3.13 Example (Functions on Polydisks)

Consider the return difference matrix relevant to the complex structured singular value problem:

*This discrepancy was brought to our attention by Mrs. Nanaz Fathpour, who expanded upon this in her "Robust Control" class project.

$$I + kL\Delta = I + kL(j\omega) \begin{pmatrix} q_1 & 0 & \dots & 0 \\ 0 & q_2 & \dots & 0 \\ \vdots & \vdots & \ddots & \vdots \\ 0 & 0 & \dots & q_{n_q} \end{pmatrix}, \quad q_i \in \overline{\mathbb{D}},\ k > 0$$

where the loop matrix $L(s)$ is assumed to be stable. The problem is to compute the largest possible gain $k_{\max}$ such that the closed-loop system remains stable for all $q_i \in \overline{\mathbb{D}}$. To be more specific,

$$k_{\max} = \inf\{k : \det(I + kL(j\omega)\Delta) = 0 \text{ for some } \omega \in \mathbb{R},\ q_i \in \overline{\mathbb{D}}\}$$

Indeed, whenever $k < k_{\max}$, the closed-loop system remains stable $\forall q_i \in \overline{\mathbb{D}}$.

To develop a holomorphic theory of such a function as $\det(I + kL\Delta)$ defined over $\overline{\mathbb{D}}^{n_q}$, we assume that there exists a domain $\mho \supset \overline{\mathbb{D}}^{n_q}$ over which the function is still defined. With this convention, the fixed-frequency mapping

$$\begin{aligned} f : \overline{\mathbb{D}}^{n_q} \subset \mho &\to \mathbb{C} \\ (q_1, q_2, \dots) &\mapsto \det(I + kL\Delta) \end{aligned}$$

is a typical example of a **holomorphic function on a polydisk**; see [Rudin 1969] and [D'Angelo 1993]. By definition, a **polydisk** is the topological product of many copies of the unit disk of the complex plane. A continuously differentiable function $\mho \to \mathbb{C}$ defined on a domain $\mho$ of $\mathbb{C}^n$ is said to be **holomorphic** iff it satisfies the Cauchy–Riemann condition in each variable,

$$\frac{\partial}{\partial \bar{q}_i} f = 0, \quad \forall i$$

It can be shown that a function $\mho \to \mathbb{C}$ is holomorphic iff it is holomorphic in each variable separately.

There is an important boundary concept in functions on polydisks. It can be shown (see Section 5.1 or [Dugundji 1970, page 91]) that

$$\partial \Pi_{i=1}^{n_q} \mathbb{D}_i = \cup_{i=1}^{n_q} \overline{\mathbb{D}_1} \times \dots \times \overline{\mathbb{D}_{i-1}} \times (\partial \mathbb{D}_i) \times \dots \times \overline{\mathbb{D}_{n_q}}$$

Clearly, if we take the Cartesian product of the boundaries of all disks, we obtain just a *subset* of the boundary of the polydisk. This is, by definition, the **distinguished boundary of the polydisk**,

$$\partial_{\bullet} \mathbb{D}^{n_q} = (\partial \mathbb{D})^{n_q} = \mathbb{T}^{n_q}$$

Clearly, this distinguished boundary is an n_q-torus.

An important fact about functions on polydisks is that they attain their maxima *on the distinguished boundary.*

Theorem 3.22. *If* $f : \overline{\mathbb{D}}^{n_q} \subset \mho \to \mathbb{C}$ *is a holomorphic function on a polydisk,*

$$\sup_{q \in \mathbb{D}^{n_q}} |f(q)| = \sup_{q \in \partial_{\bullet} \mathbb{D}^{n_q}} |f(q)|$$

Proof. The proof hinges on Jordan's theorem for many complex variables,

$$f(q) = \left(\frac{1}{2\pi j}\right)^{n_q} \int_{\partial_\bullet \mathbb{D}^{n_q}} \frac{f(z)}{\Pi_i(z_i - q_i)} \Pi_i dz_i$$

Taking the norm of both sides yields the result. For the details, see [Rudin 1969] or [D'Angelo 1993, page 6, Theorem 4]. ■

Restated, this theorem says that the extreme radial point of the template $N = f(\overline{\mathbb{D}}^{n_q})$ can be reached from the distinguished boundary. Actually, more can be said,

Theorem 3.23. *For the holomorphic mapping* $f : \overline{\mathbb{D}}^{n_q} \subset \mho \to \mathbb{C}$ *defined on a polydisk,*

$$f^{-1}\partial N \subseteq \partial_\bullet \mathbb{D}^{n_q}$$

where $N = f(\overline{\mathbb{D}}^{n_q})$.

Proof. Take $z \in \partial N$ and let $(q_1^*, q_2^*, ...)$ be a preimage. Let us freeze all but one q and consider the mapping $q_i \mapsto f(q_1^*, q_2^*, ..., q_{i-1}^*, q_i, q_{i+1}^*, ...)$. Consider the partial template $f(q_1^*, q_2^*, ..., q_{i-1}^*, \mathbb{D}, q_{i+1}^*, ...) \subseteq N$ associated with the variable q_i. For $q_i = q_i^*$, the image reaches the common boundary point between $f(q_1^*, ..., q_{i-1}^*, \mathbb{D}, q_{i+1}^*, ...)$ and N. Therefore, from the one complex variable case of Section 3.9, $q_i^* \in \partial\mathbb{D}$. Repeating the argument for all i's yields the result. ■

3.14 Example (Holomorphic Functions of Several Complex Variables)

The preceding section can be generalized to bounded domains of $\mathbb{C}^n$ other than polydisks. The definition of holomorphy is the same. The only point to watch is the maximum principle, because indeed the maximum is achieved on the boundary, not the distinguished boundary. Here, we quickly review the boundary behavior of functions of several complex variables, and we approach the maximum principle of functions of several complex variables via the related concept of *plurisubharmonic functions.*

The fundamental result is the following:

Theorem 3.24. *Let* $f : D \supset \mho \to \mathbb{C}$ *be a holomorphic function of several complex variables defined over a domain* $\mho$ *that contains the closed subset* D*. Then the mapping is open. Consequently*

$$f^{-1}(\partial N) \subseteq \partial D$$

where $N = f(D)$.

Proof. The proof of the open mapping property is the same as that of the one variable case (again, it relies on the Weierstrass preparation theorem). For another proof, see [Grauert and Remmert 1984, page 109]. The proof that the open mapping property implies the boundary behavior is again the same as that of the one variable case. ■

The attentive reader will see a connection between the open mapping property of holomorphic functions of several complex variables and [Doyle 1982, Lemma1].

Now we can formulate a sweeping generalization of the maximum principle of elementary complex analysis:

Theorem 3.25. *Under the same conditions and notation as the previous theorem,*

$$\sup_{q\in D} |f(q)| = \sup_{q\in\partial D} |f(q)$$

Proof. See [Grauert and Remmert 1984, Chapter 5, Section 5]. ■

We show in the following that the maximum principle is closely related to the boundary behavior.

Theorem 3.26. *If $f : D \subset \mho \to \mathbb{C}$ is holomorphic in $\mho \subset \mathbb{C}^n$ and if the boundary $\partial N = \partial f(D)$ is smooth, then*

$$f^{-1}(\partial N) \subseteq \partial D$$

Proof. Let $z_0 \in \partial N$. It is claimed that for any $q^* \in f^{-1}(z_0)$ we have $q^* \in \partial D$. Shift the template by a quantity a such that $z_0 + a$ is the point on the shifted template $N + a$ the closest to $0 + j0$, to be more precise,

$$\inf_{z\in N+a} |z| = |z_0 + a| > 0$$

(This is accomplished by moving the template so that the normal to $N + a$ at $z_0 + a$ passes through $0 + j0$.) Define the holomorphic function $g = \frac{1}{f+a}$. Clearly, the maximum of this function is $|\frac{1}{z_0+a}|$. By the maximum modulus theorem, $g^{-1}(\frac{1}{z_0+a}) \subseteq \partial D$. But it is easily seen that $g^{-1}(\frac{1}{z_0+a}) = f^{-1}(z_0) \ni q^*$. Therefore, $q^* \in \partial N$. Repeating the argument for all $z_0 \in \partial N$ and all $q^* \in f^{-1}(z_0)$ yields $f^{-1}(\partial N) \subseteq \partial D$. ■

3.15 Example (Plurisubharmonic functions)

Another way to deal with the maximum principle is by making use of the following concept:

Definition 3.27. *A function $F : \mho \to \mathbb{R}$ is said to be* **plurisubharmonic** *iff*

$$\begin{pmatrix} \frac{\partial^2 F}{\partial q_1 \partial \bar{q}_1} & \frac{\partial^2 F}{\partial q_1 \partial \bar{q}_2} & \cdots & \frac{\partial^2 F}{\partial q_1 \partial \bar{q}_n} \\ \frac{\partial^2 F}{\partial q_2 \partial \bar{q}_1} & \frac{\partial^2 F}{\partial q_2 \partial \bar{q}_2} & \cdots & \frac{\partial^2 F}{\partial q_2 \partial \bar{q}_n} \\ \vdots & \vdots & \ddots & \vdots \\ \frac{\partial^2 F}{\partial q_n \partial \bar{q}_1} & \frac{\partial^2 F}{\partial q_n \partial \bar{q}_2} & \cdots & \frac{\partial^2 F}{\partial q_n \partial \bar{q}_n} \end{pmatrix} \geq 0, \quad \forall q \in \mho$$

The reader can easily check that for a function of one complex variable $q = x + jy$,

$$\frac{\partial^2}{\partial q \partial \bar{q}} = \frac{1}{4}(\frac{\partial}{\partial x} - j\frac{\partial}{\partial y})(\frac{\partial}{\partial x} + j\frac{\partial}{\partial y}) = \frac{1}{4}(\frac{\partial^2}{\partial x^2} + \frac{\partial^2}{\partial y^2})$$

so that the concept of plurisubharmonic function reduces to the elementary concept of subharmonic function. The usefulness of the concept is in the following:

Theorem 3.28. *If $f : \mho \to \mathbb{C}$ is holomorphic, then $|f(q)|^2$ is plurisubharmonic.*

Proof. First, in the case where $f(q)$ has real coefficients,

$$\begin{aligned} \left\{ \frac{\partial^2 |f(q)|^2}{\partial q_i \partial \bar{q}_j} \right\} &= \left\{ \frac{\partial^2 f(q) f(\bar{q})}{\partial q_i \partial \bar{q}_j} \right\} \\ &= \left\{ \frac{\partial f(q)}{\partial q_i} \frac{\partial f(\bar{q})}{\partial \bar{q}_j} \right\} \geq 0 \end{aligned}$$

In the case where $f(q)$ has complex coefficients, the proof is the same and omitted. ∎

Theorem 3.29. (Maximum Principle) *If $F : D \subset \mho \to \mathbb{R}$ is a plurisubharmonic function, then*

$$\sup_{q \in D} F(q) = \sup_{q \in \partial D} F(q)$$

Proof. See [D'Angelo 1993]. ∎

3.16 Example (Proper Holomorphic and Biholomorphic Maps)

A holomorphic function $f : \mho_1 \to \mho_2$ between domains of $\mathbb{C}^n$ is said to be **biholomorphic** iff it is bijective with a holomorphic inverse. (The concepts of "conformal" and "biholomorphic" maps coincide for $n = 1$; however, they become quite different for $n > 1$, as exemplified by the failure of the Riemann mapping theorem for $n > 1$.) It is easily seen that biholomorphic functions are proper. As seen in Section 3.10, for a proper map f, the image of a sequence converging to $\partial\mho_1$ converges to $\partial\mho_2$. The question is whether a proper holomorphic or biholomorphic function extends to the boundary.

A most traditional result is the following:

Theorem 3.30. (Poincaré) *There is no biholomorphic mapping $\mathbb{D}^n \to B(\mathbb{C}^n)$ in dimension $n \geq 2$.*

Proof. The main idea of the modern proof, quite different from Poincaré's, is as follows: We proceed from the fact that the unit ball does not have parameterized complex-analytic sets in its boundary (on the other hand, the polydisk does). Assume there is a biholomorphic mapping $f : \mathbb{D} \times \mathbb{D}^{n-1} \to B(\mathbb{C}^n)$. Take a converging sequence in the first factor, $z_{11}^k \to \partial\mathbb{D}$ and define $g(z_{12}) = \lim f(z_{11}^{k_i}, z_{12})$ for some converging subsequence $\{f(z_{11}^{k_i}, z_{12}) : i\}$. By properness, g maps into $\partial B(\mathbb{C}^n)$. Clearly, $g(\mathbb{D}^{n-1})$ is a parameterized analytic set in $\partial B(\mathbb{C}^n)$. A contradiction. For the details, see [D'Angelo 1993, Section 1.2.2]. ■

Other results along the same line require the following concept:

Definition 3.31. *Given a domain* $\mho = \{z \in \mathbb{C}^n : r(z, \overline{z}) < 0\}$ *where* $r(x, y)$ *has a converging power expansion and* $r(z, \overline{z})$ *is real, the* **Levi form** *of the domain is the quadratic form*

$$\begin{pmatrix} \overline{w}_1 & \dots & \overline{w}_n \end{pmatrix} \begin{pmatrix} \frac{\partial^2 r(z,\overline{z})}{\partial z_1 \partial \overline{z}_1} & \cdots & \frac{\partial^2 r(z,\overline{z})}{\partial z_1 \partial \overline{z}_n} \\ \vdots & \ddots & \vdots \\ \frac{\partial^2 r(z,\overline{z})}{\partial z_n \partial \overline{z}_1} & \cdots & \frac{\partial^2 r(z,\overline{z})}{\partial z_n \partial \overline{z}_n} \end{pmatrix} \begin{pmatrix} w_1 \\ \vdots \\ w_n \end{pmatrix}, \quad z \in \mho$$

where the quadratic form is restricted to the subspace of complex codimension 1 *defined by*

$$\begin{pmatrix} \frac{\partial r}{\partial z_1} & \dots & \frac{\partial r}{\partial z_n} \end{pmatrix} \begin{pmatrix} w_1 \\ \vdots \\ w_n \end{pmatrix} = 0$$

The domain $\mho$ *is* **strictly pseudoconvex** *iff its Levi form is* > 0 *all over* $\partial\mho$. *The domain is* **pseudoconvex** *iff the Levi form is* ≥ 0.

(As we will see in Section 21.10, a generalization of the underlying real structure of the Levi form, the so-called *quadratic differential of the Nyquist map*, will play a crucial role.)

Theorem 3.32. (Bell) *Consider a biholomorphic map* $f : \mho_1 \to \mho_2$ *between smoothly bounded, strictly pseudoconvex domains* $\mho_1, \mho_2$ *of* $\mathbb{C}^n$. *Then* f *extends to a smooth diffeomorphism of the boundaries.*

Proof. See [Beals, Fefferman, and Grossman 1983, Theorem 1]. If the domains are just pseudoconvex, then much more algebraic work involving higher-order derivatives of r in the spirit of Appendix D is needed; see [D'Angelo 1993, Chapter 7, Theorem 2 and Corollary 1] for the extension of this result to the case of pseudoconvex domains. ■

3.17 Example (Whitney's Root System)

Here we introduce an entirely new procedure, based on the intuitive idea of mapping the uncertainty to the set of closed-loop poles. By *closed-loop poles*, $w_1, w_2, \ldots, w_m$, we mean the roots of

$$\det(I + L(s, q)) = 0$$

In the lumped parameter case, this amounts to

$$a_m(q)s^m + a_{m-1}(q)s^{m-1} + ... + a_1(q)s + a_0(q) = 0$$

where the coefficients $a_i : D \to \mathbb{C}$ are holomorphic functions defined over the bounded domain $D \subset \mathbb{C}^n$. The problem is that there are some mathematical difficulties at defining such a "mapping" as

$$q \mapsto \begin{pmatrix} w_1(q) \\ w_2(q) \\ \vdots \\ w_m(q) \end{pmatrix}$$

since it would require some way to order the (complex) roots $w_1, w_2, ...$ To make the above a *mapping*, we resort to the concept of *root system* introduced by [Whitney 1972] in this specific context, although the concept was already known in algebraic geometry (see [Griffiths and Harris 1994, page 236]). The idea is to map the uncertainty into the *symmetrized version* of $\mathbb{C}^m$, in which two points, say $(w_1, w_2, w_3, ..., w_m)$, $(w_2, w_1, w_3, ..., w_m)$, are identified if their coordinates differ by a permutation. By definition, the **root system** of a polynomial is its set of roots, considered as a "point" in the symmetrized power of $\mathbb{C}$. With this concept, we can safely define the *mapping* from the uncertainty to the root system.

On top of this, there is the question as to whether the leading coefficient $a_m(q)$ could vanish, in which case one of the roots goes to ∞, and we need to use the Riemann sphere S^2 rather than the complex plane $\mathbb{C}$. Since there are some discrepancies between the complex-analytic structures of the two cases, we treat them separately.

3.17.1 No Degree Uncertainty

If $a_m(q) \neq 0, \forall q \in D$ and if the $a_i(q)$'s are holomorphic in, say, a polydisk, all roots remain bounded and we use $\mathbb{C}$ as basic space in which every single root is running.

The problem as to whether the roots are holomorphic functions is revealed by the following:

Lemma 3.33. *If the coefficients of the polynomial are holomorphic functions of the complex variables $q_1, ..., q_n$, then every single root of the polynomial is a holomorphic function of q as long as it remains simple.*

Proof. This is a corollary of the Weierstrass preparation theorem. Formulate the problem as the implicit equation $F(s,q) \equiv a_m(q)s^m + a_{m-1}s^{m-1} + ... + a_0(q) = 0$. Let $w_1(q)$ be the simple root of interest where q runs in a neighborhood of 0. Assume without loss of generality that $w_1(0) = 0$. It follows that $a_0(0) = 0$, so that $F(s, q = 0) = s(a_m(0)s^{m-1} + ... + a_1(0)) \equiv sg(s)$. If the root $w_1 = 0$ for $q = 0$ is simple, we clearly have $g(0) \neq 0$.

Therefore, by the Weierstrass preparation theorem (see Appendix D), it follows that the implicit equation $F(s,q) = 0$ is equivalent to $s + r_0(q) = 0$ for some holomorphic function r_0. The theorem is proved. ■

This result, compounded by the lack of objective way to order the roots, indicates that if we map the uncertainty in the root *system*, we might get a *holomorphic* mapping with good boundary behavior properties.

To be formal, define over $\mathbb{C}^m$ the equivalence relation $\sim$ as follows:

$$(w_1, w_2, ..., w_m) \sim (v_1, v_2, ..., v_m)$$

iff there exists a permutation

$$\pi : \{1, 2, ..., m\} \to \{1, 2, ..., m\}$$

such that

$$(v_1, v_2, ..., v_m) = (w_{\pi(1)}, w_{\pi(2)}, ..., w_{\pi(m)})$$

The quotient space

$$\mathbb{C}^m / \sim$$

is called **symmetric power of** $\mathbb{C}$. Associated with this is the natural projection

$$\pi_\sim : \mathbb{C}^m \to \mathbb{C}^m / \sim$$

By definition the **root system** is the equivalence class of $(w_1, ..., w_m)$ in $\mathbb{C}^m / \sim$; that is, $\pi_\sim(w_1, ..., w_m)$. With these concepts, we define the *mapping of the uncertainty to the root system:*

$$\begin{array}{rcl} f : D & \to & \mathbb{C}^m / \sim \\ q & \mapsto & \pi_\sim(w_1(q), w_2(q), ..., w_m(q)) \end{array}$$

where $(w_1, ..., w_m)$ are the roots of $\det(I + L(s,q))$, the ordering being irrelevant.

Before we can claim that f is holomorphic, we need to understand what the space $\mathbb{C}^m / \sim$ is and endow it with a complex-analytic structure, from which the holomorphic property of f could be defined.

Without loss of generality, we assume that the polynomial is monic; that is, $a_m(q) \equiv 1$. Consider the roots-to-coefficients mapping:

$$\begin{array}{rcl} \gamma : \mathbb{C}^m & \to & \mathbb{C}^m \\ (w_1, ..., w_m) & \mapsto & \begin{pmatrix} a_{m-1} = -\sum_i w_i \\ a_{m-2} = \sum_{i \neq j} w_i w_j \\ \vdots \\ a_0 = (-1)^m \Pi_i w_i \end{pmatrix} \end{array}$$

Observe that the components of $\gamma(w)$ are the so-called *symmetric functions* of the roots, themselves the coefficients of the monic polynomial that has $w_1, ..., w_m$ as roots. By the fundamental theorem of algebra, this mapping is surjective. Since the coefficients of a monic polynomial define the roots only up to permutation, the above mapping is many-to-one. It is easily seen that there exists a surjective, one-to-one mapping:

$$\gamma_\sim : \mathbb{C}^m / \sim \to \mathbb{C}^m$$

such that the following diagram commutes:

$$\begin{array}{ccc} \mathbb{C}^m & \xrightarrow{\gamma} & \mathbb{C}^m \\ & & \gamma_\sim \\ \downarrow \pi_\sim & \nearrow & \\ \mathbb{C}^m / \sim & & \end{array}$$

Next we want to make $\gamma_\sim$ a homeomorphism, by correctly choosing the topology of $\mathbb{C}^m / \sim$.

The symmetric power of $\mathbb{C}$ is topologized by defining basis sets as

$$\{\pi_\sim(z_1, ..., z_m) : z_i \in O_i\}$$

where the O_i's are arbitrary open sets of $\mathbb{C}$ for its usual topology. An equivalent way is to define, on the symmetrized power, the metric

$$d(\pi_\sim(a), \pi_\sim(b)) = \inf_\pi \sup_i |a_i - b_{\pi(i)}|$$

where π runs over all permutations of the set of indices. For these topologies, the mapping $\gamma_\sim$ is easily seen to be a homeomorphism. In fact, the above is also a complex-analytic structure for the symmetric power of $\mathbb{C}$.

Actually, we can go even further and make $\mathbb{C}^m / \sim$ a complex-analytic manifold; see [Griffiths and Harris 1994, page 236].

Now it is clear how to define the holomorphic property of $f : D \to \mathbb{C}^m / \sim$. Since there exists a homeomorphism $\gamma_\sim : \mathbb{C}^m / \sim \to \mathbb{C}^m$, the obvious way to go is as follows:

Definition 3.34. *$f : D \to \mathbb{C}^m / \sim$ is* **holomorphic** *iff the m components of the map $\gamma_\sim \circ f : D \to \mathbb{C}^m$ are holomorphic in the sense of Section* 3.13.

Now, it is trivial to observe that

$$\begin{array}{rcl} \gamma_\sim \circ f : D & \to & \mathbb{C}^m \\ q & \mapsto & \begin{pmatrix} a_{m-1}(q) \\ \vdots \\ a_0(q) \end{pmatrix} \end{array}$$

This is *clearly* holomorphic. Therefore, we have the following theorem.

Theorem 3.35. *If the mapping from the uncertainty to the coefficients of the closed-loop polynomial is holomorphic, the mapping from q to the root system is holomorphic.*

Proof. (comments): In this particular context of a *polynomial* equation with coefficients holomorphic in q, the proof is trivial. The result remains valid when the w_i's are the s-roots of a transcendental, holomorphic equation, $F(q,s) = 0$, which would typically occur in a distributed parameter problem. The general proof is, however, much more complicated; it hinges on the Weierstrass preparation theorem; see [Whitney 1972, Theorem 7F, page 22]. ■

This theorem has far-reaching consequences for the boundary of the "template" N of closed-loop poles in the symmetric power of $\mathbb{C}$.

Lemma 3.36. *If $f : \mathbb{C}^n \to \mathbb{C}^m$ is a holomorphic function, then $||f||^2 = \sum_i |f_i|^2 : \mathbb{C}^n \to \mathbb{R}$ is plurisubharmonic.*

Proof. Each $|f_i(q)|^2$ is plurisubharmonic and hence so is $\sum_i |f_i(q)|^2$. ■

Theorem 3.37. *Let the uncertainty q run in the closed, bounded subset $D \subset \mathbb{C}^n$ and let the coefficients $a_i : D \subset \mho \to \mathbb{C}$ of the monic closed-loop polynomial be holomorphic functions in several complex variables defined over the domain $\mho$. If $f : D \subset \mho \to \mathbb{C}^m / \sim$ is the mapping of the uncertainty to the root system and if $\partial N = \partial f(D)$ is smooth,*

$$f^{-1}(\partial N) \subseteq \partial D$$

Proof. Consider the mapping f as defined from the parameter space D into the space of symmetric polynomials of the coefficients—that is, $\mathbb{C}^m$. The function $|f|$ is plurisubharmonic and hence satisfies the maximum principle. The proof then follows the same lines as that of Theorem 3.26. ■

3.17.2 Uncertain Degree

If the leading coefficient $a_m(q)$ vanishes somewhere in the domain of uncertainty, at least one root goes to infinity and, in this situation, it is natural to think of the roots as running in the Riemann sphere.

The symmetric power of the Riemann sphere,

$$(S^2)^m / \sim$$

is defined the same way as in the $\mathbb{C}$-case. With this concept, we define the *mapping* from the uncertainty to the root system,

$$f : D \to (S^2)^m / \sim$$

and it is expected that this mapping will be "holomorphic" with good boundary behavior property.

To define the holomorphic property of f, we need to understand the topological structure of the symmetric power of the sphere. As for the $\mathbb{C}$ case, the key is the roots-to-coefficients mapping, except for some extra technicalities that need to be dealt with. If none of the roots $w_1, w_2, ..., w_m$ is ∞, define the mapping as

$$\gamma : (S^2)^m \quad \to \quad \mathbb{C}^{m+1}$$

$$(w_1, \ldots, w_m) \quad \mapsto \quad \begin{pmatrix} 1 \\ -\sum_i w_i \\ \sum_{i \neq j} w_i w_j \\ \vdots \\ (-1)^m \Pi_i w_i \end{pmatrix}$$

Should some of the roots be infinity, then the image column starts with as many zeros as they are vanishing roots; for example,

$$(w_1, w_2, \ldots, w_{m-1}, \infty) \mapsto \begin{pmatrix} 0 \\ 1 \\ -\sum_i w_i \\ \sum_{i \neq j} w_i w_j \\ \vdots \\ (-1)^{m-1} \Pi_i w_i \end{pmatrix}$$

This mapping is certainly not onto. To make it onto, observe that the roots of a polynomial are unaffected by multiplication of all coefficients by a nonvanishing complex number. This suggests to map the roots to the complex projective space $\mathbb{CP}^m$. Remember that $\mathbb{CP}^m$ is the collection of all analytic lines like $\{\lambda(a_m, \ldots, a_0) : \lambda \in \mathbb{C}\}$ in $\mathbb{C}^{m+1}$; such an analytic line is usually written in homogeneous coordinates as $[a_m, \ldots, a_0]$, where the a's are not all vanishing. Next, the coefficients determine the roots, up to permutation, so that to make the mapping one-to-one it is necessary to define it over the symmetrized power of the Riemann sphere. To sum up, the γ mapping induces a one-to-one mapping:

$$\gamma_\sim : (S^2)^m / \sim \xrightarrow{\cong} \mathbb{CP}^m$$

The next step is to choose topologies for both the domain and the image so that $\gamma_\sim$ becomes a homeomorphism.

The symmetric power of the Riemann sphere can be topologized the same way we topologized the symmetric power of the complex plane.

The complex projective space can be topologized, actually it can be metrized, as follows: Consider two points written in homogeneous coordinates as $[a_m, \ldots, a_0], [b_m, \ldots, b_0]$. The lines $\{\lambda(a_m, a_{m-1}, \ldots, a_0) : \lambda \in \mathbb{C}\}, \{\lambda(b_m, b_{m-1}, \ldots, b_0) : \lambda \in \mathbb{C}\}$ in $\mathbb{C}^{m+1}$ intersect the sphere $S^{2m+1} \subseteq \mathbb{C}^{m+1}$ along two circles, Γ_a, Γ_b, respectively. Therefore, the distance between the two points is defined as $d([a_m, \ldots, a_0], [b_m, \ldots, b_0]) = \inf\{||x - y|| : x \in \Gamma_a, y \in \Gamma_b\}$, the minimum Euclidean distance between the two circles. Since the complex projective space is a Grassmannian, it can also be metrized using the sine of the angle between the two lines $\{\lambda a : \lambda \in \mathbb{C}\}, \{\lambda b : \lambda \in \mathbb{C}\} \subset \mathbb{C}^{m+1}$.

Relative to these topologies, the mapping $\gamma_\sim : (S^2)^m / \sim \to \mathbb{CP}^m$ is a homeomorphism.

The complex projective space $\mathbb{CP}^m$ can also be endowed with a complex-analytic structure by defining a system of open sets $O_i = \{(a_m, ..., a_i = 1, ..., a_0) : a_k \in \mathbb{C}, k \neq i\} \cong \mathbb{C}^m$. This system of open set covers the complex projective space and it is readily checked that the gluing maps are biholomorphic. It follows that the complex projective space is a compact complex-analytic manifold.

Now, it is obvious how to define the holomorphic property of the map from the uncertainty to the root system:

Definition 3.38. *The map* $f : D \to (S^2)^m/\sim$ *is* **holomorphic** *iff the map* $\gamma_\sim \circ f : D \to \mathbb{CP}^m$ *is holomorphic.*

Again, a trivial exercise shows that

$$\gamma_\sim \circ f : q \mapsto [a_m(q), ..., a_0(q)]$$

This mapping is certainly holomorphic. Therefore, we have the following theorem.

Theorem 3.39. *The mapping* $f : D \to (S^2)^m/\sim$ *from the uncertainty to the root system on the symmetrized power of the Riemann sphere is holomorphic.*

Theorem 3.40. *Consider the map* $f : D \to \mathbb{CP}^n$ *from the coefficients of the closed-loop polynomial to the symmetric functions of the coefficients. Then,*

$$f^{-1}(\partial N) = \partial D$$

Proof. This map is obviously a homeomorphism and the result follows from the Brouwer domain invariance. ■

BIBLIOGRAPHICAL AND HISTORICAL NOTES

The robust stability problem subject to variation of *many* parameters was probably first formulated in [Horowitz 1963], where the essence of the boundary behavior of the Nyquist map is already formulated. The retrieval of [Kharitonov 1979] revived the problem and created a substantial amount of work, among other results the "edge test" milestone of [Bartlett, Hollot, and Lin 1988]. The state-of-the-art of the classical parametric robust stability problem is summarized in [Ackerman 1993], [Barmish 1994], and [Bhattacharyya, Chapellat, and Keel 1995].

It is around the turn of this century that Poincaré formulated the new line of problems in several complex variables that include the boundary behavior and related issues. The boundary behavior of conformal maps in one variable was taken up by Carathéodory. Pursuing this line of investigation, Poincaré, in a paper dealing with Riemann surfaces, makes, without proof, the statement that the one-to-one continuous image of an open n-D set is open in an n-D manifold. Brouwer, in a correspondence with Poincaré, made the point that the statement is correct, but not obvious, and this prompted Brouwer to formulate his celebrated invariance of the domain; see [Alexandroff 1972] for a deep historical insight on this subject.

4

SIMPLICIAL APPROXIMATION

Why should this particular notion, simplicial decomposition of polyhedra, constitute the central point of all topology?

P. Alexandroff,
Elementary Concepts of Topology,
Dover, New York, 1961.

SUMMARY AND PROSPECTS

From the examples of the previous chapter, it should be clear that a crucial issue is whether the Nyquist mapping preserves the boundary. In the simple Kharitonov case, f preserves the boundary. In the overwhelming majority of other cases, it does not. Therefore, what do we do when the map f does not preserve the boundary?

A basic result of algebraic/combinatorial topology states that, given a map $f : D \times \Omega \to N$, which does not necessarily preserve the boundary, and given a tolerance level $\epsilon > 0$, there exists a decomposition of $D \times \Omega$ into tetrahedra, triangles, line segments, and so on, a decomposition of the template into triangles, line segments, and so on, and a *simplicial approximation* $\tilde{f}$—that is, a piecewise-linear, decomposition-preserving map such that

$$\tilde{f}\partial\sigma \supseteq \partial\tilde{f}\sigma$$

for any "tetrahedron" σ of the decomposition of the domain, and furthermore the approximation matches the original mapping up to the preassigned tolerance level,

$$|f(p) - \tilde{f}(p)| < \epsilon, \ \forall p \in D \times \Omega$$

The price we have to pay for the nice boundary-preserving property is that $D \times \Omega$ and N have to be broken up into small simplexes (tetrahedra, triangles, etc.). The guaranteed boundary preserving property is meant to apply to all constituting simplexes. Inside each simplex, we have a Kharitonov-like situation. Somehow, the amount of dissection needed to achieve a prescribed tolerance level ϵ is a measure of the complexity of the problem, as well as a measure of how much it departs from the simple Kharitonov case.

Because the idea of decomposing the uncertainty space into simplexes becomes so highly motivated, we found that this chapter is the right place to introduce the basic algebraic topological concepts that underlie the decomposition of polyhedra into simplexes and the construction of decomposition preserving maps between polyhedra. Among such concepts developed in this chapter are *simplicial complexes*, *chain groups*, *simplicial approximations*, *algebraic boundary*, *chain maps*, and so on. These concepts are the forerunners of *homology theory*—that is, the formalization of such global pathologies as "loops," "twists," "holes," and so on, of a manifold as inferred from the combinatorics of its decomposition; see Chapter 10.

In this chapter, the simplicial approximation theorem is presented more as a *conceptual* than computational result. It will take us quite a lot of extra effort to make the simplicial approximation a computational tool. In Chapter 6, a rather faithful implementation of the simplicial approximation theorem is proposed using emerging tools from computational geometry. In Chapter 9, we develop the so-called *simplicial algorithms*, where, for the sake of its easy implementation, the simplicial approximation theorem is modified almost beyond recognition.

We are not limited to decomposing a topological space into simplexes. We could as well decompose it into *cubes*, as done by [Massey 1980]. We can even depart from rectilinear objects and decompose the domain and the template into *cells*; see Section 21.16. This latter decomposition does not require a dissection as fine as the simplicial one, but at the expense of allowing curvilinear cells that might not have the easiest possible shape.

4.1 Simplexes, Complexes, and Polyhedra

A set of $n+1$ points $a^0, a^1, ..., a^n$ in infinite-dimensional Hilbert space are said to be **affinely independent** iff the vectors $\vec{a^0a^1}, \vec{a^0a^2}, ..., \vec{a^0a^n}$ are linearly independent. It is easily seen that this linear independence does not depend on the particular choice of a^0 as common origin of the vectors.

Given a (not necessarily affinely independent) set of vertices $a^0, a^1, ..., a^n$, their **convex hull**, written $\text{conv}(a^0, a^1, ..., a^n)$, is defined to be the set of points p such that

$$\vec{a^0p} = \sum_{i=1}^{n} p_i \, \vec{a^0a^i},$$

$$\sum_{i=1}^{n} p_i \leq 1, \;\; p_i \geq 0$$

The above linear combination can be rewritten as

$$\vec{op} = (1 - \sum_{i=1}^{n} p_i) \, \vec{oa^0} + \sum_{i=1}^{n} p_i \, \vec{oa^i}$$

where o is the origin of the Hilbert space. If we define the so-called

barycentric coordinates of the point p

$$\lambda_0 = 1 - \sum_{i=1}^{n} p_i$$
$$\lambda_i = p_i, \ i = 1, ..., n$$

it follows that $p \in \text{conv}(a^0, a^1, ..., a^n)$ is equivalent to

$$\vec{op} = \sum_{i=0}^{n} \lambda_i \, \vec{oa^i}$$

$$\sum_{i=0}^{n} \lambda_i = 1, \ \lambda_i \geq 0$$

Clearly, the barycentric coordinates are unique iff the points $a^0, a^1, ..., a^n$ are affinely independent. It is easily seen that the barycentric coordinates are unaffected by a change of the reference point o. Therefore, the above linear combination is rewritten, more intrinsically, as

$$p = \sum_{i=0}^{n} \lambda_i a^i$$

The above allows for an easy interpretation of the barycentric coordinates. Clearly, if a mass λ_i is lumped in the vertex a^i, $i = 0, 1, ..., n$, the *barycenter* or *center of gravity* of the resulting system of masses is precisely p.

The geometrical simplex generated by the set of vertices $a^0, ..., a^n$, $(a^0, ..., a^n)$, is defined as *interior* of $\text{conv}(\{a^i\})$; clearly,

$$p \in (a^0, a^1, ..., a^n)$$

iff

$$p = \sum_{i=0}^{n} \lambda_i a^i$$

$$\sum_{i=0}^{n} \lambda_i = 1; \ \lambda_i > 0, \ i = 0, 1, ..., n$$

If the vertices $a^0, a^1, ..., a^n$ are affinely independent, then the interior of $\text{conv}(a^0, a^1, ..., a^n)$ is called ***n*-simplex** and is written

$$\sigma_n = (a^0, ..., a^n)$$

Clearly, σ_1 is an open line segment, σ_2 is an open triangle, σ_3 is an open tetrahedron, and so on.

A **face** of the simplex $\sigma_n = (a^0, ..., a^n)$ is the simplex generated by a subset of vertices. In particular, the simplex $(a^0, ..., a^{i-1}, a^{i+1}, ..., a^n)$ is called **face opposed to** a^i. Clearly, edges are 1-faces and vertices are

0-faces in a higher-dimensional simplex.

The **diameter** of a simplex is defined as

$$\text{diameter}(\sigma_n) = \sup_{x,y\in\sigma_n} ||x - y||$$

where $||\cdot||$ denotes the Euclidean norm of the underlying Hilbert space. It can be shown that the diameter of a simplex is achieved on one of its edges.

It is sometimes convenient to think the n-simplex as embedded in the the Euclidean space $\mathbb{R}^{n+1}$. Let $e^0, e^1, ..., e^n$ be the usual basis of the Euclidean space $\mathbb{R}^{n+1}$, namely,

$$e^0 = \begin{pmatrix} 1 \\ 0 \\ 0 \\ \vdots \end{pmatrix}, \; e^1 = \begin{pmatrix} 0 \\ 1 \\ 0 \\ \vdots \end{pmatrix}, \; \dots$$

$e^0, e^1, ..., e^n$ can be thought of as vertices on the coordinate axes at a distance of 1 from the origin. Clearly, $e^0, e^1, ..., e^n$ are affinely independent. The simplex they generate, $(e^0, e^1, ..., e^n)$, is called **standard n-simplex** and is written Σ_n. It is embedded in the hyperplane

$$x_0 + x_1 + ... + x_n = 1$$

and the standard n-simplex is the subset of the hyperplane subject to

$$x_i > 0, \; i = 0, 1, ..., n$$

Therefore, the barycentric coordinates of the standard simplex $(e^0, e^1, ...)$ in Euclidean space $\mathbb{R}^{n+1}$ are simply the Euclidean coordinates of the underlying space $\mathbb{R}^{n+1}$.

The standard 2-simplex is shown in Figure 4.1.

Like the triangle in Greek geometry, the simplex is a basic building block with which more complicated spaces are constructed. *"Gluing"* simplexes is a basic procedure to construct more *complex* objects:

Definition 4.1. *A* **(geometric) simplicial complex** K *is a collection of simplexes of varying dimensions, all embedded in a Euclidean space of adequate dimension or in a Hilbert space, such that*

- *The closures of two simplexes either are disjoint or intersect in a common face of each other.*
- *If $\sigma \in K$ and if τ is a face of σ, then $\tau \in K$.*

Definition 4.2. *The* **n-skeleton** *of the simplicial complex K, K^n, is the simplicial complex made up with all simplexes of K of dimension $\leq n$.*

Definition 4.3. *A* **polyhedron** $|K|$ *is the (closed) subset of Euclidean space that underlies the simplexes of a complex K,*

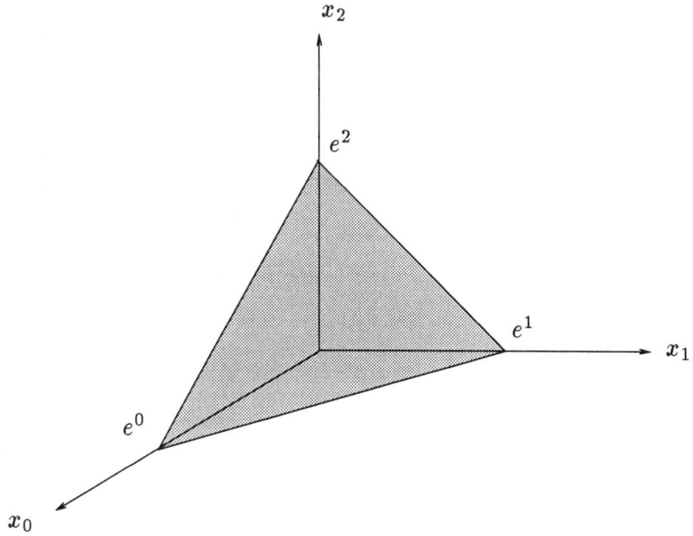

Fig. 4.1. The standard 2-simplex.

$$|K| = \{p : \exists \sigma \in K \textit{ such that } p \in \sigma\}$$

The reader should understand the difference between the *complex* and the *polyhedron*. The complex refers to a *specific* assembly of simplexes, while the polyhedron refers only to the underlying subspace of the Euclidean space. In particular, a given polyhedron may be decomposed into simplexes in many different ways; this yields *many* complexes all of them having the *same* polyhedron.

A simple example of the "gluing" process is to assemble three 1-simplexes and their vertices into a closed chain that provides a combinatorial model of the circle S^1. The resulting simplicial complex is

$$K = \{(a^0, a^1), (a^1, a^2), (a^2, a^0), (a^0), (a^1), (a^2)\}$$

and, to be precise, $|K| \cong S^1$. Likewise, consider four affinely independent vertices and define the complex

$$\begin{aligned} K = \quad & \{(a^0, a^1, a^2), (a^1, a^2, a^3), (a^2, a^3, a^0), (a^3, a^0, a^1), \\ & (a^0, a^1), (a^1, a^2), (a^2, a^3), (a^3, a^0), \\ & (a^0), (a^1), (a^2), (a^3)\} \end{aligned}$$

Clearly, this is the complex of the boundary of a tetrahedron and $|K| \cong S^2$.

Very often, attempting to construct a complicated geometrical object by gluing simplexes requires going to higher-dimensional space (to have enough space to maneuver), with difficulty to visualize. It is sometimes more convenient to do some partial gluing in low dimension and, rather

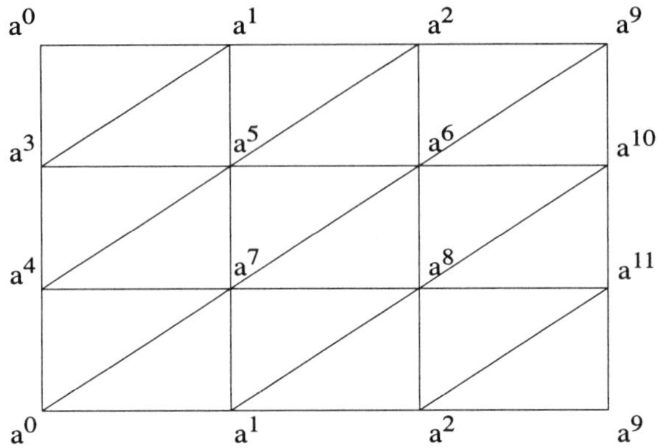

Fig. 4.2. Simplicial cylinder.

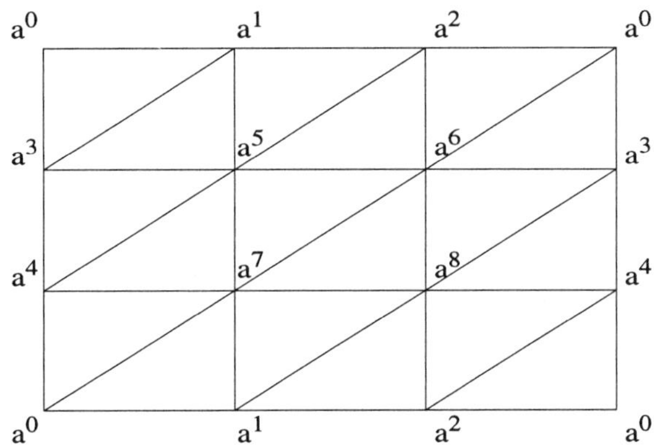

Fig. 4.3. Simplicial 2-torus.

than doing the final gluing in higher dimension, specify vertices to be identified. Consider for example the simplicial cylinder of Figure 4.2; partial gluing has been done in 2-D, resulting in a rectangle, and rather than doing the final gluing of the top and bottom edges of the rectangle in 3-D we rather identify the vertices of the top and the bottom edges. The same procedure is shown in the simplicial torus of Figure 4.3; partial gluing is done up to 2-D, resulting in a rectangle, and rather than gluing opposite faces we identify their vertices.

The technique of identifying vertices is also very useful as a mean of specifying a "twist." Probably the best example is the simplicial definition of the Möbius strip shown in Figure 4.4. From the way the vertices a^0 and a^3 are positioned, it is clear that the strip must be "twisted" once before "gluing" the right and left sides.

Some caution needs to be exercised with the trick of identifying ver-

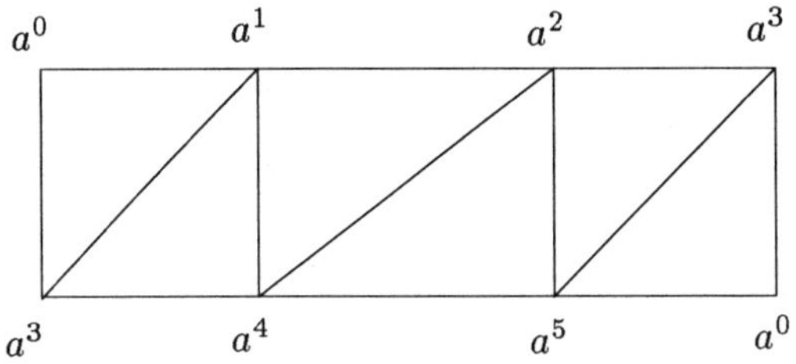

Fig. 4.4. The simplicial Möbius strip.

tices in low dimension. Indeed, identification of vertices could result in a "snowball" effect, and the resulting low-dimensional simplicial complex might collapse to one single point! For example, construct a rectangle by gluing two triangles, identify opposite faces in an attempt to generate a 2-torus, and the resulting complex is one single vertex! To avoid this collapse under identification of vertices, the rectangle has to be decomposed in a *minimum* number of triangles, so that identification of opposite edges does not result in the "snowball" effect; see [Hilton and Wylie 1965, Section 1.11] and [Munkres 1984].

The converse process of decomposing a polyhedron into many simplexes properly glued together is called **triangulation**, **(simplicial) dissection**, or **(simplicial) decomposition** of a polyhedron. This process can be extended to any *compact set* by invoking Alexandroff's theorem.

It will prove convenient at many different places to consider the **facing relation** of a simplicial complex—that is,

$$\tau \preceq \sigma \Leftrightarrow \tau \text{ is a face of } \sigma$$

This is clearly a partial ordering relation defined on the set of simplexes. Recall that a **partial ordering** is a relation that is

- **reflexive**: $\sigma \preceq \sigma$
- **transitive**: $\upsilon \preceq \tau$ and $\tau \preceq \sigma \Rightarrow \upsilon \preceq \sigma$
- **antisymmetric**: $\tau \preceq \sigma$ and $\sigma \preceq \tau \Rightarrow \tau = \sigma$

From a more conceptual point of view, observe that the facing relation in the polyhedron of a simplicial complex can be rewritten

$$\tau \preceq \sigma \Leftrightarrow \tau \subseteq \overline{\sigma}$$

Facing relations like $\tau \subseteq \overline{\sigma}$ are not limited to simplexes of a polyhedron. They exist among more complicated objects—typically, the strata of a stratified space; see Subsection 21.6.4, Subsection 21.6.7, and Definition 22.4.

4.2 Abstract Complexes

It is possible to strip the concepts of simplex and complex from their geometric substratum and define them as algebraic objects. Consider a collection of (abstract) vertices, $a^0, a^1, \ldots$. We define a set K of subsets of $\{a^0, a^1, \ldots\}$ subject to the rule that whenever $\sigma \in K$ and $\tau \subseteq \sigma$, then $\tau \in K$. A subset of vertices in K is called (abstract) simplex. K is called **(abstract) simplicial complex**. It is also called **vertex scheme**, in the sense that it specifies *what* subsets of $\{a^0, a^1, \ldots\}$ constitute an admissible (abstract) simplex. The **dimension** of an abstract simplicial complex K is the maximal cardinality of the subsets of K minus 1.

It can be shown that an abstract simplicial complex of dimension n has a geometric realization in Euclidean space $\mathbb{R}^{2n+1}$; see [Hilton and Wylie 1965]. This result is in the same spirit as Whitney's theorem asserting that a real n-manifold can be embedded in $\mathbb{R}^{2n+1}$.

It is possible to go one level up in the abstraction scale and define an abstract simplicial complex to be a collection of abstract simplexes, partially ordered by a facing relation $\preceq$.

4.3 Alexandroff Theorem

For technical reasons that will become clearer later, existence of a simplicial approximation requires the domain of definition $D \times \Omega$ to be a *polyhedron*. Therefore, the natural question is, To what extent can it be assumed that $D \times \Omega$ is a polyhedron?

The "gluing" process developed in the previous section allows us to find simplicial complexes topologically equivalent, more precisely homeomorphic, to such spaces as toris, Möbius strips, spheres, and so on. The problem is that these homeomorphisms are not easily computable, so that it is desirable, for computational purposes, to have a polyhedral approximation of the compactum, *in the same underlying space.*

As long as *all* uncertain parameters, including frequency, are confined within compact intervals, the uncertainty space is a parallelepiped. As we shall see in Chapter 5, a parallelepiped can be decomposed into simplexes and is hence a polyhedron. A first difficulty appears when we compactify the imaginary axis which results in Ω being the unit circle. The unit circle is *not* a polyhedron. However, it has been known ever since the Greek geometers that the circle can be approximated arbitrarily closely by a regular polygonal arc. Likewise, the unit disk of the complex plane, the latter playing a crucial role in the structured singular values, can be approximated with a regular polygon. *It is with this kind of approximations that Alexandroff's theorem deals with.*

Alexandroff theorem states that any compactum can be arbitrarily closely approximated by a polyhedron. To get the feeling for this theorem, con-

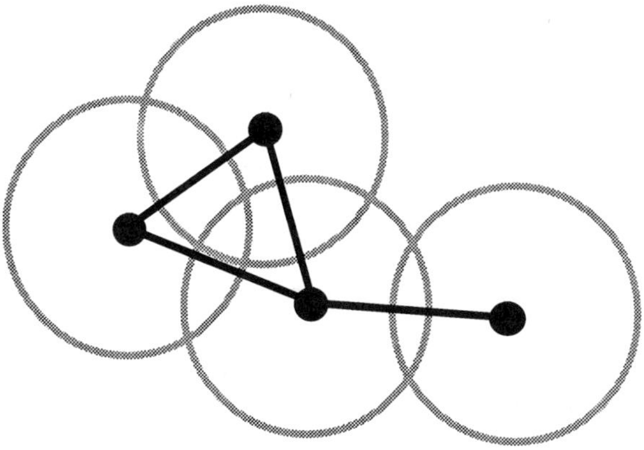

Fig. 4.5. Illustration of the nerve of a covering.

sider a compact $D\times\Omega$. For any $p \in D\times\Omega$, define $B_p(\epsilon)$ to be the open ball with its center at p and with radius ϵ. Clearly, $\cup_{p\in D\times\Omega}B_p(\epsilon)$ is an open covering of $D\times\Omega$. Since $D\times\Omega$ is compact, there exists a finite subcover, $\cup_i B_{a^i}(\epsilon)$, where i runs in a *finite* set. We define an abstract simplicial complex as follows: Its vertices are the $B_{a^i}(\epsilon)$'s, and $\{B_{a^{i_0}}(\epsilon), ..., B_{a^{i_n}}(\epsilon)\}$ is a simplex iff

$$\cap_{j=0}^{n} B_{a^{i_j}}(\epsilon) \neq \emptyset \tag{4.1}$$

This abstract simplicial complex is called **nerve of the covering** $\cup_i B_{a^i}(\epsilon)$. This abstract simplicial complex can be geometrically realized as a polyhedron $\hat{P}$ constructed as follows: The vertices of $\hat{P}$ are the a^i's, the centers of the open balls of the covering, and $(a^{i_0}, a^{i_2}, ..., a^{i_n})$ is a simplex iff Equation 4.1 is satisfied. These concepts are illustrated in Figure 4.5. Clearly, the polyhedron $\hat{P}$ is an approximation of $D\times\Omega$ in the sense that $\forall p \in D\times\Omega$ there exists a point a of the polyhedron such that $||p-a|| < \epsilon$.

Actually, much more can be said:

Theorem 4.4. (Alexandroff) *If $D\times\Omega$ is a compact set, for any $\epsilon_0 > 0$, there exist a polyhedron P and a continuous map $h : D\times\Omega \to P$ such that $\forall p \in D\times\Omega \Rightarrow ||p-h(p)|| < \epsilon_0$.*

Proof. Enclose $D\times\Omega$ together with its ϵ-neighborhood in a compact set C. Cover C with (finitely) many balls of radius $\epsilon/2$. Let $\cup_i B_{a^i}(\epsilon/2)$ be this covering. The geometric realization of the nerve of the latter covering yields a polyhedron $\tilde{P}$. Let $\hat{P}$ be the subpolyhedron consisting of those simplexes that intersect $D\times\Omega$. Clearly, $\hat{P} \supseteq D\times\Omega$ and $\forall\sigma \in \hat{P} \Rightarrow$ diameter$(\sigma) < \epsilon$. Define P to be the subpolyhedron of $\hat{P}$ obtained after removal of those vertices of $\hat{P}$ not in $D\times\Omega$. We now construct the continuous mapping $h : D\times\Omega \to P$. Take $p \in D\times\Omega$ and let $\sigma \ni p$. If σ has all of its vertices in $D\times\Omega$, $h(p) = p$. If σ has vertices outside $D\times\Omega$, write this simplex as $a^0a^1....a^{n'}b^{n'+1}...b^n$, where the vertices $b^{n'+1},...,b^n$ are not in $D \times \Omega$. Write p in barycentric coordinates,

$$p = \sum_{i=0}^{n'} \lambda_i a^i + \sum_{j=n'+1}^{n} \lambda_j b^j$$

Then we define

$$h(p) = \frac{\sum_{i=0}^{n'} \lambda_i a^i}{\sum_{i=0}^{n'} \lambda_i}$$

The motion from p to $h(p)$ is along a radial entirely embedded in a simplex of $\hat{P}$ that has a diameter $< \epsilon$. Hence $||h(p) - p|| < \epsilon$. It is also easily seen that the mapping is continuous. Hence the proof. ■

In the remainder of this chapter, it is tacitly assumed that $D \times \Omega$ is a polyhedron. (We postpone a more "manifold" approach to Chapter 21.)

The same procedure applies to N, except that we take a shorter approach than invoking Alexandroff's theorem. Clearly, there exists a polyhedron $\hat{N}$ that includes N, and we further agree that $f^{-1}(\hat{N} - N) = \emptyset$.

4.4 Simplicial Approximation—Point Set Topology

The requirement that the original map f and its simplicial approximation $\tilde{f}$ be ϵ-close is dealt with by decomposing the image space $\hat{N}$ into "small" simplexes. By "small," we mean that the supremum of the diameter of all simplexes (the **mesh**) should be less than ϵ. Therefore, the tolerance requirement is achieved by asking $f(p)$ and $\tilde{f}(p)$ to lie in the same simplex (also called a **carrier** [Hilton and Wylie 1965, 1.2, 1.7.4]).

The above might appear too strong a requirement to ensure that the tolerance requirement is met. The reward is that whenever both $f(p)$ and $\tilde{f}(p)$ lie within the same carrier the maps $\tilde{f}$ and f are homotopic [Hilton Wylie 1965, 1.7.5].

Besides the simplex, there are several other basic building blocks that can be used to decompose a compact set: cubes, cells, and so on. Here we follow modern computational geometry [Preparata and Shamos 1985], [Edelsbrunner 1987] and choose the *simplex* as the basic geometrical object.

Now, we proceed to the construction of the simplicial approximation. Remember, we have a triangulation of $\hat{N}$ such that for any of its constituting simplexes we have

$$\text{diameter}(b^0b^1...) < \epsilon$$

To capture the neighboring relation among vertices, we define the **star** of a vertex b^j, star(b^j), to be the set of simplexes having b^j as a vertex. Clearly, $\cup_j$star(b^j) is an open covering of N.

To map the tolerance relation in $D \times \Omega$, observe that $\cup_j f^{-1}(\text{star}(b^j))$ is an open covering of $D \times \Omega$. Let δ be the Lebesgue number of this

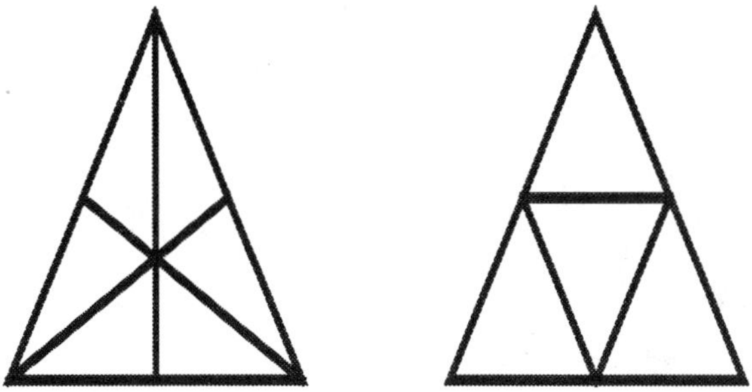

Fig. 4.6. The barycentric and Q-subdivisions of the simplex.

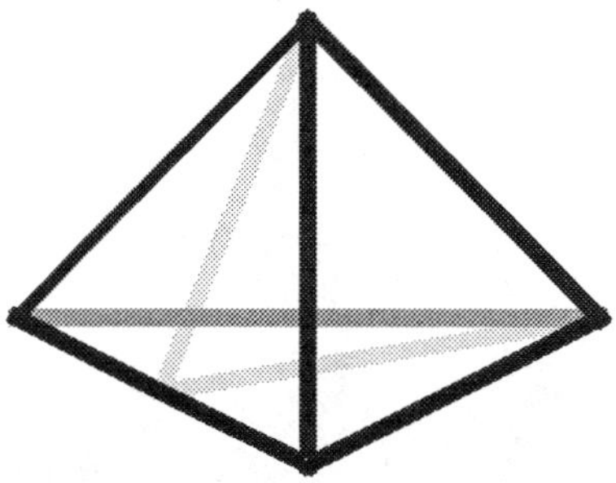

Fig. 4.7. The stellar subdivision of the simplex.

covering [Hilton and Wylie 1965, 1.8]. The *crucial step* is to subdivide the polyhedron $D \times \Omega$ fine enough such that for any simplex $a^0 a^1 ...$ of the triangulation we have

$$\text{diameter}(a^0 a^1 ...) < \delta/2$$

There are many subdivision rules that can be used. The most traditional one is the so-called *barycentric subdivision* shown in Figure 4.6. Here, however, we prefer the so-called Q-subdivision, also shown in Figure 4.6, and formally defined in Section 5.4. It is important to observe that the refinement rule must have the property that iterative application of the refinement rule yields a triangulation of mesh decreasing to 0. Observe that the so-called *stellar subdivision* illustrated in Figure 4.7 does not guarantee a mesh decreasing to zero. Whatever the refinement rule, if the mesh of the triangulation of $D \times \Omega$ is $< \delta/2$, we have

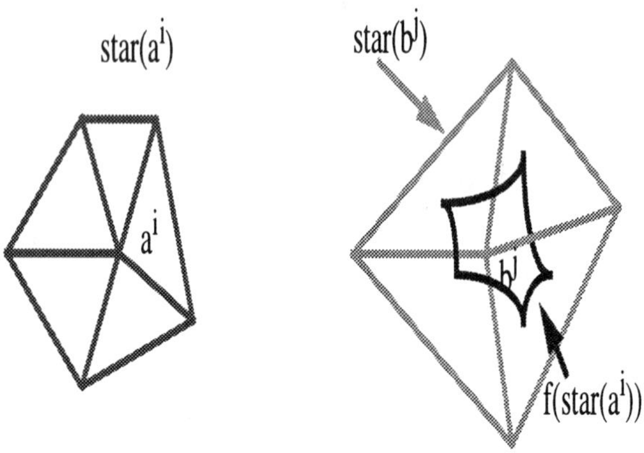

Fig. 4.8. The basic "star condition" in the construction of a simplicial approximation.

$$\text{diameter}(\text{star}(a^i)) < \delta, \ \forall i$$

Therefore, by the Lebesgue covering theorem, for any a^i there exists a b^j such that

$$\text{star}(a^i) \subseteq f^{-1}(\text{star}(b^j))$$

or, equivalently,

$$f(\text{star}(a^i)) \subseteq \text{star}(b^j)$$

There might be many b^j's satisfying the above, but if we retain one of them for each a^i, we define a vertex transformation

$$\varphi_0 : a^i \mapsto b^j$$

This basic "star relation" between triangulations of $D \times \Omega$ and N is illustrated in Figure 4.8.

Next, we prove that the geometry of the stars implies that this vertex transformation maps a *simplex* of $D \times \Omega$ into a *simplex* of $\hat{N}$; see [Hilton and Wylie 1965, 1.8.1]. Such a vertex transformation, mapping any *simplex* into a *simplex*, is said to be a **simplicial map**. The simplicial property is closely related to the targeted "boundary preserving" property.

To prove that φ_0 maps simplexes into simplexes, let $a^0, ..., a^n$ span a simplex of $D \times \Omega$. We have to show that $\varphi_0(a^0), ..., \varphi_0(a^n)$ span a simplex of $\hat{N}$. Since $a^0, ..., a^n$ span a simplex, it is easily seen that $\cap_{i=0}^{n}\text{star}(a^i) \neq \emptyset$. From the star inclusion condition, it follows that $\emptyset \neq \cap f(\text{star}(a^i)) \subseteq \cap\text{star}(\varphi(a^i))$ Therefore, $\cap\text{star}(\varphi(a^i)) \neq \emptyset$. By definition, $\text{star}(\varphi(a^i))$ is the collection of simplexes of the form $\varphi(a^i)...$ together with the permutations of the latter. Therefore, $\cap\text{star}(\varphi(a^i))$ is the collection of simplexes of the form $\varphi(a^0)...\varphi(a^n)....$ Remember, the face of a simplex is a simplex; in particular, if we retain the face $\varphi(a^0)...\varphi(a^n)$, it is a simplex. Therefore, φ maps a simplex into a simplex, as claimed.

With this property, we can define the **piecewise-linear extension** of φ_0. This is done as follows: Take a point p in the simplex $(a^0, a^1, ..., a^n)$ with barycentric coordinates λ_i,

$$p = \lambda_0 a^0 + \lambda_1 a^1 + ... + \lambda_n a^n$$

$$\sum_i \lambda_i = 1$$

$$\lambda_i > 0$$

Therefore, the image of p is defined as

$$\tilde{f}(p) = \sum_i \lambda_i \varphi_0(a^i)$$

Finally, another argument on the geometry of the stars shows that $\tilde{f}(p)$ is in the closure of the carrier of $f(p)$, so that the tolerance requirement is matched. This is proved as follows (see [Spanier 1989, Chapter 3, Section 4]): Take an arbitrary $p \in D \times \Omega$. Since $D \times \Omega$ is decomposed into *disjoint* simplexes, there exists a *unique* simplex σ that contains p. This simplex is called **carrier** of p. Likewise, let τ be the carrier of $f(p)$. We have to show that $\tilde{f}(p) \in \overline{\tau}$. Let a be a vertex of σ. Therefore, $p \in \text{star}(a)$. It follows that $f(p) \in f(\text{star}(a)) \subseteq \text{star}(\tilde{f}(a))$. It follows that $\tilde{f}(a)$ is a vertex of τ, and the latter is true for any vertex a of σ. Therefore, $\tilde{f}(\overline{\sigma}) \subseteq \overline{\tau}$, as claimed.

Therefore, the map $\tilde{f}$ is the announced simplicial approximation. We summarize this most important result in the following theorem:

Theorem 4.5. (Simplicial Approximation) *Given the continuous Nyquist map $f : D \times \Omega \to \hat{N}$ defined over the triangulated, compact polyhedron $D \times \Omega$ and mapping into the triangulated polygon $\hat{N}$ and given an arbitrary tolerance level $\epsilon > 0$, the map f admits a* **simplicial approximation** *$\tilde{f}$; that is,*

- *For some refinements $(D \times \Omega)'$ and N', with mesh$(N') < \epsilon$, the map $\tilde{f} : (D \times \Omega)' \to \hat{N}'$ is simplicial—that is, $\tilde{f}$ maps every simplex of $(D \times \Omega)'$ to a simplex of N'.*
- *$\tilde{f}(p) \in \overline{carrier}\,(f(p)), \forall p \in D \times \Omega$; consequently, f and $\tilde{f}$ are homotopic ([Hilton and Wylie 1965, 1.7]).*

Furthermore, the simplicial approximation matches the Nyquist map with the required tolerance, namely,

- $\|\tilde{f}(p) - f(p)\| < \epsilon, \forall p \in D \times \Omega.$

The basic fact that the map $\tilde{f}$ is *simplicial*—that is, $\tilde{f}$ maps simplexes to simplexes—is enough to guarantee the desirable commutativity properties identified in Chapter 3. We start with the most intuitive commutativity property, which is interesting from the computational standpoint.

Theorem 4.6. *$\tilde{f}(\partial\sigma_{n_p}) \supseteq \partial\tilde{f}(\sigma_{n_p})$, where n_p is the dimension of $D \times \Omega$.*

Proof.

$$\begin{aligned}\tilde{f}(\partial\sigma_{n_p}) &= \tilde{f}(\partial(a^0, a^1, ..., a^{n_p}))\\ &= \tilde{f}(\cup_i \overline{(a^0, ..., a^{i-1}, a^{i+1}, ..., a^{n_p})})\\ &= \cup_i \tilde{f}(\overline{(a^0, ..., a^{i-1}, a^{i+1}, ..., a^{n_p})})\\ &= \cup_i \overline{(\tilde{f}(a^0), ..., \tilde{f}(a^{i-1}), \tilde{f}(a^{i+1}), ..., \tilde{f}(a^n))}\end{aligned}$$

(Observe that the string $(\tilde{f}(a^0), ..., \tilde{f}(a^{i-1}), \tilde{f}(a^{i+1}), ..., \tilde{f}(a^{n_p}))$ has repetition among its vertices; what we mean by this notation is the usual geometrical simplex obtained after retaining only one representative in each set of repeated vertices.)

Now, consider $\partial\tilde{f}(\sigma_{n_p})$. Let us assume that $\tilde{f}(\sigma_{n_p})$ is a 2-simplex and let the (pairwise distinct) vertices of $\tilde{f}(\sigma_{n_p})$ be b^0, b^1, b^2. In general, there are many vertices in $\varphi_0^{-1}(b^j)$. Let us choose for each b^j a representative a^j of its inverse image (this might require relabeling). Therefore, we have

$$\begin{aligned}\partial\tilde{f}(\sigma_{n_p}) &= \partial(b^0, b^1, b^2)\\ &= \overline{(b^1, b^2)} \cup \overline{(b^0, b^2)} \cup \overline{(b^0, b^1)}\\ &= \overline{(\tilde{f}(a^1), \tilde{f}(a^2))} \cup \overline{(\tilde{f}(a^0), \tilde{f}(a^2))} \cup \overline{(\tilde{f}(a^0), \tilde{f}(a^1))}\\ &\subseteq \cup_i \overline{(\tilde{f}(a^0), ..., \tilde{f}(a^{i-1}), \tilde{f}(a^{i+1}), ..., \tilde{f}(a^{n_p})}\end{aligned}$$

Combining the above strings, the result follows for the case where $\tilde{f}(\sigma_{n_p})$ is a 2-simplex. If $\tilde{f}(\sigma_{n_p})$ is a 1-simplex or a 0-simplex, the proof is the same and is omitted. ■

In control terminology, the above means that, in order to get the boundary of the Nyquist template N, it suffices to check the boundary of all constituting simplexes of $D \times \Omega$.

The simplicial property of $\tilde{f}$ also guarantees the *conceptually* interesting commutativity between the inverse image and the boundary:

Theorem 4.7. *Let τ_2 be a 2-simplex of N that is the image of a n_p-simplex σ_{n_p} of $D \times \Omega$. Let $\overline{f} : \overline{\sigma_{n_p}} \to \overline{\tau_2}$ be the restriction of the simplicial approximation to the closure of σ_{n_p}. Then we have*

$$\partial\overline{f}^{-1}\tau_2 \supseteq \overline{f}^{-1}\partial\tau_2$$

Proof. Since the n_p-simplex σ_{n_p} is mapped *onto* the 2-simplex τ_2, it follows that every vertex of $\tau_2 = (b^0, b^1, b^2)$ has a nontrivial (i.e., nonempty) inverse image. Take an arbitrary edge, say (b^0, b^1), of $\partial\overline{\tau_2}$. The inverse images of b^0 and b^1 together contain *at most* n_p vertices because the inverse image of b^2 has *at least* one vertex. Therefore, $\overline{f}^{-1}(\overline{(b^0, b^1)})$ has *at most* n_p

vertices and hence is at most a $(n_p - 1)$-dimensional face of σ_{n_p}—that is, is contained in $\partial\sigma_{n_p} = \partial\overline{f}^{-1}\tau_2$. Repeating the same argument for the edges $(b^1, b^2), (b^0, b^2)$ yields the proof. ■

The above commutativity properties of the topological map $\tilde{f}$ lacks symmetry. Clearly, they would be more aesthetically appealing if they were *equality* relations. The reason why equality is not achieved is that the topological map $\tilde{f}$ maps simplexes of varying dimensions *indiscriminately*. Being more careful with the dimension results in *equality* relations. This is formally accomplished by "algebraization" of the above ... .

4.5 Simplicial Map—Algebra

The *point set topological* concepts developed in the previous sections all have an underlying *combinatorial* structure. The combinatorial structure provides a complete description of the topological structure: A geometrical simplex is uniquely specified by its vertices; the topological map $\tilde{f}$ is uniquely specified by the vertex transformation φ_0. In this section, we strip the problem from its topological substratum and develop an algebraization of its combinatorial structure.

4.5.1 Simplicial Theory

Algebraically, an n-simplex σ_n is an *ordered* set of (affinely independent) vertices, $(a^0, a^1, ..., a^n)$. For all practical algebraic purposes, this ordered set can be considered a skew-commutative product written as $a^0 a^1 ... a^n$. By **skew-commutative product**, we mean that for any permutation $\pi : \{0, 1, ..., n\} \to \{0, 1, ..., n\}$ we have

$$a^{\pi(0)} a^{\pi(1)} ... a^{\pi(n)} = \text{sign}(\pi) a^0 a^1 ... a^n$$

where $\text{sign}(\pi)$ is the **signature** of the permutation—that is, $(-1)^{\#}$ where $\#$ is the number of elementary permutations of pairs of contiguous vertices needed to obtain π. A linear combination with integral coefficients of n-simplexes (e.g., $\sigma_n + 2\tau_n$) is called a **chain**. The best intuitive picture of a *chain* is the integration domain of a differential form; see [Hilton and Wylie 1965, Section 2.7] and Section 12.2. The set of all chains of a given simplicial complex together with formal addition is a commutative (Abelian) group C_n, called an n**-chain group**. The direct sum $\oplus_n C_n$ is called a **graded group** and n is called **graduation**.

To avoid the kind of intricacies of the topological boundary identified in Section 3.2, it is convenient to algebraize the boundary as well. The boundary of a simplex is *defined* as the collection of its faces together with an algebraic trick to ensure that common face of adjacent simplexes in a simplicial decomposition cancel. From this, the boundary of any polyhedron can be computed combinatorially. Formally, the **(algebraic) boundary**

operator is defined for any simplex as

$$\partial_n a^0 ... a^n = a^1 a^2 ... a^n - a^0 a^2 ... a^n + ... (-1)^n a^0 a^1 ... a^{n-1}$$

Observe that the right-hand side is an $(n-1)$-chain. The boundary of a chain c_n is defined by linearity from the boundary of its constituting simplexes. Therefore, more formally, the boundary appears to be a group homomorphism

$$\partial_n : C_n \to C_{n-1}$$

The *fundamental* property of the algebraic boundary operator is the following:

Theorem 4.8.

$$\partial_{n-1} \partial_n a^0 a^1 ... a^n = 0$$

Proof.

$$\begin{aligned}
&\partial_{n-1} \partial_n a^0 a^1 ... a^n \\
&= \partial_{n-1} \sum_i (-1)^i a^0 ... a^{i-1} a^{i+1} ... a^n \\
&= \sum_i (-1)^i \partial_{n-1} a^0 ... a^{i-1} a^{i+1} ... a^n \\
&= \sum_i (-1)^i \Big(\sum_{j=0}^{i-1} (-1)^j a^0 ... a^{j-1} a^{j+1} ... a^{i-1} a^{i+1} ... a^n \\
&\quad + \sum_{j=i+1}^{n} (-1)^{j-1} a^0 ... a^{i-1} a^{i+1} ... a^{j-1} a^{j+1} ... a^n \Big) \\
&= \sum_{j<i} (-1)^{i+j} a^0 ... a^{j-1} a^{j+1} ... a^{i-1} a^{i+1} ... a^n \\
&\quad + \sum_{i<j} (-1)^{i+j-1} a^0 ... a^{i-1} a^{i+1} .. a^{j-1} a^{+1} ... a^n \\
&= 0
\end{aligned}$$

■

By linearity it follows from the above theorem that

$$\partial_{n-1} \partial_n c_n = 0, \quad \forall c_n \in C_n$$

A graded chain group $\oplus_n C_n$ together with a collection of homomorphisms $\partial_n : C_n \to C_{n-1}$ such that $\partial_{n-1} \partial_n = 0$ is called a **chain complex**. A chain complex is denoted as $\{C_*, \partial_*\}$.

Likewise, the vertex transformation φ_0 that underlies the simplicial map is a group homomorphism:

$$\varphi_0 : C_0(D \times \Omega) \to C_0(N)$$

Remember, the vertex transformation φ_0 maps simplexes to simplexes. Therefore, if $a^0a^1...a^n$ spans a simplex of $D \times \Omega$, it follows that

$$\varphi_0(a^0)\varphi_0(a^1)...\varphi_0(a^n)$$

spans a simplex of $\hat{N}$. Therefore, φ_0 induces, for every n, a group homomorphism:

$$\begin{array}{rcl} \varphi_n : C_n(D \times \Omega) & \rightarrow & C_n(N) \\ a^0a^1...a^n & \mapsto & \varphi_0(a^0)\varphi_0(a^1)...\varphi_0(a^n) \end{array}$$

Here, we face an important issue. Typical in this Nyquist mapping problem is that a *high-dimensional* uncertainty simplex is mapped into a *low-dimensional* simplex of the Nyquist template. More precisely,

$$\varphi_0(a^0)\varphi_0(a^1)...\varphi_0(a^n)$$

spans a simplex of dimension *at most* 2. Therefore, for $n > 2$, there *have to be* some repetitions among these vertices, and if we follow the rules of the skew-commutative product, we have $\varphi_n = 0$ for $n > 2$. This results in substantial simplification, but the question is whether *in this robust stability context* we can afford disregarding all simplexes of dimension greater than 2 in the uncertainty space? It turns out that, if our objective is solely to use φ for robust stability *test*—that is, answer by "yes" or "no" the question as to whether some instability develops *somewhere* in $D \times \Omega$ without bothering as to *where exactly* the instability develops—we can safely disregard simplexes of dimension $n > 2$ and rely on φ. This is precisely stated in the following theorem:

Theorem 4.9. *The equation $\tilde{f}(p) = 0$ has a solution iff $\tilde{f}|(D \times \Omega)^2(p) = 0$ has a solution.*

Proof. Let p be a solution to $\tilde{f}(p) = 0$. There exists a unique simplex $\sigma_n = (a^0, a^1, ..., a^n)$ in the complex of $D \times \Omega$ such that $\sigma_n \ni p$. Let $(\lambda_0, \lambda_1, ..., \lambda_n)$ be the barycentric coordinates of p in the simplex σ_n. Existence of the solution p means that there exists a solution λ to the system of equations

$$\begin{pmatrix} \tilde{f}(a^0) & \tilde{f}(a^1) & \cdots & \tilde{f}(a^n) \\ 1 & 1 & \cdots & 1 \end{pmatrix} \begin{pmatrix} \lambda_0 \\ \lambda_1 \\ \vdots \\ \lambda_n \end{pmatrix} = \begin{pmatrix} 0 \\ 0 \\ 1 \end{pmatrix}$$

In other words, $\begin{pmatrix} 0 \\ 0 \\ 1 \end{pmatrix}$ is in the column range of the matrix. Therefore, there exists another solution λ' that involves at most three columns of the matrix. In other words, at most three barycentric coordinates λ' are nonvanishing. λ' are the barycentric coordinates of another solution p' to $\tilde{f}(p') = 0$; this solution is in a $2, 1$, or even 0-face of σ_n, depending on whether λ' has $3, 2$, or 1 nonvanishing component, respectively. In either case, $p' \in (D \times \Omega)^2$. ■

It goes without saying that φ is inadequate to capture the approximate crossover $\tilde{f}^{-1}(0 + j0)$.

With the homomorphism φ_n we are getting close to a symmetric commutativity relation between the mapping and the boundary. In fact, φ commutes with the boundary. To be more precise, we have the following theorem.

Theorem 4.10. *The collection of homomorphisms* $\{\varphi_n\}$ *induced by the simplicial map* $\tilde{f}$ *is a* **chain map**—*that is,*

$$\varphi_{n-1}\partial_n = \partial_n\varphi_n$$

Proof. The proof is *implicitly* contained in [Hilton and Wylie 1965, 3.2]. Since this result plays a central role, it deserves a self-contained proof.

By linearity, it suffices to prove this result for *all* simplexes of $D \times \Omega$. Assume first that the simplex $\sigma_n = a^0...a^n$ has none of its faces collapsed under the simplicial map. In other words, because σ_n is a simplex of the standard complex, it has no repetitions among its vertices, and after mapping those vertices no repetitions appear either. Under those circumstances, we have

$$\begin{aligned}
\varphi_{n-1}\partial_n\sigma_n &= \varphi_{n-1}\sum(-1)^k a^0...a^{k-1}a^{k+1}...a^n \\
&= \sum(-1)^k\varphi(a^0)...\varphi(a^{k-1})\varphi(a^{k+1})...\varphi(a^n) \\
&= \partial\varphi(a^0)...\varphi(a^n) \\
&= \partial_n\varphi_n(a^0...a^n)
\end{aligned}$$

and the result is proved for this first case.

We now assume that some faces of σ_n are collapsed. Write $\sigma = \gamma\rho$ where $\gamma = g^0g^1...$ is collapsed to a vertex not in the image of ρ, and no faces of ρ are collapsed. It is easily seen that $\varphi(\gamma\rho) = \varphi(\gamma)\varphi(\rho) = 0$, because $\varphi(\gamma)$ is just the repetition of the same vertex and by the skew commutative product rule it vanishes. Therefore, it remains to be shown that

$$\varphi_{n-1}\partial_n\gamma\rho = 0$$

Assume first that the dimension of γ is one, namely, $\gamma = g^0g^1$. We have

$$\begin{aligned}
\varphi_{n-1}\partial_n\gamma\rho &= \varphi_{n-1}(g^1\rho - g^0\rho + \gamma(\partial\rho)) \\
&= \varphi(g^1)\varphi(\rho) - \varphi(g^0)\varphi(\rho) + \varphi(\gamma)\varphi(\partial\rho) \\
&= 0
\end{aligned}$$

because $\varphi(g^0) = \varphi(g^1)$ and $\varphi(\gamma)$ is the repetition of the same vertex.

Now, if we assume that the dimension of $\gamma = g^0 g^1 g^2 ...$ is greater than or equal to 2, we have

$$\begin{aligned}\varphi_{n-1}\partial_n \gamma\rho &= \varphi_{n-1}(g^1 g^2 ...\rho - g^0 g^2 ...\rho \pm \gamma(\partial\rho)) \\ &= \varphi(g^1)\varphi(g^2)...\varphi(\rho) - \varphi(g^0)\varphi(g^2)...\varphi(\rho) \pm \varphi(\gamma)\varphi(\partial\rho) \\ &= 0\end{aligned}$$

because there are repetitions among the vertices of all of those simplexes. The theorem is proved. ■

Observe that only the simplicial *map* property of $\tilde{f}$ is needed to prove the result; we do not need the stronger property that $\tilde{f}$ is a simplicial *approximation*. In some computational implementations of the simplicial approximation theorem, we will trade the "star condition" and the resulting homotopic property for speed of implementation, but at *no stage* are we going to relinquish the simplicial *map* property.

The *chain map* property is best illustrated by commutativity of the diagram

$$\begin{array}{ccccc} \ldots \to & C_n(D \times \Omega) & \xrightarrow{\partial_n} & C_{n-1}(D \times \Omega) & \to \ldots \\ & \downarrow \varphi_n & & \downarrow \varphi_{n-1} & \\ \ldots \to & C_n(N) & \xrightarrow{\partial_n} & C_{n-1}(N) & \to \ldots \end{array}$$

It is important to understand that the top row depicts the topological structure of the domain of uncertainty, that the bottom row depicts the topological structure of the Nyquist template, and, finally, that the "down arrows" indicate how the topological structure of the uncertainty is mapped into the topological structure of the Nyquist template.

With this algebraic formulation, we are able to write a symmetric commutativity relation between the inverse image and the boundary. This, however, requires a little preliminary lemma:

Lemma 4.11. *If* $\partial_n \sigma_n^1 = \partial_n \sigma_n^2$ *then* $\sigma_n^1 = \sigma_n^2$.

Proof. Since $\partial_n \sigma_n^1, \partial_n \sigma_n^2$ are made up with vertices of σ_n^1, σ_n^2, these latter simplexes must have *some* vertices in common, because otherwise the boundaries could not be equal. Let ρ be the $(m-1)$-simplex made up of *all* vertices that σ_n^1, σ_n^2 have in common. Let this set of vertices be denoted as R and assume that R contains m vertices. We have

$$\sigma^i = \rho\tau^i$$

where $\tau^i \in T$, the set of vertices of σ^1, σ^2 that are *not* shared by both of them. To be more formal, T is the *symmetric difference* between the set of vertices of σ^1 and the set of vertices of σ^2. Clearly, ρ has m vertices in R while τ^i has $(n-m+1)$ vertices in T, and furthermore T and R are disjoint. From elementary algebra, we have

$$\partial\sigma^i = (\partial\rho)\tau^i \pm \rho(\partial\tau^i)$$

Therefore, the equality between the boundaries yields

$$(\partial\rho)\tau^1 \pm \rho(\partial\tau^1) = (\partial\rho)\tau^2 \pm \rho(\partial\tau^2)$$

Because the vertex sets R and T are disjoint, it follows that the above *chain equality* involves two set of simplexes: those that have exactly $(m-1)$ vertices in R, $(\partial\rho)\tau^i$, and those that have exactly m vertices in R, $\rho(\partial\tau^i)$. The above chain equality can therefore be split into two chain equalities: one involving simplexes with $(m-1)$ vertices in R and another involving simplexes with m vertices in R. This yields, respectively,

$$\begin{aligned} (\partial\rho)\tau^1 &= (\partial\rho)\tau^2 \\ \rho(\partial\tau^1) &= \rho(\partial\tau^2) \end{aligned}$$

Formal division of the first one by $\partial\rho$ yields $\tau^1 = \tau^2$. Hence $\sigma^1 = \sigma^2$. ■

Theorem 4.12. *Let $\varphi_*: C_*(D \times \Omega) \to C_*(N)$ be the chain map induced by the simplicial map $\tilde{f}: D \times \Omega \to N$. For any $\tau_n \in im(\varphi_n)$, $n \leq 2$, define $\varphi_n^{-1}(\tau_n)$ to be the set of simplexes σ_n such that $\varphi_n(\sigma_n) = \tau_n$. For an arbitrary chain c_{n-1}, define $\varphi_{n-1}^{-1}(c_{n-1})$ to be the set of boundary chains b_{n-1} such that $\varphi_{n-1}(b_{n-1}) = c_{n-1}$. We have*

$$\varphi_{n-1}^{-1}\partial_n\tau_n = \partial_n\varphi_n^{-1}\tau_n$$

Proof.

$$\begin{aligned} \varphi_n^{-1}\tau_n &= \{\sigma_n : \varphi_n(\sigma_n) = \tau_n\} \\ \partial_n\varphi_n^{-1}\tau_n &= \{\partial_n\sigma_n : \varphi_n(\sigma_n) = \tau_n\} \\ &= \{\partial_n\sigma_n : \partial_n\varphi_n\sigma_n = \partial_n\tau_n\} \\ &= \{\partial_n\sigma_n : \varphi_{n-1}\partial_n\sigma_n = \partial_n\tau_n\} \\ &= \{c_{n-1} : \varphi_{n-1}c_{n-1} = \partial_n\tau_n\} \cap \text{ im } \partial_n \\ &= \varphi_{n-1}^{-1}\partial_n\tau_n \end{aligned}$$

(Observe that the lemma is invoked at the third equality.) ■

This theorem provides an algebraic formulation of the interplay between the inverse image and the boundary. This theorem is not easily formulated and requires many conditions, the motivations for which are not clear. The reason for this clumsiness is that this theorem hides a very deep concept: the *twisting of the fiber*. Unfortunately, the twisting problem requires a long string of concepts, definitions, and results, which is relegated to Chapter 13. The interplay between the preimage and the boundary will be reformulated in Subsection 13.2.6, where we will show that the twisting of the fiber prevents commutativity.

4.5.2 Semisimplicial Theory

While the *simplicial* approach is adequate for *checking* robust stability, it becomes inadequate to capture the approximate crossover $\tilde{f}^{-1}(0+j0)$. If the latter is our objective, it is necessary to keep $\varphi_0(a^0)...\varphi_0(a^n)$ as a *non-vanishing* mathematical object, even though there are repetitions among the vertices. To make a "simplex" with repetition among vertices *non-vanishing*, we have to drop the skew-commutative product property; in other words, we no longer identify simplexes differing by an even permutation of vertices. To make this distinction clear, we write this "simplex" as $[\varphi_0(a^0)...\varphi_0(a^n)]$, we *define* it as an *ordered* array of vertices, spanning a geometrical simplex, with repetition allowed, and we call it **total simplex**. The set of all linear combinations of those total n-simplexes is called the **total or semisimplicial chain group**, C_n^{total}, and decomposing a polyhedron into such simplexes results in the so-called **total or semisimplicial complex**. The boundary operator on the total complex is defined as

$$\partial[b^{i_0}...b^{i_n}] = \sum_{j=0}^{n}(-1)^j[b^{i_0}...b^{i_{j-1}}b^{i_{j+1}}...b^{i_n}]$$

and we still have the fundamental relationship

$$\partial\partial = 0$$

In other words, $\{C_*^{\text{total}}, \partial_*^{\text{total}}\}$ is a chain complex.

The chain map property has been proved for the usual complex, characterized by cancellation of ordered arrays containing repetition of vertices. Actually, the chain map property also holds for the total complex:

Theorem 4.13. *$\varphi: C^{total}(D\times\Omega)\to C^{total}(N)$ is a chain map.*

Proof. Actually, the proof of the chain map property is even more straightforward for the total complex than for the regular complex. Indeed

$$\begin{aligned}
\varphi_{n-1}\partial_n[a^{i_0}...a^{i_n}] &= \varphi_{n-1}\sum(-1)^k[a^{i_0}...a^{i_{k-1}}a^{i_{k+1}}...a^{i_n}] \\
&= \sum(-1)^k\varphi_{n-1}[a^{i_0}...a^{i_{k-1}}a^{i_{k+1}}...a^{i_n}] \\
&= \sum(-1)^k[\varphi(a^{i_0})...\varphi(a^{i_{k-1}})\varphi(a^{i_{k+1}})...\varphi(a^{i_n})] \\
&= \partial[\varphi(a^{i_0})...\varphi(a^{i_n})] \\
&= \partial\varphi_n[a^{i_0}...a^{i_n}]
\end{aligned}$$

■

Observe that, from a purely formal point of view, we could introduce *face operators* F^j on the total complex,

$$F^j : C_n^{\text{total}} \to C_{n-1}^{\text{total}}$$

$$[a^{i_0}\ldots a^{i_n}] \quad \mapsto \quad [a^{i_0}\ldots a^{i_{j-1}}a^{i_{j+1}}\ldots a^{i_n}]$$

With this notation, the boundary operator on the total complex can be rewritten

$$\partial = \sum_{j=0}^{n}(-1)^j F^j$$

Also, for reasons that at this stage appear *purely formal*, we could define *degeneracy operators* D^j

$$\begin{aligned} D^j : C_n^{\text{total}} &\rightarrow C_{n+1}^{\text{total}} \\ [a^{i_0}\ldots a^{i_n}] &\mapsto [a^{i_0}\ldots a^{i_j}a^{i_j}\ldots a^{i_n}] \end{aligned}$$

With this concept, we could define a *degenerate simplex*—that is, a simplex that has repetitions among its vertices—as a simplex in the image of a D^j.

It is easy, although a little tortuous, to show that the face and degeneracy operators defined on the total complex satisfy the following relations:

$$\begin{aligned} F^iF^j &= F^{j-1}F^i, \ i < j \\ D^iD^j &= D^jD^{i-1}, \ i > j \\ F^iD^j &= D^{j-1}F^i, \ i < j \\ F^iD^i &= F^{i+1}D^i = 1 \\ F^iD^j &= D^jF^{i-1}, \ i > j+1 \end{aligned}$$

Now, we can somewhat conceptualize the notion of total complex:

Definition 4.14. *A* **semisimplicial complex** *is a graded set,* $\cup_{n\geq 0}C_n$*, together with face and degeneracy operators,* $F^i : C_n \rightarrow C_{n-1}$*,* $D^i : C_n \rightarrow C_{n+1}$*, respectively, satisfying the above five relations. The graduation is called* **dimension**. *The elements of* C_n *are called "n-simplexes."*

If we are given an *abstract* semisimplicial complex together with face operators F^i, we could form a boundary operator $\partial = \sum_i(-1)^iF^i$ and it is easily checked that $\partial\partial$ satisfies the usual condition, because

$$\begin{aligned} \partial\partial &= (\sum_i(-1)^iF^i)(\sum_j(-1)^jF^j) \\ &= \sum_{i<j}(-1)^{i+j}F^iF^j + \sum_{i\geq j}(-1)^{i+j}F^iF^j \\ &= \sum_{i<j}(-1)^{i+j}F^{j-1}F^i + \sum_{i\geq j}(-1)^{i+j}F^iF^j \\ &\quad (i' = j-1, j' = i) \\ &= \sum_{j'\leq i'}(-1)^{i'+j'+1}F^{i'}F^{j'} + \sum_{i\geq j}(-1)^{i+j}F^iF^j \\ &= 0 \end{aligned}$$

Clearly, $C_*^{\text{total}}(D \times \Omega)$ and $C_*^{\text{total}}(N)$, endowed with the face and degeneracy operators, are semisimplicial complexes.

We shall sometimes drops the superscript "total" to alleviate the notational burden.

Finally, we need a concept of what we would call a "simplicial map," except that it should fit within the *semi*-simplicial setting:

Definition 4.15. *A* **semisimplicial map** $f_n : C_*(D \times \Omega) \to C_*(N)$ *is a map that maps simplexes to simplexes and that commutes with the face and degeneracy operators.*

Besides its intuitive motivation as a theory that allows for repetition of vertices in the simplexes of the template, the semisimplicial theory has a much deeper motivation: It is apparently the only conceptual setup to formulate the *twisting of the fiber* in the uncertainty space that prevents commutativity of the inverse chain map and the boundary; see Chapter 13.

4.6 Computational Issues

In this section, we take a hard look at the computational issues germane to the construction of a simplicial approximation as it is practiced in algebraic topology. It turns out that, in order to make the simplicial approximation theorem computationally implementable and more importantly to mesh it with simplicial algorithms for searching the stability boundary, the simplicial approximation theorem will have to be modified almost beyond recognition.

The very first step in the implementation of the simplicial approximation theorem is to construct a triangulated polygon of $\mathbb{C}$ into which the domain of uncertainty maps. This is already a nontrivial step, because it is not too clear how big a polygon we need. However, there exists an easy way to go about this. Let $\{a^i\}$ be the vertex set of some triangulation of $D \times \Omega$. Map those vertices, $b^i = f(a^i)$, construct the convex hull of $\{b^i\}$ and triangulate it using the b^i's as vertices. Constructing a triangulation that hinges on a set of points $\{b^i\}$ is a nontrivial problem, but it is a well-understood, fundamental construction of *computational geometry* that has received many fast implementations. Chapter 6 is specially dedicated to this aspect of the problem. The next problem is that it is not, in general, true that $f(D \times \Omega) \subseteq \text{conv}(\{b^i\})$. The latter is a well-known convexity problem. However, if we decide to go for a simplicial approximation of the piecewise-linear extension f_{PL} of the vertex transformation $a^i \mapsto f(a^i)$, then we have $f_{PL}(D \times \Omega) \subseteq \text{conv}(\{b^i\})$. This point of view is developed in Chapter 7.

A second computationally burdensome step in the simplicial approximation theorem is the evaluation of the Lebesgue number δ of the covering of $D \times \Omega$ by the inverse images of the stars of the vertices of N. δ can be estimated via the Jacobian as follows:

$$\delta \approx (\sigma_{\max}(J))^{-1} \times \text{ Lebesgue number of } \cup_i \text{star}(b^i)$$

where J is the Jacobian of the Nyquist map:

$$J = \begin{pmatrix} \frac{\partial \Re f}{\partial p_1} & \frac{\partial \Re f}{\partial p_2} & \cdots \\ \frac{\partial \Im f}{\partial p_1} & \frac{\partial \Im f}{\partial p_2} & \cdots \end{pmatrix}$$

Third, we need a refined triangulation of $D \times \Omega$ with mesh $< \delta/2$, and this is achieved by repeated barycentric or Q-subdivision of the grid. Unfortunately, this procedure tends to be overly conservative, because the barycentric and Q-subdivisions are *isotropic* procedures, refining in all possible directions, even though refinement might not be necessary in directions along which the Nyquist map is not very sensitive. Clearly, a fast, computationally implementable construction of the simplicial approximation that requires a minimum amount of triangulation for a given accuracy is warranted. A tentative solution to this problem is proposed in Chapter 6.

Another problem is that repeated barycentric subdivisions create "long and skinny," "flat" simplexes. In general, "flat" simplexes are to be avoided because they negatively affect the accuracy of the simplicial procedure. As we will see in Chapter 5, repeated Q-triangulation keeps the flatness bounded.

Finally, once $D \times \Omega$ and N are triangulated consistently with the required accuracy, the simplicial approximation $\tilde{f}$ is defined by the basic "star condition"

$$f(\text{star}(a^i)) \subseteq \text{star}(\tilde{f}(a^i))$$

The difficulty with the above is that we need to check the image under f of the *whole star* of a^i, which has in general the same dimension as $D \times \Omega$. The irony is that our primary objective is precisely to assess $f(D \times \Omega)$ from the images of some *lower-dimensional* objects. There are several ways to go about this: In Chapter 6 we propose to assign $f(a^i)$ to the nearest vertex of $\{b^i\}$, which is efficiently implemented as the *point location problem* of computational geometry. Another approach to make the "star" condition more manageable is to approximate simplicially the f_{PL} map, in which case the star condition

$$f_{PL}(\text{star}(a^i)) \subseteq \text{star}(\tilde{f}_{PL}(a^i))$$

can be implemented using some concepts from either computational geometry or linear programming. In any event, since this basic "star condition" is too difficult to check in practice, we will sometimes have to renounce the objective of constructing a *genuine* simplicial approximation to f. However, *at no stage* are we going to renounce the simplicial map property of $\tilde{f}$, since the latter property ensures commutativity between the inverse image and the boundary.

4.7 Relative Simplicial Approximation

For the purpose of efficient implementation, it would be interesting to make the simplicial approximation *recursive* over bigger and bigger subpolyhedra of P. To be specific, assume the map has been made simplicial over some subpolyhedron $Q \subseteq P$ (this is a trivial exercise in the extreme case when Q is a 1-simplex). The question is whether the simplicial map over Q can be extended to a simplicial map over P by refining the simplexes of $P \setminus Q$.

Let S_Q be the subdivision operator of P that barycentrically subdivides all simplexes of $P \setminus Q$ but that does not subdivide the simplexes of Q. Inspiring oneself from the absolute simplicial approximation, it is hoped that, for some finite k, any $\text{star}(a'^i), a'^i \in S_Q^k P$ would be included in some $f^{-1}(\text{star}(b^j))$, in which case the inclusion $\text{star}(a'^i) \subseteq f^{-1}(\text{star}(b^j))$ would define a vertex transformation $a'^i \mapsto b^j$, itself defining the relative simplicial approximation. Unfortunately, there are counterexamples where for no finite k does the covering $\cup_{a'^i \in S_Q^k P} \text{star}(a'^i)$ refine the covering $\cup_j f^{-1}(\text{star}(b^j))$. It turns out that a relative simplicial approximation *does* exist, but it has to be constructed by departing from the absolute case.

Theorem 4.16. (Zeeman's Relative Simplicial Approximation) *Assume we are given a topological map $f : P \to N$ such that over some subpolyhedron $Q \subseteq P$ the restriction $f|Q$ is simplicial. Then there exists a refined polyhedron P' and a simplicial map $\tilde{f} : P' \to N$ such that $\tilde{f}|Q = f|Q$ and furthermore f and $\tilde{f}$ are homotopic with the extra requirement that for every $p \in Q$ the image of p remains fixed during the entire homotopy from f to $\tilde{f}$.*

Proof. Consider the simplexes σ of $P \setminus Q$ such that $\bar{\sigma} \cap Q \neq \emptyset$, and let S be the barycentric subdivision of the faces of σ not in Q. Define Δ to be the simplicial neighborhood of Q in $SS_Q^k P$; that is, $\Delta = \cup_{a'^i \in Q} \bar{\text{star}}(a'^i)$. We define a simplicial retracting map $h : \Delta \to Q$ by the following vertex transformation: if $a \in \Delta \setminus Q$, then a is the barycenter of a simplex σ of S_Q^k such that $\bar{\sigma} \cap L \neq \emptyset$ and $h(a)$ is defined to be a vertex of $\bar{\sigma} \cup Q$; otherwise, $h(a) = a$. The crucial idea is to construct an absolute simplicial approximation $\tilde{f}$ of $fh : SS_Q^k P \to N$, which is easily seen to be the required relative simplicial approximation to f. For more details see [Zeeman 1964] or [Glaser 1970]. ■

Therefore, if we already have a simplicial map over the subpolyhedron Q, it is possible to get a simplicial map over P, without refining Q, but some caution needs to be exercised in the vertex assignment in the simplicial neighborhood of Q. What we have to remember is that the vertex assignment is directed by the star conditions on fh, and *not* on f.

4.8 Cell Complexes and Cellular Maps

The concept of "cellular complex" is somewhere in between geometry and algebra.

We are given a collection of (abstract) "cells," $\{e_n^i\}$, where the subscript n stands for the "dimension" and the superscript i runs in some index set for the n-dimensional cells. On the collection of cells, we define a partial ordering, **facing relation** $\preceq$. The facing relation and the "dimension" are compatible in the following sense: If $e_n^i \preceq e_m^j$ and $e_n^i \neq e_m^j$ then $n < m$. Given two cells e_n^i, e_{n-1}^j, we define their **incidence number**, $[e_n^i : e_{n-1}^j] = 0, \pm 1$, subject to the following constraints:

$$[e_n^i : e_{n-1}^j] \neq 0 \Leftrightarrow e_{n-1}^j \preceq e_n^i$$

$$\sum_i [e_n : e_{n-1}^i][e_{n-1}^i : e_{n-2}] = 0, \ \ \forall e_n, e_{n-2}$$

A collection of cells satisfying the above requirements is called **(abstract) cell complex**. The boundary of a cell is defined as

$$\partial e_n = \sum_i [e_n : e_{n-1}^i] e_{n-1}^i$$

and it is easily seen that $\partial\partial = 0$. We define *cellular chains* as formal linear combinations of cells. This leads to the concept of *cellular chain groups.* The boundary can be extended to a cellular chain group homomorphism satisfying $\partial\partial = 0$. Finally, we define a cellular chain map as a cellular chain group homomorphism commuting with the boundary.

There are many decompositions of manifolds and other topological spaces into (geometric) cells of varying dimensions satisfying the axioms of cellular complex. Cell decomposition is in general much more economical than decomposition into rectilinear simplexes.

Finally, we come to what can be loosely referred to as the cellular counterpart of a simplicial map. The topological map $f : X \to Y$, where X and Y are (geometric) cellular complexes, is said to be a **cellular map** if for every cell e of X, $f(e)$ is included in a cell of Y. Although a cellular approximation theorem can be devised, as far as the examples of this book are concerned, one can make the original map cellular because of the freedom involved in defining the cells.

For an application of these concepts to the Nyquist map, see Section 21.16.

4.9 HISTORICAL NOTES

The simplicial approximation theorem, in its original version, is due to Brouwer. The modern construction of the simplicial approximation, based on the star covering argument, is due to Alexander.

It is fair to say that the simplicial approximation theorem, as formulated by Brouwer around 1910–1912, was a breakthrough in the history of mathematics. The simplicial approximation, coupled with the concept of degree (see Section 18.12), allowed Brouwer to prove such famous theorems as the invariance of the dimension and the Jordan–Brouwer separation theorem, the correct proofs of which had eluded mathematicians of the pre-Brouwer era. Furthermore, the Brouwer simplicial *methods* were instrumental in Alexander's and Lefschetz's program aimed at providing more solid foundation to Poincaré's simplicial topology.

Probably the most incisive exposition of the simplicial approximation theorem and its impact on topology, very much in the spirit of the work of the pioneers, is [Alexandroff 1961].

5

CARTESIAN PRODUCT OF MANY UNCERTAINTIES

SUMMARY

In the previous chapters, we have decomposed the space of uncertainty $D \times \Omega$ as a *whole*, ignoring the fact that it is, actually, the topological product of two spaces, namely D and Ω. In this section, we investigate what can be gained by exploiting the Cartesian product structure of the generalized domain of uncertainty.

It should be intuitively clear that the higher the dimension of a space, the more difficult it is to triangulate. Therefore, the natural question is whether, given a triangulation of D and a triangulation of Ω, we can construct a triangulation of $D \times \Omega$. This question is not trivial, because the Cartesian product of two simplexes is, unfortunately, not a simplex. Inspiring ourselves from the low-dimensional example of a 2-simplex topologically multiplied by a line segment, the Cartesian product of two simplexes is called *prism*. The crux of the matter is that a prism of arbitrary dimension can be decomposed into simplexes by a simple combinatorial rule, without the need to "visualize" the situation. This yields the so-called *prismatic decomposition* of $D \times \Omega$, constructed from simplicial decompositions of D and Ω, that will be very useful at several different places.

Next, we observe that the combinatorics of the prismatic triangulation is the same as that of the *Hex board*. The *Game of Hex* is a tic-tac-toe game that is played on a hexagonal tile board, rather than the usual square tile board. In view of the equivalence between prismatic decomposition and Hex board, some simplicial algorithms for finding the stability boundary will receive, in later chapters, an interpretation in terms of strategies for finding a winning path in the game of Hex.

There is yet another interpretation of the prismatic triangulation and the Hex board: It turns out that their combinatorics is the same as that of the so-called *Q-triangulation*. The *Q-triangulation* is an ad hoc decomposition that underlies many simplicial algorithms that have appeared in the context of operations research. In view of the equivalence between the two triangulations, we will be able to run simplicial algorithms devised for Q-triangulation on the underlying prismatic triangulation of the uncertainty space.

An advantage of these combinatorially equivalent triangulations is that, if we use them as iterative refinement rule, no "long and skinny" simplexes will appear at any refinement level of the triangulation. The latter is essential for *accuracy* of the simplicial algorithms that recursively narrow down the neutral stability region within smaller and smaller simplexes.

Finally, we propose yet another interpretation of the equivalent triangulation rules in terms of the concept of *sphere packing* borrowed from lattice theory and error correcting codes.

In presenting the material of this chapter, we have strived to achieve unification among results borrowed from combinatorial topology (prismatic triangulation), operations research (Q-triangulation), game theory (Game of Hex), and some concepts of communications theory (sphere packing).

5.1 Prismatic Decomposition

5.1.1 Simplicial Uncertainty-Frequency Product

The difficulty we are facing is that the Cartesian (or topological) product of two simplexes is *not* a simplex. The good news, however, is that the Cartesian product of two simplexes, called **prism** by topologists and **simplotope** by operations researchers, can be decomposed into simplexes. Consider an arbitrary simplex of D,

$$\sigma_n = a^0 ... a^n \in C_n(D)$$

together with an arbitrary simplex of Ω,

$$\omega^l \omega^{l+1} = \tau_1 \in C_1(\Omega)$$

The decomposition of the Cartesian product into simplexes is given by

$$\sigma_n \times \tau_1 = \sum_{k=0}^{n} (-1)^k a^0_{\omega^l} ... a^k_{\omega^l} a^k_{\omega^{l+1}} ... a^n_{\omega^{l+1}} \tag{5.1}$$

In the above, $a^i_{\omega^l}$ is the vertex a^i elevated at height ω^l in the product space $D \times \Omega$.

An easy way to come to grips with this formula is to consider the decomposition of a product like $a^0 a^1 \times \omega^l \omega^{l+1}$. This is shown in Figure 5.1.

The reader is also invited to get the geometric feeling for this construction by considering a 2-D simplex $a^0 a^1 a^2$ of D and computing its Cartesian product with $\omega^l \omega^{l+1}$. This Cartesian product is a **prism**, in the elementary geometry definition of the term. The above decomposition of the Cartesian product, although somewhat combinatorial in nature, corresponds *exactly* to the three-dimensional situation of a prism decomposed into tetrahedra, a construction well-known in elementary geometry and shown in Figure 5.2. For some examples of topological applications of this useful recipe, see [Hocking and Young 1961] and [Glaser 1970].

The above intuitive geometric arguments justify the "unsigned" version of Equation 5.1 where $(-1)^k$ is dropped and "$\sum$" is replaced with "$\cup$". To

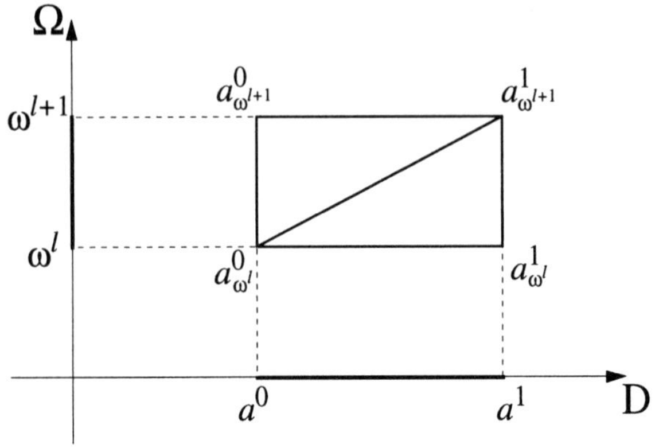

Fig. 5.1. Prismatic decomposition of a rectangle.

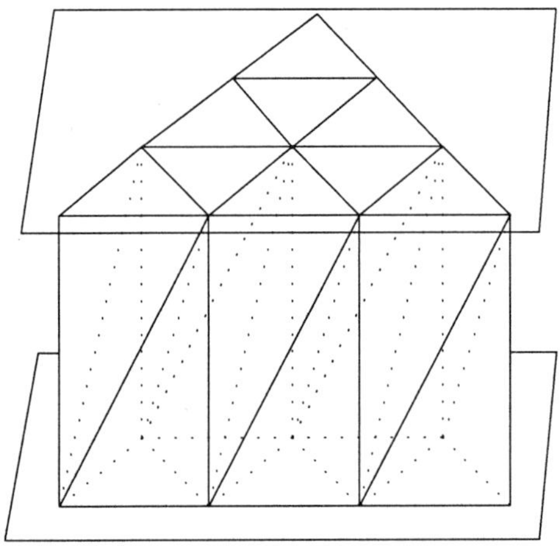

Fig. 5.2. Prismatic triangulation.

justify the "signed" formula, observe that the signature is chosen such that, in the boundary of the Cartesian product, $\partial\sigma_n \times \tau_1$, two n-faces common to adjacent simplexes of the prismatic decomposition cancel, as expected. We shall come back to this fact in more generality later.

This prismatic decomposition formula is extended to any n-chain (i.e., linear combination of n-simplexes) of D and any 1-chain of Ω by bilinearity. It follows that the prismatic decomposition is, formally, a homomorphism

$$C_n(D) \otimes C_1(\Omega) \to C_{n+1}(D \times \Omega)$$

to which we shall return later.

In an attempt to generalize the above recipe, consider the rather trivial case of the Cartesian product of an n-dimensional simplex of D and a

0-dimensional simplex of Ω,

$$\omega^l = \tau_0 \in C_0(\Omega)$$

Observe that this Cartesian product can be written as follows:

$$\sigma_n \times \tau_0 = a^0_{\omega^l} a^1_{\omega^l} ... a^n_{\omega^l}$$

5.1.2 Semisimplicial Conceptualization

In the above approach to the Cartesian product, we have strived to proceed from geometric intuition. A deeper conceptualization of the notion of Cartesian product of complexes is available in [May 1967, Definition 6.1]. This conceptualization, however, requires the semisimplicial theory of Subsection 4.5.2. In essence, the simplex $a^0_{\omega^l} ... a^k_{\omega^l} a^k_{\omega^{l+1}} ... a^n_{\omega^{l+1}}$ is, more formally, the juxtaposition, or product, of the (total) simplexes $[a^0 ... a^k a^k ... a^n]$ and $[\omega^l ... \omega^l \omega^{l+1} ... \omega^{l+1}]$ both of length $(n+2)$. Adopting the same notation as in [May 1967, Definition 6.1], this juxtaposition or product is written as

$$([a^0 ... a^k a^k ... a^n], [\omega^l ... \omega^l \omega^{l+1} ... \omega^{l+1}])$$

It is convenient to rewrite this juxtaposition in matrix form as

$$\begin{bmatrix} a^0 & \dots & a^k & a^k & \dots & a^n \\ \omega^l & \dots & \omega^l & \omega^{l+1} & \dots & \omega^{l+1} \end{bmatrix}$$

The advantage of this notation is that the first column, for example, of this matrix, namely,

$$\begin{pmatrix} a^0 \\ \omega^l \end{pmatrix}$$

can be viewed as the coordinates of the first vertex $a^0_{\omega^l}$ of the juxtaposition simplex

$$([a^0 ... a^k a^k ... a^n], [\omega^l ... \omega^l \omega^{l+1} ... \omega^{l+1}])$$

Indeed, a^0 can be viewed as the coordinates of the projection of the vertex $a^0_{\omega_l}$ on the simplex $a^0 ... a^n$ while ω^l is the coordinate of the projection of the same vertex on the line segment $\omega^l \omega^{l+1}$.

This conceptualization, which appears to be purely formal, is unfortunately the only way to go beyond elementary 3-D geometry. With the formal apparatus of [May 1967], we can define the prismatic decomposition of two arbitrary simplexes $a^0 ... a^n$ and $b^0 ... b^p$ as

$$a^0 ... a^n \times b^0 ... b^p = \sum (-1)^{p(k)} ([a^{k_0} a^{k_1} ... a^{k_{n+p}}], [b^{l_0} b^{l_1} ... b^{l_{n+p}}]) \qquad (5.2)$$

In the above, the square-bracketed simplexes are total $(n+p)$-simplexes. The first total $(n+p)$-simplex is made up with the $(n+1)$ pairwise distinct vertices of $a^0 ... a^n$, some of them being repeated in order to achieve a square-bracketed array of length $(n+p+1)$. Likewise, the second total $(n+p)$-simplex is made up with the pairwise distinct vertices of $b^0 ... b^p$. The sum is extended over all increasing sequences of k's and all increasing sequences of l's subject to the restrictions

$$0 = k_0 \leq \ldots \leq k_i \leq k_{i+1} \leq \ldots \leq k_{n+p} = n$$

$$0 = l_0 \leq \ldots \leq l_j \leq l_{j+1} \leq \ldots \leq l_{n+p} = p$$

$$k_i = k_{i+1} \Leftrightarrow l_i < l_{i+1}$$

$$l_j = l_{j+1} \Leftrightarrow k_j < k_{j+1}$$

It is easily seen that, once an admissible sequence of k's is specified, the above restrictions specify a unique sequence of l's. Actually, we have

$$k_i + l_i = i$$

Therefore, the sum can be rewritten as

$$\sum_{0=k_0 \ldots \leq k_i \leq k_{i+1} \ldots \leq k_{n+p}=n}$$

With these concepts, we can define $p(k)$, the **parity** of the sequence k, as

$$p(k) = \sum_{k_{i+1}=k_i} i$$

The reader can easily verify for himself that writing Equation 5.2 for $p = 1$ and translating the resulting equation from the semisimplicial back to the simplicial theory yields Equation 5.1.

In the decomposition of $a^0 \ldots a^n \times b^0 \ldots b^p$, it might not be completely clear why the decomposition is signed. The casual reader might want to suppress the signs and replace $\sum$ by $\cup$, and the formula interpreted in the elementary geometric sense would remain correct. However, the signed version of the formula makes its potential impact bigger, because the signature is such that the common boundary faces of adjacent simplexes of the decomposition cancel. Actually, this observation is enough to prove that Equation 5.2 is the correct generalization of the prismatic decomposition of low-dimensional geometry. This is the topic of the next section.

5.2 Boundary of Cartesian Product

5.2.1 Simplicial Approach

We start with the following "warm-up" exercise:

Theorem 5.1.

$$\partial \sigma_n \times \tau_p = (-1)^p (\partial \sigma_n) \times \tau_p + (-1)^{p+1} \sigma_n \times (\partial \tau_p), \quad p = 0, 1$$

Proof. Consider first a simplex of the form

$$\sigma_n \times \tau_1 \quad = \quad a^0 a^1 \ldots a^n \times \omega^l \omega^{l+1}$$

$$= \sum_{k=0}^{n}(-1)^k a^0_{\omega^l}...a^k_{\omega^l}a^k_{\omega^{l+1}}...a^n_{\omega^{l+1}}$$

By straightforward computation, we get

$$\begin{aligned}
&\partial\sigma_n \times \tau_1 \\
&= \sum_{m=0}^{k-1}(-1)^m \sum_{k=0}^{n} a^0_{\omega^l}...a^{m-1}_{\omega^l}a^{m+1}_{\omega^l}...a^k_{\omega^l}a^k_{\omega^{l+1}}...a^n_{\omega^{l+1}} \\
&\quad+\sum_{k=0}^{n}(-1)^k(-1)^k a^0_{\omega^l}...a^{k-1}_{\omega^l}a^k_{\omega^{l+1}}...a^n_{\omega^{l+1}} \\
&\quad+\sum_{k=1}^{n}(-1)^k(-1)^{k+1} a^0_{\omega^l}...a^k_{\omega^l}a^{k+1}_{\omega^{l+1}}...a^n_{\omega^{l+1}} \\
&\quad+\sum_{m=k+1}(-1)^{m+1}\sum_{k=0}^{n} a^0_{\omega^l}...a^k_{\omega^l}a^k_{\omega^{l+1}}...a^{m-1}_{\omega^{l+1}}a^{m+1}_{\omega^{l+1}}...a^n_{\omega^{l+1}}
\end{aligned}$$

It is easily seen that the second and third terms amount to

$$a^0_{\omega^{l+1}}...a^n_{\omega^{l+1}} - a^0_{\omega^l}...a^n_{\omega^l} = a^0...a^n \times \partial\omega^l\omega^{l+1}$$

On the other hand, it is not hard to see that the first and fourth terms reduce to

$$-(\partial a^0 a^1...a^n) \times \omega^l\omega^{l+1}$$

Hence the theorem is proved for $k = 1$. The result for $\partial\sigma_n \times \tau_0$ is straightforward and left to the reader as an exercise. ■

5.2.2 Semisimplicial Approach

The above theorem can be reformulated, in the semisimplicial setting, as follows:

Theorem 5.2.

$$\begin{aligned}
&\partial_{n+p}([a^{k_0}...a^{k_{n+p}}],[b^{l_0}...b^{l_{n+p}}]) \\
&= \sum_i(-1)^i F^i([a^{k_0}...a^{k_{n+p}}],[b^{l_0}...b^{l_{n+p}}]) \\
&= \sum_i(-1)^i (F^i[a^{k_0}...a^{k_{n+p}}],F^i[b^{l_0}...b^{l_{n+p}}]) \\
&= \sum_i(-1)^i ([a^{k_0}...a^{k_{i-1}}a^{k_{i+1}}...a^{k_{n+p}}],[b^{l_0}...b^{l_{i-1}}b^{l_{i+1}}...b^{l_{n+p}}])
\end{aligned}$$

where F^i is the face operator of the semisimplicial theory.

Proof. To prove this formula, translate it to the simplicial setting and observe that it becomes just the classical formula for the boundary of a simplex. (The above formula is essentially the same as that of [May 1967, Definition 6.1].) ■

Now we are in a position to prove our claim:

Theorem 5.3. *The common simplicial $(n+p-1)$-face of two adjacent simplexes of the prismatic decomposition of Equation 5.2 cancels in*

$$\partial_{n+p}(a^0 \ldots a^n \times b^0 \ldots b^p)$$

Proof. Let k and k' be the increasing sequences specifying two adjacent simplexes of the prismatic decomposition of $a^0 \ldots a^n \times b^0 \ldots b^p$. The condition that these two simplexes share $(n+p)$ vertices is reflected by the fact that k and k' differ by *one* at *exactly one* single index, say i_0. More precisely,

$$k_i - k'_i = \pm \delta_{i-i_0}$$

where δ is the usual Kronecker symbol. To be more illustrative, the two simplexes share the vertices

$$\begin{pmatrix} a^{k_i} \\ b^{l_i} \end{pmatrix}, \quad i \neq i_0$$

while the vertices

$$\begin{pmatrix} a^{k_{i_0}} \\ b^{l_{i_0}} \end{pmatrix}, \begin{pmatrix} a^{k'_{i_0}} \\ b^{l'_{i_0}} \end{pmatrix}$$

are opposite to the common face. It is also easily seen that the sequences k and k' have different parities:

$$(-1)^{p(k)} = -(-1)^{p(k')}$$

In the prismatic decomposition the two adjacent simplexes appear as

$$\begin{array}{rl} & (-1)^{p(k)}([a^{k_0} \ldots a^{k_{n+p}}], [b^{l_0} \ldots b^{l_{n+p}}]) \\ + & (-1)^{p(k')}([a^{k'_0} \ldots a^{k'_{n+p}}], [b^{l'_0} \ldots b^{l'_{n+p}}]) \end{array}$$

Consider the boundary of the above using the recipe provided by Theorem 5.2. To retain the two faces common to the two adjacent simplexes we drop in the first simplex the index k_{i_0} while in the second simplex we drop the index k'_{i_0}. This yields

$$\begin{array}{rl} & (-1)^{p(k)}(-1)^{i_0}([a^{k_0} \ldots a^{k_{i_0-1}} a^{k_{i_0+1}} \ldots a^{k_{n+p}}], [b^{l_0} \ldots b^{l_{i_0-1}} b^{l_{i_0+1}} \ldots b^{l_{n+p}}]) \\ + & (-1)^{p(k')}(-1)^{i_0}([a^{k'_0} \ldots a^{k'_{i_0-1}} a^{k'_{i_0+1}} \ldots a^{k'_{n+p}}], [b^{l'_0} \ldots b^{l'_{i_0-1}} b^{l'_{i_0+1}} \ldots b^{l'_{n+p}}]) \end{array}$$

Clearly, these two faces cancel, as claimed. ■

5.3 Simplicial Combinatorics of Cube

Here, we develop the prismatic decomposition of the Cartesian product of *many* factors. Going to higher dimensions entails some further conceptualization effort. The reward will be the unveiling of a fundamental combinatorial structure that will also appear in the so-called Q-triangulation and Hex game, leading to a sweeping unification of many concepts.

Consider the cube, $[0,1]^n$—that is, the Cartesian product of n copies of the unit interval. Let $x_1, ..., x_n$ be a system of Euclidean coordinates. Let us put $m+1$ equally spaced vertices on each coordinate axis interval $[0,1]$; the first vertex is placed at 0 and the $(m+1)$th vertex is placed at 1. The vertices on the first coordinate axis are written as $x_1^0, x_1^1, ..., x_1^m$. The superscripts can be viewed as first Euclidean coordinates in integer multiples of $1/m$. The vertices on the second coordinate axis are written $x_2^0, x_2^1, x_2^2, ..., x_2^m$. Etc. By recursion on the general rule, we find that an n-simplex of the prismatic decomposition of the Cartesian product

$$x_1^{i_1} x_1^{i_1+1} \times x_2^{i_2} x_2^{i_2+1} \times ...$$

is a juxtaposition of the form

$$([x_1^{i_1} ... x_1^{i_1} x_1^{i_1+1} ... x_1^{i_1+1}], [x_2^{i_2} ... x_2^{i_2} x_2^{i_2+1} ... x_2^{i_2+1}], ...) = \sigma_n$$

It is convenient to rewrite the above in matrix representation, by putting one string on top of the other,

$$\begin{array}{cccccccccc}
x_1^{i_1} & \cdots & x_1^{i_1} & x_1^{i_1+1} & \cdots & \cdots & \cdots & \cdots & x_1^{i_1+1} \\
x_2^{i_2} & \cdots & \cdots & \cdots & \cdots & x_2^{i_2} & x_2^{i_2+1} & \cdots & x_2^{i_2+1} \\
\vdots & \cdots & \cdots & \cdots & \cdots & \cdots & \cdots & \cdots & \vdots
\end{array}$$

The columns of this matrix representation are closely related to the Euclidean coordinates of the vertices that form the simplex σ_n. To be more precise, the superscripts appearing in a column of the above matrix are the Euclidean coordinates, in integer multiples of $\frac{1}{m}$, of the corresponding vertex of the simplex σ_n. For example,

$$a^0 = \frac{1}{m} \begin{pmatrix} i_1 \\ i_2 \\ \vdots \end{pmatrix}$$

Now, consider the matrix of superscripts of the above tableau:

$$\begin{array}{cccccccccc}
i_1 & \cdots & i_1 & i_1+1 & \cdots & i_1+1 & i_1+1 & \cdots & i_1+1 \\
i_2 & \cdots & i_2 & i_2 & \cdots & i_2 & i_2+1 & \cdots & i_2+1 \\
\vdots & \cdots & \cdots & \cdots & \cdots & \cdots & \cdots & \cdots & \vdots
\end{array}$$

The columns of the above matrix are the Euclidean coordinates, in integer multiples of $1/m$, of the vertices of an arbitrary simplex of the decomposition. The fundamental combinatorial rule can easily be understood if one observes that as we go from one column to the right adjacent one, *one and only one* coordinate *increases* by *exactly* 1. The fundamental combinatorial rule can be formalized as follows: The vertices $\{a^0, ..., a^n\}$ span a simplex of the prismatic decomposition of $[0,1]^n$ iff the vertices can be ordered so that

$$a^{i-1}a^i = \frac{1}{m}\vec{e}_{\pi_i}$$

where $\{e_1, ..., e_i, ..., e_n\}$ is a basis of $\mathbb{R}^n$ with each basis vector aligned with an edge of the cube $[0,1]^n$ and the ordered set $(\pi_1, ..., \pi_n)$ is a permutation of the ordered set $(1, 2, ..., n)$.

The prismatic decomposition, as applied to the parallelogram constructed on a *nonorthonormal* basis $\{\vec{e}_1, \vec{e}_2\}$, is illustrated in Figure 5.3.

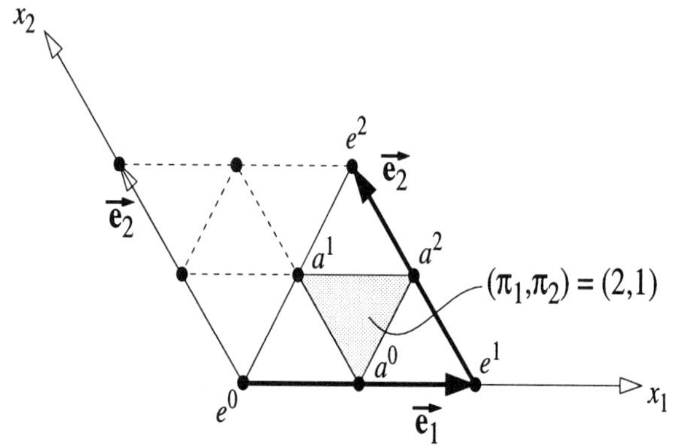

Fig. 5.3. The prismatic versus the Q-triangulation. The Q-triangulation is in solid lines; the dashed lines complete the Q-triangulation of the triangle into the prismatic triangulation of the parallelogram.

5.4 Q-Triangulation

The so-called *Q-triangulation* is a decomposition of the of the domain over which a zero point or fixed point equation is defined. It underlies many algorithms that have appeared in the context of operations research (see [Doup 1988]). The Q-triangulation is more specifically devised for domains that have the geometrical structure of a *simplex*; that is, the Q-triangulation is best suited for refining the accuracy of the solution once the solution has been narrowed down to a simplex of the polyhedron.

Consider the low-resolution simplex $e^0e^1...e^n$. There are two ways to view this simplex:

- This simplex could viewed as lying in an n-dimensional Euclidean space with basis vectors

$$\vec{e}_{i+1} = e^ie^{i+1}, \quad i = 0, ..., n-1$$

Observe that this Euclidean basis need not be orthonormal. The only requirement on the vertices is that they be affinely independent; they are not required to be arranged in some "nice" pattern. This formulation is useful to derive the *purely combinatorial* properties of the triangulation.

- Another possibility is to view $e^0e^1...e^n$ as the *standard n-simplex* embedded in $\mathbb{R}^{n+1}$, where $\{\vec{oe^0}, \vec{oe^1}, ..., \vec{oe^n}\}$ is an orthonormal basis of $\mathbb{R}^{n+1}$. This formulation is preferred to study the regular, lattice pattern properties of the triangulation.

A point a in this simplex can be charted using either Euclidean coordinates relative to the Euclidean basis,

$$\vec{e^0a} = \sum_{i=1}^{n} x_i \vec{e^{i-1}e^i}$$

or the barycentric coordinates relative to the system of vertices $e^0, e^1, ..., e^n$,

$$\vec{oa} = \sum_{i=0}^{n} \lambda_i \vec{oa^i}$$

It is an easy exercise to derive the following relationship between Euclidean and barycentric coordinates:

$$\vec{oa} = \underbrace{(1 - x_1)}_{\lambda_0} \vec{oe^0} + \sum_{i=1}^{n-1} \underbrace{(x_i - x_{i+1})}_{\lambda_i} \vec{oe^i} + \underbrace{x_n}_{\lambda_n} \vec{oe^n}$$

Definition 5.4. *The Q-decomposition of size m of the simplex $e^0e^1...e^n$ has its vertices characterized by barycentric coordinates that are integer multiples $0, 1, 2, ..., m$ of $1/m$. Equivalently, the vertices have Euclidean coordinates that are integer multiples of $1/m$. A collection of $n+1$ vertices, $\{a^0, a^1, ..., a^n\}$, form a simplex whenever there exists an ordering so that*

$$\vec{a^i a^{i+1}} = \frac{1}{m}\vec{e}_{\pi_i}, \quad i = 1, ..., n$$

where $(\pi_1, ..., \pi_n)$ is a permutation of the ordered set $(1, 2, ..., n)$.

The geometric intuition behind this definition is illustrated in Figure 5.4. Also, Figure 5.3 shows the Q-triangulation of a regular 2-simplex embedded in a parallelogram.

5.5 Combinatorial Equivalence

We are now in a position to formulate the combinatorial equivalence between the above two triangulation rules:

Theorem 5.5. *The*

- *prismatic triangulation of the cube $[0, 1]^n$ with grid size m and*

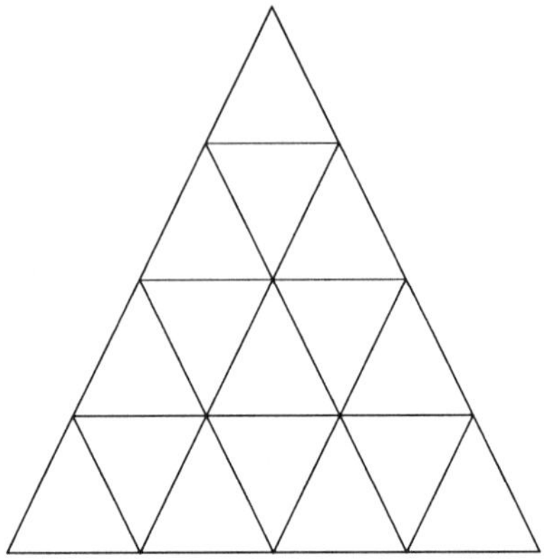

Fig. 5.4. Q-triangulation of a 2-D simplex with $m = 4$.

- *the Q-triangulation of the simplex* $e^0 e^1 ... e^n$ *with size* m

are combinatorially equivalent in the sense that the Euclidean coordinates of their vertices are integer multiples of $1/m$ *and* $a^0...a^n$ *is a simplex whenever, after reordering,* $a^{i-1}a^i = \frac{1}{m}\vec{e}_{\pi_i}; i = 1, ..., n$, *where* $(\pi_1, ..., \pi_n)$ *is a permutation of* $(1, ..., n)$ *and* $\{\vec{e}_i\}$ *is a Euclidean basis.*

This combinatorial equivalence between prismatic and Q-triangulations is illustrated in Figure 5.3.

If we are solely interested in the *combinatorial* properties of the triangulation, we could choose *any* Euclidean basis $\{\vec{e}_i\}$. If, in addition, we would like to provide the triangulation with some regular pattern properties, we choose $e_i = e^i e^{i+1}$, where $\{\vec{oe^i}\}$ is an orthonormal basis of the underlying $\mathbb{R}^{n+1}$ Euclidean space.

5.6 Flatness

For many reasons, a simplicial decomposition of $D \times \Omega$ that has "flat" simplexes should be avoided. For example, the point location problem—that is, the problem of finding *on a digital machine* what simplex σ of $D \times \Omega$ contains a given point p—is made difficult, even ambiguous, when the point p happens to be within or near a "flat" simplex. Flat simplexes in $D \times \Omega$ could create clusters of aligned points in the template, which in turn create problems to triangulate the template. The accuracy of the simplicial computation of the crossover ultimately depends on the flatness of the simplex $b^0 b^1 b^2$ of the template that contains $0 + j0$. We will come back to these issues in Chapter 6.

The flatness problem is twofold. First, there is the need to define a flatness measure for a simplex. Second, and more importantly, if this simplex is iteratively refined in the course of implementing the simplicial approximation theorem, what kind of flatness can be expected from the smaller simplexes?

As in [Freudenthal 1942], we define the "flatness" of a Euclidean n-simplex, Σ_n, as follows:

$$\text{flatness} = \frac{(\text{diameter})^n}{\text{volume}}$$

Recall that the diameter is defined as

$$\text{diameter}(\Sigma_n) = \sup_{x,y \in \Sigma_n} ||x - y||$$

Also recall that this diameter is achieved on an edge. The volume is defined in the usual way: Let

$$\begin{pmatrix} x_1^i \\ x_2^i \\ \vdots \\ x_n^i \end{pmatrix}$$

be the Euclidean coordinates of the vector $\vec{a^0 a^i}$. Then integrating the exterior volume form $dx_1 \wedge dx_2 \wedge \ldots \wedge dx_n$ (see Subsection 19.5.3) over the simplex Σ_n yields the **volume** formula

$$\text{volume}(\Sigma_n) = \frac{1}{n!} \det \begin{pmatrix} x_1^1 & x_1^2 & \ldots & x_1^n \\ x_2^1 & x_2^2 & \ldots & x_2^n \\ \vdots & \vdots & \ddots & \vdots \\ x_n^1 & x_n^2 & \ldots & x_n^n \end{pmatrix}$$

For details, see [Chu 1996].

With this flatness measure, we conjecture that the lowest flatness n-simplex is the standard n-simplex $a^0 \ldots a^n$ embedded in $\mathbb{R}^{n+1}$. Remember, the vertices of this simplex are chosen so that $\{\vec{oa^0}, \ldots, \vec{oa^i}, \ldots, \vec{oa^n}\}$ is an orthonormal basis of $\mathbb{R}^{n+1}$. It is possible to reformulate this lowest flatness conjecture more intrinsically, without reference to the $\mathbb{R}^{n+1}$ space in which the simplex is embedded. This, however, requires some concepts of roots lattice and root system of Lie algebra, which we will not go through here. The interested reader is referred to the Ph.D. dissertation of [Chu 1996].

Regarding the iterative refinement of an n-simplex Σ_n, the Q-triangulation guarantees bounded flatness of the smaller simplexes, as shown by Freudenthal:

Theorem 5.6. (Freudenthal) *Let the n-dimensional simplex Σ_n be iteratively decomposed using the Q-triangulation. Let S_Q be the operation of Q-refinement with* **mesh ratio** $\frac{1}{2}$*; that is, every simplex is Q-refined by breaking all of its edges into two subedges of equal length. Let $\sigma_n^i(S_Q^m(\Sigma_n))$ be an arbitrary n-simplex of the m-fold refinement $S_Q^m(\Sigma_n)$. Then there exists a bound, M, such that*

$$\textit{flatness}(\sigma_n^i(S_Q^m(\Sigma_n))) \le M$$

for all i's and for all m's.

Proof. See [Freudenthal 1942] for a high-level, conceptual proof; see [Chu 1996] for a more explicit, geometrical proof based on lattice theory. ■

The theorem can be extended in a trivial matter to the case where S_Q is the Q-refinement with an arbitrary mesh ratio.

HISTORICAL NOTES

The simplicial decomposition of the Cartesian product of two simplexes has roots tracing back to [Freudenthal 1937b]. At that time, this decomposition did not receive any specific name. Only later did it become known under *prismatic decomposition.* The prismatic decomposition has ever since been extensively used, under that name, by topologists (see, e.g., [Hocking and Young 1961]).

L. E. J. Brouwer initiated the problem as to whether there exists a simplicial refinement rule such that, if it is iteratively used to produce finer and finer triangulations of a given polyhedron, no “long and skinny” simplexes would appear at any stage. We can only conjecture that Brouwer formulated this problem after he observed that the barycentric subdivision, which he invented along with his simplicial approximation theorem, *does* produce simplexes with of flat aspect ratio. Whether Brouwer initiated this problem as a pure combinatorial–geometrical puzzle or whether he had in his mind a simplicial algorithm for solving his famous fixed-point theorem is a fascinating question for mathematical historians.

Freudenthal responded to Brouwer's “flatness” challenge. Using the fundamental combinatorial rules of the prismatic triangulation, which he had invented a few years earlier, Freudenthal derived a refinement rule for the simplex—a refinement rule that became known under *Q-triangulation*—and showed that it answered Brouwer's question in the affirmative.

The combinatorial formulation of the Q-triangulation in terms of permutation of indices is already in the work of [Freudenthal 1942] and has ever since been extensively exploited by operation researchers (see, e.g., [Doup 1988]).

6

COMPUTATIONAL GEOMETRY

SUMMARY

In this chapter, we develop an implementation of the simplicial approximation theorem based on *computational geometry*, with the objective of developing a combinatorial algorithm for computing an approximate crossover in the uncertainty space.

The starting point is an initial, coarse triangulation—for example, Q-triangulation—of the uncertainty space $D \times \Omega$ with vertex set $\{a^i\}$. Next, we map those vertices to the complex plane. Contrary to the regular vertex pattern of a Q-triangulated uncertainty space, the vertex set $\{b^i = f(a^i)\}$ is a disorderly constellation of points in the complex plane, and finding a triangulation that hinges on the $\{b^i\}$ is not a trivial problem. Intuitively, we would connect each vertex to its closest neighbor. Although nontrivial, this is a *well-understood* problem of *computational geometry*. We construct the *Voronoi diagram* of this set of points [Preparata and Shamos 1985], [Edelsbrunner 1987]. Essentially, the Voronoi diagram is a partition of the plane into polygonal cells; each cell encompasses one and only one vertex b^j; the cell of b^j is defined as the set of points $b \in \mathbb{C}$ that have b^j as closest vertex (think about an air traffic, collision avoidance problem). The (straight edge) graph-theoretic dual of the Voronoi diagram provides a triangulation, called the *Delaunay triangulation* [Preparata and Shamos 1985], [Edelsbrunner 1987].

From there on, we apply the techniques developed in the proof of the simplicial approximation theorem: We *freeze* the Delaunay triangulation of the Nyquist template, we *refine* the triangulation of $D \times \Omega$, and the new vertices are assigned to existing vertices of the Delaunay triangulation using some ad hoc implementation of the "star condition" until a simplicial map $(D \times \Omega)' \to N$ is obtained. Then, through efficient search, we chase all of those simplexes of $(D \times \Omega)'$ mapped simplicially to the unique simplex $b^0 b^1 b^2$ that contains $0 + j0$. This yields an assembly of simplexes that is an approximate crossover.

As a byproduct of this construction, some generic combinatorial properties of simplicial maps from a *high-dimensional* uncertainty polyhedron to the *two-dimensional* plane are derived.

Finally, some variants of this construction, departing from the strict lines of the simplicial approximation theorem, are also proposed.

6.1 Delaunay Triangulation of Template

The starting point of a robust stability crossover computation is an initial, "coarse" triangulation of $D \times \Omega$. The only requirement on this triangulation is that it be fine enough to reveal the topological subtleties of $D \times \Omega$. As such, the triangulation is subject to a minimum number of vertices, edges, and so on. (Remember the "simplicial torus" example of Chapter 4.) In addition to this minimum requirement, we might somewhat refine the triangulation for better accuracy of the algorithm. In principle, *any* refinement rule is acceptable. However, for reasons already explained, it is desirable, although not absolutely necessary, to have a *prismatic* or Q-refinement of $D \times \Omega$. Indeed, this triangulation respects the Cartesian product structure of the uncertainty space and more importantly has bounded flatness. Although a finer initial triangulation improves accuracy, going for a very fine initial triangulation is not necessary, even not recommended, because indeed there are other factors involved in the accuracy of the algorithm. One such critical factor is the flatness of the simplexes of N; and should the simplex containing $0 + j0$ have minimum flatness (i.e., be a regular triangle), we obtain extremely accurate results even with a coarse initial triangulation.

Next, we need a triangulation of $f(D \times \Omega)$. Since the very purpose of a fast robust stability check is precisely to avoid computing the whole image, the most natural way to proceed is to map the vertices a^i of the triangulation of $D \times \Omega$ using the exact Nyquist map and to use the resulting points $b^i = f(a^i)$ as vertices of a triangulation of the Nyquist template, yet to be determined.

To triangulate N, given that its vertices are a priori specified, the most natural procedure is to connect each vertex b^i to its closest neighbors. This is a nontrivial but well-understood problem of computational geometry (see [Edelsbrunner 1987, Chapter 13]). Given a constellation of vertices b^i, we first partition $\mathbb{C}$ into *cells*; each cell contains *exactly one* vertex, and all points within the cell of b^i have b^i as closest vertex. More formally, we have:

Definition 6.1. (Voronoi Diagram) *The* **Voronoi cell** $VC(b^i)$ *of a vertex* b^i *is defined as* $\{s \in \mathbb{C} : |b^i - s| < |b^j - s|, \forall j \neq i\}$. *The* **Voronoi diagram** $VD(\{b^i\})$ *of a set of points is the collection of all Voronoi cells together with their boundaries. The Voronoi diagram is said to be* **simple** *whenever every vertex of* VD *has exactly three edges incident upon it.*

The Voronoi diagram can be formalized as a *cell complex* in the sense of Section 4.8. The cells e_n^i are defined as follows:

- The 2-cells are $e_2^i = VC(b^i)$.
- The 1-cells are defined by $|e_1^{ij}| = \text{int}(\overline{VC(b^i)} \cap \overline{VC(b^j)})$, provided that

$\overline{VC(b^i)} \cap \overline{VC(b^j)} \neq \emptyset, \neq \{point\}$; $i \neq j$; furthermore, the cells are oriented so that $e_1^{ij} = -e_1^{ji}$.

- The 0-cells are defined by $|e_0^{ijk}| = \overline{VC(b^i)} \cap \overline{VC(b^j)} \cap \overline{VC(b^k)}$, provided that the latter intersection is $\neq \emptyset$; furthermore, the cells are oriented so that $e_0^{\pi_i \pi_j \pi_k} = \text{sign}(\pi) e_0^{ijk}$, where (π_i, π_j, π_k) is a permutation of the ordered set (i, j, k).

The facing relation $e_{n-1} \preceq e_n$ is defined by $e_{n-1} \subseteq \overline{e_n}$ and the incidence numbers are defined as follows:

- The 2-1 incidence numbers are defined by $[e_2^i : e_1^{ij}] = +1$ and $[e_2^i : e_1^{jk}] = 0$ when $i \notin \{j, k\}$.
- The 1-0 incidence numbers are defined by $[e_1^{ij} : e_0^{ikl}] = +1$ and $[e_1^{ij} : e_0^{klm}] = 0$ whenever $\{i, j\} \cap \{k, l, m\} = \emptyset$.

Furthermore, the incidence relation is bilinear in the sense that $[-e_n : e_{n-1}] = [e_n : -e_{n-1}] = -[e_n : e_{n-1}]$. It is easily seen that the fundamental relation $\sum_{e_{n-1}} [e_n : e_{n-1}][e_{n-1} : e_{n-2}] = 0$ is satisfied.

From now on we assume that the Voronoi diagram is simple, because the opposite situation is rather pathological.

The straight-edge graph-theoretic dual of the Voronoi diagram provides the desired triangulation:

Definition 6.2. (Delaunay Triangulation) *The* **(completed) Delaunay triangulation** *of* $\{b^i\}$, $DT(\{b^i\})$, *is the simplicial complex, the vertices of which are the* b^i*'s, and the vertex scheme of which is the following:* $b^i b^j$ *is an edge iff* $\overline{VC(b^i)} \cap \overline{VC(b^j)} \neq \emptyset$; *and* $b^i b^j b^k$ *is a 2-simplex whenever* $\overline{VC(b^i)} \cap \overline{VC(b^j)} \cap \overline{VC(b^k)} \neq \emptyset$.

An illustration of the Voronoi diagram/Delaunay triangulation is shown in Figure 6.1. A control-theoretic illustration is provided by Figure 6.3 and Figure 6.4 which show the Voronoi diagram and Delaunay triangulation, respectively, of the Horowitz supertemplate of the example of Section 21.15.

The interesting fact is that, precisely in dimension 2, the Voronoi and Delaunay constructions admit fast implementations:

Theorem 6.3. *The Voronoi diagram and the Delaunay triangulation of* $\{b^i : i = 1, ..., n_u\}$ *can be constructed in* $0(n_u \log n_u)$ *operations.*

Proof. See [Edelsbrunner 1987, Corollary 13.6]. ■

At this stage, we have an initial, coarse triangulation of $D \times \Omega$ and the Delaunay triangulation of N. The next step is to construct a simplicial map between the two, by refining the triangulation of $D \times \Omega$, *without altering the Delaunay triangulation of* N. From the formal proof of Section 4.4, the accuracy of the simplicial approximation is limited by the mesh of $DT(N)$. However, under some "flatness" conditions, it turns out

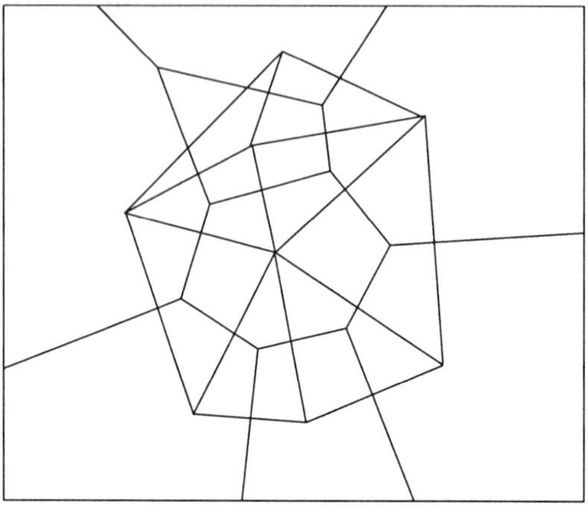

Fig. 6.1. Voronoi diagram and Delaunay triangulation.

that preimage $f^{-1}(0+j0)$ can be reconstructed from the simplicial *approximation* with *infinite precision*, despite the finite refinement of $DT(N)$; see Section 6.4.

6.2 Simplicial Edge Mapping

If we want to construct a simplicial map between the triangulation of $D\times\Omega$ and the Delaunay triangulation of $\{b^i\}$, the problem we will face is that some simplexes of $D \times \Omega$ will *not* mapped to simplexes of the Delaunay triangulation of N. One of the factors that compounds this difficulty is that $D \times \Omega$ is in general of a much higher dimension than N. Clearly, an n-simplex, $n > 2$, of $D\times\Omega$ could not possibly be mapped to a 2-simplex, a 1-simplex, or a 0-simplex of N, unless at least $n-1$ vertices a^i are mapped to the same vertex b. Even the 1-simplexes of $D \times \Omega$ could not be all mapped to edges of the Delaunay triangulation, because $D \times \Omega$ has many more edges than N. To see this, let $D \times \Omega$ be an n-simplex, $n \geq 2$. The number of edges of an n-simplex of $D \times \Omega$ is the binomial coefficient

$$\#(\text{edges of an } n\text{-simplex of } D \times \Omega) = \binom{n+1}{2} = \frac{n(n+1)}{2}$$

On the other hand, the number of edges of the Delaunay triangulation equals the number of edges of the Voronoi diagram, because the Delaunay triangulation is the dual of the Voronoi diagram. The number of edges of the Voronoi diagram with $n+1$ sites is at most $3(n+1)-6$, and the bound is tight (see [Edelsbrunner 1987, Corollary 13.7]). Therefore we have the following corollary:

$$\#(\text{ edges of Delaunay triangulation}) \leq 3n-3$$

Therefore

Corollary 6.1. *The number of excess edges—that is, the number of edges of an n-simplex* $a^0a^1...a^n$, $n \geq 2$, *of* $D \times \Omega$ *minus the number of edges of the Delaunay triangulation of* $\{f(a^i)\}$*—is greater than or equal to*

$$\binom{n+1}{2} - (3n-3) = \frac{n^2 - 5n + 6}{2}$$

This lower bound on the excess edge is tight, vanishes for $n = 2, 3$ *and then grows as* $0(n^2)$ *for* $n > 3$.

From the above corollary, we see the discrepancy between number of edges of $D \times \Omega$ and number of edges in the Delaunay triangulation. This *excess edge* is somehow a measure of the difficulty of the problem, which remains polynomial.

We now look at the general situation: The problem is to guarantee that *all* simplexes of *all* dimensions of $D \times \Omega$ are mapped to simplexes of $DT(N)$. We proceed to show that, generically, it suffices to check that all *edges* are mapped simplicially. The proof is based on a recursion on the dimension. The starting point is the following:

Theorem 6.4. *Let* $n \geq 3$. *If all simplexes* σ_n *of* $D \times \Omega$ *are mapped to simplexes of* $DT(N)$, *then all simplexes* σ_{n+1} *of* $D \times \Omega$ *are also mapped to simplexes of* $DT(N)$.

Proof. Consider an arbitrary simplex $\sigma_{n+1} = a^0a^1...a^na^{n+1}$ of $D \times \Omega$. Consider its n-face $a^0...a^n$. This n-face is an n-simplex, and therefore, by induction hypothesis, this n-face must be mapped to a simplex of $DT(N)$. Let $b^0b^1b^2$ be this image, where, without loss of generality, the indices have been relabeled such that $\varphi(a^0) = b^0, \varphi(a^1) = b^1, \varphi(a^2) = b^2$. Observe that, because $n \geq 3$, there are at least two vertices of $a^0...a^n$ that are mapped to one single vertex of $b^0b^1b^2$. Let a^k, a^l be two such vertices mapped to b^0, after yet another relabeling. Clearly, the removal of the vertex a^k from $a^0...a^{n+1}$ does not affect the image under the simplicial map $\tilde{f}$, because $\tilde{f}(a^k)$ is already represented by $\tilde{f}(a^l)$. To be more specific, we have $\tilde{f}(a^0...a^{n+1}) = \tilde{f}(a^0...a^{k-1}a^{k+1}...a^{n+1})$. But $a^0...a^{k-1}a^{k+1}...a^{n+1}$ is an n-simplex. Therefore, by induction hypothesis, it must be mapped to a simplex of $DT(N)$. Thus $\tilde{f}(a^0...a^{k-1}a^{k+1}...a^{n+1})$ is a simplex of $DT(N)$ and so is $\tilde{f}(a^0...a^{n+1})$, as claimed. ■

The following theorem is telling us that we can actually start the recursion on the dimension at $n = 2$. The reason why we have not incorporated this case in the previous theorem is that the argument is different:

Theorem 6.5. *If all* 1*-simplexes and all* 2*-simplexes of* $D \times \Omega$ *are mapped to simplexes of* $DT(N)$, *then all* 3*-simplexes of* $D \times \Omega$ *are also mapped to simplexes of* $DT(N)$.

Proof. Consider the 3-simplex, the tetrahedron, $a^0a^1a^2a^3$ of $D \times \Omega$. We have to prove that $a^0a^1a^2a^3$ is mapped to a simplex of $DT(N)$. Consider the face $a^0a^1a^2$. Since it is a 2-simplex, it is mapped to a simplex of $DT(N)$. Let $b^0b^1b^2$ be this simplex. To begin, we assume that $b^0b^1b^2$ is a 2-simplex. Thus $b^0b^1b^2$ is a 2-simplex and as such it does not contain any other vertices in its closure. Now, consider the face $a^3a^1a^2$ resulting from a trivial substitution of a vertex of the face $a^0a^1a^2$. Let the image be $b^3b^1b^2$. Observe that b^3 is not allowed to lay on any of the open simplexes b^0b^1, b^0b^2, b^1b^2, because otherwise some edges of the tetrahedron would not be mapped to simplexes. It is claimed that $b^3 = b^0, b^1$, or b^2. Indeed, in the opposite situation, since b^3 is not allowed anywhere on the perimeter of the triangle $b^0b^1b^2$, by the Jordan curve theorem, b^3 would have to be either inside $b^0b^1b^2$ or outside $b^0b^1b^2$. But the vertex b^3 could not be inside the triangle $b^0b^1b^2$ because $b^0b^1b^2$ is a simplex. Therefore, b^3 would have to be outside the triangle $b^0b^1b^2$. It follows that $\text{conv}(b^0, b^1, b^2, b^3)$ is either a quadrilatere or a triangle. If the convex hull were a quadrilatere, some edges of $a^0a^1a^2a^3$ would be mapped to the *intersecting* diagonals of the quadrilatere, thereby violating the fact that all edges are mapped simplicially. On the other hand, if the convex hull were a triangle, a face of the simplex $a^0a^1a^2a^3$ would be mapped to the triangular hull, which itself consists of three triangles, thereby violating the fact that all 2-faces are mapped simplicially. Therefore, $b^3 = b^0, b^1$, or b^2. It follows that the image of the 3-simplex $a^0a^1a^2a^3$ is $b^0b^1b^2$.

We finally look at the case where $b^0b^1b^2$ is not a 2-simplex. If $b^0b^1b^2$ happens to be a 0-simplex, it follows that $\tilde{f}(a^0a^1a^2a^3) = \tilde{f}(a^0a^3)$. By induction hypothesis, $\tilde{f}(a^0a^3)$ is a simplex of $DT(N)$ and therefore so is $\tilde{f}(a^0a^1a^2a^3)$, as claimed. Finally, assume $b^0b^1b^2$ is a 1-simplex—that is, $b^0 = b^1$—after possible relabeling. In this case, we have $\tilde{f}(a^0a^1a^2a^3) = \tilde{f}(a^0a^2a^3)$ and the result follows again from the induction hypothesis. The theorem is proved. ■

We summarize the previous two theorems in the following:

Corollary 6.2. *If all 1-simplexes and all 2-simplexes of $D \times \Omega$ are mapped to simplexes of $DT(N)$, then all simplexes of all dimensions of $D \times \Omega$ are mapped to simplexes of $DT(N)$.*

From this corollary, we learn that all we have to do is to make sure that edges and triangles of $D \times \Omega$ are mapped to simplexes of $DT(N)$. Actually, we can go even further... .

Theorem 6.6. *Generically, if all edges of $D \times \Omega$ are mapped simplicially, then all 2-faces are mapped simplicially.*

Proof. Since the edges are already mapped to simplexes, the only case of nonsimplicial mapping is a 2-face of $D \times \Omega$ mapped to a triangle of $DT(N)$

that contains a vertex b^i. This is, however, a rather exceptional situation. This corresponds to the Voronoi cell of b^i being a triangle, and Voronoi cells are usually polygons, much more complicated than mere triangles (see [Edelsbrunner 1987, Figure 13.3]). ■

Even when a nongeneric case occurs, the situation is easily rectified. Should a 2-face of $D \times \Omega$ be not properly mapped, compute the barycenter of that face, decompose the complex $D \times \Omega$ consistently with this added vertex, and assign this new barycenter to the center of the Voronoi cell in which its image falls. Observe that we are not restricted to choosing the barycenter as extra vertex to decompose the 2-face not simplicially mapped; actually, any point of that face would do. For more insight into this situation, see that part of Section 7.3 relevant to Figure 7.1.

Finally we put all of the pieces together:

Corollary 6.3. *Generically, the map from a subdivision of $D \times \Omega$ to the Delaunay triangulation $DT(\{b^i\})$ is simplicial whenever all edges of the subdivision of $D \times \Omega$ are mapped to simplexes of $DT(\{b^i\})$.*

6.3 The SimplicialVIEW Software

Here, we briefly explain how all of the ideas of the previous sections are "put together" in the SimplicialVIEW package, developed by Mr. Coutinho. Essentially, this package provides a software implementation of the simplicial approximation theorem. These are the key steps of the algorithm:

1. Obtain coarse Q-triangulation of $D \times \Omega$ and map vertices $\{a^i\}$ to the complex plane, $a^i \mapsto b^i = f(a^i)$.
2. Compute Voronoi diagram and Delaunay triangulation of $\{b^i\}$.
3. Refine the triangulation of $D\times\Omega$ and assign the new vertices of $(D\times\Omega)'$ to the *previously computed* b^j's, keeping the error to a minimum; this is referred to as *point location problem.*
4. Check the simplicial property; if it fails, go back to the refinement step.
5. Identify the unique simplex $b^0b^1b^2$ of $DT(\{b^i\})$ that contains $0 + j0$.
6. Search those simplexes of $D \times \Omega$ mapped simplicially to $b^0b^1b^2$. This assembly of simplexes is the approximate crossover.

6.3.1 Coarse Initial Refinement

For the initial refinement, systematic use is made of the Q-triangulation, prismatic decomposition.

6.3.2 Voronoi Diagram and Delaunay Triangulation

To compute the Voronoi diagram of $\{b^i\}$, the sweepline technique is used.

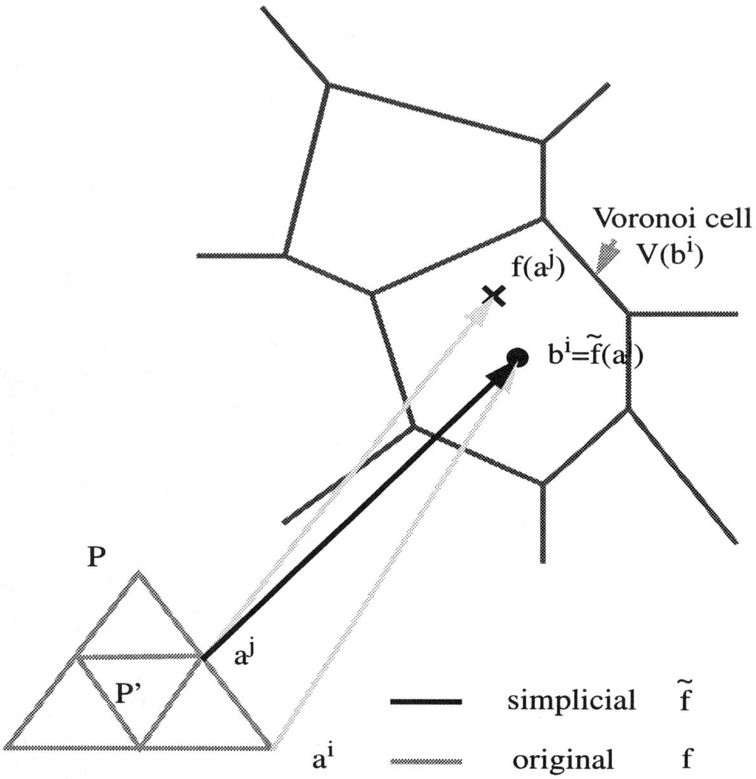

Fig. 6.2. Point location problem.

6.3.3 Refinement and Point Location

The newly created vertices should be assigned to vertices of $DT(\{b^i\})$. Rather than implementing the star conditions, we assign $\tilde{f}(a'^i)$ to the nearest vertex of $DT(N)$. In other words, we have to identify the Voronoi cell in which $f(a'^i)$ falls. This is the **point location problem**, illustrated in Figure 6.2.

6.3.4 Checking Simplicial Property

The procedure is to check all edges, making sure that they are simplicially mapped.

6.3.5 Identifying Simplex Containing the Origin

This is again a point location problem.

6.3.6 Inverse Image

Once the map has been made simplicial, we have to find those simplexes mapped into $b^0b^1b^2$.

6.4 Numerical Stability, Flatness, and Conditioning

The fundamental result is the following:

Theorem 6.7. *Consider a sequence of simplicial maps* $\tilde{f}^k : (D\times\Omega)^k \to N$, *where* $(D \times \Omega)^k$ *is a refinement of* $(D \times \Omega)^{k-1}$, *and such that*

$$\lim_{k\to\infty} mesh((D \times \Omega)^k) = 0$$

Let $(\tilde{f}^k)^{-1}(b^0b^1b^2)$ *be the collection of all simplexes of* $(D \times \Omega)^k$ *mapped simplicially to the whole triangle* $b^0b^1b^2$ *under* $\tilde{f}^k$. *Then*

$$\lim_{k\to\infty} (\tilde{f}^k)^{-1}(b^0b^1b^2) \subseteq f^{-1}(\cap_{i=0}^{2}\overline{VC}(b^i))$$

Proof. See [Coutinho and Jonckheere 1995]. ■

It follows that the computational geometry algorithm is **numerically stable**, in the sense that the computed solution is included in the *exact* solution to a *nearby* problem. The nearby problem is the preimage of the intersection of the closed Voronoi cells, rather than the preimage of $0 + j0$. Therefore, numerical stability depends on the mismatch

$$|(0 + j0) - \cap_{i=0}^{2}\overline{VC}(b^i)|$$

This mismatch is easily seen to be related to the flatness (see Section 5.6) of the simplex $b^0b^1b^2$. If this flatness is deemed unacceptable, the situation can be corrected by means of a projective transformation mapping $b^0b^1b^2$ to a regular triangle with its barycenter at $0 + j0$.

The issue of **conditioning** of the problem is, by definition, the mismatch between $f^{-1}(0 + j0)$ and $f^{-1}(\cap_{i=0}^{2}\overline{VC}(b^i))$—that is, the sensitivity of the (exact) solution to data perturbation. Numerical conditioning is crucially related to the differential topology of the map and is relegated to Chapter 23.

6.5 Making Map (Locally) Simplicial

The previous scheme was based on checking *all* edges of $D \times \Omega$ and making sure that they were mapped simplicially. This procedure might be deemed too long, but it provides the *global* simplicial map. If, however, our only interest is the *crossover*, then it would be enough to construct a *local* version of the simplicial map—that is, a simplicial map restricted to that part of $D \times \Omega$ approximately mapped to $0 + j0$.

The idea is the following: Let $b^0b^1b^2$ be the unique simplex of the Delaunay triangulation of N that contains $0 + j0$. If we trace back $b^0b^1b^2$

into $D \times \Omega$, all we will find are three points, say a^0, a^1, a^2, located at remote corners of $D \times \Omega$, so that the computation of the simplicial map can be thought of as the procedure to connect these three points be an assembly of simplexes approximating the crossover.

The procedure is as follows: Take, for example the inverse image of b^0, a^0. Since a^0 is approximately mapped to $0 + j0$, some other points around a^0 will also be mapped close to b^0, b^1, or b^2. Therefore, do a decomposition of $\mathrm{star}(a^0)$ and do a stellar decomposition of the interface between the refinement $(\mathrm{star}(a^0))'$ and $(D \times \Omega) \backslash \mathrm{star}(a^0)$. Map the vertices of $(\mathrm{star}(a^0))'$. At least one of these vertices, say a'^i, will fall in the Voronoi cell of either b^1 or b^2. Assume $f(a'^i) \in VC(b^1)$. (We say that $a'i$ has *label* 1; see Chapter 9.) At this stage, we have a 1-simplex $a^0 a'^i$ of a refinement of $D \times \Omega$ mapped to $b^0 b^1$. Now, do a decomposition of the $\mathrm{star}(a^0 a'^i)$ together with the necessary refinement of $(D \times \Omega)'$. From there on, the procedure is iterated until it gives simplexes of $D \times \Omega$, all mapped into $b^0 b^1 b^2$.

6.6 procedure

We illustrate another anisotropic gridding procedure on a simple example. Consider a 3-simplex $a^0 a^1 a^2 a^3$ of $D \times \Omega$, and let the Delaunay triangulation of $\{b^i = f(a^i)\}$ have the following vertex scheme

$$\begin{array}{l} b^0, b^1, b^2, b^3 \\ b^0 b^1, b^1 b^2, b^2 b^3, b^3 b^0, b^0 b^2 \\ b^0 b^1 b^2, b^0 b^2 b^3 \end{array}$$

The problem is that the edge $a^1 a^3$ is not mapped into an edge of the Delaunay triangulation. Since *the whole* edge $a^1 a^3$ cannot be mapped in a simplicial manner, the only way to proceed is to decompose the edge $a^1 a^3$ with the hope that the resulting pieces could be mapped in a simplicial manner. We choose the most trivial decomposition of $a^1 a^3$, the *barycentric* subdivision

$$a^1 a^3 = a^1 a^{1,3} + a^{1,3} a^3$$

where $a^{1,3}$ is the *barycenter*, i.e., center of gravity of $a^1 a^3$. Once this vertex has been added, to preserve the simplicial complex property, $D \times \Omega$ has to be subdivided into a refined simplicial complex with vertex scheme

$$\begin{array}{l} a^0, a^1, a^2, a^3, a^{1,3} \\ a^0 a^1, a^0 a^{1,3}, a^0 a^3, a^2 a^1, a^2 a^{1,3}, a^2 a^3 \\ a^0 a^1 a^2, a^0 a^{1,3} a^2, a^0 a^2 a^3, a^0 a^1 a^{1,3}, a^0 a^{1,3} a^3, a^1 a^{1,3} a^2, a^3 a^2 a^{1,3} \\ a^0 a^1 a^2 a^{1,3}, a^0 a^3 a^2 a^{1,3} \end{array}$$

The extra vertex $a^{1,3}$ that has been added to the simplicial complex of $D \times \Omega$ *must* be given an image under $\tilde{f}$, by the very definition of the simplicial map. Assigning $\tilde{f}(a^{1,3})$ its *exact* value $f(a^{1,3})$ would create another vertex in N, $f(a^{1,3})$, and this would require re-triangulating N, resulting in an endless cycle. Therefore, we renounce to add another vertex to N, and we rather assign $\tilde{f}(a^{1,3})$ an already existing vertex of N, without of course jeopardizing the simplicial property of $\tilde{f}$. To preserve the simplicial property, the only way that $a^1 a^{1,3} + a^{1,3} a^3$ could possibly be mapped is into either $b^1 b^0 + b^0 b^3$ or $b^1 b^2 + b^2 b^3$. This means that $a^{1,3}$ must be mapped into either b^0 or b^2. This assignment is made so as to minimize the error

$$\min_{i=0,1} |f(a^{1,3}) - f(a^i)|$$

Let a^i be the solution. Then we complete the definition of $\tilde{f}$ by

$$\tilde{f}(a^{1,3}) = f(a^i)$$

We first propose a procedure to fix the edges: The idea is to take all edges of $D \times \Omega$ that are not properly mapped, do a barycentric subdivision of each of them, and assign each barycenter to the center of the Voronoi cell in which its image under f falls.

Procedure: edge mapping

$E = \{a^k a^l \in C_1(K) : f(a^k)f(a^l)$ is not a simplex of $DT(N)\}$

while $E \neq \emptyset$ **do**

for each $a^k a^l \in E$

a^{kl} = barycenter of $a^k a^l$

Find $VC(b^i) \ni f(a^{kl})$

$\tilde{f}(a^{kl}) = b^i$

endfor

Find refined complex K' consistent with barycentric subdivision of selected edges

$K \leftarrow K'$

$E = \{a^k a^l \in C_1(K) : \tilde{f}(a^k)\tilde{f}(a^l)$ is not a simplex of $DT(N)\}$

endwhile

Observe that the above is *not* the usual, isotropic barycentric subdivision that indiscriminately subdivides *all* simplexes. Here, we decompose an edge *only if* it is not properly mapped.

BIBLIOGRAPHICAL AND HISTORICAL NOTES

The historical roots of the Voronoi diagram / Delaunay triangulation is the classical *tessellation* problem of Kepler—that is, the problem of distributing in some regular pattern the celestial bodies on the great sphere of the firmament.

The Voronoi diagram appeared around the turn of this century while the Delaunay triangulation was developed later in the 1930s.

The first systematic treatment of what became known as *computational geometry* was [Preparata and Shamos 1985]; for a more up-to-date book, see [Edelsbrunner 1987].

For generalizations of the Voronoi diagram to other spaces that $\mathbb{R}^2$ or other distances that the Euclidean distance, see [Klein 1987].

The applications of Voronoi diagrams to error correcting codes are tracing back to [Conway and Sloane 1982]; for a more up-to-date, comprehensive exposition, see [Conway and Sloane 1991].

More recently, the applications of Voronoi diagrams/Delaunay triangulations have exploded and have even been applied to cortical mappings; see [Martinetz and Schulten 1994].

A 3-D generalization of Voronoi diagram was developed by our student, Mr. Coutinho. In addition to providing the 3-D extension, this code allows for fast dynamic update of the diagram when a new site is added, with in view Air Traffic Control applications.

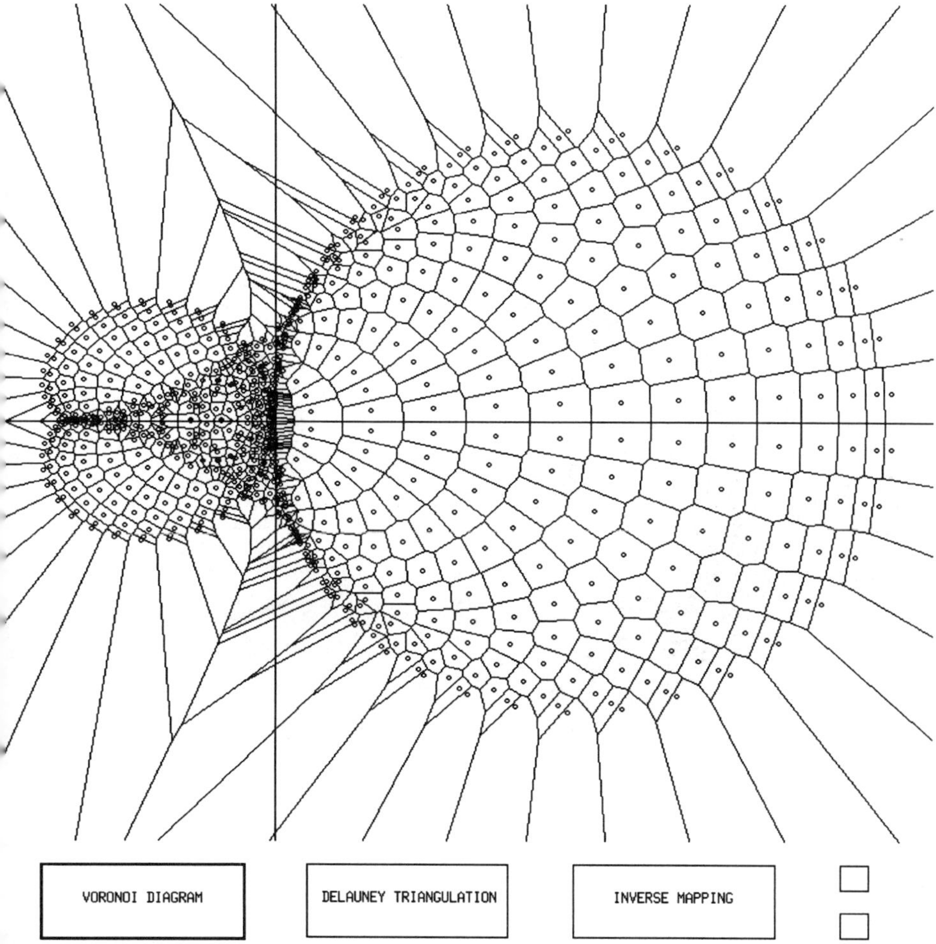

Fig. 6.3. Voronoi diagram of Horowitz supertemplate of a robust stability problem. The number of vertices has been deliberately inflated to reveal the "clustering" of the sample vertices around the so-called critical value curves; see Chapter 21. The underlying robust stability problem is that of Section 21.15. This pictures was generated by an early version of the SimplicialVIEW software developed by Dr. Coutinho.

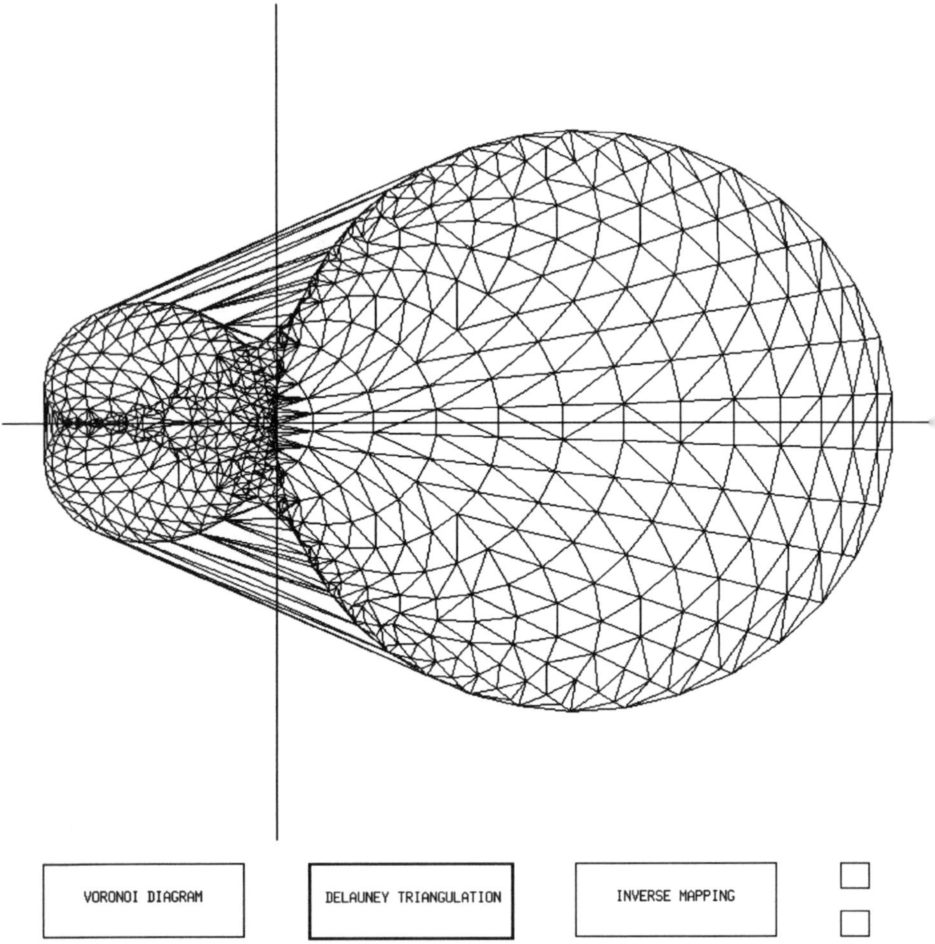

Fig. 6.4. The Delaunay triangulation, dual of the Voronoi diagram of Figure 6.3.

7

PIECEWISE-LINEAR NYQUIST MAP

SUMMARY

Computing the crossover amounts to solving the Nyquist equation $f(p) = 0 + j0$. Assuming this equation is defined over the *polyhedron* $D \times \Omega$, the natural approach to implement an ad hoc solution in the realm of finite computation is to map the vertex set $\{a^i\}$ of the domain $D \times \Omega$ to the complex plane, $a^i \mapsto b^i = f(a^i)$, and then work with the *piecewise-linear extension* of the vertex transformation $a^i \mapsto b^i$. A typical feature of the resulting map is that the image of a simplex is a convex polygon, so that this piecewise-linear extension is not, in general, simplicial. However, an attractive feature of the piecewise-linear extension is that computation of the approximate crossover can be formulated as a linear program. Next, since the piecewise-linear extension lacks simplicial property, the standard fixup would be to compute a simplicial *approximation* to the piecewise-linear map. However, a conceptually more elegant approach is to make the piecewise-linear extension *itself* simplicial relative to *refined* subdivisions of *both* $D \times \Omega$ and N. The reward is that crossover computation on a simplicial piecewise-linear map amounts to pure combinatorics, and the need for a linear program is obviated. It follows that, in a certain sense, simplicial approximation and linear programming are equivalent.

7.1 Piecewise-Linear Nyquist Map

We start with a simplicial dissection of the polyhedron $D \times \Omega$ with vertex set $\{a^i\}$. Let $b^i = f(a^i)$. To define the piecewise-linear extension of $a^i \mapsto b^i$, take an arbitrary $p \in D \times \Omega$. There exists a unique simplex $\sigma_n = a^0 a^1 ... a^n$ such that $\sigma_n \ni p$. Write p in terms of barycentric coordinates

$$p = \sum_{i=0}^{n} \lambda_i a^i$$

Then the **piecewise-linear extension** of the vertex transformation is defined as

$$f_{PL}(p) = \sum_{i=0}^{n} \lambda_i f(a^i)$$

Observe that this map is not, in general, simplicial. Indeed, the image of the simplex $a^0a^1...a^n$ under this map is the convex hull of

$$\{f(a^0), f(a^1), ..., f(a^n)\}$$

which does not, in general, degenerate in a simplex.

Suppose we want to compute the approximate crossover $f_{PL}^{-1}(0 + j0)$. This involves two interrelated problems:

- Find all σ's such that $f_{PL}(\sigma) \ni 0 + j0$.
- For each such σ, compute $(f_{PL}|\sigma)^{-1}(0 + j0)$.

7.1.1 A Linear Program

We develop a benchmark linear program of good conceptual, but of poor computational, value. It addresses the problem of deciding whether a given simplex σ_n contains some solution to $f_{PL}(p) = 0 + j0$.

Let p be a point in the arbitrary simplex σ_n represented by the vector of barycentric coordinates,

$$\lambda = \begin{pmatrix} \lambda_0 \\ \lambda_1 \\ \vdots \\ \lambda_n \end{pmatrix}$$

The condition $\sum_{i=0}^{n} \lambda_i = 1$ is written, in matrix representation, as

$$e^T \lambda = 1$$

where $e = (1, 1, ...)^T$. Let F be the matrix of two-dimensional Euclidean coordinates of the images under f of the vertices of the polyhedron of uncertainties

$$F = \begin{pmatrix} f(a^0) & f(a^1) & f(a^2) & ... \end{pmatrix}$$

The linear program for finding the crossover can be written as

$$\begin{aligned} \begin{pmatrix} F \\ e^T \end{pmatrix} \lambda &= \begin{pmatrix} 0 \\ 1 \end{pmatrix} \\ \lambda &\geq 0 \end{aligned}$$

To recover the standard linear programming formulation in terms of a linear performance index, introduce the slack variables $\mu \geq 0$. After introduction of the slack variables the linear program becomes

$$\min e^T \mu$$

subject to

$$\begin{pmatrix} I & 0 & F \\ 0 & 1 & e^T \end{pmatrix} \begin{pmatrix} \mu_1 \\ \mu_2 \\ \lambda \end{pmatrix} = \begin{pmatrix} 0 \\ 1 \end{pmatrix} \tag{7.1}$$

$$\begin{pmatrix} \mu \\ \lambda \end{pmatrix} \geq 0 \qquad (7.2)$$

In other words, the linear programming "tableau" is

$$\begin{matrix} I & 0 & F & 0 \\ 0 & 1 & e^T & 1 \\ e^T & 1 & 0 & 0 \end{matrix}$$

We immediately see that $\mu = \begin{pmatrix} \mu_1 \\ \mu_2 \end{pmatrix}$ are *basic* variables, except for the fact that they have some associated cost. To get a *canonical* linear program, we remove by row operation on the tableau the cost associated with the basic variables; this yields the canonical tableau

$$\begin{matrix} I & 0 & F & 0 \\ 0 & 1 & e^T & 1 \\ 0 & 0 & (-e^T - e^T F) & -1 \end{matrix}$$

In this latest formulation, we immediately have a so-called *basic feasible solution*, namely,

$$\mu = \begin{pmatrix} 0 \\ 1 \end{pmatrix}$$

$$\lambda = 0$$

From this basic feasible solution, we can, in principle, start pivoting until the optimal solution is obtained. Clearly, the simplex σ contains solutions to $f_{PL}(p) = 0 + j0$ iff, at optimality, $\mu = 0$.

There are, however, several problems we are likely to encounter if we attempt to implement this scheme. Besides the curse of dimensionality, the basic feasible solution is **degenerate**, in the sense that it has vanishing components. The problem with a degenerate solution is that it is not clear what the pivot should be once we have decided what barycentric coordinate should be activated. To make things worse, the linear programming tableau is very sparse, and, as argued by [Borgwardt 1988, page 43], other degenerate solutions are likely to occur in the course of the search process. Under those circumstances, the simplex methods could "cycle;" see [Zornig 1991].

Less fundamental but also annoying is the fact that the *primal* pivot step by *row* operation does not have an obvious interpretation in the geometric setup of the template approach to robust control. The usual geometrical interpretation of the primal pivot step as the transition from a vertex to an adjacent one along an edge of the polyhedron 7.1 hardly has any control interpretation. To be more explicit, the activation of a barycentric coordinate, which reduces the corresponding column of $\begin{pmatrix} F \\ e^T \end{pmatrix}$ to a Euclidean basis vector by row operations, immediately destroys the geometrical interpretation of the columns of $\begin{pmatrix} F \\ e^T \end{pmatrix}$ as coordinate vectors of the vertices of the template.

Last but not least, if it is observed that σ fails to contain a solution, it is not clear in what simplex a solution could be expected.

These defects are remedied by the "simplicial algorithms with vector labeling," to be introduced in Chapter 9 and inspired from [Doup 1988]. Essentially, the simplicial algorithms proceed from the above linear programming formulation, except that the solution is sought by *dual* pivoting, by *column* operation. The dual pivot has an easy geometric interpretation as addition/removal of vertices in the template. Furthermore, if σ fails to have a solution, the dual linear program provides clues as to where to expect one.

7.1.2 Polyhedral Crossover

Using the above linear program we investigate properties of the polyhedral crossover.

Theorem 7.1. *Let $f_{PL} : P \to N$ be the piecewise-linear extension of the Nyquist map. Then $f_{PL}^{-1}(0 + j0)$ is a polyhedron embedded in P, but not in general as a subpolyhedron of P.*

Proof. Run the linear program in each (closed) simplex of P. The solutions are consistent within the intersection $\overline{\sigma} \cap \overline{\tau}$ of two closed simplexes, as can be seen by running the linear program in the closed simplex $\overline{\sigma} \cap \overline{\tau}$. ∎

7.2 From Piecewise-Linear to Simplicial Map

The issue addressed in this section is how to make f_{PL} simplicial, and what benefit can be expected from this extra effort. Essentially, there are two ways to go:

- Refine $D \times \Omega$ *only* to get a simplicial *approximation* to f_{PL}.
- Refine *both* $D \times \Omega$ and N so as to make f_{PL} simplicial relative to the *refined* subdivisions (as we shall see, this is always possible).

It is important to understand the difference between the two approaches. In the first approach, *only* P is refined with the drawback that the map f_{PL} is affected in the process of making it simplicial. In the second one, *both* P and N are refined with the reward that the map f_{PL} is not affected.

7.2.1 Simplicial Approximation to Piecewise-Linear Map

The simplicial approximation $\tilde{f}_{PL}$ could be constructed by implementing the "star" conditions. (For a construction more specifically devised for piecewise-linear maps, see Section 7.3.) Since in this simplicial approximation process the Nyquist map is perturbed, there are some questions as to how the solution set is affected. We address the issue of $\tilde{f}_{PL}^{-1}(0+j0)$ versus $f_{PL}^{-1}(0+j0)$ in this piecewise-linear setup, since it appears hard to deal with outside the PL setup.

The fundamental problem is that it is not, in general, true that the simplicial approximation $\tilde{f}$ maps an arbitrary n_p-simplex of $D \times \Omega$ onto a 2-simplex of the template. This is easy to see. Assume the dissection of $D \times \Omega$ is so fine that, for some n_p-simplex σ_{n_p}, we have $f(\cup_i \text{star}(a^i))$ completely contained within one single simplex, say $b^0b^1b^2$, of N. The key to constructing a simplicial approximation is the star condition $f(\text{star}(a^i)) \subseteq \text{star}(\tilde{f}(a^i))$ and if $f(\text{star}(a^i)) \subseteq b^0b^1b^2, \forall i$, then $\tilde{f}$ could certainly assign all n_p+1 vertices $a^0, ..., a^{n_p}$ to the same vertex of N, say b^0. In this case, the n_p-simplex σ_{n_p} is mapped to a single point, b^0! The reader can easily verify for himself that we could have a situation where σ_{n_p} is mapped to a line. If there is a simplex σ_{n_p} that contains an exact solution to $f(p)=0$ but that is mapped to either a line or a point under $\tilde{f}$, there won't be an exact solution to $\tilde{f}(p)=0$ and an algorithm based on $\tilde{f}$ will fail to detect the solution.

To go about this difficulty, there are two solutions. The first is to make sure that every n_p-simplex of $D \times \Omega$ is mapped onto a 2-simplex of N. Some indications as to how this could be done will be given in Section 7.3. However, this is hard to manage and the reader should realize that, in this case, we no longer have the luxury to dissect $D \times \Omega$ in an arbitrary fashion. The second possibility is to make sure that in the situation where σ_{n_p} contains an exact solution to $f_{PL}(p)=0$, $\overline{\sigma_{n_p}}$ contains an exact solution to $\tilde{f}_{PL}=0$. We show how this can be managed.

Theorem 7.2. *Let $D \times \Omega$ be a polyhedron P with vertex set $\{a^i\}$. Define $b^i = f(a^i)$ and let N be the polyhedron that has $DT(\{b^i\})$ as underlying triangulation. Let $f_{PL}: P \to N$ be the piecewise linear extension of the vertex transformation $a^i \mapsto b^i$. Let $\tilde{f}_{PL}: P' \to N$ be a simplicial approximation of f_{PL} such that the diameter of any star of P' is less than $\frac{1}{2}$ of the mesh of $DT(\{b^i\})$. If $f_{PL}(p)=0$ has an exact solution in the simplex σ_n, $\tilde{f}_{PL}(p)=0$ has an exact solution in $\overline{\sigma_n}$.*

Proof. Let an exact solution to $f_{PL}(p)=0+j0$ be in the simplex $\sigma_n = a^{j_0}a^{j_1}...a^{j_n}$. Define $b^{j_i} = f_{PL}(a^{j_i})$, $i=0,1,...,n$. Clearly,

$$f_{PL}(\overline{\sigma_n}) = \text{conv}(b^{j_0}, b^{j_1}, ..., b^{j_n}) \ni 0+j0$$

Let us triangulate $\text{conv}(\{b^{j_i} : i=0,...,n\})$; since this convex hull covers $0+j0$, there exists a simplex, say $b^{j_0}b^{j_1}b^{j_2}$ (possibly after some relabeling), of the triangulation of the convex hull such that $b^{j_0}b^{j_1}b^{j_2} \ni 0+j0$; in other words, the face $\overline{\sigma_2} = \overline{a^{j_0}a^{j_1}a^{j_2}}$ of σ_n contains an *exact* solution.

Let the Delaunay triangulation of $\{b^i\}$ be given and let $b^0b^1b^2$ be the unique triangle of $DT(\{b^i\})$ that contains $0 + j0$. To construct the simplicial approximation $\tilde{f}_{PL}$, the complex P and, in particular, the simplexes $\overline{\sigma_n}$, $\overline{\sigma_2}$ are refined. Both the barycentric and the Q-subdivisions have the advantage that the subdivision of $\overline{\sigma_n}$ induces the same subdivision of the face $\overline{\sigma_2}$. Refine P, enough so that a simplicial approximation exists, and, if necessary, do some further refinement so that

$$\text{diameter}(f_{PL}(\text{star}(a^i))) < \frac{1}{2}\text{mesh}(DT(\{b^i\})), \quad \forall i$$

Under those conditions, there exists a vertex a'^{i_0} of $\overline{\sigma_2}'$ such that

$$f_{PL}(\text{star}(a'^{i_0})) \ni b^0$$

and, because of the mesh requirement, the image of this star contains neither b^1 nor b^2. Likewise, there exists a vertex a'^{i_1} of $\overline{\sigma_2}'$ such that $f_{PL}(\text{star}(a'^{i_1}))$ contains b^1 but does not contain b^0, b^2. Finally, there exists a vertex a'^{i_2} such that the image of its star contains b^2 but does not contain b^0, b^1. From these star properties, it follows that $\tilde{f}_{PL}(a'^{i_j}) = b^j$, $j = 0, 1, 2$. Now, connect the vertices a'^{i_0},a'^{i_1} by a path of edges in $\overline{\sigma_2}'$ such that, for any vertex a' of this edge-path, we have $f_{PL}(\text{star}(a')) \cap b^0b^1 \neq \emptyset$; since the only stars of N that contain the edge b^0b^1 are star(b^0), star(b^1), it follows that $\tilde{f}_{PL}(a') = b^0$ or b^1. We do the same for an edge-path connecting a'^{i_1} to a'^{i_2} and an edge-path connecting a'^{i_2} to a'^{i_0}. Finally, for any a' in the face σ_2, the simplicial approximation $\tilde{f}_{PL}$ has to assign either b^0, b^1, or b^2 to a'. Indeed, $f_{PL}(\text{star}(a')) \cap b^0b^1b^2 \neq \emptyset$ and only star(b^0), star(b^1), and star(b^2) contain $b^0b^1b^2$.

To sum up, every vertex of $\overline{\sigma_2}'$ within the paths of edges joining a'^{i_0}, a'^{i_1}, a'^{i_2} is assigned, by the simplicial approximation $\tilde{f}_{PL}$, a vertex b^0, b^1, or b^2. This b-vertex is considered a *label* attached to the corresponding vertex of $\overline{\sigma_2}'$. The property that

$$\tilde{f}_{PL}(a'^{i_j}) = b^j, \quad j = 0, 1, 2$$

is called *complete labeling.* Furthermore, the property that

$$\tilde{f}_{PL}(a' \in \text{edge-path } a'^{i_j}a'^{i_k}) \in \{b^j, b^k\}$$

is called *Sperner properness* of the labeling. By a slight generalization of *Sperner's lemma*, the proof of which requires only a slight modification of the argument of Lemma 9.7, it follows that there exists a *completely labeled* simplex of $\overline{\sigma_2}'$ within the path of edges connecting a'^{i_0}, a'^{i_1}, a'^{i_2}; that is, there exists a simplex $a'^ia'^ja'^k$ of $\overline{\sigma_2}'$ that carries all three labels b^0,b^1,b^2 at its vertices; more specifically,

$$\{\tilde{f}_{PL}(a'^i), \tilde{f}_{PL}(a'^j), \tilde{f}_{PL}(a'^k)\} = \{b^0, b^1, b^2\}$$

Since the simplex $b^0b^1b^2$ contains $0 + j0$, the equation $\tilde{f}_{PL}(p) = 0$ contains an exact solution in $a'^i a'^j a'^k$. ■

7.2.2 Making Piecewise-Linear Map Simplicial

The basic result is the following:

Theorem 7.3. *Given the piecewise-linear Nyquist map $f_{PL} : P \to N$ between polyhedra P and N, there exist refined polyhedra P', N' relative to which the map $f_{PL} : P' \to N'$ is simplicial.*

Proof. A quite general, detailed proof can be found in [Glaser 1970, Proposition 1.5]. Here, we sketch a proof specialized to the case of a two-dimensional range space.

Every simplex $\sigma = a^0 a^1 ... a^n$ of $D \times \Omega$ is mapped by f_{PL} into the convex hull of $\{b^0 = f(a^0), b^1 = f(a^1), ..., b^n = f(a^n)\}$. Taking all simplexes of $D \times \Omega$ into consideration, the template N is obtained as the finite union of convex hulls; any two overlapping convex hulls intersect in the convex hull of the vertices they share in common.

The first step of the proof is to give a simplicial complex structure to the polygonal template N (which is not in general convex). The construction goes as follows: Map all edges of $D \times \Omega$. The images of these edges are line segments joining selected pairs of b^i's. These line segments remain entirely within N. To see this, observe that any edge $a^i a^{i+1}$ must belong to a simplex $\sigma = a^0 a^1 ... a^i a^{i+1} ... a^n$ of $D \times \Omega$; hence the edge $a^i a^{i+1}$ maps into $\text{conv}(f(a^0), f(a^1), ..., f(a^i), f(a^{i+1}), ..., f(a^n)) \subseteq N$. Since the map f_{PL} is not simplicial, some of these line segments, typically $b^i b^{i+1}$, will intersect at points that are not vertices. Assign a vertex to each of these intersection points. At this stage, we have a convex polygonal cell decomposition for N. Some of the cells of this decomposition might be simplexes, and no further construction is needed. If a cell is a polygon, decompose the polygon into simplexes by joining selected pairs of vertices of the polygon. After doing so, *the polygonal template N has a simplicial complex structure.*

The second step of the proof is to map every simplex σ of $D \times \Omega$ and to refine *both* σ and $f(\sigma)$ so that $f|\sigma$ will be simplicial. Let $T = f(\sigma)$. Clearly, T is a convex subpolygon in N. Furthermore, the vertices of T are among the vertices of the simplicial complex of N. As such, T has a simplicial complex structure. We write this as $T = \cup\tau$, the disjoint union of the constituting simplexes of T. Taking the preimage of this disjoint union, it follows from a linear programming argument that $(f|\sigma)^{-1}(\cup\tau)$ is a convex polygonal cell decomposition of σ. Let C be one such cell that maps into a simplex τ, $f(C) = \tau$. Here we are at the crucial juncture of the proof: Take a point

$$a \in \text{int}(C) \cap (f|\sigma)^{-1}(\text{barycenter}(\tau))$$

With this interior point, we construct a *derived subdivision* of C; it is constructed by first decomposing the faces of the polygonal cell C into simplexes and then forming all simplexes made up with the interior point a together with a simplex of the decomposition of a face of the polygonal cell; call this derived subdivision C'. Furthermore, in the image, we perform a barycentric subdivision of τ; call τ' the simplicial complex resulting from this barycentric subdivision. It is easily seen that $(f|\sigma)$ maps C' onto τ' *simplicially*. Finally, we "piece things together" in both $D \times \Omega$ and N. In N, every simplex of the original decomposition is barycentrically subdivided, and therefore things match correctly at the common faces between two simplexes. Write N' to be this barycentric subdivision. In $D \times \Omega$, a point is taken within the interior of every simplex. The simplexes of the original decomposition are glued along common faces of each other, and this guarantees the same for the derived subdivision. Write this one as $(D\times\Omega)'$. With everything matching up correctly, it follows that $f : (D \times \Omega)' \to N'$ is simplicial. ■

Since the map has not changed, the crossover could be defined as $f_{PL}^{-1}(0+j0)$, which amounts to feasible solutions to a linear program. However, since $0 + j0$ is not in general a vertex of N', remaining within the realm of pure combinatorics would instead require defining the crossover as $f_{PL}^{-1}(b'^0b'^1b'^2)$, where $b'^0b'^1b'^2$ is the unique simplex of N' containing $0+j0$. The major advantage of working with the refined dissections of P and N is the following:

Theorem 7.4. *$f_{PL}^{-1}(\overline{b'^0b'^1b'^2})$ is a (closed) subpolyhedron of P'.*

Proof. Since $\overline{b'^0b'^1b'^2}$ is closed and since f_{PL} is continuous, the inverse image $f_{PL}^{-1}(\overline{b'^0b'^1b'^2})$ is closed. Next, to prove that the crossover is a subpolyhedron, remember that

$$f_{PL}^{-1}(\overline{b'^0b'^1b'^2}) = \{p : f_{PL}(p) \in \overline{b'^0b'^1b'^2}\}$$

A point p in the right-hand side of the above equation is in a unique simplex σ. Since f_{PL} is simplicial, it follows that $f_{PL}(\sigma)$ is a simplex. Since $\sigma \ni p$, $f_{PL}(\sigma) \ni f_{PL}(p) \in \overline{b'^0b'^1b'^2}$, so that the simplex $f_{PL}(\sigma)$ intersects $\overline{b'^0b'^1b'^2}$ and is therefore contained in $\overline{b'^0b'^1b'^2}$. It follows that

$$f_{PL}^{-1}(\overline{b'^0b'^1b'^2}) = \{\sigma : f_{PL}(\sigma) \subseteq \overline{b'^0b'^1b'^2}\}$$

By the simplicial property, $f_{PL}(\sigma)$ is a simplex, so that it is either $b'^0b'^1b'^2$, $b'^0b'^1$, $b'^1b'^2$, $b'^2b'^0$, b'^0, b'^1, or b'^2. Furthermore, if $\tau \preceq \sigma$, it follows that $f_{PL}(\tau) \subseteq \overline{b'^0b'^1b'^2}$. Therefore, if σ is in $f_{PL}^{-1}(\overline{b'^0b'^1b'^2})$, so is its face τ. It follows that $f_{PL}^{-1}(\overline{b'^0b'^1b'^2})$ is a collection of simplexes *along with their faces*. Therefore $f_{PL}^{-1}(\overline{b'^0b'^1b'^2})$ has a simplicial complex structure defining a polyhedron. ■

7.3 Strict Linear Complementarity

Here, we illustrate the procedure constructing $\tilde{f}_{PL}$ or making f_{PL} simplicial on the simple example of a tetrahedron mapped to $\mathbb{R}^2$. A deeper aim of this exercise is to show that the linear programming interpretation of this procedure is the so-called *(strict) linear complementarity problem.*

Consider a tetrahedron $a^0a^1a^2a^3$ in three-dimensional uncertainty space and let us map its vertices into the two-dimensional space, $b^i = f(a^i)$. This is shown in Figure 7.1 and Figure 7.2. Clearly, the piecewise-linear extension of the vertex transformation $a^i \mapsto b^i$ maps onto the convex hull of $\{b^i\}$. There are two generic cases: Either conv($\{b^i\}$) is a triangle (Figure 7.1) or conv($\{b^i\}$) is a quadrilatere (Figure 7.2).

Consider first the case of a triangle. In this case, the template is Delaunay-triangulated as shown in the right-hand side of Figure 7.1. Let us check the simplicial property of f_{PL}. Clearly all edges of $a^0a^1a^2a^3$ are mapped simplicially. On the other hand, the face $a^1a^2a^3$ does not map simplicially because the vertex b^0 lies in its image. Some refinement of $a^0a^1a^2a^3$ is needed. Clearly, $f_{PL}^{-1}(b^0)$ is a line segment, say a^0x for some $x \in a^1a^2a^3$. Define x to be a vertex of the refined triangulation of $a^0a^1a^2a^3$. In addition to $a^i \mapsto b^i$, let $x \mapsto b^0$; the resulting vertex transformation induces a map $f_{PL} : (a^0a^1a^2a^3)' \to N$ that is simplicial and that coincides with f_{PL}. Observe that, in this simplicial map, every 3-simplex of the refined triangulation is mapped to a 2-simplex of the template.

Consider now the case where the convex hull of $\{b^i\}$ is a quadrilatere. Do a Delaunay triangulation of $\{b^i\}$ and suppose b^1b^3 is an edge of the triangulation. Let us check the simplicial property. Since $f_{PL}(a^1a^3) = b^1b^3$ and $f_{PL}(a^0a^2) = b^0b^2$ intersect, the simplicial property fails. To rectify the situation, the line segments b^1b^3 and b^0b^2 are broken down into edges by assigning a new vertex, say z, to their intersection. This yields the *refined* triangulation N'. Back into the uncertainty space, a^1a^3 and a^0a^2 are broken down so that the resulting edges map to any of the four edges b^1z, b^3z, b^0z, b^2z. This is done by computing $f_{PL}^{-1}(z)$. This yields points $x \in a^1a^3$ and $y \in a^0a^2$ mapped to the same point z of the two-dimensional plane. Clearly, $f_{PL}^{-1}(z) = xy$. By declaring x and y vertices, do a double stellar decomposition of the tetrahedron. The resulting vertex transformation $a^i \mapsto b^i; x, y \mapsto z$ induces a simplicial map $f_{PL} : (a^0a^1a^2a^3)' \to N'$ that agrees with the original map on all vertices. Also observe that all four tetrahedra of the refined domain of uncertainty are mapped into 2-simplexes in the two-dimensional plane. It is, in general, hard to obtain a simplicial map where all maximal dimensional simplexes in the domain of definition are mapped onto maximal-dimensional simplexes in the image.

The above construction has a linear programming interpretation. To compute x, y, form the matrix

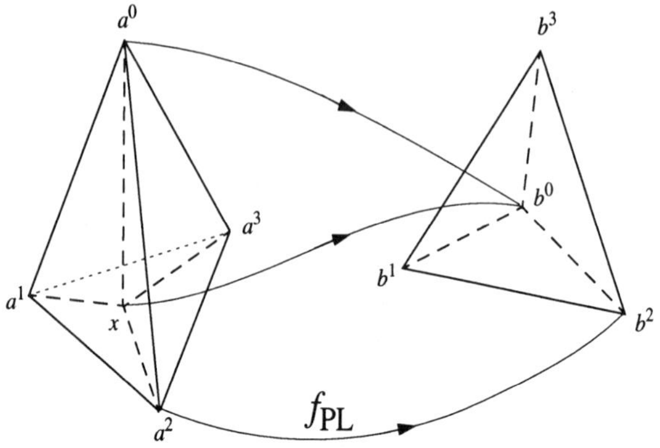

Fig. 7.1. Illustrative example of making a piecewise-linear map simplicial in the case where the convex hull of the sample image vertices is a triangle.

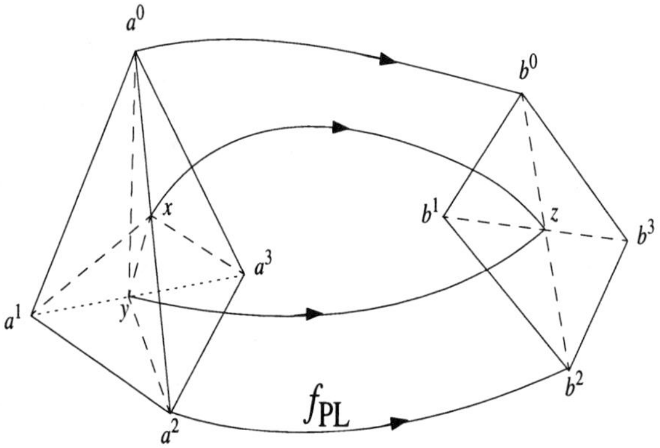

Fig. 7.2. Illustrative example of making a piecewise-linear map simplicial in the case where the convex hull of the sample image vertices is a quadrilatere.

$$A = \begin{pmatrix} f(a^0) & f(a^1) & f(a^2) & f(a^3) \\ 1 & 1 & 1 & 1 \end{pmatrix}$$

Clearly, the barycentric coordinates λ, μ of x, y are given by

$$\begin{aligned} A\lambda &= A\mu \\ e^T\lambda &= 1 \\ e^T\mu &= 1 \\ \lambda &\geq 0 \\ \mu &\geq 0 \end{aligned}$$

together with a condition that stipulates that x, y are on edges *that have no vertices in common*

$$\lambda_i = 0 \quad \Leftrightarrow \mu_i > 0$$
$$\lambda_i > 0 \quad \Leftrightarrow \mu_i = 0$$

Computation of λ, μ would be a linear programming problem if it weren't for the last condition. This last condition implies, but is not implied by, $\lambda^T \mu = 0$. A linear equation and inequality problem in the variables λ, μ together with $\lambda^T \mu = 0$ is the so-called **linear complementary problem**. This problem has roots tracing back to the *complementary slackness* theorem of [Dantzig 1963] and duality in linear programming. $\lambda^T \mu = 0$ can also be interpreted as a Kuhn–Tucker condition. In the complementarity problem, the situation where $\lambda_i = 0, \mu_i = 0$ for some i is allowed. This latter situation is, however, not allowed in our problem, and we would therefore call the computation of λ, μ *strict linear complementary problem.*

To solve the strict complementarity problem, find a vector in the null space of A—that is, $A(\lambda - \mu) = 0$. Put the absolute values of the positive, negative components of the null vector in λ, μ, respectively, and renormalize so that $\sum_i \lambda_i = 1, \sum_i \mu_i = 1$. Observe that the *strict* complementarity condition requires *all* components of the null vector to be nonvanishing.

The strict linear complementarity interpretation of the case of Figure 7.1 is essentially the same and is left to the reader.

Finally, another reward brought about by the construction of Figure 7.2 is that it reveals the *fiber structure* of the f_{PL} map—that is, the set of $f_{PL}^{-1}(b)$'s as b runs across the template. Clearly, for any b in the interior of conv (b^0, b^1, b^2, b^3), $f_{PL}^{-1}(b)$ is easily seen to be a line segment parallel to xy. In this case, the fiber structure is pretty easy—it is actually a *trivial bundle* in the sense of Chapter 13. The fiber structure of the case of Figure 7.1 is left to the reader.

8

GAME OF HEX ALGORITHM

... it was no doubt the influence of game theory, and associated fixed point theorems that gradually reduced the dependence on calculus.

S. Smale,
The Mathematics of Time,
Springer-Verlag, New York, 1980,
page 109.

SUMMARY

In this chapter, we introduce a tic-tac-toe game, called *game of Hex*, that is played on a board consisting of a gapless assembly of regular hexagons. The purpose of this "fun chapter" is twofold: First, we establish the combinatorial equivalence between the (dual of the) Hex board and the Q and prismatic triangulations introduced earlier. Next, and most importantly, the strategy for finding a winning path, developed in the highly "visual" environment of the Hex game, reveals the fundamental combinatorial algorithm that underlies the "simplicial algorithms" for finding the crossover. It is hoped that the game of Hex will provide the reader with a visual aid to understand the simplicial algorithms on Q-triangulation, to come in the next chapter.

8.1 2-D Hex Board

The two-dimensional Hex board of size m consists of m^2 regular hexagonal "tiles" assembled without gaps to form a diamond-shaped board; see Figure 8.1. A first player marks the tiles with O's and attempts to establish a North–West $\leftrightarrow$ South–East path while the other player marks the tiles with X's attempting to establish a South–West $\leftrightarrow$ North–East path. Contrary to the well-known tic-tac-toe game played on a *square* board, there is *always* a winner on the *hexagonal* board. [Gale 1979] has an amusing "proof" of this so-called *Hex theorem* in a model of the game where the O player is a river attempting to flow North–West $\rightarrow$ South–East while the X player is attempting to build a dam by laying X stones. There are *exactly* two possibilities: Either the river makes its way around the stones and the

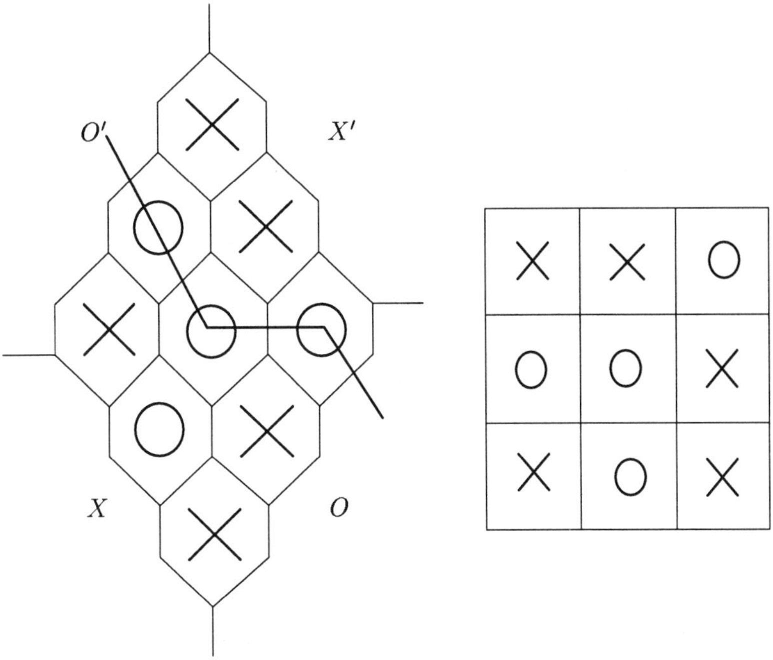

Fig. 8.1. The game of Hex versus the traditional tic-tac-toe game.

O player wins or the X player manages to connect both banks of the river with X stones, thereby interrupting the flow of the river. (We leave it to the reader to find out what is breaking down with this argument as applied to the square-tile problem.)

Observe that the decomposition of the board into regular hexagons can be formalized as a *cell complex* in the sense of Section 4.8.

Before attempting to define the more general n-dimensional Hex board, it is useful to consider the graph-theoretic dual of the Hex board as illustrated in Figure 8.2. Formally, a *vertex* is placed within each tile. Two vertices are connected by an *edge* if the tiles they represent are neighbors. We can go one step further and say that three vertices (tiles) form a *2-simplex* whenever the corresponding three tiles have a point in common. By this process, the dual of the Hex board is endowed with an *abstract simplicial complex structure*. This abstract simplicial complex has an easy geometric realization: The vertices of the geometric realization are the centers of the regular hexagons, the edges of the geometric realization are the median lines to the edges of the hexagons, and so on. As easily seen from Figure 8.2, and as easily proved using elementary geometry, the geometric realization of the graph-theoretic dual of the Hex board is the regular Q-triangulation.

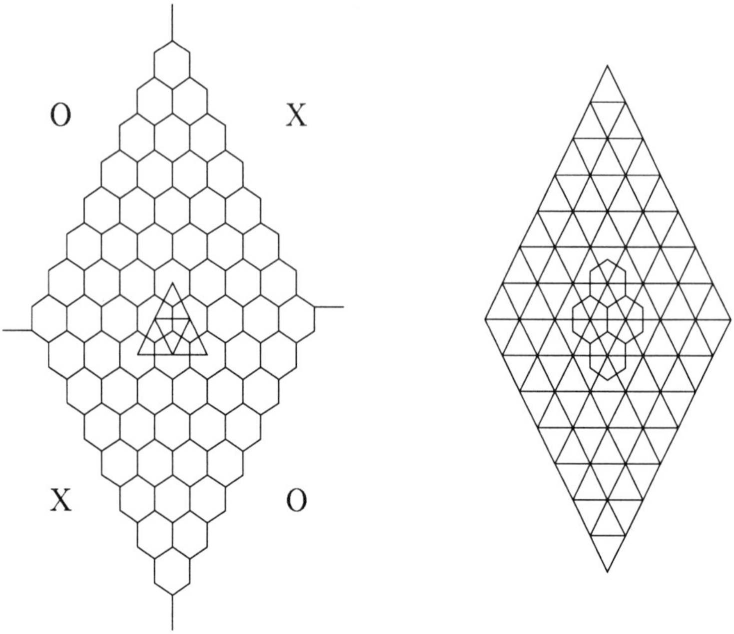

Fig. 8.2. 8×8 Hex board and its dual graph.

To go from dual back to primal, observe that the Voronoi cells of the Q-triangulation are regular hexagons. In other words, *the regular Hex board is the Voronoi diagram of the dual Q-triangulation.*

With the above insight, we are in a position to formalize the notion of the two-dimensional Hex board of size m.

Definition 8.1. *The (dual of the) two-dimensional Hex board of size m is the set of vertices having Euclidean coordinates that are integer multiples 0,1,...,m of 1/m in a basis* $\{\vec{e_0}, \vec{e_1}\}$, *where* $\vec{e_0}$ *and* $\vec{e_1}$ *are at a 120-degree angle. Two vertices* a^0, a^1 *form and edge (along which a player is allowed to move) whenever the vertices can be ordered so that*

$$(a^0)_j \leq (a^1)_j, \quad \textit{for } j = 1, 2$$

$$\sup_j |(a^0)_j - (a^1)_j| = 1$$

Three vertices a^0, a^1, *and* a^2 *form a 2-simplex whenever they can be ordered so that*

$$(a^i)_j \leq (a^{i+1})_j, \ \forall i, \forall j$$

$$a^0 a^1 = \frac{1}{m} \vec{e}_{\pi_0}$$

$$a^1a^2 = \frac{1}{m}\vec{e}_{\pi_1}$$

where (π_0, π_1) *is a permutation of the ordered set* $(0, 1)$.

We could renounce to the aesthetic appeal of the *regular* hexagonal tile pattern and "straighten out" the graph-theoretic dual by taking the basis $\{\vec{e_0}, \vec{e_1}\}$ to be *orthonormal.* In doing so, the combinatorial features of the resulting Hex board and its dual remain unchanged, but the "flatness" of the resulting geometric simplexes increases.

8.2 n-D Hex Board

Now, it should be clear what the n-dimensional hex board is

Definition 8.2. *The (dual of the) n-dimensional Hex board of size m is the set of vertices that have Euclidean coordinates that are integer multiples (0,1,2,...,m) of 1/m relative to some Euclidean basis* $\{\vec{e_0}, ..., \vec{e_{n-1}}\}$. *A collection of n+1 vertices* $a^0, ..., a^n$ *form a simplex whenever the vertices can be ordered so that*

$$(a^i)_j \le (a^{i+1})_j, \ \forall i, \forall j$$

$$a^i a^{i+1} = \frac{1}{m}\vec{e}_{\pi_i}, \ \forall i$$

where $(\pi_0, \pi_1, ..., \pi_{n-1})$ *is a permutation of the ordered set* $(0, 1, 2, ..., n-1)$.

If we choose an orthonormal basis, we cannot expect the n-D board to have the "nice-looking" features of the regular 2-D Hex board. To obtain a "regular" n-D board, the basis vectors have to be carefully chosen and, even if we do so, we cannot get *all* of the regular features of the 2-D case. This is because "symmetry" is much harder to achieve in higher dimensions. The formulation of this requires some concepts from lattice theory and root system of Lie algebras that are not reproduced here; the interested reader is referred to [Chu 1996]. Nevertheless, regardless of the way the basis is chosen, the *combinatorial* features of the resulting board are the same.

8.3 Combinatorial Equivalence

The above combinatorial formalization of the (dual of the) Hex board is closely related to the the combinatorial equivalence between the Q and prismatic triangulations developed in Chapter 5. Actually, they are all equivalent:

Theorem 8.3. *The*

- *prismatic triangulation of the cube* $[0, 1]^n$ *with grid size m,*
- *the (dual of the) n-dimensional Hex board of size m,*
- *the Q-triangulation of the simplex* $e^0e^1...e^n$ *with size m*

are combinatorially equivalent in the sense that the Euclidean coordinates of their vertices are integer multiples of $1/m$ *and* $a^0 \dots a^n$ *is a simplex whenever, after reordering,* $a^i a^{i+1} = \frac{1}{m}\vec{e}_{\pi_i}$, $i = 0, \dots, n-1$, *where* $(\pi_0, \dots, \pi_{n-1})$ *is a permutation of* $(0, 1, \dots, n-1)$ *and* $\{\vec{e}_i\}$ *is a Euclidean basis.*

For the *purely combinatorial* features of the above triangulation rule, we could take an orthonormal, or any other, Euclidean basis. If we want the aesthetic appeal of a regular pattern triangulation, or more importantly if we want simplexes of minimal flatness, the basis has to be carefully chosen. To be more specific, the basis vectors have to be the roots of a Lie algebra [Chu 1996].

8.4 Two-Dimensional Hex Game Algorithm

We first develop the Hex algorithm on the *primal* hex board consisting of an assembly of hexagonal tiles and then reformulate the algorithm on the *dual* Q or prismatic triangulation.

8.4.1 Primal

In the primal Hex game, the assembly of hexagonal tiles is embedded into four regions of constant symbols: the North–West and South–East regions with symbol O and the South–West and North–East regions with symbol X. This is shown in Figure 8.3. These extra layers are quite natural in the primal game, since they tell the players what should be connected to what. As we shall see later, these regions do not appear quite naturally in the dual Q- or prismatic triangulation; for that reason, the dual Q- or prismatic triangulation is said to have *artificial layers* and the algorithms are sometimes referred to as *artificial start algorithms*, a concept due to [Kuhn 1968]; see also [Kuhn and MacKinnon 1975].

The *crucial* observation for finding the winning path is that, with the extra layers correctly labeled, there is a XO transition *somewhere* on any of the four boundaries interfacing the board and the extra layers. Therefore, the idea is to start at that XO transition on the interface, move inside the board by following the *unique* path that has, say, O-tiles to its right and X-tiles to its left. Clearly, the path cannot terminate inside the board; it must terminate on a boundary; it is easily seen that the path could return to the starting boundary; however, since the number of XO transitions on this boundary is *odd*, there is *at least* one path that will *never* return to the starting boundary; it is also easily seen that this path will go to the *opposite* boundary. Therefore, we have a path between tiles, O tiles to the right, X tiles to the left, joining a boundary to the opposite boundary. To the right of this path, there is a string of adjacent O tiles and to the left there is a string of adjacent X tiles. If the path joins the external O layer to the opposite external O layer, the string of O tiles to the right is the

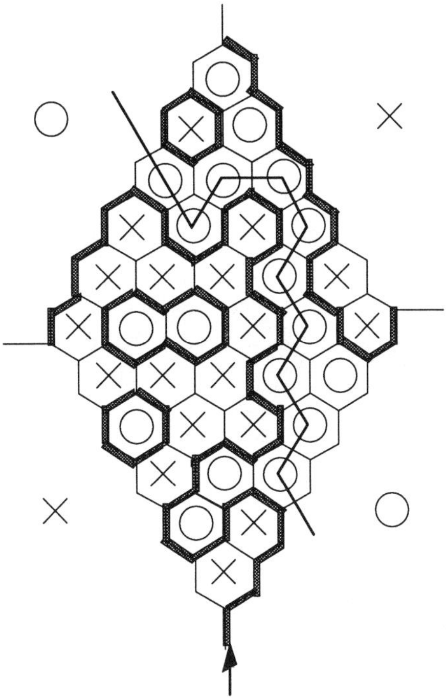

Fig. 8.3. Finding winning path of the Hex game.

winner.

A conceptualization of the strategy for finding a winning path—the so-called *fundamental graph lemma*—will be instrumental in formulating the general simplicial algorithm for finding the crossover. The same strategy for finding a winning path is also instrumental in the constructive proof of Sperner's lemma. Also, the correct marking of the extra layers is, at a more conceptual level, what is usually referred to as *Sperner properness* of the labeling.

Also, observe that the idea of walking on the board with the O tiles to the right and the X tiles to the left is somewhat similar to the strategy for finding its way in a maze: Walk by always keeping the wall to the right and the hallway to the left.

8.4.2 Dual

The dual Hex game is played on the prismatic or Q-triangulation of a square board, as shown in Figure 8.4. The vertices are marked ("labeled") with X's and O's. The vertices on the boundary are marked consistently with the "artificial layer" concept. The algorithm consists in finding a string of O vertices or a string of X vertices, all connected by edges of

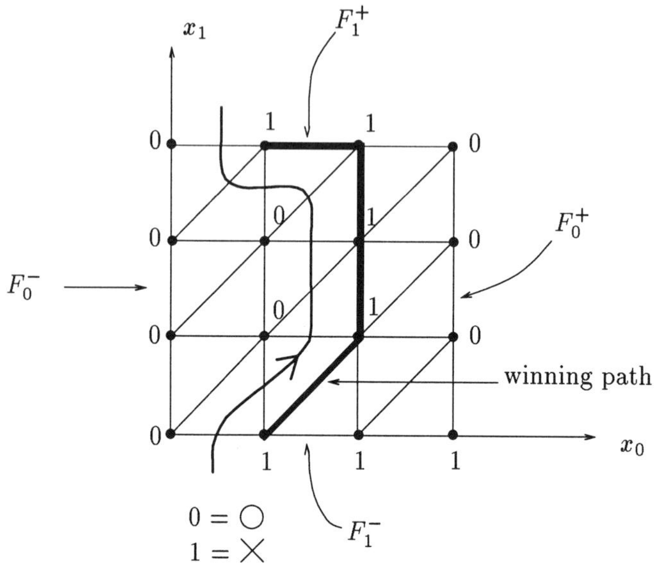

Fig. 8.4. The 2-D Hex game played on the dual Hex board.

the triangulation, and joining two opposite boundaries of the square. The algorithm consists of two steps: Find a XO transition on a boundary and consider the unique triangle having XO as an edge. From there, move inside the board, by going from one simplex to an adjacent one, through their common XO edge. The algorithm cannot terminate inside the board. It terminates on the boundary of the board; from there it is easy to figure out whether X or O is the winner.

8.4.3 Complexity

The game of Hex has been developed in the *static, noncompetitive* setting: The board is first all marked with X's and O's and then an algorithm determines who the winner (X or O) is.

The game can also be formulated in a *dynamic, competitive* fashion: The tiles are labeled one step at a time; at a given step a player labels a tile with, say, an X, then the other player labels another tile with an O, and so on, until one of the players manages to connect opposite faces.

The competitive Hex game can be thought of as the worst-case complexity of the noncompetitive game. Indeed, in complexity theory, the worst-case problem is often generated by a player, opposed to the algorithm, attempting to fool the algorithm.

[Reisch 1981] and [Even Tarjan 1976] showed that the competitive Hex game—that is, the problem of deciding at a certain stage who the winner will be—is PSPACE-complete. This means that any PSPACE problem can be reduced via polynomial space manipulation to the Hex game. In other words, the Hex game is at least as hard as any PSPACE problem. PSPACE means "polynomial space"—that is, the class of problems that require polynomial memory storage for their solution. As argued by [Even and Tarjan 1976], it appears very unlikely that a PSPACE complete problem can be solved in polynomial run time, so that we will conjecture that the game of Hex requires exponential run time.

From these considerations, we conjecture that the worst-case Hex game has exponential run time; see [Chu 1996] for some simulation results.

A manifestation of this is that the winning path could "wiggle" across the whole board, somewhat like a space curve. This phenomenon reflects intrinsic *worst-case* difficulty of the game—complicated winning path—rather than poor performance of the algorithm. Besides, a "wiggling" winning path is, statistically, unlikely to occur, so that the real issue is the mathematical expectation of the length of the winning path for a stochastic geometry model of the universe of games. Early results show a substantial gap between worst case and stochastic complexities of the static game; see [Chu 1996]. These results appear parallel to the ground-breaking results of [Borgwardt 1988] in linear programming: While the number of pivot steps is exponential in the worst case, the number of pivot steps becomes polynomial when averaged over the universe of linear programs endowed with a certain probability measure.

8.5 Three-Dimensional Hex Game Algorithm

8.5.1 Dual

We argue on the Q-triangulation, or the equivalent prismatic triangulation, of the cube $[0,1]^3$. In this case, there are three players and we use $0, 1,$ and 2 to "label" the vertices. We assume that the cube $[0,1]^3$ has the extra layers, telling the players which face should be connected to which. Opposite faces of the cube are given the same label, indicating that a potential winning path is a path connecting the two opposite faces with a string of adjacent, uniformly labeled vertices. (Two vertices are *adjacent* iff they form an edge of the triangulation.) To be more precise, let (x_0, x_1, x_2) be a system of Euclidean coordinates for the cube, and let F_i^+, F_i^- be the front and back faces, respectively, perpendicular to the i axis; that is,

$$\begin{aligned} F_i^+ &= \{x : x_i = 1\} \\ F_i^- &= \{x : x_i = 0\} \end{aligned}$$

(We note that these concepts of *faces* are central in cubical homology theory [Massey 1980].) Next, we define the labels on the boundary of the cube as follows:

$$\ell(x) = \min_i \{i : x \in F_i^-\} \text{ if } x \in \cup_i F_i^-$$

$$= \min_i \{i : x \in F_i^+\} \text{ otherwise}$$

The above is a 3-D conceptualized version of the artificial layers of the 2-D Hex board. (This is closely related to labeling rules for the simplotope [Doup 1988] and some attempt to formulate a "cubical" version of Sperner's lemma [Freund 1986].) The above is telling the ith player to attempt to connect F_i^+ and F_i^- with a string of adjacent vertices bearing the label i. Figure 8.4 attempts to illustrate these labeling concepts in a 2-D situation.

With this labeling, it is easy to show that there *exists* a 012 triangle in some face of the cube. To find this boundary 012 triangle, we start from the origin, proceed along an edge until the first label transition, then move in the face using the 2-D Hex algorithm, until we hit the 012 triangle. The basic idea of the three-dimensional game of hex algorithm is to start with this 012 triangle on the boundary, then move inside the cube by following a sequence of adjacent tetrahedra, two adjacent tetrahedra intersecting each other in a 012 face. Clearly, if we start from the 012 face $a^0a^1a^2$ ($\ell(a^i) = i$) on the boundary, there is a *unique* vertex a such that $aa^0a^1a^2$ is a tetrahedron. Let us assume, for clarity of the exposition, that $\ell(a) = 0$. Therefore, the face $\sigma_2 = aa^1a^2$ is also completely labeled. This completely labeled face σ_2 identifies a *unique* tetrahedron aba^1a^2 adjacent to $aa^0a^1a^2$ and intersecting $aa^0a^1a^2$ in the common face σ_2. (In the language of linear programming, running from one simplex to an adjacent one through a common face is a dual pivot operation.) In the new tetrahedron aba^1a^2, there is exactly one completely labeled face in addition to aa^1a^2 which itself defines a new adjacent tetrahedron. Following this recursion, the sequence of tetrahedra makes its way inside the cube, and the recursion can only terminate when a newly identified tetrahedron hits a face of the cube. Depending on whether the algorithm terminates in $F_0^{\pm}, F_1^{\pm}$, or $F_2^{\pm}$, the $0, 1$, or 2 player, respectively, wins.

8.5.2 Primal

First, we make the dual more aesthetically appealing by giving it the regular pattern of the Q-triangulation. The vertices of the Q-triangulation are arranged in a regular lattice pattern. The lattice of the vertices of the Q-triangulation is called *face-centered cubic (FCC)* or A_3 lattice; see [Chu 1996].

This lattice pattern is known in error correcting codes. Any vertex of the lattice is a nominal code word. The communications problem is that the code words can be distorted in the channel and the decoding problem is to find the lattice vertex the closest to the corrupted code word. This naturally leads to the Voronoi cells of the vertices of the lattice. From [Conway and Sloane 1991], it follows that the Voronoi cells of the 3-D Q-triangulation are *rhombic dodecahedra* (see Figure 8.5). Contrary to the regular hexagon of the 3-D case, the rhombic dodecahedron is just a

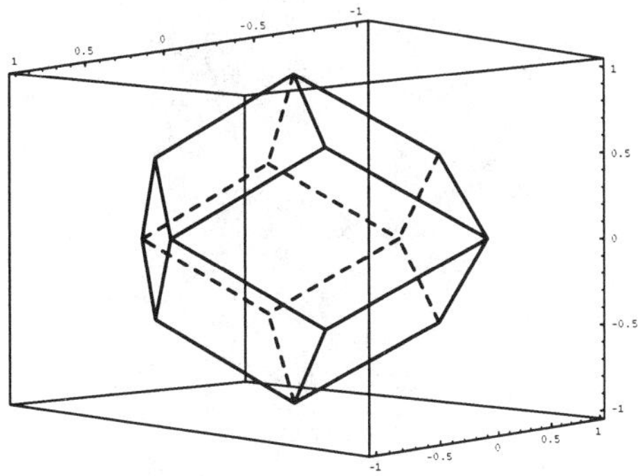

Fig. 8.5. Rhombic dodecahedron.

*semi*regular solid; that is, its faces are all the same but its vertices are not all look-alike; see [Coxeter 1973]. This *semi*regularity in the 3-D case is a manifestation of the fact that symmetry is harder to achieve in higher dimension.

Rather than embedding each vertex of the Q-triangulation within its Voronoi cell, we could also embed each vertex a^i in a sphere with its center at a^i. The common radius to all of these spheres can be adjusted so that the spheres are tangent to the faces of the rhombic dodecahedra. This results in a FCC sphere packing; see Figure 8.6,8.7. The sphere packing argument is another way to capture the symmetry properties of the lattice.

Thus the building blocks of the (primal) 3-D Hex game are rhombic dodecahedra. Each rhombic dodecahedron is marked with a 0, a 1, or a 2. A winning path is a string of adjacent dodecahedra bearing the same label.

Coming back to the dual, the crossover is a string of adjacent tetrahedra intersecting in common 012 faces. Translating this into the primal, the crossover is within the common edge between 0-dodecahedra, 1-dodecahedra, and 2-dodecahedra. This is exactly the 3-D version of the elementary 2-D strategy.

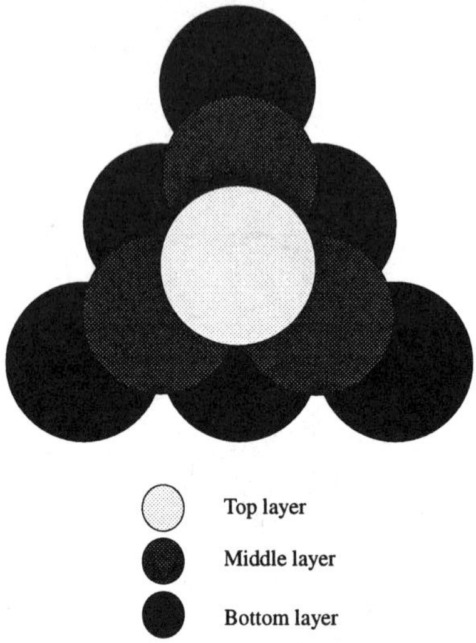

Fig. 8.6. FCC sphere packing I.

8.6 Higher-Dimensional Hex Games

The reader will have realized that probably the only approach to the higher-dimensional Hex games is to proceed via the *dual*, since the dual is the well-understood Q-triangulation together with its well-understood A_n lattice structure; see [Chu 1996].

If, however, we want to reinterpret the results in the primal, it seems that we hit quite a lot of as yet unanswered questions. Following [Conway and Sloane 1991], there are some results, as well as some unanswered questions, regarding the corresponding sphere packing in dimension 4. However, from dimension 5 on, it seems that we are venturing in totally uncharted territory.

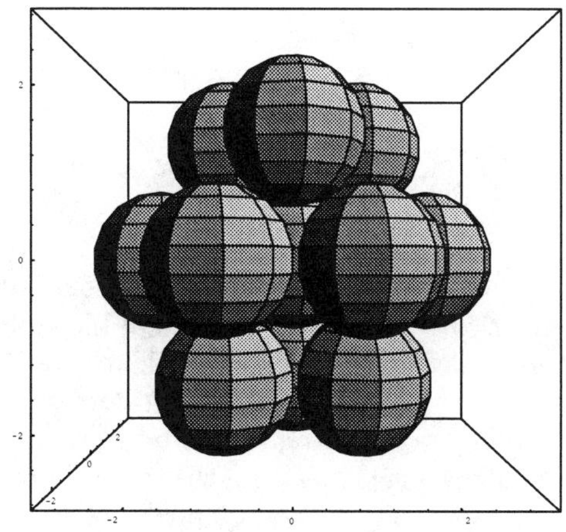

Fig. 8.7. FCC sphere packing II.

9

SIMPLICIAL ALGORITHMS OVER LABELED UNCERTAINTY SPACE

It is also of interest to analyze whether the piecewise linear path of a simplicial algorithm can be interpreted as the path of an economically meaningful adjustment mechanism...

T. Doup,
Simplicial Algorithms on the Simplotope,
Springer-Verlag, New York, 1988,
page 12.

SUMMARY

This chapter is in the same *spirit* as the computational geometry algorithm for searching the crossover, as developed in Chapter 6. Both chapters share in common the concept of polyhedral dissection of the space of uncertainty and the idea of approximating the crossover with an assembly of simplexes. However, contrary to Chapter 6, the approach developed in the present chapter is meant to be approximative, with a bit of heuristics, but has the definite advantage of obviating the need for the rather formidable apparatus of Delaunay triangulation, point location, and simplicial approximation.

In the new, somewhat heuristic, algorithm a *label* 0, 1, or 2 is assigned to each vertex of the polyhedron of uncertainty. This label is computed from the value taken by the (not necessarily simplicial) Nyquist map at that vertex. The *central result* is the fact that a simplex $a^0 \ldots a^n$ contains some approximate crossover points iff it is *maximally labeled*—that is, its vertices carry all three labels $0, 1, 2$. The algorithm for searching the crossover therefore becomes the *purely combinatorial* problem of chasing maximally labeled simplexes, a problem that has the same combinatorial structure as searching the winning path in the game of Hex.

Next, an attractive feature of the new simplicial algorithm is that it allows iterative improvement of the accuracy of the solution, provided that a technical condition—*Sperner properness* of the labeling—is satisfied. Should the simplex $a^0 \ldots a^n$ be maximally labeled, this simplex is refined, labels are assigned to the refined triangulation, and a new maximally labeled simplex is found, thereby narrowing down the solution to smaller and smaller simplexes.

To stress the connection between the computational geometry and the present approaches, the difficulty of ensuring Sperner properness and the heuristics associated with the labeling procedure completely evaporate in case the Nyquist map is simplicial.

9.1 Simplicial Algorithms Over 2-D Uncertainty Space

To grasp the concept of "simplicial algorithms" in a simple setting, we start with a 2-D uncertainty space $D \times \Omega$, as it typically happens in the SISO phase and gain margins. What simplifies the problem in this case is that the Nyquist map is **equidimensional**; that is, the domain of definition and the target have the same dimension. In the rational case and under generic conditions, the crossover consists of finitely many points.

We hasten to say that the *concepts* developed in this section apply to higher-dimensional uncertainty spaces. Besides, we could certainly run the 2-D simplicial algorithm on several 2-D slices through a higher-dimensional uncertainty space and then assemble * the crossover points in X_ω; we hasten to say, however, that this is not the fastest way to go.

The idea of generating the crossover by a 2-D sweep plane Γ through a higher-dimensional $D \times \Omega$ has a deeper interpretation. The number of solution points to $f|\Gamma(p) = 0 + j0$, as generated by the 2-D simplicial algorithm, could change as Γ sweeps $D \times \Omega$; however, the *topological degree* of the map $f|\Gamma : \Gamma \to \mathbb{C}$ remains invariant; roughly speaking, the degree is the *signed* number of solutions; as such the invariance of the degree provides some guidelines as to how the crossover points should be connected as the number of solution points changes. For the details, the reader is referred to Section 18.3 and Subsection 18.5.4.

9.1.1 Integer Labeling

Simplicial algorithms work as follows. Each vertex a of the 2-D simplicial complex of uncertain parameters $D \times \Omega$ is assigned a **label**,

$$\ell(a) = 0, 1, \text{or } 2.$$

This label is computed from the value that the Nyquist map f or its simplicial approximation $\tilde{f}$ takes at that vertex. A simplex $a^0a^1a^2$ is said to be **completely labeled** iff

$$\{\ell(a^0), \ell(a^1), \ell(a^2)\} = \{0, 1, 2\}$$

The crux of the matter is to define the labeling such that whenever the simplex $a^0a^1a^2$ is completely labeled, then $a^0a^1a^2$ contains an approximate solution to $\det(I + L\Delta) = 0$, thereby revealing a crossover point.

*A video animation of this concept on an example developed by Dr. R. Colgren, of the Lockheed-Martin Skunk Works, is available.

In such applications as the computational solution to the Brouwer fixed-point theorem [Doup 1988], [Todd 1976], or the computation of equilibrium prices in a free exchange economy, where the domain and the image are the same n-simplex, this n-simplex is labeled by embedding it into an $(n+1)$-D Euclidean space. Indeed the (standard) n-simplex can be defined by $x_0 + x_1 + \ldots + x_n = 1, 0 < x_i < 1$. A vertex is labeled according to what component of its image has maximum (or minimum) value. In this robust control problem, we have a natural way to embed the 2-simplexes of the complex plane in a 3-D space by using Riemann's stereographic projection. Therefore, let a be a vertex of (the triangulation of) $D \times \Omega$. Map it to the complex plane using the Nyquist map or its simplicial approximation

$$f(a) = \Re + j\Im$$

Consider the Riemann sphere embedded in a Euclidean space with coordinates (x_0, x_1, x_2). The center of the sphere has Euclidean coordinates $(0, 0, 0)$. The point of contact between the complex plane and the Riemann sphere is $0 + j0 \in \mathbb{C}$ and has Euclidean coordinates $(1/\sqrt{3}, 1/\sqrt{3}, 1/\sqrt{3})$. The pole of the stereographic projection is $(-1/\sqrt{3}, -1/\sqrt{3}, -1/\sqrt{3})$. Map the complex number $\Re + j\Im$ into the Riemann sphere using the stereographic projection

$$\begin{pmatrix} x_0 \\ x_1 \\ x_2 \end{pmatrix} = \begin{pmatrix} 1/\sqrt{3} & 0 & -\sqrt{2/3} \\ 1/\sqrt{3} & \sqrt{2}/2 & 1/\sqrt{6} \\ 1/\sqrt{3} & -\sqrt{2}/2 & 1/\sqrt{6} \end{pmatrix} \begin{pmatrix} \frac{1-\frac{1}{4}(\Re^2+\Im^2)}{1+\frac{1}{4}(\Re^2+\Im^2)} \\ \frac{\Re}{1+\frac{1}{4}(\Re^2+\Im^2)} \\ \frac{\Im}{1+\frac{1}{4}(\Re^2+\Im^2)} \end{pmatrix}$$

The label of the vertex a is defined by

$$\ell(a) = \min\{i \in \{0, 1, 2\} : x_i = \min_j \{x_j\}\}$$

In the above, we have defined the labeling function by following the historical path of operations research; see [Doup 1988]. Actually, in the present control context, the labeling function takes an easy interpretation in terms of the argument of $f(a)$. Indeed, it is easily seen that

$$\begin{aligned} 0 < \arg f(a) \leq 2\pi/3 &\Leftrightarrow \ell(a) = 1 \\ 2\pi/3 < \arg f(a) < 4\pi/3 &\Leftrightarrow \ell(a) = 2 \\ 4\pi/3 \leq \arg f(a) \leq 2\pi &\Leftrightarrow \ell(a) = 0 \end{aligned}$$

The butterfly-shaped regions of constant label in the complex plane are shown in Figure 9.1.

The *labeling function* $\ell(\cdot)$ has been defined over the vertices of the polyhedron of uncertainty. To be precise, we also define a labeling function $l(\cdot)$ over the vertices of the Nyquist template N. The *label* of a vertex b^i of N is simply defined via its position in the complex plane relative to the $0, \pm 2\pi/3$ butterfly (see Figure 9.1). We say that a simplex $b^0 b^1 b^2$ of N is

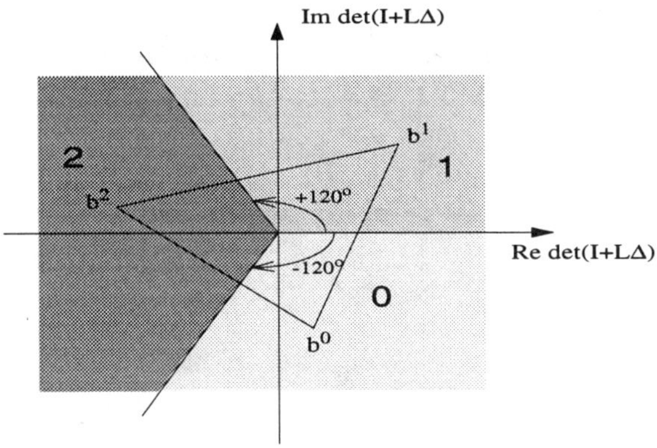

Fig. 9.1. Illustration of the labeling. The label of a vertex depends on where its image b^i falls relative to the 120-degree butterfly in the complex plane.

completely labeled whenever $\{l(b^0), l(b^1), l(b^2)\} = \{0, 1, 2\}$.

The crux of the matter is that the labeling function is able to spot crossover points in the following precise sense.

Theorem 9.1. *Consider a completely labeled simplex* $b^0b^1b^2$ *of the Nyquist template* N *with mesh*$(b^0b^1b^2) < \sqrt{3}\epsilon/2$. *Then* $\forall z \in b^0b^1b^2$ *we have* $|z| < \epsilon$.

Theorem 9.2. *Assume the mapping* $f : D \times \Omega \to N$ *is uniformly continuous. Choose* δ *such that* $\forall p, p', ||p - p'|| < \delta \Rightarrow |f(p) - f(p')| < \sqrt{3}\epsilon/2$. *Let* $\sigma = a^0a^1a^2$ *be a completely labeled simplex of* $D \times \Omega$ *with mesh*$(\sigma) < \delta$. *Then* $\forall p = (q, \omega) \in \sigma$ *we have* $|(\det(I + L(j\omega)\Delta(q)))_{PL}| < \epsilon$, *where* $(\cdot)_{PL}$ *denotes the piecewise linear extension of the vertex transformation* : $a^i \mapsto f(a^i)$, $i = 0, 1, 2$.

Now, the problem of spotting crossover points is reduced to chasing completely labeled simplexes. This becomes a purely combinatorial problem for which many fast algorithms have been devised [Gale 1979], [Todd 1976], [Doup 1988]. All of these algorithms are based on the simple combinatorial idea illustrated in Figure 9.2. Clearly, the search algorithm goes from one simplex to an adjacent one through a 01 face. Interestingly, these combinatorial search algorithms are application-independent, in the sense that they only require a simplicial complex and a rule for labeling the vertices, *independently of the problem that underlies the labeling.* Therefore, many combinatorial search algorithms, devised for such problems as computational solution to Brouwer's fixed-point theorem, computation of equilibrium prices in a free exchange economy, and so on, can be used in this robust control problem.

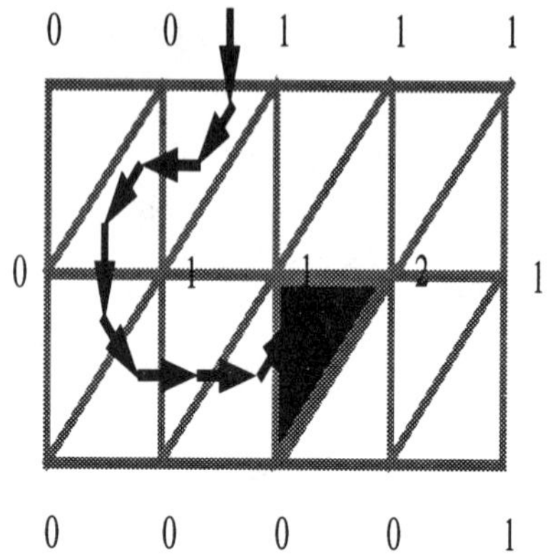

Fig. 9.2. The intuitive idea behind the algorithms for chasing completely labeled simplexes.

9.1.2 Searching—Fundamental Graph Lemma

Searching completely labeled simplexes is heavily dependent on the domain. However, regardless of whether the domain is a simplex or a simplotope (the Cartesian product of simplexes), all of the search algorithms rely on a Fundamental Graph Lemma. Given a 2-D labeled simplicial complex, we define a graph Γ_{ij} as follows: The nodes of the graph are the *almost completely labeled simplexes*—that is, simplexes $a^0a^1a^2$ such that

$$\{\ell(a^0), \ell(a^1), \ell(a^2)\} = \{i, j\};\ \ i, j = 0, 1,\ \text{or}\ 2;\ \ i \neq j$$

Two nodes are connected by a graph edge whenever the (closed) simplexes intersect in a ij edge.

Definition 9.3. *The* **degree** *of a graph* Γ *is the number of graph edges emanating from a node.*

Clearly, the degree of the graph Γ_{01} is the number of 2-simplexes that share a 01 edge. It is intuitively clear from Figure 9.2 that, for the search algorithm to work, at most two simplexes could share a common 01 edge, because otherwise there would not be a unique search path. In other words, $D \times \Omega$ should be a *pseudomanifold*; that is, every 1-simplex of $D \times \Omega$ is the face of at most two 2-simplexes (see Definition 18.1 for a precise statement).

Lemma 9.4. *If* $D \times \Omega$ *is a pseudomanifold, the degree of the graph* Γ_{ij} *is at most two.*

Lemma 9.5. (Fundamental Graph Lemma) *A graph of degree not exceeding two consists of single nodes, simple paths, or loops.*

Observe that, once an ij $(i \neq j)$ edge is identified, a very simple combinatorial algorithm can generate the Γ_{ij} graph. Essentially, the algorithm goes from one almost completely labeled simplex to the adjacent one, through the common ij edge. This is illustrated in Figure 9.3.

Now we can envision how the search algorithms work. Let 012 be a completely labeled simplex. Run the Γ_{ij} algorithm backward starting from an edge of the 012 triangle. By the fundamental graph lemma, the algorithm either terminates on another 012 triangle or on the boundary. Therefore, we can envision two kinds of algorithms:

- *Boundary Starting Algorithm*: Sweep the boundary until an ij $(i \neq j)$ edge is identified. Run the Γ_{ij} algorithm from the boundary; from there on there are two possibilities: either the algorithm returns to the boundary, and the procedure has to be restarted somewhere else, or the algorithm terminates at a targeted 012 triangle.
- *Interior Starting Algorithm*: There are cases where the boundary of $D \times \Omega$ is uniformly labeled and the algorithm cannot be started on the boundary. Besides, there are cases where $D \times \Omega$ has no boundary. In either case, we need to start inside: Search an ij edge; from this edge, run the Γ_{ij} algorithm that could only terminate at a targeted 012 triangle.

Observe that these algorithms search for a completely labeled simplex without having to check all vertices. The label of a vertex is computed only if it appears on the path of the Γ_{ij} algorithm. This clearly alleviates the computational burden in some stochastic complexity sense; see [Chu 1996].

9.1.3 Grid Refinement and Sperner's Lemma

Once a completely labeled simplex containing an approximate solution is identified, this simplex, and this simplex only, is triangulated. Inside this simplex, the search algorithm is restarted in order to identify a new completely labeled simplex which is smaller in diameter and which will provide a more accurate solution. This is the *variable grid refinement restart algorithm*. For this procedure to work, one has to make sure that a completely labeled simplex *exists* at the higher resolution level. This is where *Sperner's lemma* [Sperner 1928] is instrumental.

Definition 9.6. *A labeling function ℓ is* **Sperner proper** *over the simplex $a^0a^1a^2$, iff for any edge a^ka^l of the simplex we have*

$$\ell(\lambda a^k + (1-\lambda)a^l) \in \{\ell(a^k), \ell(a^l)\}, \ \forall \lambda \in [0,1]$$

The motivation for Sperner's lemma is that it ensures consistency of the variable resolution scheme:

Lemma 9.7. (Sperner) *Let the coarse resolution simplex $a^0a^1a^2$ be completely labeled for some Sperner proper labeling function. Let this simplex* *be triangulated, and let labels be assigned to all vertices of the refined triangulation. Then there exists a completely labeled simplex at the higher resolution level.*

Proof. We provide a *constructive* proof to this lemma, because in addition to proving this important lemma, this constructive proof reveals one of the many combinatorial algorithms that can be devised for searching completely labeled simplexes. Consider a 2-simplex, completely labeled for some Sperner proper labeling function (see Figure 9.3). Consider the graph Γ_{01}, the nodes of which are the *almost completely labeled* simplexes—that is, simplexes that carry the labels 0,1 at their vertices. Two nodes of the graph Γ_{01} are connected by a graph edge if the simplexes they represent intersect in a 01 edge. The degree of Γ_{01} is ≤ 2. By the Fundamental Graph Lemma, a graph of degree ≤ 2 consists of disjoint subgraphs that are either isolated nodes, simple paths, or simple loops. Consider the 01 edge of the low-resolution completely labeled simplex. By properness, the edge 01 does not contain the label 2, and therefore the 01 edge contains an *odd* number of 0–1 transitions. Therefore, some components of the graph Γ_{01} start at such a transition and go back to another 0–1 transition on the same edge. Because the number of such transitions is *odd*, there is at least one component of the graph that leaves an 0–1 transition and never comes back to the 01 edge. By the Fundamental Graph Lemma, this component is either an isolated node or a simple path. If it is an isolated node, it means that the simplex corresponding to this isolated node is completely labeled. If the subgraph of Γ_{01} leaving a 0–1 transition is a simple path, this path must terminate somewhere. Either it terminates inside, in which case it terminates with a completely labeled simplex, or the component of the graph Γ_{01} terminates on another edge, say the 12 edge, in which case by Sperner properness it terminates with a completely labeled simplex. ■

It turns out that more can be said about the existence of completely labeled simplexes:

Lemma 9.8. (Sperner—Strong Version) *Under the same conditions as the preceding lemma, there exists an odd number of completely labeled high-resolution simplexes.*

Proof. This strong version is a corollary of the above constructive proof. Remember that the subgraphs of Γ_{01} that leave the 01 edge and never return to it are *odd* in number and terminate in completely labeled simplexes. The problem is that this constructive procedure is not guaranteed to find *all* completely labeled simplexes. However, the theorem will be proved if we can show that the simplexes missed by the constructive procedure are *even* in number. Consider an *arbitrary* completely labeled, high-resolution simplex. Consider the graph Γ_{01} in its reverse direction and consider the subgraph of Γ_{01} that departs from this completely labeled simplex. A first

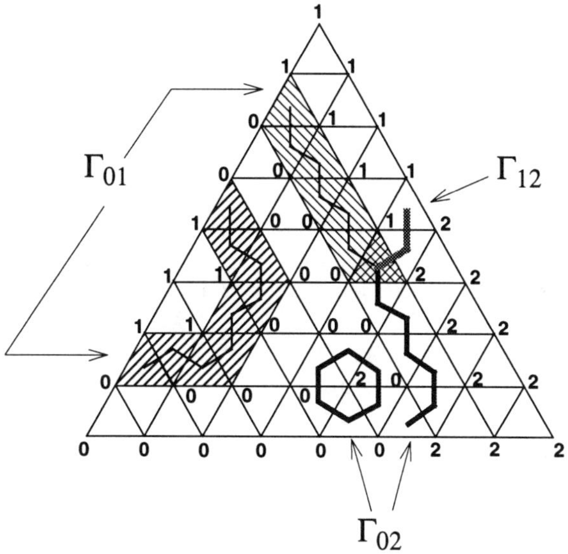

Fig. 9.3. Constructive proof of Sperner's lemma.

possibility is that this subgraph terminates on the 01 edge, in which case the simplex can be reached by the constructive procedure. The other possibility is that this subgraph goes to another completely labeled simplex, in which case both completely labeled simplexes cannot be reached by the constructive procedure. It follows that the completely labeled simplexes missed by the constructive procedure appear in pairs, and are hence *even* in number. ■

Lemma 9.9. (Sperner—Superstrong Version) *Under the same conditions as the preceding lemma, let n_+ be the number of high-resolution simplexes coherently oriented with $a^0a^1a^2$ and let n_- be the number of high-resolution simplexes of opposite orientation. Then*

$$n_+ - n_- = 1$$

Proof. To forge the connection between algorithms and topology, we defer the proof of this result to Chapter 18, where we will develop a proof based on degree theory. ■

If the labeling function is Sperner proper, then we can strengthen the result of Theorem 9.2.

Corollary 9.1. *Let the labeling function ℓ be Sperner proper over the simplex $a^0a^1a^2$ and let this simplex be completely labeled. Then the simplex $a^0a^1a^2$ contains an exact solution to $f(p) = 0$.*

Proof. Refine the simplex in an iterative, nested fashion so that the mesh of the refined triangulation decreases to zero. (The barycentric and Q-triangulations will do.) Invoking Sperner's lemma at every single resolution level, there exists at least one completely labeled simplex, and this simplex can be chosen of arbitrary small mesh. Therefore, invoking Theorem 9.2 yields the result. ■

The way the definitions have been brought about indicates that Sperner properness is a property of the labeling function *and* the triangulation on which it operates. From the definition of Sperner properness, we have the freedom to take *any* simplex $a^k a^l$ and check $\ell(\lambda a^k + (1-\lambda)a^l)$ for $\lambda \in [0,1]$. If we think the labeling function as partitioning $D \times \Omega$ into regions of constant label, unless all separating boundaries are on the same plane, it is trivial to position a simplex $a^k a^l$ so as to have more than one change of label along it. This is to say that it is not easy to get a Sperner proper labeling function by our definition. For all practical purposes, however, it suffices that ℓ be Sperner proper over all simplexes of the decomposition of $D \times \Omega$ and all simplexes resulting from refinements of the grid.

It is clear that, if Sperner properness cannot be guaranteed, we could have the pathology that at a coarse resolution level we have a completely labeled simplex and that at a finer resolution level we lose it, indicating that we have to move to an adjacent simplex; this is accomplished by the "vector labeling" algorithm.

9.1.4 Vector Labeling

The vector labeling technique is some sort of a dual implementation of the linear programming technique of Subsection 7.1.1.

In the integer labeling scheme, we aim at a completely labeled simplex containing an approximate solution in a purely *combinatorial* fashion without *numerical* data as to how close we are to the targeted simplex. The remedy for this is the technique of *vector labeling* introduced by [Eaves 1971]. In the present context, the *vector label* of a vertex a is defined to be the set of Euclidean coordinates of the stereographic projection of $f(a)$:

$$\ell(a) = \begin{pmatrix} x_0(f(a)) \\ x_1(f(a)) \\ x_2(f(a)) \end{pmatrix}$$

The simplex $a^0 a^1 a^2$ is said to be *completely vector-labeled* iff there exist solutions $\lambda_i \geq 0, \beta \geq 0$ to

$$\begin{bmatrix} \ell(a^0) & \ell(a^1) & \ell(a^2) \\ 1 & 1 & 1 \end{bmatrix} \begin{bmatrix} \lambda_0 \\ \lambda_1 \\ \lambda_2 \end{bmatrix} = \begin{bmatrix} 0 \\ 0 \\ 0 \\ 1 \end{bmatrix} + \beta \begin{bmatrix} 1 \\ 1 \\ 1 \\ 0 \end{bmatrix}$$

We have the following result:

Theorem 9.10. *The simplex $a^0a^1a^2$ contains an exact solution to*

$$f_{PL}(p) = 0$$

if and only if it is completely vector-labeled.

Proof. Consider the curvilinear triangle $\ell(a^0)\ell(a^1)\ell(a^2)$ on the Riemann sphere. With the vertices of this curvilinear triangle, we also construct the affine, rectilinear triangle $\ell(a^0)\ell(a^1)\ell(a^2)$. The stereographic projection establishes a one-to-one correspondence between the curvilinear and the rectilinear triangles. Under the stereographic projection, the point $0+j0$ on the Riemann sphere is mapped to the intersection of the $\ell(a^0), \ell(a^1), \ell(a^2)$ plane and the line with Euclidean equations $x_0 = x_1 = x_2$. Clearly, $0 + j0$ is in the curvilinear triangle iff the line $x_0 = x_1 = x_2$ intersects the plane $\ell(a^0), \ell(a^1), \ell(a^2)$ inside the rectilinear triangle $\ell(a^0)\ell(a^1)\ell(a^2)$. Let $\lambda_0, \lambda_1, \lambda_2$ be tentative barycentric coordinates of this intersection and β be the distance along $x_0 = x_1 = x_2$ between the origin O of the Euclidean coordinate system and the intersection point. This distance is measured positively or negatively depending on whether the intersection occurs in the foreground or the background of O, respectively. The linear program simply establishes that $\lambda_0, \lambda_1, \lambda_2$ are genuine barycentric coordinates of an intersection point *affinely dependent* of the vertices $\ell(a^0), \ell(a^1), \ell(a^2)$. ■

Using a dual pivot rule, it is possible to go from one simplex to an adjacent simplex until a completely vector-labeled simplex is identified [Doup 1988, 4.4]. The most elementary pivot rule is as follows: If $\lambda_i < 0, \lambda_j > 0, \lambda_k > 0$ we proceed from the simplex $a^ia^ja^k$ to the adjacent simplex $a^{i'}a^ja^k$ through their common face a^ka^l. In other words, the vertex a^i is replaced by the vertex $a^{i'}$. In case of several negative λ's, remove the vertex that has the lowest barycentric coordinate.

In practice, the above idea is implemented by introducing "slack" variables μ_j,

$$\sum_i \lambda_i \begin{pmatrix} \ell(a^i) \\ 1 \end{pmatrix} + \sum_j \mu_j \begin{pmatrix} e^j \\ 0 \end{pmatrix} - \beta \begin{pmatrix} e \\ 0 \end{pmatrix} = \begin{pmatrix} 0 \\ 1 \end{pmatrix}$$

where

$$(e^0 \; e^1 \; e^2) = \begin{pmatrix} 1 & 0 & 0 \\ 0 & 1 & 0 \\ 0 & 0 & 1 \end{pmatrix}$$

and

$$e = \begin{pmatrix} 1 \\ 1 \\ 1 \end{pmatrix}$$

The slack variables can be considered affine coordinates of new vertices

e^0, e^1, e^2 on the Riemann sphere. Clearly, (the stereographic projection of) $0 + j0$ can always be expressed as an affine combination of $\ell(a^i), e^j$, so that there is always a starting solution to the above system. The "pivoting" problem is trying to substitute vertices in such a way as to obtain $\mu_j = 0$, in which case a solution to $f_{PL}(p) = 0$ is found. During the process of interchanging the vertices, at most three among the λ, μ, variables are energized.

A starting solution can be obtained as follows: Define i^* by

$$\ell_{i^*}(a^0) = \max_i \ell_i(a^0)$$

Therefore a so-called *basic feasible solution* is

$$\begin{aligned} &\lambda_0 = 1; \ \lambda_i = 0, \ i \neq 0; \\ &\mu_i = \ell_{i^*}(a^0) - \ell_i(a^0); \\ &\beta = \ell_{i^*}(a^0) \end{aligned}$$

The next step is to introduce a new vertex $\ell(a^1)$. Either one of the μ's is set to zero, and progress is made toward the solution, or λ_0 is set to zero. In the latter case, no real progress is made toward the solution, a second new vertex $\ell(a^2)$ needs to be introduced, with the hope that it will result in de-energizing one of the μ's.

If, in the process, we have obtained a (λ, μ, β) solution, with at most three λ, μ variables energized, there is actually a line segment of (λ, μ, β) solutions. The extreme points can be found by (dual) linear programming pivoting. The (λ, μ, β) line segment in turn induces a line segment in the polyhedron of uncertainty via $\sum \lambda_i a^i$. This line segment, along the direction of decreasing μ, indicates through what face we have to move from a simplex to an adjacent one.

Here is the algorithm:

1. Choose a vertex a^{i_0} at random. Write βe as a (unique) affine combination of $\ell(a^{i_0}), e^{j_0}, e^{j_1}$.
2. Choose vertex a^{i_1} in star (a^{i_0}), and introduce the vertex $\ell(a^{i_1})$ in the linear system:

$$\begin{aligned} &\begin{pmatrix} \ell(a^{i_0}) \\ 1 \end{pmatrix} \lambda_{i_0} + \begin{pmatrix} \ell(a^{i_1}) \\ 1 \end{pmatrix} \lambda_{i_1} \\ &+ \begin{pmatrix} e^{j_0} \\ 0 \end{pmatrix} \mu_{j_0} + \begin{pmatrix} e^{j_1} \\ 0 \end{pmatrix} \mu_{j_1} - \begin{pmatrix} e \\ 0 \end{pmatrix} \beta = \begin{pmatrix} 0 \\ 1 \end{pmatrix} \end{aligned}$$

By dual pivoting, compute the line of (λ, μ, β) solutions and consider the induced line $a^{i_0}\lambda_{i_0} + a^{i_1}\lambda_{i_1}$ in P.

   (a) Either the induced line in P is $[a^{i_0}a^{i_1}]$, in which case no μ cancels. Set $a^{i_0} \leftarrow a^{i_1}$ and go back to step 2.

   (b) Or in the line of solution a μ, say μ_{j_*}, cancels, in which case set $\{e^{j_0}\} \leftarrow \{e^{j_0}, e^{j_1}\} \setminus \{e^{j_*}\}$. Hence βe is an affine combination of

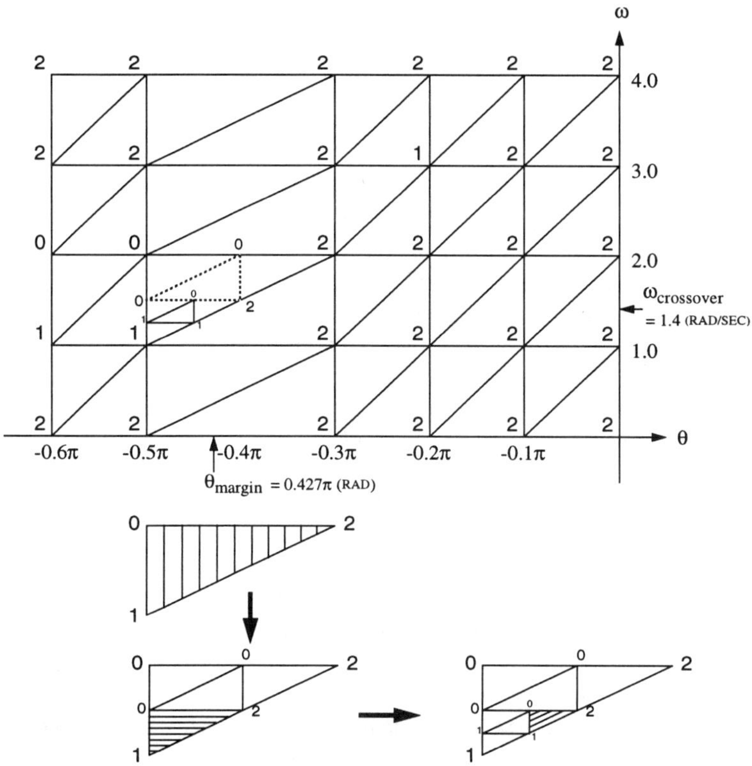

Fig. 9.4. The "variable grid refinement" concept, illustrated on a simple phase margin example.

$\ell(a^{i_0}), \ell(a^{i_1}), e^{j_0}$.

3. Choose a^{i_2} in star $(a^{i_0}a^{i_1})$ and introduce $\ell(a^{i_2})$ in the linear system:

$$\begin{pmatrix} \ell(a^{i_0}) \\ 1 \end{pmatrix} \lambda_{i_0} + \begin{pmatrix} \ell(a^{i_1}) \\ 1 \end{pmatrix} \lambda_{i_1} + \begin{pmatrix} \ell(a^{i_2}) \\ 1 \end{pmatrix} \lambda_{i_2}$$

$$+ \begin{pmatrix} e^{j_0} \\ 0 \end{pmatrix} \mu_{j_0} - \begin{pmatrix} e \\ 0 \end{pmatrix} \beta = \begin{pmatrix} 0 \\ 1 \end{pmatrix}$$

Compute the line of (λ, μ, β) solutions and the induced line $a^{i_0}\lambda_{i_0} + a^{i_1}\lambda_{i_1} + a^{i_2}\lambda_{i_2}$ in P.

(a) If μ_{j_0} cancels, the algorithm terminates with an exact solution.
(b) If a λ, say λ_{i_*}, cancels, then set $\{a^{i_0}, a^{i_1}\} \leftarrow \{a^{i_0}, a^{i_1}, a^{i_2}\} \setminus \{a^{j_*}\}$ and return to step 3.

9.1.5 Textbook Example

Consider the loop function

$$L(s) = \frac{2(s+1)}{s^3 + 2s^2 + 3s + 1}$$

In the feedback path, we insert the usual phase delay element

$$\Delta = \exp j\theta$$

It is easily found that we have a phase margin of 77 degrees—that is, $0.4277\pi = 1.3439023$ rad occurring at a crossover of $\omega = 1.4563$ rad/sec. For the purpose of illustration, we somewhat narrow down the search region to a frequency between 2 and 4 rad/sec and a phase angle between -0.7π and 0. In this case, $D \times \Omega$ is the Cartesian product $[-0.7\pi, 0] \times [2, 4]$.

The search process is illustrated in Figure 9.4. We start from an arbitrary dissection of the rectangle of uncertain parameters. This dissection need not be isotropic. We compute the labels, and a completely labeled simplex immediately comes to our attention. From there on, we repeatedly refine the completely labeled simplexes until an acceptable level of accuracy is reached. We have adopted the *Q-triangulation* (see [Doup 1988]) to refine the simplexes, although any kind of refinement scheme—for example, the barycentric subdivision—is allowed at this stage. Observe that the labeling *appears* Sperner proper, because indeed at every time we refine a triangle a new completely labeled simplex appears at the higher resolution level.

Actually, much more information is gained by looking at the vector labels. We write the linear programs for the nested sequence of completely labeled simplexes.

$$\begin{bmatrix} 0.1524 & 0.8334 & 0.4791 & -1 \\ 0.9523 & 0.2322 & 0.8263 & -1 \\ 0.2610 & 0.5015 & 0.2960 & -1 \\ 1 & 1 & 1 & 0 \end{bmatrix} \begin{bmatrix} \lambda_0 \\ \lambda_1 \\ \lambda_2 \\ \beta \end{bmatrix} = \begin{bmatrix} 0 \\ 0 \\ 0 \\ 1 \end{bmatrix}$$

$$\Rightarrow \begin{bmatrix} \lambda_0 \\ \lambda_1 \\ \lambda_2 \\ \beta \end{bmatrix} = \begin{bmatrix} 1.0766 \\ 0.8800 \\ -0.9566 \\ 0.4392 \end{bmatrix}$$

$$\begin{bmatrix} 0.3116 & 0.8334 & 0.9785 & -1 \\ 0.8976 & 0.2322 & 0.1414 & -1 \\ 0.3118 & 0.5015 & 0.1501 & -1 \\ 1 & 1 & 1 & 0 \end{bmatrix} \begin{bmatrix} \lambda_0 \\ \lambda_1 \\ \lambda_2 \\ \beta \end{bmatrix} = \begin{bmatrix} 0 \\ 0 \\ 0 \\ 1 \end{bmatrix}$$

$$\Rightarrow \begin{bmatrix} \lambda_0 \\ \lambda_1 \\ \lambda_2 \\ \beta \end{bmatrix} = \begin{bmatrix} 0.4619 \\ 0.7622 \\ -0.2241 \\ 0.5599 \end{bmatrix}$$

$$\begin{bmatrix} 0.4777 & 0.9204 & 0.6066 & -1 \\ 0.6395 & 0.1851 & 0.5761 & -1 \\ 0.6024 & 0.3444 & 0.5479 & -1 \\ 1 & 1 & 1 & 0 \end{bmatrix} \begin{bmatrix} \lambda_0 \\ \lambda_1 \\ \lambda_2 \\ \beta \end{bmatrix} = \begin{bmatrix} 0 \\ 0 \\ 0 \\ 1 \end{bmatrix}$$

$$\Rightarrow \begin{bmatrix} \lambda_0 \\ \lambda_1 \\ \lambda_2 \\ \beta \end{bmatrix} = \begin{bmatrix} 0.8593 \\ 0.1912 \\ -0.0505 \\ 0.5558 \end{bmatrix}$$

The intriguing feature is that the completely labeled triangles in the nested sequence contain no *exact* solutions! It is, however, clear from the sequence of solutions to the linear program that we are converging to an *approximate* solution. The reason for this glitch is that a closer look would reveal that the labeling is *not* Sperner proper. A completely labeled simplex at the low-resolution level does not necessarily contain an exact solution. It only contains an approximate solution. If we dissect the completely labeled simplex and if we still find completely labeled simplexes at the higher resolution level—which is the case in this example—then we can expect to further narrow down the solution. If we are less lucky, we would lose the completely labeled simplex somewhere up the resolution scale, at which level we will not be able to improve the accuracy of our estimate of the crossover.

9.2 Simplicial Algorithms Over 3-D Uncertainty Space

Consider the case of a 3-D uncertain parameter vector, $p = (q_1, q_2, \omega)$, running in the space $D \times \Omega$, properly triangulated. Under the extra assumption that $L(j\omega)$ and $\Delta(q_1, q_2)$ are rational, the crossover,

$$\begin{aligned} \Re \det(I + L(j\omega)\Delta(q_1, q_2)) &= 0 \\ \Im \det(I + L(j\omega)\Delta(q_1, q_2)) &= 0 \end{aligned}$$

is an *algebraic (space) curve* in the most traditional sense (see [Walker 1978]). In this finite resolution approach, the curve is approximated by a string of adjacent tetrahedra. By the fundamental 012 rule, each 012 face contains an approximate solution to $\det(I + L\Delta) = 0$. However, a tetrahedron with a completely labeled face has, in fact, *two* completely labeled faces. It follows that the crossover curve pierces in the tetrahedron through a 012 face and comes out of the tetrahedron through the other 012 face. Therefore, *the crossover curve is approximated by a string of adjacent tetrahedra intersecting in completely labeled faces; this is exactly the strategy for finding a winning path in the 3-D game of Hex.* This simplicial generation of an algebraic crossover curve is illustrated in Figure 9.5. The same figure also shows how, combinatorially, two "pieces" of curves (these are called **places** of curve; see [Walker 1978]) interconnect at a so-called

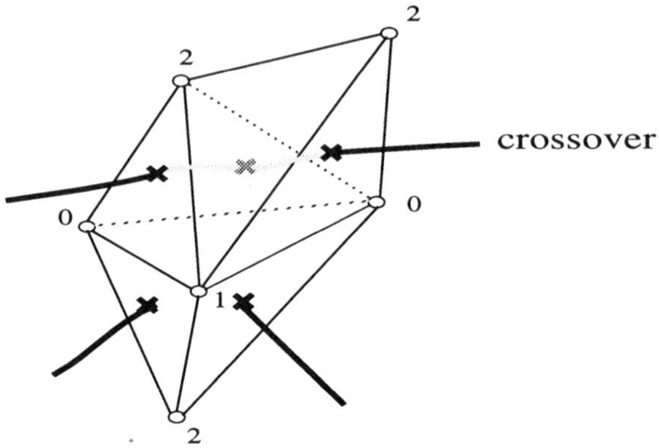

Fig. 9.5. Simplicial generation of algebraic crossover curve.

singular point of the curve; we will come back to the singular point problem later.

The only remaining difficulty with the above scheme is that we need to start with a tetrahedron that has a completely labeled face. Remember, in the 3-D game of Hex, the starting tetrahedron was found by running from a vertex, along an edge, then across a face of the cube until, invoking Sperner's lemma, a completely labeled face was identified, from which we ran inside the cube. In this robust control context, there are several ways to find a starting tetrahedron:

- Either use the simplicial algorithm over the absolute or relative uncertainty complex until it hits a tetrahedron with a completely labeled face.
- Or take, for example, a constant frequency slice through the cube of uncertainties and run the basic two-dimensional algorithm until it hits a 012 face.

It should be clear that the algorithm consists of two stages:

- Find an initial tetrahedron that has a 012 face.
- Once the algorithm "hits" a piece of crossover, the simplicial algorithm "locks" on the crossover curve and runs across it, as directed by the game of hex algorithm.

Observe that the above procedure is not limited to 3-D uncertainty spaces. Indeed, if the uncertainty space is higher-dimensional, we could run the game of hex algorithm on a varying 3-D section through $D \times \Omega$ and then assemble the pieces of crossover. In the particular case where $\dim(D \times \Omega) = 4$, the crossover is a surface. Taking varying 3-D slices through $D \times \Omega$, computing the crossover curve in each 3-D slice, and then assembling the curves together is the combinatorial analogue of generation of a surface by a *foliation*.

9.2.1 Algorithm

Game of Hex Algorithm. The algorithm we use to find the crossover is as follows:

STEP 0 : Set initial point $v_w = (v, \omega_i)^T$ and terminal frequency ω_t. Run Doup's integer labeling algorithm (Section 9.5) to find a completely labeled n-simplex $\sigma = \sigma(y^1, \pi)$ on a fixed frequency level.

STEP 1 : Set $y^1_\omega = (y^1, \omega_i)^T, \pi_\omega = (\pi, n+2), \sigma_\omega = \sigma(y^1_\omega, \pi_\omega)$ and $\overline{y}_\omega =$ new vertex of σ_ω.

STEP 2 : Calculate $l(\overline{y}_\omega)$. There is exactly one vertex $y^p_\omega \neq \overline{y}_\omega$ such that $l(y^p_\omega) = l(\overline{y}_\omega)$.

STEP 3 (Pivoting):
If last entry $y^p_{\omega(n+2)} > \omega_t$, then algorithm terminates. Otherwise, adapt y^1_ω and π_ω according to Table 9.1 (with y^1, q^0 substituted by y^1_ω, q^0_ω respectively) by replacing y^p_ω and go to Step 2 with the new simplex.

9.2.2 A 2-Torus Example

As an example of labeling a tetrahedron, consider the transfer matrix (see [Jonckheere and Bar-on 1991]),

$$L(s) = \frac{1}{s^4 + 1.75s^3 + 7.5s^2 + 4s + 8} \begin{bmatrix} 0.0625s + 0.25 & 0.25s^2 + 0.1875s + 0.75 \\ s^2 + s + 4 & s + 4 \end{bmatrix} \tag{9.1}$$

from the force actuator input (F_1, F_2) to the position sensor output (x_2, x_1) of a double spring–mass–dashpot system. Observe that the two sensor channels have deliberately been crossed. Let us close the loop with the identity matrix,

$$\begin{pmatrix} F_1 \\ F_2 \end{pmatrix} = -I \cdot \begin{pmatrix} x_2 \\ x_1 \end{pmatrix}$$

This *nominal* feedback system is stable. Let us now replace the nominal identity matrix in the feedback path with the perturbation

$$\Delta = \begin{pmatrix} \cos\theta \exp(j\phi) & \sin\theta \\ -\sin\theta & \cos\theta \exp(-j\phi) \end{pmatrix} \tag{9.2}$$

The mapping $(\theta, \phi) \mapsto \Delta$ defined on $[0, 2\pi]^2$ clearly induces a mapping

$$\begin{aligned} \mathbb{T}^2 &\rightarrow \{\Delta\} \\ (\theta, \phi) &\mapsto \Delta \end{aligned}$$

Despite the identification of 0 and 2π, the above mapping is still many-to-one. It is, however, convenient to think the collection of perturbations to be $\mathbb{T}^2$, with some of its points identified.

The generalized crossover frequency interval (see [Bar-on and Jonckheere 1990])—that is, the set of frequencies for which a destabilizing perturbation exists—is $[.643, 1.613]$, in rad/sec. Hence we take Ω to be this interval. Figure 9.6 shows the relevant part of the space of uncertainty displayed as an interval product. We consider D at two sample frequencies, 0.65 rad/sec and 1 rad/sec. The search algorithm yields the tetrahedron $a^0a^1a^2a^3$,

$$\begin{aligned}
a^0 &= (\theta = 25 \text{ deg} \quad \phi_1 = 150 \text{ deg} \quad \omega = 1 \text{ rad/sec})\\
a^1 &= (\theta = 25 \text{ deg} \quad \phi_1 = 125 \text{ deg} \quad \omega = 1 \text{ rad/sec})\\
a^2 &= (\theta = 50 \text{ deg} \quad \phi_1 = 150 \text{ deg} \quad \omega = 1 \text{ rad/sec})\\
a^2 &= (\theta = 50 \text{ deg} \quad \phi_1 = 125 \text{ deg} \quad \omega = 0.65 \text{ rad/sec})
\end{aligned}$$

This is shown in Figure 9.6. The *integer* labels are

$$\ell(a^0) = 0,\ \ell(a^1) = 1,\ \ell(a^2) = 2,\ \ell(a^3) = 2$$

This tetrahedron of $D \times \Omega$ is *maximally labeled.* The edge a^2a^3 is *collapsed* under the simplicial map. The triangle $a^0a^1a^2$ of $D\times\Omega$ mod G is *completely labeled.* Hence we know that within the tetrahedron we have an *approximate* solution to a simplicial approximation of $\det(I + L\Delta) = 0$, a simplicial approximation that collapses simplexes in a manner consistent with ℓ.

We now compute the vector labels, put them into the linear system of equations, compute the solution, and check *complete vector labeling.* The result is

$$\begin{pmatrix} 0.4031 & 0.8437 & 0.5800 & -1 \\ 0.5693 & 0.3322 & 0.8105 & -1 \\ 0.7166 & 0.4217 & 0.0821 & -1 \\ 1.000 & 1.000 & 1.000 & 0 \end{pmatrix} \begin{pmatrix} \lambda_0 \\ \lambda_1 \\ \lambda_2 \\ \beta \end{pmatrix} = \begin{pmatrix} 0 \\ 0 \\ 0 \\ 1 \end{pmatrix}$$

$$\begin{pmatrix} \lambda_0 \\ \lambda_1 \\ \lambda_2 \end{pmatrix} = \begin{pmatrix} 0.5893 \\ 0.2596 \\ 0.1511 \end{pmatrix}$$

$$\beta = 0.5442$$

Therefore the triangle $a^0a^1a^2$ is *completely vector-labeled,* from which we learn that it contains an *exact* solution.

9.2.3 Example

As an example of simplicial generation of an algebraic crossover curve, consider (see [Colgren 1993])

$$L(s)\Delta(q_1, q_2) = \frac{44(s + 0.1 + 0.00277q_2)}{\pi q_1(s + 0.1)(s - 1)}$$

The result of the game of hex algorithm is shown in Figure 9.7.

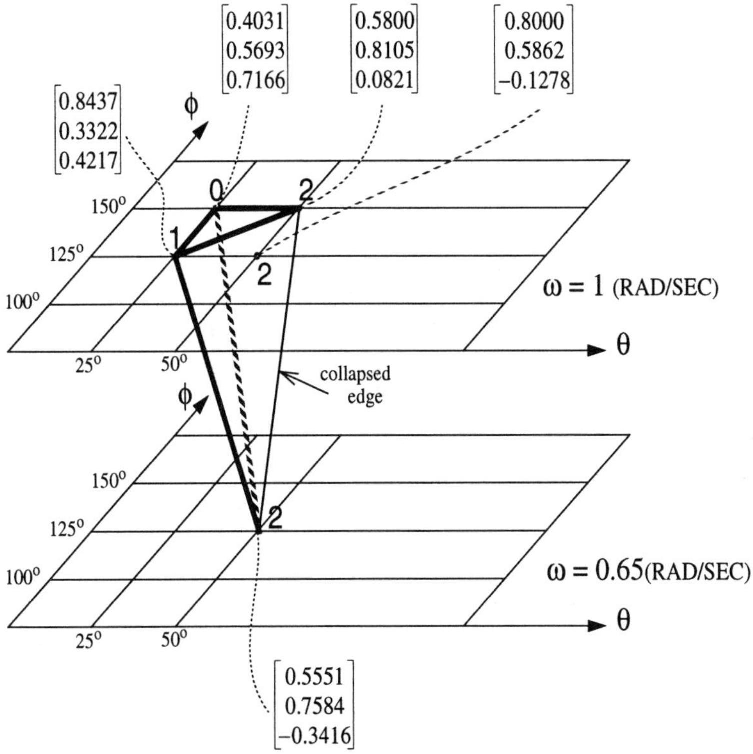

Fig. 9.6. The 2-torus example.

A top view of the same diagram, along the direction of the ω-axis, is shown in Figure 9.8. Projecting the neutral stability curve on the (q_1, q_2)-plane yields the boundary between the regions of constant number of LHP poles, as shown in Figure 9.9.

For this particular example, let

$$1 + L\Delta = \frac{\Re(q_1, q_2, \omega) + j\Im(q_1, q_2, \omega)}{\pi q_1(j\omega + 0.1)(j\omega - 1)}$$

and it is readily found that

$$\begin{pmatrix} \Re \\ \Im \end{pmatrix} = \begin{pmatrix} -(\omega^2 + 0.1)\pi q_1 + 44(0.1 + 0.00277 q_2) \\ -\pi 0.9 \omega q_1 + 44\omega \end{pmatrix}$$

Setting $\omega = 0$ yields $\Im = 0$ and

$$\Re(q_1, q_2, \omega = 0) \equiv -0.314159 q_1 + 4.4 + 0.1219 q_2 = 0$$

The above line of solutions in the $\omega = 0$ section of the cube of parameters projects as is on the (q_1, q_2)-plane, and this projection can be easily seen in

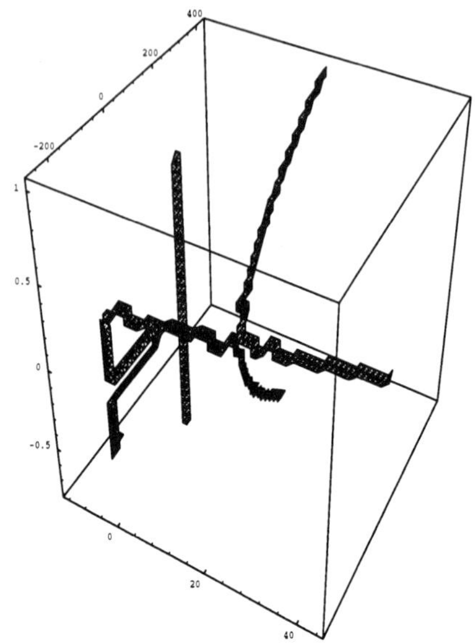

Fig. 9.7. Stability crossover curve in $D \times \Omega$; the vertical axis is the frequency.

Figure 9.9. Next, because $\omega^2 + 0.1 \neq 0$, the equation $\Re = 0$ can be solved for q_1, namely,

$$q_1 = \frac{44(0.1 + 0.00277q_2)}{(\omega^2 + 0.1)\pi}$$

Substituting for the above in $\Im = 0$ yields

$$-\pi 0.9\omega 44(0.1 + 0.00277q_2) + 44\omega\pi(\omega^2 + 0.1) = 0$$

Factoring ω out yields the equation of the parabola in the $(q_1 = \frac{44}{\pi 0.9} = 15.5618, q_2, \omega)$-plane that can be seen in Figure 9.7. This parabola has a zero-order contact with the crossover line in the $\omega = 0$ plane at the point

$$(q_1 = \frac{44}{\pi 0.9} = 15.5618,\, q_2 = \frac{\frac{44}{9} - 4.4}{0.00277 \times 44} = 4.0112,\, \omega = 0)$$

As we shall see in the next subsection, this is a *multiple point of an algebraic crossover curve.* This parabola projects onto a vertical, semi-infinite line that connects to the line $-0.314159q_1 + 4.4 + 0.1219q_2 = 0$ of the diagram of Figure 9.9 at the point $(q_1 = 15.5618,\, q_2 = 4.0112)$.

9.2.4 Algebraic Curve Interpretation

As already hinted at in the preceding subsection, in the case of three uncertain parameters together with a rational loop function, the crossover is an algebraic (space) curve. An **algebraic (space) curve** in $\mathbb{R}^3$ is the solution set of a system of two polynomial equations in three real variables like

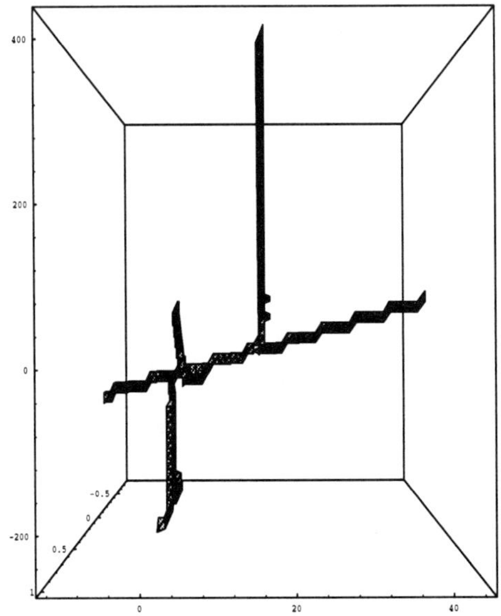

Fig. 9.8. Top view of stability crossover curve, equivalent to projecting the crossover curve on the space of parameters D and getting the boundary between the regions in which the number of LHP closed-loop poles is constant.

$$\Re(q_1, q_2, \omega) = 0, \ \Im(q_1, q_2, \omega) = 0$$

A **singular point** p^* on an algebraic curve is a point around which the curve has more than one (equivalence classes of) parameterization. A **parameterization** of the curve $\Re = 0, \Im = 0$ in a neighborhood O_{p^*} of p^* is a homeomorphism:

$$\begin{array}{ccc} O_0 & \rightarrow & O_{p^*} \\ t & \mapsto & (\gamma_1(t), \gamma_2(t), \gamma_3(t)) \end{array}$$

such that

$$\Re(\gamma_1(t), \gamma_2(t), \gamma_3(t)) = 0, \ \Im(\gamma_1(t), \gamma_2(t), \gamma_3(t)) = 0, \ \ \forall t \in O_0$$

where O_0 is some neighborhood of 0 on the real line. Two parameterizations $(O_0, \gamma), (O_0', \gamma')$ are **equivalent** iff there exists a homeomorphism h such that the following diagram commutes:

$$\begin{array}{ccc} O_0 & \xrightarrow{\gamma} & O_{p^*} \\ h\downarrow & & \downarrow = \\ O_0' & \xrightarrow{\gamma'} & O_{p^*} \end{array}$$

We now proceed to show that the point $p^* = (15.5618, 4.0112, 0)$ on the algebraic space curve of Figure 9.7 where two pieces of curves are welded

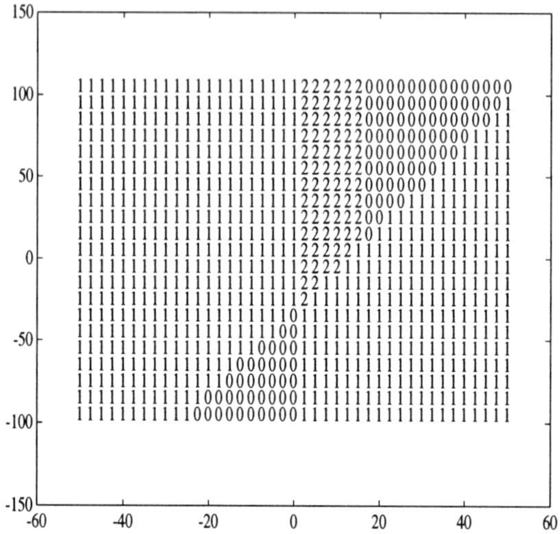

Fig. 9.9. Number of LHP closed-loop poles diagram. The horizontal axis is q_1; the vertical axis is q_2.

together is a *singular point* in the sense of the preceding definition.

The source of the difficulty can be seen from the Jacobian, evaluated at $p = p^*$,

$$\begin{pmatrix} \frac{\partial \Re}{\partial q_1} & \frac{\partial \Re}{\partial q_2} & \frac{\partial \Re}{\partial \omega} \\ \frac{\partial \Im}{\partial q_1} & \frac{\partial \Im}{\partial q_2} & \frac{\partial \Re}{\partial \omega} \end{pmatrix} = \begin{pmatrix} -0.1 \times \pi & 44 \times 0.00277 & 0 \\ 0 & 0 & 0 \end{pmatrix}$$

If this Jacobian were of full rank, then by the implicit function theorem we could find an explicit form of two variables as functions of the remaining variable, which in turn yields a parameterization. Clearly, here the Jacobian is rank-deficient and therefore we cannot expect a single parameterization.

The reader is referred to Chapter 24 and Section 24.5 for the deeper algebraic geometry interpretation of the above.

9.3 Simplicial Algorithms Over Relative Uncertainty Complex

For the sake of the argument, we look at the case where the simplicial algorithm runs on a grid with labels computed from the *simplicial* map $\tilde{f} : (D \times \Omega)' \to N$. The working triangulation of N is chosen to be $DT(\{b^i = f(a^i)\})$, where the a^i's the vertices of $D \times \Omega$.

First, in case of a simplicial map, Sperner properness is easily formulated:

Theorem 9.11. *The labeling function ℓ defined via the simplicial map $\tilde{f}$ is Sperner proper iff any edge of the Delaunay triangulation $DT(\{b^i\})$ intersects at most one radial of the 120-degree butterfly.*

The simplicial algorithm could be run on a varying 2-D slice through $D \times \Omega$. However, when the underlying map is simplicial, there is another way to go.

A typical property of a simplicial map from a *high dimensional* simplex σ of the space of uncertainties to the *low dimensional* complex plane is that a great many vertices of σ are mapped into one single point in the Nyquist template. We say that a simplex τ is **collapsed** whenever all of its vertices are mapped to the same point (see [Hocking and Young 1961, 6.10] and [Hilton and Wylie 1965, 8.2]). Clearly, quite a few simplexes of $D \times \Omega$ are collapsed. Since the simplicial map does not see the difference among vertices of a collapsed simplex, the complex $D \times \Omega$, as it stands, appears too big. It is therefore desirable to reduce the complex in such a way as to avoid unnecessary searches across collapsed simplexes. The reduced or collapsed complex has received the name of *relative complex* by Lefschetz.

We now proceed more formally. Every simplex σ_n, $n > 2$, of $D \times \Omega$ has faces, say τ^i, of maximal dimensions that are collapsed to a point, namely, $\tilde{f}(\tau^i) = b^i$. Define the subcomplex of $D \times \Omega$ formed by those τ^i's together with their faces. Let us denote this subcomplex by G and call it **degeneracy complex**. In terms of chain groups, we have the subgroup property

$$C_*(G) \subseteq C_*(D \times \Omega)$$

The **relative chain group** $C_*(D \times \Omega \bmod G)$ is defined as the quotient group of the chain group of $D \times \Omega$ and the chain group of G:

$$C_n(D \times \Omega \bmod G) = C_n(D \times \Omega)/C_n(G)$$

Observe that the boundary operator maps as follows:

$$\partial_n : C_n(D \times \Omega), C_n(G) \rightarrow C_{n-1}(D \times \Omega), C_{n-1}(G)$$

We therefore define $\bar{\partial}$ from commutativity of the diagram

$$\begin{array}{ccc} C_n(D \times \Omega) & \stackrel{\partial}{\rightarrow} & C_{n-1}(D \times \Omega) \\ \downarrow & & \downarrow \\ C_n(D \times \Omega \bmod G) & \stackrel{\bar{\partial}}{\rightarrow} & C_{n-1}(D \times \Omega \bmod G) \end{array}$$

where the down-arrows denote the natural projections $\pi_* : C_*(D \times \Omega) \rightarrow C_*(D \times \Omega \bmod G)$. It is easily verified that $\bar{\partial}\bar{\partial} = 0$. Therefore, the relative chain groups together with $\bar{\partial}$ form a chain complex, $\{C_*(D \times \Omega \bmod G), \bar{\partial}\}$ called **relative chain complex**.

By well-known universal results, there exists a unique (simplicial) map $\overline{\varphi}$ such that the following diagram commutes:

$$\begin{array}{ccc} C_*(D \times \Omega) & \overset{\varphi}{\to} & C_*(N) \\ \pi \downarrow & \overset{\overline{\varphi}}{\nearrow} & \\ C_*(D \times \Omega \bmod G) & & \end{array}$$

The above reveals an equidimensional, simplicial map $\overline{\varphi}$ from the 2-D relative uncertainty complex to the 2-D complex of the template; furthermore, $\tilde{f}$ and $\overline{\varphi}$ induce the same labeling. Therefore, we could run the simplicial algorithm of Section 9.1 on the relative complex. A problem with this is that the relative complex might fail to be a pseudomanifold (see Definition 18.1); more specifically, there may be more than one 2-simplexes sharing in common the edge $a^i a^j$ of $a^i a^j a^k$. To overcome this difficulty, the algorithm could be run on a cross section assuming that π be a semisimplicial bundle (see Chapter 13).

9.4 Simplicial Labeling Map

In this section, we reinterpret the labeling function as a simplicial map from $D \times \Omega$ to some "labels complex." First the "labels complex" is considered as an abstract complex. Second, to link the results with computational geometry, we show how the abstract "labels complex" can be geometrically realized.

9.4.1 Abstract Label Complex

The labeling function $l(\cdot)$ defined over the vertices of N can be viewed as a *simplicial* map for some abstract *"labels complex."*

Definition 9.12. *The* **labels complex** L *is the abstract simplicial complex with vertices* $0, 1, 2$ *and with the following vertex scheme:* $01, 02, 12, 012$.

Clearly, the map $l : N \to L$ is simplicial. It is also clear that

$$\ell = l \circ f : D \times \Omega \to L$$

Furthermore, the map $\ell(\cdot)$ is simplicial, because it is the composition of two simplicial maps.

Now, we treat the labeling function as the simplicial map that underlies the problem; we follow the same lines of ideas as in Section 9.3 using ℓ instead of $\tilde{f}$ and proceed to define a new, easier-to-compute degeneracy complex. Contrary to the previous chapters, no triangulation of N and refined triangulation of $D \times \Omega$ is required.

The *label* of a vertex somehow provides an idea as to where its image is in the Nyquist template. If we agree that two edge-connected vertices with the same label are somewhat close, then we could collapse the edge joining them; in more formal terms, we could assume that these two vertices form a simplex of some *degeneracy subcomplex.* This leads to the following definition:

Definition 9.13. *The* **degeneracy complex** G_ℓ *of* $D \times \Omega$ *is the subcomplex of (the complex of)* $D \times \Omega$ *with the same vertices as* $D \times \Omega$ *and the following vertex scheme:* $a^0a^1...a^n$ *forms a simplex of* G_ℓ *iff* $\ell(a^0) = \ell(a^1) = ... = \ell(a^n)$.

We can somewhat conceptualize a little further by saying that the new degeneracy complex is the kernel of the labeling function, $\ker(\ell)$. The domain of definition of ℓ is clearly the complex $D \times \Omega$. The image of ℓ is the labels complex. Since ℓ is a simplicial map, it has an underlying algebraic structure of group homomorphism, from which $\ker(\ell)$ receives its precise definition.

Therefore, the situation is summarized by the following commutative diagram:

$$\begin{array}{ccc} D \times \Omega & \rightarrow & N \\ & \ell & \\ \downarrow & \searrow & \downarrow \\ D \times \Omega \bmod G_\ell & \rightarrow & L \end{array}$$

9.4.2 (Strong) Deformation Retract of Template

In the previous subsection we treated the labels complex as abstract. Here we show how the abstract labels complex can be geometrically realized by deformation retract of the Delaunay triangulation on its triangle that contains $0 + j0$. A deeper aim of this subsection is to link up the labeling and the computational geometry approaches.

Let $\{a^i\}$ be the vertex set of $D \times \Omega$ and consider the Delaunay triangulation $DT(\{b^i = f(a^i)\})$ of the template. To make the exposition by far more accurate, we work with the piecewise-linear extension of the vertex transformation $a^i \mapsto f(a^i)$,

$$f_{PL} : D \times \Omega \rightarrow N$$

rather than the original map. In this case, the template $N = f_{PL}(D \times \Omega)$ is a polyhedron that has its underlying Delaunay triangulation.

There are two ways to go: Either we use the usual labeling function defined from the 120-degree butterfly, or, to stress the connection with computational geometry, we define a new labeling function based on the position of the vertex relative to three Voronoi cells.

9.4.2.1 *120-degree Butterfly*

The starting result is the following:

Theorem 9.14. *There is at most one completely labeled triangle of the Delaunay triangulation of* N.

Proof. Simple geometric exercise. ■

This completely labeled triangle is denoted as $b^0b^1b^2$, with the convention that $l(b^i) = i$, $i = 0, 1, 2$. Depending on whether or not the triangulation is Sperner proper, this completely labeled triangle either contains $0+j0$ or has a point z of modulus less than a certain tolerance level.

The basic idea is to manipulate the map f_{PL} in such a way that its image shrinks from N to $\overline{b^0b^1b^2}$, yet the set of completely labeled simplexes of $D \times \Omega$ remains unchanged. (However, the solution set to $f_{PL}(p) \in \overline{b^0b^1b^2}$ could change.) The new map $D \times \Omega \to \overline{b^0b^1b^2}$ would therefore be the composition of the original map f_{PL} and the map $N \to \overline{b^0b^1b^2}$ that shrinks the image set. There are several ways to implement the map $N \to \overline{b^0b^1b^2}$...

.

The most natural way is to define a so-called **retracting map** (see [Hilton and Wylie 1965, page 31]),

$$r : N \to \overline{b^0b^1b^2}$$

that is, a continuous map such that $r|\overline{b^0b^1b^2} = 1$. To define the retracting map, take $z \in N$. Therefore, z belongs to a unique simplex, say $b^{i_0}b^{i_1}b^{i_2}$, with respect to which z has barycentric coordinates $\lambda_{i_0}, \lambda_{i_1}, \lambda_{i_2}$. Therefore the retracting map is defined as

$$\begin{array}{rcl} z & \mapsto & r(z) \\ \lambda_{i_0}b^{i_0} + \lambda_{i_1}b^{i_1} + \lambda_{i_2}b^{i_2} & \mapsto & \lambda_{i_0}b^{\ell(b^{i_0})} + \lambda_{i_1}b^{\ell(b^{i_1})} + \lambda_{i_2}b^{\ell(b^{i_2})} \end{array}$$

It is easily seen that the retracting map is *simplicial*; more specifically, it is the simplicial map induced by the vertex transformation:

$$b^i \mapsto b^{\ell(b^i)}$$

Now, consider the composition map $r \circ f_{PL}$ in the following commutative diagram:

$$\begin{array}{ccc} D \times \Omega & \xrightarrow{f_{PL}} & N \\ & \overset{r \circ f_{PL}}{\searrow} & \downarrow r \\ & & \overline{b^0b^1b^2} \end{array}$$

The following result is also easily proved:

Theorem 9.15. *The map $r \circ f_{PL} : D \times \Omega \to L$ is simplicial, where the geometrically realized labels complex L is defined to be the complex of $\overline{b^0b^1b^2}$.*

The relationship between the labeling function ℓ and the composition map $r \circ f_{PL}$ should be clear; to be precise,

$$\ell = r \circ f_{PL}|(D \times \Omega)^0$$

At this stage we could extend the definition of ℓ and define it to be $r \circ f_{PL}$ over the entire polyhedron $D \times \Omega$, rather than its 0th skeleton. This leads to the following definition (see [Munkres 1984, Chapter 1, Section 3]):

Definition 9.16. *The labeling function $\ell : D \times \Omega \to L$ over the entire polyhedron of uncertainty is defined as the simplicial map $r \circ f_{PL} : D \times \Omega \to L$.*

Since we are not interested in the details of images of simplexes that do not contain $0+j0$, all such simplexes are mapped by ℓ to an edge or a vertex of the triangle $b^0b^1b^2$. Only those simplexes of $D \times \Omega$ that contain some crossover elements are mapped to the whole triangle $b^0b^1b^2$. In a certain sense, ℓ can be viewed as a very crude approximation of f, yet adequate enough insofar as f is only a recipe to capture the crossover.

The next question is whether there exists a homotopy that connects f_{PL} and $r \circ f_{PL}$. This issue is related to whether is it possible to reduce N to $\overline{b^0b^1b^2}$ by means of a *(strong) deformation retract* (see [Hilton and Wylie 1965, page 31]). By definition, a **(strong) deformation retract** is a homotopy

$$R : N \times [0,1] \to N$$

that keeps $\overline{b^0b^1b^2}$ fixed during the entire homotopy; that is,

$$\forall z \in \overline{b^0b^1b^2} \Rightarrow R(z,t) = z, \forall t$$

and such that

$$\begin{aligned} R(\cdot, 0) &= 1_N \\ R(N, 1) &= \overline{b^0b^1b^2} \end{aligned}$$

The strong deformation retract R is defined as follows: Take an arbitrary $z \in N$; assume z in the simplex $b^{i_0}b^{i_1}b^{i_2}$, relative to which z has barycentric coordinates $\lambda_{i_0}, \lambda_{i_1}, \lambda_{i_2}$. The strong deformation retract is defined by

$$\begin{aligned} &(\lambda_{i_0}, \lambda_{i_1}, \lambda_{i_2}, t) \mapsto \\ &\qquad \lambda_{i_0}((1-t)b^{i_0} + tb^{\ell(b^{i_0})}) + \lambda_{i_1}((1-t)b^{i_1} + tb^{\ell(b^{i_1})}) \\ &\qquad +\lambda_{i_2}((1-t)b^{i_2} + tb^{\ell(b^{i_2})}) \end{aligned}$$

Should the point z belong to a 1-simplex or a 0-simplex of N, the strong deformation retract is defined via a trivial modification of the above.

Now, consider the composition map

$$R(f_{PL}(\cdot), \cdot) : (D \times \Omega) \times [0,1] \to N$$

It is easily checked that

$$\begin{aligned} R(f_{PL}(p), 0) &= f_{PL}(p) \\ R(f_{PL}(p), 1) &= r \circ f_{PL}(p) \end{aligned}$$

Clearly, $R(f_{PL}(\cdot), \cdot)$ provides the homotopy between f_{PL} and $r \circ f_{PL}$.

The attentive reader will probably have realized that the deformation retraction of N onto $\overline{b^0b^1b^2}$ destroys the simplicial decomposition of N. To be more specific, take τ to be a simplex of N; clearly, $R(\tau, t)$ is not a simplex of N, nor is it in general included in a simplex of N. In Section 18.10 we will see how to remedy this situation by resorting to the concept of *elementary collapses*; we will by the same token reinterpret the material of this subsection within the context of the *Brouwer degree*.

9.4.2.2 *Voronoi Cell Butterfly*

Let $b^0b^1b^2$ be the unique simplex of $DT(\{b^i\})$ that contains $0 + j0$. The three vertices b^0, b^1, and b^2 induce a Voronoi decomposition of $\mathbb{C}$. Let $VC(b^0)$, $VC(b^1)$, and $VC(b^2)$ be the Voronoi cells of the latter decomposition. The edges of the Voronoi cells, $\partial VC(b^i)$, intersect at the center of the circle circumscribed to the triangle $b^0b^1b^2$. Clearly, the edges of the Voronoi cells define a new "butterfly" relative to which new "labels" can be defined, generically, as $l(b^k) = i \Leftrightarrow b^k \in VC(b^i)$, $i = 0, 1, 2$. From here on, we follow the same procedure as in the previous subsection: We define a retraction of N onto $\overline{b^0b^1b^2}$, a retraction during which the vertices move within one Voronoi cell, without crossing borders between Voronoi cells. The resulting map is the labeling defined from the Voronoi decomposition $\cup_{i=0}^{2} VC(b^i)$. The details are left to the reader.

9.5 Algorithm—Integer Search

The *constructive* proof of Sperner's lemma has revealed an algorithm that *starts on the boundary* and aims at a completely labeled simplex. There are a few problems with this algorithm:

- A first problem is that a *boundary starting* algorithm might miss some completely labeled simplexes. To see this, consider two interior, disjoint 012 simplexes; assume that if we run the Γ_{ij} algorithm from the ij edge of the first simplex we end up on the ij edge of the second triangle, for $ij = 01, 12$, and 02; since no Γ_{ij} graphs connect an ij edge of the interior 012 triangles to the boundary, no boundary starting algorithm can reach these 012 triangles.
- The boundary starting algorithm uses the combinatorial idiosyncrasies of the boundary to find a boundary edge from which the algorithm enters the simplex. As such the algorithm relies crucially on the Sperner properness of the labeling, without which the algorithm would not work.
- The computational implementation of the algorithm sometimes requires adding an artificial layer around the simplex, leading to the so-called "artificial start" algorithm of [Kuhn 1968] and [Kuhn and MacKinnon 1975]. This artificial layer is a bit cumbersome, although it can be better motivated in the Game of Hex (see Chapter 8).
- Finally, and most importantly, if we have some a priori knowledge as to where to solution is, we cannot start the algorithm inside, near the guessed solution, in an attempt to speed up the search.

It is clear than an *interior starting* algorithm would have many attractive features, among other things, less reliance of the combinatorics of the simplex. The following is an interior starting, variable dimension algorithm in the sense that it generates a sequences of simplexes of varying dimension

(0, 1, or 2) until a completely labeled simplex is hit.

Doup's Algorithm for Sperner Proper Integer Labeling on Q-Triangulation [Doup 1988, pages 75–76]
The notation is as follows: $e^i = (0,0,...,0,1,0,...,0), 1 \leq i \leq n+1$ is the $(n+1)$-dimensional Euclidean basis vector with the 1 in the ith position. Define $\vec{e}_i = e^{i+1} - e^i, i = 1,...,n$, and $\vec{e}^{n+1} = e^1 - e^{n+1}$. T is a subset of $\{1,2,...,n+1\}$ and $\pi = (\pi_1, \pi_2, ..., \pi_t)$ is a permutation of T. $\sigma(a^1, \pi(T)$ denotes the simplex $a^1a^2...a^t$, where $a^{i+1} = a^i + (1/m)\vec{e}_{\pi_i}$. The *variable dimension restart algorithm* for Sperner proper integer labeling rules is as follows:

STEP 0 : Set $T = \emptyset, t = 0, y^1 = v$, where v is the initial point, $\sigma = \sigma(y^1, \pi(\emptyset)), \overline{y} = y^1$, and $R_i = 0, i = 1, \ldots, n+1$.
STEP 1 : Calculate $\ell(\overline{y})$. Set $l(\overline{y}) = 1 + \ell(\overline{y})$. If $l(\overline{y})$ is not an element in T, then go to Step 3. Otherwise, there is exactly one vertex $y^p \neq \overline{y}$ such that $l(y^p) = l(\overline{y})$.
STEP 2 : If $p = t+1$ and $R_{\pi_t} = 0$, then go to Step 4. Otherwise, $\sigma(y^1, \pi(T))$ and R are adapted according to Table 9.1 by replacing y^p. Return to Step 1 with $\overline{y}$ equal to the new vertex of σ.
STEP 3 (Increase Dimension):
If $t = n$, then σ is a completely labeled simplex and the algorithm terminates. Otherwise, set $T = T \cup \{l(\overline{y})\}, \pi(T) = (\pi_1, \ldots, \pi_t, l(\overline{y}))$, and $t = t+1$. Return to Step 1 with $\overline{y} = y^{t+1}$.
STEP 4 (Decrease Dimension):
Exactly one vertex of σ, say $y^{p\prime}, p\prime \leq t$, has label π_t. Set $T = T \setminus \{\pi_t\}, \pi(T) = (\pi_1, \ldots, \pi_{t-1}), \sigma = \sigma(y^1, \pi(T))$, and $t = t-1$. Return to Step 2 with $p = p\prime$.

Table 9.1 p is the Index of the Vertex to Be Replaced

	$\overline{y}^1$	$\overline{\pi}(T)$	$\overline{R}$
$p = 1$	$y^1 + m^{-1}\vec{e}_{\pi_1}$	$(\pi_2, \ldots, \pi_t, \pi_1)$	$R + e^{\pi_1}$
$1 < p < t+1$	y^1	$(\pi_1, \ldots, \pi_{p-2}, \pi_p, \pi_{p-1}, \pi_{p+1}, \ldots, \pi_t)$	R
$p = t+1$	$y^1 - m^{-1}\vec{e}_{\pi_t}$	$(\pi_t, \pi_1, \ldots, \pi_{t-1})$	$R - e^{\pi_t}$

The above algorithm, due to [Doup 1988], was successfully implemented to get a starting point on the crossover curve of Figure 9.7 from which the Hex algorithm generated the whole curve.

9.6 Algorithm—Vector Labeling Search

For the vector labeling version of the integer labeling algorithm, the reader is referred to [Doup 1988].

HISTORICAL AND BIBLIOGRAPHICAL NOTES

The historical root of the simplicial algorithms is probably iterative adjustment of prices towards equilibrium in a free exchange economy; see [Smale 1975] for an excellent survey of the fundamental economics principles and their mathematically rigorous models.

What became known as *Sperner's lemma* was introduced in [Sperner 1928] as a combinatorial tool for proving the Brouwer invariance of the dimension and the domain.

The basic paper that introduced for the first time the concept of *simplicial algorithm*—as a tool for computing the Brouwer fixed-points of a mapping $f : \overline{\Sigma_n} \to \overline{\Sigma_n}$—is [Scarf 1967]. Although Scarf used *primitive subsets* instead of a triangulation of Σ_n, the fundamental ideas of [Scarf 1967] are unmistakingly the same as those of the modern simplicial algorithms running on a triangulation. Given a cluster of vertices $\{a^{i_j}\}$, each carrying a label ℓ such that $f_\ell(a) \leq a_\ell$, a primitive subset is constructed out of $n+1$ vertices by drawing the hyperplanes, parallel to the faces of Σ_n, passing through the selected vertices. Whenever all vertices $a^{i_0}, ..., a^{i_n}$ of a primitive set are labeled differently, the primitive set contains an approximate fixed point. Among the novel ideas of [Scarf 1967] is the utilization of Sperner's lemma to prove, in a constructive fashion, that there exists at least one completely labeled primitive subset.

The landmark paper that introduced simplicial algorithms running on a *triangulation* for computing fixed points, in the same spirit as the robust control simplicial algorithms, is [Kuhn 1968].

The method of *vector labeling*, also referred to as *complementary pivoting*, was introduced by [Eaves 1971], again in the context of computing Brouwer fixed-points.

The potential existence of fixed points for problems defined over an *unbounded* region—as it happens in robust control—motivated the so-called *homotopy method* of [Eaves and Saigal 1972]. The idea is to deform a fixed-point problem over an unbounded set to a problem over a bounded set, find the fixed points over the bounded set, and then trace the fixed points backwards to the unbounded region.

Another "homotopy" method due to [Eaves 1972] was motivated by efficient implementation of the variable grid refinement restart method. Essentially, the iteratively refined grids for the simplex Σ_n are all embedded in the prism $\Sigma_n \times [0,1]$, the region between two consecutive refinements of Σ_n is triangulated, and the iterative refinement of the solution is accomplished by moving from the $t = 0$ section (coarse resolution) to the $t = 1$ section (fine resolution) following a homotopy path.

These homotopy methods, more specifically that of [Eaves and Saigal 1972], can safely be credited to be the forerunners of the so-called *continuation methods* as described in [Alexander 1978].

An excellent survey of both integer and vector labeling methods for computing fixed points is [Todd 1976]. For a more modern exposition of the simplicial algorithms, see [Doup 1988].

Part II

HOMOLOGY OF ROBUST STABILITY

10

HOMOLOGY OF UNCERTAINTY AND OTHER SPACES

Anyone who wishes to learn topology with an interest in its applications must begin with the Betti groups,...

P. Alexandroff,
Elementary Concepts of Topology,
Dover, New York, 1961.

SUMMARY

Central in the previous chapters was the decomposition of the space of uncertainties into simplexes. Here, we show how the combinatorics of the assembly of simplexes leads to the formalization of such pathologies as "loops," "holes," "twists," and so on, of which we can acquire an intuitive, visual picture in low-dimension. Such pathologies are algebraically characterized in terms of *homology groups*. The homology groups can be loosely referred to as the "*global* topology," to contrast it with the *local*, point set topological properties.

Computing the global topology (i.e., loops, holes, twists, etc.) of such a space as $D \times \Omega$ requires manipulating a number of simplexes that grows exponentially with the dimension. Since this rate of growth is more than linear, it would be beneficial to break the space into the Cartesian product of two spaces, typically D and Ω, compute the global topology of each space, and then infer the topology of the product space from the topology of the factor spaces. This last nontrivial step is precisely the objective of two celebrated results of topology: the Eilenberg–Zilber theorem and the Künneth theorem (see [Munkres 1984, Sections 58 and 59] and [Hilton and Wylie 1965, Sections 5.7 and 8.7]). This approach is not limited to the Cartesian product of D and Ω. Indeed, D may itself be the Cartesian product of many uncertainties, requiring iterative utilization of the Eilenberg–Zilber–Künneth theorem.

In many instances, despite fast algorithms for implementing the combinatorial computations on an assembly of simplexes and clever utilization of the Eilenberg–Zilber–Künneth theorem, the homology computations remain hard, even intractable. To overcome these difficulties, an arsenal of computational tools is available. As a prelude to this, we introduce the relative homology sequence and the Mayer–Vietoris sequence at the end of this chapter and relegate the more powerful, algebraically oriented computational tools to forthcoming chapters.

10.1 Simplicial Homology

10.1.1 Homology Groups

As a byproduct of the simplicial approximation theorem we have a decomposition of $D \times \Omega$ into a great many simplexes of varying dimensions. All n-dimensional simplexes of $D \times \Omega$ are lumped into the Abelian group $C_n(D \times \Omega)$. The Abelian groups of varying dimensions are linked by a collection of boundary homomorphisms

$$\ldots \to C_{n+1}(D \times \Omega) \stackrel{\partial_{n+1}}{\to} C_n(D \times \Omega) \stackrel{\partial_n}{\to} C_{n-1}(D \times \Omega) \to \ldots$$

with the basic property that

$$\partial_n \partial_{n+1} = 0$$

A crucial pathology that a topological space can exhibit is a "loop"—that is, a subset of the space that closes on itself, which has no boundary. A circle is an example of a one-dimensional object that has no boundary. The sphere S^2 is an example of a two-dimensional object that closes on itself. We approximate such an n-D object by an assembly of n-simplexes, formalize it as a chain $z_n \in C_n(D \times \Omega)$ such that $\partial_n z_n = 0$, and call it a **cycle**. In the example of the simplicial cylinder of Figure 4.2 we have

$$\partial_1 \underbrace{(a^1a^5 + a^5a^7 + a^7a^1)}_{z_1} = 0$$

so that z_1 is indeed a 1-cycle. We define the **group of n-cycles** as

$$Z_n(D \times \Omega) = \ker \partial_n$$

It is easily seen that Z_n is a *subgroup* of C_n.

Another important concept is that of an n-dimensional *boundary*—that is, an n-D object that "bounds" an $(n+1)$-D object. The unit circle is the boundary of the unit disk. The unit sphere is the boundary of the unit ball. We formalize an n-D object bounding an $(n+1)$-D object as a chain $b_n \in C_n$ such that $b_n = \partial_{n+1} c_{n+1}$ for some $(n+1)$-D chain c_{n+1}. We call b_n an n-**boundary**. For example, in the simplicial cylinder of Figure 4.2, it is easily seen that

$$\underbrace{a^0a^3 + a^3a^4 + a^4a^0 - (a^9a^{10} + a^{10}a^{11} + a^{11}a^9)}_{b_1} =$$
$$\partial_2(a^0a^3a^1 + a^1a^3a^5 + a^5a^3a^4 + \ldots)$$

so that b_1 is indeed a boundary. We define the **subgroup of n-boundaries** as

$$B_n(D \times \Omega) = \text{im } \partial_{n+1}$$

Since $\partial_n \partial_{n+1} = 0$, it follows that

$$B_n(D \times \Omega) \subseteq Z_n(D \times \Omega)$$

The problem is that a topological space has too many cycles, not all of them being equally important. Take a cylindrical surface. The two base circles together bound the cylindrical surface; hence they are considered equivalent; they are said to be *homologous.* Formally, we say that two cycles $z_n, z'_n \in Z_n$ are **homologous** if they differ by a boundary, $z_n - z'_n \in B_n$. Coming back to the simplicial example of Figure 4.2 it is easily seen that the cycles

$$\begin{aligned} z_1 &= a^0a^3 + a^3a^4 + a^4a^0 \\ z'_1 &= a^9a^{10} + a^{10}a^{11} + a^{11}a^9 \end{aligned}$$

are homologous. The reader can easily verify for himself that the property of being *homologous* defines an *equivalence relation* on Z_n. Not all cycles are homologous, though. Take the large (poloidal) and the small (toroidal) circles of a torus. They do not bound the same surface. Hence they are not homologous. More precisely, looking at the simplicial torus of Figure 4.3 it is easily seen that the cycles

$$a^3a^5 + a^5a^6 + a^6a^3$$

$$a^1a^5 + a^5a^7 + a^7a^1$$

are not homologous. We formalize these concepts by exploiting the fact that $B_n \subseteq Z_n$ and agreeing that what is relevant in a topological space is not the set of cycles, but the *equivalent classes*, the *cosets* of cycles. The resulting *quotient group* is called **homology group**

$$H_n(D \times \Omega) = Z_n(D \times \Omega)/B_n(D \times \Omega)$$

Clearly, this group is commutative, Abelian, since Z_n, B_n are. We will use the notation $\{z^i\}$ to denote the *coset* or *homology class* of the cycle z^i. If $\{z^i\} \neq 0$, any cycle in $\{z^i\}$ is said to be **nonbounding**, because it is not the boundary of any $(n+1)$-D object. Two arbitrary cycles in $\{z^i\}$ together bound some $(n+1)$-D object.

In general, in such a homology group as $H_n(D \times \Omega)$, if we can find finitely many elements $\{z^i : i = 1, ..., m\}$ such that

$$H_n(D \times \Omega) = \{\alpha_1\{z^1\} + ... + \alpha_m\{z^m\} : \alpha_i \in \mathbb{Z}\}$$

the group is said to be **finitely generated** and the $\{z^i\}$'s are called **generators**. Contrary to the case of a vector space, the coefficients α_i of a finitely generated Abelian group do not form a *field.* Consequently, it is possible that, for some $\{z^i\} \neq 0$, we get $\{\alpha z_i\} = 0$ for some $\alpha \in \mathbb{Z}, \neq 0$. Such cycles are called **torsion cycles** and they typically happen in the Möbius strip and in the projective plane. Removing the torsion cycles from H_n yields the **free part** of H_n. This yields the next important concept:

Definition 10.1. *The* **Betti number** b_n *of the homology group* H_n *is the number of generators in the free part of* H_n.

Next, we could do the same analysis for the template N. In case of the usual Nyquist stability test, N is a subset of $\mathbb{C}$ so that its homology can be expected to be very simple. To understand $H_0(N)$, it is necessary to be a little more specific as to how the chain complex terminates:

$$\dots \to C_1(N) \xrightarrow{\partial_1} C_0(N) \xrightarrow{\partial_0=0} 0$$

From the above, it follows that

$$\begin{aligned} H_0(N) &= \ker \partial_0 / \mathrm{im} \partial_1 \\ &= C_0(N)/B_0(N) \end{aligned}$$

Clearly, $B_0(N)$ contains all elements of the form $\partial(b^{i_0}b^{i_1} + b^{i_1}b^{i_2} + \dots + b^{i_{m-1}}b^{i_m}) = b^{i_m} - b^{i_0}$. In other words, in C_0/B_0 are identified those vertices that can be joined by a connected path of 1-simplexes. Therefore, *the generators of* $H_0(N)$ *are the connected components of the template.* In general, the template has one connected component so that $H_0(N)$ is on one generator. A homology group on one generator is isomorphic to the additive group of the integers, which we write as $H_0(N) \cong \mathbb{Z}$. Next, we look at $H_1(N) = Z_1(N)/B_1(N)$. Take the example of the simplicial template with a "hole" shown in Figure 10.1. The outer boundary of the "hole" is, simplicially,

$$\begin{aligned} z_1 = \quad & b^4b^5 + b^5b^6 \\ & + b^6b^7 + b^7b^8 \\ & + b^8b^9 \\ & + b^9b^{10} + b^{10}b^{11} \\ & + b^{11}b^{12} + b^{12}b^{13} \\ & + b^{13}b^4 \end{aligned}$$

The inner boundary of the hole is

$$z_1' = b^0b^1 + b^1b^2 + b^2b^3 + b^3b^0$$

Observe that the cycles z_1, z_1' are homologous because, indeed,

$$z_1 - z_1' = \partial_1(b^4b^5b^0 + b^0b^5b^1 + \dots)$$

From there on, it is easily seen that any cycle encompassing the hole is in the homology class $\{z_1\} = \{z_1'\}$. Therefore, the homology group $H_1(N)$ is on one generator. *This single generator can be identified with the "hole" in the template.* We write the latter as $H_1(N) \cong \mathbb{Z}$. In general, the number of generators of $H_1(N)$ is the number of "holes" in the template. Remember

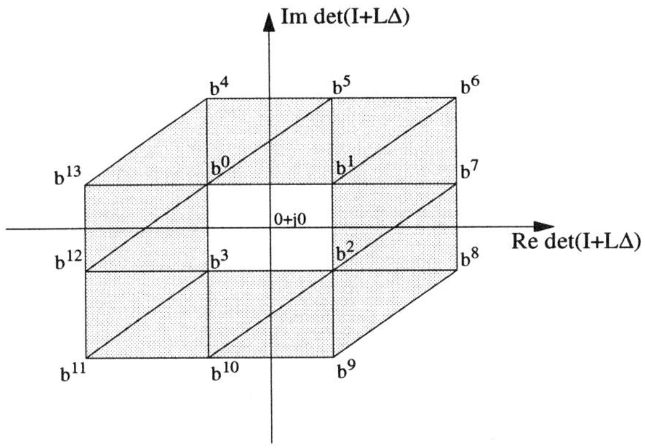

Fig. 10.1. Simplicial hole in the supertemplate.

that we need at least one hole in the template to stabilize robustly an open-loop unstable system. Finally, insofar as N is a 2-D space, we clearly have $H_n(N) = 0$, $\forall n > 2$. In the *exceptional* case where the template happens to be the whole Riemann sphere, $H_2(S^2) = \mathbb{Z}$. In all other cases where the template is a (closed) domain that can be approximated with a polyhedron, we have $H_2(N) = 0$.

As said in our review of stability criteria, it is sometimes convenient, even necessary, to map the uncertainty in a space more complicated than $\mathbb{C}$, in which case the template should be expected to have more complicated homology.

Here, we have developed the relevant Abelian group concepts in the intuitive geometrical setting. The reader should consult Appendix A, and probably Appendix C as well, for the deeper algebraic insights.

10.1.2 Homology Group Homomorphism

The next question is, How can we link the homology of $D \times \Omega$ and the homology of N? The simplicial approximation theorem yields a chain map $\varphi_n : C_n(D \times \Omega) \to C_n(N)$. Since this map commutes with the boundary, it maps cycles to cycles and boundaries to boundaries. Therefore, it induces the homology group homomorphism $\varphi_* : H_*(D \times \Omega) \to H_*(N)$. To be more specific, take $\{z_n\} \in H_n(D \times \Omega)$, where z_n is a representative of the homology class. Clearly, $\varphi(z_n)$ is a cycle, because $\partial\varphi(z_n) = \varphi(\partial z_n) = \varphi(0) = 0$. Since $\varphi(z_n)$ is a cycle, it defines a unique homology class of $H_n(N)$. It remains to show that the homology class $\{\varphi(z_n)\}$ does not depend on the representative of the homology class of $H_n(D \times \Omega)$. Take another representative, $z_n + b_n$, where $b_n = \partial c_n$ is a boundary. Clearly, the homology class induced by the representative $c_n + b_n$ is

$$\{\varphi(z_n + b_n)\} \quad = \quad \{\varphi(z_n) + \varphi(b_n)\}$$

$$
\begin{aligned}
&= \{\varphi(z_n) + \varphi(\partial c_n)\} \\
&= \{\varphi(z_n) + \partial\varphi(c_n)\} \\
&= \{\varphi(z_n)\}
\end{aligned}
$$

The fundamental concept of going from the chain map to the homology group homomorphism is depicted in the following diagram:

$$
\begin{array}{ccc}
C(D\times\Omega) & \stackrel{\varphi}{\rightarrow} & C(N) \\
\downarrow & & \downarrow \\
H(D\times\Omega) & \stackrel{\varphi_\sharp}{\rightarrow} & H(N)
\end{array}
$$

Now, we would like to formalize the "downarrow" process. Remember that the chain map itself comes from a simplicial map $\tilde{f} : D \times \Omega \to N$, where $D \times \Omega$ and N are meant to be triangulated polyhedra. Therefore, we could put together everything in the diagram

$$
\begin{array}{ccc}
D\times\Omega & \stackrel{\tilde{f}}{\rightarrow} & N \\
\downarrow & & \downarrow \\
C(D\times\Omega) & \stackrel{\varphi}{\rightarrow} & C(N) \\
\downarrow & & \downarrow \\
H(D\times\Omega) & \stackrel{\varphi_\sharp}{\rightarrow} & H(N)
\end{array}
$$

Clearly, we can collapse the above diagram into

$$
\begin{array}{ccc}
D\times\Omega & \stackrel{\tilde{f}}{\rightarrow} & N \\
\downarrow & & \downarrow \\
H(D\times\Omega) & \stackrel{\varphi_\sharp}{\rightarrow} & H(N)
\end{array}
$$

The downarrows of the above diagram—the formal process of going from the simplicial map to the homology group homomorphism—is an example of a *covariant functor*. The specific functor involved here is the *homology functor*. The concept of "functor" is formalized in Appendix A, and here we will just develop the intuition for it. The **homology functor** maps the polyhedra $D \times \Omega$, N to their homology groups, and it maps the simplicial Nyquist map to the homology group homomorphism. In other words, the homology functor provides an "algebraic picture" of the simplicial Nyquist map. **Covariance** of the functor means that in the algebraic picture the arrow flows in the *same* direction. Now we could go one step further and say that the homology functor could map *any* string of simplicial maps among polyhedra, including their compositions, into a string of Abelian groups and group homomorphisms, including their compositions. There is thus a need to formalize *what* the functor maps into *what*. This leads to the concept of *category*. A **category** is a collection of objects (polyhedra, resp. Abelian groups) and arrows (simplicial maps, resp. group homomorphisms) subject to some composition rules (see Appendix A). The top rows of the diagram is in the *category of polyhedra and simplicial maps* and bottom row is in the *category of Abelian groups and group homomorphisms.* In formal language, the homology functor maps the category of polyhedra and simplicial maps into the category of Abelian groups and homomorphisms.

10.1.3 Computation

Computation of simplicial homology groups is a long process. First, the space has to be triangulated. Triangulation of a space is not the easiest possible process (How would we triangulate the compact space of $n_l \times n_l$ unitary matrices?); nevertheless, we assume here that it can be done. The reader should realize that triangulation is a monstrous machinery to produce a *chain complex*

$$\ldots \to C_{n+1} \stackrel{\partial_{n+1}}{\to} C_n \stackrel{\partial_n}{\to} C_{n-1} \to \ldots$$

$$\partial_n \partial_{n+1} = 0 \Leftrightarrow \text{im } \partial_{n+1} \subseteq \ker \partial_n$$

from which the homology groups are *defined* and computed as

$$H_n = \ker \partial_n / \text{im } \partial_{n+1}$$

Clearly, choosing bases for the Abelian groups C_{n+1}, C_n and C_{n-1} yields matrix representations for ∂_{n+1} and ∂_n. These matrices have large size in case of a finely triangulated polyhedron; however, they are sparse and their entries are integer. Computing the kernel and the image of ∂_n and ∂_{n+1}, respectively, yields the homology groups. This procedure is implemented in an algorithm developed in Appendix B. Despite the fact that this algorithm involves only *finitely* many arithmetic operations, it is still long and involved, and presupposes the existence of a triangulation. Clearly, faster procedures for computing homology groups are in order. One such procedure, which we will go through in this chapter, consists in breaking the topological space into many factors and computing the homology of the product from the homology of the factors.

10.2 Semisimplicial Homology

Besides the standard simplicial theory, where a "simplex" is a skew commutative product of vertices (no repetition allowed) spanning a geometric simplex of a polyhedron, we could also use "total simplexes" defined as ordered arrays of vertices (with repetition allowed) spanning geometric simplexes. As seen in Section 4.5.2, we can define total chain groups and total boundary operators, leading to the total, or semisimplicial, chain complexes

$$\{C_*^{\text{total}}(D \times \Omega), \partial_*^{\text{total}}\}, \ \{C_*^{\text{total}}(N), \partial_*^{\text{total}}\}$$

Since we have $\partial^{\text{total}} \partial^{\text{total}} = 0$, we can compute total, or semisimplicial, homology groups as

$$H_n^{\text{total}}(D \times \Omega) = \ker(\partial_n^{\text{total}}) / \operatorname{im}(\partial_{n+1}^{\text{total}})$$

In Section 4.5.2, we also defined the total chain map

$$C_*^{\text{total}}(D \times \Omega) \xrightarrow{\varphi_*^{\text{total}}} C_*^{\text{total}}(N)$$

By the same process as in the simplicial theory, this chain map induces a total, or semisimplicial, homology group homomorphism:

$$H_n^{\text{total}}(D \times \Omega) \xrightarrow{\varphi_n^{\text{total}}} H_n^{\text{total}}(N)$$

The *total complex* has by far more simplexes that the regular complex, and computations on the total complex are likely to be prohibitively large. However, the reason why sometimes the total complex is useful is that some proofs appear more transparent in the total complex than they would appear in the usual complex. Of course, jumping to the total complex to shed light on some issues requires the assurance that the homology groups and homology group homomorphisms are not affected by this jump. It turns out that there is no discrepancy if we restrict ourselves to the category of polyhedra and simplicial maps. It, however, takes a substantial amount of tedious combinatorial analysis [Hilton and Wylie 1965, Chapter 3] to prove that we have the flexibility of using either the regular or the total complex. From now on we will take this fact for granted, and we will drop the superscript "total."

10.3 Homology of a Chain Complex

The above material has a *clear* geometric significance. Now, we somewhat conceptualize. Assume we have a collection of Abelian groups $\{C_*\}$ together with a collection of homomorphisms $\{\partial_*\}$, $\partial_n : C_n \to C_{n-1}$, such that $\partial_n \partial_{n+1} = 0$. This algebraic conceptualization is called **chain complex** and is usually written as $\{C_*, \partial_*\}$. This *chain complex* could be a purely algebraic object that need not come from the triangulation of a polyhedron. Nevertheless, the *algebraic structure* of a *chain complex* is enough to define homology groups. Whether an algebraic chain complex has some geometric significance is the so-called *geometric realization problem.*

10.4 Homotopy Invariance

In this section, we aim at the fundamental result that two homotopic maps induce the same homology group homomorphisms. The *fundamental tool* is the concept of *chain homotopy.* The latter is a chain map version of the usual concept of homotopy. There are several important corollaries to these homotopy concepts—among other things, the fact that homology groups are triangulation independent. To be general and to broaden the impact of the results, we make systematic use of chain complexes.

10.4.1 Chain Homotopy

The fundamental concept is the following:

Definition 10.2. *Two chain maps $\varphi_*, \gamma_* : C_* \to D_*$ are said to be* **chain homotopic**, *$\varphi_* \simeq \gamma_*$, iff there exist homeomorphisms $d_n : C_n \to D_{n+1}$ such that $\varphi_n - \gamma_n = \partial_{n+1} d_n + d_{n-1}\partial_n$. The collection of homeomorphisms d_* is called* **chain homotopy** *between φ_* and γ_*.*

The situation can be visualized with the aid of the following *noncommutative* diagram:

$$\begin{array}{ccccccc}
\cdots & C_{n+1} & \rightarrow & C_n & \stackrel{\partial_n}{\rightarrow} & C_{n-1} & \cdots \\
& & d_n & & d_{n-1} & & \\
\varphi_{n+1} - \gamma_{n+1} \downarrow & & \swarrow & \downarrow & \swarrow & \downarrow \varphi_{n-1} - \gamma_{n-1} & \\
\cdots & D_{n+1} & \stackrel{\partial_{n+1}}{\rightarrow} & D_n & \rightarrow & D_{n-1} & \cdots
\end{array}$$

Theorem 10.3. *Chain homotopic maps, $\varphi_* \simeq \gamma_*$, induce the same homology group homomorphism.*

Proof. Take a cycle z_n in C_n. We must show that

$$\{\varphi_n(\{z_n\})\} = \{\gamma_n(\{z_n\})\}$$

Equivalently, we must show that $\varphi_n(z_n) - \gamma_n(z_n)$ is a boundary. By the chain homotopy property, we have $\varphi_n(z_n) - \gamma_n(z_n) = \partial_{n+1} d_n(z_n) + d_{n-1}\partial_n z_n = \partial_{n+1} d_n(z_n)$, so that $\varphi(z_n) - \gamma(z_n)$ is indeed a boundary. Therefore, $\varphi(z_n)$ and $\gamma(z_n)$ induce the same homology class. ■

Definition 10.4. *A chain map $\varphi : C \to D$ is said to be a* **chain equivalence** *iff it has a* **chain homotopic inverse**; *that is, a chain map $\varphi^\dagger : D \to C$ such that $\varphi\varphi^\dagger \simeq 1 : D \to D$ and $\varphi^\dagger\varphi \simeq 1 : C \to C$ (see [Hilton and Wylie 1965, Section 3.3].)*

Theorem 10.5. *A chain equivalence induces a homology group isomorphism.*

Proof. Since $\varphi\varphi^\dagger \simeq 1 : D \to D$ it follows that $\varphi\varphi^\dagger$ and 1 induce the same homology group isomorphism. Therefore $\varphi_*\varphi_*^\dagger : H_*(D) \to H_*(D)$ is the identity. By the same argument, $\varphi_*^\dagger\varphi_* : H_*(C) \to H(C)$ is also the identity. It follows that $\varphi_* : H_*(C) \to H_*(D)$ has an inverse $\varphi_*^\dagger : H_*(D) \to H_*(C)$. Clearly, $H_*(C) \cong H_*(D)$ (see [Hilton and Wylie 1965, Theorem 3.3.8]). ■

10.4.2 Acyclic Carriers

In general, it is not easy to find the chain homotopy between two chain maps. To prove that two maps are chain homotopic (and hence induce the same homology group homomorphism) it is easier to show that the two maps have a common carrier.

Definition 10.6. *The chain maps $\varphi_*, \gamma_* : C_* \to D_*$ are said to be* **carried by the same acyclic carrier** *if there exists a map Γ defined on the basis elements $\{\sigma\}$ of C_* such that $\Gamma(\sigma)$ is a subcomplex of D_*, and subject to the following conditions:*

- *If σ_{n-1} is a face of σ_n then $\Gamma(\sigma_{n-1})$ is a subcomplex of $\Gamma(\sigma_n)$.*
- *$\forall \sigma \in C_*$, $\varphi(\sigma)$ and $\gamma(\sigma)$ are chains of $\Gamma(\sigma)$.*
- *$H_n(\Gamma(\sigma)) = 0$ for $n \neq 0$ and $H_0(\Gamma(\sigma)) \cong \mathbb{Z}$; such a space is said to be* **acyclic**.

Theorem 10.7. *Two maps are chain homotopic if their images are within the same acyclic carrier.*

Proof. See [Hilton and Wylie 1965, 3.4.4, 3.4.13]. ∎

10.4.3 Invariance Under Homotopy

We now proceed to show the *fundamental result* that whenever two topological maps $f, g : |K| \to |L|$ are homotopic, they induce the same homology group homomorphism. The first step is to construct simplicial approximations $\tilde{f}, \tilde{g}$ defined on *the same refinement* of K (it is easy to see that this can be done). The next concept is the following:

Definition 10.8. *Two simplicial maps $\tilde{f}, \tilde{g} : |K| \to |L|$ are said to be* **simplicially homotopic** *if there exists a simplicial map $\tilde{F} : (K \times [0,1])' \to L$ such that $\tilde{F}(\cdot\,, 0) = \tilde{f}$ and $\tilde{F}(\cdot\,, 1) = \tilde{g}$.*

Theorem 10.9. *Given two homotopic topological maps, $f \simeq g : |K| \to |L|$, there exist simplicially homotopic simplicial approximations $\tilde{f}, \tilde{g}$.*

Proof. Given two homotopic topological maps, it is easy to construct simplicial approximations $\tilde{f}, \tilde{g}$ defined on the *same* refinement of K. Since f, g are homotopic, so are $\tilde{f}, \tilde{g}$ and therefore there exists a topological map $F : K \times [0,1] \to L$ such that $F(\cdot, 0) = \tilde{f}$ and $F(\cdot, 1) = \tilde{g}$. We define a simplicial approximation to the topological map by refining K, by refining $[0,1]$ as $\cup_j (t^j, t^{j+1}) \cup_j t^j$, and by prismatically decomposing $\sigma \times (t^i, t^{j+1})$, where σ is a simplex of K. It is easily seen that during this refinement, the mesh decreases to zero and hence there exists such a refinement that allows for a simplicial approximation $\tilde{F}$ to the topological map $K \times [0,1] \to L$. In the construction of the latter simplicial approximation, we have to ensure that it agrees with $\tilde{f}$ on $K \times \{0\}$ and with $\tilde{g}$ on $K \times \{1\}$. It is easily seen that this can be done. ∎

Theorem 10.10. *If two maps $\tilde{f}, \tilde{g} : K \to L$ are simplicially homotopic, they induce chain homotopic chain maps.*

Proof. It suffices to construct a carrier function Γ. Take a simplex σ_n of K. This simplex is the basis of a unique prism $\sigma_n \times [0,1]$, itself decomposed into many prisms, $\cup_j \sigma \times \overline{(t^j, t^{j+1})}$, each prism itself decomposed into simplexes of the complex of $K \times [0,1]$. Take $\Gamma(\sigma_n)$ to be the latter decomposition of $\sigma_n \times [0,1]$ in $(n+1)$-simplexes of $K \times [0,1]$.

■

Corollary 10.1. *Two homotopic topological maps $f, g : |K| \to |L|$ between polyhedra induce the same homology group homomorphism.*

10.4.4 Homotopy Equivalence

Definition 10.11. *The map $f : |K| \to |L|$ is said to be a* **homotopy equivalence** *iff it has a* **homotopy inverse***; that is, a map $f^\dagger : |L| \to |K|$ such that $ff^\dagger \simeq 1 : |L| \to |L|$ and $f^\dagger f \simeq 1 : |K| \to |K|$. In this case, the polyhedra $|K|$ and $|L|$ are said to be* **homotopically equivalent** *or to have the same* **homotopy type**.

We leave it up to the reader to extend the above definition to arbitrary topological spaces X and Y.

Corollary 10.2. *If the polyhedra $|K|$ and $|L|$ are homotopically equivalent, then $H_*(|K|) = H_*(|L|)$.*

Consider two spaces $X_0 \subseteq X$ and assume that there exists a (strong) deformation retract $R(\cdot, t) : X \times [0,1] \to X$ of X onto X_0 (see Subsection 9.4.2). It is easily seen that the inclusion map $i : X_0 \to X$ has a homotopic inverse, $R(\cdot, 1) : X \to X_0$. Therefore, if there exists a (strong) deformation retract of X onto X_0, the spaces X and X_0 are homotopically equivalent. The space X is said to be **contractible** if there exists a strong deformation retract of X onto one of its points; equivalently, if X is homotopically equivalent to a point.

Corollary 10.3. *If $|K|$ is a contractible polyhedron, then $|K|$ is* **acyclic***; that is, $H_n(|K|) = 0$ for $n \neq 0$ and $H_0(|K|) \cong \mathbb{Z}$.*

10.5 Homology of Product of Uncertainty

The higher the dimension of the set of uncertainties, the more complicated the triangulation, the more difficult the computation of the homology groups. Therefore, the question is whether it is possible, from triangulations of D and Ω and their homology groups, to compute the homology groups of $D \times \Omega$?

We already know that the *prism construction* allows us to obtain, in a straightforward manner, a triangulation of $D \times \Omega$ from a triangulation of D and a triangulation of Ω. It remains to develop an efficient way to compute the homology groups of $D \times \Omega$ in terms of the homology groups of D and Ω. This is precisely what the Eilenberg–Zilber and the Künneth theorems do.

10.5.1 Eilenberg–Zilber Theorem

Define the *prismatic decomposition map*

$$\begin{aligned}\psi : C_*(D) \oplus C_*(\Omega) &\rightarrow C_*(D \times \Omega)\\ (\sigma, \tau) &\mapsto \sum_i (-1)^i a^0_{\omega^l} \ldots a^i_{\omega^l} a^i_{\omega^{l+1}} \ldots a^n_{\omega^{l+1}}\end{aligned}$$

Geometrically, (σ, τ) can be thought of as the Cartesian product $\sigma \times \tau$ and ψ takes the prism $\sigma \times \tau$ and decomposes it, prismatically, into simplexes of $C_*(D \times \Omega)$. Algebraically, (σ, τ) is a bilinear pairing and hence a tensor product, $\sigma \otimes \tau$. (See Appendix A.) Therefore, the prismatic decomposition map can be redefined as

$$\begin{aligned}\psi : C_*(D) \otimes C_*(\Omega) &\rightarrow C_*(D \times \Omega)\\ \sigma \otimes \tau &\mapsto \sum_i (-1)^i a^0_{\omega^l} \ldots a^i_{\omega^l} a^i_{\omega^{l+1}} \ldots a^n_{\omega^{l+1}}\end{aligned}$$

In the above, the right-hand side is the usual simplicial chain group of $D \times \Omega$ which, when endowed with ∂, becomes the usual simplicial chain complex. In the left-hand side of the above, the **tensor product chain group** is defined as

$$C_*(D) \otimes C_*(\Omega) = \{\sigma \otimes \tau : \sigma \in C_*(D), \tau \in C_*(\Omega)\}$$

The intuitive geometric interpretation of its generators are prisms, each made up with a simplex of D and a simplex of Ω (without allowing any decomposition of the prism). It turns out that the left-hand side can also be made a chain *complex*. Remember that in the geometric chain group $C_*(D \times \Omega)$ the boundary of (the decomposition of) a prism is

$$\begin{aligned}&\partial \text{ prismatic decomposition of } \sigma_n \times \tau_k\\ =\;& (-1)^k \text{ prismatic decomposition of } (\partial\sigma_n) \times \tau_k\\ &+(-1)^{k+1} \text{ prismatic decomposition of } \sigma_n \times (\partial\tau_k)\end{aligned}$$

Rather than prismatically decomposing a prism like $\sigma \times \tau$, we now keep it as a single entity and consider it as the algebraic object $\sigma \otimes \tau$. This defines an homomorphism $\partial^{\otimes}$ on $C_*(D) \otimes C_*(\Omega)$:

$$\partial^{\otimes} \sigma_n \otimes \tau_k = (-1)^k (\partial\sigma_n) \otimes \tau_k + (-1)^{k+1} \sigma_n \otimes (\partial\tau_k) \tag{10.1}$$

The right-hand side boundaries are those defined on the factor chain groups. With the latter, we have the following theorem:

Theorem 10.12. *$\{C_*(D) \otimes C_*(\Omega), \partial^{\otimes}\}$ is a chain complex, called* **tensor product chain complex.**

Proof. All that remains to be proved is $\partial^{\otimes}\partial^{\otimes} = 0$, which is trivial from the definition of the boundary operator on the tensor product. ■

The link ψ between the two chain complexes is actually a chain map,

Theorem 10.13. *ψ is a chain map; that is, $\psi\partial^{\otimes} = \partial\psi$.*

Proof. Straightforward from the various definitions. ■

A very vivid way to display what has been achieved so far is by means of the following commutative diagram:

$$\begin{array}{ccc} C_*(D) \otimes C_*(\Omega) & \stackrel{\partial^{\otimes}}{\rightarrow} & C_*(D) \otimes C_*(\Omega) \\ \downarrow \psi & & \downarrow \psi \\ C_*(D \times \Omega) & \stackrel{\partial}{\rightarrow} & C_*(D \times \Omega) \end{array} \tag{10.2}$$

The bottom row is the original chain complex. The top row is the tensor product chain complex. The prismatic decomposition map ψ maps the top row into the bottom row. The fact that the diagram commutes means that ψ is a *chain map*.

As we have seen (see [Hilton and Wylie 1965, Theorem 3.1.9]), a chain map induces homology group homomorphisms:

$$\psi_* : H_*(C_*(D \otimes \Omega)) \to H_*(D \times \Omega)$$

Now we come to the very motivation of this section—that is, the problem of splitting the high-dimensional problem $D \times \Omega$ into two (or more) lower-dimensional problems, namely, D and Ω. We have a chain complex for D, $\{C_*(D), \partial_*\}$ and a chain complex for Ω, $\{C_*(\Omega), \partial_*\}$. These are "standard" chain complexes in the sense that they decompose the spaces into simplexes. Since the tensor product $\{C_*(D) \otimes C_*(\Omega), \partial^{\otimes}\}$ has the chain complex structure, it follows that one can compute homology groups $H_*^{\otimes} = H_*(C_*(D \otimes \Omega))$ using a purely algebraic procedure. Now the question is, How does the homology $H_*^{\otimes}$ compare with the original one $H_*(D \times \Omega)$? The celebrated Eilenberg–Zilber theorem asserts that they are isomorphic, in other words, that ψ_* is an *isomorphism*. It follows that the homology groups of the chain complex $C_*(D \times \Omega)$ can, in fact, be computed from the chain complex $C_*(D) \otimes C_*(\Omega)$. Besides, the latter approach is *faster* because $C_*(D) \otimes C_*(\Omega)$ has by far less generators than $C_*(D \times \Omega)$.

Theorem 10.14. (Eilenberg–Zilber) *$\psi_* : H_*(C_*(D \otimes \Omega)) \to H_*(D \times \Omega)$ is an isomorphism.*

Proof. We use the method of chain homotopy and acyclic carriers. To show that ψ_* is an isomorphism, it suffices to show that the chain map ψ has a chain inverse [Hilton and Wylie 1965, Theorem 3.3.8]. Remember, a *chain inverse* is a chain map $\psi^{\dagger} : C_*(D \times \Omega) \to C_*(D) \otimes C_*(\Omega)$ such that both $\psi\psi^{\dagger}$ and $\psi^{\dagger}\psi$ are *chain homotopic* to the identity map 1 [Hilton and Wylie 1965, Section 3.3]. To define the chain homotopic inverse $\psi^\dagger : C_*(D \times \Omega) \to C_*(D) \otimes C_*(\Omega)$, let us start with an arbitrary simplex σ_{n_q+1} of dimension $n_p = n_q + 1$. By definition of the triangulation of $D \times \Omega$, this simplex is of the form

$$\sigma_{n_q+1} = a^0_{\omega^l} \ldots a^m_{\omega^l} a^m_{\omega^{l+1}} \ldots a^{n_q}_{\omega^{l+1}}$$

It is claimed that the chain homotopic inverse is

$$\begin{aligned} \psi^\dagger(a^0_{\omega^l} \ldots a^m_{\omega^l} a^m_{\omega^{l+1}} \ldots a^{n_q}_{\omega^{l+1}}) &= a^0 \ldots a^{n_q} \times \omega^l \omega^{l+1} \quad \text{for } m = n_q \\ &= 0 \quad \text{otherwise} \end{aligned}$$

It is easily verified that $\psi^\dagger \psi = 1$. To prove that $\psi \psi^\dagger \simeq 1$ we show that their images are within the same *acyclic carrier* [Hilton and Wylie 1965, 3.4.4]. It is easily seen that the following is the acyclic carrier. Take any simplex $a^0_{\omega^l} \ldots a^m_{\omega^l} a^m_{\omega^{l+1}} \ldots a^{n_q}_{\omega^{l+1}}$ of $C_*(D \times \Omega)$. Define the following:

$$\Gamma(a^0_{\omega^l} \ldots a^m_{\omega^l} a^m_{\omega^{l+1}} \ldots a^{n_q}_{\omega^{l+1}}) = \sum_k a^0_{\omega^l} \ldots a^k_{\omega^l} a^k_{\omega^{l+1}} \ldots a^{n_q}_{\omega^{l+1}}$$

It is easily seen that the above is an acyclic carrier for $\psi\psi^\dagger$ and 1. This completes the proof in dimension $n_q + 1$. The proof in lower dimension follows exactly the same lines and is omitted. Therefore, the theorem is proved. ■

10.5.2 Künneth Theorem

From the Eilenberg–Zilber theorem, we learn that one can use either the top or the bottom chain complex of the commutative diagram 10.2 in order to compute the homology of $D \times \Omega$. It turns out that using the top chain complex—the tensor product complex—is faster, because it involves triangulation of spaces lower-dimensional than the bottom chain complex. Furthermore, the homology of the tensor complex can be computed *purely algebraically* from the homology of the factors. This is the celebrated *Künneth formula*:

Theorem 10.15. (Künneth) *Assume that neither D nor Ω has torsion. Then the homology of $D \times \Omega$ is given by the "convolution product"*

$$H_n(C_*(D \otimes \Omega)) = \bigoplus_k H_k(D) \otimes H_{n-k}(\Omega)$$

Proof. Again, this is a purely algebraic result that need not invoke any elements of point set topology. It relies on tensor algebra. For the details, see [Hilton and Wylie 1965, 5.7.17] or [Munkres 1984, Section 58]. ■

Corollary 10.4. *Under the same conditions as the Künneth theorem,*

$$H_n(D \times \Omega) = \bigoplus_k H_k(D) \otimes H_{n-k}(\Omega)$$

10.5.3 Remark

Instead of the expression 10.1, the formula for the boundary of the tensor product is usually written in the alternative form:

$$\partial^{\otimes}\sigma_n \otimes \tau_p = (\partial\sigma_n) \otimes \tau_p + (-1)^n \sigma_n \otimes (\partial\tau_p) \tag{10.3}$$

Following [Munkres 1984, page 341, Exercise 3], it turns out that there are *many* ways to define the boundary of the tensor product of two chain complexes in such a way that the homology of $\{C_*(D) \otimes C_*(\Omega), \partial^{\otimes}\}$ coincides with the homology of $D \times \Omega$. The reason why we obtained the rather unusual form 10.1 is that we proceeded from a precise prismatic decomposition of $D \times \Omega$ and then went on to algebraize it to obtain 10.1. It can be shown from [Munkres 1984, page 341, Exercise 3] that the complex 10.1 and the complex 10.3 are isomorphic and hence yield the same homology groups for the Cartesian product $D \times \Omega$. In general, with some cellular homology concepts, the technical details of the prismatic decomposition can be bypassed, so that the expression 10.3 is preferred because it simplifies the notation.

10.5.4 Application—Uncertainty-Frequency Product

As an illustration of the Künneth formula, assume that Ω is the unit circle $\mathbb{T}$. The unit circle has no torsion and its homology is given by

$$H_0(\Omega) = \mathbb{Z} \Rightarrow b_0(\Omega) = 1$$
$$H_1(\Omega) = \mathbb{Z} \Rightarrow b_1(\Omega) = 1$$

Therefore, an application of the Künneth formula yields

$$\begin{aligned} H_n(D \times \Omega) &= (H_{n-1}(D) \otimes H_1(\Omega)) \oplus (H_n(D) \otimes H_0(\Omega)) \\ &= (H_{n-1}(D) \otimes \mathbb{Z}) \oplus (H_n(D) \otimes \mathbb{Z}) \\ &= H_{n-1}(D) \oplus H_n(D) \end{aligned}$$

Therefore, in case Ω is the unit circle and D has no torsion, we find the Betti numbers of $D \times \Omega$ in terms of the Betti numbers of D,

$$b_n(D \times \Omega) = b_{n-1}(D) + b_n(D) \tag{10.4}$$

10.5.5 Application—Uncertainty Torus

We now illustrate how to recursively use the Künneth theorem in case D is itself the topological product of many uncertainties. Consider the case of two uncertain phase angles—that is, $D = \mathbb{T}^2$—the latter denoting the 2-torus. Applying the Künneth theorem to $D = \mathbb{T}^2 = \mathbb{T} \times \mathbb{T}$ yields

$$H_0(\mathbb{T} \times \mathbb{T}) = H_0(\mathbb{T}) \otimes H_0(\mathbb{T}) = \mathbb{Z} \otimes \mathbb{Z} = \mathbb{Z}$$

Therefore,

$$b_0(\mathbb{T}^2) = 1$$

Next, we consider the first homology group of $D = \mathbb{T}^2$:

$$\begin{aligned} H_1(\mathbb{T} \times \mathbb{T}) &= (H_0(\mathbb{T}) \otimes H_1(\mathbb{T})) \oplus (H_1(\mathbb{T}) \otimes H_0(\mathbb{T})) \\ &= (\mathbb{Z} \otimes \mathbb{Z}) \oplus (\mathbb{Z} \otimes \mathbb{Z}) \\ &= \mathbb{Z} \oplus \mathbb{Z} \end{aligned}$$

Therefore,

$$b_1(\mathbb{T}^2) = 2$$

The second homology group of $D = \mathbb{T}^2$ is

$$\begin{aligned} & H_2(\mathbb{T} \times \mathbb{T}) \\ &= (H_2(\mathbb{T}) \otimes H_0(\mathbb{T})) \oplus (H_1(\mathbb{T}) \otimes H_1(\mathbb{T})) \oplus (H_0(\mathbb{T}) \otimes H_2(\mathbb{T})) \\ &= H_1(\mathbb{T}) \otimes H_1(\mathbb{T}) \\ &= \mathbb{Z} \otimes \mathbb{Z} \\ &= \mathbb{Z} \end{aligned}$$

Therefore,

$$b_2(\mathbb{T}^2) = 1$$

Next, we compute the Betti numbers of $D \times \Omega$ from the Betti numbers of the factor $D = \mathbb{T}^2$ using Equation 10.4

$$\begin{aligned} b_0(D \times \Omega) &= b_0(D) = 1 \\ b_1(D \times \Omega) &= b_1(D) + b_0(D) = 2 + 1 = 3 \\ b_2(D \times \Omega) &= b_2(D) + b_1(D) = 1 + 2 = 3 \\ b_3(D \times \Omega) &= b_3(D) + b_2(D) = 0 + 1 = 1 \end{aligned}$$

Finally, if we consider a phase margin problem with, say, n_q uncertain phase angles, the uncertainty space D becomes the topological product of n_q identical copies of $\mathbb{T}$—that is, $D = \mathbb{T}^{n_q}$, the n_q-torus. Applying the Künneth theorem and related machinery to this latter problem, it is not hard to prove that

$$b_n(\mathbb{T}^{n_q} \times \Omega) = \binom{n_q + 1}{n}$$

10.6 Uncertainty Manifold—Mayer–Vietoris Sequence

Suppose D, or any other topological space, can be broken down as the union $D_1 \cup D_2$ of two intersecting subsets. (If D_1 and D_2 are disjoint, then it is easily verified that $H_*(D_1 \cup D_2) = H_*(D_1) \oplus H_*(D_2)$.) The question is whether it is possible to use this structure of D to gain insight into its homology. The answer is provided by the following:

Theorem 10.16. (Mayer–Vietoris Sequence) *If $D = D_1 \cup D_2$, there exists an exact sequence*

$$\begin{array}{rll} 0 \to \ldots \to & H_n(D_1 \cap D_2) & \to H_n(D_1) \oplus H_n(D_2) \to H_n(D) \\ \to & H_{n-1}(D_1 \cap D_2) & \to \ldots \to 0 \end{array}$$

Proof. We form the following short exact sequence of chain complexes:

$$0 \to C_n(D_1 \cap D_2) \to C_n(D_1) \oplus C_n(D_2) \to C_n(D_1 \cup D_2) \to 0$$

Going to the homology, we obtain a series of sequences:

$$\begin{array}{ccccc} H_{n-1}(D_1 \cap D_2) & \to & H_{n-1}(D_1) \oplus H_{n-1}(D_2) & \to & H_{n-1}(D_1 \cup D_2) \\ \uparrow & & & & \\ & \leftarrow & \overset{\partial_*}{\longleftarrow} & \leftarrow & \\ & & & & \uparrow \\ H_n(D_1 \cap D_2) & \to & H_n(D_1) \oplus H_n(D_2) & \to & H_n(D_1 \cup D_2) \\ \uparrow & & & & \\ & \leftarrow & \overset{\partial_*}{\longleftarrow} & \leftarrow & \\ & & & & \uparrow \\ H_{n+1}(D_1 \cap D_2) & \to & H_{n+1}(D_1) \oplus C_{n+1}(D_2) & \to & H_{n+1}(D_1 \cup D_2) \end{array}$$

It is not hard to see that the rows are exact sequences. Less obvious is the fact that there are *connecting morphisms* (the "wiggling" arrows), ∂_*, connecting all of the short sequences into a long, exact sequence. For an algebraic proof of the existence of the connecting morphism, see Appendix A. ■

The Mayer–Vietoris sequence is instrumental in computing the homology of compact manifolds. Indeed, the homology of a compact manifold is revealed by the combinatorics of the intersections of the open sets of a finite subcover of the manifold. This approach is exploited in [Bott and Tu 1982].

10.6.1 Application—Homology of Special Unitary Uncertainty

Remember, $SU(2, \mathbb{C}) \cong S^3$. We therefore illustrate the Mayer–Vietoris argument on the computation of the homology groups of the spheres, defined for this purpose as

$$S^m = \{x \in \mathbb{R}^{m+1} : x_1^2 + x_2^2 + \ldots + x_{m+1}^2 = 1\}$$

We define the North and South "hemispheres," respectively, as follows:

$$\begin{array}{rcl} O_2 & = & S^m \setminus (0, 0, \ldots, 0, -1) \\ O_1 & = & S^m \setminus (0, 0, \ldots, 0, +1) \end{array}$$

By a standard stereographic projection argument, O_i is homeomorphic to the Euclidean space $\mathbb{R}^m$. Write this homeomorphism as $\pi_{si} : \mathbb{R}^m \to O_i$. Furthermore,

$$O_1 \cup O_2 = S^m$$

Next the gluing map $\pi_{s2}^{-1}\pi_{s1} : \mathbb{R}^m \setminus 0 \to \mathbb{R}^m \setminus 0$ is easily seen to be homeomorphic. Therefore, O_1,O_2 together with the stereographic projections constitute an *atlas* of the manifold S^m.

(The above is a higher-dimensional generalization of the Riemann sphere concept introduced in Subsection 2.6.2 and illustrated in Figure 2.3.)

Next consider $O_1 \cap O_2$. This is an open set that comprises the equator $x_{m+1} = 0$ of the sphere S^m. The equator is clearly the sphere S^{m-1}. Furthermore, it is easily seen that $O_1 \cap O_2$ is contractible to the equator. Therefore, $H_*(O_1 \cap O_2) = H_*(S^{m-1})$.

Now, write the Mayer–Vietoris sequence for $S^m = O_1 \cup O_2$, namely,

$$\begin{array}{lll} \ldots \to & H_n(O_1 \cup O_2) & \to H_{n-1}(O_1 \cap O_2) \to H_{n-1}(O_1) \oplus H_{n-1}(O_2) \\ \to & H_{n-1}(O_1 \cup O_2) & \to \ldots \end{array}$$

It follows that

$$\ldots \to H_n(S^m) \to H_{n-1}(S^{m-1}) \to H_{n-1}(\mathbb{R}^m) \oplus H_{n-1}(\mathbb{R}^m) \to H_{n-1}(S^m) \to \ldots$$

The sequence terminates, for $n = 1$, with

$$0 \to H_1(S^m) \to H_0(S^{m-1}) \to \mathbb{Z} \oplus \mathbb{Z} \to H_0(S^m) \to 0$$

For $n > 1$, the Mayer–Vietoris sequence breaks down into several shorter exact sequences,

$$0 \to H_n(S^m) \to H_{n-1}(S^{m-1}) \to 0, \quad n > 1$$

which imply that

$$H_n(S^m) \cong H_{n-1}(S^{m-1}), \ \ n > 1$$

From the above, the homology of the spheres, including $S^3 \cong SU(2, \mathbb{C})$, can be obtained by proceeding from the homology of the lowest-dimensional sphere S^1 (see [Bott and Tu 1982, I, Example 2.9]) and then going to higher dimensions. To be more specific, the final result is

$$\begin{array}{rcl} H_n(S^m) & = & \mathbb{Z}, \ \ \text{for } n = 0, m \\ H_n(S^m) & = & 0, \ \ \text{otherwise} \end{array}$$

10.7 Neglecting an Uncertainty—Relative Homology Sequence

Assume Q is a subpolyhedron of the polyhedron P of uncertainty. We have a usual chain complex for P, $C_*(P)$, and a subcomplex $C_*(Q)$; clearly, we have

$$C_*(Q) \subseteq C_*(P)$$

Therefore, we can construct the quotient complex

$$C_*(P)/C_*(Q)$$

The above is called **relative chain complex** of P modulo Q. It is important to understand that this concept is *fundamentally algebraic* and that it does not have a clearcut geometric interpretation; roughly speaking, the relative complex is obtained from the absolute complex by "collapsing" the subcomplex Q. We can compute a homology group for the complex P, the complex Q, and the relative complex. These homology groups are related by the relative homology sequence:

Theorem 10.17. *There exists a relative homology sequence*

$$\ldots \to H_n(Q) \to H_n(P) \to H_n(P \bmod Q) \to H_{n-1}(Q) \to \ldots$$

Proof. Putting the chain complex of P, the subcomplex of Q, and the relative complex together yields the short exact sequence:

$$0 \to C_*(Q) \to C_*(P) \to C_*(P)/C_*(Q) \to 0$$

Going to the homology we obtain a series of exact rows which we arrange as follows:

$$\begin{array}{ccccc}
H_{n-1}(Q) & \to & H_{n-1}(P) & \to & H_{n-1}(P \bmod Q) \\
\uparrow & & & & \\
& \leftarrow & \stackrel{\partial_*}{\longleftarrow} & \leftarrow & \\
& & & & \uparrow \\
H_n(Q) & \to & H_n(P) & \to & H_n(P \bmod Q) \\
\uparrow & & & & \\
& \leftarrow & \stackrel{\partial_*}{\longleftarrow} & \leftarrow & \\
& & & & \uparrow \\
H_{n+1}(Q) & \to & H_{n+1}(P) & \to & H_{n+1}(P \bmod Q)
\end{array}$$

Using the connecting morphism ∂_* of Appendix A yields the relative homology sequence. ■

10.8 More Sophisticated Homology Computation

Despite the impressive arsenal of tools that we have developed to compute homology groups, there are still some topological spaces, the homology of which escapes the aforementioned analysis. Think about the homology of the unitary or orthogonal group. For those—and many other spaces that are beyond the realm of robust control—we need even more sophisticated tools,

- Fiber bundles
- Spectral sequences

Each of these topics deserves a chapter of its own (see Chapters 13 and 14, respectively). Roughly speaking, assume we have to compute the homology of a space X and that this space can be linked to a *known* space A via of a *well-behaved* map f (typically, a *fibration*). The space X can be either the domain or the image of the map; that is, we could have either $f : X \to A$ or $f : A \to X$, although the former is the most common situation. Fiber bundles and its generalization, spectral sequences, allow us to compute the homology of X from the homology of A and the properties of the map f.

11

COMBINATORIAL HOMOLOGY OF CROSSOVER AND STABILITY BOUNDARY

SUMMARY

In this chapter we study the connectivity properties of the crossover $f^{-1}(0+j0)$ and the stability boundary in D using *simple* homology tools. We hasten to say that simple homology tools do not lead us very far. More information about the crossover—or more information about the *relationship* between the topology of the domain of uncertainty, the template, and the crossover—can be obtained using the following tools, to come in later chapters,

- Spectral sequences (Chapter 14)
- Exact homotopy sequences (Chapter 15)
- Algebraic geometry (Part IV), assuming the Nyquist map is rational

11.1 Combinatorial Homology of Crossover

There are several ways to define the *crossover region*. The most natural one is to use the original Nyquist map $f : D \times \Omega \to N$ and define

$$X_\omega = f^{-1}(0 + j0) \subseteq D \times \Omega$$

This kind of crossover is best analyzed using differential topology or algebraic geometry as we will do in Chapters 23 and 24, respectively. In the present simplicial/combinatorial setup, we would of course use the simplicial approximation $\tilde{f} : D \times \Omega \to N$ and define the crossover as

$$\tilde{X}_\omega = \tilde{f}^{-1}(0 + j0)$$

A first problem is that $\tilde{X}_\omega$ and X_ω might not be homotopically equivalent because indeed the mere homotopic property of $f, \tilde{f}$ does not guarantee homotopically equivalent preimages without some further assumptions. However, in the realm of finite computation, the simplicial approximation is "the best game in town." If we want to avoid the linear programming effort associated with computing $\tilde{f}^{-1}(0 + j0)$, we could argue that, in this finite resolution approach, we can only narrow down the critical point $0+j0$ within a closed simplex, say $\overline{(b^0, b^1, b^2)}$, of the triangulation of N so that the only relevant crossover is the preimage of this closed simplex,

$$X_\omega^\Delta = \tilde{f}^{-1}(\overline{(b^0, b^1, b^2)})$$

Clearly, taking the preimage of the *closed* simplex $\overline{(b^0, b^1, b^2)}$ yields a (closed) polyhedron. We call the above **solid crossover**.

Combinatorially, computing the above crossover amounts to chasing all simplexes mapped to $b^0b^1b^2$ or any of its faces. This combinatorial search is implemented efficiently in the SimplicialVIEW software. Algebraically, this search amounts to computing the preimage of the complex of $\overline{(b^0, b^1, b^2)}$ under the total chain map φ associated with $\tilde{f}$, namely,

$$X_\omega^\triangle = \varphi_*^{-1}(\{b^0, b^1, b^2, b^0b^1, b^1b^2, b^2b^0, b^0b^1b^2\})$$

The above directly yields the simplicial complex of $X_\omega^\triangle$, $\{C_*(X_\omega^\triangle), \partial_*\}$. From this simplicial complex one could, in principle, compute its homology via $H_n = \ker \partial_n / \mathrm{im}\, \partial_{n+1}$. Computing H_n *directly* from the definition would be a computationally intensive exercise. A fast implementation of the definition proceeds from a matrix representation of $\partial_n : C_n \to C_{n-1}$ in suitable bases for C_n, C_{n-1}; the bulk of the algorithm is the Smith reduction of ∂_n over the ring of integers; once the Smith form of ∂_n is derived, the homology is trivial to compute (see [Hilton and Wylie 1965, 5.1], [Munkres 1984, Section 11], or Appendix B).

As far as homology is concerned, there is no need to use $\tilde{X}_\omega$, because the following statements apply:

Theorem 11.1. *$\tilde{X}_\omega$ is a deformation retract of $\tilde{X}_\omega^\triangle$.*

Proof. Consider a simplex $\overline{\sigma_n}$ such that $\overline{\sigma_n} \cap \tilde{X}_\omega \neq \emptyset$. Clearly, $\tilde{f}(\overline{\sigma_n}) = b^0b^1b^2$ and as such $\overline{\sigma_n} \subseteq \tilde{X}_\omega^\triangle$. We show that any such $\overline{\sigma_n}$ can be retracted onto $\tilde{X}_\omega \cap \overline{\sigma_n}$. Let μ_0, μ_1, μ_2 be the barycentric coordinates of $0+j0$ relative to the vertices b^0,b^1,b^2. In the uncertainty space, the vertices are relabeled so that $\overline{\sigma_n} = \overline{a^0a^1a^2...a^n}$ and it is assumed that $\tilde{f}(a^i) = b^2$ for $i = 2, ..., n$. (Remember, the simplicial map has to collapse some faces to points.) It is easily seen that the barycentric coordinates

$$(\lambda_0^X, \lambda_1^X, ..., \lambda_n^X)$$

are those of a point on $\tilde{X}_\omega \cap \overline{\sigma_n}$ iff

$$\begin{aligned} \lambda_0^X &= \mu_0 \\ \lambda_1^X &= \mu_1 \\ \sum_{i=2}^{n} \lambda_i^X &= \mu_2 \end{aligned}$$

Let $(\lambda_0, ..., \lambda_n)$ be an arbitrary point of $\overline{\sigma_n}$. The deformation retract is defined as

$$\left(\begin{pmatrix} \lambda_0 \\ \lambda_1 \\ \lambda_2 \\ \vdots \\ \lambda_n \end{pmatrix}, t \right) \mapsto (1-t) \begin{pmatrix} \lambda_0 \\ \lambda_1 \\ \lambda_2 \\ \vdots \\ \lambda_n \end{pmatrix} + t \begin{pmatrix} \mu_0 \\ \mu_1 \\ \frac{\mu_2}{n-1} \\ \vdots \\ \frac{\mu_2}{n-1} \end{pmatrix}$$

It is easily seen that things match up correctly at those faces of σ_n intersecting the crossover. The proof requires only trivial modification to accommodate the case where the simplicial map collapses several faces of σ_n to points. The details are left out. ∎

Corollary 11.1. $H_*(\tilde{X}_\omega) = H_*(\tilde{X}_\omega^\triangle)$.

Proof. A deformation retraction is a homotopy equivalence that does not affect the homology groups. ∎

11.2 Projecting the Crossover

In the *topological* map setup, $f : D \times \Omega \to N$, the projection $D \times \Omega \to D$ is defined as the projection π_1 on the first factor of $p = (q, \omega)$, namely, $\pi_1(p) = q$. Therefore, the projection of the crossover X_ω on D is $X = \pi_1(X_\omega)$.

In the *simplicial* map context, $\tilde{f} : D \times \Omega \to N$, the projection map $\pi_1 : D \times \Omega \to D$ is not in general simplicial, unless $D \times \Omega$ is prismatically triangulated. If π_1 is not simplicial, projecting $\tilde{X}_\omega, \tilde{X}_\omega^\triangle$ destroys the affine simplicial decomposition so that $\pi_1(\tilde{X}_\omega), \pi_1(\tilde{X}_\omega^\triangle)$ will fail to be a subcomplex of D. If $D \times \Omega$ is prismatically triangulated, then π_1 is simplicial. This is easily seen as follows: Any simplex of $\tilde{X}_\omega, \tilde{X}_\omega^\triangle$ is of the form $a^0_{\omega^l} \ldots a^i_{\omega^l} a^i_{\omega^{l+1}} \ldots a^n_{\omega^{l+1}}$ where $a^0 \ldots a^i \ldots a^n$ is a simplex of D and $\omega^l \omega^{l+1}$ is a simplex of Ω. It is easily seen that π_1 maps as follows:

$$\begin{array}{rcl} \pi_1 : D \times \Omega & \to & D \\ a^0_{\omega^l} \ldots a^i_{\omega^l} a^i_{\omega^{l+1}} \ldots a^n_{\omega^{l+1}} & \mapsto & a^0 \ldots a^i \ldots a^n \end{array}$$

Clearly, π_1 maps simplexes to simplexes and is hence simplicial. It should also be clear that π_1 can be viewed as a chain map.

The chain map π_1 therefore induces a homology group homomorphism as shown by the following commutative diagram:

$$\begin{array}{ccc} \tilde{X}_\omega, \tilde{X}_\omega^\triangle & \xrightarrow{\pi_1} & X \\ \downarrow & & \downarrow \\ H_*(\tilde{X}_\omega), H_*(\tilde{X}_\omega^\triangle) & \xrightarrow{\pi_{1*}} & H_*(X) \end{array}$$

The above provides the linkage between the homology of the crossover and the homology of its projection.

As we will show in Chapter 25, the above is the so-called Tarski–Seidenberg elimination of the (real!) variable ω.

12

COHOMOLOGY

the de Rham theory ... constitutes in some sense the most perfect example of a cohomology theory.

Bott and Tu,
Differential Forms in Algebraic Topology,
Springer-Verlag, New York, 1982,
page 3.

SUMMARY

In this chapter, we "dualize" simplicial homology. This is done by introducing all linear forms: $C_n \to \mathbb{Z}$ defined over the chain group of n-simplexes of a polyhedron P and taking value in $\mathbb{Z}$ or any other value group. Going to linear forms defined over C_n yields the *adjoint* of the boundary operator, the *coboundary* operator δ^n. From there on, working on linear forms defined over the simplexes of P together with the coboundary and following the same algebraic path as in homology, one constructs the *(simplicial) cohomology groups* of the polyhedron P.

We quickly browse through simplicial cohomology and spend more time on the *de Rham cohomology*, a cohomology based on cochain groups of differential forms together with the exterior differential acting as coboundary operator. This latter cohomology theory, which is very well suited to study manifolds and Lie groups, will be instrumental to study the homotopy properties of the Nyquist map from the uncertainty to the return difference matrix viewed as an element of the Lie group $GL(n_l, \mathbb{C})$.

12.1 Simplicial Cohomology

Remember the path of approach we followed in *homology*. An *n-simplex* was defined as a skew-commutative product $\sigma_n = a^0 a^1 ... a^n$, together with the obvious geometrical interpretation. The *boundary* ∂_n of an n-simplex was defined as $\partial_n \sigma_n = \sum_{i=0}^{n} (-1)^i a^0 ... a^{i-1} a^{i+1} ... a^n$. An *n-chain* was defined as a linear combination of n-simplexes. The best intuitive picture of a chain is the domain of integration of a differential form. (This is formalized in the celebrated de Rham cohomology.) By linearity, the boundary was extended in a trivial manner to any chain, and it was easily checked that $\partial_{n-1} \partial_n = 0$.

It followed that im $\partial_{m+1} \subseteq \ker \partial_n$ and therefore we could define quotient groups, called *homology groups*, as $H_n(P) = \ker \partial_n / \text{im } \partial_{n+1}$.

12.1.1 Cohomology Groups

We now dualize the above. Let P be a polyhedron. A **n-cochain** c^n is a linear form defined on the chains of n-simplexes of P. The value taken by the cochain c^n on the n-simplex σ_n is written (σ_n, c^n), using inner product notation to emphasize the bilinear property. An intuitive picture of an n-cochain, which is formalized in the de Rham theory, is an exterior differential form, homogeneous of degree n. The **n-cochain group** $C^n(P, G_n)$ is the group of linear combination of n-cochains defined over the chains of n-simplexes of P and taking value in the group G_n. The **coboundary** operator

$$\delta^n : C^n(P) \to C^{n+1}(P)$$

is defined as the dual of ∂_{n+1}, namely,

$$(\sigma^{n+1}, \delta^n c^n) = (\partial_{n+1}\sigma_{n+1}, c^n)$$

The best intuitive picture of the coboundary operator, again formalized in the de Rham theory, is the exterior differential—that is, the conceptualization of such elementary operators as curl, divergence, and so on. Using duality, it is trivial to prove that $\delta^{n+1}\delta^n = 0$.

To exhibit the duality a little more formally, define the cochain group as

$$C^n(P, G_n) = \text{hom}(C_n(P), G_n)$$

and consider the following diagram:

$$\begin{array}{ccc} C_n(P) & \xrightarrow{\partial_n} & C_{n-1}(P) \\ \downarrow & & \downarrow \\ \text{hom}(C_n(P), G_n) & \xleftarrow{\delta^{n-1}} & \text{hom}(C_{n-1}(P), G_n) \end{array}$$

In the above, the "down-arrows" denote the contravariant functor $\mathcal{F}$ of going from the chains to the linear forms defined on the chains. The functor is *contravariant* in the sense that it *reverses* the arrows. The arrow at the bottom level is $\delta^* = \mathcal{F}(\partial_*)$.

A collection of groups $\{C^n\}$ together with a collection of group homomorphisms $\{\delta^n : C^n \to C^{n+1}\}$ satisfying the fundamental property $\delta^{n+1}\delta^n = 0$ is called a **cochain complex**. From here on, we follow the same algebraic path as in homology. In a cochain complex, we define the **cocycle group** as

$$Z^n = \ker(\delta^n)$$

Likewise, the group of **coboundaries** is defined as

$$B^n = \text{im}(\delta^{n-1})$$

The best intuitive picture of a coboundary is an exact differential form. Since

$$B^n \subseteq Z^n,$$

we can define quotient groups, referred to as **cohomology groups**

$$H^n = Z^n / B^n$$

If z^n is a cocycle (i.e., $\delta^n z^n = 0$) we write its coset in H^n as $\{z^n\}$.

12.1.2 Cohomology Group Homomorphism

Given a simplicial map $f : P \to N$ between triangulated polyhedra, it is easily seen that a cochain of N induces a cochain of P. Indeed, if c^n is a cochain of N given by the rule $\tau_n \mapsto (c^n, \tau_n)$, it induces a cochain $f^* c^n$ given by the rule $\sigma_n \mapsto (c^n, f(\sigma_n))$. Next, it is easily seen that if c^n is a cocycle, $f^* c^n$ is a cocycle; likewise, if c^n is a coboundary, $f^* c^n$ is a coboundary. Therefore, f^* maps the cohomology class $\{z^n\}$ to a cohomology class of P. This situation is depicted in the following diagram:

$$\begin{array}{ccc} P & \xrightarrow{f} & N \\ \downarrow & & \downarrow \\ H^*(P) & \xleftarrow{f^*} & H^*(N) \end{array}$$

The "down-arrow" is the process of going from the category of polyhedra and simplicial maps to the category of Abelian groups and homomorphisms. The functor is *contravariant* because the cohomology group homomorphism flows in a direction *opposite* to the topological map.

12.1.3 Cup Product

A definite advantage of cohomology over homology is that, in the former, it is possible to define the so-called *cup product* $\smile$ that makes $(\oplus_n H^n, +, \smile)$ a *ring*, the so-called *cohomology ring*. In many instances, the extra ring structure ensures the definite superiority of cohomology over homology.

It is convenient to use the *total* complex of the underlying topological space which is assumed to be a polyhedron. The value group needs to have the extra *ring* structure, $(R, +, \times_R)$. Consider two (total) cochains c^n, c^m of the polyhedron. Remember, c^n assigns to each total simplex $[a^0 ... a^n]$ an element of R and c^n is extended to the (semisimplicial) chain complex by linearity. The **cup product** $c^n \smile c^m$ is defined as the $(n+m)$-cochain that takes the following value on an arbitrary, admissible $(n+m)$-array:

$$c^n \smile c^m([a^0 ... a^n ... a^{n+m}]) = c^n([a^0 ... a^n]) \times_R c^m([a^n ... a^{n+m}])$$

The cochain is further extended to the total chain group by linearity.

More formally, observe that the cup product induces a homomorphism:

$$\smile : C^n \otimes_R C^m \to C^{n+m}$$

Clearly, $(\oplus_n C^n, +, \smile)$ is a ring. The ring is not commutative, because indeed,

$$c^n \smile c^m = (-1)^{n+m} c^m \smile c^n$$

Furthermore, it is found, by long but elementary direct computation, that the coboundary satisfies the **Leibniz rule**,

$$\delta(c^n \smile c^m) = (\delta c^n) \smile c^m + (-1)^n c^n \smile (\delta c^m)$$

It follows from the Leibniz rule that the cup product of two cocycles is a cocycle. Hence $(Z^* = \oplus_n Z^n, +, \smile)$ is a subring of $(\oplus_n C^n, +, \smile)$. It is also easily seen from the Leibniz rule that $(B^* = \oplus_n B^n, +, \smile)$ is an *ideal* in $(\oplus_n Z^n, +, \smile)$. Therefore the quotient ring, Z^*/B^*, exists and is called **cohomology ring**.

12.1.4 Cohomology of Product Space

It is reasonable to expect that the cohomology of a Cartesian product should be expressible in terms of the cohomology of the factors.

Theorem 12.1. (Künneth's Theorem for Cohomology) *If X and Y are topological spaces and if the cohomology groups $H^*(X)$ and $H^*(Y)$ are finitely generated, there exists a split exact sequence*

$$0 \to \bigoplus_{p+q=n} H^p(X) \otimes H^q(Y) \to H^n(X \times Y, Z) \to \bigoplus_{p+q=n} H^{p+1}(X) * H^q(Y) \to 0$$

In the above, $\bigoplus$, $\otimes$, and $*$ denote the direct sum, the tensor product, and the torsion product, respectively, of Abelian groups (see Appendix A). (A sequence of groups and group homomorphisms is said to be exact whenever at every group the image of the incoming arrow equals the kernel of the departing arrow (see Appendix A).)

Proof. See [Weibel 1994, 6.1.14, pages 165–166]. ∎

12.2 de Rham Cohomology

Roughly speaking, the de Rham theory is the ultimate conceptualization of such results as Green's theorem in the plane, Stokes' theorem, and Gauss–Ostrogradsky's theorem. Central in these elementary results is the concept of *differential form*—that is, line element, surface element, volume element, and so on. Intuitively, a differential form can be viewed as an assignment of a real number—the "length," the "area," or the "volume,"—to each infinitesimal line, surface, or volume element. In other words, a differential form can be viewed as a cochain. Next, the elementary operations of curl, divergence, and so on can be generalized under the unifying concept of *exterior differential*. The exterior differential acts as coboundary operator on the cochain of differential forms. From there on, one perceives the possibility of building a cohomology theory entirely based on differential forms. This is precisely what the *de Rham cohomology* is all about.

12.2.1 Cohomology of Differential Forms

Let P be an n_p-manifold and consider the neighborhood O_p of $p \in P$ charted with Riemann coordinates $x_1, x_2, ..., x_{n_p}$. An **exterior differential form** of degree n defined in the neighborhood O_p is an expression like

$$\varpi = \sum_{i_1=1}^{n_p} \sum_{i_2=1}^{n_p} \cdots \sum_{i_n=1}^{n_p} g_{i_1 i_2 \ldots i_n}(x_1, ..., x_{n_p}) dx_{i_1} dx_{i_2} \ldots dx_{i_n}$$

The product

$$dx_{i_1} dx_{i_2} ... dx_{i_n}$$

is associative and subject to the skew-commutativity rules,

$$\begin{aligned} dx_i dx_i &= 0 \\ dx_i dx_j &= -dx_j dx_i, \forall i, j \end{aligned}$$

Furthermore, the functions $g_{i_1 i_2 \ldots i_n} : \mathbb{R}^{n_p} \to \mathbb{R}$ are smooth.

From the skew-commutativity rule, it follows that

$$dx_{\pi(1)} dx_{\pi(2)} ... dx_{\pi(n)} = \text{sign}(\pi) dx_1 dx_2 ... dx_n,$$

for any permutation $\pi : \{1, 2, ..., n\} \to \{1, 2, ..., n\}$ where $\text{sign}(\pi)$ denotes the signature of the permutation. Under those circumstances, the differential form $dx_1 dx_2 ... dx_n$ is said to be **alternating**.

A *differential form* on the manifold P is a collection of local differential forms as above, one for each chart, subject to the restriction that the local differential forms of overlapping neighborhoods O_{p_1} and O_{p_2} agree on $O_{p_1} \cap O_{p_2}$.

Clearly, the collection of all differential n-forms on P form an additive group, written $\Lambda^n(P)$. Clearly, $\Lambda^0(P)$ is the Abelian group of all smooth functions $P \to \mathbb{R}$. By the skew-commutativity rule, it is also easily seen that $\Lambda^n(P) = 0$ for $n > n_p$.

The **exterior differential** of a smooth function $g : \mathbb{R}^{n_p} \to \mathbb{R}$ is defined as

$$d^0 g = \sum_i \frac{\partial g}{\partial x_i} dx_i$$

The **exterior differential** of the n-form ϖ is defined as

$$d^n \varpi = \sum_{i_1 i_2 \ldots i_n} dg_{i_1 i_2 \ldots i_n} dx_{i_1} dx_{i_2} ... dx_{i_n}$$

The right-hand side of the above is of course subject to the skew-commutativity rules. Clearly, the exterior differential raises the degree by one; hence,

$$d^n : \Lambda^n \to \Lambda^{n+1}$$

Furthermore, it is not hard to see that

$$d^n d^{n-1} = 0$$

so that d is a coboundary. The de Rham cohomology is simply the cohomology of the cochain complex $\{\Lambda, d\}$. A differential form ϖ such that $d\varpi = 0$ is said to be **closed**. A differential form ϖ that can be written as $d\theta$ for some form θ is said to be **exact**. Clearly, the set of exact forms is included in the set of closed forms,

$$\operatorname{im}(d^{n-1}) \subseteq \ker(d^n)$$

The **de Rham cohomology** of P is the quotient of the closed forms by the exact forms,

$$H^n(P, \mathbb{R}) = \ker(d^n)/\operatorname{im}(d^{n-1})$$

The value group of the de Rham cohomology is the real line $\mathbb{R}$. As such, the de Rham cohomology misses the torsion phenomena.

One of the most important motivations of the theory of exterior differential forms is the generalization of the Gauss–Green–Ostrogradsky theorem:

Theorem 12.2. *Let P be an n_p-manifold and let ϖ be an $(n_p - 1)$-form. Then*

$$\int_P d\varpi = \int_{\partial P} \varpi$$

Proof. See [de Rham 1984]. ■

12.2.2 Pull-Back

Let $f : X \to Y$ be a smooth map between smooth manifolds and let $\varpi(y) = \sum g_{i_1,i_2,\ldots}(y_1, y_2, \ldots)dy_{i_1}dy_{i_2}\ldots$ be an exterior differential form defined on Y in terms of local Riemann coordinates y_i relative to which the map is written $y_i = f_i(x)$. The **pull-back** of the differential form ϖ by f is $f^*\varpi = \sum g_{i_1,i_2,\ldots}(f_1(x), f_2(x), \ldots)df_1(x)df_2(x)\ldots$. If ϖ is a closed form representing a cohomology class, it is clear that $\varpi \mapsto f^*\varpi$ is a concretization of the de Rham cohomology group homomorphism induced by f.

12.2.3 Wedge Product

To define a **de Rham cohomology ring**, we introduce the **exterior or wedge product** of differential forms. For

$$\begin{aligned} \varpi &= \sum g_{i_1 i_2 \ldots i_n} dx_{i_1} dx_{i_2} \ldots dx_{i_n} \\ \theta &= \sum h_{j_1 j_2 \ldots j_m} dx_{j_1} dx_{j_2} \ldots dx_{j_m} \end{aligned}$$

the wedge product is defined as

$$\varpi \wedge \theta = \sum \sum g_{i_1 i_2 \ldots i_n} h_{j_1 j_2 \ldots j_m} (dx_{i_1} dx_{i_2} \ldots dx_{i_n})(dx_{j_1} dx_{j_2} \ldots dx_{j_m})$$

Furthermore, the wedge product is skew commutative,

$$\varpi \wedge \theta = (-1)^{nm} \theta \wedge \varpi$$

From the above, it follows that Λ^* together with the wedge product $\wedge$ form an **exterior algebra**. The interplay between the differential and the wedge product is given by the following:

$$d(\varpi \wedge \theta) = (d\varpi) \wedge \theta + (-1)^n \varpi \wedge (d\theta)$$

Clearly, the wedge product of differential forms has the same structure as the cup product of semisimplicial cochains. From here on, we follow the same algebraic path as in the semisimplicial case, resulting in the de Rham cohomology ring.

13

TWISTED CARTESIAN PRODUCT OF UNCERTAINTIES

SUMMARY

In previous chapters, a central idea was the decomposition of the uncertainty space as a Cartesian product. This revealed quite a lot of structure: the prismatic decomposition of combinatorial topology, the equivalent Q-triangulation of operations research together with the extra insight provided by the underlying Hex board structure, the Eilenberg–Zilber theorem, and so on. This Cartesian product decomposition, however, is not tailored to the particular role $D \times \Omega$ plays as the *domain of definition* of the Nyquist map $f : D \times \Omega \rightarrow N$. In this specific role, more insight is gained by viewing $D \times \Omega$ as a "bundle of fibers" $f^{-1}(s), s \in N$. Under some conditions—among others, that all fibers be homeomorphic—the Nyquist map $D \times \Omega \rightarrow N$ can be viewed as a *fiber bundle* in the sense of [Steenrod 1951]. The specific feature of a fiber bundle is that the preimage $f^{-1}(O_s)$, where O_s is a neighborhood of $s \in N$, is (locally) homeomorphic to the Cartesian product $O_s \times f^{-1}(0 + j0)$. What distinguishes a fiber bundle from an ordinary Cartesian product is that it is not possible to go from local to global without "twisting" the fiber $f^{-1}(0+j0)$ as s runs across the base space N. To forge the connection with controls, we already mention that the twisting of the fiber typically reveals an open-loop unstable plant.

The idea of decomposing the domain of definition of a function as the Cartesian product of a "fiber" and the image is also the essence of Thom's first isotopy lemma. This result says that if $f : X \rightarrow \mathbb{R}^2$ is a proper stratified submersion (i.e., the matrix of partial derivatives of f has rank 2 everywhere), then there exists a stratum preserving homeomorphism $X \rightarrow f^{-1}(0) \times \mathbb{R}^2$. As we shall see in Chapter 21 of the "Differential Topology" part, the submersion property implies the absence of critical points which is itself equivalent to a Nyquist template without boundary; that is, a Nyquist template that would spread across the whole complex plane. Therefore, Thom's isotopy lemma, while conceptually worth mentioning, is of very limited applicability in this control context.

Consistently with the overall "simplicial" flavor of this book, the second half of this chapter is devoted to a *semisimplicial* version of the concept of fiber bundle. The major advantage of *semisimplicial* bundles is that they are plentiful. In fact, we show that any continuous Nyquist map $D \times \Omega \rightarrow N$, after some "preprocessing" and discretization, becomes a semisimplicial bundle. The culmination of semisimplicial bundles as applied to the Nyquist problem is that the total complex of $D \times \Omega$ is chain isomorphic to the so-called *twisted* complex defined over the Cartesian product of N and the fiber $f^{-1}(0+j0)$,

$$\{C_*(D \times \Omega), \partial_*\} \xrightarrow{\cong} \{C_*(N \times f^{-1}(0+j0)), \partial_*^{\text{twisted}}\}$$

The "simplexes" of the right-hand side complex are ordered pairs of total simplexes of the same length of N and $f^{-1}(0+j0)$, respectively. The essence of the twisting phenomenon is that to make the above a *chain map*, one has to endow the complex $C_*(N \times f^{-1}(0+j0))$ with a modified boundary operator, the *twisted* boundary operator. With this new decomposition of the domain of uncertainty, we are able to dispose of the issue of commutativity between the inverse image and the boundary that was left strictly local. It turns out that there are two factors that prevent the inverse image of a semisimplicial map to commute—globally—with the face operator: the "twisting" of the fiber and any "hole" in the fiber complex.

Invoking the above isomorphism, we could use the "template times crossover" complex to compute the homology of $D \times \Omega$. The distinguishing feature of doing so (rather than computing the homology *directly* from the complex of $D \times \Omega$) is that the former reveals the linkage between the topology of the uncertainty, the crossover and the template. This is a prelude to the chapter to come, dealing with spectral sequences.

13.1 Fiber Bundle

As already said, the motivation for bringing the concept of fiber bundle stems from our desire to view $D \times \Omega$ as a bundle of fibers $f^{-1}(s)$ as s runs in the Nyquist template. The problem is that there is a great many fibers, not all of them being necessarily homeomorphic, not even of the same homotopy type. At this stage, it is already necessary to put some restrictions on the fibers: We assume that all fibers are homeomorphic. Later, we will investigate how this condition can be relaxed by introducing the concept of *fibration* in which all fibers are of the same homotopy type. With all fibers homeomorphic, the idea is to pick a privileged fiber, say $f^{-1}(0+j0)$. We could also take the privileged fiber to be $f^{-1}(s)$ for some $s \in N$; we could even take the "fiber" to be another topological space F, provided that it be homeomorphic to $f^{-1}(0+j0)$. It turns out that, under some singularity conditions on the map f (among other conditions, that $0+j0$ be a regular value in the sense of Chapter 21), there exists O_0, an open neighborhood of $0+j0$, such that $f^{-1}(O_0)$ is homeomorphic to $O_0 \times F$. Thus, subject to some conditions, $D \times \Omega$ is locally homeomorphic to a Cartesian product. This is formalized in the concept of fiber bundle.

13.1.1 Basic Definitions and Concepts

Definition 13.1. *A space E, called* **total space**, *is said to be endowed with a* **fiber bundle** *structure over the* **base space** B *and with* **fiber space** F *if there exists a map $\pi : E \to B$, called* **bundle projection**, *such that*

- *For any $b \in B$, the fiber over b, $E_b = \pi^{-1}(b)$, is homeomorphic to F*
- *For some open covering of B, $\cup_i O_i$, and for any open set O_i of the covering, there exists a homeomorphism*

$$h_i : \pi^{-1}(O_i) \to O_i \times F$$

- *Furthermore, the composite*

$$\pi \circ h_i^{-1} : O_i \times F \to O_i$$

is the projection on the first factor; in other words, the bundle projection is locally equivalent to the projection on the first factor, as revealed by the following commutative diagram:

$$\begin{array}{ccc} O_i \times F & \xrightarrow{h_i^{-1}} & \pi^{-1}(O_i) \\ \pi_1 \downarrow & & \downarrow \pi \\ O_i & \xrightarrow{=} & O_i \end{array}$$

The homeomorphism h_i provides local coordinates and is called **chart**. *The set of all homeomorphisms $\{h_i\}$ is called* **atlas** *of the bundle.*

A convenient notation for this structure is

$$\begin{array}{ccc} F & \xrightarrow{i} & E \\ & & \downarrow \pi \\ & & B \end{array}$$

where i is the inclusion map of the fiber in the total space and π is the bundle projection.

Probably the best illustrative example of a fiber bundle is the Möbius strip. The total space E is the strip itself, the base space is a circle drawn around the strip, the fiber is a compact line segment that might be thought of as moving around the base circle with a "twist" after one lap around the base, and π is the projection onto the base circle along the fiber.

Of course what is of interest to us is to think $D \times \Omega$ as the total space, the Nyquist template N as the base space, the Nyquist map as the bundle projection, and the crossover $f^{-1}(0+j0)$ as the fiber. It should be stressed that this is not always possible.

The local versus global Cartesian product structure of the total space E can probably be better understood from

$$E = \cup_i h_i^{-1}(O_i \times F)$$

The *fact* is that it is not possible, in general, to "stitch" the local homeomorphisms h_i into one single, global homeomorphism h such that $E = h^{-1}(B \times F)$, so that the bundle fails to be, globally, a Cartesian product. The opposite case motivates the following:

Definition 13.2. *A bundle is said to be* **trivial** *if there exists a homeomorphism $h : E \to B \times F$ such that $\pi \circ h^{-1} = \pi_1$.*

It should be clear that an important bundle problem is to decide whether a given bundle is trivial.

We pursue somewhat further the coordinate approach to bundles. Consider two open sets O_i, O_j of B together with the homeomorphisms h_i, h_j. The composite

$$h_i \circ h_j^{-1} : (O_i \cap O_j) \times F \to (O_j \cap O_i) \times F$$

induces, for any $b \in O_i \cap O_j$, a homeomorphism of the fiber

$$h_i \circ h_j^{-1}(b) : F \to F$$

The homeomorphism $h_i \circ h_j^{-1}(b)$ is called **transition function** and is written more compactly as $h_{ij}(b)$. This leads to the concept of *structure group* of a bundle: The group generated by all homeomorphisms $h_i \circ h_j^{-1}(b)$ for all i, j and all $b \in B$ is called **(minimum) structure group** G. Furthermore, we require the structure group to be a **topological group**—that is, G is at the same time a group and a topological space and the two structures are compatible in the sense that the mapping $(g_1, g_2) \mapsto g_1 g_2^{-1}$ is continuous. (Roughly speaking, a topological group is a Lie group without analytic structure; formally, this is Hilbert's fifth problem.) With this concept, we can define continuity of the transition functions.

Definition 13.3. *The topological group G is said to be a* **structure group** *for the bundle $(E, \pi, B; F)$ whenever the mapping*

$$\begin{array}{rcl} O_i \cap O_j & \to & G \\ b & \mapsto & h_i \circ h_j^{-1}(b) \end{array}$$

is continuous for all i, j.

Observe that, given a bundle (e.g., the Möbius strip), there are *many* ways to set up the homeomorphisms h_i. To each such choice, there corresponds a *unique* minimum structure group G, by our definition. In the *simplest* possible set up for the Möbius strip, the structure group consists of the identity and the reflection across the midpoint of the fiber. It is sometimes convenient to somewhat enlarge this minimum group to include a broader class of homeomorphisms of the fiber. For example, for a bundle that has $\mathbb{R}^k$ as fiber it is convenient to take $GL(k, \mathbb{R})$ to be the structure group.

We will say that the fibers are **"twisted"** whenever the minimum structure group G is nontrivial.

The motivation for the structure group is that, very often, the details of the coordinate data of the bundle are not quite relevant. All we really need to know is the structure group. For example, in order to classify a bundle—that is, to compare the bundle with the so-called *universal bundles*—the knowledge of the structure group is essential.

Clearly, the transition functions and the structure group are very useful to check triviality of a bundle.

13.1.2 Bundle Morphisms

Bundles have their morphisms and isomorphisms, but the latter become relevant when applied to bundles with the same fiber and structure group.

Definition 13.4. *A* **bundle morphism** *between two bundles* (E, π, B), (E', π', B') *is a pair of maps* $g_e : E \to E'$, $g_b : B \to B'$ *such that:*

- *The morphism respects the fibers; that is, the following diagram commutes:*

$$\begin{array}{ccc} E & \xrightarrow{g_e} & E' \\ \pi \downarrow & & \downarrow \pi' \\ B & \xrightarrow{g_b} & B' \end{array}$$

The two bundles, both with the same fiber F *and the same structure group* G, *are* **equivalent** *or* **isomorphic** *iff, in addition,* g_e, g_b *are homeomorphisms and*

- *Over each* $b \in B$, *the map* $g_e|E_b$ *is a transformation of the fiber that belongs to the group* G.

With these concepts, we can redefine a bundle (E, π, B) to be **trivial** if it is isomorphic to the bundle $(B \times F, \pi_1, B)$.

Clearly, the transition functions are useful to check whether two bundles are isomorphic.

13.1.3 Clutching

They are many constructions that can be done on bundles, and we will make extensive use of the so-called *Whitney sum* of two vector bundles in Chapter 20. Here, we quickly review a generic procedure, called *clutching*, that is widely used to "glue" several bundles to construct a more complicated one. Let $B = \sqcup_{i=1}^{2} O_i$ be a base space covered with just two open sets for simplicity. Consider (possibly trivial) bundles (E_i, π_i, O_i). Define a homeomorphism $\phi : \pi_1^{-1}(O_1 \cap O_2) \to \pi_2^{-1}(O_1 \cap O_2)$. ϕ is called gluing or clutching data. Over the disjoint union $\sqcup_i E_i$, define the relation $e_1 \sim e_2$ iff $e_2 = \phi(e_1)$. Define the quotient $E = \sqcup_i E_i / \sim$. In other words, in the disjoint union we identify points e_1, e_2 linked by $e_2 = \phi(e_1)$. It can be shown ([Karoubi 1978, Chapter I, Theorem 3.2]) that the projections π_i induce a unique bundle projection $\pi : E \to B$ such that

$$\pi|\pi_1^{-1}(O_1 \cap O_2) = (\pi|\pi_2^{-1}(O_1 \cap O_2)) \circ \phi$$

The resulting bundle (E, π, B) is said to be obtained by **clutching** the bundles (E_i, π_i, O_i).

13.1.4 Cross Section

In the concept of fiber bundle, the base space does not physically reside in the total space. One might attempt to move the base space into the total space by means of a continuous map $B \to E$. This leads to the following:

Definition 13.5. *A* **cross section** *in a bundle* $(E, \pi, B; F)$ *is a continuous map* $c : B \to E$ *such that* $\pi \circ c = 1_B$.

Without much danger, we will take the freedom to identify the cross section with the image of the map $B \to E$.

The concept of a cross section allows for a better visualization of the bundle, since a cross section can be thought of as a "slice" that cuts through all fibers. Unfortunately, the visualization convenience afforded by the cross section is not a universal helper, since a cross section does not always exists! Construction of a cross section is a procedure, recursive on the dimension of the simplexes of the base, that requires some "obstruction" to vanish at every step of the procedure. We do not as yet have the obstruction theory tools necessary to construct a cross section and we therefore have to postpone this to Section 16.8 (see also [Steenrod 1951] or [Mimura and Toda 1991, Postscript, page 231] for the obstruction to cross sectioning).

Despite the fact that we cannot as yet determine whether a cross section exists, it should be clear that *existence* of a cross section is an important bundle problem.

If we make the bold assumption that the Nyquist map $D \times \Omega \to N$ is a bundle, the cross-sectioning problem is whether there exists some 2-D subspace of $D \times \Omega$ that maps *homeomorphically* to the template N. In the affirmative, it would suffice to sweep the cross section through the uncertainty to check robust stability for *all* parameter values. We will see in Subsection 13.2.5 how the cross section problem can be extended to an arbitrary simplicial Nyquist map.

13.1.5 Bundle Interpretation of Kharitonov's Theorem

Kharitonov's theorem provides us with an example of fiber bundle structure in control. Consider the Kharitonov polynomial,

$$p(s) = s^n + a_{n-1}s^{n-1} + ... + a_2 s^2 + a_1 s + a_0$$

For reasons that will become clearer later, the coefficients a_i are in the *open* cube

$$D = \Pi_{i=0}^{n-1}(\underline{a}_i, \overline{a}_i)$$

We look at the *fixed-frequency* map

$$\begin{aligned} f_\omega : D &\to \mathbb{C} \\ (a_0, ..., a_{n-1}) &\mapsto p(j\omega) \end{aligned}$$

It is easily seen that the Horowitz template $f_\omega(D) = N_\omega$ is an *open rectangle* in the complex plane. Our claim is that, for $\omega \neq 0$, (D, f_ω, N_ω) is a fiber bundle. The essential point is to look at the preimages of the points in the template N_ω. Observe that

$$\begin{pmatrix} \Re f_\omega \\ \Im f_\omega \end{pmatrix} = \begin{pmatrix} 1 & 0 & -\omega^2 & \cdots \\ 0 & \omega & 0 & \cdots \end{pmatrix} \begin{pmatrix} a_0 \\ a_1 \\ a_2 \\ \vdots \end{pmatrix} + \begin{pmatrix} \Re(j\omega)^n \\ \Im(j\omega)^n \end{pmatrix}$$

Take an arbitrary $z \in N_\omega$. Let a^* be a preimage—that is, a selected solution to the above equation. Then the "fiber" over z is

$$f_\omega^{-1}(z) = \left(a^* + \ker \begin{pmatrix} 1 & 0 & -\omega^2 & \cdots \\ 0 & \omega & 0 & \cdots \end{pmatrix} \right) \cap D$$

The term between parentheses is clearly an $(n-2)$-dimensional (affine) subspace. Its intersection with D is clearly homeomorphic to $\mathbb{R}^{n-2}$. Therefore all fibers are homeomorphic. From there follows the local triviality and, hence, the fiber bundle property of the mapping.

Another remarkable property is that this fiber bundle has a *cross section*. Indeed, consider the open rectangle

$$\Gamma = (\underline{a}_0, \underline{a}_1, \overline{a}_2, \overline{a}_3, ...)(\overline{a}_0, \underline{a}_1, \underline{a}_2, \overline{a}_3, ...)(\overline{a}_0, \overline{a}_1, \underline{a}_2, \underline{a}_3, ...)(\underline{a}_0, \overline{a}_1, \overline{a}_2, \underline{a}_3, ...)$$

It is easily seen that this is a rectangle; that is, consecutive edges are making 90-degree angles. It is also easily seen that $f_\omega|\Gamma$ is a homeomorphism $\Gamma \to N_\omega$. The inverse of that homeomorphism provides a cross section $\gamma : N_\omega \to \Gamma$. The cross section Γ provides a two-dimensional subspace of D such that $f_\omega(\Gamma) = N_\omega$.

13.1.6 Principal Bundle

Another typical bundle problem is to determine whether a bundle is trivial. The following is a key concept in dealing with bundle problems of this kind:

Definition 13.6. *A bundle $(E, \pi, B; F, G)$ is said to be* **principal** *whenever $F = G$.*

Theorem 13.7. *Any bundle $(E, \pi, B; F, G)$ has an associated principal bundle $(E^p, \pi^p, B; G, G)$ with the same base space and transition functions.*

Proof. Let $\{O_i\}$ be an open covering of B. Define the space

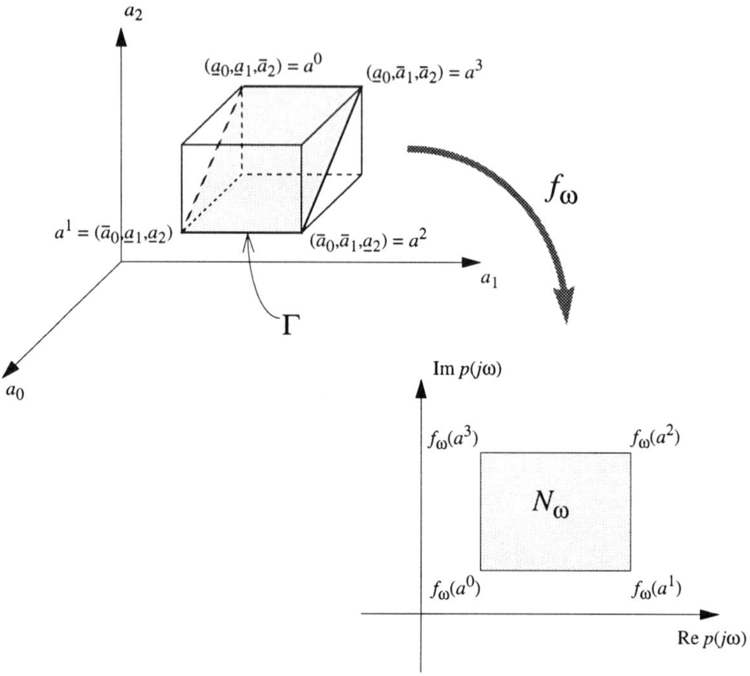

Fig. 13.1. Cross section in Kharitonov's cube.

$$\sqcup_i O_i \times G$$

An element of this space is written as (b, g), where $b \in O_i \subseteq B$ and $g \in G$. We define on this space an equivalence relation as follows: Let

$$(b, g) \in O_i \times G, \quad (b', g') \in O_j \times G$$

Then

$$(b, g) \sim (b', g')$$

iff

$$b = b' \text{ and } g = h_{ji} g'$$

The total space of the principal bundle is defined as

$$E^p = (\sqcup_i O_i \times G) / \sim$$

The principal bundle projection is just the projection on the first factor,

$$\begin{aligned} \pi^p : E^p &\rightarrow B \\ \{(b, g)\} &\mapsto b \end{aligned}$$

The required properties for (E^p, π^p, B) to be a bundle are easy to check and this is left to the reader. ∎

Observe that in the above the principal bundle is obtained by "clutching" the trivial bundles $(O_i \times G, \pi_1, O_i)$.

Theorem 13.8. *The bundle* (E, π, B) *is trivial iff the associated principal bundle* (E^p, π^p, B) *is trivial.*

Proof. Triviality of a bundle can be decided from the transition functions. Since (E, π, B) and (E^p, π^p, B) have the same transition functions, they are simultaneously trivial. ■

A fundamental result is the following:

Theorem 13.9. *There exists a cross section* $B \to E^p$ *iff* (E, π, B) *and* (E^p, π^p, B) *are trivial.*

Proof. See [Steenrod 1951, Theorems 8.2 and 8.3 and Corollary 8.4], or [Nash and Sen 1983, Section 7.4]. ■

13.1.7 Tangent Bundle

To illustrate the previously defined concepts, we introduce the *tangent bundle* to such a manifold as $D \times \Omega$, along with its *principal bundle*, called *frame bundle.*

Consider the tangent space to the manifold $D \times \Omega$ at p, namely,

$$T_p(D \times \Omega) \cong \mathbb{R}^{n_p}$$

If $D \times \Omega$ is embedded in an Euclidean space of adequate dimension (Whitney's embedding theorem), we can use the intuitive definition of the tangent space; otherwise, we have to use the more formal definition (see Subsection 21.2.1).

Stack all of the tangent spaces together as a disjoint union in the bigger space $(D \times \Omega) \times \mathbb{R}^{n_p}$,

$$T(D \times \Omega) = \sqcup_{p \in D \times \Omega} T_p(D \times \Omega)$$

The **tangent bundle**

$$(T(D \times \Omega), \pi_1, D \times \Omega)$$

of $D \times \Omega$ is defined as follows: Its total space is the disjoint union of all tangent spaces; the base space is the manifold $D \times \Omega$; to define the bundle projection, observe that, since $T(D \times \Omega) \subseteq (D \times \Omega) \times \mathbb{R}^{n_p}$, any element of $T(D \times \Omega)$ can be written (p, x), where $x \in T_p(D \times \Omega)$, $p \in D \times \Omega$; therefore, the bundle projection is defined as the projection on the first factor:

$$\begin{aligned} \pi_1 : T(D \times \Omega) &\to D \times \Omega \\ (p, x) &\mapsto p \end{aligned}$$

The fiber F is the vector space $\mathbb{R}^{n_p}$; the transition functions are easily seen to be the Jacobians of the transition functions of the manifold $D \times \Omega$; and the structure group is $GL(n_p, \mathbb{R})$.

The choice of the *general* linear group as structure group is just for the convenience of it. Quite often, by carefully choosing the charts, the structure group can be reduced to, say, $O(n_p)$, and typical in some orientation problems is whether the structure group can be further *reduced* to $SO(n_p)$.

A **vector field** on the manifold $D \times \Omega$ is a cross section through its tangent bundle—that is, a continuous assignment of a tangent vector to any point of the manifold. Observe that such a vector field could vanish at some points of the manifold.

The *principal* bundle associated with the tangent bundle is defined as follows: Its base space is $D \times \Omega$; the fiber over p consists of all nonsingular transformations of $T_p(D \times \Omega)$; that is, $GL(n_p, \mathbb{R})$. The *principal bundle* associated with a *tangent bundle* is called **frame bundle**. A cross section in the frame bundle, $D \times \Omega \to GL(n_p, \mathbb{R})$, is a continuous assignment of a nonsingular transformation to any $p \in D \times \Omega$. Since a basis of $T_p(D \times \Omega)$ can be obtained from a *fixed* basis through a nonsingular transformation, it follows that *existence of a cross section through the frame bundle is equivalent to existence of a continuous basis for the tangent space.* For more details, see [Nash and Sen 1983, Section 7.3] and [Karoubi 1978, Chapter 1, Section 3.18, page 16].

The tangent bundle is a first example of a *vector bundle*—that is, a bundle where the fiber is such a vector space as $\mathbb{R}^n$ or $\mathbb{C}^n$. Vector bundles are instrumental in K-theory (see Chapter 20).

To address the nontrivial issue of whether a manifold has an *everywhere nonvanishing* vector field, we introduce the *unit tangent bundle.* Define the unit sphere S^{n_p-1} in every tangent space $T_p(D \times \Omega) \cong \mathbb{R}^{n_p}$. Stack all of the spheres together as a disjoint union in $(D \times \Omega) \times S^{n_p-1}$ and define the bundle projection on the first factor. This is an example of a **sphere bundle**—that is, a bundle whose fiber is a sphere. The structure group is taken as $\text{Diff}(S^{n_p-1})$ or $O(n_p)$. Clearly, existence of an everywhere nonvanishing vector field is equivalent to existence of a cross section through the unit tangent bundle.

13.1.8 Elementary Homotopy Theory of Bundles

The next concept is that of *pull-back.* Consider a bundle (E, π, B) and a continuous mapping $\phi : X \to B$. Schematically,

$$\begin{array}{ccc} & & E \\ & & \downarrow \pi \\ X & \xrightarrow{\phi} & B \end{array}$$

Define

$$\phi^* E = \{(x, e) \in X \times E : \phi(x) = \pi(e)\} \tag{13.1}$$

Let π_1, π_2 be the projections on the first and second factors, respectively. Clearly, there exists a unique map $\phi^*\pi$ ($\doteq \pi_1$) such that the following diagram commutes:

$$\begin{array}{ccc} \phi^* E & \xrightarrow{\pi_2} & E \\ \phi^*\pi = \pi_1 \downarrow & & \downarrow \pi \\ X & \xrightarrow{\phi} & B \end{array} \tag{13.2}$$

It is easily seen that $(\phi^* E, \phi^* \pi, X)$ is a bundle, called **pull-back** of (E, π, B) by $\phi : X \to B$. This pull-back bundle is also written $\phi^*(E, \pi, B)$. The fundamental homotopy property of the pull-back is the following:

Theorem 13.10. *If $\phi_1, \phi_2 : X \to B$ are homotopic, the pull-back bundles, $\phi_1^* E, \phi_2^* E$ are equivalent in the sense that they have the same transition functions.*

Proof. See [Steenrod 1951] or [Karoubi 1978, Chapter 1, 7.2] for the specific case of a vector bundle. ■

For convenience, we recall the following definition:

Definition 13.11. *A topological space X is said to be* **contractible** *if the identity map on X, 1_X, is homotopic to the map that sends any $x \in X$ to the same base point $x_* \in X$—in other words, if there exists a homotopy*

$$h_t : X \times [0, 1] \to X$$

such that

$$\begin{aligned} h_0(x) &= x \\ h_1(x) &= x_*, \ \forall x \in X \end{aligned}$$

Theorem 13.12. *If the basis B is contractible, the bundle is trivial.*

Proof. Consider the identity $1_B : B \to B$ and the injection $i : * \to B$, where $*$ is a base point in B. Because the space B is contractible, 1_B and i are homotopic. Therefore, the pull-back bundles $1_B^* E, i^* E$ are equivalent. Since $i^* E$ is a bundle over one single point, it is trivial. Therefore, so is $1_B^* E = E$. ■

Theorem 13.13. *If the fiber F is contractible, then the bundle has a cross section.*

Proof. See [Nash and Sen 1983, page 161] or [Steenrod 1951]. ■

Theorem 13.14. *If either the base space B or the structure group G, viewed as a topological space, is contractible, then the bundle is trivial.*

Proof. If B is contractible, we already know that the bundle is trivial. If G is contractible, go to the principal bundle E^p; its fiber is contractible. Therefore, E^p has a cross section. Hence E is trivial. ■

The control interpretation of the above is as follows: Assume the Nyquist template—the base space—is contractible. This is very likely to happen with an open-loop stable system. Then we can guarantee that the bundle is trivial and has a cross section. On the other hand, suppose that the system is open-loop unstable and that it is required to be closed-loop robustly stable. In this case, the Nyquist template—the base space—has the shape of an annulus around $0 + j0$, and the base space is not contractible to a point. Therefore, in that case, we could have a nontrivial bundle; in other words, the fibers could be "twisted."

13.1.9 Fiber Bundle Interpretation of Doležal's Theorem

The traditional theorem of Doležal [1964], instrumental in linear time-varying systems theory, is the following:

Theorem 13.15. (Doležal) *Consider a continuous, matrix-valued map,*

$$\begin{array}{rcl} [0,1] & \to & \mathbb{M}(m,n;\mathbb{R}) \\ t & \mapsto & M(t) \end{array}$$

where $m \leq n$, and let the kernel of $M(t)$ have constant dimension,

$$\dim \ker(M(t)) = d, \ \ \forall t \in [0,1]$$

Then there exists a continuous basis, which can be chosen orthonormal, for $\ker M(t)$*; that is, there exist continuous functions*

$$\begin{array}{rcl} [0,1] & \to & \mathbb{R}^n \\ t & \mapsto & \eta^i(t) \end{array}$$

such that

$$\textit{rank} \ \begin{pmatrix} \eta^1(t) & \eta^2(t) & \dots & \eta^d(t) \end{pmatrix} = d, \ \ \forall t \in [0,1]$$

and

$$M(t) \begin{pmatrix} \eta^1(t) & \dots & \eta^d(t) \end{pmatrix} = 0, \ \ \forall t$$

Proof. We construct the following bundle over the parameter space

$$(\sqcup_{t \in [0,1]} \ker M(t), \pi, [0,1])$$

The total space is the disjoint union, topologized as a subspace of $[0,1] \times \mathbb{R}^n$. The base space is the parameter space $[0,1]$ endowed with the usual relative topology. The bundle projection π is just the projection on the first factor. The fiber $\pi^{-1}(t) = \ker M(t)$ is clearly homeomorphic to $\mathbb{R}^d$. By choosing an $(m-d) \times (m-d)$ submatrix of $M(t)$, nonsingular over a sufficiently small open interval (t_1, t_2), it is easy to construct a *local* ($t \in (t_1, t_2)$) continuous basis of $\ker M(t)$, from which it follows that $\pi^{-1}((t_1, t_2)) = \sqcup_{t \in (t_1, t_2)} \ker M(t)$ is homeomorphic to $(t_1, t_2) \times \mathbb{R}^d$. Write this as $\pi^{-1}((t_1, t_2)) = h_i^{-1}((t_1, t_2) \times \mathbb{R}^d)$. Clearly, the homeomorphism can be taken *linear* on the fiber. It also follows that $(t_1, t_2) = \pi\pi^{-1}((t_1, t_2)) = \pi \circ h_i^{-1}((t_1, t_2) \times \mathbb{R}^d)$, so that $\pi \circ h_i^{-1}$ is indeed the identity on the first factor. The structure group is $GL(d, \mathbb{R})$.

The *crucial point* is that the base space $[0, 1]$ is contractible, so that the bundle is trivial. Therefore, the total space is *globally* isomorphic to the Cartesian product of the base space and the fiber,

$$E = \sqcup_{t\in[0,1]} \ker M(t) = h^{-1}([0,1] \times \mathbb{R}^d)$$

and the isomorphism can be taken linear on the fiber. Because the homeomorphism is linear on the fiber, it maps a basis into a basis. Therefore, take $e^1, ..., e^d$, the usual Euclidean basis of $\mathbb{R}^d$. It follows that the continuous basis for $\ker M(t)$ is

$$\eta^i(t) = h^{-1}(t, e^i), \ \ i = 1, ..., d$$

■

It is easy to generalize the proof and show that Doležal's theorem holds for any *contractible* parameter space.

13.2 Semisimplicial Bundles and Twisted Cartesian Product

In this section, we develop a simplicial version of the concept of fiber bundle. Since the vertex transformation of the simplicial map $D \times \Omega \to N$ is *many-to-one*, we have to resort to the *semi*simplicial theory in order to keep a "simplex" of N with repetition of vertices as an acceptable mathematical entity.

Let E, B be semisimplicial complexes, graduated as $\{E_n : n \geq 0\}, \{B_n : n \geq 0\}$, where the graduation index n stands for the "length"—that is, the number of vertices, with repetition allowed, minus 1. E can be thought of as the domain of uncertainty, while B should be thought of as the Horowitz supertemplate. Both complexes are endowed with the usual face and degeneracy operators, F^i, D^i, respectively, defined in Subsection 4.5.2. The two complexes are linked by the semisimplicial projection map $\pi : E \to B$. E is called **total complex**, B is called **base complex**, and π is the **bundle projection**.

The issue of whether (E, π, B) is a semisimplicial bundle has to do with whether, for an arbitrary simplex β_n of B, $\pi^{-1}(\beta_n)$ is isomorphic to $\Delta[n] \times F$, where $\Delta[n]$ is some standard semisimplicial complex that underlies all n-dimensional objects and F is the fiber complex. If we define $\pi^{-1}(\beta_n)$ to be the collection of all simplexes of E mapped to β_n, then $\pi^{-1}(\beta_n)$ fails to be a complex; therefore, $\pi^{-1}(\beta_n)$ is instead defined to be the collection of all simplexes mapped to β_n, a *face* of β_n, or a *degenerate* simplex with its vertices in β_n. Clearly, they are plenty of simplexes in $\pi^{-1}(\beta_n)$ and the best way to organize the book-keeping is to proceed very formally.

After reviewing semisimplicial bundles and related concepts following [Barratt, Gugenheim, and Moore 1959] and [May 1967,1992], we will be able to formulate the commutativity between the boundary and the inverse image of a semisimplicial Nyquist map.

13.2.1 Background

We first need to set up some standard, reference semisimplicial complex $\Delta[n]$ that will underlie all n-dimensional objects and in particular the n-dimensional objects of the base space.

Let $v^0, v^1, v^2, ...$ be affinely independent points in infinite-dimensional Hilbert space. Let

$$\Sigma_n = v^0 ... v^n$$

From a geometrical standpoint, Σ_n is called *standard n-simplex.* However, the geometric substratum in which Σ_n has been defined is not absolutely necessary. From an abstract point of view, Σ_n is called "universal generator in dimension n." The justification for this terminology will be explained later but roughly goes as follows: The semisimplicial complex $\Delta[n]$ that models dimension n is *defined* as the complex generated by Σ_n and all simplexes obtained by removal and repetition of vertices of Σ_n (see [Barratt, Gugenheim, and Moore 1959]).

For every dimension n, we construct a semisimplicial complex $\Delta[n]$, called **n-dimensional model**. An m-simplex of $\Delta[n]$ is an array

$$[v^{i_0} v^{i_1} ... v^{i_m}]$$

where the indices are subject to the condition

$$i_0 \leq i_1 \leq ... \leq i_m \leq n$$

In other words, an m-simplex of $\Delta[n]$ is made up of vertices of the geometrical n-simplex Σ_n in Hilbert space. Clearly, $\Sigma_n \in \Delta[n]$. The face and degeneracy operators of the semisimplicial complex $\Delta[n]$ are defined as usual as

$$\begin{aligned} F^j [v^{i_0} v^{i_1} ... v^{i_m}] &= [v^{i_0} v^{i_1} ... v^{i_{j-1}} v^{i_{j+1}} ... v^{i_m}] \\ D^j [v^{i_0} v^{i_1} ... v^{i_m}] &= [v^{i_0} v^{i_1} ... v^{i_{j-1}} v^{i_j} v^{i_j} v^{i_{j+1}} ... v^{i_m}] \end{aligned}$$

It is easily seen that an m-simplex of $\Delta[n]$ can be written in a *unique* fashion as

$$D^{i_l} ... D^{i_1} F^{j_1} ... F^{j_k} \Sigma_n$$

for

$$i_1 < ... < i_l, \;\; j_1 < ... < j_k$$

and

$$l - k = m - n$$

(See [Gugenheim 1957, Lemma 1].) A composition of degeneracy and face operators as above is called **semisimplicial operator**. It follows from the above that the semisimplicial complex $\Delta[n]$ is generated from Σ_n by all semisimplicial operators. This motivates the terminology of "generator in dimension n" for Σ_n.

As in singular homology (see [Hilton and Wylie 1965, Section 8.1, page 314]) define the length preserving operators

$$\begin{array}{rcl} V^i : \Delta[n-1] & \to & \Delta[n] \\ \left[v^0 \dots v^{n-1}\right] & \mapsto & [v^0 \dots v^{i-1} v^{i+1} \dots v^n] \end{array}$$

$$\begin{array}{rcl} T^i : \Delta[n] & \to & \Delta[n-1] \\ \left[v^0 \dots v^n\right] & \mapsto & [v^0 \dots v^{i-1} v^i v^i v^{i+1} \dots v^{n-1}] \end{array}$$

by linearity from the above vertex transformations. Observe the following, easily checked, relations:

$$\begin{aligned} V^i \Sigma_{n-1} &= F^i \Sigma_n \\ T^i \Sigma_{n+1} &= D^i \Sigma_n \end{aligned}$$

Not unlike singular homology, we will view a total simplex with $(n+1)$ vertices of B as a semisimplicial mapping

$$b : \Delta[n] \to B$$

This semisimplicial map induces the array $b(\Sigma_n)$. With a slight abuse, we shall sometimes confuse the array $b(\Sigma_n)$ and the map b. This identification is not harmful. Indeed, given any total simplex or array with $(n+1)$ vertices, β_n, there exists an *unique* semisimplicial map $b : \Delta[n] \to B$ such that $\beta_n = b(\Sigma_n)$. Therefore, b and β_n determine each other. Since b is defined to be a semisimplicial map, the face and degeneracy of b are defined as

$$\begin{aligned} F^i b(\sigma) &= b(F^i \sigma) \\ D^i b(\sigma) &= b(D^i \sigma) \end{aligned}$$

To forge the connection with singular homology theory, observe that

$$F^i b(\Sigma_n) = b(F^i \Sigma_n) = b(V^i \Sigma_{n-1})$$

Therefore, $F^i b(\Sigma_n) = b(V^i \Sigma_{n-1})$. The latter is the singular homology definition of the face operator; see [Hilton and Wylie 1965, Section 8.1, pages 314–315]. The same remark applies to the degeneracy operator. Observe that $D^i b(\Sigma_n) = b(D^i \Sigma_n) = b(T^i \Sigma_{n+1})$ from which the singular definition of the degeneracy operator, $D^i b(\Sigma_n) = b(T^i \Sigma_{n+1})$, emerges.

The same remarks applies to the total complex E and the fiber complex F. We shall sometimes confuse an array e_n of $n+1$ vertices of E with the semisimplicial mapping $e_n : \Delta[n] \to E$.

Next we need a precise definition of $\pi^{-1}(\beta_n)$. Define

$$E_b = \{(\sigma, e) : \sigma \in \Delta[n], e \in E, b(\sigma) = \pi(e)\}$$

Observe that E_b, the semisimplicial *fiber over* b, makes the following diagram commute:

$$\begin{array}{ccc} E_b & \xrightarrow{\pi_2} & E \\ \pi_1 \downarrow & & \downarrow \pi \\ \Delta[n] & \xrightarrow{b} & B \end{array}$$

In the above, π_1, π_2 denote the projection on the first, second component, respectively. Such a commutative diagram is called **pull-back** of π by b. It is the semisimplicial analogue of the pull-back 13.1– 13.2 of a bundle by a continuous map. The motivation for the concept of semisimplicial pull-back is that it allows precise definition of the "inverse image." Indeed, consider the set of second components of E_b associated with the first component Σ_n. Clearly, the set $\{x : (\Sigma_n, x) \in E_b\}$ can be identified with $\pi^{-1}(\beta_n) = \{\sigma \in E : \pi(\sigma) = \beta\}$, provided that we identify the map b with the image $\beta_n = b(\Sigma_n)$. However, because E_b carries the entire nested structure generated by the complex of β_n, it contains much more information than $\pi^{-1}(\beta_n)$.

13.2.2 Semisimplicial Bundle

The issue of π being a semisimplicial bundle projection is whether there exists a so-called fiber complex F such that the complex of $\pi^{-1}(\beta_n)$ is isomorphic to $\Delta[n] \times F$, where the latter is defined as

$$\Delta[n] \times F = \{(\sigma_n, f) : \sigma_n \in \Delta[n], f \in F\}$$

It should be stressed that $\Delta[n] \times F$ is merely a set of ordered pairs of simplexes of equal length, that need not be interpreted as the prismatic decomposition of the Cartesian product of the geometrical complex of $\Delta[n]$ and the geometrical complex of F.

Definition 13.16. *Let E, B, F be semisimplicial complexes, called* **total complex, base complex,** *and* **fiber complex,** *respectively, and consider a semisimplicial map $\pi : E \to B$. $(E, \pi, B; F)$ is said to be a* **semisimplicial bundle** *iff π is onto and for every n and every map (total simplex) $b : \Delta[n] \to B$ there exists a strong isomorphism $h(b)$ such that the following diagram commutes:*

$$\begin{array}{ccccc} \Delta[n] \times F & \overset{h(b)}{\to} & E_b & \to & E \\ \downarrow & & \downarrow & & \downarrow \\ \Delta[n] & \overset{1}{\to} & \Delta[n] & \overset{b}{\to} & B \end{array}$$

The isomorphism $h(b)$ is **strong** *in the sense that it is the identity on the first component.*

We also define

$$\begin{array}{rcl} \pi_2 \circ h(b) = g(b) : \Delta[n] \times F & \to & E \\ (\sigma, f) & \mapsto & g(b)(\sigma, f) \end{array}$$

The set of transformations $\{g(b) : b \in B\}$ is called the **atlas**.

We look at the effect of the degeneracy operator on the atlas, namely, $g(D^i b)$. The latter is defined by the following commutative diagram,

$$\begin{array}{ccccc} & \rightarrow & \stackrel{g(D^i b)}{\rightarrow} & \rightarrow & \\ \uparrow & & & & \downarrow \\ \Delta[n+1]\times F & \stackrel{h(D^i b)}{\rightarrow} & E_{D^i b} & \rightarrow & E \\ \downarrow & & \downarrow & & \downarrow \\ \Delta[n+1] & \rightarrow & \Delta[n+1] & \stackrel{D^i b}{\rightarrow} & B \end{array}$$

The issue is that, in general, $g(D^i b)(\sigma, f) \neq g(b)(T^i\sigma, f)$. However, the atlas can be *normalized* as

$$g(D^i b)(\sigma, f) = g(b)(T^i\sigma, f)$$

It is easily seen that this can be done. Indeed, first define $g(b)$ on the non-degenerate simplexes b's, and *then* define $g(b)$ on the degenerate simplexes by the right-hand side of the above.

The situation is quite different with the face operator. $g(F^i b)$ is defined by the following commutative diagram:

$$\begin{array}{ccccc} & \rightarrow & \stackrel{g(F^i b)}{\rightarrow} & \rightarrow & \\ \uparrow & & & & \downarrow \\ \Delta[n-1]\times F & \stackrel{h(F^i b)}{\rightarrow} & E_{F^i b} & \rightarrow & E \\ \downarrow & & \downarrow & & \downarrow \\ \Delta[n-1] & \rightarrow & \Delta[n-1] & \stackrel{F^i b}{\rightarrow} & B \end{array}$$

In an attempt to relate $g(F^i b)$ to $g(b)$ we have to relate the n-dimensional and $(n-1)$-dimensional models. This is most easily accomplished by means of the following commutative diagram:

$$\begin{array}{cccc} \Delta[n]\times F & \stackrel{h(b)}{\longrightarrow} & E_b & \rightarrow \quad E \\ \uparrow (V^i \times 1) & & \uparrow (V^i \times 1) & \uparrow \\ \Delta[n-1]\times F & \stackrel{h'(F^i b)}{\longrightarrow} & E_{F^i b} & \rightarrow \quad E \end{array}$$

It can be shown that there exists a morphism $h'(F^i b)$ that makes the above diagram commute. However, the crucial issue is that, in general,

$$h(F^i b) \neq h'(F^i b)$$

which is equivalent to

$$g(F^i b)(\sigma_{n-1}, f_{n-1}) \neq g(b)(V^i\sigma_{n-1}, f_{n-1})$$

To achieve equality, the fiber in the left-hand side of the inequality should be transformed. Define the transformation

$$t_i^*(b) = (h(F^i b))^{-1} h'(F^i b)$$

$$t_i^*(b) : \Delta[n-1] \times F \to \Delta[n-1] \times F$$

By an obvious projection on the second argument, we define the transformation:

$$t_i(b) : \Delta[n-1] \times F \to F$$

The two sets of transformations are related as

$$\begin{array}{rcl} t_i^*(b) : \Delta[n-1] \times F & \to & \Delta[n-1] \times F \\ (\sigma, f) & \mapsto & (\sigma, t_i(b)(\sigma, f)) \end{array}$$

The set $\{t_i(b) : b \in B\}$ is called **transformation set** associated with the atlas. With the transformation $t(b)$ at hand, we get

$$g(F^i b)(\sigma_{n-1}, t_i(b)(\sigma_{n-1}, f_{n-1})) = g(b)(V^i \sigma_{n-1}, f_{n-1})$$

We now proceed to the semisimplicial version of the structure group of the bundle.

Definition 13.17. *The* **group complex** *G of the bundle is the group generated by all $t_i(b)$'s such that $t_i^*(b)$ is invertible; this group receives its semisimplicial complex structure from the face and degeneracy operators $F^j t_i(b) = t_i(b)(V^j \times 1)$ and $D^j t_i(b) = t_i(b)(T^j \times 1)$.*

Like continuous bundle, the atlas is not an intrinsic feature of the bundle and is hence nonunique. However, the group G of the bundle is a much more intrinsic object.

The next idea is to carefully choose the atlas in such a way as to afford some simplifications.

Lemma 13.18. *There exists an atlas with some group G such that $t_i^*(b) = 1, \forall b \in B, i > 0$, in other words,*

$$g(F^i b)(\sigma_{n-1}, f_{n-1}) = g(b)(V^i \sigma_{n-1}, f_{n-1}), \quad i > 0$$

Proof. See [Barratt, Gugenheim, and Moore 1959, Lemma 2.5] or [May 1967, page 77, 19.2]. ■

We will assume, from now on, that this particular atlas has been chosen.

13.2.3 Twisted Cartesian Product

Now, we come to the fundamental objective of establishing a chain isomorphism between the Cartesian product $B \times F$ and the total space E.

$B \times F$ is viewed as a semisimplicial complex, graduated as $(B \times F)_n = B_n \times F_n$. Algebraically, $B_n \times F_n$ is to be understood as the set of all ordered pairs of total simplexes of length $(n+1)$ of B and F, respectively. $B \times F$ is an abstract semisimplicial complex relative to the face and degeneracy operators defined as $F^i(b, f) = (F^i b, F^i f), D^i(b, f) = (D^i b, D^i f)$, respectively. Geometrically, $B \times F$ can be thought of as a *prismatically triangulated* Cartesian product, because indeed a pair (β_n, f_n) of total simplexes can be thought of as the "juxtaposition" $([\beta_n], [f_n])$ of the two total simplexes, itself a total simplex of the prismatic decomposition of $B \times F$.

E is viewed as a semisimplicial complex graduated as E_n. Denote by $\{E, \partial\}$ the semisimplicial chain complex of E endowed with the usual face and degeneracy operators, F^i, D^i, respectively, from which the boundary operator is defined as $\partial = \sum_i (-1)^i F^i$.

Define a homomorphism

$$\iota_n : B_n \times F_n \to E_n$$

by

$$\iota(\beta_n, f_n) = g(\beta_n)(\Sigma_n, f_n)$$

We prove that this mapping is one-to-one and hence that ι is an isomorphism. Indeed, take $e \in E_n$ and let $\pi(e) = [b^{i_0}...b^{i_n}]$. Clearly, there exists a *unique* mapping $b : \Delta[n] \to B$ such that $b(\Sigma_n) = [b^{i_0}...b^{i_n}]$. With this specific b consider the following commutative diagram:

$$\begin{array}{ccc} \Delta[n] \times F & \stackrel{h(b)}{\longrightarrow} & E_b \\ \downarrow & & \downarrow \\ \Delta[n] & \stackrel{=}{\longrightarrow} & \Delta[n] \end{array}$$

where $h(b)$ is a strong isomorphism. From this isomorphism property and commutativity of the diagram, there exists a *unique* $f \in F_n$ such that $h(b)(\Sigma_n, f) = (\Sigma_n, e)$. It follows that $g(b)(\Sigma_n, f) = e$. Since $g(b)$ is also an isomorphism, it follows that f is *uniquely* determined by the equation $g(b)(\Sigma_n, f) = e$. Therefore, there exists a *unique* pair $(b, f) \in B_n \times F_n$ such that $\iota(b, f) = e$ and ι is an isomorphism.

The *crucial issue* is whether the isomorphism $\iota : B \times F \to E$ is also a semisimplicial map relative to the respective face and degeneracy operators defined on $B \times F$ and E. To check whether ι is a semisimplicial map, we first examine the interplay between the face operator and the ι map. Assuming that the atlas has been normalized, we easily get the following string:

$$\begin{aligned} D^i \iota(b_n, f_n) &= D^i g(b_n)(\Sigma_n, f_n) \\ &= g(b_n)(D^i \Sigma_n, D^i f_n) \\ &= g(b_n)(T^i \Sigma_{n+1}, D^i f_n) \\ &= g(D^i b)(\Sigma_{n+1}, D^i f_n) \\ &= \iota(D^i b_n, D^i f_n) \end{aligned}$$

Thus, with a normalized atlas, we have the required commutativity between the degeneracy operator and the map,

$$D^i \iota(\beta_n, f_n) = \iota(D^i \beta_n, D^i f_n), \quad i \geq 0$$

However, the situation is more complicated with the face operator. Observe the following string, for $i > 0$,

$$\begin{aligned} F^i \iota(b_n, f_n) &= F^i g(b_n)(\Sigma_n, f_n) \\ &= g(b_n)(F^i \Sigma_n, F^i f_n) \\ &= g(b_n)(V^i \Sigma_{n-1}, F^i f_n) \\ &= g(F^i b_n)(\Sigma_{n-1}, F^i f_n) \\ &= \iota(F^i b_n, F^i f_n) \end{aligned}$$

Thus, with a carefully chosen the atlas, we have

$$F^i \iota(\beta_n, f_n) = \iota(F^i \beta_n, F^i f_n), \;\; i > 0$$

The only thing that remains in the way to make ι a semisimplicial map is the case $i = 0$. As above, consider the string,

$$\begin{aligned} F^0 \iota(b_n, f_n) &= F^0 g(b_n)(\Sigma_n, f_n) \\ &= g(b_n)(F^0 \Sigma_n, F^0 f_n) \\ &= g(b_n)(V^0 \Sigma_{n-1}, F^0 f_n) \\ &= g(F^0 b_n)(\Sigma_{n-1}, t_0(b_n)(\Sigma_{n-1}, F^0 f_n)) \\ &= \iota(F^0 b_n, t_0(b_n)(\Sigma_{n-1}, F^0 f_n)) \end{aligned}$$

Now, define τ, the **twisting function**, as

$$\tau(\beta_n) F^0 f_n = t_0(\beta_n)(\Sigma_{n-1}, F^0 f_n)$$

With this concept the interplay between ι and F^0 becomes

$$F^0 \iota(\beta_n, f_n) = \iota(F^0 \beta_n, \tau(\beta_n) F^0 f_n)$$

The situation for $i = 0$ means that ι is *not* a semisimplicial map for the usual face operators F^i defined on $B \times F$. However, if we define the **twisted face operator** $F^0 \times \tau F^0$, keeping $F^i, i > 0$ unchanged, then ι becomes a semisimplicial map. It remains to check, however, whether $B \times F$ remains a semisimplicial complex after modification of F^0. It is easily seen that all we have to check are the following fiber complex identities:

$$\begin{aligned} \tau F^0 F^j &= F^{j-1} \tau F^0 \\ \tau F^0 D^j &= D^{j-1} \tau F^0 \\ \tau F^0 D^0 &= F^1 D^0 = 1 \end{aligned}$$

To prove, for purpose of illustration, the first equality, observe that $\iota(b, f)$ is in the semisimplicial complex E endowed with the usual face operators. Therefore, we get

$$\begin{aligned} F^{j-1} F^0 \iota(b, f) &= F^0 F^j \iota(b, f) \\ F^{j-1} \iota(F^0 b, \tau(b) F^0 f) &= F^0 \iota(F^j b, F^j f) \\ \iota(F^{j-1} F^0 b, F^{j-1} \tau(b) F^0 f) &= \iota(F^0 F^j b, \tau(F^j f) F^0 F^j f) \end{aligned}$$

Since ι is an isomorphism, it follows that $F^{j-1} \tau(b) F^0 f = \tau(F^j f) F^0 F^j f$, which is exactly what we had to prove.

With the new face operator, we define the **twisted boundary operator** on $B \times F$ as follows:

$$\partial^{\text{twisted}}([\beta],[f]) = ([F^0\beta],[\tau(\beta)F^0 f]) + \sum_{k=1}(-1)^k([F^k\beta],[F^k f])$$

Since this boundary comes from the face of a semisimplicial complex, we have

$$\partial^{\text{twisted}}\partial^{\text{twisted}} = 0$$

Relative to $\partial^{\text{twisted}}$, $B \times F$ is hence a chain complex. Furthermore ι becomes a chain map,

$$\iota\partial^{\text{twisted}} = \partial\iota$$

Putting everything together, we have a chain map isomorphism between chain complexes

$$\{B \times F, \partial^{\text{twisted}}\} \xrightarrow{\cong} \{E, \partial\}$$

The Cartesian product $B \times F$, when it is endowed with the boundary operator $\partial^{\text{twisted}}$, is called **twisted Cartesian product**.

13.2.4 Example

We illustrate the twisting phenomenon on the example of Figure 13.2 that shows a simplicial version of the *principal bundle* of the Möbius strip of Figure 4.4.

To be more explicit, the principal bundle of the Möbius strip consists of the boundary of the Möbius strip that projects onto the base space S^1 and provides a double covering of S^1. The fiber is clearly discrete and consists of two points. Therefore, the principal bundle of the Möbius strip is a covering space of S^1 in the sense of Section 3.8.

Coming back to the semisimplicial version shown in Figure 13.2, observe that the preimage of a simplex of the complex B splits into two disconnected simplexes so that the fiber complex $F = \{f^0, f^1\}$ consists of two vertices. This is an example of a semisimplicial covering map, a concept due to Reidemeister (see [Dieudonné 1989, pages 120–121]).

To see what the twisting is, we construct the isomorphism ι. The basic exercise here is to assign to every simplex of E a pair consisting of its projection on B and a vertex of F, to be determined. The following table shows one such assignment:

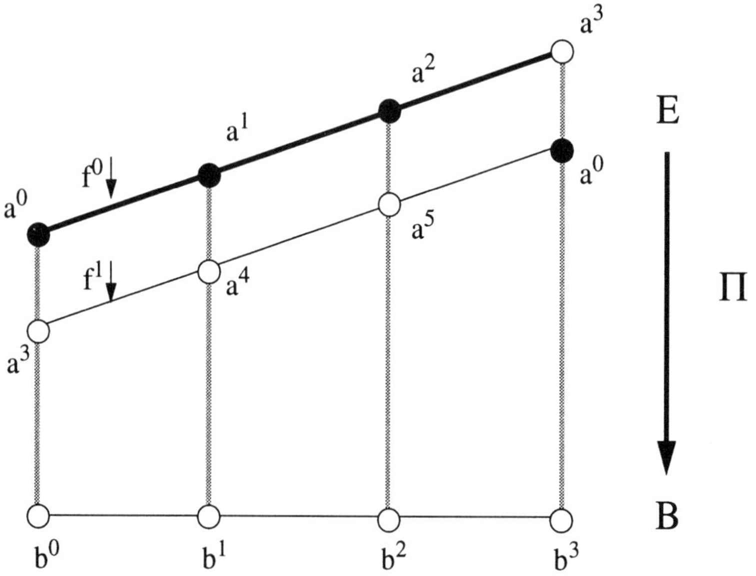

Fig. 13.2. Semisimplicial version of the principal bundle of the Möbius strip. Thick lines and dark vertices correspond to the vertex f^0 of the fiber complex; thin lines and circled vertices correspond to vertex f^1 of fiber complex $F = \{f^0, f^1\}$.

b_n	f_n	$\iota(b_n, f_n)$
$[b^0b^1]$	$[f^0f^0]$	$[a^0a^1]$
$[b^1b^2]$	$[f^0f^0]$	$[a^1a^2]$
$[b^2b^0]$	$[f^0f^0]$	$[a^2a^3]$
$[b^0b^1]$	$[f^1f^1]$	$[a^3a^4]$
$[b^1b^2]$	$[f^1f^1]$	$[a^4a^5]$
$[b^2b^0]$	$[f^1f^1]$	$[a^5a^0]$
$[b^0]$	$[f^0]$	$[a^0]$
$[b^1]$	$[f^0]$	$[a^1]$
$[b^2]$	$[f^0]$	$[a^2]$
$[b^0]$	$[f^1]$	$[a^3]$
$[b^1]$	$[f^1]$	$[a^4]$
$[b^2]$	$[f^1]$	$[a^5]$

Of course this assignment is not unique, but all such assignments share one common feature: It is not possible to do the assignment such that for an arbitrary simplex $([b^ib^j], [f^kf^k])$ of E, the pair (b^j, f^k) is in the boundary of $([b^ib^j], [f^kf^k])$. To be more specific, consider the figure and the simplex $[a^5a^0] \cong ([b^2b^0], [f^1f^1])$. b^0 is clearly in the boundary of $[b^2b^0]$; however, $(b^0, f^1) \cong a^3$ is not in the boundary of $([b^2b^0], [f^1f^1]) \cong [a^5a^0]$. The reason is that we have to connect the rightmost vertex of the *bottom* sheet (f^1) to the leftmost vertex of the *top* sheet (f^0).

To relate these features with the twisting function, observe that

$$F^0\iota([b^2b^0],[f^0f^0]) = F^0[a^2a^3] = a^3$$

while

$$\iota(F^0[b^2b^0], F^0[f^0f^0]) = \iota(b^0, f^0) = a^0$$

Hence $F^0\iota([b^2b^0],[f^0f^0]) \neq \iota(F^0[b^2b^0], F^0[f^0f^0])$ and the twisting function is easily seen to be a permutation of (f^0, f^1).

13.2.5 Cross Section

We provide a semisimplicial version of the concept of cross section:

Definition 13.19. *A* **pseudo-cross section** *in a semisimplicial map* $\pi : E \to B$ *is a map* $c : B \to E$ *such that* $\pi \circ c = 1_B$, $F^i c = cF^i, i > 0$, *and* $D^i c = cD^i, i \geq 0$. *If, in addition,* $F^0 c = cF^0$, *then* c *is called* **cross section**.

Observe that the concept of (pseudo-) cross section is not restricted to semisimplicial bundles. As such, it is directly applicable to the semisimplicial approximation to the Nyquist map. However, to *construct* a pseudo-cross section, it is convenient to proceed from the twisted Cartesian product formulation of a semisimplicial bundle (see Section 13.5 how to reduce an arbitrary continuous map to a semisimplicial bundle). Clearly, a cross section exists if there is no twisting.

13.2.6 Commutativity

Now, we are in a position to come back to the issue of commutativity between the inverse image and the boundary in the semisimplicial setting.

Clearly, ι takes a simplex β_n of the Nyquist template together with a simplex f_n of the fiber and associates with this pair a unique simplex of the polyhedron of uncertainties in the preimage of β_n. From that point of view, ι does the inverse image operation, with the fiber complex parameterizing the preimage.

To be more specific, let $\beta_{n_p} = b(\Sigma_{n_p})$ be a total n_p-simplex of the Nyquist template. Because the following diagram commutes

$$\begin{array}{ccc} \Delta[n_p] \times F & \overset{g(b)}{\to} & E \\ \downarrow & & \downarrow \pi \\ \Delta[n_p] & \overset{b}{\to} & B \end{array}$$

we have

$$\pi(g(b)(\Sigma_{n_p}, f_{n_p})) = b(\Sigma_{n_p})$$

that is,

$$\pi(\iota(\beta_{n_p}, f_{n_p})) = \beta_{n_p}$$

In other words, for any $f_{n_p} \in F_{n_p}$, $\iota(\beta_{n_p}, f_{n_p})$ is an n_p-simplex of E such that $\pi(\iota(\beta_{n_p}, f_{n_p})) = \beta_{n_p}$. Therefore, if we define $\pi^{-1}(\beta_{n_p})$ to be the set of n_p-simplexes of E mapped onto β_{n_p}, we get

$$\pi^{-1}(\beta_{n_p}) = \cup_{f_{n_p} \in F_{n_p}} \iota(\beta_{n_p}, f_{n_p})$$

Now, we can write down a result related to commutativity between face operator and inverse image.

Theorem 13.20. *Let $E \xrightarrow{\pi} B$ be a semisimplicial bundle with fiber complex F. Let $\pi^{-1}(\beta_{n_p})$ be the collection of n_p-simplexes of E mapped onto β_{n_p}. Then, for $i > 0$ and provided that any $(n_p - 1)$-simplex of F is the face of an n_p-simplex of F, we have $F^i\pi^{-1}(\beta_{n_p}) = \pi^{-1}(F^i\beta_{n_p})$.*

Proof.

$$\begin{aligned} F^i\pi^{-1}(\beta_{n_p}) &= F^i \cup_{f_{n_p} \in F_{n_p}} \iota(\beta_{n_p}, f_{n_p}) \\ &= \cup_{f_{n_p} \in F_{n_p}} F^i\iota(\beta_{n_p}, f_{n_p}) \\ &= \cup_{f_{n_p} \in F_{n_p}} \iota(F^i\beta_{n_p}, F^i f_{n_p}) \\ &= \cup_{f_{n_p-1} \in F_{n_p-1}} \iota(F^i\beta_{n_p}, f_{n_p-1}) \\ &= \pi^{-1}(F^i\beta_{n_p}) \end{aligned}$$

■

The lesson we learn from this theorem is that there are two features that limit the commutativity between the inverse image and the face: on the one hand, the *twisting of the fiber*, and, on the other hand, any "hole" in the fiber space.

13.2.7 Twisted Tensor Product

We have associated with E a twisted *Cartesian product* complex $\{C_*(B \times F), \partial^{\text{twisted}}\}$ from which the homology of E can be computed from the basis, the fiber, and the twisting. However, in the spirit of the Eilenberg–Zilber theorem, it turns out that the homology of E can also be computed from a twisted *tensor product* complex $\{C_*(B) \otimes C_*(F), \partial^{\otimes,\text{twisted}}\}$. The key result is the following theorem:

Theorem 13.21. (Brown) *The twisted Cartesian product complex $\{C_*(B \times F), \partial^{twisted}\}$ is chain homotopy equivalent to the twisted tensor product complex $\{C(B) \otimes C(F), \partial^{\otimes,twisted}\}$.*

Proof. See [May 1992, Chapter VI, Section 31]. ■

The complex $C(B) \otimes C(F)$ is smaller than the complex of $C(B \times F)$. Indeed, in the former, the formal product of two simplexes—a prism—is kept as an acceptable entity, whereas in the latter the prism is decomposed into simplexes using the prismatic triangulation. Clearly, $C(B\times F)$ contains much more objects than $C(B) \otimes C(F)$. One would hence expect some simplification from the utilization of $C(B) \otimes C(F)$ instead of $C(B \times F)$.

In Section 14.7 we will put these results into use to develop a spectral sequence revealing the relation between the uncertainty, the template, and the crossover of a Nyquist problem.

13.3 Nyquist Fibration

The concept of fiber bundle, in both its continuous and semisimplicial versions, has allowed the clarification of several issues in robust stability. However, it is limited, because fiber bundle is a very restrictive concept. The purpose of the remaining sections of this chapter is to show that, up to homotopy equivalence, the concept of *semisimplicial* bundle is, actually, very general.

Specifically, in this section, we introduce the concept of *fibration* that applies to a continuous map. We relegate its semisimplicial analogue to the following section.

We approach the concept of fibration in the spirit of the theory of fiber bundles. Let P be the polyhedron of uncertainties considered as the "total space" while the Nyquist template is considered as a "base space." The polyhedron P is considered as lying *over* N, the Nyquist map being considered as a projection:

$$\begin{array}{c} P \\ \downarrow f \\ N \end{array}$$

An important concept in homotopy theory is that of *curve lifting.* Assume that we trace a curve $[0,1] \to N$ in the Nyquist template. This curve is required to start at a preassigned point of N for $t=0$. To indicate that we are soon to vary the starting point of the curve, we write it as a mapping $g : z_0 \times [0,1] \to N$. $g_0(z_0)$ is the initial point of the curve in N, and z_0 is soon to be allowed to run in the space Z. Putting everything together, we get the diagram

$$\begin{array}{ccc} & & P \\ & & \downarrow \\ z_0 \times [0,1] & \to & N \end{array}$$

The question is whether the curve traced in N can be lifted to a curve in P starting at a certain preassigned initial point. To define the starting point in P, we define a mapping $h : z_0 \to P$, the image of which is the initial condition. Of course $f(h(z_0))$ must be the starting point of the curve in N. The situation is summarized in the following diagram:

$$\begin{array}{ccc} z_0 \times 0 & \rightarrow & P \\ i \downarrow & & \downarrow \\ z_0 \times [0,1] & \rightarrow & N \end{array}$$

The curve lifting issue is whether there exists a homotopy: $z_0 \times [0,1] \rightarrow P$ such that the following diagram commutes:

$$\begin{array}{ccccc} z_0 \times 0 & \rightarrow & P \\ \downarrow & \overset{?}{\nearrow} & \downarrow \\ z_0 \times [0,1] & \rightarrow & N \end{array}$$

The lifting may or may not exist. In case f is a *covering map*, the lifting *exists* and is *unique*.

We now vary the initial point of the curve traced in N; in other words, we allow $z_0 \in Z$, where Z is some topological space. Not only do we want liftability of a single curve, but we also want liftability of a family of curves parameterized by the initial condition $z_0 \in Z$. This is the concept of fibration.

Definition 13.22. *A map $f : P \rightarrow N$ is said to be a* **fibration** *or have the* **homotopy lifting property** *iff for any topological space Z, homotopy $g : Z \times [0,1] \rightarrow N, (z,t) \mapsto g_t(z)$, and map $h : Z \times 0 \rightarrow P$ such that $fh = g_0$, the homotopy g_t can be lifted—that is, there exists a map $Z \times [0,1] \rightarrow P$ such that the following diagram commutes:*

$$\begin{array}{ccc} Z \times 0 & \overset{h}{\rightarrow} & P \\ \downarrow & \overset{?}{\nearrow} & \downarrow f \\ Z \times [0,1] & \overset{g}{\rightarrow} & N \end{array} \tag{13.3}$$

If the map f has the homotopy lifting property whenever Z is an arbitrary polyhedron, then f is said to be a **Serre fibration**.

Theorem 13.23. *If $f : P \rightarrow N$ is a fibration, then any two fibers P_a, P_b, $a, b \in N$, have the same* **homotopy type**; *that is, there are maps $\alpha : P_a \rightarrow P_b$ and $\beta : P_b \rightarrow P_a$ such that $\alpha\beta$ is homotopic to 1_{P_b} and $\beta\alpha$ is homotopic to 1_{P_b}.*

Corollary 13.1. *If $f : P \rightarrow N$ is a fibration, then for any two fibers P_a, P_b, we have $\pi_n(P_a) = \pi_n(P_b)$.*

To get an intuitive "mental picture" of the concept of fibration, consider Figure 13.3 that shows the orthogonal projection of a circle on a line. The most direct way to convince oneself that this projection is *not* a fibration is to observe that the fibers P_{b^0}, P_{b^1} consist each of one single point, while the other fibers consist of two points. Clearly, the fibers are *not* homotopically equivalent and hence the projection is *not* a fibration. More subtle is the fact that the curve lifting is not continuous relative to z when the "starting

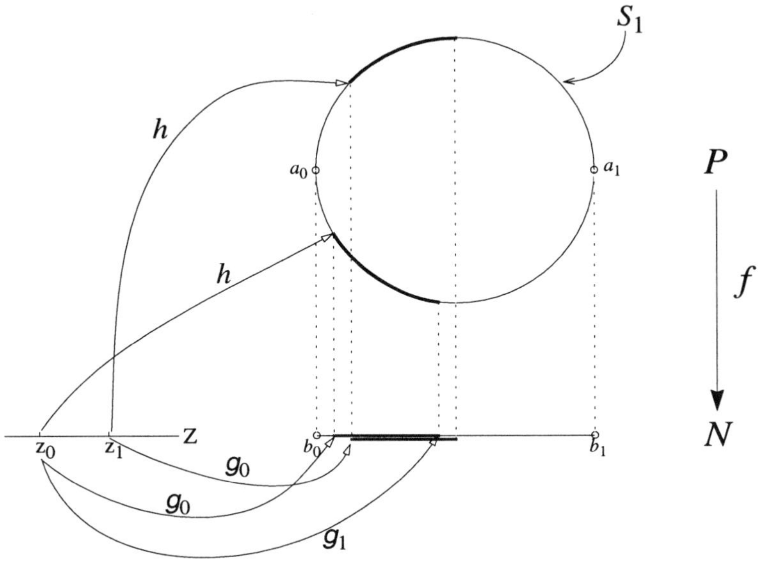

Fig. 13.3. Illustration of lack of fibration property of the orthogonal projection of a circle on a line.

point" $h(z)$ in P is close to the critical point a^0 (or a^1 for that matter). Actually, the lack of fibration property is due to the critical points a^0, a^1. Observe that after removal of the critical points/values, the resulting map $S^1\backslash\{a^0, a^1\} \to b^0b^1$ *is* a fibration (remember b^0b^1 denotes the *open* simplex).

Last but not least, we have to justify how it can be assumed that a Nyquist map is a fibration.

Theorem 13.24. *A map $f : P \to N$ is homotopically equivalent to a fibration; that is, there exists a fibration $\pi_f : E_f \to N$ and a homotopy equivalence $s : P \to E_f$ such that the following diagram commutes:*

$$\begin{array}{ccc} P & \xrightarrow{s} & E_f \\ f\downarrow & & \downarrow \pi_f \\ N & = & N \end{array} \tag{13.4}$$

Proof. We follow the traditional approach; see [Whitehead 1978, page 43], [Spanier 1989, Chapter 2, Section 8], and [McCleary 1985, page 105]; see also [Bredon 1993, page 457] for a slightly different approach. With the template N, we construct the map

$$\begin{array}{rcl} \pi^{(0)} : N^{[0,1]} & \to & N \\ \gamma & \mapsto & \gamma(0) \end{array}$$

This mapping, which assigns to a curve in N its initial point, is easily seen to be a fibration. Bringing $\pi^{(0)}$ and f together in the same diagram yields

$$\begin{array}{ccc} & & N^{[0,1]} \\ & & \downarrow \pi^{(0)} \\ P & \xrightarrow{f} & N \end{array}$$

Next, consider the pull-back of $\pi^{(0)}$ by f,

$$\begin{array}{ccc} E_f & \xrightarrow{\pi_1} & N^{[0,1]} \\ \pi_2 \downarrow & & \downarrow \pi^{(0)} \\ P & \xrightarrow{f} & N \end{array}$$

where

$$E_f = \{(\gamma, p) \in N^{[0,1]} \times P : \gamma(0) = f(p)\}$$

E_f can be thought of as the *curve lifting data* of $f : P \to N$. π_1 is the projection on the first factor. π_2, the projection on the second factor, is the **pull-back** of the fibration $\pi^{(0)}$ over the map f; it is a general feature that the pull-back of a fibration over a map is a fibration, and, as such, π_2 is a fibration. Next, we define a homotopy inverse for π_2. Define the section

$$\begin{array}{rcl} s : P & \to & E_f \\ p & \mapsto & (\gamma_{f(p)}, p) \end{array}$$

where $\gamma_{f(p)}$ is the constant curve at $f(p)$. From the definitions, it is easily seen that $\pi_2 s = 1_P$ and $s\pi_2 \simeq 1_{E_f}$ so that π_2, s are homotopic inverses of each other. Now we define the projection:

$$\begin{array}{rcl} \pi_f : E_f & \to & N \\ (\gamma, p) & \mapsto & \gamma(1) \end{array}$$

We leave it to the reader to verify that π_f is a fibration; for a proof, see [Whitehead 1978, Theorem 7.3, page 42]. Furthermore, if we define the map $\pi^{(1)}$ that assigns to a curve its end point, it is easily seen that π_f makes the following diagram commutative:

$$\begin{array}{ccc} E_f & \xrightarrow{\pi_1} & N^{[0,1]} \\ s \uparrow & \overset{\pi_f}{\searrow} & \downarrow \pi^{(1)} \\ P & \xrightarrow{f} & N \end{array}$$

The lower triangle of the above is exactly the same as Diagram 13.4; furthermore, π_f is a fibration and s has homotopic inverse π_2. Hence the theorem is proved. ■

The spaces P and E_f are homotopically equivalent; to be more specific, E_f can be retracted onto $s(P)$; indeed, it suffices to retract all curves in the first factor of E_f to their initial points. Now, let us see what can be said about the preimages. Take an arbitrary point $b \in N$; in the most typical situation we would take $b = 0 + j0$. From Diagram 13.4, we define two different preimages: the "fiber" $f^{-1}(b) = P_b$ (for $b = 0 + j0$ this is the crossover X_ω) and the fiber $\pi_f^{-1}(b) = (E_f)_b$. Because π_f is a fibration, all $(E_f)_b : b \in N$ are homotopically equivalent. However, since f need not be a fibration, the "fibers" $P_b : b \in N$ need not be all homotopically equivalent. The reader can easily supply simple examples of this fact; besides, this is the basic theme of Morse theory. The question, therefore, is whether anything can be said about the homotopy properties of P_b versus the homotopy properties of $(E_f)_b$. By commutativity of Diagram 13.4, it follows that

$$s(P_b) \subseteq (E_f)_b, \quad \forall b \in N$$

It follows that P_b is homotopically equivalent to a *subset*, $s(P_b)$, of $(E_f)_b$. Unfortunately, $(E_f)_b$ cannot in general be retracted to $s(P_b)$; indeed, the curves in the first factor of $(E_f)_b$ have b as their common end point and hence these curves cannot be retracted to their initial points without "ripping apart" their common tail. Therefore, $(E_f)_b$ can be thought of as a big space, the subsets of which generate, up to homotopy equivalence, the preimages P_b.

A dual approach using the concept of mapping cylinder and cofibration is in [Bott and Tu 1982, page 249]. (The semisimplicial counterpart of these results are in [May 1967, Theorem 10.10].)

13.4 Semisimplicial Fibration

In this section, we develop a concept, more general than that of semisimplicial bundle, the so-called *semisimplicial fibration*. It is the semisimplicial analogue of the concept of fibration introduced in the previous section. We also show how any continuous map can be reduced, up to homotopy equivalence, to a semisimplicial bundle.

Definition 13.25. *(E, π, B) is said to be a* **semisimplicial fibration** *if π is onto and if, for an arbitrary dimension n, the following condition holds: Given an integer k, $0 \leq k \leq n$, simplexes $e_i \in E_{n-1}, \forall i \neq k$, and $b \in B_n$ such that*

$$\begin{aligned} F^i e_j &= F^{j-1} e_i; \; i < j; \; i,j \neq k \\ \pi e_i &= F^i b \end{aligned}$$

there exists a simplex $e \in E_n$ such that

$$\begin{aligned} e_i &= F^i e, \; i \neq k \\ b &= \pi e \end{aligned}$$

Furthermore, if $\pi(d) = \pi(e)$ together with $F^i d = F^i e, \forall i \neq k$ implies that $d = e$, then the fibration is said to be **minimal**.

The concept of semisimplicial fibration is the combinatorial analogue of the concept of Serre fibration introduced earlier.

Semisimplicial fibrations crop up quite naturally as follows: Invoke the previous section to reduce up to a homotopy equivalence a continuous map to a Serre fibration; then it can be shown that the induced map of the singular chain complex of the total space to the singular chain complex of the base space is a semisimplicial fibration (see [Barratt, Gugenheim, and Moore 1959]).

Next, we show that by a deformation retract a semisimplicial fibration can be made minimal.

Theorem 13.26. *Let (E, π, B) be a semisimplicial fibration. Then there exists a subcomplex $E' \subseteq E$, a map $\pi' = \pi|E'$, and a minimal fibration (E', π', B) that is a strong deformation retract of (E, π, B).*

Proof. See [Barratt, Gugenheim, and Moore 1959, Theorem 4.2] or [May 1992, page 41, Theorem 10.10]. ■

Finally, this last theorem substantially enlarges the domain of applicability of semisimplicial bundles:

Theorem 13.27. *Every minimal semisimplicial fibration with a connected base is a semisimplicial fiber bundle with a Kan fiber complex.*

Proof. See [Barratt, Gugenheim, and Moore 1959, Proposition 2.2] or [May 1967, page 46, Theorem 11.11]. ■

13.5 Summary

The (long!) process of going from an arbitrary continuous map to a semisimplicial bundle is illustrated in the following diagram:

$$
\begin{array}{rcl}
D \times \Omega & \xrightarrow{f} & N \\
\downarrow & & \downarrow \text{ homotopy equivalence} \\
E_f & \xrightarrow{\text{fibration}} & N \\
\downarrow & & \downarrow \text{ go to singular complex} \\
E = C^{\text{sing}}(E_f) & \xrightarrow{\text{sing chain = s-s fibr}} & C^{\text{sing}}(N) = B \\
\downarrow & & \downarrow \text{ deformation retract} \\
E' & \xrightarrow{\text{s-s bundle}} & B
\end{array}
$$

BIBLIOGRAPHICAL AND HISTORICAL NOTES

The concept of fiber bundle took its final shape in the classic by [Steenrod 1951]. However, prior to this, the concept of covering map (a bundle with discrete fiber) had been around. Following [Dieudonné 1989, pages 120–121], it was apparently Reidemeister who in the 1930s formulated the concept of "twisting" in the framework of simplicial covering map (the twisting is in this case a permutation of the fiber).

The *semi*simplicial theory grew up out of the work by [Eilenberg and Zilber 1950] on the theorem that bears their names.

The ground-breaking work by [Kan 1955a,b] on the combinatorial reformulation of homotopy groups (see Section 15.7) clearly demonstrated how the semi*cubical* theory can be put to work.

The work of Kan was reformulated within the semi*simplicial* setting by [Gugenheim 1957] and [Gugenheim and Moore 1957].

The concept of semisimplicial bundles is apparently due to [Barratt, Gugenheim, and Moore 1959]. The recently re-edited monograph by [May 1967] brought under the same cover a variety of results that had appeared scattered through the literature. There, the semisimplicial theory is conceptualized in a categorical setup. We define the *semisimplicial category*: its objects are sets of integers $\{0, 1, ..., n\}$ and its morphisms are monotone increasing maps. Viewing V^i, T^i as maps between the set of *indices* of the vertices v^i, it is clear that V^i, T^i are morphisms of the semisimplicial category and any morphism in the semisimplicial category can be factored as a product of V^i's, T^i's. A contravariant functor from the semisimplicial category to any category is called **simplicial object** and endows the category with the semisimplicial structure. A contravariant functor from the semisimplicial category to a category of sets is called **simplicial set**. The face and degeneracy operators F^i, D^i are images of V^i, T^i under the appropriate functor. This categorical formulation of the semisimplicial theory has received widespread acceptance (see [Gillet 1992], [Weibel 1994], and [McCleary 1985]).

Fiber bundles have already received control applications; for example, the mapping $(A, B, C) \mapsto C(sI - A)^{-1}B$, where (A, B, C) is an n-D minimal realization, can be shown to be a principal bundle with $GL(n, \mathbb{R})$ as structure group (see [Delchamps and Byrnes 1982]). However, the applications of fiber bundles and semisimplicial bundles to Nyquist maps of robust stability problems, as developed in this chapter, are, to our knowledge, new.

14

SPECTRAL SEQUENCE OF NYQUIST MAP

SUMMARY AND MOTIVATION

A *spectral sequence* is an algebraic object: It is a sequence

$$..., E^r, E^{r+1}, ...$$

of bigraded groups or modules,

$$E^r = \{E^r_{u,v} : u, v \in \mathbb{Z}\}$$

Each group or module is endowed with a boundary endomorphism

$$\partial^r : E^r \to E^r, \; \partial^r \partial^r = 0$$

and the *specific* feature of a *spectral* sequence is that each group or module is the homology of its predecessor,

$$E^{r+1} = \ker(\partial^r) / \operatorname{im}(\partial^r)$$

It is further required that the sequence converges after a finite number of steps; the steady-state group or module is written, with a slight abuse, as E^∞.

Spectral sequences provide a formalization of a fairly general successive approximation procedure to compute (co)homology groups. Spectral sequences also appear in a variety of algebraic problems. Indeed, many computational algebra problems are homology computations in disguise; for example, the algebraic system theory problem of computing the strongly reachable subspace can be recast as a homology computation and the corresponding spectral sequence appears to be the well-known geometric subspace recursion or, equivalently, Silverman's dual structure algorithm (see [Kitapci, Jonckheere, and Silverman 1986]).

Closer to the robust stability problem, spectral sequences occur in a variety of topological situations and have been instrumental in the derivation of some of the most spectacular results of algebraic topology. The widespread application of spectral sequences (see [McCleary 1985]) has, however, served to obscure the fundamental motivation for which they were introduced in the first place. We owe it to [Dieudonné 1989, page 141] to have recast spectral sequences in the proper historical context in which they were introduced by Leray soon after World War II. Indeed, the original objective of spectral sequences is no more esoteric than to exhibit, in a recursive fashion, relations between the (co)homology groups of the domain of definition, the image, and the "fiber" $f^{-1}(0+j0)$ of such a map as $f : D \times \Omega \to N$.

A forerunner to Leray's spectral sequence is a 1927 result by Vietoris that reads as follows: Consider compact metric spaces P, N and let $f : P \to N$ be a continuous map. Assume the lower-order Vietoris homology groups of the fiber vanish, $H_i(f^{-1}(z)) = 0, 1 \leq i \leq n, \forall z \in N$. Then Vietoris' result asserts that f induces *isomorphisms* $f_i^* : H_i(P) \to H_i(N)$. However, when the fiber has nontrivial lower homology groups, or even worse when the homology of the fiber changes with z, the interplay between the homology of P, N and the fiber immediately reaches higher levels of complexity.

To grasp the concept of a spectral sequence, it is convenient to proceed algebraically, even though the underlying problem is topological. If we define the formal group $G := \bigoplus_n C_n(D \times \Omega)$, it appears naturally decomposed into a direct sum of disjoint subgroups. Such a decomposition is called *graduation*. The usual boundary operator ∂_n can trivially be extended to an endomorphism $\partial : G \to G$, $\partial\partial = 0$ and the homology of the group is defined as $H(G) = \ker(\partial)/\operatorname{im}(\partial)$. The *novel* concept in a spectral sequence is that, on top of the graduation of the group, we introduce a (increasing) *filtration*—that is, a (increasing) sequence of subgroups starting at 0 and ending at G. Filtrating the graduation yields a host of subgroups indexed by two integers—one for the graduation, the other for the filtration—and immediately brings the problem in the realm of *bigraded* groups. The spectral sequence can be defined as a sequence of successive approximations of $H(G)$. To be more specific, in case of a *field* of coefficients, $\bigoplus_{u+v=n} E^{\infty}_{u,v} = H_n(G)$, the short notation of which is $E^r \Rightarrow H(G)$. The approximations are constructed as follows: With the filtration, we can define a sequence of "approximate" cycles—that is, a sequence of "cycles" such that their boundaries converge to the head of the filtration; we can also define a sequence of "approximate" boundary groups that bound toward the tail of the filtration. With these "approximate" cycle and boundary groups, we define "approximate" homology groups that converge to the exact homology as $r \to \infty$.

To go from algebra to topology—that is, to make spectral sequences of any use in such a topological problem as robust stability—the whole difficulty is to correctly set up the filtration, *consistently with the underlying topological problem of unraveling the relations between the homology of* $P = D \times \Omega$, *the homology of* N, *and the properties of the map.* Furthermore, the objective of a spectral sequence is not only to secure $E^r \Rightarrow H(P)$, but to set up the spectral sequence in such a way that the early term of the spectral sequence, typically E^2, provides a *meaningful*, easily computable approximation of $H(P)$, so that we know "how to start." At this stage, there are many ways to proceed, resulting in a great many possible spectral sequences.

Probably the most natural way to grasp the filtration problem is to go back to the history of the problem—as compiled by [Dieudonné 1989]—that starts with the work of Leray on sheaf cohomology and spectral sequences as tools for investigating the cohomology properties of such a map as $f : P \to N$. First, we have to define a cohomology. Simplicial cohomology is not quite appropriate for an arbitrary continuous Nyquist map because indeed the preimage of a simplex of the template—assuming it is triangulated—is not, in general, a simplex of the uncertainty space. Leray defined a cohomology for P that requires nothing more than the underlying topological space structure: We are given an *abstract* cochain complex K^* and we link it to P by defining the *support* $\mathrm{Supp}_P(\cdot)$ that takes any cochain c_K to a subset of P. The pair (K^*, Supp_P) is called *concrete* complex. A cohomology can be computed from the abstract complex using standard algebra; however, since there are *many* such complexes, it takes some work to define a *unique* cohomology from a great many abstract cochain complexes or single out *one* such complex so that the resulting cohomology is isomorphic to the Alexander-Spanier cohomology. The essence of a spectral sequence is the following: Rather than computing the cohomology of P *directly* from the cochain complex K^*, to highlight the role of P as the domain of the Nyquist map, we prefer to compute $H^*(P)$ from the cohomology of N and the properties of f. To this effect, we choose an abstract cochain complex L^* for N such that any cochain of L^* has a support that is a subset of N. Clearly, we have a (decreasing) filtration of the cochain group for N as

$$\ldots \supseteq \bigoplus_{i \geq u} L^i \supseteq \bigoplus_{i \geq u+1} L^i \supseteq \ldots$$

Next, we define the concrete complex $(f^{-1}(L^*), \mathrm{Supp}_P)$ as follows: Its abstract cochain complex is still L^*; however, its support is defined in P; to be more specific, take $c_L \in L^*$; then $\mathrm{Supp}_P(c_L) = f^{-1}(\mathrm{Supp}_N(c_L))$. This in turn induces a filtration of the cochain complex K^* of the uncertainty,

$$\ldots \supseteq K \circ f^{-1}(\bigoplus_{i \geq u} L^i) \supseteq K \circ f^{-1}(\bigoplus_{i \geq u+1} L^i) \supseteq \ldots$$

In the above, $K \circ f^{-1}(L^*)$ is a new concrete complex; its abstract complex is $K^* \otimes f^{-1}(L^*)$; its coboundary is $\delta(c_K^i \otimes c_L^j) = (\delta c_K^i) \otimes c_L^j + (-1)^i c_K^i \otimes (\delta c_L^j)$; and the support of a cochain $c_K \otimes c_L$ is the intersection of the support of c_K and c_L in P. With the above filtration of the cochain of the uncertainty, the algebra is set up, so that we can start running the spectral sequence until it converges to the cohomology of P.

The identification of the E_2 term of the Leray spectral sequence leads, quite naturally, to the modern concept of *sheaf cohomology*. A **(pre)sheaf** on a topological space, say N, is a rule that assigns to each subset $Y \subseteq N$ an Abelian group or a module $S(Y)$, typically $H^*(f^{-1}(Y))$, subject to the following conditions:

- For any $Y_1 \subseteq Y_2$, there exists a morphism $S^{Y_1,Y_2} : S(Y_2) \to S(Y_1)$;
- $S^{Y,Y} = 1_{S^Y}$;
- For $Y_1 \subseteq Y_2 \subseteq Y_3$, $S^{Y_1,Y_3} = S^{Y_1,Y_2} \circ S^{Y_2,Y_3}$.

It turns out that the E_2 term of the spectral sequence is the cohomology of N with coefficients in the sheaf $\{S(Y) = H^*(f^{-1}(Y)) : Y \subseteq N\}$. To be more specific, define the cochain group $C^*(N, \{S(Y) : Y \subseteq N\})$ of N with coefficients in the sheaf S as the group of all linear combinations of the form

$$\sum s_n g_N^n$$

where g_N^n is a generator in dimension n of the concrete cochain complex (L^*, Supp) of N and the coefficient s_n assigns to any $Y \subseteq N$, subject to the condition that $Y \supseteq \text{Supp}(g_N^n)$, an element $s_n(Y) \in S(Y)$. The coboundary of this cochain is defined as

$$\delta^n(\sum s_n g_N^n) = \sum s_n \delta^n g_N^n$$

The cochain complex $\{C^*(N,S), \delta^*\}$ has a cohomology which is the precise definition of what we referred to as *cohomology of N with coefficients in* $\{H^*(f^{-1}(Y)) : Y \subseteq N\}$. It turns out that the E_2 term of the Leray spectral sequence is

$$E_2^{u,v} = H^u(N, \{H^v(f^{-1}(Y) : Y \subseteq N\})$$

Despite the conceptual appeal of this formula, computation of sheaf cohomology is clumsy and the cohomology of the fiber $f^{-1}(0 + j0)$ does not appear in the formula.

The breakthrough came with Serre, who managed to get the cohomology of the fiber *clearly* into the picture. To accomplish this task, Serre, however, had to restrict himself to *fibrations* (see Chapter 13) together with *singular cubical* homology. In a few words, a *singular cube* is the image of a standard cube under a continuous map. The uth subset of the filtration of P is defined as that consisting of those singular cubes that have a $(n_p - u)$-dimensional subcubes mapped into the fiber $f^{-1}(0 + j0)$. Using this approach, Serre was able to derive results, much neater than those of Leray. In particular the E^2 terms take a very simple form; since the map is a fibration, the fibers $f^{-1}(z)$ are all homotopically equivalent and their homology $H_*(f^{-1}(z)$ does not depend on z and the E^2 term is just the homology of N with coefficients in the homology of the fiber

$$E_{u,v}^2 = H_u(N, H_v(f^{-1}(z)))$$

The problem of course is that the topological Nyquist map is not always a fibration.

Consistently with our overall objective of remaining within *simplicial* theory, here we shall focus on two "simplicial" versions of the homology spectral sequence of such a continuous map as $f : D\times\Omega \to N$. In presenting these two spectral sequences, we somehow follow chronological order. The first spectral sequence—the Zeeman dihomology spectral sequence—follows in the footstep of the fundamental work of Leray. It is characterized by a clever filtration that requires no more on the Nyquist map $f : D \times \Omega \to N$ than to be *simplicial*. Because of the fairly general nature of the map f, the difficulty of the homology of the fiber depending on the selected simplex τ of the Nyquist template is present. However, the E^2 term of the spectral sequence still provides a reasonably simple approximation of $H(D \times \Omega)$ in terms of $H(N)$ and $H(f^{-1}(\tau))$. The second spectral sequence rather follows in the footstep of Serre by providing a semisimplicial version of it. The map is required to be *semisimplicial fiber bundle* in the sense of Chapter 13. The advantage is that the homology of the fiber is constant, resulting in E^2 being an even more transparent approximation of $H(D\times\Omega)$ in terms of $H(N)$ and $H(\text{fiber})$. The disadvantage of this approach is that it requires the Nyquist map $f : D\times\Omega \to N$ to be a semisimplicial fiber bundle. However, as we saw in Section 13.5, allowing for some preprocessing, this is not too much of a restriction.

14.1 Homology Spectral Sequence

14.1.1 Graduation and Filtration of Differential Group

We consider an Abelian group G that can be decomposed as the direct sum of subgroups,

$$G = \bigoplus_n G_n$$

Such a decomposition is called **graduation**, and G is called **graded group**. The index n is called **total degree** of the subgroup G_n. To get a better feeling of the concept, the reader can replace G_n by $C_n(D \times \Omega)$.

A **differential of degree** s defined on a group G is a group endomorphism

$$\partial : G \to G$$

such that

$$\partial\partial = 0$$

$$\partial G_n \subseteq G_{n+s}$$

Again, to be more concrete, ∂ can be thought of as the extension of the usual boundary $\partial_n : C_n(D \times \Omega) \to C_{n-1}(D \times \Omega)$ to $\bigoplus_n C_n(D \times \Omega)$. It is clear from this insight that most of the differentials we have dealt with so far have degree -1. However, soon we will introduce differentials with a more complicated degree structure. A graded group endowed with a differential is said to be a **differential graded group**.

Observe that the concept of differential does not really require a graduation of the group. We could certainly define on a group G an endomorphism $\partial : G \to G$ such that $\partial\partial = 0$. In this case, (G, ∂) is called **differential group**.

The next concept is that of a **(increasing) filtration** of the group G—that is, an increasing sequence of subgroups,

$$0 \subseteq A_0 \subseteq A_1 \subseteq \ldots \subseteq A_u \subseteq A_{u+1} \subseteq \ldots \subseteq A_l = G$$

The index u is called **filtration degree** and l is called **filtration length**. A group G together with a filtration is called **filtered group**.

When defining a filtration on top of a differential group, we will always assume compatibility between filtration and differential; that is,

$$\partial A_u \subseteq A_u$$

Under those circumstances, G is said to be a **differential filtered group**.

When filtering a graded group, we will always assume that each A_u is graded by G_n; that is,

$$A_u = \oplus_u (G_n \cap A_u)$$

Finally, we will for most of the time put the other restriction that each element $a \in G_n$ has filtration length not exceeding n, namely,

$$G_n \subseteq A_n$$

Definition 14.1. *A* **differential filtered graded group** *is a quadruple* (G, G_n, ∂, A_u)*, where* G *is a group,* G_n *a graduation,* ∂ *a differential, and* A_u *a filtration subject to the following compatibility conditions:* $\partial A_u \subseteq A_u$*,* $A_u = \oplus_n (G_n \cap A_u)$*, and* $G_n \subseteq A_n$*. When the last condition is lifted, the group is said to be* **unrestricted***.*

We shall always assume here that the compatibility condition $\partial A_u \subseteq A_u$ is satisfied (see [Spanier 1989, page 469] and [McCleary 1985, page 32]); some results, though, can be extended to the case where the compatibility condition is not satisfied (see [Hilton and Wylie 1965, page 397]). The condition $A_u = \oplus_n (A_u \cap G_n)$ is hard to relax and we shall always assume that it holds (see [Spanier 1989, page 468]). On the other hand, the condition $G_n \subseteq A_n$ can be lifted and we shall occasionally do so. As we shall see later, the condition $G_n \subseteq A_n$ guarantees that the spectral sequence is in the first quadrant; it also guarantees better convergence properties than in the case where it fails (see [Hilton and Wylie 1965, pages 406 and 407 and Section 10.9].

Observe that a graduation G_n of G yields a filtration of G,

$$A_u = \bigoplus_{n \leq u} G_n,$$

Conversely, a filtration A_u of G yields a graduation

$$\bigoplus_u A_{u+1}/A_u$$

The problem is that the graded group $\oplus_u(A_{u+1}/A_u)$ is not always isomorphic to G. In case of a *field* of coefficients, G is *uniquely* determined by $\oplus_u A_{u+1}/A_u \cong G$. To be more specific, if A_u and G are (the additive groups of) vector spaces, then $\oplus_u(A_{u+1}/A_u) \cong G$. In the general case, $\oplus_u A_{u+1}/A_u$ specifies G up to finitely many module extensions (see [McCleary 1985, page 30] and [Spanier 1989, pages 469 and 471]). To get the feeling for the extension problem, consider the near-trivial filtration:

$$0 = A_0 \subseteq A_1 \subseteq A_2 = G$$

This filtration clearly yields the short exact sequence

$$0 \to A_1/A_0 \to G \to A_2/A_1 \to 0$$

Clearly, the original group G fits in the middle of a short exact sequence constructed with A_{u+1}/A_u. The salient feature is that *only if the sequence splits* do we have

$$G \cong (A_1/A_0) \oplus (A_2/A_1)$$

In the general situation, given the graduation, there are many G's fitting in the middle of the short exact sequence, and finding all such G's is the *extension problem*, central in homological algebra; see Appendix A.

14.1.2 Filtration of Cycle and Boundary Groups

Since the group G comes with a filtration, it is natural to look at the effect of this filtration on cycles and boundaries of G.

With the mere differential group structure (G, ∂), we can define the cycle subgroup as

$$Z = \partial^{-1}0$$

We look at how the cycles and filtration interfere. Consider the following two strings,

$$0 \subseteq \ldots \subseteq A_u \subseteq A_{u+1} \subseteq \ldots \subseteq G$$

$$Z = \partial^{-1}0 \subseteq \ldots \subseteq \partial^{-1}A_u \subseteq \partial^{-1}A_{u+1} \subseteq \ldots \subseteq \partial^{-1}G = G$$

The second string is obtained by applying ∂^{-1} to the first string. Now, we intersect every single element of the first string with an element of the second string; that is, we define

$$Z_u^r = \partial^{-1}(A_{u-r}) \cap A_u$$

The above is the group of "cycles modulo A_{u-r}" in A_u, because indeed, if $z \in Z_u^r$, it follows that $\partial z = 0 \bmod A_{u-r}$. Clearly, as r increases, A_{u-r} becomes smaller and smaller and eventually, after a finite number of steps, vanishes. With a slight abuse, we denote as Z_u^∞ the corresponding steady-state value. Clearly,

$$Z_u^\infty = A_u \cap Z$$

so that the Z_u^∞'s provide a filtration of Z.

Now, we introduce the graduation. The index u is the filtration degree, n is called total degree, and we introduce a complementary degree v, namely, $u + v = n$,

$$Z_{u,v}^r = G_n \cap Z_u^r$$

Now, we perform a similar analysis on the boundaries. The subgroup of boundaries of the differential group (G, ∂) is defined as

$$B = \partial G$$

As before, consider the two filtrations:

$$0 = A_0 \subseteq ... \subseteq A_u \subseteq A_{u+1} \subseteq ... \subseteq G$$

$$0 = \partial 0 \subseteq ... \subseteq \partial A_u \subseteq A_{u+1} \subseteq ... \subseteq \partial G = B$$

Clearly, the second string is obtained by applying ∂ to the first one. We intersect every single element of the first string with an element of the second string. Define

$$B_u^r = \partial A_{u+r} \cap A_u$$

the elements of A_u that bound in A_{u+r}. Increasing r, there exists a steady-state value B_u^∞ of boundaries in A_u,

$$B_u^\infty = B \cap A_u$$

And clearly, the B_u^∞'s provide a filtration of B.

Next, we introduce the graduation,

$$B_{u,v}^r = G_n \cap B_u^r$$

From the definition of B_u^∞, Z_u^∞, it is trivial to observe that $B_u^\infty \subseteq Z_u^\infty$. From this, one obtains the following long string of inclusions:

$$... \subseteq B_u^r \subseteq B_u^{r+1} \subseteq ... \subseteq B_u^\infty \subseteq Z_u^\infty \subseteq ... \subseteq Z_u^{r+1} \subseteq Z_u^r \subseteq ...$$

Intersecting the above string with G_n, we get

$$... \subseteq B_{u,v}^r \subseteq B_{u,v}^{r+1} \subseteq ... \subseteq B_{u,v}^\infty \subseteq Z_{u,v}^\infty \subseteq ... \subseteq Z_{u,v}^{r+1} \subseteq Z_{u,v}^r \subseteq ...$$

14.1.3 Filtration of Homology Group

After filtering the cycle and boundary groups, the next natural question is whether the filtration of Z and B could provide a filtration of the homology group.

With the mere differential group structure (G, ∂), we define the homology of G as

$$\begin{aligned} H(G) &= \partial^{-1} 0 / \partial G \\ &= Z/B \end{aligned}$$

The compatibility condition $\partial A_u \subseteq A_u$ makes (A_u, ∂) a differential group, so that its homology $H(A_u)$ can be defined as above.

To construct a filtration of $H(G)$, we consider the following diagram:

$$\begin{array}{ccccc} A_{u-1} & \overset{j}{\rightarrow} & A_u & \overset{i}{\rightarrow} & G \\ \downarrow & & \downarrow & & \downarrow \\ H(A_{u-1}) & \overset{j_*}{\rightarrow} & H(A_u) & \overset{i_*}{\rightarrow} & H(G) \end{array}$$

In the above, j, i are injection maps. They are easily seen to be chain maps and hence they induce homology group homomorphisms j_*, i_*, respectively. We define

$$D_u = i_*(H(A_u))$$

First we show that the D_u's provide a *filtration* of $H(G)$. Indeed, observe the following:

$$\begin{aligned} D_{u-1} &= (ij)_*(H(A_{u-1})) \\ &= i_*(j_*(H(A_{u-1}))) \end{aligned}$$

Since $j_*(H(A_{u-1})) \subseteq H(A_u)$, it follows that

$$D_{u-1} \subseteq i_*(H(A_u)) = D_u$$

The reader can easily check the limiting behavior of D_u for u large and u small and derive,

$$0 \subseteq D_0 \subseteq \ldots \subseteq D_u \subseteq D_{u+1} \subseteq \ldots \subseteq H(G)$$

Now, we would like to get a concrete interpretation of D_u. Take $\{z\} \in H(A_u)$. Write the coset of the cycle z as $z + \partial A_u$. Here we have to introduce some notation that might appear pedantic. We denote as ∂_{A_u} the boundary of the differential group A_u while we denote as ∂_G the boundary of the differential group G. The chain map property of i yields

$$\begin{aligned} i(z + \partial_{A_u} A_u) &= i(z) + i(\partial_{A_u} A_u) \\ &= i(z) + \partial_G i(A_u) \\ &= z + \partial_G A_u \\ &= z + (A_u \cap \partial_G G) \end{aligned}$$

Going to the homology, it follows that

$$D_u = Z_u^\infty / B_u^\infty$$

Putting it in other words, D_u consists of those homology classes of $H(G)$ that contains cycles in A_u. (Hence we recover the already known fact that $\{D_u\}$ yields a filtration of $H(G)$.)

Remember, $H(G)$ also comes with the usual graduation, $\oplus_n H_n(G)$. This graduation in turn induces a graduation on the D_u,

$$D_{u,v} = D_u \cap H_n(G)$$

where, as usual, $u + v = n$. Retaining the elements of graduation n in $D_u = Z_u^\infty / B_\infty^u$ yields

$$D_{u,v} = Z_{u,v}^\infty / B_{u,v}^\infty$$

It turns out that the spectral sequence reveals the homology $H(G)$, through $\oplus_u D_{u+1}/D_u$.

14.1.4 Successive Approximation

The relation $D_u = Z_u^\infty / B_\infty^u$ gives us the *clue* that by judiciously assembling some terms of the long string of inclusions, we might get a meaningful approximation to $H(G)$.

It turns out that a meaningful rth-order approximation of $D_u \cap H_n(G)$ is

$$E_{u,v}^r = Z_{u,v}^r / (Z_{u-1,v+1}^{r-1} + B_{u,v}^{r-1})$$

The motivation for this choice will become clearer later; in a certain sense, "the end justifies the means."

First, for the above make sense, we have to check whether $Z_{u-1,v+1}^{r-1} \subseteq Z_{u,v}^r$ and $B_{u,v}^{r-1} \subseteq Z_{u,v}^r$. This is easily done from the definitions of the various groups and is left to the reader.

Next, we investigate the effect of the differential operator ∂ on the various groups. We have

$$\begin{aligned}
\partial Z_{u,v}^r &= \partial(G_n \cap \partial^{-1} A_{u-r} \cap A_u) \\
&\subseteq \partial G_n \cap (A_{u-r} \cap \partial A_u) \\
&\subseteq G_{n-1} \cap B_{u-r}^r \\
&= B_{u-r,v+r-1}^r \\
&\subseteq Z_{u-r,v+r-1}^r
\end{aligned}$$

Next consider

$$\begin{aligned}
\partial(Z_{u-1,v+1}^{r-1} + B_{u,v}^{r-1}) &= \partial Z_{u-1,v+1}^{r-1} + 0 \\
&= \partial(G_n \cap Z_{u-1}^{r-1}) \\
&= \partial(G_n \cap \partial^{-1} A_{u-r} \cap A_{u-1}) \\
&\subseteq \partial G_n \cap A_{u-r} \cap \partial A_{u-1} \\
&\subseteq G_{n-1} \cap B_{u-r}^{r-1} \\
&= B_{u-r,v+r-1}^{r-1} \\
&\subseteq Z_{u-r-1,v+r}^{r-1} + B_{u-r,v+r-1}^{r-1}
\end{aligned}$$

Therefore, ∂ induces an operator

$$\partial_{u,v}^r : E_{u,v}^r \to E_{u-r,v+r-1}^r$$

Further, it is trivial that

$$\partial_{u-r,v+r-1}^r \partial_{u,v}^r = 0$$

In other words, $\partial^r_{u,v}$ is a differential operator of **bidegree** $(-r, r-1)$. The latter, in turn, induces an operator $\partial^r : \oplus_{u,v} E^r_{u,v} \to E^r_{u,v}$. It follows that

$$(E^r = \oplus_{u,v} E^r_{u,v}, \partial^r)$$

is a bigraded differential group, with differential of bidegree $(-r, r-1)$ and total degree $-r + r - 1 = -1$.

From the above, we can define the homology of E^r relative to ∂^r. Like E^r, its homology $H(E^r)$ is also bigraded. $H_{u,v}(E^r, \partial^r)$ is defined as the subgroup of those homology classes that have a ∂^r-cycle in $E^r_{u,v}$; that is,

$$H_{u,v}(E^r, \partial^r) = \ker(\partial^r : E^r_{u,v} \to E^r_{u-r,v+r-1})/\partial^r E^r_{u+r,v-r+1}$$

A major result is that the E^r's form a *spectral sequence*:

Theorem 14.2. *For a differential unrestricted filtered graded group, we have* $E^{r+1}_{u,v} = H_{u,v}(E^r, \partial^r), \forall u, v$, *which we write more compactly as* $E^{r+1} = H(E^r)$.

Proof. This is not the easiest possible proof, but it is purely algebraic. See, for example, [Hilton and Wylie 1965, Theorems 10.2.5 and 10.3.6, pages 402 and 408, respectively] or [McCleary 1985, Section 2.2.2, page 33]. ∎

As a final remark, this subsection has a single index, filtration only version that can be recovered from the bigraded version by removing the complementary degree and defining $E^r_u = \oplus_v E^r_{u,v}$. It is easily checked that $E^r_u = Z^r_u/(Z^{r-1}_{u-1} + B^{r-1}_u)$.

14.1.5 Initialization

We look at the very first term of the spectral sequence,

$$E^0_{u,v} = Z^0_{u,v}/(Z^{-1}_{u-1,v+1} + B^{-1}_{u,v})$$

so as to know how to start. We have

$$Z^0_{u,v} = G_n \cap \partial^{-1} A_u \cap A_u$$

Since $\partial A_u \subseteq A_u$, it follows that $\partial^{-1} A_u \supseteq A_u$ so that $\partial^{-1} A_u \cap A_u = A_u$. Therefore,

$$Z^0_{u,v} = G_n \cap A_u$$

Following the same line of argument,

$$\begin{aligned} Z^{-1}_{u-1,v+1} &= G_n \cap \partial^{-1} A_{u-1+1} \cap A_{u-1} \\ &= G_n \cap A_{u-1} \end{aligned}$$

and

$$B^{-1}_{u,v} = G_n \cap \partial A_{u-1} \cap A_u$$

$$\subseteq \quad G_n \cap A_{u-1}$$

Therefore,

$$E^0_{u,v} = (G_n \cap A_u)/(G_n \cap A_{u-1})$$

14.1.6 Convergence

Having set up the spectral sequence, it remains to show that it converges after finitely many steps and that the steady-state value $E^\infty_{u,v}$ provides meaningful information about $H(G)$.

Define

$$E^\infty_{u,v} = Z^\infty_{u,v}/(Z^\infty_{u-1,v+1} + B^\infty_{u,v})$$

The first convergence result is the following:

Theorem 14.3. *For a differential (restricted) filtered graded group,* $E^r_{u,v} = E^\infty_{u,v}$, *for* $r > \max(u, v+1)$.

Proof. See [Hilton and Wylie 1965, Theorem 10.3.9, page 409]. ■

If the group is *unrestricted*, then E^∞ can only be equated to the direct limit of $E^r, E^{r+1}, \ldots$ (see [Hilton and Wylie 1965, Section 10.9]).

Finally, we precisely relate the limit of the spectral sequence E^∞ to the homology $H(G)$.

Theorem 14.4. $E^\infty_u = D_u/D_{u-1}$ *which implies that* $E^\infty = \bigoplus_u D_u/D_{u-1}$.

Proof. See [Hilton and Wylie 1965, Theorem 10.2.7, page 405 and Proposition 10.3.7, page 408] and [McCleary 1985, Section 2.2.2, page 33]. ■

As a final word of warning, remember that $\oplus_u D_u/D_{u-1}$ is not always isomorphic to $H(G)$. Nevertheless, the limit of the spectral sequence determines $H(G)$ up to extension.

14.2 Example (Spectral Sequence of a Matrix)

A finite-dimensional matrix $A : \mathbb{R}^n \to \mathbb{R}^n$ induces the chain complex

$$0 \to \mathbb{R}^n \xrightarrow{A} \mathbb{R}^n \to 0$$

To get a better understanding of this, compare the above with

$$G_3 \xrightarrow{\partial_3} G_2 \xrightarrow{\partial_2} G_1 \xrightarrow{\partial_1} G_0$$

where $G_3 = G_0 = 0$, $G_2 = G_1 = \mathbb{R}^n$, $\partial_3 = 0$, $\partial_1 = 0$, and $\partial_2 = A$. The chain complex condition $\partial_{n-1}\partial_n = 0$ is trivially satisfied. The nontrivial homology groups of this chain complex are clearly

$$\begin{aligned} H_2 &= \ker A \\ H_1 &= \mathbb{R}^n/A\mathbb{R}^n = \text{coker } A \end{aligned}$$

We propose to show how the spectral sequence of this chain complex—for an appropriate filtration—provides a successive approximation scheme to compute the kernel and the cokernel of a matrix A.

(A deeper motivation for this exercise is to get the reader acquainted with spectral sequences of chain complexes like $0 \to C_2 \xrightarrow{\partial} C_1 \to 0$ which are instrumental in polynomial quotient module computation for singularity analysis of the Nyquist map.)

For this example, the group G is *graduated* as

$$G = G_3 \oplus G_2 \oplus G_1 \oplus G_0$$

Given a subspace $S \subseteq \mathbb{R}^n$, we define a *filtration* $\{A_u\}$ as follows:

$$\begin{aligned} A_0 &= 0 \oplus 0 \oplus 0 \oplus 0 \\ A_1 &= 0 \oplus S \oplus \mathbb{R}^n \oplus 0 \\ A_2 &= 0 \oplus \mathbb{R}^n \oplus \mathbb{R}^n \oplus 0 \end{aligned}$$

We clearly have $0 = A_0 \subseteq A_1 \subseteq A_2 = G$. Next, we check the compatibility condition $\partial A_u \subseteq A_u$; this is trivial for $u = 0, 2$; for $u = 1$ we check that

$$\begin{aligned} \partial A_1 &= \partial(0 \oplus S \oplus \mathbb{R}^n \oplus 0) \\ &= 0 \oplus \partial_3 0 \oplus AS \oplus \partial_1 \mathbb{R}^n \\ &\subseteq 0 \oplus 0 \oplus AS \oplus 0 \\ &\subseteq 0 \oplus \mathbb{R}^n \oplus \mathbb{R}^n \oplus 0 = A_1 \end{aligned}$$

Verification of the condition $A_u = \oplus_n (A_u \cap G_n)$ is left to the reader. Finally, the filtration length condition $G_n \subseteq A_n$ is easily checked for $n = 0, 2$; for $n = 1$, we get

$$\begin{aligned} G_1 = \mathbb{R}^n &\cong 0 \oplus 0 \oplus \mathbb{R}^n \oplus 0 \\ &\subseteq 0 \oplus S \oplus \mathbb{R}^n \oplus 0 \\ &= A_1 \end{aligned}$$

After these preliminaries, we are in a position to compute the E^0 term of the spectral sequence:

$$\begin{aligned} E^0_{0,v} &= (G_v \cap A_0)/(G_v \cap A_{-1}) = 0/0 = 0 \\ E^0_{1,v} &= (G_{v+1} \cap A_1)/(G_{v+1} \cap A_0) = G_{v+1} \cap A_1 \\ E^0_{2,v} &= (G_{v+2} \cap A_2)/(G_{v+2} \cap A_1) = G_{v+2}/(G_{v+2} \cap A_1) \end{aligned}$$

It is easily seen that E^0 is a *first-quadrant bigraded group*; that is, $E^0_{u,v} = 0$ for $u < 0$ or $v < 0$. This property is a corollary of the filtration length condition. We display the subgroups of E^0 in the following two-dimensional array:

$$E^0 = \begin{array}{c|cccc}
v & & & & \\
2 & & 0 & & \\
& & \downarrow & & \\
1 & 0 & S & 0 & \\
& & \downarrow & \downarrow & \\
0 & 0 & \mathbb{R}^n & \mathbb{R}^n/S & \\
\hline
& 0 & 1 & 2 & u
\end{array}$$

The bidegree of ∂^0 is $(0,-1)$ so that the ∂^0 arrow flows down one level. Next we compute E^1 as the homology of $\{E^0, \partial^0\}$ and this yields

$$E^1 = \begin{array}{c|ccccc}
v & & & & & \\
2 & & 0 & & & \\
1 & 0 & \ker A|S & & 0 & \\
0 & 0 & \mathbb{R}^n/AS & \leftarrow & \mathbb{R}^n/S & \\
\hline
& 0 & 1 & & 2 & u
\end{array}$$

The subgroups appearing on the $u+v=2$ line, $\oplus_{u+v=2}E^1_{u,v}$, constitute an approximation of $H_1 = \ker A$. One clearly sees on that line $\ker A|S$, which is a first approximation of $\ker A$. Likewise, $\oplus_{u+v=1}E^1_{u,v}$ is a first-order approximation to $H_1 = \mathbb{R}^n/A\mathbb{R}^n$. At this stage, the latter is approximated by $\mathbb{R}^n/AS$, which is meaningful. The bidegree of E^1 is $(-1,0)$, which means that the ∂^1 arrow flows one step to the left. As a second approximation, we compute E^2 as the homology of $\{E^1, \partial^1\}$. The resulting E^2 diagram is

$$E^2 =$$

$$\begin{array}{c|cccc}
v & & & & \\
2 & & 0 & & \\
1 & 0 & \ker A|S & 0 & \\
0 & 0 & (\mathbb{R}^n/AS)/A(\mathbb{R}^n/S) & \ker(\partial^1 : \mathbb{R}^n/S \to \mathbb{R}^n/AS) & \\
\hline
& 0 & 1 & 2 & u
\end{array}$$

To understand where do we go from here, observe that ∂^2 has bidegree $(-2,+1)$, so that the ∂^2 arrow of the E^2 diagram goes two steps back, one level up. Clearly, there are no nontrivial such arrows in the E^2 diagram. It follows that $E^2 = E^3 = \ldots = E^\infty$, the spectral sequence *collapses* at E^2, and the *exact* solution is reached at that level. It remains to double check that we have the correct subspaces in the E^2 diagram. The E^2 diagram tells us that

$$H_1 = \oplus_{u+v=1} E^2_{u,v} = (\mathbb{R}^n/AS)/A(\mathbb{R}^n/S)$$

The following lemma is useful:

Lemma 14.5. (Noether Isomorphism) *For a string of subgroups $U \subseteq V \subseteq W$, we have*

$$(W/U)/(V/U) \cong W/V$$

Proof. See [Hilton and Wu 1974, Chapter 1, Theorem 4.11 and Chapter 4, Theorem 2.4]. This theorem is also valid for modules. ∎

Clearly, $AS \subseteq A\mathbb{R}^n \subseteq \mathbb{R}^n$. Therefore, applying the lemma yields

$$H_1 = \oplus_{u+v=1} E^2_{u,v} = (\mathbb{R}^n/AS)/(A\mathbb{R}^n/AS) = \mathbb{R}^n/A\mathbb{R}^n$$

and the spectral sequence argument correctly yields the cokernel. Now we check whether $\oplus_{u+v=2} E^2_{u,v}$ indeed yields the kernel,

$$H_2 = \oplus_{u+v=2} E^2_{u,v} = \ker A|S \oplus \ker(\partial^1 : \mathbb{R}^n/S \to \mathbb{R}^n/AS)$$

To sort out the extreme right-hand side of this equality, choose a basis of $\mathbb{R}^n$ where the first elements generate S. Partition the matrix A conformably

$$A = \begin{pmatrix} A_{11} & A_{12} \\ A_{21} & A_{22} \end{pmatrix}$$

By elementary linear algebra, we have

$$\ker A|S = \ker \begin{pmatrix} A_{11} \\ A_{21} \end{pmatrix}$$

and

$$\ker(\partial^1 : \mathbb{R}^n/S \to \mathbb{R}^n/AS) = \operatorname{im} \begin{pmatrix} A_{12} \\ A_{22} \end{pmatrix} \cap \operatorname{im} \begin{pmatrix} A_{11} \\ A_{21} \end{pmatrix}$$

It is easily verified that

$$\ker A = \ker \begin{pmatrix} A_{11} \\ A_{21} \end{pmatrix} \oplus \left(\operatorname{im} \begin{pmatrix} A_{12} \\ A_{22} \end{pmatrix} \cap \operatorname{im} \begin{pmatrix} A_{11} \\ A_{12} \end{pmatrix} \right)$$

so that

$$\oplus_{u+v=2} E^2_{u,v} = \ker A$$

which verifies that the spectral sequence argument yields the expected result.

14.3 Spectral Sequence of Geometric System Theory

We quickly establish the connection between spectral sequences and the so-called *geometric theory* of such a linear, time-invariant, finite-dimensional state-space system as

$$\begin{aligned} x(k+1) &= Fx(k) + Gw(k) \\ y(k) &= Hx(k) + Jw(k) \end{aligned}$$

Consider the input/output relationship,

$$\begin{pmatrix} \vdots \\ y(-2) \\ y(-1) \\ y(0) \end{pmatrix} = \underbrace{\begin{pmatrix} \ddots & \ddots & \ddots & \\ \ddots & J & 0 & 0 \\ \ddots & HG & J & 0 \\ & HFG & HG & J \end{pmatrix}}_{A=\partial} \begin{pmatrix} \vdots \\ w(-2) \\ w(-1) \\ w(0) \end{pmatrix}$$

The kernel of $A = \partial$ consists of the so-called output nulling control sequences that generate the strongly reachable subspace (at $k = 0$). This control sequence is generated by Silverman's dual structure algorithm, which amounts to computing all output nulling control sequences recursively starting at $k = 0$ and then proceeding backward. Alternatively, the corresponding increasing sequence of state subspaces is generated by the Basile–Marro geometric subspace recursion. We indicate how these traditional algorithms can be viewed as the spectral sequence to compute the kernel of the above matrix.

The chain complex is set up as above, with

$$G_2 = \begin{pmatrix} \vdots \\ \mathbb{R}^{n_w} \\ \mathbb{R}^{n_w} \\ \mathbb{R}^{n_w} \end{pmatrix}$$

G_2 is defined similarly. The filtration is defined as

$$A_u = 0 \oplus \begin{pmatrix} \vdots \\ 0 \\ \oplus_{i=0}^{u} \mathbb{R}^{n_w} \end{pmatrix} \oplus \begin{pmatrix} \vdots \\ 0 \\ \oplus_{i=0}^{u} \mathbb{R}^{n_y} \end{pmatrix} \oplus 0$$

As easily seen, causality takes care of all of the required conditions on the filtration. Therefore, following the same procedure as for the general matrix A, the spectral sequence for computing the kernel of A is easily found to be equivalent to the dual of the structure algorithm or the geometric subspace recursion to compute the strongly reachable subspace also called *minimal input-containing conditioned-invariant subspace* (see [Kitapci, Jonckheere, and Silverman 1986]).

14.4 Cohomology Spectral Sequence

The cohomology spectral sequence is obtained by just dualizing the homology spectral sequence. We strive to follow the same lines as in the homology case, restricting ourselves to just write the basic facts and results and leaving the details of the dualization to the reader.

We consider a graded group

$$G = \bigoplus_n G^n$$

together with a differential δ of degree $+1$, namely,

$$\delta : G \to G$$
$$\delta\delta = 0$$
$$\delta G^n \subseteq G^{n+1}$$

which makes (G, δ) a differential graded group. In a *cohomology* spectral sequence, we start with a **decreasing filtration** of the differential group (G, δ),

$$G = A^0 \supseteq A^1 \supseteq A^2 \supseteq \ldots \supseteq A^u \supseteq A^{u+1} \supseteq \ldots \supseteq A^l = 0$$

The filtration and graduation are compatible in the sense that

$$\delta A^u \subseteq A^u$$
$$A^u = \bigoplus_n (G^n \cap A^u)$$
$$G^n \cap A^{n+1} = 0$$

The objective is to compute the cohomology of the differential group (G, δ)

$$\begin{aligned} H(G) &= \delta^{-1}0/\delta G \\ &= Z/B \end{aligned}$$

which is revealed through some decreasing filtration

$$H(G) = D^0 \supseteq \ldots \supseteq D^u \supseteq D^{u+1} \supseteq \ldots \supseteq 0$$

To construct the filtration of $H(G)$ define

$$\begin{aligned} Z_r^u &= \delta^{-1} A^{u+r} \cap A^u \\ B_r^u &= \delta A^{u-r} \cap A^u \end{aligned}$$

These are rth-order approximations in the sense that

$$\begin{aligned} Z_\infty^u &= Z \cap A^u \\ B_\infty^u &= B \cap A^u \end{aligned}$$

Therefore, the filtration of the cohomology group $H(G)$ is given by

$$D^u = Z_\infty^u / B_\infty^u$$

Also define

$$D^{u,v} = D^u \cap G^n$$

where, as before, $u + v = n$ and

$$\begin{aligned} Z_r^{u,v} &= \delta^{-1}(A^{u+r}) \cap (A^u \cap G^n) \\ B_r^{u,v} &= \delta A^{u-r} \cap (A^u \cap G^n) \end{aligned}$$

To set up the successive approximation scheme, define

$$E_r^{u,v} = Z_r^{u,v} / (Z_{r-1}^{u+1,v-1} + B_{r-1}^{u,v})$$

The δ operator induces an operator

$$\delta_r^{u,v} : E_r^{u,v} \to E_r^{u+r,v-r+1}$$

of bidegree $(r, -r+1)$. The fundamental result is that the E_r's form a spectral sequence, in the sense that $H(E_r, \delta_r) = E_{r+1}$. The spectral sequence converges in the sense that $E_r^{u,v} = E_\infty^{u,v}$ for $r > \max(u, v+1)$. Finally,

$$\begin{aligned} E_\infty^{u,v} &= D^{u,v} / D^{u+1,v-1} \\ E_\infty^u &= D^u / D^{u+1} \\ E_\infty &= \bigoplus_u D^u / D^{u+1} \end{aligned}$$

To sum up, after running the $E_r^{u,v}$ recursion, the cohomology of G is given, *up to module extension*, as

$$H^n(G) = \bigoplus_{u+v=n} E_\infty^{u,v}$$

For more details about cohomology spectral sequences, see [Hilton and Wylie 1965, Section 10.5], [Dieudonné 1989, Part 1, Chapter IV, Section 7.D, pages 132–138], and [McCleary 1985, Chapter 2].

The reader interested in a *typical* cohomology spectral sequence argument is referred to Subsection 19.5.1. It is there shown how to set up the spectral sequence of a fiber bundle that yields the de Rham cohomology of the general linear group.

14.5 Zeeman Dihomology Spectral Sequence of Simplicial Nyquist Map

The spectral sequence developed in this section is very much in the spirit of the original work of Leray, except for three things: First, it is a *homology* rather than *cohomology* theory as in the original work by Leray. Second, it deals with *simplicial* rather than arbitrary continuous maps. Third, it deals with the concept of *covariant stacks* rather than *cohomology sheaves.* Covariant stacks formalize the fact that the homology of the fiber depends on the simplex or the point of the Nyquist template above which the fiber lies. This adaptation of the early work by Leray is due to Zeeman and was developed within the context of his dihomology theory. In a few words, *di*homology is a technique for studying a map from its graph, considered as a subcomplex of the *double* complex $C_*(D \times \Omega) \otimes C_*(N)$.

Consider decompositions of $D \times \Omega$ and N and let $f : D \times \Omega \to N$ be the Nyquist simplicial map. The difficulty is to define the "fiber" above a simplex τ of the Nyquist template. In this *simplicial* approach, the fiber above $\tau \in N$ is defined as

$$f^{-l}(\tau) = \{\sigma \in D \times \Omega : f(\text{star}(\sigma)) \cap \text{star}(\tau) \neq \emptyset\}.$$

Observe that this formal definition of the fiber is related to the topological map as follows:

$$f^{-l}(\tau) = \overline{f^{-1}(\text{star}(\tau))}$$

Clearly we have an increasing filtration of $G = \oplus_n C_n(D \times \Omega)$,

$$A_u = \cup_{\tau_n, n \leq u} f^{-l}(\tau_n)$$

It is easily seen that $\{G, G_n, \partial, A_u\}$ is a differential filtered graded group. From there on, we could run the spectral sequence.

In order to get a meaningful approximation in the lower-order terms of the sequence, we prefer to adopt the graph approach. The **closed neighborhood graph** (see [Zeeman 1962b, page 642]) of the simplicial map $f : D \times \Omega \to N$ is defined as

$$\mathcal{G} = \{\sigma \otimes \tau \in (D \times \Omega) \otimes N : f(\text{star}(\sigma)) \cap \text{star}(\tau) \neq \emptyset\}$$

Theorem 14.6. *Given $\sigma \in D \times \Omega$, the so-called* **right facet**

$$\{\tau \in N : \sigma \otimes \tau \in \mathcal{G}\} = \overline{f(star(\sigma))}$$

is acyclic; that is, $H_n(\{\tau\}) = \mathbb{Z}$ for $n = 0$ and 0 otherwise.

Proof. This is fairly trivial from the definition of the simplicial map (see [Zeeman 1962b, page 642]). ■

Observe that the so-called **left facet** of $\tau \in N$, or fiber over τ,

$$\{\sigma \in D \times \Omega : \sigma \otimes \tau \in \mathcal{G}\} = f^{-l}(\tau)$$

is not always acyclic.

Theorem 14.7. *There is an isomorphism:*

$$H_*(\mathcal{G}) \xrightarrow{\cong} H_*(D \times \Omega)$$

Proof. This result can be gotten by dualizing (right facet $\leftrightarrow$ left facet) the spectral sequence argument of [Zeeman 1962a, Theorem 1, pages 618–619]. This result is also asserted in [Zeeman 1962b, page 641]. The spectral sequence argument goes as follows: Define the augmentation of N, $\epsilon_N : N \to \{point\}$, to be the chain map from N to the point complex. This augmentation induces a chain map $\epsilon_N : (D \times \Omega) \otimes N \to (D \times \Omega) \otimes \{point\}$, which in turn induces $\epsilon_N : \mathcal{G} \to (D \times \Omega) \otimes \{point\}$. Consider filtrations of $\mathcal{G}$ and $(D \times \Omega) \otimes \{point\}$ corresponding to the dimension n of the simplexes in $D \times \Omega$. Consider the spectral sequences converging to

$H_*(\mathcal{G})$ and $H_*((D \times \Omega) \otimes \{point\})$. The crucial step is to observe that the augmentation ϵ_N induces an isomorphism of the spectral sequences, $E^r(\mathcal{G}) \xrightarrow{\cong} E^r(D \times \Omega), r \geq 1$. Therefore, the isomorphism of the E^∞ terms induces the required isomorphism $H_*(\mathcal{G}) \xrightarrow{\cong} H_*(D \times \Omega)$. ∎

Therefore, to come up with a spectral sequence converging to $H_*(D \times \Omega)$, it suffices to filter the graph of the map. The graph of the map as a subcomplex of $\{\sigma_v \otimes \tau_u\}$ is filtered relative to u. From here on, we can run the *complete* spectral sequence starting at $r = 0$. It is, however, not necessary to run the spectral sequence for the first two values of r. Indeed, without any other information than the simplicial property of f, it is possible to compute E^2. The interesting thing about the E^2 term of the spectral sequence is that it reveals a simple approximation of $H_*(D \times \Omega)$ as the homology of the Nyquist template with coefficients in the homology of the fiber.

The difficulty with the above concept is to define the homology of N with coefficients in the homology of the fiber when the latter varies as we sweep across the Nyquist template. Here, the concept of *stack* comes into play. The **covariant (homology) stack** is the rule S that assigns to each $\tau \in N$ the group

$$S(\tau) = H_*(f^{-l}(\tau))$$

Furthermore, the stack induces transition functions as follows: Consider a pair of simplexes such that τ^j is a face of τ^i,

$$\tau^i \succeq \tau^j$$

It is not hard to see that

$$\text{star}(\tau^i) \subseteq \text{star}(\tau^j)$$

Therefore, from the definition of the fiber, we have

$$f^{-l}(\tau^i) \subseteq f^{-l}(\tau^j)$$

This yields an inclusion map and this inclusion map yields a homology group homomorphism, as revealed by the following commutative diagram:

$$\begin{array}{ccc} f^{-l}(\tau^i) & \xrightarrow{i} & f^{-l}(\tau^j) \\ \downarrow & & \downarrow \\ H_*(f^{-l}(\tau^i)) & \xrightarrow{S^{\tau^j,\tau^i}} & H_*(f^{-l}(\tau^j)) \end{array}$$

It follows that the rule S comes together with transition homomorphisms between any pair of simplexes linked by the facing relation $\tau^i \succeq \tau^j$,

$$S^{\tau^j,\tau^i} : S(\tau^i) \rightarrow S(\tau^j)$$

Furthermore, it is not hard to see that the *compatibility* relation is satisfied:

$$S^{\tau^k,\tau^j} S^{\tau^j,\tau^i} = S^{\tau^k,\tau^i}$$

The above properties can be formalized by viewing the stack as a functor: The simplicial complex of $D \times \Omega$ is made a category by defining the objects to be the simplexes of $D \times \Omega$ and by defining a unique arrow or morphism $\sigma^i \to \sigma^j$ iff $\sigma^i \succeq \sigma^j$. It follows that the stack is a *covariant functor* from the category $D \times \Omega$ to the category of Abelian groups and group homomorphisms.

With this concept, we can define what we have so far loosely called homology of N with coefficients in the homology of the fiber. To define it, we must first define the chain group $C_*(N, S)$ of simplexes of N with their coefficients in the stack S. A chain is anything like

$$\sum h\tau_n$$

where

$$\begin{array}{rcl} \tau_n & \in & N \\ h & \in & S(\tau_n) \end{array}$$

Next we define a boundary operator ∂ on this chain complex. Let

$$\tau_n = b^0 b^1 \ldots b^n$$

The boundary ∂ is defined by linearity from

$$\partial(h\tau_n) = \sum_j (-1)^j S^{b^0 \ldots b^{j-1} b^{j+1} \ldots b^n, \tau_n}(h)\, b^0 \ldots b^{j-1} b^{j+1} \ldots b^n$$

It is easily verified that

$$\partial\partial = 0$$

Therefore, $H_*(N, S)$ is simply defined as the homology of the chain complex $\{C_*(N, S), \partial_*\}$.

Now we can state the final result.

Theorem 14.8. (Zeeman) *Given the simplicial map $f : D \times \Omega \to N$ between simplicial complexes $D \times \Omega$ and N together with the filtration of $\mathcal{G}$ corresponding to the dimension of the simplexes of N, there exists a spectral sequence with*

$$E^2_{u,v} = H_u(N, \{H_v(f^{-1}(\tau))\})$$

and converging to the homology of $D \times \Omega$,

$$E^r \Rightarrow H_*(D \times \Omega)$$

Proof. See [Zeeman 1962b]. ∎

14.6 Leray–Serre Spectral Sequence

The Serre spectral sequence was historically the first successful application of spectral sequences in which the topological results came out *clearly*. Given a *fibration*

$$\begin{array}{ccc} F & \xrightarrow{i} & E \\ & & \downarrow \pi \\ & & B \end{array}$$

the Serre spectral sequence constructs the homology of the total space E from the homology of the base space B and the homology of the fiber $\pi^{-1}(b_*) = F$, with an identifiable E^2 term.

The Serre spectral sequence involves several technical difficulties that are, unfortunately, beyond the scope of this book. A first problem is that, since we are dealing with a continuous map without any underlying triangulation, we need either singular cubical homology theory, as was first done by Serre, or the modern approach based on the homology of CW complexes. The next difficulty is to define a local system of coefficients for the base B taking value in $H_*(\pi^{-1}(b))$. This is somewhat similar to Zeeman's concept of *stack*. The problem is that for $b^1, b^2 \in B$, the groups $H_*(\pi^{-1}(b^1)), H_*(\pi^{-1}(b^2))$ are isomorphic, but the isomorphism depends on the path joining b^1 and b^2. For a homotopy class γ of paths joining b^1 and b^2, we need to define a transition isomorphism $H_*(\pi^{-1}(b^1)) \rightarrow H_*(\pi^{-1}(b^2))$. This in turn yields some technical difficulties in defining the chain complex $\{C_*(B, H_*(\pi^{-1}(b))), \partial\}$ of the base with coefficients in the homology of the fiber. Assuming that this chain complex has been defined, then the homology of this chain group is called *homology of the base with coefficients in the homology of the fiber* and is written $H_*(B, H_*(\pi^{-1}(b)))$. While these topics go beyond the scope of this book, for the sake of completeness we quote the following result:

Theorem 14.9. (Leray–Serre) *Given the fibration $\pi : E \rightarrow B$ with path connected base space B and fiber F, there exists a first-quadrant spectral sequence*

$$E^r_{**} \Rightarrow H_*(E)$$

with

$$E^2_{uv} \cong H_u(B, H_v(\pi^{-1}(b)))$$

where $H_*(B, H_*(\pi^{-1}(b)))$ is the homology of the base B with coefficients in the homology of the fiber $\pi^{-1}(b)$. Furthermore, if B is simply connected, then the local system of coefficients becomes one single group, $H_*(F)$, with constant transition isomorphism and

$$E^2_{uv} \cong H_u(B, H_v(F))$$

Proof. The original proof of Serre used singular cubical homology; see [Hilton and Wylie 1965, Section 10.4] and [Hu 1959, Chapter IX]. The modern proofs rather rely on the homology of CW complexes; see [McCleary 1985, Chapter 5] and [Spanier 1989]. ■

If the base is pathwise and simply connected, $E^2_{uv} = H_u(B, H_v(F))$ can be computed from the universal coefficient theorem (see Appendix A);

specifically, there exists a split short exact sequence

$$0 \to H_u(B) \otimes H_v(F) \to E^2_{uv} \to H_{u-1}(B) * H_v(F) \to 0$$

If either the base or the fiber has no torsion, the above can be further reduced to

$$E^2_{uv} = H_u(B) \otimes H_v(F)$$

Now, consider a Nyquist map $f : P \to N$ together with its "fiber" $F = f^{-1}(0 + j0) = X_\omega$. As seen in Chapter 13, this Nyquist map has a homotopically equivalent, induced fibration $\pi_f : E_f \to N$ with fiber $F_f = \pi_f^{-1}(0 + j0)$. Remember, $E_f \simeq D \times \Omega$ and X_ω is homotopically equivalent to a *subset* of F_f. As a corollary of the Serre spectral sequence, we have the following theorem.

Theorem 14.10. (Leray–Serre) *Given the fibration $\pi_f : E_f \to N$, there exists a first-quadrant spectral sequence*

$$E^r_{**} \Rightarrow H_*(E_f) \cong H_*(D \times \Omega)$$

with

$$E^2_{uv} \cong H_u(N, H_v(\pi_f^{-1}(n)))$$

where $H_*(N, H_*(\pi_f^{-1}(n)))$ is the homology of the template with coefficients in the homology of the fiber. If the template is simply connected,

$$E^2_{u,v} \cong H_u(N, H_v(F_f)) \cong H_u(N) \otimes H_v(F_f)$$

and furthermore under those circumstances the spectral sequence collapses at E^2, so that

$$H_n(D \times \Omega) \cong \oplus_{u+v=n} H_u(N) \otimes H_v(F_f) \tag{14.1}$$

Proof. Trivial corollary of the Serre spectral sequence. ∎

The discrepancy between $H_*(N, H_*(\pi_f^{-1}(n)))$ and $H_*(N) \otimes H_*(F_f)$ is the *twisting* of the fiber. In case of no twisting (Equation 14.1), the situation is basically the same as that of the Eilenberg–Zilber theorem.

There is an obvious cohomology dual of the Serre spectral sequence. We leave it to the reader.

Finally, since a fiber bundle is a fibration, the Serre spectral sequence also applies to the former.

14.7 Semisimplicial Serre Spectral Sequence of Nyquist Map

The spectral sequence of this section is in the spirit of that of Serre, except for two things: First, it deals with a semisimplicial bundle map, rather than a fibration as in Serre, and second we use the *semisimplicial*, rather than *singular cubical*, homology. This semisimplicial version of the Serre sequence is due to [May 1992].

From Subsection 13.2.7, it follows that, to compute the homology of $D \times \Omega$ in such as way as to reveal the relation between the uncertainty, the template and the crossover of a semisimplicial Nyquist bundle, we can proceed from either the twisted Cartesian product complex $\{C_*(N \times F), \partial^{\text{twisted}}\}$ or the twisted tensor product complex $\{C_*(N) \otimes C_*(F), \partial^{\otimes,\text{twisted}}\}$.

14.7.1 Twisted Cartesian Product Spectral Sequence

We filter $C(N \times F)$ as follows: The filtration A_u of $C_n(N \times F)$, $A_u C_n(N \times F)$, is defined to be the set of $(b, f) \in C_n(N \times F)$ such that $b = D^{j_m} ... D^{j_1} b'$, where b' is nondegenerate, $0 \leq j_1 < ... < j_m < n$, and $m \geq n - u$.

There results a spectral sequence E^r that converges to $H_*(N \times F) = H_*(D \times \Omega)$; see [May 1992, Chapter VI].

14.7.2 Twisted Tensor Product Spectral Sequence

The filtration of $C(N) \otimes C(F)$ is as follows:

$$A_u = \bigoplus_{n \leq u} C_n(N) \otimes C(F)$$

There results the spectral sequence $\overline{E}^r$ converging to $H_*(D \times \Omega)$. Clearly, $\overline{E}^1_{u,v} = C_u(N) \otimes H_v(F)$. However, the most interesting thing about this spectral sequence is that, if the twisting is one-trivial (see [May 1992, Chapter VI]), we have

$$\overline{E}^r_{u,v} = H_u(N, H_v(F)), \ \ r \geq 2$$

For the proof, see [May 1992, Chapter VI].

14.8 Eilenberg–Moore Spectral Sequence

So far, spectral sequences have been presented as tools for computing the homology of the domain of definition from the homology of the image and the homology of the fiber. In robust control, the problem is posed the other way around; $(D \times \Omega)$ is in general a relatively simple space and its homology can be computed using elementary techniques; since the template is two-dimensional, its homology can be computed using, for example, its Delaunay triangulation; therefore, the *real challenge* is to compute the homology of the crossover from the homology of $D \times \Omega$ and N. This problem can be tackled by turning around the Serre spectral sequence argument, but this requires great imagination and luck. Clearly, a *more systematic* procedure to get the cohomology of $f^{-1}(0 + j0)$ from that of $D \times \Omega$ and N would be welcome. This is precisely what the *Eilenberg–Moore* spectral sequence does.

We first consider a more general problem in terms of pull-back of fibration. Consider a fibration (E, π, B), another space X, and a mapping $g : X \to B$, and let (E_g, π_g, X) be the pull-back fibration. Putting everything together in diagram form yields

$$\begin{array}{ccc} E_g & \longrightarrow & E \\ \pi_g \downarrow & & \downarrow \pi \\ X & \xrightarrow{g} & B \end{array} \tag{14.2}$$

The problem is to find the cohomology of E_g from the cohomology of X, B, E and the properties of the maps π, g.

We use cohomology because the cup product induces quite a lot of extra structure. The cochain groups $C^*(E_g), C^*(X), C^*(B), C^*(E)$ are viewed as *differential graded modules* over $\mathbb{Z}$ or some other ring R (see Appendix C). The cup product makes the base cochain $C^*(B)$ a *differential graded algebra* (see Appendix C). A crucial step is to endow $C^*(E_g), C^*(X), C^*(E)$ with $C^*(B)$-module structure. This is accomplished by going to the cochain of the pull-back diagram, namely,

$$\begin{array}{ccc} C^*(E_g) & \longleftarrow & C^*(E) \\ \pi_g^* \uparrow & & \uparrow \pi^* \\ C^*(X) & \xleftarrow{g^*} & C^*(B) \end{array}$$

The left module action of $C^*(B)$ on $C^*(E)$ is given by

$$\begin{array}{rcl} C^*(B) \otimes C^*(E) & \to & C^*(E) \\ (b, e) & \mapsto & (\pi^*(b)) \smile e \end{array}$$

By a similar process, $C^*(X)$ is made a right $C^*(B)$-module.

Finally, by going to the cohomology, it is easy to endow $H^*(E)$ and $H^*(X)$ with the left and right $H^*(B)$-module structure, respectively, where $H^*(B)$ is considered a differential graded algebra.

The key result is the following theorem:

Theorem 14.11. (Eilenberg–Moore) *Referring to Diagram 14.2, and assuming that the isomorphism $H^*(\pi^{-1}(b^1)) \to H^*(\pi^{-1}(b^2))$ does not depend on the path joining b^1, b^2 in B (see Subsection 16.2.1), there exists a second-quadrant spectral sequence $(E_r^{*,*}, \delta_r)$ starting at the torsion product of the modules $H^*(X)$, $H^*(E)$ over the algebra $H^*(B)$ (see Appendix C)*

$$E_2^{u,v} = Tor_{H^*(B)}^{u,v}(H^*(X), H^*(E))$$

and converging to the cohomology of the pull-back total space

$$E_r^{*,*} \Rightarrow H^*(E_g)$$

Proof. See [McCleary 1985, Chapter 7]. ■

To recover the crossover problem, consider the Nyquist fibration $f : D \times \Omega \to N$ together with the inclusion map $i : 0 + j0 \to N$. Clearly, the total space of the pull-back bundle is the crossover $f^{-1}(0 + j0)$,

$$\begin{array}{ccc} f^{-1}(0+j0) & \rightarrow & D\times\Omega \\ \downarrow & & \downarrow \\ 0+j0 & \xrightarrow{i} & N \end{array}$$

Therefore, $H^*(X)$ of the general result becomes $H^*(\{0+j0\})$, which is fairly trivial.

The analysis of the Nyquist map, *up to homotopy equivalence*, relies crucially on whether or not the template is contractible.

14.8.1 Contractible Template—Open-Loop Stable Case

If the template N is contractible, its cohomology is concentrated in zero dimension:

$$H^*(N) = H^0(N) = \mathbb{Z}$$

Therefore, the E_2 term of the Eilenberg–Moore spectral sequence of the Nyquist map is

$$\begin{aligned} E_2 &= \text{Tor}_{\,H^\bullet(N)}(H^*(0+j0), H^*(D\times\Omega)) \\ &= \text{Tor}_{\,\mathbb{Z}}(\mathbb{Z}, H^*(D\times\Omega)) \end{aligned}$$

To make the computation of the torsion product simpler, we use the projective resolution of the first factor, $\mathbb{Z}$:

$$0 \xrightarrow{d_P} \mathbb{Z} = P^0 \xrightarrow{\epsilon} \mathbb{Z} \longrightarrow 0$$

To contrast this trivial case with the general situation where each module of the resolution sequence comes with a graduation and a coboundary operator, we write the diagram:

$$\begin{array}{ccccccc} 0 & \longrightarrow & 0 & \longrightarrow & 0 & \rightarrow & 0 \\ & & \uparrow \delta_P & & \uparrow & & \\ 0 & \xrightarrow{d_P} & \mathbb{Z} & \xrightarrow{\epsilon} & \mathbb{Z} & \rightarrow & 0 \end{array}$$

The total cochain complex $\mathcal{P}$ constructed from the projective resolution of the first factor in the torsion product is

$$\begin{aligned} \mathcal{P}^i &= 0, \ i \neq 0 \\ \mathcal{P}^0 &= \mathbb{Z}, \ \delta_{\mathcal{P}} = 0 \end{aligned}$$

Therefore, the torsion product becomes

$$\begin{aligned} &\text{Tor}_{\,H^\bullet(N)}(H^*(0+j0), H^*(D\times\Omega)) \\ &= H(\mathcal{P}^* \otimes_{H^\bullet(N)} H^*(D\times\Omega), \delta^*_{\mathcal{P}} \otimes 1 \pm 1 \otimes \delta_{H^\bullet(D\times\Omega)}) \\ &= H(\mathcal{P}^* \otimes_{\mathbb{Z}} H^*(D\times\Omega), 0) \\ &= \mathcal{P} \otimes_{\mathbb{Z}} H(D\times\Omega) \\ &= H(D\times\Omega) \end{aligned}$$

It follows that

$$E_2^{-n} = \text{Tor}\ {}_{H^\bullet(\{0+j0\})}^{-n}(H^*(N), H^*(D \times \Omega)) = H^n(D \times \Omega)$$

Furthermore, $\text{Tor}^{-n,v}$ is nontrivial for only one v so that the spectral sequence collapses at E_2. Finally, we get

$$H^n(f^{-1}(0 + j0)) = H^n(D \times \Omega)$$

14.8.2 Uncontractible Template

Consider an open-loop unstable system that is robustly stable in the closed loop. The template is an annulus around $0 + j0$. Take a base point $*_N$ in the template and let us compute the cohomology of the fiber $f^{-1}(*_N)$. Referring to the diagram,

$$\begin{array}{ccc} f^{-1}(*_N) & \rightarrow & D \times \Omega \\ \downarrow & & \downarrow \\ *_N & \rightarrow & N \end{array}$$

we construct the Eilenberg–Moore spectral sequence. The computation of the E_2 term is essentially the same as in the open-loop stable case; to be more specific,

$$\begin{aligned} E_2 &= \text{Tor}\ {}_{H^\bullet(N)}(H^*(f^{-1}(*_N)), H^*(D \times \Omega)) \\ &= H(\mathcal{P}^* \otimes_{H^\bullet(N)} H^*(D \times \Omega), 0) \\ &= \mathcal{P} \otimes_{H^\bullet(N)} H(D \times \Omega) \end{aligned}$$

The only difference with the open-loop stable case is that here the tensor product is over the algebra $H^*(N)$, which is not as trivial as in the open-loop stable case. To be more specific,

$$\mathcal{P}^* \otimes_{H^\bullet(N)} H^*(D \times \Omega) = \mathcal{P}^* \otimes_{\mathbb{Z}} H^*(D \times \Omega) / \ \text{im}\,(a_r \otimes 1 - 1 \otimes a_l)$$

In the above, a_r is the right action of $H^*(N)$ on $H^*(*_N)$ and a_l is the left action of $H^*(N)$ on $H^*(D \times \Omega)$.

Part III

HOMOTOPY OF ROBUST STABILITY

15

HOMOTOPY GROUPS AND SEQUENCES

SUMMARY

In this chapter, we "officially" start the homotopy theory of robust stabilization. "Officially" because the concept of homotopy has already pervaded the previous parts, and in particular because semisimplicial bundles and fibrations are traditionally considered to be part of homotopy theory. However, serious homotopy theory starts with the concept of *homotopy groups*, which are introduced in this chapter. Homotopy groups provide yet another algebraic picture of topological spaces, in addition to (co)homology groups. Probably the most important motivation for homotopy groups in robust stability is that they provide the *natural* value groups for the obstructions to extending the Nyquist map to higher-and-higher-dimensional skeleta of the polyhedron of uncertainty. Homotopy groups also yield exact sequences that play a role similar to homology exact sequences. In particular, in this chapter, we derive an exact homotopy sequence linking the homotopy groups of the uncertainty, the crossover, and the template of a robust stability problem.

15.1 Homotopy Groups

Let Y be a topological space with base point $y_* \in Y$. We define the *homotopy set* $\pi_n(Y, y_*)$, $n \geq 0$, to be the set of homotopy classes of maps $g : [0,1]^n \to Y$ sending the boundary of the cube $\partial[0,1]^n$ to the base point y_*. If the boundary of the n-cube is pinched to a point s_*, we get an object that is topologically equivalent to the n-sphere S^n; formally, $[0,1]^n/\partial[0,1]^n \cong S^n$; see [Karoubi 1978, II.2.38]. Therefore, one can equally define an element of $\pi_n(Y, y_*)$ as a homotopy class of all maps $f : S^n \to Y$ sending a privileged point s_* of S^n, typically its "north pole," to y_*. More precisely, consider the following commutative diagram:

$$\begin{array}{ccc} [0,1]^n, \partial[0,1]^n & \overset{g,h}{\to} & Y, y_* \\ \pi \downarrow & \overset{f}{\nearrow} & \\ S^n, s_* & & \end{array}$$

In the above, π is the obvious projection. Clearly, given a map g there exists a unique map f such that the diagram commutes. Conversely, given

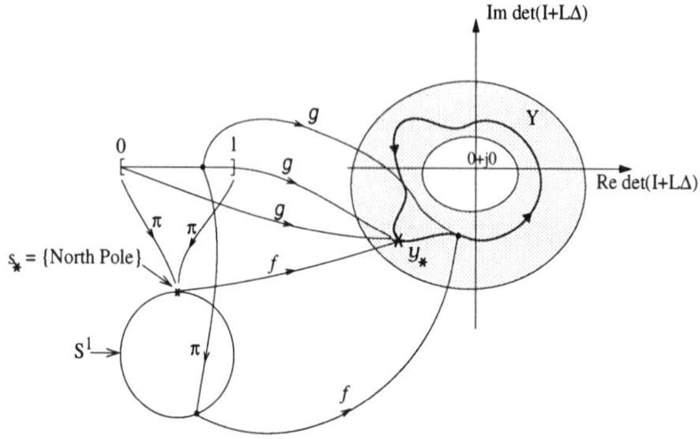

Fig. 15.1. The fundamental or Poincaré group of the template of an open-loop unstable system.

f there exists a map g, unique up to homotopy, that makes the diagram commute.

For $n = 0$, $\pi_0(Y, y_*)$ consists of the homotopy classes of maps from the 0-dimensional cube, or the 0-dimensional sphere, to Y, subject to the condition that $\partial[0,1]^0$, or $s_* \in S^0$, maps to y_*. Clearly, $\pi_0(Y, y_*)$ can be identified with the unique connected component of Y that contains y_*. π_0 is called **zeroth-order homotopy set**.

Next, for $n > 0$, we define a *group operation* on the homotopy set $\pi_n(Y, y_*)$. While mapping the *sphere* might look more aesthetically appealing than mapping the *cube* to the space Y, we prefer the latter formulation to define the group operation. Let $\{g\}$,$\{h\}$ be two homotopy classes of maps. The composition or product $\{g\} * \{h\}$ is defined as the homotopy class of the map

$$\begin{aligned} g * h(x_1, ..., x_n) &= g(2x_1, x_2, ..., x_n), & 0 \leq x_1 \leq 1/2 \\ &= h(2x_1 - 1, x_2, ..., x_n), & 1/2 \leq x_1 \leq 1 \end{aligned}$$

It is easy, although tedious, to show that the homotopy class of the above map does not depend on the representatives g and h. The $*$ operation can be shown to be associative, to have a unit element and an inverse. The homotopy set $\pi_n(Y, y_*)$, $n > 0$, together with the $*$ operation, is called ***n*th-order homotopy group** of Y with base point y_*.

There is no group structure that can be defined on the zeroth homotopy set.

The first homotopy group $\pi_1(Y, y_*)$, referred to as **fundamental group** or **Poincaré group**, is in general non-Abelian—that is, noncommutative. It is illustrated in Figure 15.1. The homotopy groups $\pi_n(Y, y_*)$, $n \geq 2$, referred to as **higher homotopy groups**, *are* Abelian.

A space Y is said to be **simply connected** if $\pi_1(Y, y_*)$ is trivial.

In the same spirit as the Eilenberg–Zilber theorem of homology, the fundamental group is well-behaved under the Cartesian product,

$$\pi_1(Y \times Z) \cong \pi_1(Y) \times \pi_1(Z)$$

However, the fundamental group does not behave well under the union operation (Seifert–Van Kampen theorem; see [Bredon 1993, Chapter III, Section 9]).

For $n = 1$, and if Y is arcwise connected, there is a very simple relation between the homology and homotopy groups. To be more specific, the homology group is the "Abelianized" version of the homotopy group,

$$H_1(Y) = \pi_1(Y)/[\pi_1(Y), \pi_1(Y)]$$

In the above, $[A, A]$ denotes the **commutator subgroup**—that is, the subgroup of A generated by elements of the form $(a*b)*(b*a)^{-1}$, $\forall a, b \in A$.

15.2 Homotopy Group Homomorphism

Given a continuous map $f : X \to Y$, given a homotopy class $\{g : S^n \to X\} \in \pi_n(X)$, there is clearly an induced homotopy class $\{f \circ g : S^n \to Y\} \in \pi_n(Y)$. We *define* $f_n(\{g\}) = \{f \circ g\} \in \pi_n(Y)$, and call $f_* : \pi_*(X) \to \pi_*(Y)$ the **homotopy group homomorphism induced by** f. This fact is depicted by the following commutative diagram:

$$\begin{array}{ccc} X & \xrightarrow{f} & Y \\ \downarrow & & \downarrow \\ \pi_*(X) & \xrightarrow{f_*} & \pi_*(Y) \end{array}$$

In other words, the homotopy functor is *covariant*. Observe that, contrary to homology, we can define the induced homotopy group homomorphism directly from the continuous map, without the need to resort to a simplicial approximation.

15.3 Homotopy Groups of Spheres

One of the most challenging problems is to compute the homotopy groups of the spheres. Since this would take us far beyond the objective of this book and since we will come back to this problem when there will be the need for some specific results, in this section we briefly review some basic facts. It is not hard to show that $\pi_1(S^1) = \mathbb{Z}$. However, it is already more difficult to show that $\pi_n(S^1) = 0$, $n > 1$. Regarding the higher-dimensional spheres, a general feature is that $\pi_n(S^n) = \mathbb{Z}$, $n \geq 1$. Next, $\pi_n(S^m) = 0$ for $1 \leq n < m$. However, quite contrary to homology, $\pi_n(S^m)$ does not in general vanish for $n > m$.

15.4 Basic Obstruction Result

As a prelude to Chapter 16, consider a map

$$S^n, s_* \xrightarrow{f} Y, y_*$$

and let us ask the question as to whether the map can be extended to B^{n+1}; to be more specific, does there exist a map $f^{n+1} : B^{n+1} \to Y$ such that $f^{n+1}|S^n = f$? The fundamental result is the following:

Theorem 15.1. *The map $f : S^n \to Y$ can be extended to B^{n+1} iff $\{f\}$, viewed as an element of $\pi_n(Y)$, vanishes.*

Proof. If the map can be extended to $f^{n+1} : B^{n+1} \to Y$, it is easy to deform it to a constant map. Restricting the deformation to S^n yields a deformation of f to the constant map. Hence $\{f\} = 0$, as claimed. Conversely, assume $\{f\} = 0$. Let $F : S^n \times [0,1] \to Y$ be a homotopy such that $F(x,1) = f(x)$ and $F(x,0) = y_*$. Assume the sphere is embedded in the Euclidean space $\mathbb{R}^{n+1}$ with its center at the origin of the Euclidean space. Then any point of B^{n+1} can be written as xt for some $x \in S^n$ and some $t \in [0,1]$. Furthermore, the xt factorization is continuous and unique, except at the center of the ball B^{n+1} where $t = 0$ and x is undefined. It is claimed that an extension is $f^{n+1}(xt) = F(x,t)$. Indeed, except possibly at the origin, this extension is continuous and agrees with f on S^n because indeed $f^{n+1}(x1) = F(x,1) = f(x)$. At the origin, $t = 0$ and x is undefined, but this does not matter, because indeed $f^{n+1}(x0) = F(x,0) = y_*$, $\forall x \in S^n$. ■

At this stage, the reader can perceive the ultimate motivation for the homotopy groups. Indeed, $\{f\}$ can be thought of as the "obstruction" to extending the map and takes value in $\pi_n(Y)$. *The homotopy groups can therefore be viewed as value groups for obstruction to extending maps.* In particular, if $\pi_n(Y)$ vanishes, there are no obstructions to extending.

15.5 Homotopy Sequence of Nyquist Fibration

In Part II, the *spectral sequence* of the Nyquist map was presented as a recursive procedure to establish the connections between the *homology* groups of the uncertainty space, the fiber $f^{-1}(0+j0)$, and the template N. Needless to say, these homology relations are shrouded in a complicated algebraic machinery. This section is in the same spirit of investigation, except that we simplify the problem by resorting to the *homotopy*, rather than homology, groups of the uncertainty, the crossover, and the template. It turns out that one can achieve a fairly simple relationship—in terms of an exact sequence between *homotopy* groups of uncertainty, template, and crossover—provided that we assume that the Nyquist map is a *fibration*. The latter condition is somehow equivalent to relinquishing some of the fine structure of the problem and study the Nyquist map *up to homotopy equivalence.*

15.5.1 Exact Homotopy Sequence

Let $f : P \to N$ be a base point preserving fibration. A base point $*$ is chosen in N, the fiber is chosen as $F = f^{-1}(*)$, and the base point $*$ of P is chosen in F. The base point preserving map f induces the homotopy groups homomorphism $f_* : \pi_n(P,*) \to \pi_n(N,*)$. Let $i : F \to P$ be the inclusion map, and let $i_* : \pi_n(F) \to \pi_n(P)$ be the homotopy group homomorphism induced by i. From the fibration property, it is not hard to show (see [McCleary 1985, Theorem 4.19]) that the following diagram has all of its rows exact:

$$\begin{array}{ccccc} \pi_{n-1}(F) & \xrightarrow{i_*} & \pi_{n-1}(P) & \xrightarrow{f_*} & \pi_{n-1}(N) \\ \uparrow & & & & \\ & \longleftarrow & \xleftarrow{\partial_*} & \longleftarrow & \\ & & & & \uparrow \\ \pi_n(F) & \xrightarrow{i_*} & \pi_n(P) & \xrightarrow{f_*} & \pi_n(N) \end{array}$$

To construct the long exact sequence, we need a "connecting" boundary homomorphism $\partial_* : \pi_n(N) \to \pi_{n-1}(F)$ which is defined as follows: Take a representative

$$\alpha : [0,1]^n, \partial[0,1]^n \to N, *$$

of an arbitrary $\{\alpha\} \in \pi_n(N)$. This representative induces a homotopy:

$$\begin{array}{lcl} [0,1]^{n-1} \times [0,1] & \to & N \\ {[0,1]^{n-1}} \times \{0\} & \to & * \end{array}$$

Let $* : [0,1]^{n-1} \to *$ be the constant map into the base point of P. We assemble everything in the following commutative diagram:

$$\begin{array}{lcl} [0,1]^{n-1} \times \{0\} & \xrightarrow{*} & P \\ & & \downarrow f \\ {[0,1]^{n-1}} \times [0,1] & \xrightarrow{\alpha} & N \end{array}$$

By the fibration property, this homotopy can be lifted to P; that is, there exists $A : [0,1]^{n-1} \times [0,1] \to P$ such that the above diagram commutes. Therefore, $\partial_*\alpha$ is defined as the homotopy class of the map $A|[0,1]^{n-1} \times \{1\} : [0,1]^{n-1} \times \{1\} \to F$ and we have

Theorem 15.2. *If $f : P \to N$ is a fibration with typical fiber $F = P_a$, then there exists an exact sequence*

$$\ldots \to \pi_n(F) \xrightarrow{i_*} \pi_n(P) \xrightarrow{f_*} \pi_n(N) \xrightarrow{\partial_*} \pi_{n-1}(F) \to \ldots \to \pi_0(P) \to \pi_0(N) \to 0$$

Proof. We have already proved existence of ∂_*; exactness at the domain and the target of ∂_* is easily proved; for the details, see [Hu 1959, page

152]. ■

Clearly, the Nyquist template has the property that $\pi_n(N) = 0, n \geq 2$. Therefore, the long exact sequence splits into several short exact sequences.

Corollary 15.1. *Whenever* $\pi_n(N) = 0, n \geq 2$, *then*

$$0 \to \pi_n(F) \to \pi_n(P) \to 0, n \geq 2$$

In other words, $\pi_n(F)$ *and* $\pi_n(P)$ *are isomorphic.*

15.6 Corollaries of Exact Homotopy Sequence

As we have seen, the fundamental group is well-behaved under Cartesian product; actually, the higher homotopy groups are well-behaved as well:

Corollary 15.2.

$$\pi_n(Y \times Z) \cong \pi_n(Y) \oplus \pi_n(Z), \quad n \geq 2$$

Proof. Consider the natural projection $\pi : Y \times Z \to Y$. It is clearly a fibration with fiber Z. The exact homotopy sequence therefore yields

$$\ldots \to \pi_n(Z) \to \pi_n(Y \times Z) \xrightarrow{\pi_*} \pi_n(Y) \xrightarrow{\partial_*} \pi_{n-1}(Z) \to \ldots$$

The fibration $\pi : Y \times Z \to Y$ clearly has a cross section $c : Y \to Y \times Z$, $y \mapsto (y, z_*)$ such that $\pi c = 1_Y$. The latter in turn induces a homomorphism $c_* : \pi_n(Y) \to \pi_n(Y \times Z)$ such that $\pi_* c_* = 1_{\pi_n(Y)}$. It follows that π_* is an epimorphism so that $\partial_* = 0$. The long exat sequence therefore splits into several short exact sequences

$$0 \to \pi_n(Z) \to \pi_n(Y \times Z) \underset{\pi_*}{\overset{c_*}{\rightleftarrows}} \pi_n(Y) \to 0$$

This short exact sequence splits for $n \geq 2$, see Theorem A.11, so that $\pi_n(Y \times Z) = \pi_n(Y) \oplus \pi_n(Y)$, as claimed. ■

As we did for homology, we can compute the homotopy groups of the product $D \times \Omega$ in terms of the homotopy groups of the factors. From the above, we get

$$\pi_n(D \times S^1) \cong \pi_n(D) \oplus \pi_n(S^1)$$

In particular, for $n = 1$, we get

$$\pi_1(D \times S^1) \cong \pi_1(D) \oplus \mathbb{Z}$$

For $n > 1$, using the nontrivial result that the higher homotopy groups of S^1 vanish, we get

$$\pi_n(D \times S^1) \cong \pi_n(D), \quad n > 1$$

Finally, for illustrative purposes, we focus on the homotopy groups of $D = \mathbb{T}^{n_q}$ relevant to the multichannel phase margin problem. Not unlike the homology result, we get

$$\pi_1(\mathbb{T}^{n_q}) \cong \underbrace{\mathbb{Z} \oplus \ldots \oplus \mathbb{Z}}_{n_q \text{ terms}}$$

However, *contrary to homology,* we have

$$\pi_n(\mathbb{T}^{n_q}) = 0, \quad n > 1$$

15.7 HISTORICAL NOTES

The fundamental group π_1 was developed by Poincaré around the turn of this century. The higher homotopy groups were formulated by Hurewicz in the 1930s.

The difficulty of computing the homotopy groups motivated the purely combinatorial approach to homotopy groups due to [Kan 1955a,b]. To be slightly more specific, given a Kan (semisimplicial) complex K with base point k_*, define $L_n = \{\sigma \in K_n : F^i\sigma = k_*, \, i = 0, ..., n\}$; two simplexes are said to be homotopic, $\sigma_n \simeq \tau_n$, iff there exists a simplex σ_{n+1} such that

$$\begin{aligned} F^i\sigma_{n+1} &= k_*, \quad \text{for} \quad i < n \\ &= \sigma_n, \quad \text{for} \quad i = n \\ &= \tau_n, \quad \text{for} \quad i = n+1 \end{aligned}$$

The reader can easily verify that $\simeq$ is an equivalent relation whenever K is a Kan complex. Hence, the homotopy groups can be defined by $\pi_n(K, k_*) = L_n / \simeq$.

The reader can easily verify that, for $n = 1$, the above definition is a formalization of the *group of edge-loops* over a vertex of a simplicial complex proposed by [Hilton and Wylie 1965, Section 6.3] for computing the fundamental group of a polyhedron. The edge-loop group is a combinatorial, computationally tractable version of the fundamental group. To be specific, given a simplicial complex K together with a base-point vertex a^*, define an *edge-loop* to be a polygonal arc consisting of 1-simplexes of the complex interconnected at their common vertices starting at a^* and ending at a^*. An equivalence relation among edge-loops is defined—Two edge-loops are *equivalent* iff one can be derived from the other using such allowable operations as replacing $\overline{a^0a^1} \cup \overline{a^1a^2}$ by $\overline{a^0a^2}$ whenever $a^0a^1a^2$ is a simplex. The set of (equivalence classes of) edge-loops together with the obvious composition operation form a group, the group of edge-loops of the complex, isomorphic to the fundamental group.

The reader can also verify that the above construction bears some striking similarity with the so-called *"computation of monotone simple circuits in the plane,"* a construction of computational geometry (see [Toussaint 1988]).

16

OBSTRUCTION TO EXTENDING THE NYQUIST MAP

SUMMARY

In this chapter, as well as in the subsequent chapters of Part III, we take a radically new approach, to the same problem though. Rather than focusing on the crossover $f^{-1}(0+j0)$, we *a priori* remove $0+j0$ from the complex plane, and our primary concern is *existence* of the Nyquist map $f : D\times\Omega \to \mathbb{C}\setminus\{0+j0\}$. Should the Nyquist map exist, then the closed-loop system is stable (provided the open-loop system is stable).

If, however, the Nyquist map $D\times\Omega \to \mathbb{C}\setminus\{0+j0\}$ does not exist, there are some questions as to how this mathematical fact has to be interpreted in the proper control engineering context. What is typically happening is that the Nyquist map exists over some low-dimensional skeleton $(D\times\Omega)^n$, while the Nyquist map is not *extendable* to the entire polyhedron.

The problem of extension of continuous maps, typically extension from $(D\times\Omega)^n$ to $D\times\Omega$, is a central problem in topology. Extension may or may not exist. The approach, due to Eilenberg, consists in a recursive test on the dimension n. Assume the map over $(D\times\Omega)^n$ exists; we attempt to construct a map over $(D\times\Omega)^{n+1}$ that agrees with the map over $(D\times\Omega)^n$; sometimes in this process, an "obstruction" develops. Mathematically, the obstruction takes value in the homotopy group π_n of the range space. This probably provides *one of the most important motivations of homotopy groups*, as already argued in Chapter 15. The extension exists iff the *obstruction* vanishes.

It should be intuitively clear at this stage that Kharitonov's theorem and the "edge test" can be reformulated within the setup of obstruction theory: We perform a test on a low-dimensional skeleton of the uncertainty space and, if the test passes, there are no further obstructions to extending the Nyquist to the entire uncertainty space and the system is robustly stable. Gleaning on this insight, it should be clear that obstruction theory provides the conceptual framework to dispose of such issues as to *when*, *why* and *how* a test on a low-dimensional subset of uncertainty is enough to guarantee robust stability.

16.1 Statement of Nyquist Extension Problem

As usual, define a generalized space of uncertainties where the frequency is treated as yet another uncertain parameter

$$P = D \times \Omega$$

It is essential to assume that P is a *polyhedron*, with an underlying simplicial complex structure. To justify this assumption, we invoke Alexandroff's theorem together with the prismatic triangulation of the product space $D \times \Omega$. The central issue is the Nyquist mapping

$$\begin{aligned} f : P &\rightarrow S^2 \\ (q, \omega) &\mapsto f(q, \omega) = \det(I + L(j\omega)\Delta(q)) \end{aligned}$$

Here, for a change, we plot the Nyquist curve on the Riemann sphere S^2. We also assume that the open-loop system $L(s)$ is stable. For homotopy reasons, we remove the north pole ∞ of the Riemann sphere. Indeed, if we leave the north pole as is, a Nyquist curve circling around $0 + j0$ could be continuously deformed through the north pole into a point and hence appear nullhomotopic, although in this control context it has to be considered of a nonvanishing homotopy class. For most obvious stability reasons, we also remove the south pole $0 + j0$, or a small disk around the south pole. This yields the doubly punctured Riemann sphere, S^2_{**}, which is clearly homeomorphic to the punctured complex plane,

$$S^2_{**} \cong \mathbb{C} \setminus \{0 + j0\}$$

The robust stability problem is to *decide* whether

$$f(P) \subseteq S^2_{**}$$

We hasten to say that in addition to the above there is an encirclement condition, which will be taken care of by the *homotopy* extension formulation.

Computing the entire image $f(P)$ and checking whether it is included in S^2_{**} might entail too much computation so that we have to take some shortcut. Let P^n be the set of m-simplexes, $m \leq n$, of P, also called the n-**skeleton**. The idea is to compute, say $f(P^n)$, check whether it is included in S^2_{**}, and, if the answer is yes, try to determine whether it is possible to conclude that $f(P) \subseteq S^2_{**}$. Kharitonov's theorem clearly falls within this category of problems: Since testing Hurwitzness is equivalent to a frequency sweep, it follows that the Kharitonov Hurwitzness test on selected vertices is equivalent to a test on (a subset of) $(\text{Cube} \times \Omega)^1$.

To reformulate the problem in a precise topological set up, assume we know that the restriction of f to P^n

$$f|P^n = f^n$$

passes the robust stability test

$$f^n(P^n) \subseteq S^2_{**}$$

In other words, the map

$$f^n: P^n \to S^2_{**}$$

exists. Therefore, the problem is to figure out whether the map f^n can be extended to the entire polyhedron P. It is convenient to represent this **extension** problem by the commutative diagram [Whitehead 1978]:

$$\begin{array}{ccc} (D \times \Omega)^n & \stackrel{f^n}{\rightarrow} & \mathbb{C} \setminus \{0 + j0\} \\ i \downarrow & \stackrel{?}{\nearrow} & \\ (D \times \Omega) & & \end{array}$$

where i denotes the inclusion and the arrow $\stackrel{?}{\rightarrow}$ represents the map the existence of which is in question.

Extension of continuous maps is a central problem in topology. Eilenberg proposed an approach to extending continuous maps. It involves recursively extending the map to higher-dimensional simplicial complexes. At each step—that is, when jumping from n-simplexes to $(n+1)$-simplexes—the extension is possible iff the so-called *obstruction cocycle* vanishes. Intuitively, the obstruction cocycle measures the obstruction that one will encounter when extending the map over n-simplexes to a map over $(n+1)$-simplexes. To be somewhat more specific, there is an obstruction whenever for some simplex σ_{n+1} the image $f^n(\partial\sigma_{n+1})$ "wraps around" $0 + j0$. This "wrapping" can be formalized as a nontrivial homotopy class of $f|\partial\sigma_{n+1}$; since $\partial\sigma_{n+1} \cong S^n$, this "wrapping" can equivalently be formalized as a cochain taking value in the *homotopy group* $\pi_n(\mathbb{C} \setminus \{0 + j0\})$. Next, to deal with the nesting properties of the extension across two dimensional jumps, typically from $(D \times \Omega)^{n-1}$ to $(D \times \Omega)^{n+1}$, we have to consider the *cohomology class* of the obstruction in $H^{n+1}(P, \pi_n(S^2_{**}))$.

In the above, we have treated ω as a mere uncertain parameter, temporarily ignoring the privileged role it plays. As such, we have lost the important concept that there is a Nyquist *curve* for every $q \in D$. To restore this concept, we have to formulate the problem of extending *the whole Nyquist curve* $f(q, \omega \in \Omega)$ from the n-skeleton of D, D^n, to the entire space of uncertainty D. The Nyquist curve can be viewed as a homotopy

$$\begin{aligned} f_t(q) : D \times [0,1] &\rightarrow S^2_{**} \\ (q, t) &\mapsto \det(I + L(\frac{\exp(2\pi jt) + 1}{\exp(2\pi jt) - 1})\Delta(q)) \end{aligned}$$

In other words, homotopy is used as a mathematical formulation of the intuitive idea of continuous deformation of the Horowitz template under frequency sweep. From this point of view, the new extension problem is to determine whether the partial homotopy f_t^n, or Nyquist curve, over D^n can be extended over to D. This is the **homotopy extension** problem.

It is convenient to represent it by the following diagram:

$$\begin{array}{ccc} D \times \partial[0,1] \cup D^n \times [0,1] & \stackrel{f_t^n}{\rightarrow} & \mathbb{C} \setminus \{0 + j0\} \\ \downarrow i & \stackrel{?}{\nearrow} & \\ D \times [0,1] & & \end{array} \tag{16.1}$$

The above *homotopy extension problem* is the dual of the *homotopy lifting problem* of Chapter 13. To get the feeling for this duality, it suffices to compare Diagram 16.1 and Diagram 13.3. This duality is developed in [James 1984, Chapter 6]. A map (typically Nyquist map) that has the homotopy lifting property is said to be a *fibration*. In view of the duality, a map (typically the inclusion map) that has the homotopy extension property is said to be a **cofibration**.

It is important to understand the difference between the extension problem and the homotopy extension problem. The extension problem does not address such an issue as the homotopy type of the curve obtained by varying ω keeping all other parameters fixed. The homotopy extension problem, on the other hand, does address this issue. In the homotopy extension problem, if the Nyquist curve is nullhomotopic on D^0, all Nyquist curves resulting from the homotopy extension will remain null homotopic over D^n.

The reader will already have realized that Kharitonov's theorem involves a homotopy extension from the 0-th skeleton D^0 to D. Likewise, the "edge test" of [Bartlett, Hollot, and Lin 1988] can be interpreted as a homotopy extension from the 1-skeleton D^1. In both cases, if the test on the low-dimensional skeleton passes, then, because of the topological nature of the map, no obstruction develops, the homotopy extension exists, and the system is robustly stable.

16.2 Obstruction to Extending a General Map

We quickly review the obstruction to extending maps—a very general topological problem. The standard Nyquist map is into S^2_{**}. However, as we alluded to in Section 2.8, and as several examples in Chapter 3 clearly showed, there are robust stability problems that are more naturally characterized by maps into *other* spaces. In particular, in Chapter 19, we will map the uncertainty into the return difference *matrix*. Therefore, for the sake of generality, and also for the sake of clarity, in this section we consider a map

$$f: P \to Y$$

from a polyhedron to a *general* topological space Y. In the next section, we will specialize the results to $Y = S^2_{**}$. The reader anxious to get to the point might want to already replace Y by S^2_{**}.

16.2.1 Path Connectedness of Range

The extension problem depends *crucially* on the *homotopy groups* $\pi_n(Y, y_0)$ of the range space Y relative to some *base point* $y_0 \in Y$.

The homotopy groups $\pi_n(Y, y_0)$ depend on the base point y_0. Clearly, this base point makes a crucial difference in case Y has more than one path connected components. However, if Y is path connected, the dependency of π_n on y_0 becomes somewhat looser. More precisely, let Y be path connected, let y_0 and y_1 be two base points and let w be a path connecting them. Then there exists an isomorphism

$$i(y_0, y_1, w) : \pi_n(Y, y_0) \rightarrow \pi_n(Y, y_1)$$

Unfortunately, this isomorphism depends on the path that joins y_0 to y_1, so that even when the space Y is path connected its homotopy groups still depend on the base point. To remove the dependency of the homotopy groups on the base point, not only do we have to require Y to be path connected, but in addition we have to require the isomorphism: $\pi_n(Y, y_0) \rightarrow \pi_n(Y, y_1)$ to be independent on the path that joins y_0 to y_1. Under those circumstances, the space Y is said to be **simple**. From now on, we assume Y to be *simple*, and drop the dependency of the homotopy group on the base point.

16.2.2 Absolute Obstruction to Extension

Assume we know, or have computed, the map

$$f^n : P^n \rightarrow Y$$

and we want to extend the map to $P^{(n+1)}$. Let σ_{n+1} be an oriented $(n+1)$-simplex of P. We define a cochain c_f of

$$C^{n+1}(P, \pi_n(Y))$$

by the rule

$$(\sigma_{n+1}, c_f) = \{f|\partial\sigma_{n+1}\}$$

In the above, $C^{n+1}(P, \pi_n(Y))$ is the Abelian group of $(n+1)$-cochains defined on the $(n+1)$-simplexes of P and taking value in $\pi_n(Y)$, the n-homotopy group of Y. The notation $\{f|\bullet\}$ refers to the homotopy class of f restricted to $\bullet$. $\partial\sigma_{n+1}$ is the frontier of σ_{n+1}; it is the closure of σ_{n+1} minus σ_{n+1}; it is an n-dimensional simplicial complex homeomorphic to S^n. The cochain c_f is referred to as **obstruction cochain**. Regardless of whether or not an extension exists, the obstruction has the following property:

Theorem 16.1. *The obstruction cochain c_f is a cocycle of the cochain complex $C^*(P, \pi_n(Y))$.*

Proof. See [Hilton and Wylie 1965, Theorem 7.2.3, page 292]. ∎

c_f is called **obstruction cocycle**. The motivation for the terminology of *obstruction* cocycle stems from the following:

Theorem 16.2. *f^n can be extended over to P^{n+1} iff $c_f = 0$.*

Proof. This follows from Theorem 15.1. See also [Hilton and Wylie 1965, Proposition 7.2.1, page 291]. ∎

Clearly, the problem of extending f^n from dimension n to dimension $(n+1)$ is taken care of by the above theorem. Consider, however, the recursion on the dimension more carefully. Assume f^{n-1} can be extended to f^n in a nonunique way. Therefore, we choose one f^n and seek an extension f^{n+1}. If such an extension exists—that is, $c_{f^n} = 0$—fine. If, however, the extension f^{n+1} does not exist, we might have to come back to the previous extension from $(n-1)$ to n and revise our choice of f^n. Obstruction theory provides us with a procedure to correct the initial guess at the level of the extension from $(n-1)$ to n. We skip the detail (see [Whitehead 1965, Section 2.7]) and just write out the key results.

Theorem 16.3. *If f^n and f'^n are two different extensions of f^{n-1}, then the cohomology classes of their obstructions $\{c_{f^n}\}$ and $\{c_{f'^n}\}$ are equal in the cohomology group $H^{n+1}(P, \pi_n(Y))$.*

Proof. See [Hilton and Wylie 1965, Theorem 7.2.6, page 293]. ∎

Theorem 16.4. *Let f^{n-1} have an extension f'^n with obstruction $c_{f'^n}$ cohomologous to the cocycle z in $C^{n+1}(P, \pi_n(Y))$. Then, there exists an extension f^n such that $c_{f^n} = z$.*

Proof. See [Hilton and Wylie 1965, Theorem 7.2.7, page 293]. ∎

Theorem 16.5. *If f^{n-1} has an extension f^n, then f^{n-1} has an extension over to P^{n+1} iff c_{f^n} is cohomologous to 0.*

Proof. See [Hilton and Wylie 1965, Theorem 7.2.8, page 293]. ∎

Combining the above three theorems, the problem of the extension from $(n-1)$ to $(n+1)$ should be clear: Start with f^{n-1}. Assume $c_{f^{n-1}} = 0$. Find a provisional extension f'^n such that its obstruction $c_{f'^n}$ is a coboundary—that is, $\{c_{f'^n}\} = 0$. Then use this data to revise the choice and find an extension f^n such that $c_{f^n} = 0$. This last choice of f^n can be extended over to f^{n+1}.

The above procedure has limitations: Assume we know f^0. The above allows us to check existence of f^2. Assume f^2 exists. If we try to extend to f^3, an extension from f^2 may not exist, although an extension of f^0 exists. This is to say that obstruction theory has limitations. As we shall see soon, almost by miracle, these difficulties evaporate in the robust control problem.

16.2.3 Relative Obstruction to Extension

Let Q be a subcomplex of P. Assume the partial Nyquist map $Q \to S^2_{**}$ exists. We want to extend the Nyquist map over P. Let P^n be the n-skeleton of P. We start by extending the map over $P^0 \cup Q$ by assigning arbitrary images to the 0-simplexes—that is, vertices—of $P \bmod Q$. We then extend it over to $P^1 \cup Q$, and if this not possible, we revise our choice of extension over to $P^0 \cup Q$. We therefore proceed recursively as in the absolute case, with the same limitation. In this context, to define the obstruction cocycle, we have to consider the *relative* cochain complex

$$C^{n+1}(P \bmod Q, \pi_n(Y))$$

By definition $C^{n+1}(P \bmod Q, \pi_n(Y))$ is the group of linear forms defined on the $(n+1)$-simplexes of P, taking values in $\pi_n(Y)$, and "vanishing" on the $(n+1)$-simplexes of Q. By "vanish" we mean that the linear form takes as value the identity or null element of the group $\pi_n(Y)$.

Theorem 16.6. *Assume the map $f : P^n \cup Q \to Y$ exists for a subcomplex $Q \subseteq P$. The map can be extended to $P^{n+1} \cup Q$ iff the obstruction cocycle c_{f^1} of the cochain complex $C^{n+1}(P \bmod Q, \pi_n(Y))$ vanishes.*

16.3 Obstruction to Extending Nyquist Map

The first issue is the homotopy characteristic of S^2_{**}. Observe that the doubly punctured Riemann sphere, the image of the map f, is a *path connected* space. It follows that its homotopy groups $\pi_n(S^2_{**}, s_0)$ do not depend on the base point s_0, in the sense that given two base points s_0 and s_1 and a path joining them there exists an isomorphism $\pi_n(S^2_{**}, s_0) \to \pi_n(S^2_{**}, s_1)$. The problem is that this isomorphism might depend on the path that joins s_0 to s_1. Therefore, we shall, at least temporarily, keep the based point dependency in the notation, $\pi_n(S^2_{**}, s_0)$.

Let us first look at $\pi_0(S^2_{**}, s_0)$. Since S^2_{**} is path-connected, the *set* π_0 is in one-to-one correspondence with a singleton, the unique path connected component of S^2_{**}. Therefore, π_0 can safely be considered as null.

We now look at the first and higher homotopy groups. The key result is that the doubly punctured sphere S^2_{**} is *homotopically equivalent* to the unit circle, or 1-sphere, S^1.

Theorem 16.7. *The spaces S^2_{**} and S^1 are* **homotopically equivalent***; that is, there exist mappings*

$$\begin{aligned} g : S^2_{**} &\to S^1 \\ h : S^1 &\to S^2_{**} \end{aligned}$$

such that $g \circ h$ and $h \circ g$ are homotopic to the identity mapping.

Proof. It suffices to take g to be the orthogonal projection along an arc of great circle onto the "equator" which is homeomorphic to S^1. (The mapping $g : S^2_{**} \to S^1$ is a *retract* of S^2_{**} into S^1.) Furthermore, h is defined to be the inclusion map Equator($= S^1$) $\to S^2_{**}$. ■

Theorem 16.8. *Since S^2_{**} and S^1 are homotopically equivalent, they have isomorphic homotopy groups, namely, $\pi_n(S^2_{**}, s_0) = \pi_n(S^1, t_0)$.*

Proof. It is a fairly general result that homotopically equivalent spaces have isomorphic homotopy groups (see [Hocking and Young 1961, page 185]). ■

Theorem 16.9. *The fundamental group $\pi_1(S^2_{**}, s_0) = \pi_1(S^1, t_0)$ is on one generator, is Abelian, and is simple; that is, it does not depend on the base point.*

Proof. The fact that the first homotopy group of the circle is on one generator is obvious and well-known. Clearly, the generator of $\pi_1(S^2_{**}, s_0)$ could be taken to be the equator whenever s_0 lies on the equator. Furthermore, it is a purely algebraic fact that a group on one generator is Abelian. Finally, it follows from [Spanier 1989, page 51] that whenever $\pi_1(\cdot, s_0)$ is Abelian for a base point (and hence for all base points), the isomorphism $\pi_1(\cdot, s_0) \to \pi_1(\cdot, s_1)$ is unique—that is, does not depend on the path that joins s_0 to s_1. Therefore, S^1 and S^2_{**} are simple. ■

Regarding the higher homotopy groups, here is the key result:

Theorem 16.10. $\pi_n(S^2_{**}, s_0) = \pi_n(S^1, s_0) = 0, n \geq 2$.

Proof. The fact that the higher homotopy groups of the circle S^1 are null is a known, although nonobvious, result. See [Freudenthal 1937a] and [Hilton and Wylie 1965, 6.615]. ■

Since $\pi_n(S^2_{**}, s_0), n \geq 2$, are null, they are of course Abelian and they do not depend on the base point.

To sum up, the range space S^2_{**} is path-connected, is simple (the homotopy groups do not depend on the base point and we therefore drop this dependency in the notation), and, furthermore, $\pi_0 = 0, \pi_1 = \mathbb{Z}$, and $\pi_n = 0, n \geq 2$.

Now, we look more carefully at the recursion in this particular case of the robust stability problem.

16.3.1 Absolute Results

Clearly, $C^1(P, \pi_0(S^2_{**})) = 0$, so that there are no obstructions to be worried about when jumping from P^0 (vertices) to P^1 (edges).

Likewise, for $n \geq 2$, $C^{n+1}(P, \pi_n(S^2_{**})) = 0$, so that no obstruction could possibly develop from P^n to P^{n+1}.

Therefore, the only obstruction is while jumping from P^1 (edges) to P^2 (faces), since the cochain group $C^2(P, \pi_1(S^2_{**}))$ does not in general vanish.

Theorem 16.11. *A necessary condition for robust stability over the entire polyhedron $P = D \times \Omega$ is that the obstruction cocycle c_{f^1} of the Nyquist map restricted to the edges of the polyhedron vanishes.*

16.3.2 Relative Results

A relative version of the above also exists. To be more precise, let Q be a subcomplex of P and let P^n be the set of m-simplexes, $m \leq n$, of P. Assume the map f is known over Q, and we want to extend it over to P. We start by extending it over to $P^0 \cup Q$ by assigning arbitrary images to the 0-simplexes—that is, vertices—of $P \bmod Q$. We then extend it over to $P^1 \cup Q$, and if this not possible, we revise our choice of extension over to $P^0 \cup Q$. We therefore proceed recursively as above, with the same limitation. In this context, to define the obstruction cocycle, we have to consider the relative cochain complex

$$C^{n+1}(P \bmod Q, \pi_n(S^2_{**}))$$

By definition, $C^{n+1}(P \bmod Q, \pi_n(S^2_{**}))$ is the group of linear forms defined on the $(n+1)$-simplexes of P, taking values in $\pi_n(S^2_{**})$, and "vanishing" on the $(n+1)$-simplexes of Q. By "vanish" we mean that the linear form takes as value the identity or null element of the group $\pi_n(S^2_{**})$.

Theorem 16.12. *Assume the feedback system is robustly stable for $Q \subseteq P$. Necessary for the system to be robustly stable all over P is that the obstruction cocycle c_{f^1} of the cochain complex $C^2(P \bmod Q, \pi_1(S^2_{**}))$ vanishes.*

Robust control applications of this *relative* obstruction theory are easy to conceive. The typical example is whether stability of a feedback system with an uncertainty of the form

$$\begin{pmatrix} q_1 & \dots & 0 \\ \vdots & \ddots & \vdots \\ 0 & \dots & q_n \end{pmatrix}, \quad |q_i| \leq 1$$

is equivalent to stability of a system with uncertainty

$$\begin{pmatrix} \exp(-j\theta_1) & \dots & 0 \\ \vdots & \ddots & \vdots \\ 0 & \dots & \exp(-j\theta_n) \end{pmatrix}$$

In this case, $Q = (\mathbb{T} \times \ldots \times \mathbb{T}) \times \Omega$, while $P = (\overline{\mathbb{D}} \times \ldots \times \overline{\mathbb{D}}) \times \Omega$. The problem is hence to figure out whether checking stability over Q is enough to guarantee stability over P. Again, the answer is obstruction theory.

16.4 Weak Converse

Clearly, from Theorem 16.11, the absence of obstruction is *necessary* for robust stability. Now, assume that no obstruction develops all the way to the entire polyhedron P. Is this *sufficient* for robust stability? The difficulty of the problem is compounded by the fact that, if no obstruction exists, the extension is far from unique (although all extensions are somewhat related).

An arbitrary extension from P^0 to P^1 of course agrees with f on the vertices of P. Subject to this restriction, there are several possible extensions. The idea is to choose as extension the simplicial approximation, *any* simplicial approximation, that agrees with the Nyquist map f on the vertices of P. We prove that such a simplicial approximation exists.

Theorem 16.13. *Consider the polyhedron P with vertices $\{a^i\}$, together with the Nyquist map $f : P \to \mathbb{C}$. Let $b^i = f(a^i)$ be the images of the vertices of P. Let the set of points $\{b^i\}$ be triangulated using, for example, the Delaunay triangulation of computational geometry, resulting in the complex N. Then there exists a refined polyhedron P' and a simplicial approximation $\tilde{f} : P' \to N$ that agrees with f on the vertices of P.*

Proof. This requires just a slight amendment of the proof of the simplicial approximation theorem. When refining the polyhedron, one has to refine it small enough such that $f(\text{star}(a^i)) \subseteq \text{star}(b^i)$ so that the simplicial approximation is allowed to transform the vertex a^i into $b^i = f(a^i)$. ■

If $\tilde{f}$ passes the crucial obstruction test $c_{\tilde{f}1} = 0$, it is easily seen that the simplicial approximation passes the robust stability test in the sense that $\tilde{f}(P) \subseteq S^2_{**}$.

The mismatch between the original map and the simplicial approximation can be evaluated using the Hessian of f. To be more precise, let $\mu(P)$ be the mesh of P—that is, the supremum of the diameters of all simplexes of P. Define

$$\epsilon = \frac{\mu^2}{2}\sqrt{||H_r||^2 + ||H_i||^2}$$

where H_r and H_i are the Hessian of the real and imaginary parts, respectively, of f. Then an upper bound on the mismatch is given by

$$|f(p) - \tilde{f}(p)| \leq \epsilon, \forall p$$

From there on, the weak converse is as follows:

Theorem 16.14. (Weak Converse) *Let $\tilde{f} : P' \to N$ be a simplicial approximation that agrees with the Nyquist map $f : P \to N$ over the vertices of P. If the obstruction cocycle $c_{\tilde{f}1}$ vanishes, then the robust stability test $\tilde{f}(P) \subseteq S^2_{**}$ passes for the simplicial approximation of the Nyquist map, which is an ϵ-perturbation of the original Nyquist map.*

16.5 Computation of Homotopy Class

The preceding section has left us with the central problem of computing the homotopy class $\{f^1|\partial\sigma_2\}$ of the Nyquist map restricted to the boundary of an arbitrary 2-simplex σ_2 of the polyhedron P.

In this simple case of a map to the unit circle, it is easy to "visualize" the situation. We plot the image $f^1(\partial\sigma_2)$ and one can safely identify the obstruction cocycle with the number of times the image of $f|\partial\sigma_2$ circles around $0 + j0$,

$$\{f^1|\partial\sigma_2\} = \frac{1}{2\pi j}\oint_{\partial\sigma_2} d\log f^1(p)$$

It is because the punctured complex plane has a simple homotopy structure that there is such an easy formulation of the obstruction. In case of a map into a more general space Y, the generalization of the above is not easy; it essentially involves the algebraic topological concept of *degree*, which itself relies on invariant differential forms on Y; we will come back to this in Chapter 19.

Coming back to the above recipe, by (bi)linearity, the contour of integration could be any closed path along the edges of the polyhedron of uncertainties $P = D \times \Omega$. If the polyhedron P is prismatically triangulated, there exists a path along the edges of P along which $\Delta =$ constant. Then the above obstruction test boils down to the usual necessary and sufficient Nyquist stability criterion for open-loop stable feedback systems.

As an example, let us come back to the "2-torus" example of Equations 9.1–9.2 with the relevant part of the uncertainty space shown in Figure 9.6. Consider the completely labeled simplex σ_2 on the $\omega = 1$ rad/sec slice. The image of the boundary $\partial\sigma_2$ under the Nyquist map is shown in Figure 16.1. Clearly,

$$\{f^1|\partial\sigma_2\} = \frac{1}{2\pi j}\oint_{\partial\sigma_2} d\log\det(I + L\Delta) = -1$$

Therefore, it is impossible to extend the Nyquist map from $\partial\sigma_2$ to σ_2. In other words, the simplex σ_2 contains a neutral stability point.

16.5.1 Piecewise-Linear Nyquist Map

To ensure that the obstruction cocycle vanishes, and hence that the Nyquist map can be extended, the homotopy class $\{f|\partial\sigma_2\}$ must vanish for *all* 2-simplexes of P. This might end up being a computationally intensive effort, but the simplicial techniques of Part I come to the rescue.

Rather than computing the *whole* images of *all* edges, we would rather compute the images of the vertices of P and then use a piecewise-linear extension of f, f_{PL}. Remember the definition of the piecewise-linear extension. Take a point $p \in \sigma_2 = a^0a^1a^2$. The point p can be written as an affine combination of the vertices, namely,

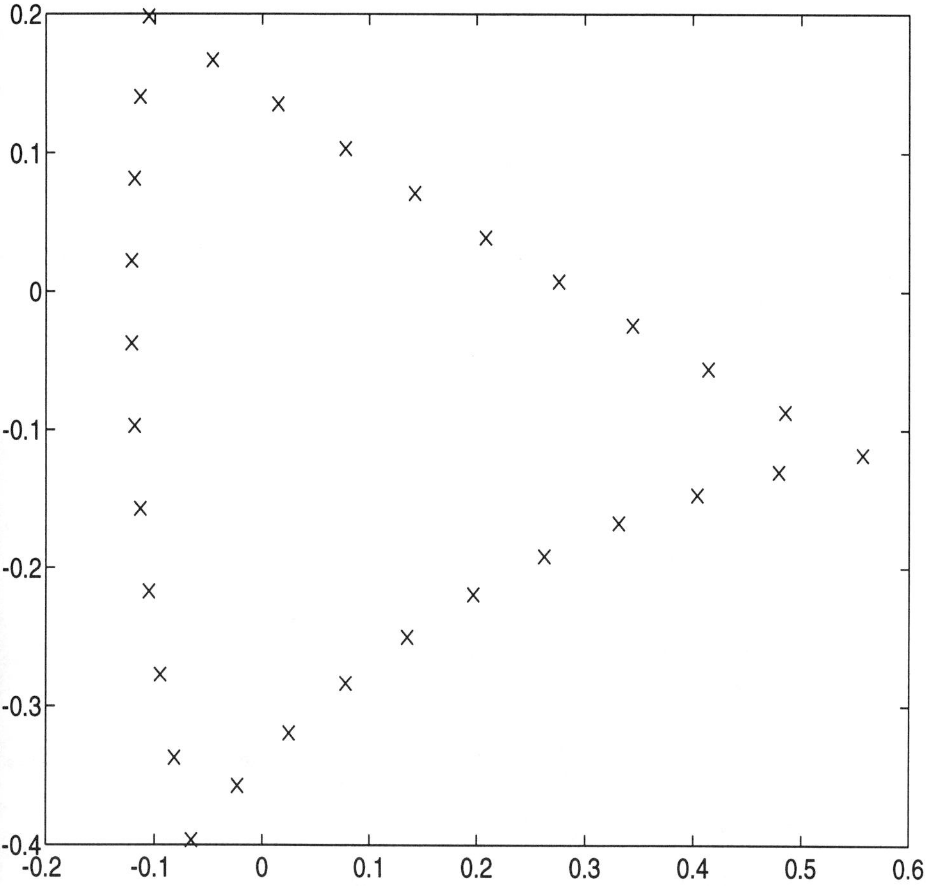

Fig. 16.1. Image of $\partial\sigma_2$ under Nyquist map.

$$p = \lambda_0 a^0 + \lambda_1 a^1 + \lambda_2 a^2, \sum_i \lambda_i = 1, 0 < \lambda_i < 1$$

The λ_i's are the barycentric coordinates of p. The piecewise-linear extension for f, f_{PL}, is defined as

$$f_{PL}(p) = \sum_i \lambda_i f(a^i)$$

Of course, to give sense to the affine combination of the right-hand side, it has to be carried over in the complex plane that has the correct linear structure. This affine structure in the complex plane is mapped into curvilinear simplexes on the Riemann sphere S^2 by the stereographic projection.

Remember, the celebrated *simplicial approximation theorem* asserts that there exists a piecewise linear, simplicial map $\tilde{f} : (D \times \Omega)^1 \to S^2_{**}$ *that is homotopic to* f, so that $\{f|\partial\sigma_2\} = \{\tilde{f}|\partial\sigma_2\}$. Once a simplicial map $\tilde{f}$ is obtained, we propose to compute the homotopy class $\{\tilde{f}|\partial\sigma_2\}$ using the technique of *labeling* of vertices developed in Chapter 9.

16.5.1.1 *Integer Labeling*

The connection between the labeling technique of Chapter 9 and the obstruction test is easy to establish. Remember, the labeling $\ell(\cdot)$ is said to be Sperner proper if for any edge a^0a^1 of P, the straight edge $f(a^0)f(a^1)$ in the complex plane intersects the Butterfly pattern at at most one point. Now we are in a position to establish the connection between labeling and homotopy class.

Theorem 16.15. *Let* $\ell(\cdot)$ *be Sperner proper. Then the homotopy class* $\{\tilde{f}|\partial a^0a^1a^2\}$ *of the Nyquist map restricted to the boundary of the simplex* $a^0a^1a^2$ *is nonvanishing iff the simplex is completely labeled—that is,* $\{\ell(a^0), \ell(a^1), \ell(a^2)\} = \{0, 1, 2\}$.

16.5.1.2 *Vector Labeling*

In case Sperner properness cannot be guaranteed, there exists an *exact* homotopy test, but it is slightly more computationally intensive that the integer label test. Remember, the *vector label* of the vertex a of the polyhedron P is defined as the Euclidean coordinates of the image of $f(a)$ under stereographic projection

$$l(a) = \begin{pmatrix} x_0(\tilde{f}(a)) \\ x_1(\tilde{f}(a)) \\ x_2(\tilde{f}(a)) \end{pmatrix}$$

The simplex $a^0a^1a^2$ is said to be *completely vector labeled* iff there exist solutions $\lambda_i \geq 0, \beta \geq 0$ to

$$\begin{bmatrix} l(a^0) & l(a^1) & l(a^2) \\ 1 & 1 & 1 \end{bmatrix} \begin{bmatrix} \lambda_0 \\ \lambda_1 \\ \lambda_2 \end{bmatrix} = \begin{bmatrix} 0 \\ 0 \\ 0 \\ 1 \end{bmatrix} + \beta \begin{bmatrix} 1 \\ 1 \\ 1 \\ 0 \end{bmatrix}$$

We have the following result:

Theorem 16.16. *The homotopy class* $\{\tilde{f}|\partial a^0a^1a^2\}$ *is nonvanishing iff the simplex* $a^0a^1a^2$ *is completely vector-labeled.*

16.5.2 Comparison with Part I

From the above, it follows that the obstruction approach to robust stability amounts to checking that there does not exist completely (integer or vector) labeled 2-faces of the polyhedron $P = D \times \Omega$. This is quite similar to the results of Part I, obtained via a different avenue of approach. Remember, in the latter, the instability was related to existence of a completely labeled simplex of the relative 2-complex, which is clearly equivalent to a 2-face of a simplex of $D \times \Omega$ being completely labeled. This establishes the link between Part I and Part III.

16.5.3 Relative Results

In the relative case, the obstruction test hinges on the computation of $\{f|\partial\sigma_2 \bmod Q\}$. If none of the simplexes of σ_2 are in Q, then $\{f|\partial\sigma_2 \bmod Q\} = \{f|\partial\sigma_2\}$. If σ_2 belongs to the subcomplex Q, then $\{f|\partial\sigma_2 \bmod Q\} = 0$. The novelty here appears when σ_2 has an edge, say a^0a^1, in the subcomplex Q. In this case, $\partial\sigma_2 \bmod Q$ can be thought of as a "lace over Q", to use the terminology of [Cerf 1970]. A "lace over Q" is a path, starting in Q, going anywhere in P, and eventually ending up in Q. The obstruction test is thus to make sure that any lace over Q can be deformed with its extremities remaining in Q such that the image of the lace under f can be shrunken to a point.

If we want to implement this idea in the realm of piecewise linear maps, the first step is to construct a simplicial approximation of $f|Q$. The original map f and its simplicial approximation, both restricted to Q, are homotopic. The difficulty, however, appears when attempting to construct a simplicial approximation *over* P that is consistent with the simplicial approximation over Q. The solution is given by Zeeman's relative simplicial approximation. (Remember, constructing the *relative* simplicial approximation cannot, in general, be done using the absolute simplicial approximation algorithm.) If we compute the homotopy class of the Nyquist map using with Zeeman's relative simplicial approximation rather than the original map, no discrepancy will occur.

16.6 Homotopy Extension

Here, we combine *within one single section*, the absolute case, the relative case, the general case of an arbitrary range space Y, and the Nyquist problem $Y = S^2_{**}$.

The general *homotopy extension* problem is the following: Assume the extreme functions f_0, f_1 of a homotopy are known over the complete domain of definition D,

$$f_0, f_1 : D \to Y$$

On the other hand, assume a *partial* homotopy f_t^E from $t = 0$ to $t = 1$ is known over a *subspace* $E \subseteq D$ of the domain of definition D,

$$f_t^E : E \times [0,1] \to Y.$$

The homotopy extension problem is to determine whether the partial homotopy f_t^E can be extended over the complete domain of definition,

$$f_t : D \times [0,1] \to Y$$

The problem can be described by the following diagram:

$$\begin{array}{ccc} D \times \partial[0,1] \cup E \times [0,1] & \to & Y \\ \downarrow i & & \overset{?}{\nearrow} \\ D \times [0,1] & & \end{array}$$

In the case of interest to us, $Y = S^2_{**}$. Furthermore, if the Nyquist curve is correctly set up as a homotopy, as we did in Section 16.1, the extreme values $t = 0, 1$ of the homotopy parameter correspond to $\omega = -\infty, +\infty$. Therefore, under mild roll-off conditions at $\omega = \pm\infty$, the complete uncertainty domain D is mapped to $1 + j0$,

$$f_0, f_1 : D \to \{1 + j0\} \subseteq S^2_{**}$$

Therefore, in the robust stability problem, we do know the extreme functions of the homotopy over the complete domain of uncertainty. Furthermore, we assume that closed-loop stability is guaranteed for some subspace $E \subseteq D$, which could collapse to the vertex of nominal parameters; in other words, we have the partial homotopy,

$$f_t^E : E \times [0,1] \to S^2_{**}$$

Robust stability for all parameters values in D amounts to extending the partial homotopy to the complete homotopy

$$f_t : D \times [0,1] \to S^2_{**}$$

The general homotopy extension problem can be reduced to an extension problem. In fact, in the homotopy extension problem, we know a map

$$F^0 : (D \times \partial[0,1]) \cup (E \times [0,1]) \to Y$$

and we seek for an extension over

$$K^n = (D \times \partial[0,1]) \cup ((E \cup D^{n-1}) \times [0,1])$$

Let F^n be the extension to K^n. Consider the extension from K^n to K^{n+1}. This corresponds to the extension from D^{n-1} to D^n. Any simplex σ_n of D could potentially create problems. The homotopy obstruction cochain $c_F \in C^n(D \bmod E, \pi_n(Y))$ is defined as

$$(\sigma_n, c_F) = \{F^n | \partial(\overline{\sigma_n} \times [0,1])\}$$

As before, $\{F|\bullet\}$ stands for the homotopy class of F restricted to $\bullet$. It is easily checked that c_F is a cochain of $D \bmod E$, in the sense that c_F vanishes on the simplexes of E.

Theorem 16.17. (Homotopy Extension) *$F^n : K^n \to Y$ is extendable over K^{n+1} iff $c_F = 0$.*

Proof. See [Hilton and Wylie 1965, Proposition 7.3.1, page 295]. ■

Theorem 16.18. *c_F is a cocycle of the cochain complex*

$$C^n(D \bmod E, \pi_n(Y))$$

Proof. See [Hilton and Wylie 1965, Theorem 7.3.4, page 295]. ■

Theorem 16.19. *If F and F' are two different extensions of the map $K^n \to Y$ over K^{n+1}, then $\{c_F\} = \{c_{F'}\}$.*

Proof. See [Hilton and Wylie 1965, Theorem 7.3.5, page 296]. ■

Theorem 16.20. *Let $F^{n-1} : K^{n-1} \to Y$. Let F^n be an extension over K^n. Then F^{n-1} is extendable over K^{n+1} iff $\{c_F\} = 0$.*

Proof. See [Hilton and Wylie 1965, Theorem 7.3.6, page 296]. ■

Now we specialize the results to $Y = S^2_{**}$. The only nontrivial cochain group is $C^1(D \bmod E, \pi_1(S^2_{**}))$. Therefore, the only problem that could possibly arise is at the level of the extension from K^1 to K^2—that is, the extension from D^0 to D^1. This is the case $n = 1$. Therefore, the only test is to make sure that the homotopy class

$$(\sigma_1, c_F) = \{F|\partial(\overline{\sigma_1} \times [0,1])\}$$

is trivial for all simplexes $\sigma_1 \in D^1$. A slight problem is that the Cartesian product of two simplexes, typically $\sigma_1 \times [0,1]$, is not a simplex. However, this Cartesian product can be decomposed into simplexes using the prismatic triangulation. From there on, the techniques of piecewise-linear topology and labeling of the previous section apply.

The advantage of this *homotopy* extension formulation is that the obstruction, if any, takes value in $C^1(D \bmod E, \pi_1(S^2_{**}))$. The basic obstruction test therefore appears to be a test on the *edges* of D. As such, our obstruction test begins to resemble Bartlett's edge test.

16.7 Comparison Between Homotopy Extension and Edge Tests

At this juncture, we can explain in more details how the obstruction to homotopy extension and the "edge test" of [Bartlett, Hollot, and Lin 1988] compare. The obstruction test is

$$\{F|\partial(\overline{\sigma_1} \times [0,1])\} = 0, \forall \sigma_1 \in D^1$$

The edge test is

$$\text{polynomial is Hurwitz } \forall q \in \sigma_1, \forall \text{ exposed edge } \sigma_1 \text{ of } D^1$$

To draw a comparison, we have to treat the Hurwitzness test as a frequency sweep. (The latter is a fair comparison only for very high degree polynomials.) The Hurwitzness test on an edge clearly amounts to a *continuum* of frequency sweeps. Regarding the obstruction test, remember that the Nyquist mapping maps D to *one single point* $(1 + j0)$ at the boundary of the homotopy interval $(\omega = \pm\infty)$. Therefore the obstruction to homotopy extension amount to a homotopy test on the Nyquist map restricted to

$$\partial(\overline{\sigma_1} \times [0,1]) = ((\partial\sigma_1) \times [0,1]) \cup \underbrace{(\sigma_1 \times \partial[0,1])}_{\text{point}}$$

Therefore, the obstruction test amounts to *two* frequency sweeps—one for each boundary point of σ_1. *Here we are at the crucial point.* The edge test verifies that the Nyquist map exists over D^1 while the obstruction test captures the problem at the more fundamental level of deciding whether there exists an extension to D^1.

If the obstruction vanishes, there are many extensions. If the Nyquist map is affine, then choosing the affine extension will result in an extension that agrees with the Nyquist map.

However, if we take into consideration the nonlinearities of the Nyquist map, it is possible that, even though there are no obstructions, the nonlinear Nyquist map will fail the test on P. The reason for this is that the nonlinearities of the Nyquist map drives it outside the set of extensions of f^2. The latter is the explanation of the counterexamples to the "edge test" that have been reported. In these cases, while no obstructions exist (the edge test passes), the Nyquist map fails to be within the set of extensions of f^2.

16.8 Appendix—Obstruction to Cross Sectioning

Consider a fiber bundle

$$\begin{array}{ccc} F & \rightarrow & E \\ & & \downarrow \pi \\ & & B \end{array}$$

where the base space B is a triangulated polyhedron. Clearly, for any simplex σ_n of B, $\overline{\sigma_n}$ is contractible and therefore $\pi^{-1}(\overline{\sigma_n}) \cong \overline{\sigma_n} \times F$.

We attempt to construct a cross section—that is, a map $\gamma : B \rightarrow E$ such that $\pi\gamma = 1_B$. The procedure is recursive over higher-and-higher-skeleta of B. First, $\forall b^i \in B$, choose $\gamma(b^i) \in \pi^{-1}(b^i)$. This yields a map $\gamma^0 : B^0 \rightarrow E$ such that $\pi\gamma^0 = 1_{B^0}$. The problem is to extent this map to higher-and-higher-skeleta of B. We proceed inductively. We assume we have a map $\gamma^n : B^n \rightarrow E$, $\pi\gamma^n = 1_{B^n}$ and we attempt to extend it to B^{n+1}. Take an arbitrary simplex σ_{n+1} of B. By induction hypothesis, the cross section exists over $\partial\sigma_{n+1}$; in other words, we have a map :

$$\partial\sigma_{n+1} \cong S^n \rightarrow \pi^{-1}(\overline{\sigma_{n+1}}) \cong \overline{\sigma_{n+1}} \times F$$

Clearly, the obstruction to extending this map to $\overline{\sigma_{n+1}}$ is

$$\{\gamma^n | \partial\sigma_{n+1}\} \in \pi_n(\overline{\sigma_{n+1}} \times F) = \pi_n(\overline{\sigma_{n+1}}) \oplus \pi_n(F) = \pi_n(F)$$

The obstruction cocycle c_γ^n is defined by

$$(\sigma_{n+1}, c_\gamma^n) = \{\gamma^n | \partial\sigma_{n+1}\} \in C^{n+1}(B, \pi_n(F))$$

Clearly, the cross section can be extended from B^n to B^{n+1} iff the primary obstruction vanishes, namely, $c_\gamma^n = 0$. The secondary obstruction to extending from B^{n-1} to B^{n+1} is the cohomology class $\{c_\gamma^n\} \in H^{n+1}(B, \pi_n(F))$ and *the cross section can be extended from B^{n-1} to B^{n+1} iff $\{c_\gamma^n\} = 0$.*

NOTE

The obstruction approach to the robust Nyquist stability criterion developed in this chapter is believed to be novel. The only reference that we are aware of and that might come close to the point of view adopted here is [DeCarlo and Seaks 1977], where the idea of "removing $0 + j0$ from $\mathbb{C}$" is also instrumental.

17

HOMOTOPY CLASSIFICATION OF NYQUIST MAPS

SUMMARY

Since we have removed $0 + j0$ from the complex plane, the Nyquist map $D \times \Omega \to \mathbb{C} \setminus \{0 + j0\}$, *if it exists*, could have a nontrivial homotopy class. The set of all possible homotopy classes of all such maps depends essentially on D. This chapter addresses the question as to what homotopy classes can be expected for a given uncertainty space. In case of such uncontractible uncertainty spaces as toris—multichannel phase margin problems—the set of all homotopy classes is quite large. The next question is to determine which of these homotopy classes yield closed-loop stability. Finally, given an uncertainty structure D, it is shown how the *loop function* $L(s)$ induces a homotopy class of the map.

17.1 Fundamental Classification Result

The doubly punctured sphere S^2_{**}, on which the Nyquist template is plotted, has the remarkable property that all but one of its homotopy groups vanish. Such a space is called *Eilenberg–Mac Lane space.*

Definition 17.1. *Y is said to be an* **Eilenberg–Mac Lane space** *iff it is path-connected, simple, and such that $\pi_n(Y) = 0$ for all $n \neq m$. In this case, Y is written as $K(\pi_m(Y), m)$. Conversely, if π is a group, $K(\pi, m)$ is any path-connected, simple space Y such that $\pi_n(Y) = 0$, $n \neq m$, and $\pi_m(Y) = \pi$.*

Remember, for $n \neq 0$, $H_n(S^m)$ vanishes except for $m = n$ in which case $H_n(S^n) = \mathbb{Z}$. Therefore, Eilenberg–Mac Lane spaces can be viewed as those spaces which play in homotopy theory the same role as spheres play in homology theory. Clearly, S^1 is a $K(\mathbb{Z}, 1)$ space. However, the latter is the exception rather than the rule. Indeed, only the 1-D sphere behaves consistently under the homotopy and homology functors. Eilenberg–Mac Lane spaces other than the trivial $K(\mathbb{Z}, 1)$ space exist but are much harder to characterize. Nevertheless, they play a fundamental role in homotopy theory.

There is a straightforward classification of all maps from a polyhedron to an Eilenberg–Mac Lane space:

Theorem 17.2. *The homotopy classes of maps $f : P \to K(\pi_m, m)$ from a polyhedron P to the Eilenberg–Mac Lane space $K(\pi_m, m)$ are in one-to-one correspondence with the elements of the cohomology group $H^m(P, \pi_m)$.*

Proof. This result is not easy to prove; see [Hilton and Wylie 1965, Theorem 7.4.1, page 298 and Corollary 7.4.4, page 302]. For a generalized version of the same theorem, see [Spanier 1989, Chapter 8, Section 1, Theorem 10, page 428]. ∎

In the robust stability problem, the doubly punctured Riemann sphere S^2_{**} is homotopically equivalent to S^1 and is therefore an Eilenberg–Mac Lane space of type $K(\pi_1(S^2_{**}), 1) = K(\pi_1(S^1), 1) = K(\mathbb{Z}, 1)$. Therefore, we have the following corollary:

Corollary 17.1. (Hopf) *The homotopy classes of maps $f : D \times \Omega \to S^2_{**}$ are in one-to-one correspondence with the cohomology group $H^1(D \times \Omega, \mathbb{Z})$.*

In robust stability, closed-loop stability is determined by the homotopy class of the Nyquist map, which is itself an element of $H^1(D \times \Omega, \mathbb{Z})$. Therefore, the typical robust stability exercise of "guessing" the homotopy class of a partial Nyquist map over a subspace $E \subseteq D$ consists in picking the correct element from $H^1(D \times \Omega, \mathbb{Z})$ so that the richer this cohomology group, the more complex the problem is.

17.2 Classification of Maps to Spheres

There is another path of approach to Corollary 17.1. The starting point is the following:

Theorem 17.3. (Hopf) *If P is a polyhedron of dimension $m \geq 1$, the set of homotopy classes of maps $P \to S^m$ is in a one-to-one correspondence with the cohomology group $H^m(P, \mathbb{Z})$.*

Proof. See [Hilton and Wylie 1965, Theorem 7.4.1, page 298 and Corollary 7.4.2, page 301]. ∎

Taking $m = 1$ yields Corollary 17.1. However, it should be kept in mind that Theorem 17.2 and Theorem 17.3 are two different paths of approach. Indeed, only for $m = 1$ is S^m an Eilenberg–Mac Lane space.

17.3 Elementary Proof of Main Result

The preceding theorem and corollary have a long history of complicated proofs. Nevertheless, an elementary proof, specially devised for the case $m = 1$, can be devised along the lines of [Hilton and Wylie 1965, Section 2.7]. In addition to proving the result, it reveals the mechanism through which a map $f : P \to S^2_{**}$ induces a cohomology class. The latter is of crucial importance to get the feeling for the homotopy classification problem for Nyquist maps.

We now proceed to proving the corollary. Since S^2_{**} and S^1 are homotopically equivalent, the problem reduces to proving that a map $f : P \to S^1$ induces a unique cohomology class and vice versa. Choose a base point $s_* \in S^1$. If we have a map $f : P \to S^1$, we naturally have a unique restriction $f : P^1 \to S^1$, and we write this restriction as f_0 to indicate that it is soon to be the beginning of a homotopy. By the same token, we have yet another restriction $g_0 : P^0 \to S^1$ and of course $f|P^0 = g_0$. It is easy to see that the map $g_0 : P^0 \to S^1$ can be homotopically deformed into a map $g_1 : P^0 \to s_*$. Let g_t be this homotopy deformation. Now, we have reached a situation relevant to the homotopy extension problem for polyhedra:

Lemma 17.4. (Homotopy Extension for Polyhedra) *Assume we are given a partial homotopy over some subpolyhedron $L \subseteq K$,*

$$g_t : L \times [0,1] \to Y$$

Assume the g_0 map in the homotopy can be extended to the full polyhedron; that is, there exists a map f_0,

$$\begin{aligned} f_0 : K &\to Y \\ f_0|L = g_0 : L &\to Y \end{aligned}$$

Then there exists a full homotopy

$$f_t : K \times [0,1] \to Y$$

that extends the partial homotopy, namely,

$$f_t|L = g_t$$

In other words, the following diagram commutes:

$$\begin{array}{ccc} K \times \{0\} \cup L \times [0,1] & \stackrel{f_0, g_t}{\longrightarrow} & Y \\ i \downarrow & & \stackrel{f_t}{\nearrow} \\ K \times [0,1] & & \end{array}$$

Before going through the proof, a few remarks are in order. It is important to understand the difference between this result and Theorem 16.17. Here, there is no obstruction to extension because, contrary to Theorem 16.17, the end function f_1 of the full homotopy is not specified. Actually, the proof proceeds by recursion over higher-and-higher-dimensional skeleta of K, ensuring at every step that there is no obstruction.

Proof. The proof is inspired from [Hilton and Wylie 1965, 1.6.10], and consists in recursively constructing extensions over higher-and-higher-dimensional skeleta of K. Clearly, from the data of the problem, there exists a map:

$$(K \times \{0\}) \cup (L \times [0,1]) \to Y$$

The initial step of the recursion is to extend the above to a map

$$f_t^0 : (K \times \{0\}) \cup ((L \cup K^0) \times [0,1]) \to Y$$

such that $f_t^0|(K \times \{0\}) \cup (L \times [0,1]) = g_t$. To construct such an extension, all we have to do is define f_t^0 over the vertices of K not in L. If a^0 is one such vertex, we define $f_t^0(a^0) = f_0(a^0)$, and it is easily seen that this provides the required extension. The recursive step of the proof is to show that if we have a map

$$f_t^n : (K \times \{0\}) \cup ((L \cup K^n) \times [0,1]) \to Y$$

then one can construct a map f_t^{n+1} such that $f_t^{n+1}|(K \times \{0\}) \cup ((K \cup L^n) \times [0,1]) = f_t^n$. Take a closed simplex $\overline{\sigma_{n+1}}$ in K. Clearly, $\partial\sigma_{n+1} \in K \cup L^n$. By recursive hypothesis, we have an extension over $(\overline{\sigma_{n+1}} \times \{0\}) \cup (\partial\sigma \times [0,1])$, and we seek an extension over $\overline{\sigma_{n+1}} \times [0,1]$. The key point in the proof is to observe that there exists a deformation retract:

$$r : \overline{\sigma_{n+1}} \times [0,1] \to (\overline{\sigma_{n+1}} \times \{0\}) \cup (\partial\sigma_{n+1} \times [0,1])$$

defined as follows: Let $\overline{\sigma_{n+1}} \times [0,1]$ be embedded in Euclidean space $\mathbb{R}^{n+1} \times \mathbb{R}$. Let b be the Euclidean coordinates of the barycenter of σ in $\mathbb{R}^{n+1}$. It is easily seen that the radial projection from the point with Euclidean coordinates $(b, 2)$ in $\mathbb{R}^{n+2}$ provides the required deformation retract. With this deformation retract, we define

$$f_t^{n+1} = f_t^n \circ r : \overline{\sigma_{n+1}} \times [0,1] \to Y$$

By this process, all functions in the homotopy f_t^{n+1} are defined over all closed simplexes $\overline{\sigma_{n+1}}$ and it is easily seen that these partial extensions can be "stitched together" to give an extension over $(K \times \{0\}) \cup ((K \cup L^{n+1}) \times [0,1])$. The proof is completed. ■

We apply the homotopy extension to the Nyquist problem by setting $K = P^1$, $L = P^0$, and $Y = S^1$. Let f_t be the full homotopy. Let the end function f_1 of this homotopy be written as h. Clearly, f and h are homotopic as maps from P^1 to S^1,

$$f \simeq h : P^1 \to S^1$$

Furthermore, h maps P^0 into s_*, which we write as

$$h : P^1, P^0 \to S^1, s_*$$

Since f and h are in the same homotopy class, it suffices to indicate how h determines a cohomology class. The crucial point is that h defines a cochain c_h as follows: Take a simplex σ_1 of P^1 and define the value taken by the cochain c_h on the simplex σ_1, (σ_1, c_h), as the number of times, counted trigonometrically, that the image of σ_1 under h winds around S^1,

$$(\sigma_1, c_h) = \text{ winding number of } h(\sigma^1) \text{ around } S^1$$

More formally, (σ_1, c_h) is an element of $\pi_1(S^1, s_*)$.

Next, we prove that c_h is a cocycle; that is, $\delta c_h = 0$. Take a 2-simplex $\sigma_2 = a^0 a^1 a^2$. By duality, we have to show that $(\sigma_2, \delta c_h) = (\partial\sigma_2, c_h) = 0$. Equivalently, expanding $\partial\sigma_2$, we have to show that $(a^0a^1 + a^1a^2 + a^2a^0, c_h) = 0$, which means that the image under h of the closed loop $a^0a^1 + a^1a^2 + a^2a^0$ does not wind around the unit circle S^1. As we have seen in the previous chapter, the basic obstruction fact is that the latter is equivalent to h having an extension from $\partial\sigma_2$ to σ_2. Another appeal to the homotopy extension theorem for polyhedra reveals that this extension exists; hence c_h is indeed a cocycle.

Clearly, h determines a unique cocycle c_h, but the original map f does not uniquely determine h. Indeed, g_t is nonunique, and hence the extension f_t along with $f_1 = h$ are nonunique also. Let h and h' be two different extensions obtained from the same original map f. Clearly, h and h' are homotopic

$$h \simeq h' : P^1 \to S^1$$

and both h and h' map P^0 into s_*

$$h, h' : P^1, P^0 \to S^1, s_*$$

In the *particular case* where P^0 is mapped into s_* *during the entire homotopy* from h to h', which we write as

$$h \simeq h' : P^1, P^0 \to S^1, s_*$$

then it is trivial to see that

$$c_h = c_{h'}$$

However, it is not always true that during the homotopy from h to h' the images of the vertices P^0 do not move. Define a 0-cochain $c_{h,h'}$ by the rule that $(a^0, c_{h,h'})$ is the number of times the image of the vertex a^0 wanders around the circle during the homotopy from h to h'. It is claimed that

$$\delta c_{h,h'} = c_h - c_{h'}$$

so that if h and h' are two extensions from f they determine cohomologous cocycles and hence a *unique* cohomology class. To prove the above, we evaluate both the right-hand side and the left-hand side cochains for an arbitrary simplex a^0a^1 and we get

$$(a^0, c_{h,h'}) - (a^1, c_{h,h'}) = (a^0a^1, c_h) - (a^0a^1, c_{h'})$$

The above is nothing other than the geometrically obvious fact that the increase in the number of times the image of a^0a^1 circles around S^1 as we go from h to h' is equal to the number of rounds of a^1 minus the number of rounds of a^0 under the homotopy.

The result is proved. ■

17.4 Cohomology of Product of Uncertainty

Now, we look more carefully at the cohomology group $H^1(D \times \Omega, \mathbb{Z})$. Let us look at the second factor—the frequency domain Ω that is homeomorphic to the unit circle. It is well-known that the homology groups of the circle are finitely generated. Furthermore, the cohomology of the circle is given by (see [Munkres 1984, page 254])

$$\begin{aligned} H^0(\Omega, \mathbb{Z}) &= \mathbb{Z} \\ H^1(\Omega, \mathbb{Z}) &= \mathbb{Z} \end{aligned}$$

Obviously, the above cohomology groups have no torsion, so that the torsion product $H^{p+1}(D) * H^q(\Omega)$ in Künneth's theorem for cohomology drops. Therefore, we get the short exact sequence

$$0 \rightarrow \bigoplus_{p+q=n} H^p(D) \otimes H^q(\Omega) \rightarrow H^n(D \times \Omega, \mathbb{Z}) \rightarrow 0$$

In other words,

$$H^n(D \times \Omega, \mathbb{Z}) = \bigoplus_{p+q=n} H^p(D) \otimes H^q(\Omega)$$

Finally, for $n = 1$ the above becomes

$$\begin{aligned} H^1(D \times \Omega, \mathbb{Z}) &= (H^1(D) \otimes H^0(\Omega)) \oplus (H^0(D) \otimes H^1(\Omega)) \\ &= (H^1(D) \otimes \mathbb{Z}) \oplus (H^0(D) \otimes \mathbb{Z}) \\ &= H^1(D) \oplus H^0(D) \end{aligned}$$

Corollary 17.2. *Assume the homology groups of D are finitely generated. The homotopy classes of maps $f : D \times \Omega \rightarrow S^2_{**}$ are in one-to-one correspondence with the group $H^1(D) \oplus H^0(D)$.*

It is reasonable to assume that D is path connected, in which case (see [Munkres 1984, page 255])

$$H^0(D) = \mathbb{Z}$$

Intuitively, the above component of the group of homotopy classes of Nyquist maps represents the possibility that the frequency circle maps into a curve circling around the south pole $0 + j0$ of the Riemann sphere. In addition, the uncertainty space D might contain loops that have the potential of creating the same situation on the Riemann sphere. The likelihood of this last situation to occur depends on the number of generators of the group $H^1(D)$. We look at a couple of examples.

17.4.1 Multivariable Phase Margin

Assume D is a diagonal matrix of n phase delays. Therefore, D is the topological product of n copies of the unit circle—that is, $\mathbb{T}^n$, the n-torus. Repeated application of Künneth's theorem for cohomology tells us that the cohomology group $H^1(\mathbb{T}^n)$ is on $\binom{n}{1} = n$ generators. Therefore, in addition to the the frequency circle, there are in the uncertainty space many loops that could potentially map into a loop around $0 + j0$. The homotopy classification of the Nyquist maps consists in sorting out the situation: Figure out whether the Nyquist template circles around $0 + j0$; if it does, find out where the circle around $0 + j0$ is coming from, whether it comes from the frequency circle, in which case we have an "encirclement condition," or whether it is instead due to loops in the parameter space.

17.4.2 Special Orthogonal Perturbation

The physical motivation for $D = SO(3)$ is an uncertain rotation in a feedback path containing three covariant quantities. A typical example is the misalignment between body and inertial axes in a strapped-down navigation system. The issue is the cohomology of $SO(3)$. The singular cohomology with integer coefficients of $SO(3)$ is (see [Bott and Tu 1982, page 195])

$$\begin{aligned} H^0(SO(3)) &= \mathbb{Z} \\ H^1(SO(3)) &= 0 \\ H^2(SO(3)) &= \mathbb{Z} \bmod 2 \\ H^3(SO(3)) &= \mathbb{Z} \end{aligned}$$

Therefore, since $H^1(SO(3)) = 0$, the $SO(3)$ problem appears less complicated than the multivariable phase margin problem since the possibility of D producing loops around $0 + j0$ is nonexistent.

17.5 Formal Classification

We now take a more global look the classification problem. We are given an a priori set D of structured perturbation. Any zero excluding loop function $L(s) = K(s)G(s)$ wrapped around the structured perturbation $\Delta \in D$ induces a Nyquist map that is just one among the many possible maps

$$f : D \times \Omega \to S^2_{**}$$

This Nyquist map has a homotopy class, which, by Hopf's theorem, is an element of $H^1(D \times \Omega, \pi) = H^1(D) \oplus H^1(\Omega)$. By this procedure, we have defined a "classification" mapping

$$\gamma : \{\text{zero excluding } L(s)\} \to H^1(D \times \Omega, \pi)$$

It is not as yet clear how the factorization structure $L(s) = K(s)G(s)$ of the loop function affects the map γ.

As an example, take the a priori set of structured perturbations

$$D = \{\begin{bmatrix} \exp(-j\theta) & 0 \\ 0 & q \end{bmatrix} : \theta \in [0, 2\pi), q \in \mathbb{C}, |q| \leq 1\}$$

In this case, the set D of structured perturbations has the topological structure of a "donut" and it is easily seen that its cohomology group $H^1(D)$ is on one generator. Consider a loop function

$$L(s) = \begin{bmatrix} 100 & 1 \\ 0 & \frac{0.1}{s+1} \end{bmatrix}$$

It is easily checked that

$$\det(I + L\Delta) = (1 + \frac{0.1q}{s+1})(1 + 100\exp(-j\theta))$$

and therefore, the map γ sends $L(s)$ to an element of $H^1(D) \oplus 0$.

18

BROUWER DEGREE OF NYQUIST MAP

It is quite remarkable that Brouwer proved all his big theorems by a skillful (and sometimes quite tortuous) use of a single concept he discovered and studied in the first days of 1910, the degree of a map.

J. Dieudonné
A History of Algebraic and Differential Topology 1900–1960,
Birkhäuser, Boston, 1989,
page 169.

SUMMARY

From a *conceptual* point of view, the obstruction test for robust stability is a test on the *homotopy class* of the map $f^1 : (D \times \Omega)^1 \to S^1$. More specifically, f^1 should be nullhomotopic for the extension to exist and the robust stability test to pass. An important homotopy invariant of a map is its Brouwer degree. Roughly speaking, the *Brouwer degree* of a map can be defined as an integer that is associated with the map, that is a homotopy invariant, and that vanishes iff the map is nullhomotopic. In this chapter, we reformulate the obstruction problem, and more broadly the robust stability problem, in terms of the *Brouwer degree* of the Nyquist map. A key result is that f^1 is nullhomotopic iff its Brouwer degree vanishes.

With this motivation in mind, we embark on a more comprehensive overview of the various degree concepts with potential applications to many Nyquist-related problems. For example, the degree of the restriction of the Nyquist map $D \times \Omega \to N$ to a 2-D section through the domain is interpreted as the "signed" number of crossover points in the section.

From another point of view, this chapter provides some new insight into the simplicial algorithms introduced in Part I. By exploiting the fundamental homotopy invariance of the degree, we manipulate the Nyquist map by means of homotopies, retract the template to the labels complex, and show that the labeling function is homotopic to the original Nyquist map. In other words, the simplicial algorithms of Chapter 9 appear to be tests on the degree of the map.

The material of this chapter is deliberately kept elementary. From a certain point of view, this chapter is just a "warm-up exercise" for the chapter to come, where a far less elementary degree concept will be developed.

18.1 Orientation

Before we can jump into degree theory, it is necessary to formalize the concept of *orientation.*

An **orientation** of a simplex σ_n is just an ordering of its vertices, namely, $a^0, ..., a^1$. Two simplexes differing by an *even* permutation of vertices, namely,

$$a^0 a^1 a^2 a^3 ... a^n, \quad a^1 a^2 a^0 a^3 ... a^n$$

are said to be **coherently oriented**. Two simplexes differing by an *odd* permutation of their vertices, namely,

$$a^0 a^1 a^2 ... a^n, \quad a^1 a^0 a^2 ... a^n$$

are said to be of **opposite orientation**. Consistently with the skew-commutative product rule, the fact that the above simplexes are of *opposite* orientation is written

$$a^0 a^1 a^2 ... a^n = -a^1 a^0 a^2 ... a^n$$

An orientation of an n-simplex induces an orientation of any of its $(n-1)$-faces. To be specific, the orientation induced by $a^0 ... a^n$ on the face opposed to a^i is

$$(-1)^i a^0 ... a^{i-1} a^{i+1} ... a^n$$

Conversely, an orientation of an $(n-1)$-face, σ_{n-1}, induces an orientation of any n-simplex that has σ_{n-1} as a face. To be specific, let a be the vertex of σ_n not in the face σ_{n-1}. The orientation of the n-simplex induced by σ_{n-1} is $a\sigma_{n-1}$. Indeed, the orientation induced by $a\sigma_{n-1}$ on the face opposed to a is σ_{n-1}.

The next step is to define the concept of orientation for an assembly of simplexes meant to be a combinatorial model of such a geometric object as a manifold. For the purpose of developing a degree theory for maps between manifolds, it is useful to introduce the concept of *pseudomanifold.*

Definition 18.1. (Spanier) *An n-D* **pseudomanifold** *K is an n-D simplicial complex subject to the following restrictions:*

- *K is homogeneous of dimension n; that is, every simplex is the face some n-simplex.*
-
  - *Either every (n−1)-D simplex is the face of exactly two n-simplexes, in which case the complex is said to be a* **pseudomanifold without boundary***;*
  - *Or every (n−1)-D simplex is the face of at most two n-simplexes, in which case the complex is said to be a* **pseudomanifold with boundary***.*

- *If σ_n,σ'_n are simplexes of K, there exists a sequence $\sigma_n = \sigma_n^1, ..., \sigma_n^m = \sigma'_n$ of simplexes such that σ_n^i and σ_n^{i+1} share a common $(n-1)$-face.*

Furthermore, the boundary of the pseudomanifold, ∂K, is the complex generated by the collection of $(n-1)$-simplexes that are faces of exactly one n-simplex. For the details, see [Spanier 1989, page 150].

As an example, the complex generated by $\partial\sigma_{n+1}$ is a pseudomanifold without boundary. The polyhedron of this complex is homeomorphic to the sphere S^n.

Consider an n-D pseudomanifold without boundary. An **orientation** of the pseudomanifold is, *if it exists*, a choice of orientation of all its n-simplexes such that

$$\partial \sum \sigma_n = 0$$

If such a choice exists, the pseudomanifold is said to be **orientable**. Clearly, $\sum \sigma_n$ is a cycle. Furthermore, there are no n-boundaries because the dimension of the underlying complex is n. *Therefore, an orientation of an n-D pseudomanifold without boundary is a choice of a preferred generator of $H_n(K)$.*

For example, the pseudomanifold $\partial\sigma_3$, a combinatorial model of S^2, is orientable. Indeed, an orientation is

$$z_2 = a^0a^1a^2 - a^1a^2a^3 + a^2a^3a^0 - a^3a^0a^1$$

Let K be an n-D pseudomanifold with boundary. An **orientation** is a choice of orientation of the n-simplexes of K such that

$$\overline{\partial \sum \sigma_n} = \partial K$$

If such a choice exists, the pseudomanifold is said to be **orientable**. Phrased another way, an orientation is a choice of a preferred generator of the relative homology group $H_n(K, \partial K)$.

For degree theory, it is important to observe that two adjacent n-simplexes of an n-D pseudomanifold induce *opposite* orientations on their common face.

18.2 Combinatorial Degree

Consider a simplicial map $f : X \to Y$ between triangulated n-dimensional pseudomanifolds without boundary. Consider a point $y \in Y$, embedded in an n-simplex τ_n. Let $f^{-1}(y) = \{x^1, ..., x^k, ...\}$ be the (finite) preimage set of y. Since $y \in \tau_n$, each preimage point belongs to an n-simplex, $x^k \in \sigma_n^k$. The simplex σ_n^k contains *exactly one* preimage point, x^k, because if there were more than one such preimage point the simplicial restriction $f|\sigma_n^k : \sigma_n^k \to \tau_n$ could not be onto. Clearly, f maps every σ_n^k onto τ_n. Let n_+ be the number of n-simplexes σ_n^k mapped onto τ_n with *coherent* orientation and let n_- be the number of n-simplexes mapped to τ_n with *reverse* orientation. The **(Brouwer) degree** of f is defined as

$$\text{degree}(f) = n_+ - n_-$$

See [Spanier 1989, Chapter 4, Exercise F.1, page 207].

To make the above a property of the *map* f rather than the *point* y, observe that the degree is unchanged as y sweeps τ_n. Assume now that y follows an arc γ crossing the $(n-1)$-simplex τ_{n-1} common to τ_n and τ_n' at the point y_0. As $y \to y_0$, the x^k's approach common faces between the σ_n^k's and adjacent simplexes $\sigma_n'^k$'s. At this stage, there are several possibilities.

If $\sigma_n'^k$ does not contain any point in $f^{-1}(y)$, x^k crosses the common face between σ_n^k and $\sigma_n'^k$. This situation is illustrated in Figure 18.1.

If $\sigma_n'^k$ is mapped *onto* τ_n', $y' \in \tau_n'$ has a unique preimage in $\sigma_n'^k$. From the definition of orientation, σ_n^k is mapped to τ_n with coherent (opposite, resp.) orientation iff $\sigma_n'^k$ is mapped to τ_n' with coherent (opposite, resp.) orientation. Therefore, as we proceed from y to y', both n_+ and n_- remain unchanged and the degree remains unchanged.

If $\sigma_n'^k$ is mapped onto the face τ_{n-1}, $f^{-1}(y_0)$ has a line of solutions in $\sigma_n'^k$ going from the face σ_{n-1}^k to another face, say σ_{n-1}', common between $\sigma_n'^k$ and $\sigma_n''^k$. If $\sigma_n''^k$ is mapped onto τ_n', we are back to the previous case. If not, follow the line of solution $f^{-1}(y_0)$ in $\sigma_n''^k$ until a simplex mapped onto τ_n' is found.

Now, assume $\sigma_n'^k$ contains a preimage x^l of y. As $y \to y_0$, both x^k and x^l converge to the face σ_{n-1}^k and annihilate each other. It follows that both n_+ and n_- decrease by 1 so that the degree, $n_+ - n_-$, remains unchanged. This situation is illustrated in Figure 18.2.

The opposite situation also exists: Let $y' \in \tau_n'$ converge to y_0. As y' crosses the face τ_{n-1}, a pair of preimage points is born in the face σ_{n-1}^k. As y gets in σ_n^k, the two preimage points immediately move to adjacent simplexes. It follows that n_+ and n_- both increase by 1, and the degree $n_+ - n_-$ still remains unchanged.

This concept of degree is sometimes referred to as **combinatorial degree**.

An important observation is that this concept of degree, although defined *locally*, is actually a *global* invariant.

A *fundamental* property of the degree is its invariance under homotopy. To prove this, dissect finely the homotopy interval $[0, 1]$ and refine the triangulation of X finely enough so that two adjacent maps in the homotopy of maps differ on one vertex only of every simplex of the refined triangulation of X. From there, the invariance of the degree under homotopy is easily seen.

It is also possible to extend the concept of degree for maps between pseudomanifolds *with boundaries*. All of the properties of the degree remain valid, as long as y stays away from the boundary of Y and as long as no

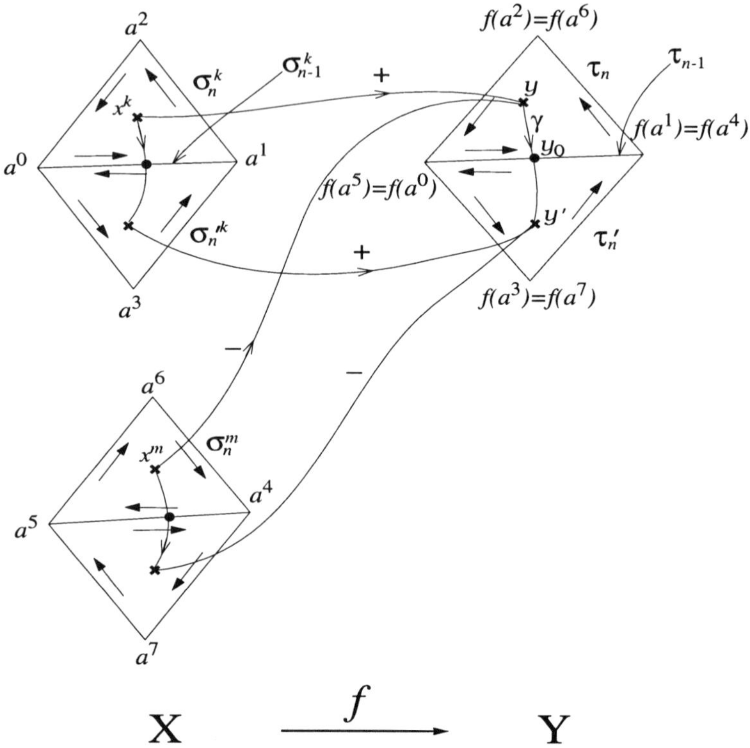

Fig. 18.1. Illustration of the Brouwer combinatorial degree in the regular case. The degree of the map of this picture is 0.

preimage points hit the boundary of X.

Finally, for an arbitrary map $f : X \to Y$, its degree can be defined combinatorially as the degree of its simplicial approximation.

18.3 Analytical Degree

It is also possible to define the same degree *concept* for maps that are *not* simplicial without resorting to a simplicial approximation. The fundamental idea is to embed $x^k \in f^{-1}(y)$ in either a ball B^n, its homeomorph σ_n, or a cube, and the *fundamental* degree concept is related to whether these objects have their orientation *preserved* or *reversed* under the map f. In particular, if the map $f : X \to Y$ is differentiable, the issue as to whether the orientation is preserved or reversed can be decided by computing the sign of the determinant of the Jacobian matrix of partial derivatives of f at the preimage points.

To be more specific, let the map $f : X \to Y$ be differentiable, between

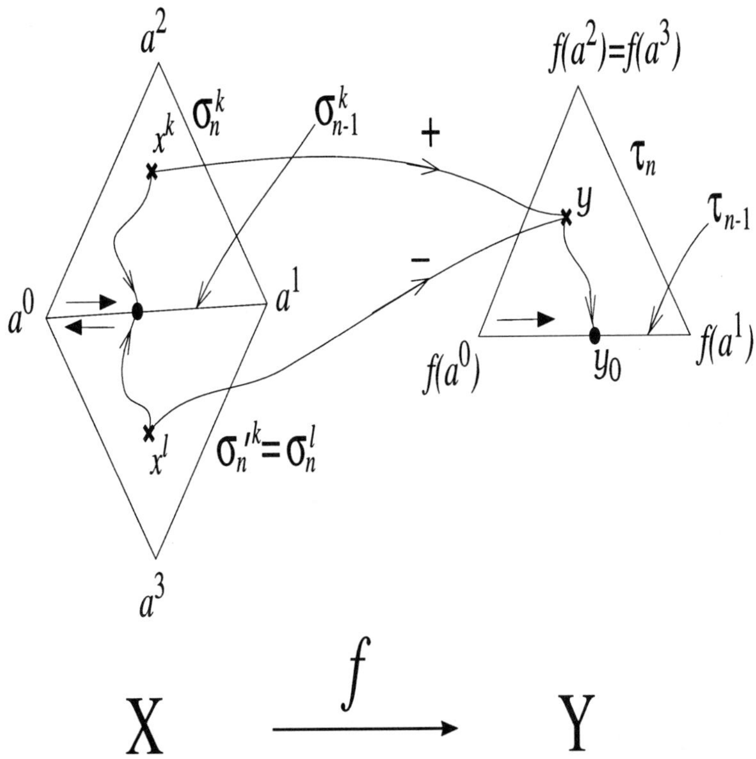

Fig. 18.2. Illustration of the combinatorial Brouwer degree when two preimage points annihilate each other.

compact differentiable, orientable manifolds of the same dimension n. Pick a regular value $y \in Y$. It is not hard to show that $f^{-1}(y)$ contains at most finitely many points (see [Bott and Tu 1982, page 41]). Let $J(x)$ denote the Jacobian,

$$J(x) = \begin{pmatrix} \frac{\partial f_1}{\partial x_1} & \cdots & \frac{\partial f_1}{\partial x_n} \\ \vdots & \ddots & \vdots \\ \frac{\partial f_n}{\partial x_1} & \cdots & \frac{\partial f_n}{\partial x_n} \end{pmatrix}$$

evaluated at x. The map f preserves or reverses the orientation around x iff $\det J(x)$ is positive or negative, respectively. Therefore,

$$\text{degree } f = \sum_{x \in f^{-1}(y)} \text{sign } \det J(x)$$

This is the so-called **analytical degree**, also referred to as the **Leray–Schauder degree** in the infinite-dimensional situation. As we will see in Subsection 19.5.8, this concept can be derived from the machinery of differential, "volume" forms, an interpretation that has roots tracing back to Brouwer.

18.4 Homological Degree of Maps Between Spheres

Traditionally, the *Brouwer degree* is defined for a map $f : S^n \to S^n$ from the n-sphere to the n-sphere. Let g be a generator of the homology group $H_n(S^n) = \mathbb{Z}$. To define how the homology class g is mapped, remember that the map f induces a homology group homomorphism $f_* : H_n(S^n) \to H_n(S^n)$. Clearly, $f_*(g)$ is an element of $H_n(S^n)$, and as such there exists an integer d such that $f_*(g) = dg$. This integer is called **(Brouwer) homological degree of the map**. Intuitively, the degree is the number of times $f(S^n)$ "wraps around" S^n.

It can be shown that Brouwer's concept of the degree encompasses the usual notion of "degree" of a rational transfer function $S^2 \to S^2$, where the "degree" is meant to be the maximum of the algebraic degrees of the numerator and denominator polynomials (see [Dieudonné 1989, page 169]). A simple illustration of this fact will be developed in Section 18.5.

To link the homological definition of the degree with the intuitive combinatorial definition, we have to resort to simplicial approximation. Remember that S^n is homeomorphic to the boundary of the standard $(n+1)$-simplex. Therefore, the map $f : S^n \to S^n$ induces a map $\partial\sigma_{n+1} \to \partial\sigma_{n+1}$. Let $\tilde{f} : Sd\partial\sigma_{n+1} \to \partial\sigma_{n+1}$ be a simplicial approximation of the latter map for some subdivision of $\partial\sigma_{n+1}$. By arguing on the simplicial map, it is not hard to show that the homological degree and the combinatorial degree match.

A fundamental property of the homological degree is that it is an invariant under homotopy. This is easily proved from the general homotopy invariance of homology. Actually much more can be said.

Theorem 18.2. (Brouwer) *Two maps $f, g : S^n \to S^n$ are homotopic if and only if they have the same degree.*

Proof. The "only if" part has already been proved. The "if" part is much harder (see [Dieudonné 1989, Part 3, Chapter II]). ■

18.5 Simple Examples

18.5.1 Degree of a Linear Map

Consider an $n \times n$ nonsingular matrix A. It can be viewed as mapping the Euclidean space $\mathbb{R}^n$ onto itself. To take into consideration what is happening at infinity, we use the one-point compactification of $\mathbb{R}^n$, $\mathbb{R}^n \cup \{\infty\}$, which is homeomorphic to S^n (see Section 2.5). Clearly, the map $A : \mathbb{R}^n \to \mathbb{R}^n$ is proper (see Section 2.5) and as such the map can be extended to the one-point compactification, as shown in the following diagram:

$$\begin{array}{ccc} \mathbb{R}^n & \xrightarrow{A} & \mathbb{R}^n \\ \downarrow & & \downarrow \\ S^n & \xrightarrow{A^+} & S^n \end{array}$$

Therefore, the degree of the linear map A is defined as the Brouwer degree of the extended map A^+. Intuitively, $A^+(S^n)$ "wraps" *once* around S^n, so that the degree is ± 1, depending on whether the map preserves or reverses the orientation. The latter clearly depends on the sign of the determinant of A, so that

$$\text{degree}(A) = \text{sign}\det A$$

For a formal exposition of this intuitive argument, see [Bredon 1993].

18.5.2 Degree of a Holomorphic Function

A polynomial map $p : \mathbb{C} \to \mathbb{C}$ can clearly be extended to the one-point compactification of the complex plane, $\mathbb{C} \cup \{\infty\} \cong S^2$, with the understanding that $\{\infty\}$ is mapped to $\{\infty\}$,

$$\begin{array}{ccc} \mathbb{C} & \xrightarrow{p} & \mathbb{C} \\ \downarrow & & \downarrow \\ S^2 & \xrightarrow{p^+} & S^2 \end{array}$$

Therefore, the degree of a polynomial, holomorphic map is defined as the Brouwer degree of the extended map $p^+ : S^2 \to S^2$. As an example, let us evaluate the degree of $z \mapsto z^2$, homologically. Define the following vertices:

$$\begin{aligned} o &= 0 + j0 \\ a^0 &= \exp(j0(\pi/2)) \\ a^1 &= \exp(j(\pi/2)) \\ a^2 &= \exp(j2(\pi/2)) \\ a^3 &= \exp(j3(\pi/2)) \\ \infty &= \infty \end{aligned}$$

With these vertices we construct the following generator of $H_2(S^2)$:

$$\begin{aligned} z_2 &= oa^0a^1 + oa^1a^2 + oa^2a^3 + oa^3a^0 \\ &\quad + \infty a^1a^0 + \infty a^2a^1 + \infty a^3a^2 + \infty a^0a^3 \end{aligned}$$

The crux of the matter is to observe the following:

$$\begin{aligned} p^+(oa^0a^1) &= oa^0a^1 + oa^1a^2 \\ p^+(oa^1a^2) &= oa^2a^3 + oa^3a^0 \\ p^+(oa^2a^3) &= oa^0a^1 + oa^1a^2 \\ p^+(oa^3a^0) &= oa^2a^3 + oa^3a^0 \\ p^+(\infty a^1a^0) &= \infty a^1a^0 + \infty a^2a^1 \end{aligned}$$

$$\begin{aligned} p^+(\infty a^2 a^1) &= \infty a^3 a^2 + \infty a^0 a^3 \\ p^+(\infty a^3 a^2) &= \infty a^1 a^0 + \infty a^2 a^1 \\ p^+(\infty a^0 a^3) &= \infty a^3 a^2 + \infty a^0 a^3 \end{aligned}$$

Adding the above equalities yields

$$p^+(z_2) = 2z_2$$

so that the Brouwer degree is 2, as expected.

18.5.3 Degree of Real Polynomial Map

The reader should not fall in the trap of believing that all polynomial maps have a Brouwer degree equal to the largest exponent. As a counterexample, consider the map $p : \mathbb{R} \to \mathbb{R}, x \mapsto x^2$. The Jacobian is clearly $J(x) = 2x$; next, take $y > 0$; the analytic degree evaluated for y is $\text{sign}J(\sqrt{y}) + \text{sign}J(-\sqrt{y}) = +1 + (-1) = 0$, so that, *contrary to the complex case*, the topological degree vanishes! The same result can be gotten from a Brouwer kind of argument. Compactify the real line to S^1 and consider the induced map $p^+ : S^1 \to S^1$. Clearly, $p^+(S^1)$ is that part of S^1 mapped stereographically onto $[0, \infty)$. Clearly, $p^+(S^1)$ can be contracted to a point, the map is nullhomotopic, and its degree vanishes.

The reader should realize that we have come across a *fundamental discrepancy* between real and complex analysis.

18.5.4 Application to Robust Stability

The concept of analytical degree becomes very useful when studying the restriction of the Nyquist map $f|\Gamma$ to a 2-D cross section through $D \times \Omega$. Since the map $f|\Gamma$ is equidimensional, the inverse image of $s \in N$ is generically a finite set of points, so that the analytical degree that can be evaluated from the Jacobian. Furthermore, as the section Γ sweeps across $D \times \Omega$, the restricted map $f|\Gamma$ is homotopically deformed. By the homotopy invariance, the degree does not change. From there we can infer how the solution set to $f|\Gamma = 0 + j0$ is changed and gain information as to how these solution sets should be pieced together to reconstruct X_ω. This is illustrated in Figure 18.3.

As a particular case, take $\dim D = 2$ and let Γ_ω be a constant frequency slice. The solution of $f|\Gamma_\omega = 0 + j0$ is, generically, a finite set of points. Furthermore, as ω changes, $f|\Gamma_\omega$ is changed by no more than a homotopy. From the homotopy invariance it is possible to infer how all of these solution points should be pieced together to reconstruct the crossover curve X_ω.

18.6 Application (Index of Vector Field)

A useful application of the Brouwer degree of maps between spheres is the concept of *index of a vector field*. Given an n-manifold X, a **vector**

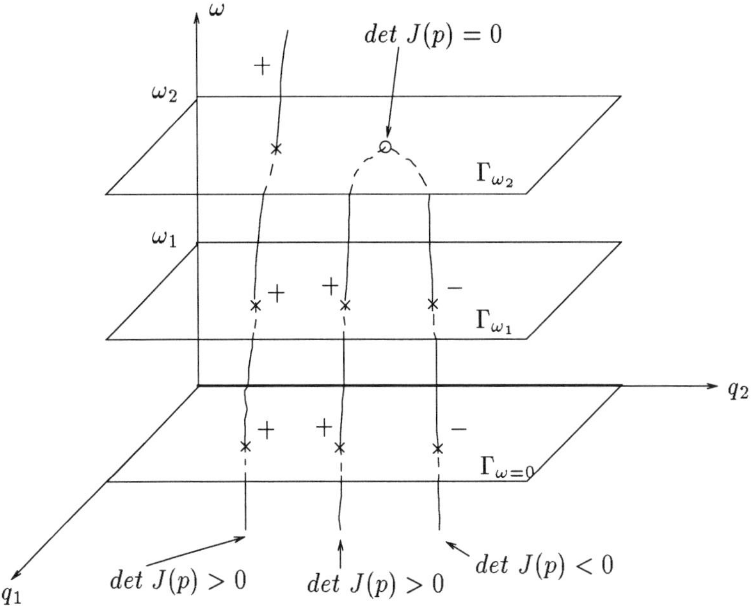

Fig. 18.3. Illustration of homotopy invariance of degree of fixed-frequency Nyquist mapping.

field V is a rule that assigns to each $x \in X$ a vector $v(x) \in T_x X$. (See Subsection 13.1.7 for the more formal definition.) Take a point x^* and consider a ball $B_{x^*}(r)$, small enough so that its closure does not contain any other zero point of the field than x^*; clearly, $\partial B_{x^*}(r) \cong S^{n-1}$. For any $x \in \partial B_{x^*}(r)$, the vector $\frac{v(x)}{\|v(x)\|}$, with its origin translated to the origin of the Euclidean space $\mathbb{R}^n$, defines a point on the unit sphere S^{n-1} of the Euclidean space. This, in turn, induces the so-called **Gauss map** $S^{n-1} \to S^{n-1}$ of the field $v(x)$ around x^*. The **index of the vector field** V at x^*, written $\text{ind}(x^*, V)$, is defined to be the Brouwer degree of the Gauss map.

18.7 (Co)homological Degree of Maps from Manifolds to Spheres

When dealing with maps that are *not* from S^n to itself, several difficulties appear as to how the concept of homological degree should be extended and the extent to which its fundamental properties remain valid. Probably the best account as to how the concept of the Brouwer homological degree should be defined for maps that are *not* from S^n to itself is to be found in [Hu 1959]. Essentially, the original concept of the degree extends without too much hurdle to maps $S^n \to M_n$ or maps $M_n \to S^n$, where M_n is some

n-dimensional manifold.

18.7.1 Degree of Nyquist-Related Map

At this juncture, we make the connection between the degree and the obstruction to extending the Nyquist map. Remember the map $(D \times \Omega)^1 \to S^2_{**}$. To be complete, we show that this map can be homotopically deformed to the map $(D \times \Omega)^1 \to S^1$. This is done as follows: Let $e(p)$ and $a(p)$ be the elevation and the azimuth, respectively, of $f(p)$ on the doubly punctured Riemann sphere—that is,

$$f(p) = \begin{pmatrix} e(p) \\ a(p) \end{pmatrix}$$

It is easily seen that the homotopy deformation is

$$F(p,t) = \begin{pmatrix} (1-t)e(p) \\ a(p) \end{pmatrix}$$

This homotopy yields a map $(D \times \Omega)^1 \to S^1$ from the one-skeleton of the uncertainty space, to the circle *of the same dimension*. This is clearly the $n = 1$ case of maps of the form $M_n \to S^n$, the Brouwer degree of which is formalized in [Hu 1959]. Because the map is *to a sphere*, we have to use *cohomology*, rather than homology, groups to capture the degree. The topological map clearly induces a cohomology group homomorphism f^*,

$$\begin{array}{ccc} (D \times \Omega)^1 & \xrightarrow{f} & S^1 \\ \downarrow & & \downarrow \\ H^*((D \times \Omega)^1) & \xleftarrow{f^*} & H^*(S^1) \end{array}$$

Observe that the bottom arrow is *reversed*, which in the language of categories means that the "downarrows"—that is, the operation of going to the cohomology groups—is a *contravariant functor*. Let g be a preferred generator of $H^1(S^1)$. Therefore, [Hu 1959] defines the **Brouwer (cohomological) degree** to be $f^*(g)$, viewed as an element of $H^*((D \times \Omega)^1)$. Clearly, it is a homotopy invariant.

Because the domain is not a sphere, there is no way to associate an integer with $f^*(g)$ and call it "degree" without loosing some crucial property that any degree concept should enjoy.

To grasp this concept of the degree and to link it with the early definition of Brouwer, the first thing is to clarify what the generator g of $H^1(S^1)$ is. Observe that any cochain c^1 that associates with a 1-simplex σ_1 of S^1 a scalar (σ_1, c^1) is a cocycle. To define a generator g of $H^1(S^1)$, it suffices to make sure that the cocyle c^1 is not a coboundary. Let the 1-simplexes of S^1 be assembled so as to form a cycle (e.g., $\gamma = a^0a^1 + a^1a^2 + a^2a^0$). It is easily seen that c^1 is a coboundary iff $(\gamma, c^1) = 0$. Therefore, a generator g of $H^1(S^1)$ is a cochain such that $(\gamma, g) \neq 0$. Likewise, a cohomology class of $(D \times \Omega)^1$ is a cochain that associates a nonvanishing scalar to at least one cycle of $(D \times \Omega)^1$. The next thing is to understand how the cohomology group homomorphism goes in a direction *opposite* to the topological map. Let φ be the chain map induced by f. An arbitrary cochain c^1 of S^1 induces a cochain d^1 on $D \times \Omega$ by the rule $(\sigma_1, d^1) = (\varphi(\sigma_1), c^1)$, where σ_1 is an arbitrary simplex of $(D \times \Omega)^1$. This defines a mapping $c^1 \mapsto d^1, \hom(S^1, G) \to \hom((D \times \Omega, G)^1, G)$ that is easily seen to map cocycle into cocycle and coboundary into coboundary, so that it induces the required homomorphism $H^1(S^1) \to H^1((D \times \Omega)^1)$. To get a little more specific as to what this latter homomorphism is, let z_i be the cycles of $(D \times \Omega)^1$ that are mapped into γ—that is, $\varphi(z_i) = \gamma$. By the basic rule, the cochain g assigns to the cycles z_i's the scalar $(\varphi(z_i), g) = (\gamma, g) \neq 0$. It follows that the rule $z_i \mapsto (\varphi(z_i), g)$ is a cohomology class that could be safely identified with $f^*(g)$. Equivalently, the degree $f^*(g)$ can be defined by $(z_i, f^*(g)) = (\varphi(z_i), g)$.

It follows from the above that this concept of the degree relates to *whether* and *how* cycles of $(D \times \Omega)^1$ are mapped to S^1. As such, this concept begins to resemble the original combinatorial definition of Brouwer (see [Dieudonné 1989, page 178]) and the related concept of *analytical* degree.

With this degree concept, we can formulate the obstruction problem in terms of the degree.

Theorem 18.3. (Brouwer–Hopf) $f^1 : (D \times \Omega)^1 \to S^1$ *is nullhomotopic iff its degree vanishes. Consequently, there is no obstruction to extending* f^1 *iff the degree of* $f^1 : (D \times \Omega)^1 \to S^1$ *vanishes.*

Proof. Clearly, a vanishing degree is equivalent to all the cycles of $(D \times \Omega)^1$ being mapped to S^1 in a way homotopically equivalent to a constant map. ∎

18.7.2 Degree of Maps from Manifolds to Spheres

The previous subsection is merely a manifestation of a general feature of the degree of maps from manifolds to spheres. The **degree** of a map $f : M_n \to S^n$ is defined as $f^*(g)$, where g is a generator of $H^n(S^n)$.

With this concept, we can formulate the question as to whether or not a vanishing degree is equivalent to nullhomotopy.

Theorem 18.4. (Brouwer–Hopf) *If* M_n *is a compact, connected, orientable manifold (in the sense of Brouwer), then two maps* $M_n \to S^n$ *are homotopic iff they have the same degree. In particular* $f : M_n \to S^n$ *is nullhomotopic iff its degree vanishes.*

Proof. This is a very complicated proof. For the history of this problem and a sketch of the proof, see [Dieudonné 1989, page 311]. ∎

The reader is warned that this result in the particular case $n = 1$ cannot, in general, be applied to the Nyquist problem, for indeed $(D \times \Omega)^1$ is not, in general, a manifold.

18.7.3 A Counterexample

We show that for such a map as $f^1 : (D \times \Omega)^1 \to S^1$, there is a discrepancy between the cohomological degree $f^*(g), g \in H^1(S^1)$ and the combinatorial degree in the general case where $(D \times \Omega)^1$ fails to be a manifold.

Assume $(D \times \Omega)^1$ is **orientable**—that is, the 1-simplexes of $(D \times \Omega)^1$ can be oriented so that the sum of *all* oriented simplexes, $a^0a^1 + a^1a^2 + ...$, is a cycle, z. This cycle is a homology class of $H_1((D \times \Omega)^1)$ as well. Therefore, z, viewed as an element of $H_1((D \times \Omega)^1)$, is mapped to an element of $H_1(S^1)$. In terms of a preferred generator γ of $H_1(S^1)$, we have

$$f_*(z) = (n_+ - n_-)\gamma$$

where $n_+ - n_-$ is the number of cycles of $(D \times \Omega)^1$ mapped to γ with *coherent* orientation minus the number of cycles of $(D \times \Omega)^1$ mapped to γ with *opposite* orientation. The difference, $n_+ - n_-$, is yet another concept of the degree that is formulated in, among other places, [Spanier 1989, page 207, F1]. It is clearly equivalent to the combinatorial degree.

The important question is whether the cohomological degree $f^*(g)$ is equal to the combinatorial degree $n_+ - n_-$. Unfortunately this is not, in general, the case. A counterexample is given by the one-dimensional complex of Figure 18.4 mapped simplicially to the labels complex. In this case, the orientation of $(D \times \Omega)^1$ is

$$\begin{aligned} z &= a^0a^1 + a^1a^2 + a^2a^0 \\ &\quad + a^0a^3 + a^3a^4 + a^4a^1 + a^1a^3 + a^3a^5 + a^5a^0 \\ &\quad + a^4a^5 + a^5a^6 + a^6a^4 \end{aligned}$$

It is indeed easily checked that $\partial_1 z = 0$. Furthermore, the simplicial approximation maps as follows:

$$\begin{aligned} a^0 &\mapsto 0 \\ a^1 &\mapsto 1 \\ a^2 &\mapsto 2 \\ a^3 &\mapsto 0 \\ a^4 &\mapsto 1 \\ a^5 &\mapsto 0 \\ a^6 &\mapsto 2 \end{aligned}$$

It is easily seen that the only cycles in $(D \times \Omega)^1$ that map to γ are $a^0a^1 + a^1a^2 + a^2a^0 = z_1$ and $a^4a^5 + a^5a^6 + a^6a^4 = z_2$. The first one maps with *coherent* orientation while the second one maps with *opposite* orientation.

Therefore, $n_+ - n_- = 0$. Finally, it is easily seen that

$$\begin{aligned} f^*(g) \quad &: \quad z_1 \mapsto (\gamma, g) \neq 0 \\ &: \quad z_2 \mapsto -(\gamma, g) \neq 0 \end{aligned}$$

Clearly, $f^*(g)$ is not trivial and this establishes the required contradiction.

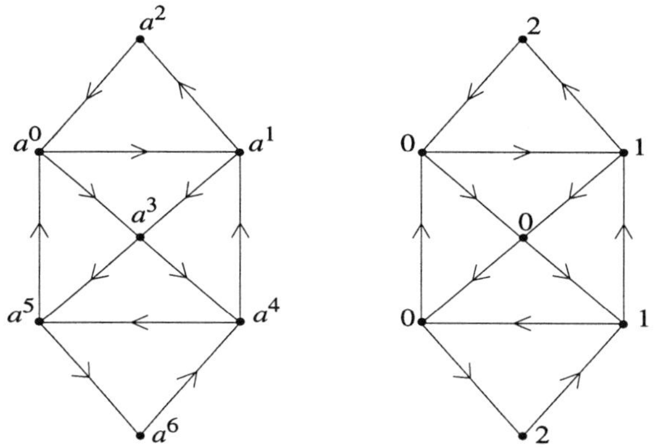

Fig. 18.4. Left diagram: The one-dimensional simplicial complex of the degree counterexample and its orientation. Right diagram: The labeling of the complex.

Despite this counterexample, there is, however, at least one important particular case where the equivalence $f^*(g) = 0 \Leftrightarrow n_+ - n_- = 0$ can be guaranteed. It is the case where $(D \times \Omega)^1$ is combinatorially equivalent to the 1-skeleton of the Q-triangulation of the standard 2-simplex. The equivalence is easily seen to be a trivial corollary of the superstrong version of Sperner's lemma.

18.8 Example (Brouwer Degree Proof of Superstrong Sperner Lemma)

As an example of application of the combinatorial degree and its homotopy invariance, we develop a proof of the superstrong version of Sperner's lemma.

Let $\sigma_n = a^0 ... a^n$ be a completely labeled simplex—that is,

$$\ell(a^i) = i, \quad i = 0, 1, ..., n$$

Let $S_Q \sigma_n$ be the polyhedron obtained after Q-refinement of this simplex. The vertices of the refined simplex are labeled, subject to the conditions of Sperner properness. To be specific, for any vertex a of the refined triangulation in a face of the original simplex,

$$a \in a^{i_0} a^{i_1} ... a^{i_m}, \quad i_0, i_1, ..., i_m \in \{0, 1, ..., n\}$$

its label is restricted to be in the set of indices of the face,

$$\ell(a) \in \{i_0, i_1, ..., i_m\}$$

By Sperner's lemma, we know that there are completely labeled simplexes at the higher resolution level. The labeling of a completely labeled high-resolution simplex defines an orientation of this simplex that may be coherent with, or opposed to, the orientation $a^0...a^n$ of the big simplex. Let n_+ be the number of high-resolution completely labeled simplexes with orientation *coherent* with that of $a^0...a^n$. Let n_- be the number of completely labeled high-resolution simplexes with orientation *opposite* to $a^0...a^n$.

Theorem 18.5. (Sperner—Superstrong Version) $n_+ - n_- = 1$.

Proof. Consider the identity map

$$1 : S_Q\sigma_n \to S_Q\sigma_n$$

Next, the labeling ℓ induces the simplicial map $l(\cdot)$ defined by the following vertex transformation:

$$\begin{array}{rcl} l(\cdot) : S_Q\sigma_n & \to & S_Q\sigma_n \\ a & \mapsto & a^{\ell(a)} \end{array}$$

Both of these maps have a degree. The (combinatorial) degree of the identity map is 1. The combinatorial degree of $l(\cdot)$ is easily seen to be $n_+ - n_-$. Clearly the two maps are homotopic. The problem we are facing in applying the homotopy invariance of the degree theory in this context is that the maps are defined over polyhedra with boundary. However, the *crucial point* is to observe that during the homotopy the boundary remains *fixed*. From there, it is not hard to prove that the degree is invariant under homotopy. Therefore,

$$\text{degree}(l(\cdot)) = n_+ - n_- = \text{degree}(1) = 1$$

as claimed. ∎

18.9 Degree of Maps Between Pseudomanifolds

Let X, Y be two orientable pseudomanifolds with boundaries and consider a map f together with its induced homology group homomorphism, as depicted in the following diagram:

$$\begin{array}{ccc} X, \partial X & \xrightarrow{f} & Y, \partial Y \\ \downarrow & & \downarrow \\ H_*(X, \partial X) & \xrightarrow{f_*} & H_*(Y, \partial Y) \end{array}$$

Let g_x, g_y be orientations of X, Y, respectively; in other words, g_x, g_y are generators of $H_n(X, \partial X), H_n(Y, \partial Y)$, respectively. The **(homological)**

degree of the map f between pseudomanifolds is the unique integer such that

$$g_y = \text{degree}(f) f_*(g_x)$$

18.10 Homotopy Collapse of Template

In this section, we do a deformation retract of the template onto the unique triangle that contains $0 + j0$ and show how a degree test on the resulting map is equivalent to the labeling test of Chapter 9. The retraction of the template is very much in the spirit of Subsection 9.4.2, except that here the strong deformation retract is made "stronger" by "realizing" it as a sequence of *elementary collapses.*

Consider the simplex $b^k b^i b^j$ of the template N. An onto mapping

$$c : b^k b^i b^j \to b^k b^i \cup b^k b^j$$

is called **elementary collapse** (see [Glaser 1970]). The intuitive idea is to "push in" the face $b^i b^j$ of the simplex until it hits the other faces. The elementary collapse is written

$$b^k b^i b^j \searrow b^k b^i \cup b^k b^j$$

We also define a collapsing homotopy $C(\cdot, t)$ such that $C(\cdot, 0)$ is the identity and $C(\cdot, 1) = c(\cdot)$.

There are several ways to define the collapsing map $c(\cdot)$ and the collapsing homotopy $C(\cdot, t)$. Here we will adopt the following definition: Consider the parallelogram with vertices b^k, b^i, b^j, v constructed on the vectors $\vec{b^k b^i}, \vec{b^k b^j}$. Let b be a vertex of $b^k b^i b^j$. $c(b)$ is defined to be the intersection of the radial line bv with $b^i b^k \cup b^j b^k$. It is possible to define this collapsing map in terms of barycentric coordinates. Let b^{ij} be the barycenter of $b^i b^j$. Subdivide the simplex $b^k b^i b^j$ consistently with the barycentric subdivision of the face $b^i b^j$. Let b be a point of $b^k b^i b^j$, assumed to be in the simplex $b^k b^i b^{ij}$, with respect to which b has barycentric coordinates $(\lambda_k, \lambda_i, \lambda_{ij})$. The collapsing homotopy is defined as

$$\left(\begin{pmatrix} \lambda_k \\ \lambda_i \\ \lambda_{ij} \end{pmatrix}, t \right) \mapsto \begin{pmatrix} \lambda_k + t\lambda_{ij} \\ \lambda_i \\ \lambda_{ij}(1-t) \end{pmatrix}$$

In case of a 1-simplex $b^k b^i$, the collapse

$$b^k b^i \searrow b^k$$

is defined similarly. The details are left to the reader.

Now, consider the *simply connected* template $N = f_{PL}(D \times \Omega)$ together with its underlying Delaunay triangulation. Assume that the labeling ℓ is Sperner proper and let $b^0 b^1 b^2$, $\ell(b^i) = i$ be the unique simplex that contains $0 + j0$. Consider a simplex $b^k b^i b^j$ that has a "free" face $b^i b^j$;

that is, $b^i b^j$ is the face of *exactly one* simplex of N. Consider the collapse $b^k b^i b^j \searrow b^i b^j$ and let $C_1(\cdot, t)$ be the corresponding collapsing homotopy. Consider the composition map $C_1(\cdot, 1) \circ f_{PL}$. This map is homotopic to f_{PL} and furthermore its image $C_1(\cdot, 1) \circ f_{PL}(D \times \Omega)$ is a proper subcomplex of N. Start collapsing those simplexes that have a "free" face that intersects one of the radials of the butterfly. Then collapse the simplexes that do not intersect the butterfly. Compose all of the collapsing homotopies into the composite

$$C = \ldots \circ C_2 \circ C_1$$

Clearly,

$$C(\cdot, 1) \circ f_{PL} : D \times \Omega \to \overline{b^0 b^1 b^2}$$

is homotopic to f_{PL} and maps to $\overline{b^0 b^1 b^2}$.

Clearly, during this homotopy, the preimage remains unchanged, $f_{PL}^{-1}(0+j0) = (C(\cdot, 1) \circ f_{PL})^{-1}(0 + j0)$, and the degree remains unchanged.

Also, observe that during the collapsing homotopy, the labels of the vertices do not change.

Now, we perform an obstruction analysis on the map $C(\cdot, 1) \circ f_{PL}$. The issue becomes the degree of the map between 1-skeleta:

$$\ell = (C(\cdot, 1) \circ f_{PL})^1 : (D \times \Omega)^1 \to (b^0 b^1 b^2)^1$$

Clearly, this is nothing other than the labeling function as defined in Chapter 9. The reader can easily see that checking the degree of the latter map is equivalent to the labeling technique of Chapter 9.

The reader can easily understand some of the obvious advantages of realizing, when possible, a homotopy equivalence $N \simeq \overline{b^0 b^1 b^2}$ as a series of elementary collapses, $N \searrow \ldots \searrow \overline{b^0 b^1 b^2}$. The converse of an elementary collapse, called **elementary expansion**, is the process of attaching a simplex to a polyhedron along a common face. A homotopy equivalence that is a series of elementary collapses and expansions is called **simple homotopy equivalence.** Simple homotopy equivalence allows for a finer classification than homotopy equivalence and is codified in the language of algebraic K-theory; see [Weinberger 1994, Section 1.2] for these insights.

18.11 Continuation or Embedding Methods

The invariance of the degree of the Nyquist map $f|\Gamma_\omega : \Gamma_\omega \to \mathbb{C}$ under frequency sweep and its corollary indicating how the pieces of crossover should be assembled, see Figure 18.3, have already provided us with a prelude to the so-called *continuation* or *embedding* methods (see [Wacker 1978]). Roughly speaking, the essence of the continuation or embedding methods for computing zero-points of maps, fixed-points of maps, and so on is to exploit the homotopy invariance of the degree to reduce the problem to a more tractable one. To be slightly more explicit, given a "difficult" zero-point or fixed-point problem, construct a homotopy that deforms the problem into a "simple" one, solve the "simple" problem, and then backtrack along the homotopy path all the way to the solution of the initial, "difficult" problem.

More formally, consider the zero-point problem of computing $f^{-1}(0)$ for a map $f : X \to \mathbb{R}^n$ where X is an n-D manifold-with-boundary (e.g., a closed disk; see Section 22.1 for the precise definition). It is further assumed that $f(\partial X) \not\ni 0$. The rationale for this formulation is that we should have some a priori knowledge as to where the roots are or restrict ourselves to those roots in a bounded region. Next, we construct a map $g : X \to \mathbb{R}^n$ for which $g^{-1}(0)$ is easy to compute or already known and $g(\partial X) \not\ni 0$. For example, if $X = \mathbb{R}^n$, we could choose $g(x) = (g_1(x), ..., g_n(x))$, where the g_i's are polynomials constructed so that they have finitely many real roots in common, none of which is in ∂X. The crucial step is to construct a homotopy

$$\begin{aligned} F : X \times [0,1] &\to \mathbb{R}^n \\ (x,t) &\mapsto F(x,t) \end{aligned}$$

such that

$$\begin{aligned} F(x,0) &= f(x) \\ F(x,1) &= g(x) \end{aligned}$$

Typically, we would choose,

$$F(x,t) = (1-t)f(x) + tg(x)$$

Under some conditions, $F^{-1}(0)$ is a bundle of curves that provides some paths from the trivial solution $g^{-1}(0)$ to the sought solution $f^{-1}(0)$. This is illustrated in Figure 18.5.

To put these ideas into gear, there are three questions to be answered:

- Are there some curves joining some (all?) solutions $g^{-1}(0)$ to some (all?) solutions $f^{-1}(0)$?
- Are the curves smooth enough so that it is possible to backtrack on them?
- How to devise an algorithm to backtrack?

To be able to exploit the invariance of the degree under homotopy, which is instrumental in answering the above questions, we have to ensure that, $\forall t$, the x-roots of $F(x,t)$ do not intersect ∂X,

$$F(\partial X, t) \not\ni 0, \forall t \tag{18.1}$$

Clearly, a necessary condition for this to happen is that the maps

$$f, g : (X, \partial X) \to (\mathbb{R}^n, \mathbb{R}^n \setminus 0)$$

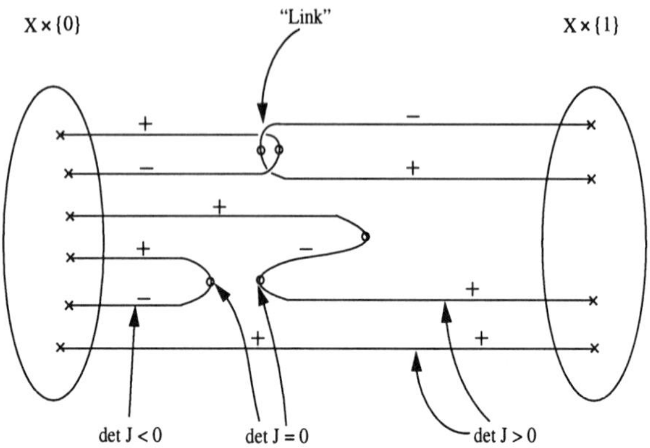

Fig. 18.5. Illustration of the "homotopy" or "continuation" method to go from the solution of a trivial problem to the sought solution.

have the same degree. (Observe that $H_n(X, \partial X) \cong \mathbb{Z}$ and $H_n(\mathbb{R}^n, \mathbb{R}^n \setminus 0) \cong H_n(S^n) \cong \mathbb{Z}$, so that the homology group homomorphism $f_n : H_n(X, \partial X) \to H_n(\mathbb{R}^n, \mathbb{R}^n \setminus 0)$ is multiplication by an integer which is the degree.)

Here is the answer to the first question:

Theorem 18.6. *Let X be an n-D manifold-with-boundary. Assume the maps*

$$f, g : (X, \partial X) \to (\mathbb{R}^n, \mathbb{R}^n \setminus 0)$$

have the same degree and that Condition 18.1 is satisfied. Then there exists at least one curve joining a solution in $g^{-1}(0) \subseteq X \times \{1\}$ to a solution in $f^{-1}(0) \subseteq X \times \{0\}$ whenever the common degree of the maps f, g is nonvanishing.

Proof. The only way $f^{-1}(0)$ and $g^{-1}(0)$ could not be connected by at least one curve is when all of the tracks leaving $f^{-1}(0)$ *return* to $f^{-1}(0)$. But the latter assertion means that the degree of the map is vanishing. Hence a contradiction. ■

A more general version of this theorem is available in [Alexander 1978]. Regarding the second question, here is a quick answer:

Theorem 18.7. *Under the same conditions as the preceding theorem and if the Jacobian of the map $x \mapsto F(x,t)$ is nonsingular $\forall t \in [0,1]$, then $F^{-1}(0)$ consists of nonintersecting curves joining every root of g to every root of f.*

Proof. If the Jacobian of the map $x \mapsto F(x,t)$ is nonsingular, the map $(x,t) \mapsto F(x,t)$ is a submersion—that is, the matrix of partial derivative has full column rank. It follows that every value, in particular 0, is regular.

It follows from the implicit function theorem that $F^{-1}(0)$ is a manifold. This implies that the curves are nonintersecting. From the preceding theorem, none of the curves goes backward. Hence the only remaining possibility is the curves joining every single root of g to every single root of f. ∎

Observe that when the maps $f(x)$, $g(x)$ are affine, namely, $f(x) = A_0x + b_0$, $g(x) = A_1x + b_1$, the condition that the maps have the same degree reads $\operatorname{sign}\det A_0 = \operatorname{sign}\det A_1$; further, the condition that the Jacobian of $F(\cdot, t)$ is nonsingular amounts to checking that the pencil $A_0 + tA_1$ has no generalized eigenvalues in the *real* interval $[0, 1]$. This can be checked, using finitely many arithmetic operations, by running the Sturm sequence of the polynomial $\det(A_0 + tA_1)$; see [Bochnak, Coste, and Roy 1987, Chapter 1].

See [Alexander 1978] for another, topologically more involved, condition guaranteeing enough smoothness of the curves.

In case of a zero-point *gradient* problem—that is, $f = \nabla F = 0$—we would take the "easy" problem to be another gradient zero-point problem—that is, $g = \nabla G = 0$. In this case, the homotopy is

$$(1-t)\nabla F + t\nabla G = \nabla((1-t)F + tG) = 0$$

The tracks going from the "easy" to the "difficult" problem solutions are the locis of the critical points of the one-parameter family of functions $(1-t)F + tG$. The latter problem is far from trivial, but fairly well-understood (see [Cerf 1970]). In Chapter 21, we will be confronted with a similar problem.

Regarding algorithms for backtracking along the homotopy path, the idea is fairly simple. Assume the current situation along the homotopy path is $F(x, t) = 0$. Perturb the homotopy parameter and solve the perturbed equation $F(x - \epsilon y, t - \epsilon) = 0$ up to first order for y, which yields the backtracking direction

$$y = -\left(\frac{\partial F}{\partial x}\right)^{-1} \frac{\partial F}{\partial t}$$

Chosing the correct step size ϵ involves some numerical stability issues which might call for predictor-corrector steps, but we will not discuss this here.

Finally, we also mention without any further discussions that the game of Hex algorithm for locking on the crossover curve (see Section 9.2) can be viewed as a *"simplicial continuation"* algorithm.

18.12 HISTORICAL NOTES

The concept of *degree of a map* is due to Brouwer. It was originally developed in the spirit of the combinatorial approach of Section 18.2. By skillful application of that *single concept*, coupled with simplicial approximation (see Section 4.9), Brouwer was able to prove the celebrated theorems that bear his name: the Brouwer fixed-point theorem, the Brouwer invariance of the domain, the Jordan–Brouwer theorem, and so on.

Later, after the concept of homology became mature, the combinatorial degree was reformulated in terms of the (co)homology group homomorphism induced by the map.

While, historically, degree theory was the first proof of the fixed-point theorem, later, many other proofs, among other proofs the so-called *constructive proofs* based on the simplicial techniques developed in Part I, came about. Degree theory was not only instrumental in the first proofs of the Brouwer theorems, but degree theory provided new proofs of, and provided new insights into, known theorems; see [Dieudonné 1989, Part 2, Chapters I, II, and III]. The lesson we learn from history is that degree theory is just *one way* to tackle a class of topological problems. This chapter has been organized in this spirit; we did not derive new results; we merely rephrased many previously known results within the framework of degree theory.

An excellent monograph, specially emphasizing the *analytic* degree, is [Lloyd 1978]. Probably the best exposition of the *(co)homological* degree of maps between manifolds and spheres is [Hu 1959]. For the history of the problem, and the connection between the analytic and homological degrees, see [Dieudonné 1989].

The continuation methods—that is, "putting degree theory to work" to solve zero-point and other problems—have achieved a certain level of popularity in control. The first such application of continuation methods was probably [Richter and DeCarlo 1984].

19

HOMOTOPY OF MATRIX RETURN DIFFERENCE MAP

It [Bott's theorem] opens up the possibility of developing large parts of algebraic topology on the basis of linear algebra rather than on the orthodox theory of simplicial complexes and homology.

Michael F. Atiyah,
"Algebraic topology and elliptic operators,"
Communications on Pure and Applied Mathematics,
vol. 20, pp. 237–249, 1967.

SUMMARY

We consider the Nyquist map that takes a combination of uncertain parameters to the corresponding *matrix* return difference $I + L\Delta$. This is motivated, in part, by the counterexamples of Section 2.2, pointing out some pitfall in the usual determinant criterion when the number of open-loop RHP poles varies.

We approach the problem in the spirit of obstruction theory as we did for the determinant map into S^1. We attempt to recursively construct a Nyquist map into $GL(n_l, \mathbb{C})$, where n_l is the number of feedback paths. The construction is recursive over higher-and-higher-dimensional subpolyhedra of the polyhedron of uncertainties. The *central* problem is to determine whether the map $\partial\sigma_n \to GL(n_l, \mathbb{C})$ can be extended to a map $\sigma_n \to GL(n_l, \mathbb{C})$. The obstruction problem immediately leads us to studying $\pi_{n-1}(GL(n_l, \mathbb{C}))$. The Bott periodicity theorem reveals an interplay between the number of uncertain parameters and the number of feedback loops. Loosely speaking, doing a loop shifting so as to make $2n_l \geq n$ results in a problem more mathematically tractable. It turns out that the homotopy groups of $GL(n_l, \mathbb{C})$ are by far less trivial than those of S^1 encountered in the determinant map. To put it simply, there are more obstructions to extending the map in $GL(n_l, \mathbb{C})$ than the map in S^1. It follows that a vanishing obstruction cocycle is getting closer to a sufficient condition for robust stability than in the determinant case. This constitutes the second motivation for looking at the map into the *matrix* return difference.

The main body of this chapter is devoted to developing a degree concept, specially devised for the map $\partial\sigma_n \to GL(n_l, \mathbb{C})$. Somehow, the degree of the map is a measure of the amount of obstruction to extending it to σ_n. Another, somewhat loose interpretation is that the degree is the number of times the image of $\partial\sigma_n$ "wraps around" the subset of singular matrices and as such the degree generalizes the intuitive concept of "winding number" around $0+j0$. Finally, we develop a computational approach to the degree based on integration of differential forms on Lie groups.

19.1 Matrix Return Difference

The central issue in the traditional multivariable Nyquist stability criterion is to ensure that $\det(I + L\Delta) \neq 0, \forall\omega, \forall\Delta \in D$. However, at a more *fundamental* level, the issue is rather to ensure that the *matrix* $I + L(j\omega)\Delta$ is *nonsingular* at all frequencies and for all uncertain parameters. To be more precise, we have the following theorem:

Theorem 19.1. *Assume $(I + L(s)\Delta_0)^{-1}$ is stable for some nominal $\Delta_0 \in D$, where D is a connected set of uncertainties. Assume the mapping $(\omega, \Delta) \mapsto (I+L(j\omega)\Delta)$ is continuous. Then $(I+L(s)\Delta)^{-1}$ is stable $\forall\Delta \in D$ iff $(I + L(j\omega)\Delta)$ is nonsingular $\forall\omega$ and $\forall\Delta \in D$.*

These considerations lead to the question of *existence* of the map

$$\begin{array}{ccc} (\Delta, \omega) & \mapsto & (I + L(j\omega)\Delta) \\ D \times \Omega & \to & GL(n_l, \mathbb{C}) \end{array}$$

where n_l is the number of feedback paths. Existence of this map appears equivalent to existence of the map $D \times \Omega \to \mathbb{C} \setminus \{0 + j0\}$ of the previous chapters.

However, if we formulate robust stability as an obstruction problem—that is, recursive robust stability check over an increasing sequence of nested uncertainty subsets—it turns out that there are *fundamental* differences between the map into $\mathbb{C} \setminus \{0 + j0\}$ and the map into $GL(n_l, \mathbb{C})$.

The map into $\mathbb{C} \setminus \{0 + j0\} \simeq S^1$ has a relatively easy obstruction structure because of the relatively simple homotopy structure of S^1. The problem is that the homotopy structure of $GL(n_l, \mathbb{C})$ is quite another story!

To elucidate the algebraic topology of the general linear group, it is convenient to retract $GL(n_l, \mathbb{C})$ into the *compact* subgroup of $n_l \times n_l$ unitary matrices, $U(n_l)$. The latter is referred to as the **unitary group**.

19.2 General Linear versus Unitary Groups

For the determinant map, we exploited the homotopy equivalence $\mathbb{C} \setminus \{0 + j0\} \simeq S^1$, from which it followed that $\mathbb{C} \setminus \{0 + j0\}$ and S^1 have the same homotopy groups, the same cohomology groups, and so on. As far as obstruction goes, the determinant Nyquist map could be viewed as a map either into $\mathbb{C} \setminus \{0 + j0\}$ or into S^1. The matrix extension of this is the following:

Theorem 19.2.

$$GL(n_l, \mathbb{C}) \simeq U(n_l)$$

Consequently,

$$\begin{aligned}\pi_{n-1}(GL(n_l, \mathbb{C})) &= \pi_{n-1}(U(n_l)) \\ H^*(GL(n_l, \mathbb{C})) &= H^*(U(n_l))\end{aligned}$$

Proof. The key point in this whole proof is that $GL(n_l, \mathbb{C})$ can be retracted to $U(n_l, \mathbb{C})$. Clearly, any $A \in GL(n_l, \mathbb{C})$ can be factored *in a unique way* (see [Marcus and Minc, page 74]) as $A = QH$, where $Q \in U(n_l)$ and H is a positive definite Hermitian matrix. Therefore, *as a set*, $GL(n_l, \mathbb{C})$ can be written

$$GL(n_l, \mathbb{C}) = U(n_l) \times \text{PDH}$$

We leave it to the reader to verify that this factorization is continuous. Therefore, we write a nonsingular matrix as (U, H), where U, H are the unitary, positive definite Hermitian, respectively, matrices of the polar factorization. The relevant fact is that the cone of positive definite Hermitian matrices is contractible to the identity matrix I, and therefore $GL(n_l, \mathbb{C})$ is contractible to $U(n_l)$. To be more specific, consider the inclusion map

$$i : U(n_l) \to GL(n_l, \mathbb{C})$$

and the projection map

$$\begin{aligned}\pi : GL(n_l, \mathbb{C}) &\to U(n_l) \\ (U, H) &\mapsto U\end{aligned}$$

Clearly, $\pi \circ i = 1_{U(n_l)}$. Furthermore, it is claimed that $i \circ \pi \simeq 1_{GL}$. Indeed, the following provides the required homotopy from $i \circ \pi$ to 1_{GL}:

$$\begin{aligned}GL(n_l, \mathbb{C}) \times [0, 1] &\to GL(n_l, \mathbb{C}) \\ ((U, H), t) &\mapsto (U, (1-t)I + tH)\end{aligned}$$

(Observe that this map is the identity on U during the entire homotopy; this is called (strong) deformation retraction.) Therefore, $GL(n_l, \mathbb{C})$ and $U(n_l)$ are homotopically equivalent, and it is a standard result that homotopically equivalent spaces have the same homotopy and cohomology groups. (For more details about these kinds of arguments, see [Bott and Tu 1982, page 36, Corollaries 4.1.2.1 and 4.1.2.2].) ■

A problem with the polar factorization argument is that when we have a GL-valued map, it is hard to perform the polar factorization for several values of the parameter and then stitch together the great many factorizations. In case of a GL-valued map, like the Nyquist map, we prefer to use the Gram–Schmidt orthogonalization procedure on the columns of the matrix, resulting in a unitary matrix. It is easily seen that the Gram–Schmidt orthogonalization is continuous (see [Atiyah 1966c]). It provides

$A(p) = U(p)R(p)$, where $R(p)$ is an upper-triangular matrix with real, non-vanishing numbers on the diagonal. Therefore, by means of a homotopy, the diagonal entries can be made equal to 1 whereas another homotopy wipes out the strictly upper-triangular part. Therefore, we have the following theorem:

Theorem 19.3. *If $A(p)$ is a continuous GL-valued map and if $A(p) = U(p)R(p)$ is obtained by application of the Gram–Schmidt orthogonalization process, then the maps*

$$\begin{aligned} p &\mapsto A(p) \\ p &\mapsto U(p) \end{aligned}$$

are homotopic.

19.3 Homotopy Groups of GL

The homotopy groups $\pi_{n-1}(GL(n_l, \mathbb{C}))$ reveal a fundamental difference between the case $2n_l \geq n$ and $2n_l < n$. The former case (large number of feedback paths) is referred to as *stable homotopy case*. The latter case (large number uncertain parameters) is by far more complicated and referred to as *unstable homotopy*.

19.3.1 Stable Homotopy ($2n_l \geq n$)

The fundamental result is the following:

Theorem 19.4. (Bott) *For $2n_l \geq n$,*

$$\begin{aligned} \pi_{n-1}(GL(n_l, \mathbb{C})) &= 0, \ n \text{ odd} \\ &= \mathbb{Z}, \ n \text{ even} \end{aligned}$$

Proof. This result has a long history of complicated proofs (see [Bott 1957, 1958, 1959, 1970]). For some simplified proofs, see [Atiyah and Bott 1964] and [Atiyah 1968]. Probably the most elementary proof, in the K-theoretic context of Chapter 20, is in [Atiyah 1969]. ■

The above is a manifestation of a very broad result, usually referred to as *Bott periodicity theorem*. This result is a cornerstone of the so-called *topological K-theory* to which we will return later. The Bott periodicity theorem reveals that robust stability has some generic features depending essentially on how the numbers n_l of feedback paths and the number n of uncertain parameters are related. It is a general trend that increasing the number of feedback paths simplifies the problem. For that reason, the domain of validity of Bott's theorem, $2n_l \geq n$, is referred to as *stable homotopy*. In the *unstable homotopy* case, $2n_l < n$, the homotopy classification of maps $S^{n-1} \to GL(n_l, \mathbb{C})$ is extraordinarily complicated and not, as yet, completely elucidated.

The well-known "loop-shifting" procedure of putting the uncertainty in "diagonal perturbation" form is an attempt to increase the number of feedback paths so as to make the problem more manageable. To illustrate this, consider the fixed-frequency mapping of the complex μ-function:

$$\begin{array}{rcl} D = (\mathbb{D})^{n_l} & \rightarrow & GL(n_l, \mathbb{C}) \\ (q_1, \ldots, q_{n_l}) & \mapsto & I + L(j\omega) \begin{pmatrix} q_1 & \ldots & 0 \\ \vdots & \ddots & \vdots \\ 0 & \ldots & q_{n_l} \end{pmatrix} \end{array}$$

Clearly, we have $2n_l = n_q \geq n$. Therefore, the obstruction to extending the partial fixed-frequency map from the skeleton D^{n-1} to the skeleton D^n is a stable homotopy problem. The last extension from the skeleton D^{n_q-1} to D is the borderline of the cases manageable via the Bott periodicity theorem. Taking the frequency into consideration, the obstruction to the extension from P^{n-1} to P^n is a stable homotopy problem as long as $n \leq n_q$; it, however, becomes an unstable homotopy problem when $n = n_p$.

Restricting the q_i's to be real—typically, $q_i \in [-1, +1]$—yields $2n_l = 2n_q \geq 2n > n$, so that the problem remains within the realm of the Bott periodicity theorem. It is easily seen that even with the additional frequency parameter we remain in the stable homotopy case.

19.3.2 Unstable Homotopy ($2n_l < n$)

In the case $2n_l < n$, not much can be said, in general, about the homotopy groups $\pi_{n-1}(GL(n_l, \mathbb{C}))$.

For the case of two feedback loops ($n_l = 2$), we can reduce the unstable homotopy of the classical group $GL(2, \mathbb{C})$ to the more orthodox problem of the homotopy groups of spheres. First observe that $\pi_{n-1}(GL(2, \mathbb{C})) = \pi_{n-1}(U(2))$, where $U(2)$ is the Lie group of 2×2 unitary matrices. (This last result is trivial from the polar factorization of an arbitrary element of $GL(2, \mathbb{C})$ and contraction of the positive definite Hermitian part.) Next we want to show that $\pi_{n-1}(U(2)) = \pi_{n-1}(SU(2, \mathbb{C}))$, $n - 1 \geq 2$, where $SU(2, \mathbb{C})$ denotes as usual the special unitary group; that is, the group of unitary matrices with determinant equals to 1. (Actually, we have the general result that $\pi_{n-1}(U(n_l)) = \pi_{n-1}(SU(n_l))$, $n-1 \geq 2$; see [Husemoller 1994, Section 8.12]). To provide an elementary proof of this fact, observe that any map $S^{n-1} \rightarrow U(2)$ induces a "determinant" map $S^{n-1} \rightarrow S^1$. However, by Freudenthal's theorem, $\pi_{n-1}(S^1) = 0$, $n - 1 \geq 2$. Therefore, the generators of $\pi_{n-1}(U(2))$ never circle around S^1 so that these maps can be deformed to maps of determinant identically equal to 1—that is, maps $S^{n-1} \rightarrow SU(2)$. Therefore, $\pi_{n-1}(U(2)) = \pi_{n-1}(SU(2, \mathbb{C}))$. Next, it is a well-known fact (see [Bar-on and Jonckheere 1992]) that $SU(2, \mathbb{C})$ is homeomorphic to S^3. To summarize,

$$\pi_{n-1}(GL(2,\mathbb{C})) = \pi_{n-1}(U(2)) = \pi_{n-1}(SU(2,\mathbb{C})) = \pi_{n-1}(S^3), n-1 \geq 2$$

As claimed, in the case of two feedback loops, we recover the problem of homotopy groups of spheres. For an account of the first unstable homotopy groups of S^3, see [Fuks and Rokhlin 1984, page 437]. To be more specific, we list here below the last two stable and the first unstable homotopy groups of $GL(2,\mathbb{C})$:

$$\begin{aligned}
\pi_2(GL(2,\mathbb{C})) &= 0 \\
\pi_3(GL(2,\mathbb{C})) &= \mathbb{Z} \\
\pi_4(GL(2,\mathbb{C})) &= \mathbb{Z} \bmod 2 \\
\pi_5(GL(2,\mathbb{C})) &= \mathbb{Z} \bmod 2 \\
\pi_6(GL(2,\mathbb{C})) &= \mathbb{Z} \bmod 12 \\
\pi_7(GL(2,\mathbb{C})) &= \mathbb{Z} \bmod 2 \\
\pi_8(GL(2,\mathbb{C})) &= \mathbb{Z} \bmod 2 \\
\pi_9(GL(2,\mathbb{C})) &= \mathbb{Z} \bmod 3 \\
\pi_{10}(GL(2,\mathbb{C})) &= \mathbb{Z} \bmod 15
\end{aligned}$$

Following [Fuks and Rokhlin 1984, page 436], the homotopy groups $\pi_n(S^3)$ have been *computed* up to $n = 25$. They also enjoy the following general feature:

Theorem 19.5. (Serre) *$\pi_i(S^3)$ is finite for $i \neq 3$.*

Proof. See [Spanier 1989]. Note that this is just a particular case of Serre's result asserting that $\pi_i(S^n)$ is finite for n odd and $i > n$ (see [Dieudonné 1989, page 487]). ■

Observe that the case $\pi_5(GL(2,\mathbb{C})) = \pi_5(S^3)$ is the particular case $n = 3$ of the map $S^{2n-1} \to S^n$. To any such map, one can attach an integer called *Hopf invariant* (see [Dieudonné 1989, page 317]). Since the Hopf invariant is a homotopy invariant, it is *part* of the structure of $\pi_{2n-1}(S^n)$. It can be shown that the Hopf invariant of $S^{2n-1} \to S^n$ vanishes for n odd, but this does not preclude the homotopy group $\pi_{2n-1}(S^n)$ to be nontrivial, as shown by the above table.

We cannot pursue this venture any further, for indeed the only spheres that have a Lie group structure are S^0, S^1, and S^3 (see [Bredon 1993, page 310]).

For $n_l > 2$, the only general facts that are known as of this date are the very first unstable homotopy groups:

Theorem 19.6. (Bott, Mimura, and Toda) *The first four unstable homotopy groups of $GL(n_l,\mathbb{C})$ are*

$$\begin{aligned}
\pi_{2n_l}(GL(n_l,\mathbb{C})) &= \mathbb{Z} \bmod (n_l)! \\
\pi_{2n_l+1}(GL(n_l,\mathbb{C})) &= \mathbb{Z} \bmod 2, \quad \textit{for } n_l \textit{ even}
\end{aligned}$$

$$
\begin{aligned}
&= 0, \quad \textit{for } n_l \textit{ odd} \\
\pi_{2n_l+2}(GL(n_l,\mathbb{C})) &= (\mathbb{Z} \bmod (n+1)!) \oplus (\mathbb{Z} \bmod 2), \quad \textit{for } n_l \textit{ even} \\
&= \mathbb{Z} \bmod (n+1)!/2, \quad \textit{for } n_l \textit{ odd} \\
\pi_{2n_l+3}(GL(n_l,\mathbb{C})) &= \mathbb{Z} \bmod \gcd(n_l(3n_l+1), 24), \quad \textit{for } n_l \textit{ even} \\
&= \mathbb{Z} \bmod \gcd(n_l(3n_l+1), 24), \quad \textit{for } n_l \textit{ odd}
\end{aligned}
$$

Proof. The first result is probably the earliest result on unstable homotopy and is already contained in the early work of [Bott 1958]. The next three unstable homotopy groups are more recent, and far less trivial, results (see [Mimura and Toda 1991, page 218]). ■

To be somewhat more specific, we specialize the above result to the case of three feedback paths (see [Mimura and Toda 1991, pages 218–219]):

$$
\begin{aligned}
\pi_2(GL(3,\mathbb{C})) &= 0 \\
\pi_3(GL(3,\mathbb{C})) &= \mathbb{Z} \\
\pi_4(GL(3,\mathbb{C})) &= 0 \\
\pi_5(GL(3,\mathbb{C})) &= \mathbb{Z} \\
\pi_6(GL(3,\mathbb{C})) &= \mathbb{Z} \bmod 6 \\
\pi_7(GL(3,\mathbb{C})) &= 0 \\
\pi_8(GL(3,\mathbb{C})) &= \mathbb{Z} \bmod 12 \\
\pi_9(GL(3,\mathbb{C})) &= \mathbb{Z} \bmod 3
\end{aligned}
$$

In case of four feedback paths, we have the following relevant homotopy groups (see [Mimura and Toda 1991, pages 218–219]):

$$
\begin{aligned}
\pi_2(GL(4,\mathbb{C})) &= 0 \\
\pi_3(GL(4,\mathbb{C})) &= \mathbb{Z} \\
\pi_4(GL(4,\mathbb{C})) &= 0 \\
\pi_5(GL(4,\mathbb{C})) &= \mathbb{Z} \\
\pi_6(GL(4,\mathbb{C})) &= 0 \\
\pi_7(GL(4,\mathbb{C})) &= \mathbb{Z} \\
\pi_8(GL(4,\mathbb{C})) &= \mathbb{Z} \bmod 24 \\
\pi_9(GL(4,\mathbb{C})) &= \mathbb{Z} \bmod 2 \\
\pi_{10}(GL(4,\mathbb{C})) &= (\mathbb{Z} \bmod 5!) \oplus (\mathbb{Z} \bmod 2) \\
\pi_{11}(GL(4,\mathbb{C})) &= \mathbb{Z} \bmod 6
\end{aligned}
$$

19.4 Degree (Stable Homotopy Case)

Remember the obstruction approach of recursive robust stability checks over higher-and-higher-dimensional simplexes of the polyhedron of uncertainty. Consider the problem of determining whether the Nyquist map exists over an n-simplex σ_n of P. Its boundary $\partial\sigma_n$ is homeomorphic to S^{n-1}. If the map over $\partial\sigma_n$ exists, we have a map

$$S^{n-1} \to GL(n_l, \mathbb{C})$$

Whether the Nyquist map can be extended to σ_n is equivalent to whether the above map is homotopic to a constant map (this is the basic obstruction result (see Theorem 15.1 or [Hilton and Wylie 1965, page 276, Proposition II.1.12])). A corollary of the Bott theorem can be worded as follows:

Theorem 19.7. (Atiyah) *Consider the stable homotopy case $2n_l \geq n$. When n is odd, every map $S^{n-1} \to GL(n_l, \mathbb{C})$ is homotopic to a constant map. If n is even, there exists an integer, called* **degree**, *such that two maps are homotopic iff they share the same degree. In particular, a map $S^{n-1} \to GL(n_l, \mathbb{C})$ is homotopic to a constant map iff its degree vanishes.*

Proof. See [Atiyah 1967b]. ∎

In the case of the determinant map into $S^1 \simeq \mathbb{C} \setminus 0 + j0$, it is easy to visualize the Nyquist template circling around $0 + j0$. Formally, the number of times of template circles around $0 + j0$ is the homotopy class of the map into S^1. It is, however, hard, even impossible, to visualize the template in $GL(n_l, \mathbb{C})$ "wrapping around" the "holes" of GL—that is, the singular matrices. The only way to survive is to proceed algebraically and use the *degree* as the only way to come to grips with what is loosely called the "wrapping."

Now, let us see how the degree can be defined.

19.4.1 $2n_l = n$

First, we discuss the borderline case, $2n_l = n$. In this case, since $f(p) \in GL(n_l, \mathbb{C})$ the first column of $f(p)$, $f_1(p)$, is nonvanishing. We therefore get a map

$$\begin{array}{ccc} p & \mapsto & \dfrac{f_1(p)}{||f_1(p)||} \\ S^{2n_l-1} & \to & S^{2n_l-1} \end{array}$$

Clearly, this map has a very classical degree, in the sense of the Brouwer degree. Furthermore, it can be shown (see [Atiyah 1967b]) that the degree does not depend on the chosen column of $f(p)$. Therefore, we define

$$\text{degree}(f) = \frac{1}{(n_l - 1)!}\text{degree}\left(\frac{f_1(p)}{||f_1(p)||}\right)$$

19.4.2 $2n_l > n$

Finally, when $2n_l > n$, as a consequence of the stable homotopy groups of GL it can be shown (see [Atiyah 1967b]) that the map $f : S^{n-1} \to GL(n_l, \mathbb{C})$ can be deformed to

$$g(p) = \begin{pmatrix} h(p) & 0 \\ 0 & I \end{pmatrix}$$

where $h(p) \in GL(\frac{n}{2}, \mathbb{C})$. Then the degree of $f(p)$ is defined to be the degree of $h(p)$, as defined in the first, the "borderline," case.

19.5 Differential Degree (Stable and Unstable Homotopy)

The case $2n_l < n$ is the unstable homotopy case. Since there is no clear repetitive pattern in $\pi_{n-1}(GL(n_l, \mathbb{C}))$, it is hard to come up with a general, streamlined, easy-to-use definition of the degree. In this section, we develop a definition based on the machinery of integration of differential forms. It is computationally involved, but it can be applied to both the stable and the unstable homotopy cases. In particular, in some unstable homotopy situations, we show that the degree can only be defined as a vector of integers.

Since the degree is defined as a quantity that should vanish iff the map $f : S^{n-1} \to GL(n_l, \mathbb{C})$ is homotopic to a constant map, we could *define* the degree as the homotopy class of the map. In other words, the degree takes value in $\pi_{n-1}(GL(n_l, \mathbb{C}))$. The problem is that it is difficult to evaluate the homotopy class of a map. Inspiring ourselves from the simple case of $f : S^1 \to \mathbb{C} \setminus \{0 + j0\}$, where the degree of the map is given by $\frac{1}{2\pi} \oint_{S^1} \frac{df}{f}$, we could think of setting up a general differential approach to evaluate the homotopy class of the map.

The starting point is the following commutative diagram

$$\begin{array}{ccc} S^{n-1} & \xrightarrow{f} & GL(n_l, \mathbb{C}) \\ \downarrow & & \downarrow \\ H^{n-1}(S^{n-1}) & \xleftarrow{f^*} & H^{n-1}(GL(n_l, \mathbb{C})) \end{array}$$

By a fundamental theorem, the cohomology group homomorphism f^* depends only on the homotopy class of f, so that f^* could be taken as the definition of the degree.

To accomplish this program, we should first compute the cohomology of $GL(n_l, \mathbb{C})$. As we already saw, $GL(n_l, \mathbb{C})$ can be retracted onto the unitary group $U(n_l)$. The cohomology of $U(n_l)$ is computed by means of the spectral sequence of a fiber bundle that has $U(n_l)$ as total space. Another way is to identify cohomology classes with bi-invariant differential forms on the compact Lie group $U(n_l)$. Next, the cohomology group homomorphism f^* is materialized as the *pull-back* of the bi-invariant differential form on the sphere S^{n-1}. Finally, we show how to associate with f^* an integer, the degree, by integration of the differential form on the sphere.

19.5.1 Cohomology of General Linear Group

Since $GL(n_l, \mathbb{C})$ can be retracted onto $U(n_l)$ they have the same cohomology and we rather focus on the latter group.

The specific feature of $U(n_l)$ is that it *acts* on S^{2n_l-1}. The *action* of the Lie group $U(n_l)$ on the manifold S^{2n_l-1} is the continuous map:

$$\begin{array}{rcl} U(n_l) \times S^{2n_l-1} & \to & S^{2n_l-1} \\ (U, s) & \mapsto & Us \end{array}$$

which clearly satisfies the required conditions of group action on manifold:

$$\begin{array}{rcl} (U_1U_2)s & = & U_1(U_2s), \ \forall U_1, U_2 \in U(n_l), \forall s \in S^{2n_l-1} \\ 1_{U(n_l)}s & = & s, \ \forall s \in S^{2n_l-1} \end{array}$$

Furthermore, take the point $s_0 = (1 + j0, 0, ..., 0)$ on the sphere. It is clear that

$$S^{2n_l-1} = \{Us_0 : U \in U(n_l)\}$$

With this extra condition, the action of $U(n_l)$ on S^{2n_l-1} is said to be *transitive*. Next, the **stabilizer** or **isotropy group** of $U(n_l)$ at a point s_0 of the manifold is defined as

$$\{U : Us_0 = s_0\}$$

Elementary matrix algebra shows that the stabilizer is the subgroup of matrices structured as

$$U = \begin{pmatrix} 1 & 0 \\ 0 & U_{22} \end{pmatrix}, \quad U_{22} \in U(n_l - 1)$$

Clearly the stabilizer is $\begin{pmatrix} 1 & 0 \\ 0 & U(n_l - 1) \end{pmatrix}$. It follows that

$$U(n_l)/\begin{pmatrix} 1 & 0 \\ 0 & U(n_l - 1) \end{pmatrix} = U(n_l)/\text{stabilizer at } s_0 = U(n_l)s_0 = S^{2n_l-1}$$

The above result is written more compactly as $U(n_l)/U(n_l - 1) = S^{2n_l-1}$. Using some general Lie group results, it follows that the projection $U(n_l) \to S^{2n_l-1}$ is a fiber bundle with fiber $U(n_l - 1)$.

To compute the de Rham cohomology of the unitary group $U(n_l)$, we use the spectral sequence argument of Section 14.4 on the following fiber bundle (see [Bott and Tu 1982, page 196]):

$$\begin{array}{ccc} U(n_l - 1) & \to & U(n_l) \\ & & \downarrow \\ & & S^{2n_l-1} \end{array}$$

$U(n_l)$ is the total space, S^{2n_l-1} is the base space, while $U(n_l - 1)$ is the fiber. The horizontal arrow denotes the inclusion map, while the vertical arrow denotes the bundle projection.

We start with the case $n_l = 2$:

$$\begin{array}{ccc} U(1) = S^1 & \rightarrow & U(2) \\ & & \downarrow \\ & & S^3 \end{array}$$

We set up the cohomology spectral sequence of this specific fiber bundle. The E_2 term is

$$E_2^{u,v} = H^u(S^3, H^v(S^1))$$

Since S^3 is simply connected ($\pi_1(S^3) = 0$), the isomorphism $H^v(S^1) \rightarrow H^v(S^1)$ between the cohomology of the fibers above two points s_1, s_2 of S^3 does not depend on the path joining s_1, s_2. Furthermore, since the de Rham theory uses the *field* $\mathbb{R}$ of coefficients, we need not bother with torsion products. Therefore, from [Bott and Tu 1982, pages 169–170] and Section 14.6, it follows that

$$E_2^{u,v} = H^u(S^3) \otimes H^v(S^1)$$

We display the E_2 term in the following diagram:

$E_2 =$

v					
1	$\mathbb{R}$			$\mathbb{R}$	
0	$\mathbb{R}$			$\mathbb{R}$	
	0	1	2	3	u

Since the differential $d_2 : E_2^{u,v} \rightarrow E_2^{u+2,v-1}$ is an arrow that goes one row down and two columns left, it vanishes. The spectral sequence collapses at E_2. Therefore, $H^*(U(2)) = E_2$, in other words,

$$H^k(U(2)) = \bigoplus_{u+v=k} E_2^{u,v}$$

so that

$$\begin{aligned} H^0(GL(2,\mathbb{C})) &= \mathbb{R} \\ H^1(GL(2,\mathbb{C})) &= \mathbb{R} \\ H^2(GL(2,\mathbb{C})) &= 0 \\ H^3(GL(2,\mathbb{C})) &= \mathbb{R} \\ H^4(GL(2,\mathbb{C})) &= \mathbb{R} \end{aligned}$$

The case $n_l = 3$ is based on the fiber bundle

$$\begin{array}{ccc} U(2) & \rightarrow & U(3) \\ & & \downarrow \\ & & S^5 \end{array}$$

We follow the same procedure as in the $n_l = 2$ case. Since S^5 is simply connected and since we are using a *field* of values, it follows that

$$E_2^{u,v} = H^u(S^5, H^v(U(2))) = H^v(S^5) \otimes H^u(U(2))$$

The relevant information is displayed in the E_2 tableau:

$E_2 =$

v						
4	$\mathbb{R}$					$\mathbb{R}$
3	$\mathbb{R}$					$\mathbb{R}$
2	0					0
1	$\mathbb{R}$					$\mathbb{R}$
0	$\mathbb{R}$					$\mathbb{R}$
	0	1	2	3	4	5	u

From this tableau it follows that the spectral sequence again collapses at E_2. Hence we get

$$\begin{aligned}
H^0(GL(3,\mathbb{C})) &= \mathbb{R} \\
H^1(GL(3,\mathbb{C})) &= \mathbb{R} \\
H^2(GL(3,\mathbb{C})) &= 0 \\
H^3(GL(3,\mathbb{C})) &= \mathbb{R} \\
H^4(GL(3,\mathbb{C})) &= \mathbb{R} \\
H^5(GL(3,\mathbb{C})) &= \mathbb{R} \\
H^6(GL(3,\mathbb{C})) &= \mathbb{R} \\
H^7(GL(3,\mathbb{C})) &= 0 \\
H^8(GL(3,\mathbb{C})) &= \mathbb{R} \\
H^9(GL(3,\mathbb{C})) &= \mathbb{R}
\end{aligned}$$

Since $GL(3,\mathbb{C})$ has the same cohomology as $U(3)$ and since $U(3)$ is a compact manifold, we have the Poincaré duality $H^k = H^{9-k}$, as is easily seen from the above tableau.

Observe that so far all cohomology groups are on *one* generator. In control-theoretic words, this means that when the number of feedback paths does not exceed 3, there is only one possibility, up to homotopy, for the template in $GL(n_l,\mathbb{C})$ to circle around singular matrices. However, beyond three feedback paths, the situation becomes more complicated.

For $n_l = 4$ we proceed as usual from the bundle

$$\begin{array}{ccc} U(3) & \to & U(4) \\ & & \downarrow \\ & & S^7 \end{array}$$

As before, we set up the cohomology spectral sequence starting at

$$E_2^{uv} = H^u(S^7) \otimes H^v(U(3))$$

which yields the tableau

$E_2 =$

v									
9	$\mathbb{R}$							$\mathbb{R}$	
8	$\mathbb{R}$							$\mathbb{R}$	
7	0	$\ddots$						0	
6	$\mathbb{R}$		$\ddots$					$\mathbb{R}$	
5	$\mathbb{R}$			$\ddots$				$\mathbb{R}$	
4	$\mathbb{R}$				$\ddots$			$\mathbb{R}$	
3	$\mathbb{R}$					$\ddots$		$\mathbb{R}$	
2	0						$\ddots$	0	
1	$\mathbb{R}$							$\mathbb{R}$	
0	$\mathbb{R}$							$\mathbb{R}$	
	0	1	2	3	4	5	6	7	u

As easily seen, the spectral sequence collapses at E_2. Next, when we perform the operation

$$H^k(GL(4,\mathbb{C})) = \oplus_{u+v=k} E_2^{u,v}$$

we get *two* generators for $k = 8$; this is indicated by the dotted $u+v = 8$ line in the E_2 diagram. By sweeping the above diagram with $u + v =$ constant lines, we get

$$\begin{aligned}
H^0(GL(4,\mathbb{C})) &= \mathbb{R} \\
H^1(GL(4,\mathbb{C})) &= \mathbb{R} \\
H^2(GL(4,\mathbb{C})) &= 0 \\
H^3(GL(4,\mathbb{C})) &= \mathbb{R} \\
H^4(GL(4,\mathbb{C})) &= \mathbb{R} \\
H^5(GL(4,\mathbb{C})) &= \mathbb{R} \\
H^6(GL(4,\mathbb{C})) &= \mathbb{R} \\
H^7(GL(4,\mathbb{C})) &= \mathbb{R} \\
H^8(GL(4,\mathbb{C})) &= \mathbb{R} \oplus \mathbb{R} \\
H^9(GL(4,\mathbb{C})) &= \mathbb{R} \\
H^{10}(GL(4,\mathbb{C})) &= \mathbb{R} \\
H^{11}(GL(4,\mathbb{C})) &= \mathbb{R} \\
H^{12}(GL(4,\mathbb{C})) &= \mathbb{R} \\
H^{13}(GL(4,\mathbb{C})) &= \mathbb{R} \\
H^{14}(GL(4,\mathbb{C})) &= 0 \\
H^{15}(GL(4,\mathbb{C})) &= \mathbb{R} \\
H^{16}(GL(4,\mathbb{C})) &= \mathbb{R}
\end{aligned}$$

Observe that the above table provides yet another manifestation of the Poincaré duality, $H^k = H^{16-k}$. The fact that $H^8(GL(4))$ is on *two* generators means that the *degree* of the matrix return difference map $\partial\sigma_9 \to GL(n_l, \mathbb{C})$ can only be defined as a $\mathbb{Z}^2$-valued function, one component per generator. In the broader context of recursive robust stability test over higher-and-higher-dimensional skeleta of uncertainty, this means that there is *accrued difficulty* at extending robust stability from the 8-D skeleton of $D \times \Omega$ to the 9-D skeleton of uncertainty in a multivariable system with four feedback paths.

We leave it as an exercise to the reader to pursue this investigation beyond $n_l = 4$.

In addition to the above spectral sequence argument, we can exploit the multiplicative structure of the cohomology groups to gain some deeper insights.

Theorem 19.8.

$$H^*(U(n_l)) = \Lambda(g_1, g_3, ..., g_{2n_l-1})$$

where $g_i \in H^i(U(n_l))$ is a generator of degree i and Λ denotes the Hopf algebra generated by the g_i's.

Proof. See [Mimura and Toda 1991, page 119]. ∎

From the cohomology algebra, we can provide another explanation of the fact that $H^8(GL(4))$ is on two generators. Indeed, $H^*(GL(4))$ is generated *as an algebra* by $\{g_1, g_3, g_5, g_7\}$, where $g_i \in H^i(GL(4))$. There are *two* ways to get an element of $H^8(GL(4))$ from these generators—namely, g_1g_7 and g_5g_3. The reader can verify that in all the other cases we have considered the Hopf algebra reveals only one generator per cohomology group.

19.5.2 de Rham Cohomology of Differential Forms

To give f^* a more computationally manageable interpretation, we use the differential form approach to cohomology. This is essentially the *de Rham cohomology*. $GL(n_l, \mathbb{C})$ is easily seen to be a smooth manifold, and around $A = \{a_{ij}\} \in GL(n_l, \mathbb{C})$ we could take $(a_{11}, a_{12}, ...)$ as local Riemann coordinates. An exterior differential $(n-1)$-form on the manifold $GL(n_l, \mathbb{C})$ is an expression like

$$\varpi(A) = \sum g_{i_1 j_1 i_2 j_2 \ldots}(a_{11}, a_{12}, ...) da_{i_1 j_1} da_{i_2 j_2} \ldots$$

The sum runs over all $i_1, j_1, i_2, j_2, ...,$ each running in $\{1, 2, ..., n_l\}$. The g's are smooth functions; there are $(n-1)$ factors in the product of differentials; and the product of differentials is skew-commutative. Define $\Lambda^{n-1}(GL)$ as the (cochain) group of exterior $(n-1)$-differential forms defined on $GL(n_l, \mathbb{C})$. We introduce the exterior differential operator $d : \Lambda^{n-1}(GL) \to$

$\Lambda^n(GL)$ and it is easily verified that $dd = 0$. The resulting cohomology theory is called *de Rham cohomology* of the general linear group.

Since $GL(n_l, \mathbb{C})$ is a smooth manifold and since the de Rham cohomology coincides with singular cohomology over smooth paracompact manifolds, there is no danger at fusing the two concepts.

The de Rham cohomology classes of $GL(n_l, \mathbb{C})$ have a simple interpretation: They are the so-called *left- and right-invariant differential forms* defined on $GL(n_l, \mathbb{C})$. The concept of *invariant* differential forms relies on the stronger *Lie group* structure that $GL(n_l, \mathbb{C})$ has in addition to being a smooth manifold.

Definition 19.9. *The $(n-1)$-differential form ϖ defined on $GL(n_l, \mathbb{C})$ is said to be* **left-invariant** *iff $\varpi(A) = \varpi(GA), \forall G \in GL(n_l, \mathbb{C})$. It is said to be* **right-invariant** *iff $\varpi(A) = \varpi(AG), \forall G \in GL(n_l, \mathbb{C})$. It is said to be* **bi-invariant** *if it is left- and right-invariant.*

We now expand on the concept of *invariant differential forms*; we start from the noncompact case of $GL(n_l, \mathbb{C})$ and then proceed to the compact case of $U(n_l)$.

19.5.3 Invariant Differential Forms on GL

The starting point of the general construction of invariant differential forms on matrix Lie groups is the expression

$$A^{-1}dA$$

where $A \in GL(n_l, \mathbb{C})$ (or $GL(n_l, \mathbb{R})$) and dA denotes the exterior differential of A; that is,

$$dA = \begin{pmatrix} da_{11} & da_{12} & \dots & da_{1n_l} \\ da_{21} & da_{22} & \dots & da_{2,n_l} \\ \vdots & \vdots & \ddots & \vdots \\ da_{n_l1} & da_{n_l2} & \dots & da_{n_ln_l} \end{pmatrix}$$

In case of a *real* matrix, we choose the a_{ij}'s as local Riemann coordinates, and da_{ij} is self-explanatory. If the matrix is complex, write $a_{i,j} = x_{ij} + jy_{ij}$ and use the x_{ij}, y_{ij}'s as local Riemann coordinates of the Lie group $GL(n_l, \mathbb{C})$, in which case

$$da_{ij} = dx_{ij} + j dy_{ij}$$

Another way to deal with complex matrices is to use the **realification** * of A; that is, if

$$A = X + jY, \quad X, Y \in \mathbb{M}(n_l, \mathbb{R})$$

the realification of A is the matrix

*This somewhat unusual terminology is borrowed from [Karoubi 1978]; the realification of a complex space is its underlying real structure.

$$\begin{pmatrix} X & Y \\ -Y & X \end{pmatrix}$$

Matrices structured as above form a subgroup of $GL(2n_l, \mathbb{R})$, isomorphic to $GL(n_l, \mathbb{C})$. Using realification, the fundamental expression becomes

$$\begin{pmatrix} X & Y \\ -Y & X \end{pmatrix}^{-1} d \begin{pmatrix} X & Y \\ -Y & X \end{pmatrix}$$

However, to keep the notation more compact, we will mainly work with the $A^{-1}dA$ representation.

As a historical digression, we will just mention that the expression $A^{-1}dA$ is instrumental in deriving a general right-invariant metric ds^2 for matrix Lie groups (see [Barut and Raczka 1986, page 110, Proposition 1]).

Invariance of $A^{-1}dA$ under left translation is trivial. On the other hand, under right translation $A \mapsto AG$, we get

$$A^{-1}dA \mapsto G^{-1}A^{-1}dAG$$

Therefore, the (left and right) invariant differential forms are the "similarity invariants" of $A^{-1}dA$. Extreme caution should be exercised here, because the matrix $A^{-1}dA$ is defined over the *skew-commutative* ring of differential 1-forms, whereas classical invariant theory requires the matrix to be defined over a *commutative* ring. As far as the trace is concerned, it is linear in the entries of $A^{-1}dA$ and hence does not involve the intricacies of skew-commutativity. Therefore, we have the following theorem:

Theorem 19.10.

$$\varpi^1(A) = \mathit{Trace}(A^{-1}dA)$$

is a invariant 1-form on $GL(n_l, \mathbb{C})$.

Proof. This is an apparently elementary result, but we need to devise a proof that does not rely on commutativity of the ground ring. Let $G \in GL(n_l, \mathbb{C})$, define $H = G^{-1}$, and let $W = A^{-1}dA$. We have

$$\begin{aligned} \varpi^1(AG) &= \mathrm{Trace}(G^{-1}A^{-1}dAG) \\ &= \sum_i (H(A^{-1}dA)G)_{ii} \\ &= \sum_{i,j,k} H_{ij}W_{jk}G_{ki} \\ &= \sum_{j,k} W_{jk}(\sum_i H_{ij}G_{ki}) \\ &= \sum_{j,k} W_{jk}\delta_{jk} \\ &= \sum_j W_{jj} \end{aligned}$$

$$= \text{Trace}(A^{-1}dA)$$
$$= \varpi^1(A)$$

■

In an attempt to construct more bi-invariant forms, consider the following easily proved lemma:

Lemma 19.11. *The wedge product of bi-invariant forms is bi-invariant.*

Unfortunately, because of skew-commutativity of 1-forms, this lemma does not lead us very far, because indeed

$$\text{Trace}(A^{-1}dA) \wedge \text{Trace}(A^{-1}dA) = 0$$

On the other hand, if the matrix A is complex, it is easily seen that $\overline{\text{Trace}(A^{-1}dA)}$ is a new bi-invariant 1-form. Therefore, wedging the latter with the former, one obtains the *nontrivial* 2-form:

$$\text{Trace}(A^{-1}dA) \wedge \overline{\text{Trace}(A^{-1}dA)}$$

At this stage, there is nothing more that can be accomplished with the fundamental invariant form $\text{Trace}(A^{-1}dA)$, because indeed, any more wedging would produce trivial invariant forms.

Another traditional invariant is the determinant, and we would be tempted to define the invariant n_l-form: $\varpi^{n_l}(A) = \det(A^{-1}dA)$. Extreme caution needs to be exercised when computing $\det(A^{-1}dA)$, since the elements of the matrix do not form a commutative ring.

Classical determinant theory requires a *commutative* ring structure for the entries. Dieudonné developed a determinant theory for matrices over a *noncommutative field* (see [Artin 1988, Chapter IV] and [Rosenberg 1994, Section 2.2]). Dieudonné's determinant theory is motivated by *left invertibility* of matrices defined over a noncommutative field. Here, the entries are in an exterior, skew-commutative algebra, without multiplicative inverse. Therefore, neither classical determinant nor Dieudonné's theory applies directly to our problem. It is possible, though, to define a determinant over a noncommutative ring, provided that we agree on the *ordering* of the factors in the products of the entries. Tracing back to the very early definition of the determinant, we *define*

$$\det_r(W) = \sum_\pi \text{sign}(\pi) w_{1\pi(1)} w_{2\pi(2)} \ldots w_{n_l\pi(n_l)}$$

where π is a permutation of $(1, 2, ..., n_l)$ and $sign(\pi)$ is the signature of the permutation. In other words, we expand the determinant using the cofactor rule on the *first row* of every matrix. We could as well define a determinant by cofactor expansion of the *first column*:

$$\det_c(W) = \sum_\pi \text{sign}(\pi) w_{\pi(1)1} w_{\pi(2)2} \ldots w_{\pi(n_l)n_l}$$

In the above, W denotes $A^{-1}dA$, or any other matrix defined over the exterior algebra of differential forms. A *crucial fact* is that the row and column expansion formulas *do not*, in general, yield the same result.

The motivation for bringing the determinant concept in the first place was to obtain invariant forms. As such, the crucial issue is whether $\det(G^{-1}(A^{-1}dA)G) = \det(G^{-1})\det(A^{-1}dA)\det G = \det(A^{-1}dA)$. Since G is a matrix defined over a commutative field, the only suspicious equality is the first one. Part of the answer is given by the following theorems, which indicate that *some* linear algebra can be done with the row and column determinants:

Theorem 19.12. *Let W be a matrix defined over the exterior algebra of differential forms; let B, C be matrices defined over $\mathbb{C}$. Then we have*

$$\begin{aligned} det_c(BW) &= \det B \, det_c W \\ det_r(WC) &= det_r W \det C \end{aligned}$$

Proof. Let us prove the second equality. If C is singular, then the *columns* of WC are linearly dependent over $\mathbb{C}$ and its *row* expansion determinant vanishes; on the other hand, $\det C = 0$; therefore, the result is proved in ca e C is singular. If C is nonsingular, we use the result of [Artin 1988, Chapter IV], asserting that any matrix (over a, possibly noncommutative, field) can be factored as $C = SD$, where $S \in SL(n_l, \mathbb{C})$, the *special linear group* of nonsingular matrices with determinant $= 1$, also called *unimodular group*; D is a matrix differing from the identity by its (n_l, n_l) term which equals

$$\mu = \det(C)$$

The matrix S can itself be factored as a finite product of elementary matrices, $S = \Pi E_{ij}(\lambda)$, $i \neq j$, where $E_{ij}(\lambda)$ differs from the identity matrix by its (i,j) element equal to λ. Clearly,

$$\det_r(WC) = \det_r((WS)D)$$

Since postmultiplication by D amounts to multiplying the last column of WS by μ, it is easily seen that

$$\det_r((WS)D) = \mu \det_r(WS)$$

Next, observe that $WS = W\Pi E_{ij}(\lambda)$. The matrix product $WE_{ij}(\lambda)$ amounts to replacing the jth column of W by the jth column plus λ times the ith column. From the definition of the *row* expansion determinant, it is easy to see that $\det_r(WE_{ij}(\lambda)) = \det_r(W)$. Therefore,

$$\det_r(WS) = \det_r(W)$$

Combing all of the above yields $\det_r(WC) = \det_r W \det C$. The proof of the first equality is the same and omitted. ■

Now we can come back to the invariance question. Consider the slightly more general problem of whether $\det(BWC)$ can be factored. Choose, for example, the column expansion of the determinant. By the theorem, we have $\det_c(BWC) = \det(B)\det_c(WC)$. Now, the problem is that nothing can be said about $\det_c(WC)$; specifically, it cannot in general be factored as $\det(W)\det(C)$ in either the column or the row version of $\det(W)$. Therefore, no invariant forms can be obtained from $\det(A^{-1}dA)$.

Besides the trace and the determinant, there are many other "similarity invariants"—namely, each coefficient of s^n in the characteristic polynomial of $A^{-1}dA$, itself equal to the sum of all $n \times n$ principal submatrices of $A^{-1}dA$. Written compactly,

$$\det(sI - A^{-1}dA) = s^{n_l} + \varpi^1(A)s^{n_l-1} + \varpi^2(A)s^{n_l-2} + \ldots + \varpi^n(A)s^{n_l-n} + \ldots + \varpi^{n_l}(A)$$

Again, this effort is bound to fail, because the determinant is not invariant.

Fortunately, some of the pitfalls observed above can be rectified and the above intuitive ideas can be made to work by considering "compounded" expressions like

$$A^{-1}dAA^{-1}dA...A^{-1}dA$$

$$\overline{A^{-1}dAA^{-1}dA}...\overline{A^{-1}dA}$$

Again, left translation invariance is obvious. Right translation invariance requires us to take similarity invariants of $(A^{-1}dA)^n$. The trace of this expression immediately provides an n-form, provided that the rules of exterior differentiation do not act in such a way as to cancel the form. Actually, by the Bott periodicity theorem, many of these forms will cancel. In particular, in the case of complex matrices, for $n > 2n_l^2$, all similarity invariants of $(A^{-1}dA)^n$ cancel, since any such similarity invariant would be an invariant form of degree higher than the dimension of the manifold.

Now, consider $(A^{-1}dA)^n = W$ where the power n is *even*. In this case, W is a matrix defined over the *commutative subring* of exterior differential forms of *even* degree. Therefore, we can take all of the traditional similarity invariants of matrices and get bi-invariant differential forms, of *even* degree. In particular, we could define the characteristic polynomial of $A^{-1}dAA^{-1}dA$,

$$\det(sI - A^{-1}dAA^{-1}dA) = s^{n_l} + \varpi^2(A)s^{n_l-1} + \varpi^4(A)s^{n_l-2} + ... + \varpi^{2n_l}(A)$$

The coefficients of this characteristic polynomials are bi-invariant forms, of even degree.

For $A \in GL(n_l, \mathbb{R})$, the bi-invariant form of highest degree is

$$\varpi^{n_l^2}(A) = \frac{da_{11}da_{12}...da_{1n_l}da_{21}...da_{n_ln_l}}{\det(A)^{n_l}}$$

Observe that this bi-invariant form cannot be derived from $\text{Trace}(A^{-1}dA)^{n_l^2}$ because indeed this trace vanishes. To prove that $\varpi^{n_l^2}(A) = \varpi^{n_l^2}(GA) = \varpi^{n_l^2}(AG)$, the easiest way is to write the Artin factorization of G as

$$\Pi E_{ij}(\lambda)D, \quad D = \text{diag}(1, 1, ..., 1, \mu)$$

where $\mu = \det G$ and $E_{ij}(\lambda)$ differs from the identity by the quantity λ in the (i, j), $i \neq j$, position. From there, elementary algebra yields $\varpi^{n_l^2}(E_{ij}(\lambda)A) = \varpi^{n_l^2}(AE_{ij}(\lambda)) = \varpi^{n_l^2}(A)$ and $\varpi^{n_l^2}(AD) = \varpi^{n_l^2}(DA) = \varpi^{n_l^2}(A)\mu^{n_l}$. From these manipulations, bi-invariance follows.

The form $|\varpi^{n_l^2}(A)|$ defines a positive measure, sometimes intuitively referred to as *volume element* or *(unsigned) volume form* (see [Murnaghan 1938, Chapter 8]). To be more precise, this form is the *Haar measure* on the Lie group $GL(n_l, \mathbb{R})$. By definition, the *Haar measure* is a (positive) measure that is defined on a topological group or a Lie group and that is left and/or right translation invariant. The Haar measure is instrumental for integration on topological and Lie groups (see [Barut and Raczka 1986, Section 3.9]).

For $A \in GL(n_l, \mathbb{C})$, the (signed) volume form is the bi-invariant $2n_l^2$-form,

$$\varpi^{2n_l^2}(A) = \frac{da_{11}da_{12}...da_{n_l n_l}}{\det(A)^{n_l}} \overline{\frac{da_{11}da_{12}...da_{n_l n_l}}{\det(A)^{n_l}}}$$

(In general, a volume form on an n-dimensional manifold charted with local Riemann coordinates $x_1, x_2, ..., x_n$ is an exterior differential form of degree n that is invariant under transformations in some group acting on the manifold.)

As an example, suppose we want to compute the invariant 3-form on $GL(2, \mathbb{C})$. As indicated above, we compute the trace of $(A^{-1}dA)^3$. Let

$$A^{-1}dA = \begin{pmatrix} \varpi_{11} & \varpi_{12} \\ \varpi_{21} & \varpi_{22} \end{pmatrix}$$

As an aside, observe that

$$(A^{-1}dA)^2 = \begin{pmatrix} \varpi_{12}\varpi_{21} & \varpi_{11}\varpi_{12} + \varpi_{12}\varpi_{22} \\ \varpi_{21}\varpi_{11} + \varpi_{22}\varpi_{21} & \varpi_{21}\varpi_{12} \end{pmatrix}$$

so that $\text{Trace}((A^{-1}dA)^2) = \varpi_{12}\varpi_{21} + \varpi_{21}\varpi_{12} = 0$. Therefore, the 2-form on $GL(2, \mathbb{C})$ vanishes, as predicted by the Bott periodicity theorem. Next we get

$$\begin{aligned} \varpi^3(A) &= \text{Trace}((A^{-1}dA)^3) \\ &= \frac{1}{(\det(A))^2}(a_{11}da_{12}da_{21}da_{22} - a_{12}da_{11}da_{21}da_{22} \\ &\quad + a_{21}da_{11}da_{12}da_{22} - a_{22}da_{11}da_{12}da_{21}) \end{aligned}$$

19.5.4 Invariant Differential Forms on U

From the invariant differential forms on $GL(n_l, \mathbb{C})$, we can derive invariant differential forms on $U(n_l)$. Formally, consider the inclusion map and its induced cohomology group isomorphism:

$$\begin{array}{ccc} U(n_l) & \xrightarrow{i} & GL(n_l, \mathbb{C}) \\ \downarrow & & \downarrow \\ H^*(U(n_l)) & \xleftarrow{i^*} & H^*(GL(n_l, \mathbb{C})) \end{array}$$

Since $U(n_l)$ is a deformation retract of $GL(n_l, \mathbb{C})$, the inclusion map is a homotopy equivalence so that i^* is an isomorphism. Therefore, *the pull-back of any invariant differential form on $GL(n_l, \mathbb{C})$ is an invariant differential form on $U(n_l)$.*

To write an *explicit* invariant differential form on $U(n_l)$, we first need Riemann coordinates on $U(n_l)$. Consider the well-known mapping from skew-Hermitian to unitary matrices:

$$U = (I+K)^{-1}(I-K)$$

It is easily seen that this mapping is smoothly invertible iff $U \neq -I$. Therefore, the above provides a diffeomorphism between the (Euclidean) space of skew Hermitian matrices and $U(n_l) \setminus \{-I\}$. In other words, the entries of K, properly constrained, provide Riemann coordinates for $U(n_l)$ around $U = I$. Following the same argument, the formula

$$U = (K+I)^{-1}(K-I)$$

provides Riemann coordinates around $U = -I$. Finally, it is easily seen that these two systems form an *atlas* for $U(n_l)$.

The pull-back of the fundamental differential form $A^{-1}dA$ on $U(n_l)$ around $U = I$ is

$$\begin{aligned} U^{-1}dU &= ((I+K)^{-1}(I-K))^{-1}d((I+K)^{-1}(I-K)) \\ &= (I-K)^{-1}(I+K)((d(I+K)^{-1})(I-K) - (I+K)^{-1}dK) \end{aligned}$$

So far, we have been using standard matrix algebra combined with exterior differentiation. Things are getting a bit more tricky when it comes to $d(I+K)^{-1}$. By definition, dA^{-1} has to be interpreted as follows: A is a matrix, the entries of which depend smoothly on Riemann coordinates. Since A is nonsingular, it follows that the entries of A^{-1} are smooth functions of the Riemann coordinates as well. dA^{-1} is therefore defined as the matrix obtained by applying the exterior differential entrywise to A^{-1}. The following lemma might look quite obvious if dA is interpreted as a "small" variation; however, the result deserves more caution in the realm of exterior differentiation.

Lemma 19.13.

$$dA^{-1} = -A^{-1}dAA^{-1}$$

Proof. We argue by induction on the size n_l of the matrix. For $n_l = 1$, the result is obvious. Now, we assume the result is correct up to order $n_l - 1$, and we proceed to show that it is valid up to order n_l. Since A is nonsingular, it has an $(n_l-1)\times(n_l-1)$ nonsingular submatrix. By row and column permutation, this submatrix can be brought to the top left-hand corner,

$$A = \begin{pmatrix} M & c \\ r & s \end{pmatrix}$$

In the above, M is a $(n_l-1)\times(n_l-1)$ nonsingular matrix, c is a column, r is a row, and s is a scalar. By the well-known *matrix inversion lemma*,

$$\begin{pmatrix} M & c \\ r & s \end{pmatrix}^{-1} = \begin{pmatrix} M^{-1} + M^{-1}ce^{-1}rM^{-1} & -M^{-1}ce^{-1} \\ -e^{-1}rM^{-1} & e^{-1} \end{pmatrix}$$

where e is the *Schur complement*

$$e = s - rM^{-1}c$$

By definition of dA^{-1}, we have

$$d\begin{pmatrix} M & c \\ r & s \end{pmatrix}^{-1} = \begin{pmatrix} d(M^{-1}+M^{-1}ce^{-1}rM^{-1}) & -d(M^{-1}ce^{-1}) \\ -d(e^{-1}rM^{-1}) & de^{-1} \end{pmatrix}$$

By induction hypothesis, we can apply the rule $dA^{-1} = -A^{-1}dAA^{-1}$ to each block, and eventually we will find out $d\begin{pmatrix} M & c \\ r & s \end{pmatrix}^{-1}$ as a linear function of dM, dr, dc, ds. After some long but elementary manipulation, it is easily found that this linear function is the same as

$$\begin{aligned}\begin{pmatrix} d(M^{-1}+M^{-1}ce^{-1}rM^{-1}) & -d(M^{-1}ce^{-1}) \\ -d(e^{-1}rM^{-1}) & de^{-1} \end{pmatrix} \\ = -\begin{pmatrix} M^{-1}+M^{-1}ce^{-1}rM^{-1} & -M^{-1}ce^{-1} \\ -e^{-1}rM^{-1} & e^{-1} \end{pmatrix}\begin{pmatrix} dM & dc \\ dr & ds \end{pmatrix} \\ \cdot\begin{pmatrix} M^{-1}+M^{-1}ce^{-1}rM^{-1} & -M^{-1}ce^{-1} \\ -e^{-1}rM^{-1} & e^{-1} \end{pmatrix}\end{aligned}$$

Therefore, $dA^{-1} = -A^{-1}dAA^{-1}$, as claimed. ■

Applying the rule $dA^{-1} = -A^{-1}dAA^{-1}$ to $U^{-1}dU$, we get the invariant differential form on the unitary group in Riemann coordinates around $U = I$ as

$$U^{-1}dU = -2(I-K)^{-1}dK(I+K)^{-1}, \ (K = -K^*)$$

To construct bi-invariant forms on the unitary group from the fundamental expression $U^{-1}dU$, we proceed as in the $GL(n_l, \mathbb{C})$ case. A 1-form is

$$\text{Trace}(U^{-1}dU)$$

A 2-form is

$$\text{Trace}(U^{-1}dU) \wedge \overline{\text{Trace}(U^{-1}dU)}$$

To construct higher forms, we have to consider "compounded" expressions like

$$U^{-1}dUU^{-1}dU...U^{-1}dU$$

A specific feature of the *unitary* case is that compounded expressions like

$$U^{-1}dU(U^{-1}dU)^*U^{-1}dU(U^{-1}dU)^*...U^{-1}dU(U^{-1}dU)^*$$

have the correct invariance property. Therefore, taking, for example, the trace of the above yields a bi-invariant form on $U(n_l)$.

Eventually, we will find an n_l^2 bi-invariant volume form on $U(n_l)$, which is normalized as

$$\int_{U(n_l)} \varpi^v(U) = 1$$

Now we can precisely relate the bi-invariant differential forms on the compact Lie group $U(n_l)$ with its de Rham cohomology.

Theorem 19.14. *The de Rham cohomology vector space $H^n(U(n_l), \mathbb{R})$ of the unitary group is isomorphic to the vector space of bi-invariant differential n-forms defined on $U(n_l)$.*

Proof. Clearly, $U(n_l, \mathbb{C})$ is a compact, connected Lie group, and it is a standard result (see [Bredon 1993, page 308]) that the cohomology with real coefficients of a compact, connected Lie group is isomorphic to the space of its bi-invariant differential forms. We sketch a proof, specially devised for the $U(n_l)$ compact group.

The key point is to prove that any bi-invariant form is closed. Unfortunately, the proof of this result involves concepts from Lie groups, Lie algebras, and Lie brackets, and is a bit involved so that we will omit this proof and refer the reader to [Bredon 1993, V.12]. It is, however, easy to show that those bi-invariant forms generated as $\text{trace}((A^{-1}dA)^n)$ are closed. Indeed, for $n = 1$, invoking Lemma 19.13, it is easily found that

$$\begin{aligned}
&d\text{Trace}(A^{-1}dA)\\
&= \text{Trace}((dA^{-1})dA)\\
&= \text{Trace}(-A^{-1}dAA^{-1}dA)\\
&= -\sum_{i,j}(A^{-1}dA)_{i,j}(A^{-1}dA)_{j,i}\\
&= -\sum_{i<j}(A^{-1}dA)_{i,j}(A^{-1}dA)_{j,i} - \sum_{i>j}(A^{-1}dA)_{i,j}(A^{-1}dA)_{j,i}\\
&= -\sum_{i<j}(A^{-1}dA)_{i,j}(A^{-1}dA)_{j,i} - \sum_{j>i}(A^{-1}dA)_{j,i}(A^{-1}dA)_{i,j}\\
&= 0
\end{aligned}$$

A similar proof shows that $d\text{Trace}((A^{-1}dA)^n) = 0$.

Next, we show that any bi-invariant form ϖ defines a *nontrivial* de Rham cohomology class; in other words, we show that ϖ could not be an exact form. We argue by contradiction and assume that

$$\varpi = d\alpha$$

for some differential form α. We make α invariant by Haar integration:

$$\beta(U) = \int \alpha(GU)\varpi^v(G)$$

It is easily seen that β is left-invariant; indeed,

$$\begin{aligned}
\beta(G_0U) &= \int \alpha(GG_0U)\varpi^v(G)\\
&= \int \alpha(GG_0U)\varpi^v(GG_0)\\
&= \int \alpha(GU)\varpi^v(G)
\end{aligned}$$

$$= \beta(U)$$

Furthermore, since ϖ is invariant, its Haar integration leaves it unchanged, so that

$$\varpi = d\beta$$

Next, we make β right-invariant by yet another Haar integration:

$$\gamma(U) = \int \beta(UG)\varpi^v(G)$$

Checking right-invariance of γ is the same as checking left-invariance of β and is left to the reader. Let us make sure that the second Haar integration has not destroyed left-invariance,

$$\begin{aligned} \gamma(G_0U) &= \int \beta(G_0UG)\varpi^v(G) \\ &= \int \beta(UG)\varpi^v(G) \\ &= \gamma(U) \end{aligned}$$

Again, the second Haar integration does not affect ϖ, so that

$$\varpi = d\gamma$$

Since γ is bi-invariant, it is closed and hence $\varpi = d\gamma = 0$, which is a contradiction since ϖ was assumed nontrivial.

Now, we prove that in every de Rham cohomology class generated by some bi-invariant form(s), there is a *unique* such bi-invariant form. Let ϖ, ϖ' be two bi-invariant forms in the same cohomology class, from which it follows that $\varpi - \varpi' = d\alpha$. Haar integration of both sides of the previous equality yields $\varpi - \varpi' = d\gamma$ for some *bi-invariant* form γ. Since γ is bi-invariant, it is closed; hence, $\varpi - \varpi' = d\gamma = 0$.

So far we have proved that any bi-invariant form generates a unique, nontrivial de Rham cohomology class. To prove that the bi-invariant forms *exhaust* all cohomology classes, we have to go through the process of counting the cohomology classes generated by the traces of the powers of $A^{-1}dA$ and observe that the number of such bi-invariant forms is consistent with the number of generators of the cohomology groups predicted by the spectral sequence argument of Subsection 19.5.1. We leave this to the reader. ■

The above theorem is very much in the spirit of the *Hodge theory*: On an m-D orientable Riemann manifold M, consider the form

$$\varpi = g_{i_1 \ldots i_n}(x_1, \ldots, x_m)dx_{i_1}\ldots dx_{i_n},\ i_1 < \ldots < i_n$$

and define the **Hodge star operation** as

$$\star\varpi = \text{sign}(\pi)g_{i_1 \ldots i_n}(x_1, \ldots, x_m)dx_{j_1}\ldots dx_{j_{m-n}},\ j_1 < \ldots < j_{m-n}$$

where $\pi = (i_1, ..., i_n, j_1, ..., j_{m-n})$ is a permutation of $(1, ..., m)$ and sign(π) is the signature of the permutation. The form ϖ is said to be **harmonic** when $d\varpi = 0$ and $\star(d \star \varpi) = 0$. The *Hodge theorem* asserts that the space of harmonic forms is isomorphic to the de Rham cohomology of M (see [Green 1993]).

19.5.5 Example (SO Group)

For example, let us see what $U^{-1}dU$ becomes in the particular case of the $SO(2)$ group. Let

$$K = \begin{pmatrix} 0 & k \\ -k & 0 \end{pmatrix}$$

so that

$$U(k) = \begin{pmatrix} \frac{1-k^2}{1+k^2} & \frac{-2k}{1+k^2} \\ \frac{2k}{1+k^2} & \frac{1-k^2}{1+k^2} \end{pmatrix}$$

After elementary manipulation, one gets

$$U^{-1}dU = \frac{2}{1+k^2} \begin{pmatrix} 0 & -dk \\ dk & 0 \end{pmatrix}$$

Clearly, Trace$(U^{-1}dU) = 0$ so that, apparently, we are not getting a bi-invariant differential form. Note, however, that $SO(2)$ is a *commutative* group. As such, the left-invariant form $U^{-1}dU$ is automatically right-invariant. Therefore, the bi-invariant 1-form on the Lie group $SO(2)$ is

$$\varpi^1(U) = \frac{2dk}{1+k^2}$$

Now, we look at $SO(3)$, which is *not* commutative. Define

$$K = \begin{pmatrix} 0 & k_{12} & k_{13} \\ -k_{12} & 0 & k_{23} \\ -k_{13} & -k_{23} & 0 \end{pmatrix}$$

After some long but elementary manipulation, we get

$$-2(I-K)^{-1}dK(I+K)^{-1} = \frac{-2}{1+k_{12}^2+k_{13}^2+k_{23}^2} \cdot \begin{pmatrix} 0 & dk_{12}+k_{23}dk_{13}-k_{13}dk_{23} & dk_{13}-k_{23}dk_{12}+k_{12}dk_{23} \\ X & 0 & dk_{23}+k_{13}dk_{12}-k_{12}dk_{13} \\ X & X & 0 \end{pmatrix}$$

("X" in the (i,j)th position denotes the (j,i)th element after the $k_{mn} \to -k_{mn}$ substitution.) Clearly, in this case, Trace$(U^{-1}dU) = 0$, which means that the first de Rham cohomology group of $SO(3)$ vanishes. This can be recovered from the already known fact that the first singular cohomology group, with value group $\mathbb{Z}$, of $SO(3)$ vanishes; therefore, from the universal coefficient theorem, it follows that the de Rham cohomology, with value group $\mathbb{R}$, also vanishes.

19.5.6 Pull-Back

Next, we specialize the cohomology group homomorphism f^* to the de Rham cohomology of $GL(n_l, \mathbb{C})$. The relevant concept is the *pull-back* operation.

Let us write the map $f : S^{n-1} \to GL(n_l, \mathbb{C})$ in local Riemann coordinates as $A(p)$. The **pull-back** of ϖ on S^{n-1}, $f^*\varpi$, is defined as

$$f^*\varpi(p) = \sum g_{i_1 j_1 i_2 j_2 \ldots}(a_{11}(p), a_{22}(p), \ldots) da_{i_1 j_1}(p) da_{i_2 j_2}(p) \ldots$$

with the understanding that the differentials $da_{i_k j_k}(p)$ are written explicitly in terms of $dp_1 dp_2 \ldots$, subject to the exterior differential rules. The crucial feature of the pull-back operation on differential forms is the following:

Theorem 19.15. *The pull-back of differential forms commutes with the exterior differential; that is, $f^* d = df^*$.*

Proof. See [Bott and Tu 1982, page 19]. ■

Therefore the pull-back maps cocycles into cocycles, coboundaries into coboundaries, and finally cohomology classes into cohomology classes. Thus if $\varpi \in H^{n-1}(GL(n_l, \mathbb{C}))$, it follows that $f^*\varpi \in H^{n-1}(S^{n-1})$. Next the cohomology of S^{n-1} is (see [Bott and Tu 1982, page 36])

$$\begin{aligned} H^*(S^{n-1}) &= \mathbb{R}, \text{ in dimension } n-1 \\ &= 0, \text{ otherwise} \end{aligned}$$

Therefore, $f^*\varpi$ is a multiple of the unique generator of $H^{n-1}(S^{n-1})$.

19.5.7 Degree

Let us identify the degree f^* with the collection of all $f^*\varpi$'s, when ϖ runs into the set of generators of $H^{n-1}(GL(n_l, \mathbb{C}))$. Each such $f^*\varpi$ is in $H^{n-1}(S^{n-1})$ and is thus a multiple of the unique generator g of $H^{n-1}(S^{n-1})$. It is that multiple—preferably an integer—that we shall call *degree.* Comparing $f^*\varpi$ and g viewed as cohomology classes requires us to find the exact forms that make the differential forms ($f^*\varpi +$ exact form) and ($g +$ exact form) multiple of each other. It is computationally more efficient to get rid of the exact forms by integration on the sphere. It turns out that the kernel of the integration map

$$\begin{aligned} \int : H^{n-1}(S^{n-1}) &\to \mathbb{R} \\ f^*\varpi &\mapsto \int_{S^{n-1}} f^*\varpi \end{aligned}$$

is 0 so that $f^* \varpi$ is not trivial iff the integral does not vanish. The fact that the integration map has vanishing kernel is a corollary of the following lemma:

Lemma 19.16. *The kernel of the integration map*

$$\begin{aligned} \int : \Lambda^{n-1}(S^{n-1}) &\to \mathbb{R} \\ \varpi &\mapsto \int_{S^{n-1}} \varpi \end{aligned}$$

coincides with the exact forms.

Proof. See [Bott and Tu 1982, page 18]. ■

Therefore we can safely define the **degree** as

$$\text{degree}(f) = \int_{S^{n-1}} f^* \varpi$$

when ϖ is a generator of $H^{n-1}(GL(n_l, \mathbb{C}))$. This definition achieves our stated objective that the degree should vanish iff the map is homotopic to the constant map. Finally, to make the degree an *integer* it suffices to choose ϖ such that

$$\int_{f(S^{n-1})} \varpi = 1$$

For a proof, see [Bott and Tu 1982, page 40]. If the mapping f is surjective, then we normalize ϖ as $\int_{GL} \varpi = 1$. However, since GL is not compact, this scaling is sometimes difficult to achieve. It is therefore desirable to retract GL onto U, which does not change the homotopy of f, after which the normalization $\int_U \varpi = 1$ is easy since integration is over a compact group.

19.5.8 Connection with Analytical Degree

It turns out that the above concept of the degree, computed by integration of differential forms, is closely related to the *analytical* or *Leray–Schauder* degree of Section 18.3. Actually, to *prove* that this degree is an *integer*, we do need its analytical interpretation in terms of the Jacobian of the map.

Consider a map $f : X \to Y$ between *compact*, differentiable, orientable manifolds of the same dimension n. Observe that Y need not be a Lie group. Define a volume form $\varpi(y)$ on Y, pull it back on X, and compute the degree as

$$\text{degree}(f) = \frac{\int_X \varpi(f(x))}{\int_Y \varpi(y)}$$

For the details, see [Bott and Tu 1982, page 41].

To see the connection with the analytical degree and to *prove* that the differential degree is an integer, assume first that the differential form $\varpi(y)$ has compact support $\overline{O_{y^o}}$, where O_{y^o} is an open neighborhood of the regular value y^0, small enough so that $f^{-1}(O_{y^0})$ is the disjoint union of neighborhoods of the preimages in $f^{-1}(y^0)$. Write the invariant differential form, locally around the regular value $y^0 \in Y$, as

$$\varpi(y) = g(y)dy_1 dy_2 ... dy_n$$

Let $x^0 \in X$ be a preimage of y^0. Clearly, $y = f(x)$, $x \in O_{x^0}$, $y \in O_{y^0}$ and therefore the local pull-back can be written

$$\begin{aligned}\varpi(f(x)) &= g(f(x))(df_1(x))...(df_n(x)) \\ &= g(f(x))(\sum_{i_1} \frac{\partial f_1(x)}{\partial x_{i_1}} dx_{i_1})...(\sum_{i_n} \frac{\partial f_n(x)}{\partial x_{i_n}} dx_{i_n}) \\ &= g(f(x))(\sum_{\pi} \operatorname{sign}(\pi) \frac{\partial f_{\pi(1)}}{\partial x_1} ... \frac{\partial f_{\pi(n)}}{\partial x_n}) dx_1 dx_2 ... dx_n \\ &= g(f(x)) \det J(x) dx_1 dx_2 ... dx_n\end{aligned}$$

In the above π is an arbitrary permutation of the ordered set $(1, 2, ..., n)$ of indices, $\operatorname{sign}(\pi)$ is the signature of the permutation and $J(x)$ denotes the Jacobian.

Since y^0 is a regular value, it follows from the implicit function theorem that the local map $f|O_{x^0} : O_{x^0} \to O_{y^0}$ is a diffeomorphism. It is a fairly general fact about integration that a change of variables in an integral affects the integral by no more than a sign, to be specific,

$$\int_{O_{y^0}} \varpi(y) = \operatorname{sign}(J(x)) \int_{O_{x^0}} \varpi(f(x))$$

Therefore, taking into consideration all of the preimage points yields

$$\frac{\int_X f^* \varpi}{\int_Y \varpi} = \sum_{x \in f^{-1}(y)} \operatorname{sign} \det J(x)$$

Clearly, we have recovered the analytical degree in case the differential form has compact support.

Now, let $\varpi(y)$ be an arbitrary differential form. Using a partition of unity argument, write it as

$$\varpi(y) = \sum_i \varpi_i(y)$$

where $\varpi_i(y)$ is a form with compact support $\overline{O_{y^i}}$ enjoying the same properties as in the previous case and $\cup_i O_{y^i}$ is a finite, open covering of Y. Evaluate the differential degree using any of the differential forms $\varpi_i(y)$; clearly, every such differential degree matches the analytical degree and since the degree does not depend on y^i, it follows that

$$\frac{\int_X \varpi_i(f(x))}{\int_Y \varpi_i(y)} = d \in \mathbb{N}, \ \forall i$$

It follows that

$$\begin{aligned} d &= \frac{\int_X \varpi_i(f(x))}{\int_Y \varpi_i(y)} \\ &= \frac{\sum_i \int_X \varpi_i(f(x))}{\sum_i \int_Y \varpi_i(y)} \\ &= \frac{\int_X \varpi(f(x))}{\int_Y \varpi(y)} \end{aligned}$$

Clearly, we recover the analytical degree.

19.6 Example (the Principle of Argument)

The encirclement issue in the elementary Nyquist stability criterion can be formulated in terms of the degree of the Nyquist map computed by integration of differential forms. The punctured complex plane $\mathbb{C}\setminus\{0+j0\}$ is viewed as the unitary group $U(1)$ with the normalized, bi-invariant 1-form,

$$\varpi(s) = \frac{1}{2\pi j}\frac{ds}{s}$$

Define the Nyquist map

$$\begin{aligned} f:\Omega &\rightarrow \mathbb{C}\setminus\{0+j0\} \\ \omega &\mapsto \det(I+L(j\omega)) \end{aligned}$$

The pull-back of ϖ on Ω is easily seen to be

$$\begin{aligned} f^*\varpi &= \frac{d\det(I+L(j\omega))}{\det(I+L(j\omega))} \\ &= d\log\det(I+L(j\omega)) \\ &= d\log|f(j\omega)| + jd\angle f(j\omega) \end{aligned}$$

The integration of the pull-back of the invariant form yields

$$\begin{aligned} \text{degree} &= \int_\Omega f^*\varpi \\ &= \frac{1}{2\pi j}\int_\Omega (d\log|f(s)| + jd\angle f(s)) \\ &= \frac{1}{2\pi}\int_{-\infty}^{+\infty} d\angle\det(I+L(j\omega)) \end{aligned}$$

Clearly, the degree obtained by integration of the pull-back of the invariant differential form on $\mathbb{C}\setminus\{0+j0\}$ matches the intuitive concept of the number of encirclements of $0+j0$ by the Nyquist plot.

19.7 Example (Mapping into SO)

19.7.1 Degree 1 Map

Consider the mapping

$$\begin{array}{ccc} S^1 & \to & SO(2) \\ \exp(j\theta) & \mapsto & \begin{pmatrix} \cos\theta & -\sin\theta \\ \sin\theta & \cos\theta \end{pmatrix} \end{array}$$

Since $SO(2) \cong S^1$ the above map is the identity on S^1, so that by an elementary Brouwer degree argument the map is of degree 1. The purpose of this illustrative example is to recover this result from the machinery of bi-invariant differential forms.

First, a technicality. We have to chart $SO(2)$ with Riemann coordinates and we need two charts—one around $U = I$ and another one around $U = -I$. Observe that the chart around $U = I$ covers $SO(2) \setminus \{-I\}$. Since the chart is used essentially as a tool to compute integrals of differential forms, it would suffice to restrict ourselves to the chart around $U = I$ if we can prove that the integral of the invariant form around an arbitrarily small neighborhood of $-I$ is arbitrarily small. It is easily seen that the chart around $U = -I$ is

$$U(k) = \begin{pmatrix} \frac{-1+k^2}{1+k^2} & \frac{-2k}{1+k^2} \\ \frac{2k}{1+k^2} & \frac{-1+k^2}{1+k^2} \end{pmatrix}$$

and that the bi-invariant differential form is

$$\varpi^1(U) = \frac{2dk}{1+k^2}$$

Let $O^{\epsilon}_{-I} = U((-\epsilon, +\epsilon))$. Clearly,

$$\int_{O^{\epsilon}_{-I}} \varpi^1 = \int_{-\epsilon}^{+\epsilon} \frac{2dk}{1+k^2} \to 0, \text{ as } \epsilon \downarrow 0$$

A similar remark applies to the circle S^1. Technically we need two charts; however, the Riemann stereographic projection on the tangent to the north pole charts the whole sphere except the south pole; since integration around an arbitrarily small neighborhood of the south pole is vanishingly small, we only need the chart of $S^1 \setminus \{\text{south pole}\}$.

When we modulo these conventions the above map, in terms of Riemann coordinates on S^1, can be written

$$k \mapsto (I+K)^{-1}(I-K) = \frac{1}{1+k^2}\begin{pmatrix} 1-k^2 & -2k \\ 2k & 1-k^2 \end{pmatrix}$$

In terms of Riemann coordinates on both S^1 and $SO(2)$, the map becomes

$$k \mapsto k$$

Clearly,

$$\begin{aligned}\int_{SO(2)} \varpi &= \int_{-\infty}^{+\infty} \frac{2dk}{1+k^2} \\ &= 2\pi\end{aligned}$$

Next, the pull-back on the sphere is easily seen to be

$$f^*\varpi = \frac{2dk}{1+k^2}$$

Clearly,

$$\int_{S^1} f^*\varpi = 2\pi.$$

Hence,

$$\text{degree} = \frac{\int_{S^1} f^*\varpi}{\int_{SO(2)} \varpi} = 1$$

and the result is consistent with the Brouwer degree.

19.7.2 Degree 2 Map

Consider next the less trivial case,

$$\begin{aligned} f : S^1 &\to SO(2) \\ \exp(j\theta) &\mapsto \begin{pmatrix} \cos 2\theta & -\sin 2\theta \\ \sin 2\theta & \cos 2\theta \end{pmatrix}\end{aligned}$$

Again, an elementary Brouwer degree argument yields a degree of 2. To recover this result using the machinery of bi-invariant forms, observe that the mapping, in terms of the Riemann coordinate k of S^1, is

$$k \mapsto (U(k))^2$$

where $U(k)$ is the mapping of the preceding case, namely,

$$U(k) = \begin{pmatrix} \frac{1-k^2}{1+k^2} & \frac{-2k}{1+k^2} \\ \frac{2k}{1+k^2} & \frac{1-k^2}{1+k^2} \end{pmatrix}$$

The essential point is to observe that the pull-back $f^*\varpi$ is

$$\begin{aligned}(U(k)^2)^{-1}d(U(k)^2) &= (U(k)^{-1})U(k)^{-1}(U(k)dU(k) + dU(k)U(k)) \\ &= U(k)^{-1}U(k)^{-1}2U(k)dU(k) \\ &= 2U(k)^{-1}dU(k) \\ &= \frac{4dk}{1+k^2}\end{aligned}$$

(Observe that $U(k)$ and $dU(k)$ commute.) Therefore, the degree is

$$\frac{\int_{S^1} f^* \varpi}{\int_{SO(2)} \varpi} = \frac{\int_{-\infty}^{+\infty} \frac{4dk}{1+k^2}}{\int_{-\infty}^{+\infty} \frac{2dk}{1+k^2}} = 2$$

as guessed from intuition.

19.8 Example (Brouwer Degree)

Since the only spheres that have Lie group structures are S^0, S^1, S^3, only for maps $S^n \to S^n$, $n = 0, 1, 3$ can the Brouwer degree be reformulated in terms of invariant differential forms. Let ϖ^n, $n = 1, 3$ be the invariant differential form defined on S^n, $n = 1, 3$. Therefore,

$$\text{Brouwer degree of } f = \frac{\int_{S^n} f^* \varpi^n}{\int_{S^n} \varpi^n}$$

19.9 Example (McMillan Degree)

As shown by [Martin and Hermann 1978], the *system-theoretic* McMillan degree of a transfer matrix can be reformulated as the *algebraic-geometrical* degree of an induced map from the Riemann sphere to the complex Grassmannian. Next, as shown in [Bucy 1991], differential forms representing the cohomology of the Grassmannian manifold can be derived from left-invariant differential forms on the complex unitary groups. This makes the machinery developed in this chapter applicable to the McMillan degree.

Consider an $m \times n$ transfer matrix with rational coprime factorization $D^{-1}(s)N(s)$ and consider the IO relationship,

$$y(s) = D^{-1}(s)N(s)u(s)$$

Clearly, the IO relationship can be written

$$\begin{pmatrix} D(s) & -N(s) \end{pmatrix} \begin{pmatrix} y(s) \\ u(s) \end{pmatrix} = 0$$

Since $D^{-1}N$ is coprime, the rank of $\begin{pmatrix} D(s) & -N(s) \end{pmatrix}$ is constant, equal to n, $\forall s \in S^2$. Therefore,

$$\begin{pmatrix} y(s) \\ u(s) \end{pmatrix} \in G_{m+n,m}$$

where $G_{m+n,m}$ denotes the (complex) Grassmannian of all m-D subspaces of $\mathbb{C}^{m+n}$. Therefore, the transfer matrix induces a map, the so-called *Martin–Hermann* map,

$$f : S^2 \to G_{m+n,m}$$

The major result of [Martin and Hermann 1978] is that the algebraic-geometrical degree of this map is equal to the McMillan degree of $D^{-1}N$.

The degree of the Martin–Hermann map can be evaluated using the machinery of differential forms. The Grassmannian receives its homogeneous space structure via the isomorphism:

$$G_{m+n,m} \cong U(m+n)/(U(n) \times U(m))$$

As shown by [Bucy 1991], left-invariant forms on U induce cohomology forms on the Grassmannian. Let ϖ^2 be the cohomology 2-form on $G_{m+n,m}$. Therefore,

$$\text{McMillan degree} = \frac{\int_{S^2} f^* \varpi^2}{\int_{f(S^2)} \varpi^2}$$

The only problem with the above is that homotopies that deform the map f from real coefficient maps to complex coefficient maps are allowed and yield the same degree. Enforcing the real coefficient property would require us to restrict ourselves to homotopies between maps *that commute with complex conjugation.* To attack this problem we need an equivariant theory along the lines of Atiyah's KR-theory.

19.10 Obstruction to Extending GL-Valued Nyquist Map

We look at the recursion for the extension of the map to higher-and-higher-dimensional skeleta, $P^0, P^1, P^2, \ldots$ of the polyhedron P of uncertainties.

By simplicial decomposition of the polyhedron P, the general extension from P^{n-1} to P^n is reduced to the extension from $\partial\sigma_n$ to σ_n, where σ_n is an arbitrary simplex. This latter extension involves a map $S^{n-1} \to GL(n_l, \mathbb{C})$. The extension is possible iff the homotopy class of $S^{n-1} \to GL(n_l, \mathbb{C})$ considered as an element of $\pi_{n-1}(GL(n_l, \mathbb{C}))$ vanishes. *Clearly, the only nontrivial case is the case where n is even*, because otherwise the map is homotopic to a constant map and the extension is always possible. Furthermore, assuming n is even, remaining within the realm of *stable homotopy* requires $n_l \geq \frac{n}{2}$. In other words, a minimum amount of feedback paths is required. It should be clear to the reader that as we proceed to higher-and-higher-dimensional skeleta, beyond a certain point we will no longer be able to remain within the realm of stable homotopy.

From an obstruction point of view, the obstruction to extending the map from P^{n-1} to P^n is the cocycle

$$c_{f^{n-1}} \in C^n(P, \pi_{n-1}(GL(n_l, \mathbb{C})))$$

defined by

$$(c_{f^{n-1}}, \sigma_n) = \{f^{n-1} | \partial\sigma_n\} \in \pi_{n-1}(GL(n_l, \mathbb{C}))$$

If $c_{f^{n-1}} = 0$, the extension is possible. If $c_{f^{n-1}} \neq 0$, the extension of f^{n-1} is not possible. However, if we revise the choice at the level of the extension from P^{n-2} to P^{n-1} it is possible to find an extension f'^{n-1} which in turn has an extension to P^n. The deciding factor is the cohomology class $\{c_{f^{n-1}}\}$ defined as an element of $H^n(P, \pi_{n-1}(GL(n_l, \mathbb{C})))$. If $c_{f^{n-1}} \neq 0$ but $\{c_{f^{n-1}}\} = 0$, there exists an extension f'^{n-1} from P^{n-2} to P^{n-1} which in turn has an extension to P^n.

NOTES

The historical roots of exterior differential forms are Stokes' theorem and its generalization and analytical dynamics. The mathematical formalization of exterior differential forms, and more specifically the theory of invariant differential forms on Lie groups, were developed in the context of multilinear algebra by E. Cartan. Two most substantial developments occurred in the 1930s with the Hodge theory of harmonic differentials and the de Rham theory of currents (the latter took its final shape in the 1950s). The application of the de Rham theory to cohomology was popularized by [Bott and Tu 1982].

More recently, differential forms have become increasingly utilized in nonholonomic kinematics and robotics. However, probably the most spectacular engineering application of differential forms is multidirectional signal estimation (see, e.g., [Bucy 1991]).

The homotopy classification of maps from spheres to general linear groups was formulated in the context of elliptic operators by Atiyah and is one of the ingredients in the proof of the Atiyah–Singer index theorem.

20

K-THEORY OF ROBUST STABILIZATION

Roughly speaking K-theory may be described as the linear algebra of large matrices,...

Michael F. Atiyah,
"A survey of K-theory,"
Proceedings of the Conference on K-Theory and Operator Algebras,
Atlanta, GA, 1977.

SUMMARY

In this chapter, rather than mapping the uncertainty to the return difference matrix as we did in the previous chapter, we map the uncertainty to the return difference *operator*. To make it simple, assume that the system is open-loop stable and time-invariant. The crucial point is the equivalence between closed-loop stability and invertibility of the Toeplitz operator induced by the return difference, $I+T_{L\Delta}$. Mapping into an appropriate space of operators immediately leads to the so-called *topological K-theory*, which is probably the most significant development that has occurred in algebraic topology after 1960.

In this chapter, we fulfill the promise made in the very first paragraph of this book—where we boldly introduced the whole subject by claiming that one of the roots of topological K-theory is a robust stabilization problem in disguise.

An operator-theoretic concept, more fundamental than invertibility, is *Fredholmness.* It has to do with finite dimensionality of the kernel and the cokernel. It is a general fact that *Fredholmness* of a Toeplitz operator is a necessary, but not sufficient, condition for its invertibility. The deeper link between operator theory and robust stability can be perceived by observing that the *zero exclusion principle* of robust stability, $\det(I + L(j\omega)\Delta) \neq 0, \forall \omega$, is in fact the condition for *Fredholmness* of the Toeplitz operator $I + T_{L\Delta}$.

As we shall see, the connection between closed-loop stability and invertibility of the Toeplitz operator induced by the return difference is more than the casual observation that their respective criteria match. This connection is actually *very natural,* and we exploit this fact to rebuild multivariable closed-loop stability from a self-contained, operator-theoretic viewpoint. As a byproduct of some advances in Toeplitz operators, we can even prepare the ground for $\mathcal{L}^2$ stability of distributed parameter and distributed sensor/actuator feedback systems.

Not unlike the obstruction approach where we a priori removed $0 + j0$ from the complex plane, here we enforce the zero exclusion principle by mapping the uncertainty into the space of *Fredholm* Toeplitz operators. The first issue is therefore *existence* of the map

$$\begin{array}{rcl} D & \rightarrow & \{\text{Fredholm operators}\} \\ \Delta & \mapsto & I + T_{L\Delta} \end{array}$$

Not unlike the homotopy extension problem that was meant to be a trick to get rid of the frequency sweep, here, by mapping into the space of *operators*, we avoid by the same token the frequency sweep. Clearly, a mapping defined on the space D is more aesthetically appealing than a mapping defined on $D \times \Omega$.

Having taken care of the zero exclusion principle, the next step is the encirclement condition—that is, the homotopy type of the map. For *one single* $\Delta \in D$, the homotopy type of the Nyquist map is related to the concept of *index* of the Toeplitz operator $I + T_{L\Delta}$. The index can be viewed as a $\mathbb{Z}$-valued mapping. In this robust stability context, we have a *family* of Fredholm Toeplitz operators and the question is what is the index of a *family* of Fredholm Toeplitz operators? This problem immediately leads us into K-theory. The *salient* result is that the natural value group for the index of the *Fredholm family* $\{I + T_{L\Delta} : \Delta \in D\}$ is the Grothendieck group $K^0(D)$ of vector bundles defined over the uncertainty space.

It is fair to say that one of the *original* motivations for the $K^0(D)$ group was the definition of a value group for the index of a *family* of Fredholm Hilbert space operators with its parameter running in D. In this chapter, we pay tribute to history by following this path of approach, at the risk of appearing a bit archaic.

In this particular robust control application, the value group $K^0(D)$ appears too big because it does not fully exploit the *Toeplitz* property. Restricting the family to be *Toeplitz* in addition to being Fredholm immediately leads to such K-idiomatic concepts as *suspension, stabilization, reduced K-groups, and higher K-groups*. The concept of *stabilization* of K-theory—not to be confused with the control-theoretic concept—refers to going to higher-dimensional spaces so as to have "more room to maneuver." In this control context, the K-concept of *stabilization* takes the concrete aspect of going to an arbitrarily large number of feedback paths or going to distributed sensing/actuating. The *higher K-groups* make K-theory not unlike a cohomology theory in the sense that it associates Abelian groups to each (reasonably well-behaved) topological space and a K-group homomorphism to each continuous map. The topological K-theory is *contravariant*, in the sense that the arrow of the K-group homomorphism flows *opposite* to the arrow of the topological map. To indicate that they are contravariant functors, the K-groups of such a space as D are written as $K^0(D), K^{-1}(D), \ldots$. A specific feature of K-theory as opposed to ordinary cohomology is the *Bott periodicity theorem*: The sequence of K-groups is periodic with period 2, so that we only need to bother with K^0 and K^{-1}.

The K-groups defined above are values groups for the index of families of *complex* Fredholm Toeplitz operators. Restricting ourselves to *real perturbations*, then we get into *real K-theory*. It is well-known in control engineering that real perturbations are harder to cope with than complex perturbations. The same situation occurs in K-theory: A manifestation of the accrued complexity of real K-theory is the real Bott periodicity theorem indicating a periodic pattern with a period of 8 in the real K-groups as opposed to a period of 2 in the complex case.

20.1 Return Difference Operator

For clarity of the exposition, the relevant concepts are developed in the *discrete-time* setting. The translation to continuous-time is left to the reader.

20.1.1 Open-Loop Stable, Discrete-Time Systems

We consider a stable, discrete-time loop transmission $L(z)\Delta$. To be consistent with the standard Hardy space terminology, we have to reverse ($z \leftrightarrow z^{-1}$) the usual z-transform notation. In other words, a discrete-time causal signal, say $e = (e_0, e_1, e_2, \ldots)$, is transformed as

$$e(z) = e_0 + e_1 z + e_2 z^2 + \ldots$$

This signal is considered **stable** iff $e(z)$ is analytic in the disk $\mathbb{D}$ and

$$\lim_{r \uparrow 1} \int_0^{2\pi} ||e(r\exp(j\theta))||^2 d\theta < \infty$$

In other words, iff $e(\cdot)$ is in the Hardy space $\mathcal{H}^2$. The loop matrix $L(z)\Delta$ is defined to be **stable** iff it is analytic in $\mathbb{D}$ and essentially bounded on $\mathbb{T}$—that is, iff $L\Delta \in \mathcal{H}^\infty$. In this case, the loop transmission has an expansion, converging in $|z| < 1$,

$$L(z)\Delta = L_0 + L_1 z + L_2 z^2 + \ldots$$

With these input/output concepts, $e \in \mathcal{H}^2$ and $L\Delta \in \mathcal{H}^\infty$ imply that $L\Delta e \in \mathcal{H}^2$. Because of this latter property, the loop transmission is said to be **$\mathcal{L}^2$ or $\mathcal{H}^2$ stable.**

The relation between the sample data error and output signals can be written as

$$\begin{pmatrix} y_0 \\ y_1 \\ y_2 \\ \vdots \end{pmatrix} = \underbrace{\begin{pmatrix} L_0 & & & \\ L_1 & L_0 & & \\ L_2 & L_1 & L_0 & \\ \ddots & \ddots & \ddots & \ddots \end{pmatrix}}_{T_{L\Delta}} \begin{pmatrix} e_0 \\ e_1 \\ e_2 \\ \vdots \end{pmatrix}$$

Clearly, the above is the matrix representation of the Toeplitz operator induced by $L(z)\Delta$.

20.1.2 Toeplitz Operators

Let $\mathcal{L}^2$ be the Hilbert space of functions defined and square integrable on the unit circle. Let $\mathcal{H}^2 \subseteq \mathcal{L}^2$ be the Hardy space of functions analytic in the open unit disk. Let $P_{\mathcal{H}^2}$ be the orthogonal projection $\mathcal{L}^2 \to \mathcal{H}^2$. Let $\mathcal{L}^\infty$ be the Banach algebra of functions essentially bounded on the unit circle, and let $\mathcal{H}^\infty \subseteq \mathcal{L}^\infty$ be the subalgebra of functions analytic in the open unit disk.

Definition 20.1. *For $L\Delta \in \mathcal{L}^\infty$, the* **Toeplitz operator** *$T_{L\Delta}$ is defined as*

$$\begin{aligned} T_{L\Delta} : \mathcal{H}^2 &\to \mathcal{H}^2 \\ f &\mapsto P_{\mathcal{H}^2} L\Delta f \end{aligned}$$

From the above, it is trivial to see that

$$T_{I+L\Delta} = I + T_{L\Delta}$$

A less trivial, but well-known, result is the following:

Theorem 20.2. *The Toeplitz operator $I + T_{L\Delta}$ is bounded iff the symbol $I + L\Delta$ is in $\mathcal{L}^\infty$, in which case $||I + T_{L\Delta}|| = ||I + L\Delta||_\infty$.*

To develop an index theory for Toeplitz operators, it is not enough to ask that the symbol be essentially bounded (see [Douglas 1972]). We require that the symbol be *continuous* on the circle, namely, $(I + L\Delta) \in C^0(\mathbb{T}, \mathbb{M}(n_l, \mathbb{C}))$.

20.1.3 Open-Loop Unstable Systems

To illustrate the nature of the problem, consider $L\Delta = \frac{1}{z-0.5}$. The matrix representation of the open-loop system is

$$(-2)\begin{pmatrix} 1 & & & \\ 2 & 1 & & \\ 4 & 2 & 1 & \\ \ddots & \ddots & \ddots & \ddots \end{pmatrix}$$

Even though this matrix has the correct structure, it is *not* a Toeplitz operator, as the reader will readily verify from the formal definition. Besides, the above operator is not bounded.

The only way to deal with unbounded operators is to define a *domain of definition*, $\mathcal{D} \subseteq \mathcal{H}^2$, over which the operator is bounded. *The operator-theoretic way of viewing stabilization of open-loop unstable systems is to design the compensator so that* $e \in \mathcal{D}$. To illustrate the concept, consider the factorization

$$L\Delta = \frac{1}{z-a} g(z)$$

where

$$|a| < 1, \; g \in \mathcal{H}^\infty$$

In this simple case, it is readily seen that

$$\mathcal{D} = (z-a)\mathcal{H}^2$$

Since $e \in \mathcal{D}$, we parameterize it as

$$e(z) = (z-a)(e_0' + e_1' z + e_2' z^2 + \ldots)$$

and the matrix representation of $L(z)\Delta$ using the parameterization $\{e_i'\}$ of $e \in \mathcal{D}$ is

$$y = T_g e'$$

Therefore, in case of an open-loop unstable system, the operator $L(z)\Delta|\mathcal{D}$ is isomorphic to the Toeplitz operator T_g induced by the $\mathcal{H}^\infty$ factor of the loop function.

20.1.4 Closed-Loop Stability

In this subsection, we rederive the multivariable closed-loop stability criterion from a self-contained, operator-theoretic viewpoint. The basic theorem is the following:

Theorem 20.3. *If the loop transmission* $L\Delta$ *is* $\mathcal{L}^2$*-stable, then the closed-loop system is* $\mathcal{L}^2$*-stable if and only if the Toeplitz operator* $I + T_{L\Delta}$ *is invertible.*

Proof. Let r be the reference signal, y the output-to-be-controlled signal, and $e = y - r$ the error signal. Considering all of these quantities to be elements of a Hilbert space of causal signals, typically $\mathcal{H}^2 \otimes \mathbb{R}^{n_l}$, the open-loop equation can be written $y = T_{L\Delta} e$. Injecting $e = y - r$ yields $(I + T_{L\Delta})e = r$. From there, the result is trivial. ∎

Observe that the above result remains valid for a *distributed parameters* open-loop transfer matrix $L\Delta$, provided that the stability be interpreted in the $\mathcal{L}^2$ sense. This theorem also remains valid in the case of distributed sensors/actuators. In this case, the Fourier coefficients L_k of $L\Delta$ are themselves Hilbert space operators from the Hilbert space of instantaneous errors to the Hilbert space of instantaneous outputs. The Toeplitz operator $I + T_{L\Delta}$ is still defined by the projection formula and can be intuitively viewed as a block-Toeplitz matrix where each "block" is itself a Hilbert space operator. For the conceptualization of this mathematical object, see [Rabindranathan 1969].

20.2 Index of Fredholm Operators

We first review the concepts of kernel and cokernel for a general operator $A : \mathcal{H} \to \mathcal{H}$ defined over a separable Hilbert space $\mathcal{H}$. This operator is meant to be the Toeplitz operator induced by the return difference. The **kernel** of A is defined as $\ker(A) = \{x \in \mathcal{H} : Ax = 0\}$. It is easily seen to be *closed.* The **cokernel** of the operator is $\text{coker}(A) = \mathcal{H}/A\mathcal{H}$. The latter quotient is defined purely algebraically; the analysis in general requires $A\mathcal{H}$ to be closed. The following two lemmas are standard results.

Lemma 20.4. *For a Hilbert space operator $A : \mathcal{H} \to \mathcal{H}$ such that $A\mathcal{H}$ is closed, we have*

$$\textit{coker}(A) = \ker(A^*) = (A\mathcal{H})^\perp$$

where A^* denotes the adjoint and $(\cdot)^\perp$ denotes the orthogonal complement.

Lemma 20.5. *For a Hilbert space operator $A : \mathcal{H} \to \mathcal{H}$, if $\dim(\mathcal{H}/A\mathcal{H}) < \infty$, then $A\mathcal{H}$ is closed.*

The next relevant concept is that of *Calkin algebra.* Let $\mathbb{B}$ be the Banach algebra of linear, bounded operators on $\mathcal{H}$. Let $\mathbb{K}$ be the subalgebra of compact operators. $\mathbb{K}$ is also an ideal in $\mathbb{B}$. Clearly, we have the short exact sequence

$$0 \to \mathbb{K} \to \mathbb{B} \xrightarrow{\pi} \mathbb{B}/\mathbb{K} \to 0$$

The quotient algebra $\mathbb{B}/\mathbb{K}$ is called **Calkin algebra**.

We link the previous concepts in the following theorem-definition:

Theorem 20.6. (Atkinson) *For an operator $F \in \mathbb{B}$, the following statements are equivalent:*

1. *$\pi(F)$ is invertible in the Calkin algebra $\mathbb{B}/\mathbb{K}$.*
2. *There exists an operator $F^\dagger \in \mathbb{B}$ such that both $FF^\dagger - I$ and $F^\dagger F - I$ are compact.*
3. *F has closed image and both $\ker(F)$ and $\ker(F^*)$ are finite-dimensional.*
4. *Both $\ker(F)$ and coker $(F) := \mathcal{H}/F\mathcal{H}$ are finite-dimensional.*

Furthermore, if any of the above statement holds, the operator is said to be **Fredholm**.

A few remarks are in order. About statement 2: Observe that $F^\dagger$ can be taken so that $I - FF^\dagger$ and $I - F^\dagger F$ are the *finite rank* projections onto

$(F\mathcal{H})^\perp$ and $\ker(F)$, respectively. About statement 4: Observe that it is related to existence of solutions to $Fx = b$.

We now define a concept specially devised for *Fredholm* operators:

Definition 20.7. *The* **index** *(also called* **analytical or Fredholm index***) of the Fredholm operator F, typically $I + T_{L\Delta}$, is defined as*

$$\begin{aligned} index\ (F) &= \dim\ker F - \dim coker\ F \\ &= \dim\ker F - \dim\ker F^* \end{aligned}$$

As an example, the reader can easily verify that any matrix $F : \mathbb{R}^m \to \mathbb{R}^n$ is a Fredholm operator with index $n - m$, whatever the rank. Other examples are provided by the following Toeplitz operators:

$$\text{index}(T_z) = \text{index}\begin{pmatrix} 0 & 0 & 0 & \\ 1 & 0 & 0 & \ddots \\ 0 & 1 & 0 & \ddots \\ & \ddots & \ddots & \ddots \end{pmatrix} = -1$$

$$\text{index}(T_{z^{-1}}) = \text{index}\begin{pmatrix} 0 & 1 & 0 & \\ 0 & 0 & 1 & \ddots \\ 0 & 0 & 0 & \ddots \\ & \ddots & \ddots & \ddots \end{pmatrix} = +1$$

The above is easily generalized to $\text{index}(T_{z^n}) = -n$. The important corollary of the latter is that the index map from the set of Fredholm operators to $\mathbb{Z}$ is *surjective.*

We now review the classical properties of the Fredholm index. By *classical,* we mean those properties that apply to *one single* Fredholm operator, as opposed to the index for a family of operators. The following three lemmas are trivial consequences of Atkinson's theorem.

Lemma 20.8. *If F_1, F_2 are Fredholm, so is F_1F_2.*

Lemma 20.9. *The Fredholm property is invariant under compact perturbation.*

Lemma 20.10. *If F is Fredholm and R is invertible, FR and RF are Fredholm and*

$$index(FR) = index(RF) = index(F)$$

The next lemma requires a little bit of work.

Lemma 20.11. *F is Fredholm with index 0 iff there exists a finite rank operator Φ such that $F + \Phi$ is invertible.*

Proof. The "only if" part is proved by constructing an operator Φ mapping $\ker(F)$ onto $\text{coker}(F)$ and by observing that $F+\Phi$ is invertible. To prove the "if" part, define P to be the orthogonal projection on $(\Phi\mathcal{H})^{\perp}$. Using the same argument as Lemma 20.22, it is seen that the Fredholm index is invariant under premultiplication of such a projection operator. Therefore,

$$\begin{aligned} 0 &= \text{index}(F+\Phi) \\ &= \text{index}(P(F+\Phi)) \\ &= \text{index}(F) \end{aligned}$$

And the proof is completed. ■

Lemma 20.12. *If F is Fredholm with index 0, then, for any compact operator K, $F+K$ is Fredholm with index 0.*

Proof. By approximating the compact operator with a finite rank operator, we recover the case of the previous lemma. ■

To proceed further, we need a simple, yet powerful, trick which is *typical* in K-theory. It is the idea of going to a higher-dimensional space so as to have more room to maneuver. It is basically the same idea as the stable homotopy groups of GL where going to infinite matrices makes the problem more mathematically manageable. Assume we have to prove something about a Fredholm operator with arbitrary index n. It is claimed that, without loss of generality, we can assume that $\text{index}(F) = 0$. Indeed, if the index is n, define the operator

$$F \oplus T_{z^n} : \mathcal{H} \oplus \ell^2 \to \mathcal{H} \oplus \ell^2$$

Clearly, $\text{index}(F \oplus T_{z^n}) = \text{index}(F) + \text{index}(T_{z^n}) = \text{index}(F) - n = 0$. Therefore, to prove something about an arbitrary Fredholm operator F with index n, go to the operator $F \oplus T_{z^n}$ that has vanishing index.

The stabilization trick immediately yields the following classical result:

Theorem 20.13. *The Fredholm index is invariant under compact perturbation.*

Proof. Just use Lemma 20.12 together with the stabilization trick. ■

Define $\mathcal{F}$ to be the set of Fredholm operators. The index can be formally viewed as a (surjective) map

$$\text{index} : \mathcal{F} \to \mathbb{Z}$$

and this map is continuous, more precisely stated in the following theorem:

Theorem 20.14. *$\mathcal{F}_n = \text{index}^{-1}(n)$ is open.*

Proof. We have to show that for a sufficiently small (operator-norm topology) neighborhood of $F \in \mathcal{F}_n$, all operators have the same index n. We reduce the problem to the case $n = 0$ by considering the operator $F \oplus T_{z^n}$. Since the latter operator has index 0, there exists a finite rank perturbation Φ such that $F \oplus T_{z^n} \oplus \Phi$ is invertible. By an elementary argument on the spectrum of the operator, one finds that there exists a neighborhood of $F \oplus T_{z^n} \oplus \Phi$, small enough, in which all operators are invertible. Because these operators are invertible, they have index 0. Shifting these operators by Φ, of finite rank, does not change the index that remains 0. Hence a neighborhood of $F \oplus T_{z^n}$ has index 0. It follows that in a neighborhood of F all operators have index n. ∎

Theorem 20.15. *Two Fredholm operators F_1, F_2 have the same index iff they can be joined by a norm continuous path of Fredholm operators.*

Proof. Let F_1, F_2 be two Fredholm operators with the same index n. Consider first $n = 0$. Hence $F_1 \in \mathcal{F}_0$, which itself contains the identity I. Since the operator F_1 has zero index, it is the finite rank perturbation Φ of an invertible operator R, $F_1 = R + \Phi$. It is therefore easy to write a norm continuous path of operators from $F_1 = R+\Phi$ to I. Likewise, we can find a path from I to F_2. Combining the two paths yields a path from F_1 to F_2, as claimed. The case where $n \neq 0$ is recovered from the case $n = 0$ using the "stabilization" trick, standard in K-theory (see [Wegge-Olsen 1993, page 229]). Conversely, if F_1, F_2 can be joined by a path of Fredholm operators, since the index is locally constant, it is constant all along the path, and F_1, F_2 have the same index. ∎

Corollary 20.1. *The $\mathcal{F}_n$'s are the connected components of $\mathcal{F}$.*

Theorem 20.16. *If F_1, F_2 are Fredholm operators, then*

$$\textit{index}\ (F_1F_2) = \textit{index}\ (F_1) + \textit{index}\ (F_2)$$

In other words, the index is a homomorphism from the multiplicative algebra of Fredholm operators to the additive group of the integers.

Proof. We would like to take this opportunity to prove this result using a K-archetypical argument. Consider the homotopy:

$$\begin{pmatrix} F_1 & 0 \\ 0 & I \end{pmatrix}\begin{pmatrix} \cos\frac{\pi t}{2} & -\sin\frac{\pi t}{2} \\ \sin\frac{\pi t}{2} & \cos\frac{\pi t}{2} \end{pmatrix}\begin{pmatrix} F_2 & 0 \\ 0 & I \end{pmatrix}\begin{pmatrix} \cos\frac{\pi t}{2} & \sin\frac{\pi t}{2} \\ -\sin\frac{\pi t}{2} & \cos\frac{\pi t}{2} \end{pmatrix}$$

It is easily seen that this is a norm-continuous homotopy of Fredholm operators from $\begin{pmatrix} F_1 & 0 \\ 0 & F_2 \end{pmatrix}$ to $\begin{pmatrix} F_1F_2 & 0 \\ 0 & I \end{pmatrix}$. The index of the former operator ($t = 0$) is index (F_1) + index (F_2), while the index of the latter ($t = 1$) is index (F_1F_2). Since during the homotopy the index remains unchanged, the result is proved. ∎

20.3 Index of Fredholm Toeplitz Operators

Restricted to *Toeplitz* operators, the concept of Fredholmness takes an even more vivid significance.

Theorem 20.17. *Let* $(I + L\Delta) \in C^0(\mathbb{T}, \mathbb{M}(n_l, \mathbb{C}))$. *The Toeplitz operator* $I + T_{L\Delta}$ *is Fredholm iff the zero exclusion principle,* $det(I + L(\exp(j\theta))\Delta) \neq 0, \forall\theta$*, holds.*

Proof. This is a "classic" result of Toeplitz operators (see [Douglas 1972, Corollary 1.4]). Here is a quick proof: Consider the string of mappings:

$$C^0(\mathbb{T}, \mathbb{M}(n_l, \mathbb{C})) \xrightarrow{T} \mathbb{B} \xrightarrow{\pi} \mathbb{B}/\mathbb{K}$$

In the above, T is the Toeplitz mapping from the symbol to the operator and π is the natural projection. Define the composite $F = \pi \circ T$. Take $f, g \in C^0(\mathbb{T}, \mathbb{M})$ with *finite* Fourier series. It is easily seen that $T_{fg} - T_f T_g$ is a finite-dimensional hence compact operator. Going to the Calkin algebra yields $F_{fg} = F_f F_g$. Since finite Fourier series are dense in $C^0(\mathbb{T}, \mathbb{M})$ and since F is continuous, it follows that $F_{fg} = F_f F_g$ extends to the whole $C^0(\mathbb{T}, \mathbb{M})$. In other words, F is a multiplicative homomorphism. Hence T_f is Fredholm iff f is invertible in $C^0(\mathbb{T}, \mathbb{M})$ and the theorem is proved. ■

Testing Fredholmness of a Toeplitz operator is, conceptually, easy. The difficulty, however, is to jump from Fredholmness to invertibility. Here the concept of Fredholm index becomes instrumental. Clearly, invertibility should be related to a vanishing Fredholm index. However, the Fredholm index is not easy to compute. It turns out that for a Fredholm *Toeplitz* operator, there exists a more specialized concept of index, easier to compute,

Definition 20.18. *The* **winding number** *(also called* **topological index***) of the Fredholm Toeplitz operator* $I + T_{L\Delta}$ *with continuous symbol is the homotopy class of the map* $\exp(j\theta) \mapsto \det(I + L(\exp(j\theta))\Delta)$*; that is, more concretely,*

$$w(I + T_{L\Delta}) = \frac{1}{2\pi j} \oint_{\mathbb{T}} d \log \det(I + L(z)\Delta)$$

The two concepts of index for *Toeplitz* operators can easily be linked:

Theorem 20.19. *Let the Toeplitz operator* $I + T_{L\Delta}$ *be Fredholm. Then*

$$w(I + T_{L\Delta}) = -index\ (I + T_{L\Delta})$$

Proof. We provide a proof of this well-known result (see [Douglas 1972, Theorem 2]), because it is precisely this index problem that took algebraic topological methods in Toeplitz operators off the ground. By the *purely analytical* argument of the previous section, $\text{index}(I+T) = \dim\ker(I+T) - \dim\text{coker}(I+T)$ is constant under a Fredholmness preserving deformation of the operator. Such a deformation corresponds to a homotopy class of $[\mathbb{T}, GL(n_l, \mathbb{C})] \cong \mathbb{Z}$ (Bott periodicity theorem) and this homotopy class is in fact the homotopy class of the determinant map $\mathbb{T} \to \mathbb{C} \setminus \{0 + j0\}$ of the symbol. Since the mapping from the symbol to the Toeplitz operator is continuous, the analytical index and winding number must be linked by a constant. Choosing T_{z^m} as testing function shows that this constant is -1. ■

In the SISO case, a result by Coburn asserts that $\ker(I + T)$ and $\text{coker}(I + T)$ cannot be *simultaneously* nontrivial; therefore, we have the following corollary:

Corollary 20.2. (Coburn) *Let $I + T_{L\Delta}$ be a Fredholm Toeplitz operator with SISO symbol. Then $I + T_{L\Delta}$ is invertible iff its winding number or its index vanishes.*

Proof. See [Douglas 1972]. ■

In the multivariable case, the above is "almost" true.

Theorem 20.20. *The set of invertible Toeplitz operators is dense in the set of Fredholm Toeplitz operators with zero index (or winding number).*

Proof. See [Douglas 1972, page 5]. ■

In this specific control application, when the system is open-loop stable, the return difference operator is causal (lower triangular), and the Coburn theorem extends to the multivariable case.

Theorem 20.21. *Let $I + T_{L\Delta}$ be a causal Fredholm Toeplitz operator with matrix-valued symbol; that is, $I + L\Delta \in \mathcal{H}^\infty(\mathbb{T}, \mathbb{M}(n_l, \mathbb{C}))$. Then $I + T_{L\Delta}$ is invertible (equivalently, the closed-loop system is $\mathcal{H}^2$ stable) iff its winding number or Fredholm index vanishes.*

Proof. Since the symbol is in $\mathcal{H}^\infty$, the winding number is clearly the number of zeros of $I + L(z)\Delta$ in $\mathbb{D}$. It follows that $(I + L\Delta)^{-1} \in \mathcal{H}^\infty$ and

$$T_{I+L\Delta}^{-1} = T_{(I+L\Delta)^{-1}}$$

so that the return difference operator is invertible and the system is closed-loop stable. ■

A related result was formulated independently by [Sefton and Ober 1994].

20.4 Index of Fredholm Family

Having defined the *index* of a Fredholm (Toeplitz) operator and stressed its relevance to robust stability, the next question is whether we can define such a thing as the index of a *family* of Fredholm operators. Here, we closely follow the early development due to [Atiyah 1969]. We stress that this section is meant to be *intuitive*, to proceed from the conceptually motivated idea of defining the index of a whole *family*, and to lead us *quite naturally* into K-theory.

For $q \in D$, define

$$T(q) = I + T_{L\Delta(q)}$$

For every $q \in D$, the index

$$\text{index } T(q) = \dim\ker T(q) - \dim\operatorname{coker} T(q)$$

is well-defined, but there are some questions as to how this index varies with q. This leads to the issue of continuity of the composite map

$$D \to \mathcal{F} \overset{\text{index}}{\longrightarrow} \mathbb{Z}$$

Since both maps are continuous, so is the composite, $q \mapsto \text{index}(T(q))$. However, this continuity does not preclude the dimensions of the kernel and the cokernel to jump around some q_0, but in that case the jumps should cancel each other so as to preserve continuity of the index. It turns out that this phenomenon can be avoided by manipulating the family by a homotopy (see Section 20.2) so as to secure a kernel and a cokernel of constant dimension across the family.

20.4.1 Constant Cokernel Dimension

Here we provide a *concrete* proof based on matrix representation of operators, due to Atiyah [1969], of the continuity of the index and the fact that the Fredholm family can be manipulated in such as way as to secure a kernel and a cokernel of constant dimension. The deeper aim of this alternate derivation is to develop an intuitive definition of K-theory as the theory of vector spaces continuously depending on a parameter. In fact, the definition of the index of a family is closely related to the Doležal problem of finding a continuous basis for the kernel of a matrix, a problem central in time-varying linear system theory.

Since the dimension of *both* the kernel and the cokernel could potentially vary with q, we will, by a simple manipulation, fix the dimension of the cokernel so that the whole problem will be lumped into the variation of the dimension of the kernel.

Let $e^0, e^1, \ldots$ be a basis of Hilbert space. Define P_n to be the projection onto the span of $e^n, e^{n+1}, \ldots$. In matrix representation,

$$P_n = \begin{pmatrix} 0 & 0 & \dots & 0 & 0 & 0 & \dots \\ 0 & 0 & \dots & 0 & 0 & 0 & \dots \\ \vdots & \vdots & \ddots & \vdots & \vdots & \vdots & \\ 0 & 0 & \dots & 0 & 0 & 0 & \dots \\ 0 & 0 & \dots & 0 & 1 & 0 & \dots \\ 0 & 0 & \dots & 0 & 0 & 1 & \dots \\ \vdots & \vdots & & \vdots & \vdots & \vdots & \ddots \end{pmatrix}$$

P_n is clearly a Fredholm operator with index 0. Define the operator

$$F_n(q) = P_n T(q)$$

Observe that F_n is not Toeplitz, but it is Fredholm as the product of two Fredholm operators. This can be further strengthened to the following:

Lemma 20.22. *For any Fredholm operator F, index $(P_n F) =$ index (F).*

Proof. Partition $F = \begin{pmatrix} F_1 \\ F_2 \end{pmatrix}$ conformably with P_n. Define $S = \ker(F_2) \cap$ row (F_1). It is easily seen that

$$\ker(P_n F) = \ker \begin{pmatrix} 0 \\ F_2 \end{pmatrix} = \ker(F) \oplus S$$

and

$$\text{coker } (P_n F) = \mathcal{H} / \begin{pmatrix} 0 \\ F_2 \end{pmatrix} \mathcal{H} = (\mathcal{H}/F\mathcal{H}) \oplus \begin{pmatrix} F_1 S \\ 0 \end{pmatrix}$$

Since the spaces S and $\begin{pmatrix} F_1 S \\ 0 \end{pmatrix}$ have the same dimension, the result is proved. ■

The crux of the matter is to exploit compactness of D and choose n, large enough, so that

$$\mathcal{H} = \text{span } \{e^0, e^1, ..., e^{n-1}\} + T(q)\mathcal{H}, \ \forall q \in D$$

Premultiplying both sides by P_n and defining $\mathcal{H}_n = P_n \mathcal{H}$ yields

$$\mathcal{H}_n = F_n(q)\mathcal{H}, \ \forall q \in D$$

so that

$$\dim \text{coker } F_n(q) = n, \ \forall q \in D$$

Hence we get

$$\text{index } T(q) = \dim \ker F_n(q) - n$$

and the remaining problem is to assess the variation of the kernel of $F_n(q)$ as q varies. We first observe the following:

Lemma 20.23. $\dim \ker F_n(q) \leq \dim \ker F_n(q_0)$ *in some neighborhood of* q_0.

Our next major claim is that $\dim\ker F_n(q)$ is *locally constant*:

Theorem 20.24. (Atiyah–Doležal) *Given $q_0 \in D$, there exists a neighborhood O_{q_0} of q_0 such that $\dim\ker F_n(q) = \dim\ker F_n(q_0), \forall q \in O_{q_0}$, and there exists a continuous basis of $\ker F_n$ in the same neighborhood.*

Proof. Define the operator

$$\begin{array}{rcl} A(q) : \mathcal{H} & \to & \mathcal{H}_n \oplus \ker F_n(q_0) \\ x & \mapsto & \begin{pmatrix} F_n(q)x \\ P_{\ker F_n(q_0)}x \end{pmatrix} \end{array}$$

Clearly, $A(q_0)$ is an isomorphism. Hence $A(q)$ remains an isomorphism in some neighborhood, O_{q_0}, of q_0. Let $f_1, ..., f_d$ be a basis of $\ker F_n(q_0)$. Define

$$g_i(q) = (A(q))^{-1} f_i$$

and we claim that $\{g_i(q)\}$ is a basis of $\ker F_n(q)$. Indeed, for $x = (A(q))^{-1} f_i$, the definition of $A(q)$ yields

$$f_i = \begin{pmatrix} F_n(q)g_i(q) \\ P_{\ker F_n(q_0)}g_i \end{pmatrix}$$

Remember, $f_i \in \ker F_n(q_0)$ so that $f_i \in P_{\ker F_n(q_0)}g_i$, and therefore

$$F_n(q)g_i(q) = 0$$

It follows that we have d linearly independent vectors $g_1, ..., g_d$ in $\ker F_n(q)$. By the lemma, the dimension of that kernel cannot exceed the dimension of $\ker F_n(q_0)$. Therefore, $\{g_i(q) : i = 1, ..., d\}$ is a basis of $\ker F_n(q)$ for $q \in U$. ■

20.4.2 Vector Bundle Formulation

From now on, we argue on the family $F_n(q)$ that has an easier structure than the family $T(q)$. The family $F_n(q)$ keeps as much of the structure of the original family $T(q)$ as homotopy permits, because indeed, as the reader can verify for himself, the family of operators

$$((1-t)I + tP_n)T(q)$$

provides a homotopy linking $T(q)$ for $t = 0$ to $F_n(q)$ for $t = 1$.

The cokernel of F_n not only has constant dimension but is *constant* as a subspace of $\mathcal{H}$. However, the kernel of F_n, while it has constant dimension, *does vary* as a subspace of $\mathcal{H}$. In view of this, the (difference of) dimension appears inadequate in the sense that it does not reveal that some subspaces could change with q, nor does it give any clues as to *how* the subspaces change. To make the index a more accurate picture as to how things change in a *family* of operators, we define the "index" of the *family* to be the formal difference of the collection of all *subspaces*, namely,

$$\text{“index”}(\{T(q) : q \in D\}) = \sqcup_{q \in D} \ker F_n(q) - \sqcup_{q \in D} \text{coker} F_n(q)$$

(For the time being, the formal difference can be viewed as an ordered pair.)

The disjoint union of all kernels is embedded in, and topologized as, a subspace of $D \times \mathcal{H}$,

$$\sqcup_{q \in D} \ker F_n(q) \subseteq D \times \mathcal{H}$$

The disjoint union of all kernels can be made the total space E of a bundle over D with projection

$$\pi : E \to D$$

defined as follows: For $x \in E \subseteq D \times \mathcal{H}$, because of the definition of E, there exists a unique q such that $x \in \ker F_n(q)$; hence define $\pi(x) = q$. Now it is easily seen that π is a bundle projection with E as total space and D as base space. The *specific* feature of this bundle is that each fiber $\pi^{-1}(q) = \ker F_n(q)$ is a *linear* subspace of E; this is the concept of *vector bundle.* Hence the kernel of the family is defined as the whole bundle

$$(E = \sqcup_{q \in D} \ker F_n(q), \pi, D)$$

Now we conduct the same analysis for the cokernel, except that this last one turns out to be trivial. Since $F_n(q)\mathcal{H} = \mathcal{H}_n$, coker $F_n(q) = \mathcal{H}_n^{\perp}$. As before, we make the collection of the cokernels a subspace of $D \times \mathcal{H}$,

$$\sqcup_{q \in D} \text{ coker } F_n(q) = \sqcup_{q \in D} \mathcal{H}_n^{\perp} = D \times \mathcal{H}_n^{\perp} \subseteq D \times \mathcal{H}$$

The cokernel bundle is

$$(D \times \mathcal{H}_n^{\perp}, \pi_1, D)$$

where π_1 is the projection on the first factor. The cokernel bundle is clearly trivial.

Therefore, the *index of the Fredholm family* becomes

$$\begin{aligned}&\text{“index”}(\{T(q) : q \in D\}) \\ &\quad = \{(\sqcup_{q \in D} \ker F_n(q), \pi, D)\} - \{(\sqcup_{q \in D} \text{coker} F_n(q), \pi_1, D)\}\end{aligned}$$

In the above, the braces mean that we are not making distinction between two bundles that differ by a mere nonsingular transformation of the total spaces and the fibers. In the sequel, we shall leave this fact implicit and drop the braces to simplify the notation.

Coming back to the "formal difference" of bundles, first observe that we can define the addition $a + b$ of two bundles as the bundle that has as fiber over q the direct sum of the fiber over q of a and the fiber over q of b. The zero bundle over D is the bundle that has zero-dimensional spaces as fibers. This zero bundle is clearly a neutral element for $+$. The difference $a - b$ of bundles should be viewed as purely formal, in the sense that there is no single bundle assigned to $a - b$; in other words, $a - b$ should be viewed as a couple (a, b). There is, however, a minimum extra structure that one should expect from the formal difference, namely, $a - b + b = a$. We also have the addition of formal differences, $(a-b)+(c-d) = (a+c)-(b+d)$, the zero bundle that acts as neutral element, and the inverse element $-(a-b) = (b-a)$. The reader can easily verify that all of this makes the set of formal differences a *commutative group.*

Proceeding a bit informally, one could say that the K^0 group of D is precisely the group of formal differences of (homeomorphic classes of) vector bundles over D. At this stage the reader might perceive the need to rephrase the above a little more formally, which is done in the following section.

20.5 K-Group

In this section, we reformulate in a more formal fashion the preceding section. The reader who is convinced by the intuitive argument of the preceding section can certainly skip this one and proceed to the next section.

20.5.1 Complex Bundle over Uncertainty Space

Roughly speaking, the group $K^0(D)$ is an algebraic picture of the set of bundles $\pi : E \to D$ that can be constructed *over* the space D with the restriction that the fibers be complex vector spaces.

Definition 20.25. *A continuous map $\pi : E \to D$ is said to be a* **complex vector bundle** *or, more specifically, a* **complex k-bundle** *over the* **base space** D *with* **total space** E *iff*

- *for any $q \in D$, the fiber over q, $\pi^{-1}(q)$, is homeomorphic to $\mathbb{C}^k$;*
- *for some open covering of D, $\cup_i O_i$, and any open set of the covering, there exists a homeomorphism*

$$h_i : \pi^{-1}(O_i) \to O_i \times \mathbb{C}^k$$

- *the composite*

$$\pi \circ h_i^{-1} : O_i \times \mathbb{C}^k \to O_i$$

  is the projection on the first factor
- *furthermore,*

$$h_i|\pi^{-1}(q) : \pi^{-1}(q) \to \{q\} \times \mathbb{C}^k$$

  is a vector space isomorphism.

If $k = 1$, the vector bundle is said to be a **line bundle***.*

We invite the reader to compare this definition with the one provided in Chapter 13.

20.5.2 Equivalent Bundles

In many cases, there are too many bundles that can be defined over D and that do not differ in a significant fashion. To reduce the set of bundles down to a more tractable size, we define an equivalence relation among them that will subsequently be used to quotient out the set.

Definition 20.26. *Two complex k-bundles (E_1, π_1, D) and (E_2, π_2, D) with the same base space are* **equivalent** *or* **isomorphic** *iff there exists a homeomorphism $h : E_1 \to E_2$ such that*

- *the homeomorphism respects the fiber—that is, $\pi_2 h = \pi_1$—or still in other words, the following diagram commutes:*

$$\begin{array}{ccc} E_1 & \stackrel{h}{\to} & E_2 \\ \pi_1 \downarrow & & \downarrow \pi_2 \\ D & = & D \end{array}$$

- *the homeomorphism h maps fiber onto fiber in such a way that $h : \pi_1^{-1}(q) \to \pi_2^{-1}(q)$ is a linear transformation in $GL(k, \mathbb{C})$.*

We define $\mathrm{Vect}(D)$ to be the set of equivalent classes of vector bundles over the uncertainty space D.

20.5.3 trivial bundle

Definition 20.27. *A* **trivial k-bundle** *over D is a bundle (E, π, D) that is equivalent to $(D \times \mathbb{C}^k, \pi_1, D)$.*

The trivial k-bundle is somewhat loosely written as (k). A very trivial bundle is obtained by setting $k = 0$ and by defining $\pi : D \times \{0\} \to D$ to be the projection on the first factor. This is called the **zero bundle** and is loosely written as (0).

20.5.4 Whitney Sum

The next step is to define a group operation—the so-called Whitney sum $\oplus$—on $\mathrm{Vect}(D)$. Intuitively, given (equivalent classes of) bundles (E_1, π_1, D) and (E_2, π_2, D), the Whitney sum $(E_1, \pi_1, D) \oplus (E_2, \pi_2, D)$ is the bundle over D that has fiber equal to the direct sum $\pi_1^{-1}(q) \oplus \pi_2^{-1}(q)$ of the fibers $\pi^{-1}(q)$ and $\pi_2^{-1}(q)$.

More formally, given given a k_1-bundle (E_1, π_1, D) and a k_2-bundle (E_2, π_2, D), define the "fiber product" (see [Karoubi 1978, page 3]) of the total spaces

$$E_1 \times_D E_2 = \{(e_1, e_2) \in E_1 \times E_2 : \pi_1(e_1) = \pi_2(e_2)\}$$

and the "product" projection

$$\begin{array}{rcl} \pi_1 \times_D \pi_2 : E_1 \times_D E_2 & \to & D \\ (e_1, e_2) & \mapsto & \pi_1(e_1) = \pi_2(e_2) \end{array}$$

It is easily seen that $\pi_1 \times_D \pi_2$ is continuous for the product topology; next,

$$(\pi_1 \times_D \pi_2)^{-1}(q) = \pi_1^{-1}(q) \oplus \pi_2^{-1}(q)$$

where $\oplus$ in the right-hand side denotes the usual direct sum. Furthermore, it is easily seen that all of the required conditions for $(E_1 \times_D E_2, \pi_1 \times_D \pi_2, D)$ to be a k-bundle are satisfied. Hence, we have the following definition:

Definition 20.28. *The* **Whitney sum** *of* (E_1, π_1, D) *and* (E_2, π_2, D) *is defined as*

$$(E_1, \pi_1, D) \oplus (E_2, \pi_2, D) = (E_1 \times_D E_2, \pi_1 \times_D \pi_2, D)$$

If E_1, F_1 and E_2, F_2 are pairs of equivalent bundles, it is easily seen that $E_1 \oplus E_2$, $F_1 \oplus F_2$ are equivalent bundles so that the Whitney sum $\oplus$ is well-defined on the quotient space. This makes $(\text{Vect}(D), \oplus)$ a *commutative semigroup.* Furthermore, it is easy to check that $E_1 \oplus 0 = 0 \oplus E_1 = E_1$, where 0 denotes the zero bundle. Therefore, $(\text{Vect}(D), \oplus)$ becomes a *commutative semigroup with zero element*—that is, a *commutative monoid.*

With the precise definition of the Whitney sum at hand, we can formulate the following, very useful theorem:

Theorem 20.29. *Given a vector bundle* (E_1, π_1, D) *over compact space* D*, there exists a bundle* (E_2, π_2, D) *such that* $(E_1, \pi_1, D) \oplus (E_2, \pi_2, D)$ *is trivial.*

Proof. See [Karoubi 1978, page 27] or [Atiyah 1967a, Corollary 1.4.14, page 27]. ■

20.5.5 Grothendieck Construction

To make the commutative semigroup $(\text{Vect}(D), \oplus)$ an Abelian group, we have to define inverse element—that is, a "substraction" operation. The addition operation on bundles had an intuitive geometrical interpretation in terms of direct sum of the fibers; however, it is hard to see what the "substraction" of bundles might be, so that we have to proceed algebraically.

Given an arbitrary commutative semigroup S, the *Grothendieck construction* provides a commutative group $\text{Groth}(S)$ together with a semigroup homomorphism $\gamma : S \to \text{Groth}(S)$. The latter condition means that the Grothendieck construction safeguards the original addition operation defined on the semigroup. The fact that the Grothendieck group is in a certain sense "natural" emerges from the following universal property of the construction: Given an arbitrary Abelian group A and a semigroup homomorphism $\phi : S \to A$, there is a "detour" via the Grothendieck group and a group homomorphism $\overline{\phi}$ such that the following diagram commutes:

$$\begin{array}{ccc} S & \xrightarrow{\phi} & A \\ {\scriptstyle\gamma}\searrow & & \nearrow{\scriptstyle\overline{\phi}} \\ & \text{Groth}(S) & \end{array}$$

Having spelled out the required properties of the group, we now define it.

Definition 20.30. *The* **Grothendieck group** *Groth(S) associated with the commutative semigroup S is the quotient group* $(S \times S)/\sim$, *where* $\sim$ *is the equivalence relation*

$$(a,b) \sim (c,d) \Leftrightarrow \exists u \in S \ni a+d+u = b+c+u$$

If $\{(a,b)\}$ *denotes the equivalent class of* (a,b), *the addition on the Grothendieck group is defined as* $\{(a,b)\} + \{(a',b')\} = \{(a+a', b+b')\}$, *and the inverse element is*

$$-\{(a,b)\} = \{(b,a)\}$$

We leave to the reader to verify that the above is indeed a substraction. The reader can also easily verify that the couple (a,b) can be viewed as the formal difference $a-b$ introduced in the previous section.

20.5.6 K-Group

The K^0 group of the uncertainty space D is defined as the Grothendieck group of (equivalence classes of) vector bundles over D endowed with the Whitney sum operation:

$$K^0(D) = \text{Groth}(\text{Vect}(D), \oplus)$$

20.5.7 K-Group Homomorphism

Clearly, K^0 is a *contravariant* functor, as shown by the diagram

$$\begin{array}{ccc} D_1 & \overset{g,h}{\longrightarrow} & D_2 \\ \downarrow & & \downarrow \\ K^0(D_1) & \overset{g^*,h^*}{\longleftarrow} & K^0(D_2) \end{array}$$

It follows, from the homotopy theory of bundles, that K^0 is a homotopy invariant. More specifically, if $g \simeq h$, then $g^* = h^*$. In particular, if g is a homotopy equivalence, $D_1 \simeq D_2$, it follows that $K^0(D_1) = K^0(D_2)$.

20.6 Index of Uncertain Return Difference Operator

The above manipulation has revealed a mapping

$$\text{Index} \ : \{D \to \mathcal{F}\} \to K^0(D)$$

that assigns to each family of Fredholm operators indexed by D an element of $K^0(D)$. We write this index map for a *family* of operators *Index*, with a capital *I*, to distinguish it from the index of a single Fredholm operator. Next, we will proceed to show that the Index depends only on the homotopy type of the map $D \to \mathcal{F}$ and that the induced map $[D, \mathcal{F}] \to K^0(D)$ is actually an *isomorphism*. The latter is the "hard" part.

The control interpretation can already be perceived at this point: Assume we are *given* a plant transfer matrix $G(s)$ and an uncertainty set D. Each compensator $K(s)$ satisfying the zero exclusion principle induces a mapping $D \to \mathcal{F}, \Delta \mapsto I + T_{KG\Delta}$. By the above, any such mapping is mapped to an element of $K^0(D)$. Therefore, to each $K(s)$, there corresponds an element of $K^0(D)$. This element of $K^0(D)$ is related to the index of the family $I + T_{L\Delta}$, itself related to the encirclement of the Nyquist plot. There is only one encirclement condition that will make the system closed-loop stable, so that there is within $K^0(D)$ only one element that corresponds to a closed-loop stable system. Therefore, *the number of generators of $K^0(D)$ somehow gives an idea as to how constrained we are in the selection of a robustly stabilizing compensator.*

To derive the properties of the Index map, we have to extend the properties of the index of *one single* operator to a *family* of operators.

Theorem 20.31. *If $F_1(q)$ and $F_2(q)$ are homotopic as maps $D \to \mathcal{F}$, then $Index(F_1(q)) = Index(F_2(q))$.*

Proof. The homotopy between F_1 and F_2 is a map $D \times [0,1] \to \mathcal{F}$. Therefore, the entire homotopy is mapped into an element of $K^0(D \times [0,1])$. By the fundamental homotopy invariance of K-theory, it follows that $K^0(D \times [0,1]) = K^0(D)$. Therefore, during the entire homotopy from F_1 to F_2, the Index is the same element of $K^0(D)$. ■

Theorem 20.32.

$$Index(F_1(q)F_2(q)) = Index(F_1(q)) \oplus Index(F_2(q))$$

Proof. The proof relies on the K-archetypical trick of constructing a homotopy between $\begin{pmatrix} F_1(q) & 0 \\ 0 & F_2(q) \end{pmatrix}$ and $\begin{pmatrix} F_1(q)F_2(q) & 0 \\ 0 & I \end{pmatrix}$. It is easy to verify that

$$\text{Index}\left(\begin{pmatrix} F_1(q) & 0 \\ 0 & F_2(q) \end{pmatrix}\right) = \text{Index}(F_1(q)) \oplus \text{Index}(F_2(q))$$

and

$$\text{Index}\left(\begin{pmatrix} F_1(q)F_2(q) & 0 \\ 0 & I \end{pmatrix}\right) = \text{Index}(F_1(q)F_2(q))$$

Since Index is not changed during homotopy, the theorem is proved. ■

At this stage, we have reached the following result:

Corollary 20.3. *The induced index map*

$$Index : [D, \mathcal{F}] \to K^0(D)$$

is a homomorphism.

Finally, we need to show that Index is an *isomorphism*. The proof relies on exactness of the sequence

$$[D, \mathcal{F}^*] \to [D, \mathcal{F}] \overset{\text{Index}}{\to} K^0(D) \to 0$$

In the above, $\mathcal{F}^*$ denotes the set of *invertible* Fredholm operators. By a standard result of homological algebra (see Appendix A), Index is an isomorphism whenever the sequence is exact (i.e., at every point of the sequence the image of the incoming arrow coincides with the kernel of the departing arrow) and $[D, \mathcal{F}^*] = 1$ (i.e., all maps $D \to \mathcal{F}^*$ are homotopic to a single constant map).

Theorem 20.33. *The above sequence is exact.*

Proof. We first show exactness at the $[D, \mathcal{F}]$ node. Take a map $F : D \to \mathcal{F}$ with 0 Index in $K^0(D)$. If F is meant to be a representative of its homotopy class, then $F \in \ker(\text{Index})$. We must show that F is in the image of the first arrow, namely, that $F : D \to \mathcal{F}$ is homotopic to a map $D \to \mathcal{F}^*$. Since F has vanishing index, its Index in $K^0(D)$ has a representative of the form $(E) - (F)$, where both (E) and (F) are, say, k-bundles. Let (G) be a trivial bundle so that $(E) \oplus (G) \cong (F) \oplus (G)$ This isomorphism induces a Fredholm family

$$\Gamma(q) : \ker F(q) \to \operatorname{coker}(F(q))$$

Clearly, $F(q) + t\Gamma(q)$ defines a homotopy from $F(q)$ to $F(q) + \Gamma(q)$, which is trivially invertible for all q.

To prove exactness at the third node, it suffices to show that Index is surjective. Every element of $K^0(D)$ has a representation of the form $(k) - (E, \pi, D)$, where (k) is a trivial k-bundle over D while (E, π, D) is a possibly nontrivial ℓ-bundle over D. To show surjectivity, we have to construct a map $D \to \mathcal{F}$ such that its Index is $(k) - (E, \pi, D)$. We first construct a map S such that $\text{Index}(S) = -(E, \pi, D)$. Let (F) denote a complement bundle so that $(E) \oplus (F)$ is isomorphic to the trivial bundle $(D \times \mathbb{C}^n, \pi_1, D)$, where π_1 denotes the projection onto the first factor. Let $E_q = \pi^{-1}(q) \cong \mathbb{C}^\ell$ denote the fiber over $q \in D$ and let $\pi_q : \mathbb{C}^n \to \mathbb{C}^n$ denote the projection onto E_q. Consider the identity and shift operators, viewed as Toeplitz operators $\mathcal{H} \to \mathcal{H}$

$$T_{z^0} = \begin{pmatrix} 1 & 0 & 0 & \\ 0 & 1 & 0 & \ddots \\ 0 & 0 & 1 & \ddots \\ & \ddots & \ddots & \ddots \end{pmatrix}, \quad T_{z^1} = \begin{pmatrix} 0 & 0 & 0 & \\ 1 & 0 & 0 & \ddots \\ 0 & 1 & 0 & \ddots \\ & \ddots & \ddots & \ddots \end{pmatrix}$$

Define the map

$$S : D \ \to \ \mathcal{F}(\mathcal{H} \otimes \mathbb{C}^n) \cong \mathcal{F}(\mathcal{H})$$

$$q \mapsto S(q) = T_{z^1} \otimes \pi_q + T_{z^0} \otimes (1 - \pi_q)$$

It is easily seen that $\ker S(q) = 0, \forall q \in D$ and that $\sqcup_{q \in D}((\mathcal{H} \otimes \mathbb{C}^n)/S(q)(\mathcal{H} \otimes \mathbb{C}^n)) = \sqcup_{q \in D} E_q = E$. Therefore $\text{Index}(S) = -(E, \pi, D)$. Next, we have to construct a map $D \to \mathcal{H}$ that has Index (k). It is trivial to see that $\text{Index}(T_{z^{-k}}) = (k)$. Finally, using the homomorphism theorem, we obtain

$$\text{Index}(T_{z^{-k}} S) = (k) - (E, \pi, D)$$

And the theorem is proved. ∎

Theorem 20.34. (Kuiper) *The space of invertible Fredholm operators $\mathcal{F}^*$ is contractible; consequently, there is only one homotopy class of maps $D \to \mathcal{F}^*$, which is written $[D, \mathcal{F}^*] = 1$, the homotopy class of the constant map.*

Proof. See [Kuiper 1965]. ∎

Finally, we put all of the pieces together,

Theorem 20.35. (Atiyah 1967a, 1969, and Jänich 1965) *The Index map*

$$\mathit{Index} : [D, \mathcal{F}] \to K^0(D)$$

is an isomorphism between the (operator) multiplicative group of (homotopy classes of) Fredholm families and the Grothendieck group of vector bundles over the uncertainty space D.

At this stage, it is instructive to consider the K-groups of a few domain of uncertainties.

Theorem 20.36.

$$\begin{aligned} K^0(\{point\}) &= \mathbb{Z} \\ K^0(\prod_i [a_i, b_i]) &= \mathbb{Z} \end{aligned}$$

Proof. The first result is very standard in K-theory (see [Wegge-Olsen 1993, page 215]). The second result is easily obtained by deformation retraction of the cube into a point. ∎

The first result indicates that even when D is reduced to a point, there are countably infinitely many homotopy classes of maps from $D = \{\text{point}\}$ to the space of Fredholm operators. $[\{\text{point}\}, \mathcal{F}]$ is easily seen to be in a one-to-one correspondence with the connected components of $\mathcal{F}$, each connected component of $\mathcal{F}$ being uniquely characterized by the index. Clearly, we can construct a Toeplitz operator of arbitrary index; therefore, each connected component of $\mathcal{F}$ contains some Toeplitz representatives.

There is a surprising resemblance between the topologies of the space of all Fredholm operators and the space of all $m \times m$ *symmetric* transfer matrices of fixed McMillan degree d. It turns out that *both* spaces split into many connected components. As we already know, every connected component of the space of Fredholm operators is uniquely characterized by the Fredholm index; likewise, it can be shown that every connected component of the space of symmetric transfer matrices of fixed McMillan degree is uniquely characterized by the (multivariable) Cauchy index (see [Bitmead and Anderson 1977]).

Given a set D of uncertainty, to each zero excluding loop transmission $L(s)$ there corresponds an element of $K^0(D)$ defined via the "classification" mapping

$$\begin{aligned} \gamma : \{\text{zero excluding } L(s)\} &\rightarrow K^0(D) \\ L(s) &\mapsto \text{Index}(T_{(I+L\Delta)}) \end{aligned}$$

(Compare with Section 17.5.) We could therefore consider a partitioning of $\{\text{zero excluding } L(s)\}$ into equivalent classes, two loop transmissions being equivalent iff they induce the same homotopy class in $[D, \mathcal{F}]$.

If $D = \{\text{point}\}$, the classification mapping $\{L(s)\} \rightarrow [D, \mathcal{F}]$ is surjective. In other words, $K^0(D)$ gives an accurate picture of all the homotopy possibilities that can be reached with $I + T_{L\Delta}$.

The second result of the preceding theorem indicates that when D is a cube, the situation remains unchanged: The classification mapping $\{L(s)\} \rightarrow K^0(D)$ is still surjective.

In the case of more general D's, the classification mapping $\{L(s)\} \rightarrow [D, \mathcal{F}]$ need not be surjective, because $K^0(D)$ allows for all mappings to the space of *Fredholm* operators, while in the time-invariant setup we are actually restricting ourselves to *Toeplitz* operators. From this perspective, $K^0(D)$ appears "too big." However, this argument should be balanced against the fact that, in $[D, \mathcal{F}]$, homotopies are allowed to move operators across the whole space of Fredholm operators without being restricted to remain within the Toeplitz subspace. With the accrued freedom in homotopies in $\mathcal{F}$ the space $K^0(D)$ appears "too small." We shall see in the next section that we are not too far off.

We could refine the above argument by fixing the plant $G(s)$ and assigning to each zero excluding compensator $K(s)$ a homotopy class in $K^0(D)$. The space of compensators would be partitioned, with two compensators being in the same class iff they induce the same element of $K^0(D)$. Clearly, there is only one homotopy class in $K^0(D)$ that is stabilizing; in other words, there is only one equivalence class of compensators, all of them being *robustly* stabilizing.

20.7 Index of Open-Loop Unstable Return Difference Operator

If the operator $I + T_{L\Delta}$ is unbounded, it has a domain, $\mathcal{D}(q)$, such that $(I + T_{L\Delta})|\mathcal{D}(q)$ is bounded. Assume $\mathcal{D}(q)$ depends continuously on q and that

$(I+T_{L\Delta})|\mathcal{D}(q)$ is a Fredholm family. Using the argument of Section 20.4, it is easily seen that the index

$$\dim\ker((I+T_{L\Delta})|\mathcal{D}(q)) - \dim\mathcal{H}/(I+T_{L\Delta})\mathcal{D}(q)$$

is locally constant. From there on, we define an Index for the whole family, taking value in $K^0(D)$,

$$\begin{aligned}&\text{Index}((I+T_{L\Delta})|\mathcal{D}(q)) = \\ &\qquad (\sqcup_{q\in D}\ker((I+T_{L\Delta})|\mathcal{D}(q)),\pi,D) - (\sqcup_{q\in D}\mathcal{H}/(I+T_{L\Delta})\mathcal{D}(q),\pi,D)\end{aligned}$$

In the above, the disjoint union is topologized as a subspace of $D\times\mathcal{H}$.

20.8 Reduced K-Groups

It would be interesting to know the *accrued* homotopy possibilities afforded by a nontrivial space D relative to the trivial case where $D=\{\text{point}\}$. Somehow, we have to remove from $K^0(D)$ its $K^0(\{\text{point}\})$ component. This is done as follows: Consider the following diagram depicting the injection map i and its induced K-group homomorphism

$$\begin{array}{ccc}\{\text{point}\} & \stackrel{i}{\rightarrow} & D \\ \downarrow & & \downarrow \\ K^0(\{\text{point}\}) & \stackrel{i^*}{\leftarrow} & K^0(D)\end{array}$$

Define the **reduced K-group**, $\tilde{K}^0(D)$, to be the kernel of i^*—that is,

$$K^0(D) = \tilde{K}^0(D)\oplus K^0(\{\text{point}\})$$

or

$$K^0(D) = \tilde{K}^0(D)\oplus\mathbb{Z}$$

or

$$\tilde{K}^0(D) = K^0(D)/\mathbb{Z}$$

More formally, $\tilde{K}^0$ is a functor specially devised for the category of *pointed spaces* (spaces with base points) and *base point preserving* maps, and furthermore the functor $\tilde{K}^0$ is a *base point preserving* homotopy invariant. Remember, K^0 is a functor defined on the category of spaces (not necessarily pointed) and maps (not necessarily base point preserving), and furthermore the functor K^0 is a *free* homotopy invariant. A manifestation of this is $K^0(D)=[D,\mathcal{F}]$, the set of *free* homotopy classes. On the other hand, remember that $\mathcal{F}$ breaks off into many connected components indexed by $\mathbb{Z}$ so that if we choose a base point q_* in D and a base point in $\mathcal{F}$, preferably in $\mathcal{F}^*$, then $\tilde{K}^0(D)=[D,\mathcal{F}]_*$, the set of *base point preserving* homotopy classes.

In the category of pointed spaces, a key concept is that of suspension:

Definition 20.37. (Atiyah) *Given two (compact) topological spaces X,Y*

with respective base points x_*, y_*, their **smash product** *is defined as*

$$X \wedge Y = (X \times Y)/(x_* \times Y \cup X \times y_*)$$

Furthermore, the **reduced suspension** *of the space X, written as SX, is defined as*

$$SX = S^1 \wedge X$$

where the unit circle S^1 is taken with an arbitrary base point s_*.

The smash product is easily seen to be associative, up to homeomorphism. Furthermore, there exists a natural homeomorphism:

$$\underbrace{S^1 \wedge S^1 \wedge \ldots \wedge S^1}_{n \text{ factors}} \cong S^n$$

Closely related to the concept of suspension, we define a **loop** over the pointed space Y to be a map $[0,1], \partial[0,1] \to Y, y_*$. Let ΩY denote the space of loops over Y. It turns out that Ω is the "dual" of the suspension in the sense that

$$[SX, Y]_* = [X, \Omega Y]_*$$

20.9 Unitary Approach to K-Theory

K-theory deals with vector bundles over a certain topological space and it is informative to compare these vector bundles with the so-called *universal bundles*. The *universal vector bundle* is

$$\begin{array}{ccc} \mathbb{C}^k & \to & E_{m,k} = \{(V,v) \in G_{m,k} \times \mathbb{C}^k : v \in V\} \\ & & \downarrow \pi_1 \\ & & G_{m,k} = U(m)/(U(k) \times U(m-k)) \end{array}$$

The base space is the Grassmann manifold $G_{m,k}$ of all subspaces of complex dimension k in $\mathbb{C}^m$. It is well-known, and easy to show, that the Grassmannian can be written in terms of the unitary groups as shown above. The intuitive idea behind this universal bundle is that above a point $V \in G_{m,k}$ lies the subspace $\mathbb{C}^k$ that V represents. This is accomplished by defining the total space as the collection of all $\mathbb{C}^k$ spaces parameterized by the Grassmannian,

$$E_{m,k} = \{(V,v) \in G_{m,k} \times \mathbb{C}^k : v \in V\}$$

and by defining the bundle projection to be the projection π_1 on the first factor. Clearly, the fiber over V is homeomorphic to $\mathbb{C}^k$ so that the above is a vector bundle with structure group $U(k)$.

Observe that the associated principal bundle is

$$\begin{array}{ccc} U(k) & \to & V_{m,k} = U(m)/U(m-k) \\ & & \downarrow \\ & & G_{m,k} \end{array} \tag{20.1}$$

In the above, $U(m)/U(m-k)$ is the so-called *Stiefel manifold* of orthonormal k-frames in $\mathbb{C}^m$.

The bundle $(E_{m,k}, \pi_1, G_{m,k})$ is **universal** in the sense that any complex k-bundle E with structure group $U(k)$ over D, with real dimension $\leq 2(m-k)$, arises as a pull-back $g^* E_{m,k}$ of the universal bundle:

$$\begin{array}{ccc} E = g^* E_{m,k} & \longrightarrow & E_{m,k} \\ \downarrow & & \downarrow \pi_1 \\ D & \xrightarrow{g} & G_{m,k} = U(m)/(U(k) \times U(m-k)) \end{array}$$

Since the pull-back bundle depends only on the homotopy class of g, it follows that

$$\mathrm{Vect}_k(D) = [D, G_{m,k}]$$

provided that

$$m > \frac{1}{2} \text{ real } \dim(D) + k$$

This technicality that m should be large enough is taken care of by defining $G_{\infty,k} = \lim_{m\to\infty} G_{m,k}$ so that

$$\mathrm{Vect}_k(D) = [D, G_{\infty,k}]$$

Finally defining $BU = \lim_{k\to\infty} G_{\infty,k}$ yields

$$\tilde{K}^0(D) = [D, BU]_*$$

BU is called **classifying space** for the unitary group. We leave it to the reader to compare the above with $K^0(D) = [D, \mathcal{F}]$.

It follows from the above that

$$\begin{aligned} \tilde{K}^0(S^n) &= [S^n, G_{\infty,k}] \\ &= \pi_n(G_{\infty,k}) \end{aligned}$$

provided that $k > \frac{n}{2}$. The homotopy groups of the Grassmannian can be worked out from the exact homotopy sequence of the fibration 20.1, and this yields

$$\tilde{K}^0(S^n) = \pi_{n-1}(U(\infty))$$

where $U(\infty)$ is the infinite unitary group. From the above theorem, we already reach a K-theoretic manifestation of the Bott periodicity theorem:

$$\tilde{K}^0(S^n) = \tilde{K}^0(S^{n+2})$$

and

$$\begin{aligned} K^0(S^{2n}) &= \mathbb{Z} \oplus \mathbb{Z} \\ K^0(S^{2n+1}) &= \mathbb{Z} \end{aligned}$$

20.9.1 Chern Classes and Character

Any k-bundle (E,π,D) with structure group $U(k)$ can be obtained by pull-back, $g^*(E_{m,k},\pi,G_{m,k})$, of the universal vector bundle provided that m is large enough. This establishes a one-to-one correspondence between k-bundles over D and homotopy classes of maps $g : D \to G_{m,k}$. An invariant of the homotopy class of g is $g^* : H^*(G_{m,k}) \to H^*(D)$, from which it follows that some invariants of the k-bundle (E,π,D) can be picked in $H^*(D,\mathbb{Z})$. These invariants are the *Chern classes* of the bundle,

$$c_i(E,\pi,D) \in H^{2i}(D,\mathbb{Z})$$

The Chern classes of the k-bundle (E,π,D) can be defined as the obstructions to cross sectioning the associated bundle:

$$\begin{array}{ccc} V_{k,l} & \to & E_{(l)} \\ & & \downarrow \pi_1 \\ & & D \end{array}$$

In the above,

$$E_{(l)} = \{(q,T) : q \in D,\ T \in \{l\text{-frames of } \pi^{-1}(q) \cong \mathbb{C}^k\}\}$$

The set of l-frames ($l \le k$) in $\mathbb{C}^k$ is the **Stiefel manifold** $V_{k,l}$; π_1 is the projection on the first factor, so that the fiber above q is $V_{k,l}$. The obstructions to cross sectioning this bundle (see Section 16.8) are in the cohomology groups,

$$H^{j+1}(D,\pi_j(V_{m,k}))$$

The homotopy groups of the Stiefel manifold $V_{k,l}$ can be computed from the exact homotopy sequence of a fibration and this yields,

$$\begin{aligned} \pi_j(V_{k,l}) &= 0,\ j \le 2(k-l) \\ \pi_{2(k-l)+1}(V_{k,l}) &= \mathbb{Z} \end{aligned}$$

The **Chern classes** of the k-bundle (E,π,D) are defined as the first obstructions to cross sectioning the bundles $(E_{(l)},\pi_1,D)$, for $l \le k$; for fixed l, the first obstruction occurs at the extension from the skeleton $D^{2(k-l)}$ to the skeleton $D^{2(k-l)+2}$,

$$c_{k-l+1}(E,\pi,D) \in H^{2(k-l)+2}(D,\underbrace{\pi_{2(k-l)+1}(V_{k,l})}_{\mathbb{Z}})$$

Furthermore, we *define*

$$c_0(E,\pi,D) = 1_{H^*}$$

The **total Chern class** of the bundle (E,π,D) is defined as

$$c(E,\pi,D) = \sum_i c_i(E,\pi,D)$$

It is easily seen that if $h^*(E,\pi,D)$ is a pull-back of (E,π,D), we have

$$h^*(c(E, \pi, D)) = c(h^*(E, \pi, D))$$

(In the left-hand side, h^* denotes the cohomology group homomorphism, while in the right-hand side, h^* denotes the bundle pull-back.) It is also easily seen that if (E_i, π_i, D) are line bundles, we have

$$c(\oplus_i (E_i, \pi_i, D)) = \oplus_i c(E_i, \pi_i, D)$$

It turns out that it is useful to do some changes of variables. Consider,

$$\sum_i c_i x^i = \Pi_i (1 - \gamma_i x)$$

where x denotes an indeterminate. The rationale for the above manipulation is to make the c_i's symmetric functions of some coefficients that happen to be the γ_i's. The **total Chern character** is defined as

$$ch(E, \pi, D) = \sum_i \exp(\gamma_i) \in H^*(D, \mathbb{Q})$$

(Observe that the power expansion of $\exp(\gamma_i)$ stops at a finite power so that the Chern character is in the *rational* cohomology ring.) Again, it is easily proved that for any two bundles (E),(F) over D, we have

$$\begin{aligned} ch(E \oplus F) &= ch(E) + ch(F) \\ ch(E \otimes F) &= ch(E)ch(F) \end{aligned}$$

It turns out that the Chern character is a powerful tool in K-theory, for it links K-theory and cohomology:

Theorem 20.38. (Atiyah and Hirzebruch) *The total Chern character induces a ring homomorphism*

$$\begin{aligned} ch : K^0(D) &\to H^*(D, \mathbb{Q}) \\ (E) - (F) &\mapsto ch(E) - ch(F) \end{aligned}$$

Proof. See [Atiyah and Hirzebruch 1961]. ■

Clearly, the Chern character establishes the link between classification of Nyquist maps into $\mathcal{F}$ and classification of Nyquist maps into $\mathbb{C} \setminus \{0 + j0\}$ as specified by Hopf's theorem.

20.10 Higher K-Groups and Bott Periodicity

The fundamental motivation for the higher K-groups is the classification of maps into GL. Recall that the previous chapter has only addressed the classification of maps from the *sphere* to GL. The **higher K-groups** are obtained by taking the K-groups of the *suspensions* of a pointed space.

Definition 20.39. (Higher K-Groups)

$$\tilde{K}^{-n}(X) = \tilde{K}^0(S^n X)$$

$$K^{-n}(X) \quad = \quad \tilde{K}^0(S^n(X^+))$$

In the above, X^+ is obtained by adding to X a point not previously in X, $X^+ = X \cup \{point\}, \{point\} \notin X$, and this new point becomes the base point of X^+.

It should be clear that

$$K^0(X^+) = K^0(X) \oplus \mathbb{Z}$$

Now, we reach a fundamental result—the K-theoretic Bott periodicity theorem:

Theorem 20.40. (K-Theoretic Bott Periodicity Theorem)

$$\tilde{K}^{-n}(X) = \tilde{K}^{-n-2}(X)$$

Proof. It is easily seen that $\tilde{K}(X) = [X, BU \times \mathbb{Z}]_*$, where the base point of $BU \times \mathbb{Z}$ is taken in $BU \times \{0\}$. A key point is the fact, proved by Bott, that $BU \times \mathbb{Z}$ is homotopically equivalent to the space of loops over the infinite unitary group, $BU \times \mathbb{Z} \simeq \Omega U$. The next key result, again due to Bott and very much in the spirit of Theorem 19.4, is that $\Omega^2 U \simeq U$. From there on, we get $\tilde{K}^{-n}(X) = [S^n X, BU \times \mathbb{Z}]_* = [X, \Omega^{n+1}U]_*$ and the periodicity is obvious. ■

The followings are useful corollaries of the Bott periodicity:

Corollary 20.4. (Bott Periodicity)

$$K^{-2}(X) = K^0(X)$$

$$\begin{aligned} K^{-n}(X) &= \tilde{K}^{-n}(X) \oplus \mathbb{Z}, \quad n \text{ even} \\ &= \tilde{K}^{-n}(X), \quad n \text{ odd} \end{aligned}$$

$$\begin{aligned} K^{-n}(\{point\}) &= \mathbb{Z}, \quad n \text{ even} \\ &= 0, \quad n \text{ odd} \end{aligned}$$

$$\begin{aligned} K^0(S^n) &= \mathbb{Z} \oplus \mathbb{Z}, \quad n \text{ even} \\ &= \mathbb{Z}, \quad n \text{ odd} \end{aligned}$$

Proof. See [Atiyah 1969, Section 3]. ■

To perform all of the K-calculation, the following is a useful recipe:

Theorem 20.41.

$$\tilde{K}^{-n}(X \times Y) = \tilde{K}^{-n}(X \wedge Y) \oplus \tilde{K}^{-n}(X) \oplus \tilde{K}^{-n}(Y)$$

Proof. See [Atiyah 1967a, Corollary 2.4.8, page 76] or [Atiyah and Hirzebruch 1961]. ■

The whole motivation for the higher K-groups is the following theorem:

Theorem 20.42. *If X is a space without base point, we have, for free homotopies,*

$$[X, GL] = K^{-1}(X) = \tilde{K}^0(SX^+)$$

where $X^+ = X \cup \{point\}$, $point \notin X$.

On the other hand, if X is a pointed space and if we choose the identity matrix to be the base point of GL, we have, for based point preserving homotopies,

$$[X, GL]_* = \tilde{K}^0(SX)$$

Proof. The free homotopy result is in [Karoubi 1978, page 75] while the based homotopy result is in [Karoubi 1978, page 78]. The based homotopy result is easily recovered from the unitary approach:

$$\begin{aligned}\tilde{K}^0(SX) &= [SX, BU \times \mathbb{Z}]_* \\ &= [SX, \Omega U]_* \\ &= [X, \Omega^2 U]_* \\ &= [X, U]_*\end{aligned}$$

Since $U \simeq GL$ the based homotopy result is proved. ■

20.11 Index for Fredholm Toeplitz Family

The Index map, as it was defined previously, applies to an *arbitrary* family of Fredholm operators. The problem is that the Fredholm family relevant to time-invariant control is quite restricted—it is a Toeplitz family. This is to say that the value group $K^0(D)$ for the Index map is "too big." Remember that in the proof of surjectivity of the Index map $[D, \mathcal{F}] \to K^0(D)$, we showed that an arbitrary element of $K^0(D)$ is the image of an operator $\mathcal{H} \otimes \mathbb{C}^n \to \mathcal{H} \otimes \mathbb{C}^n$ of the form $T_{-1} \otimes \pi_q + T_0 \otimes (1 - \pi_q)$, which is certainly Toeplitz on the space $\mathcal{H} \otimes \mathbb{C}^n$, but this Toeplitz property was destroyed under the homeomorphism $\mathcal{F}(\mathcal{H} \otimes \mathbb{C}^n) \to \mathcal{F}(\mathcal{H})$ that we used to get back to the original Hilbert space.

What we therefore need is a value group for the index of a Fredholm *Toeplitz* family. Clearly, a family of Fredholm Toeplitz operators can be viewed as a map $D \times \Omega \to GL(n_l, \mathbb{C})$. If we take the injective limit $n_l \to \infty$—that is, if we consider *infinitely many feedback paths* or an *infinitely large size* for the symbol of the return difference Toeplitz operator $T_{I+L\Delta}$—we recover the situation of Theorem 20.42 on the stable homotopy of maps to GL.

Going to infinitely large size of matrices because of the simplification it affords is the K-archetypical concept of **stabilization**. *Going to infinitely many feedback paths or distributed sensing/actuating is the control manifestation of the K-archetypical concept of stabilization.*

Besides ensuring that the Nyquist map is into the space of *Toeplitz* operators, we also have to enforce such typical real-world features as roll-off at infinite frequency. In other words, the homotopy classes of potential return difference maps are not free—they are constrained to map $\omega = \infty$ to the identity return difference matrix I. *This is the control manifestation of the base point problem.*

The roll-off at infinite frequency clearly suggests that we choose $j\infty$ to be the base point of Ω. Since at infinite frequency the loop gain vanishes, the return difference goes to the identity matrix I, so that we choose the identity matrix I to be the base point of GL.

Next, we introduce a base point for D. It turns out that, by correctly setting up the space of uncertainty D and its base point Δ_*, it is possible to replace the *Cartesian* product $D \times \Omega$ by a *smash* product. We have

$$D \wedge \Omega = (D \times \Omega)/(D \times \{j\infty\}) \cup (\Delta_* \times \Omega)$$

That we can collapse $D \times \{j\infty\}$ to a point is obvious under roll-off conditions. Indeed, at infinite frequency, the loop transmission vanishes and the Nyquist map does not depend on D. On the other hand, it is more tricky to justify the second collapse. Assume that the system is open-loop stable and let the uncertainty incorporate a multiplicative gain; in other words, $\Delta_* = 0$ is a point in D. We choose the zero gain $\Delta_* = 0$ to be the base point of D. Clearly, at Δ_*, the loop transmission vanishes and the Nyquist map no longer depends on Ω. This justifies the second collapse. We can therefore replace $D \times \Omega$ by $D \wedge \Omega$. The base point of $D \wedge \Omega$ is $(D \times \{j\infty\}) \cup (\Delta_* \times \Omega)$ properly "smashed" to a single point.

To be more formal, there exists a Nyquist map $\wedge f$ such that the following diagram commutes:

$$\begin{array}{ccc} D \times \Omega & \xrightarrow{f} & GL(n_l, \mathbb{C}) \\ \downarrow & \nearrow_{\wedge f} & \\ D \wedge \Omega & & \end{array}$$

where $D \times \Omega \to D \wedge \Omega$ is the collapsing map. Now we look at the base point preserving homotopy classes of return difference maps. We have

$$\begin{aligned} [D \wedge \Omega, GL]_* &= [D \wedge S^1, GL]_* \\ &= [SD, GL]_* \\ &= \tilde{K}^0(SSD) \\ &= \tilde{K}^{-2}(D) \\ &= \tilde{K}^0(D) \end{aligned}$$

We therefore have the following theorem:

Theorem 20.43. *Consider a return difference Nyquist map*

$$f : D \times \Omega \quad \to \quad GL(n_l, \mathbb{C})$$

$$(\Delta, \omega) \mapsto I + L(j\omega)\Delta$$

where the loop transmission is stable. Assume the following:

$$\begin{aligned} f(\Delta, \infty) &= I, \ \forall \Delta \in D \\ f(\Delta_*, \omega) &= I, \ \forall \omega \in \Omega \end{aligned}$$

Then the stable homotopy classes of all such maps are given by an isomorphism

$$\lim_{n_l \to \infty} [D \times \Omega, GL(n_l, \mathbb{C})]_* \to \tilde{K}^0(D)$$

As an illustration of the K-calculations, we compute the homotopy classes in the cases of toris of uncertainty. For $D = \mathbb{T} = S^1$ we get

$$\lim_{n_l \to \infty} [\mathbb{T} \times \Omega, GL(n_l, \mathbb{C})]_* = \tilde{K}^0(S^1) = 0$$

For $D = \mathbb{T}^2 = \mathbb{T} \times \mathbb{T}$ we first use Theorem 20.41 to get

$$\begin{aligned} \tilde{K}^0(\mathbb{T} \times \mathbb{T}) &= \tilde{K}^0(\mathbb{T} \wedge \mathbb{T}) \oplus \tilde{K}^0(\mathbb{T}) \oplus \tilde{K}^0(\mathbb{T}) \\ &= \tilde{K}^0(S^2) \oplus 0 \oplus 0 \\ &= \mathbb{Z} \end{aligned}$$

so that

$$\lim_{n_l \to \infty} [\mathbb{T}^2 \times \Omega, GL(n_l, \mathbb{C})]_* = \tilde{K}^0(\mathbb{T}^2) = \mathbb{Z}$$

Finally, consider $D = \mathbb{T}^3 = \mathbb{T}^2 \times \mathbb{T}$. Again, proceeding from Theorem 20.41 yields

$$\begin{aligned} \tilde{K}^0(\mathbb{T}^3) &= \tilde{K}^0(\mathbb{T}^2 \wedge \mathbb{T}) \oplus \tilde{K}^0(\mathbb{T}^2) \oplus \tilde{K}^0(\mathbb{T}) \\ &= \tilde{K}^0(S\mathbb{T}^2) \oplus \mathbb{Z} \oplus 0 \\ &= \tilde{K}^{-1}(\mathbb{T}^2) \oplus \mathbb{Z} \end{aligned}$$

Again, appealing one more time to Theorem 20.41, we get

$$\begin{aligned} \tilde{K}^{-1}(\mathbb{T}^2) &= \tilde{K}^{-1}(\mathbb{T} \wedge \mathbb{T}) \oplus \tilde{K}^{-1}(\mathbb{T}) \oplus \tilde{K}^{-1}(\mathbb{T}) \\ &= \tilde{K}^{-1}(S^2) \oplus \tilde{K}^{-1}(\mathbb{T}) \oplus \tilde{K}^{-1}(\mathbb{T}) \\ &= \tilde{K}^0(SS^2) \oplus \tilde{K}^0(SS^1) \oplus \tilde{K}^0(SS^1) \\ &= \tilde{K}^0(S^3) \oplus \tilde{K}^0(S^2) \oplus \tilde{K}^0(S^2) \\ &= 0 \oplus \mathbb{Z} \oplus \mathbb{Z} \end{aligned}$$

Putting everything together, we get

$$\lim_{n_l \to \infty} [\mathbb{T}^3 \times \Omega, GL(n_l, \mathbb{C})]_* = \tilde{K}^0(\mathbb{T}^3) = \mathbb{Z} \oplus \mathbb{Z} \oplus \mathbb{Z}$$

20.12 Atiyah–Hirzebruch Spectral Sequence

The classification theorem of the preceding section in terms of the K-group of D is reminiscent of the Hopf classification theorem of Nyquist curves.

We insist that this is *no accident.* Observe that, here, we get *directly* to the classification result in terms of the K-group of D. Previously, it took us along quite a detour from the Hopf theorem, via Künneth's theorem for cohomology to get rid of the frequency sweep, eventually ending in a classification result in terms of cohomology of D.

The above observation reveals some hitherto hidden links between the cohomology and the K-groups of D. The connection between cohomology and K-groups culminates in the *Atiyah–Hirzebruch spectral sequence.*

Theorem 20.44. (Atiyah and Hirzebruch) *Let D be a simplicial complex. There exists a second-quadrant spectral sequence, of the cohomology type, starting at*

$$E_1^{u,v} = C^u(D, K^v(\{point\}))$$

where $\delta_1 : E_1^{u,v} \to E_1^{u+1,v-1+1}$ is the usual coboundary operator; furthermore,

$$E_2^{u,v} = H^u(D, K^v(\{point\}))$$

and the spectral sequence converges to the K-groups of D,

$$E_r^{u,v} \Rightarrow K^{u+v}(D)$$

Proof. See [Atiyah and Hirzebruch 1961]. The argument hinges on a carefully chosen filtration of the K-groups. Let D^u be the uth skeleton of D. Consider the following diagram, where i denotes as usual the inclusion,

$$\begin{array}{ccc} D^{u-1} & \xrightarrow{i} & D \\ \downarrow & & \downarrow \\ K^*(D^{u-1}) & \xleftarrow{i^*} & K^*(D) \end{array}$$

The filtration of $K^*(D)$ is defined as

$$A_u K^*(D) = \ker(i^* : K^*(D) \to K^*(D^{u-1}))$$

From this filtration, the spectral sequence is easy to derive. ■

This connection between K-theory and cohomology is a manifestation of the fact that K-theory can be viewed as a *generalized cohomology theory*; we refer the reader to [Hilton 1971] for this insight.

20.13 KO-Theory of Real Perturbation

So far, we have been dealing with *complex* bundles over uncertainty space. The drawback is that when D is a space of *real* uncertainties, K-theory of *complex* bundles would map D into the space of *complex* Fredholm operators. Clearly, more relevant results would be obtained by mapping D into the space of *real* Fredholm operators.

If we want to emphasize the *real* aspect of the uncertainty, we have to consider *real* vector bundles over D. The kernel and the cokernel of the (real) Toeplitz operator $I + T_{L\Delta}$ are *real* spaces. The kernel and the cokernel of the *family* $\Delta \mapsto I + T_{L\Delta}$ are *real* bundles over D—that is,

$$(\sqcup_{q\in D} \ker(I + T_{L\Delta}), \pi_1, D), \ (\sqcup_{q\in D} \operatorname{coker}(I + T_{L\Delta}), \pi_1, D)$$

Each total space E is topologized as a subspace of $D \times \mathcal{H}$, where $\mathcal{H}$ is the Hilbert space over the real field. By *real bundle*, we mean that each fiber E_q is a *real* vector space. From there on, we follow exactly the same path of approach as in the complex case. We consider all real bundles over D. Two bundles are equivalent if there is a homeomorphism between their respective total spaces. The Whitney sum is defined as in the complex case. This leads to the Grothendieck group of the category of (equivalent classes of) real bundles over D, written $KO(D)$. With the mapping $\Delta \mapsto I + T_{L\Delta}$ we associate the KO-valued index,

$$(\sqcup_{q\in D} \ker(I + T_{L\Delta}), \pi, D) - (\sqcup_{q\in D} \operatorname{coker}(I + T_{L\Delta}), \pi, D)$$

and we have the following theorem:

Theorem 20.45. *There exists an isomorphism*

$$Index : [D, \mathcal{F}] \to KO(D)$$

between the set of homotopy classes of maps from the space of real uncertainty to the space of real Fredholm operators and the Grothendieck group of the category of real bundles over D.

Proof. We have already defined the arrow from $[D, \mathcal{F}]$ to $KO(D)$. It remains to show that this arrow is an isomorphism. We proceed *exactly* along the same lines as in the complex case. It is easily seen that the arguments can be adapted to the real case. The only point to watch is Kuiper's theorem in the real case. Actually, Kuiper's theorem is very general (see [Karoubi 1978, page 46]). In particular, the space of *real*, invertible Fredholm operators is contractible (see [Kuiper 1965, Introduction]). With this last result, the proof of the complex case is trivial to adapt to the real case. ■

In general, $KO(D)$ is a more complicated algebraic object than $K(D)$, because there is less space to maneuver in the space of real Fredholm operators than in the space of complex operators.

20.14 KR-Theory of Real Perturbation

In a certain sense, mapping D into the space of *complex* Fredholm operators can be justified by considering the real space D as "embedded" into a bigger, complex space. Among some examples of such embeddings, we mention the real versus complex μ-function,

$$[-1, +1]^n \subseteq \overline{\mathbb{D}}^n$$

Of course, it is known that these kinds of embeddings create overly conservative results. Another example is borrowed from [Bar-on and Jonckheere 1990]

$$SO(2) \subseteq SU(2, \mathbb{C})$$

The $SU(2, \mathbb{C})$ perturbation yields tractable computational solution to the multivariable phase margin problem, but this only provides conservative estimate of the $SO(2)$ margin problem that occurs, typically, in case of misalignment between body and inertial axes.

A slightly different problem occurs while mapping D into $GL(n_l, \mathbb{C})$. Even when D is a space of *real* perturbation, the GL-valued symbol is *still complex* because of the frequency. What we have to secure is the "real coefficients" property of the symbol. In other words, the mapping $j\Omega \to GL(n_l, \mathbb{C})$ must *commute with complex conjugation.* This is precisely where Atiyah's KR-theory comes in.

In a few words, KR-theory is the K-theory of real topological spaces embedded in bigger, complex spaces. To formalize the fact that such a topological space as D comprises a "real" subspace, the space D is endowed with an **involution**—that is, a homeomorphism $\gamma : D \to D$ such that $\gamma^2 = 1$. The real part of D, D_R, is defined as the set of fixed points of the involution.

These concepts take a concrete interpretation in case the involution is complex conjugation. However, the concept of involution is not limited to this case. Indeed, consider a lossless, reciprocal, flexible system with colocated sensors/actuators; the compensator is a scalar matrix of constant positive gains, so that $L(s)$ is symmetric—that is, $L(s) = L^T(s)$ (see [Opdenacker and Jonckheere 1985]). Write the perturbation in additive form, $L\Delta = L + \Delta_a$. For the perturbation of the plant to respect the fundamental law of reciprocity, the additive perturbation Δ_a must remain symmetric. The involution $\Delta_a \mapsto \Delta_a^T$ in turn yields the involution $\Delta \mapsto I + L^{-1}(\Delta^T - I)L^T$ and the fixed points of this last involution are the physically meaningful perturbations.

The involution on $j\Omega$ is always complex conjugation, or, equivalently, the involution on Ω is $\omega \mapsto -\omega$. If D has involution γ, then $D \times \Omega$ has involution

$$(\Delta, \omega) \mapsto (\gamma(\Delta), -\omega)$$

We now follow the same line of approach as in the case of the complex theory. We need to define a vector bundle over D that is somehow compatible with the involution.

Definition 20.46. *An* **R-vector bundle** *over the topological space D with involution γ_D is a (complex) vector bundle (E, π, D), where the space E is equipped with an involution γ_E such that*

- *the bundle projection commutes with the involution, $\pi\gamma_E = \gamma_D\pi$;*

- *the following diagram commutes:*

$$\begin{array}{ccc} \mathbb{C} \times E_q & \rightarrow & E_q \\ \downarrow & & \downarrow \\ \mathbb{C} \times E_{\gamma(q)} & \rightarrow & E_{\gamma(q)} \end{array}$$

The horizontal arrows denote the external law of the complex vector spaces, while the vertical arrows denote the involution.

A few words of comments about this definition. To define the involution on E, take $e \in E$. There exists a unique q such that $e \in E_q$. From there, the commutativity yields $\pi\gamma(e) = \gamma\pi(e) = \gamma(q)$, and further $\gamma(e) \in E_{\gamma(q)}$. Therefore, the involution on E maps E_q into $E_{\gamma(q)}$. Besides this and commutativity of the above diagram, there are no other restrictions on the involution. In case the involution on D is complex conjugation, then the involution $E_q \rightarrow E_{q^*}$ is just complex conjugation.

The kernel and the cokernel of *one single* operator $I + T_{L\Delta}$ are complex vector spaces. If the involution on D is complex conjugation, the involution on $\ker(T(q))$ and $\mathrm{coker}(T(q))$ is also complex conjugation. Next, we stack together the $\ker(T(q)) : q \in D$ and $\mathrm{coker}(T(q)) : q \in D$ as disjoint unions in $D \times \mathcal{H}$. The kernel and the cokernel of the *family* of operators $I + T_{L\Delta}$ are

$$(\sqcup_{q \in D} \ker(T(q)), \pi, D), \ (\sqcup_{q \in D} \mathrm{coker}(T(q)), \pi, D)$$

They are easily seen to be R-vector bundles over D.

There are many R-vector bundles over D, so that they need to be quotiented out by the equivalence relation that two bundles $(E, \pi, D), (F, \pi, D)$ are equivalent whenever the total space E, F are homeomorphic. However, in this context of R-bundles, there is the additional requirement that the homeomorphisms commute with the involution.

Definition 20.47. *Two R-vector bundles* $(E, \pi, D), (F, \pi, D)$ *are* **equivalent** *whenever there exists a homeomorphism* $h : E \rightarrow F$ *such that* $h\gamma = \gamma h$.

The next step is to introduce the Whitney sum of two (equivalent classes of) R-vector bundles. This results in a commutative monoid structure. Finally, we construct the associated Grothendieck group. More formally (see [Karoubi 1978]),

Definition 20.48. *The real K-group,* $KR(D)$, *of* D *is defined as the Grothendieck group of the categories of R-bundles over* D.

Therefore, the R-index of the family $\Delta \mapsto I + T_{L\Delta}$ is an element of $KR(D)$,

$$(\sqcup_{q \in D} \ker T(q), \pi, D) - (\sqcup_{q \in D} \mathrm{coker}(T(q)), \pi, D) \in KR(D)$$

Next we formulate the following:

Theorem 20.49. *There exists an isomorphism* $[D, \mathcal{F}]_\gamma \to KR(D)$ *between the set of homotopy classes of maps* $D \to \mathcal{F}$ *that commute with the involution and* $KR(D)$.

Proof. See [Atiyah 1966a]. ■

Finally, we consider the *higher* KR-groups, themselves related to the concept of *suspension*. Define $B^{u,v}$ and $S^{u,v}$ to be the unit ball and unit sphere, respectively, of $\mathbb{R}^v \oplus j\mathbb{R}^u$. The reader is warned that in this notation $S^{u,v}$ denotes a sphere of dimension $u + v - 1$. The involution on $\mathbb{R}^v + j\mathbb{R}^u$, $B^{u,v}$, and $S^{u,v}$ is defined to be complex conjugation. The higher KR-groups are defined by

$$KR^{u,v}(P) = \tilde{KR}(P \times B^{u,v}/P \times S^{u,v})$$

In the above, $P \times B^{u,v}/P \times S^{u,v}$ is the topological space obtained after collapsing the subspace $P \times S^{u,v}$ of $P \times B^{u,v}$ to a point, this point becoming the base point of $P \times B^{u,v}/P \times S^{u,v}$. $\tilde{KR}(X)$ is the kernel of the homomorphism $KR(X) \to KR(\{\text{point}\})$, $\{\text{point}\} \in X$. The involution on $P \times B^{u,v}/P \times S^{u,v}$ is induced by the involution on $P \times B^{u,v}$, itself defined as

$$\gamma_{P \times B^{u,v}} = (\gamma_P, \text{complex conjugation})$$

It turns out that the $KR^{0,v}$'s are generalizations of the classical higher K-groups defined in terms of the classical suspensions themselves defined as smash multiplications by spheres. To be more specific, remember that $K^{-n}(P) = \tilde{K}(S^n P^+)$ where S^n denotes the classical suspension operation; that is, smash multiplication by the n-sphere, and $P^+ = P \cup \{\text{point}\}$, where point $\notin P$, this latter point becoming the base point of P^+. Let B^n, S^{n-1} denote the unit ball, unit sphere, respectively, of $\mathbb{R}^n$. The key point is to observe that $B^n/S^{n-1} \cong S^n$ (see [Karoubi 1978, II.2.38]), from which it is not hard to show that

$$S^n P^+ \cong P \times B^n/P \times S^{n-1}$$

Therefore,

$$K^{-n}(P) = \tilde{K}(P \times B^n/P \times S^{n-1})$$

From the above, the analogy between the classical higher K-groups and the higher KR-groups is obvious.

With the higher KR-groups, we can specify the homotopy classes of maps from D to the space of Fredholm Toeplitz operators that commute with the involution.

Theorem 20.50.

$$\lim_{n_l \to \infty} [D \times S^1, GL(n_l, \mathbb{C})]_\gamma = KR^{0,1}(D \times S^1)$$

Proof. See [Atiyah 1966a]. ■

20.15 Connection with Algebraic K-Theory

Roughly speaking, zeroth-order algebraic K-theory deals with the collection of all projective modules that can be constructed over some ring R. Clearly, this is an algebraic version of topological K-theory, which deals with the vector bundles that can be constructed over a topological space.

Let $(M, +, R, \cdot)$ be a projective R-module. Let N be another such module. It is easily seen that their direct sum, $M \oplus N$, is another projective R-module. Let $\{M\}$ be the set of R-modules isomorphic to M, and let $\{N\}$ be the isomorphism class of N. Let Mod_R denote the set of isomorphism classes of projective R-modules. We define over Mod_R the direct sum,

$$\{M\} \oplus \{N\} = \{M \oplus N\}$$

Clearly, $(\mathrm{Mod}_R, \oplus)$ is a commutative monoid. To make it a group, it suffices to use the Grothendieck construction. The zeroth-order algebraic K-group of the ring R, $K_0(R)$, is defined as

$$K_0(R) = \mathrm{Groth}(\mathrm{Mod}_R, \oplus)$$

Observe that K_0 is a *covariant* functor.

The connection with topological K-theory is established by the following theorem:

Theorem 20.51. (Swan–Serre) *Let D be a compact Hausdorff space. Let $C^0(D, \mathbb{C})$, $C^0(D, \mathbb{R})$ be the rings of continuous functions $D \to \mathbb{C}$, $D \to \mathbb{R}$, respectively, with pointwise addition and multiplication. Then*

$$\begin{aligned} K^0(D) &= K_0(C^0(D, \mathbb{C})) \\ KO(D) &= K_0(C^0(D, \mathbb{R})) \end{aligned}$$

Proof. We sketch a proof which, in addition, reveals that there are many ways to go about K-theory. Take an element of $K^0(D)$ represented by the k-bundle (E, π, D). Because its fiber $\mathbb{C}^k$ is contractible, this bundle has a cross section $\gamma : D \to E$. This cross section can be viewed as a map in $C^0(D, \mathbb{C}^k) = (C^0(D, \mathbb{C}))^k$. This map, in turn, can be viewed as an element of the projective module

$$((C^0(D, \mathbb{C}))^k, +, C^0(D, \mathbb{C}), \cdot)$$

Conversely, given a projective module A over $C^0(D, \mathbb{C})$, there exists a module B such that

$$A \oplus B = ((C^0(D, \mathbb{C}))^k, +, C^0(D, \mathbb{C}), \cdot)$$

(See Appendix C for details.) Define the natural projection

$$\pi_1 : (C^0(D, \mathbb{C}))^k \to A$$

It turns out that π_1 has a matrix representation in the algebra of matrices over $C^0(D, \mathbb{C})$, namely,

$$\pi_1 \in \mathbb{M}(k, C^0(D, \mathbb{C})) \tag{20.2}$$

This yields the vector bundle

$$E = \{(q, v(q)) : q \in D,\ v \in (C^0(D, \mathbb{C}))^k,\ \pi_1 v = v\}$$

The above in turn defines an element of $K^0(D)$. For more details, see [Rosenberg 1994, Theorem 1.6.3], [Karoubi 1978, Chapter I, Theorem 6.18], or [Gillet 1992]. ∎

Actually, Equation 20.2 is the beginning of *K-theory of C^*-algebras*. Given a C^*-algebra A (e.g., $C^0(D, \mathbb{C})$), the C^*-algebraic K-group, $K_0(A)$, is defined as the Grothendieck group of the semigroup of equivalent classes of projections in $\mathbb{M}(\infty, A)$ under "direct sum." A **projection** π is defined by $\pi = \pi^2 = \pi^*$. The relation is defined by unitary equivalence. Given two projections $\pi \in \mathbb{M}(k, A)$, $\pi' \in \mathbb{M}(l, A)$, embed them as follows:

$$\begin{pmatrix} \pi & 0 \\ 0 & 0 \end{pmatrix}, \begin{pmatrix} 0 & 0 \\ 0 & \pi' \end{pmatrix} \in \mathbb{M}(k + l, A)$$

The "sum" of the two projections is given by

$$\pi + \pi' = \begin{pmatrix} \pi & 0 \\ 0 & \pi' \end{pmatrix}$$

The reader can easily verify for himself that **stabilization**—that is, going to infinite matrices—is essential. For details on K-theory of C^*-algebras, see [Wegge-Olsen 1993].

BIBLIOGRAPHICAL AND HISTORICAL NOTES

K-theory, or the general *index problem* that includes the Nyquist winding problem, has roots tracing back to the traditional Riemann–Roch theorem: Consider, over $\mathbb{CP}^2$, the homogeneous polynomial $\phi(z_0, z_1, z_2)$ satisfying the condition that the (complex) rank of

$$\begin{pmatrix} \frac{\partial \phi}{\partial z_0} & \frac{\partial \phi}{\partial z_1} & \frac{\partial \phi}{\partial z_2} \end{pmatrix}$$

is everywhere equal to 1. In this case, the (complex) algebraic curve $\phi(z_0, z_1, z_2) = 0$ is in fact a *compact Riemann surface* S. An invariant of the underlying topological structure of this surface is its *genus* g—that is, the number of "handles." One way to derive invariants of the finer *complex-analytic structure* of the surface is to investigate what kind of complex-analytic functions can be defined on it. The Riemann–Roch theorem addresses the problem of the (complex) dimension of the space of meromorphic functions $S \to \mathbb{C}$ having poles of order $\leq n_i$ at $P_i \in S$. Call $d = \sum_i n_i$ the **degree** and redefine, more formally, the set of poles as $\mathcal{D} = \sum_i n_i P_i$ also called a **divisor**. One can only conjecture that Riemann, after correctly observing that the dimension $n(S, \mathcal{D})$ of the space of functions could change under small perturbation of the data, thought that an invariant *robust against data perturbation* was more relevant. If we substract from $n(S, \mathcal{D})$ the so-called **index of specialty**—that is, the dimension of the space of holomorphic differential 1-forms vanishing on $\mathcal{D}$—then we obtain an invariant independent of data perturbation, to be more specific,

$$n(S, \mathcal{D}) - i(S, \mathcal{D}) = 1 + d - g$$

The above is the famous *Riemann–Roch formula.* Clearly, the analogy with the index of Fredholm operators is striking.

Even though the Riemann–Roch formula was a remarkable achievement, considerable difficulties were encountered in attempts to extend it to higher-dimensional objects.

It was the concept of *sheaf cohomology* introduced by Leray that allowed Hirzebruch to generalize the Riemann–Roch theorem in several directions. The key point is the reformulation of $n(S, \mathcal{D}) - i(S, \mathcal{D})$ as a *sheaf cohomology Euler number*: Given the compact Riemann surface S together with the data $\mathcal{D}$, it can be shown that there exists a line bundle $L_{\mathcal{D}}$ such that

$$n(S, \mathcal{D}) = \dim H^0(S, L_{\mathcal{D}}), \quad i(S, \mathcal{D}) = \dim H^1(S, L_{\mathcal{D}})$$

where the $H^i(S, L_{\mathcal{D}})$'s denote the sheaf cohomology groups of the surface S relative to the line bundle $L_{\mathcal{D}}$. Clearly, the left-hand side of the Riemann–Roch formula is, actually, the sheaf cohomology Euler number

$$n(S, \mathcal{D}) - i(S, \mathcal{D}) = \dim H^0(S, L_{\mathcal{D}}) - \dim H^1(S, L_{\mathcal{D}}) = \chi(S, L_{\mathcal{D}})$$

More generally, given any line bundle L over an algebraic variety X of dimension n, Hirzebruch proved that

$$\chi(X, L) = \sum_{i=0}^{n} \frac{c_1(L)^{n-i} T_i(c_1(L), \ldots, c_i(L))}{(n-i)!}$$

The $c_i(L)$'s are the *Chern classes* of the line bundle L, and the T_i's are the so-called *Todd polynomials*, the first few of which are

$$\begin{aligned} T_1(c_1) &= \frac{c_1}{2} \\ T_2(c_1, c_2) &= \frac{c_1^2 + c_2}{12} \\ T_3(c_1, c_2, c_3) &= \frac{c_1 c_2}{24} \\ T_4(c_1, c_2, c_3, c_4) &= \frac{-c_4 + c_3 c_1 + 3c_2^2 + 4c_2 c_1^2 - c_1^4}{720} \end{aligned}$$

The above formula for χ is referred to as *Hirzebruch–Riemann–Roch formula.*

K-theory was introduced by Grothendieck in the problem of formulating how the invariants of two complex algebraic varieties X, Y are transformed under a holomorphic map $f : X \to Y$. This is formulated in the *Grothendieck–Riemann–Roch* formula in which appeared for the first time the groups $K(X), K(Y)$ of vector bundles over the algebraic varieties X, Y. Taking Y to be a point reduces the Grothendieck–Riemann–Roch formula to the Hirzebruch–Riemann–Roch formula.

It was Atiyah and Hirzebruch who realized that the concept of K-groups could be extended to any compact topological space using the Bott periodicity theorem as a substitute for the extra structure afforded by algebraic varieties.

Parallel to these efforts, operator theory investigators were working on the problem of the index of elliptic operators. In its simplest definition, an **elliptic operator** E acts on an appropriate space of maps $X \to \mathbb{C}^m$ defined on a compact manifold X and is of the form $e(x, -j\frac{\partial}{\partial x_1}, \ldots, -j\frac{\partial}{\partial x_n})$, where, $\forall x \in X$, $e(x, \omega_1, \ldots, \omega_n)$ is a matrix of homogeneous polynomials in the ω_i's such that $\det e(x, \omega) \neq 0$, $\forall x \in X$, $\forall \omega \neq 0$. The latter is referred to as *elliptic property*. It can be shown that elliptic operators are Fredholm so that $\dim\ker E - \dim \operatorname{coker} E$ is defined and it can further be shown that this index is invariant under perturbation of the operator. This latter property led to the problem of constructing a formula for the index in terms of the topological invariants of the manifold X and the symbol $e(x, \omega)$. This line of investigation culminated in the K-theoretic *Atiyah–Singer index theorem* for elliptic operators. Furthermore, one can also define a family of elliptic operators parameterized by D in which case the index involves $K(D)$.

To tie it up all together, observe that the traditional Riemann–Roch problem can be viewed as an index problem for the Cauchy–Riemann operator $\frac{\partial}{\partial x - jy}$.

Probably one of the most spectacular achievements of K-theory is a simple proof of the fact that only the S^1, S^3, S^7 spheres are parallelizable; that is, only for $n = 1, 3, 7$ has S^n n linearly independent vector fields in its tangent space. The latter is related to Adams' simplified proof of an outstanding Hopf invariant problem.

From there on, the subject of K-theory virtually exploded and led to such spinoffs as the algebraic K-theory, the algebraic K-theory of numbers, the K-homology of Brown, Douglas, and Fillmore, the K-theory of C^*-algebras, the KK-theory of Kasparov that combines K-homology and K-theory of C^*-algebras, and finally the E-theory which can be viewed as some sort of simplification of the KK-theory.

It is through the operator-theoretic concept of Fredholm index that K-theory intruded in control. Indeed, probably the first appearance of the Fredholm index in control was in the spectral theory of the linear-quadratic problem (see, e.g, [Jonckheere and Silverman 1981]), where the spectrum of a nonself-adjoint block-Toeplitz operator was shown to be crucially related to the connection between time-domain and frequency-domain conditions.

The pathological situation of the failure of the Coburn theorem in the multivariable case led to situations where the Fredholm index, easily computable from the frequency-domain Popov function, failed to rule out existence of a potentially troublesome discrete spectrum in case of cancellation of the dimensions of the kernel and the cokernel.

Closer to our applications, a Fredholm index formulation of closed-loop stability of control systems was formulated independently by [Sefton and Ober 1994]. Along the same operator-theoretic line of ideas, the geometric formulation of stabilization of open-loop unstable systems also appears quite naturally in the gap metric approach to robustness (see, e.g., [Georgiou and Smith 1990]).

The fundamental paper introducing the topological K-theory for arbitrary topological spaces is [Atiyah and Hirzebruch 1961]. An early comprehensive treaty of topological K-theory is [Atiyah 1967a]. A somewhat more formal, less intuitive treaty is [Karoubi 1978]. A comprehensive treaty of fiber bundles and vector bundles, including elements of K-theory, is [Husemoller 1994].

A short, excellent survey of both the topological and the algebraic K-theories, stressing their inter-relationships, is [Luis-Puebla 1992]. A more comprehensive survey of algebraic K-theory is [Rosenberg 1994].

The classic for K-theory of C^*-algebras is [Blackadar 1986]. Another excellent, highly pedagogical survey of K-theory of C^*-algebras is [Wegge-Olsen 1993].

Part IV

DIFFERENTIAL TOPOLOGY OF ROBUST STABILITY

21

SINGULARITY OVER COMPACT DIFFERENTIABLE UNCERTAINTY MANIFOLDS

The study of the topology of X by Morse theory always involves passage from local information (Morse data at a critical point $p \in X$) to global information about X.

M. Goresky and R. MacPherson
Stratified Morse Theory,
Springer-Verlag, New York, 1988,
page 17.

SUMMARY

In this chapter, we begin investigating the *differential* topology of robust stability—that is, we exploit differentiability of both the manifold of uncertainty and the Nyquist map, for the class of problems they are indeed differentiable. In the previous parts, except possibly for Chapter 19, the uncertainty was assumed to be nothing more than a topological space and the Nyquist map was just continuous. With this minimum structure, the "algebraic" aspects of the topology of the problem dominate. Now, we are adding a differentiable component, resulting in the "differential topology" of robust stability.

To make the differential topological exposition as transparent as possible, we assume in this chapter that $D \times \Omega$ is a *compact* differentiable manifold—that is, a manifold-without-boundary. This is typically the case of the phase margin problem where D is a torus or the three-sphere, depending on the context. By a twist of irony, what is "easy" as far as differential topology is concerned has been deemed "difficult" as far as robust stability test is concerned. Indeed, for a manifold-without-boundary P, there is no way either $f\partial P \supseteq \partial f P$ or $f^{-1}\partial N \subseteq \partial f^{-1} N$ could possibly be satisfied, since $\partial P = \partial f^{-1} N = \emptyset$. To put it simply, since the uncertainty manifold has no boundary, there is no conceivable robust stability test on the boundary. For these kinds of problems, we could perform a simplicial decomposition of both the manifold and the supertemplate, construct a simplicial approximation, and get an approximate robust stability check on the boundary of the simplexes. We could also decompose the manifold and the template into "cells" and construct a decomposition-preserving,

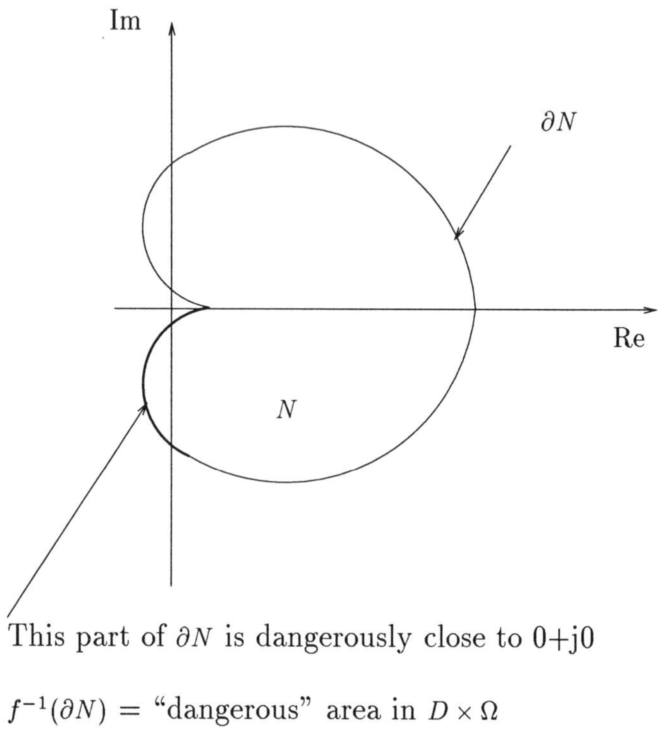

Fig. 21.1. "Dangerous" part of the supertemplate.

"cellular" map. Whatever the decomposition, "cellular" or simplicial, it is clear that the *amount* of dissection needed would be kept minimum if we choose a dissection that has $f^{-1}\partial N$ as "skeleton." To be more specific, the boundaries of the constituting pieces of P cover $f^{-1}(\partial N)$. *It is the precise purpose of this chapter to provide a differentiable topological interpretation of* $f^{-1}\partial N$.

There are plenty of other intuitive motivations for looking at $f^{-1}(\partial N)$. If the system is robustly stable, but if part of the boundary of the template goes dangerously close to zero, as shown in Figure 21.1, we might want to know what are the dangerous areas in $D \times \Omega$. Clearly, $f^{-1}(\partial N)$ covers these dangerous areas. Along the same line of arguments, it would be good to know what part of $D \times \Omega$ is mapped into ∂N. If this part can easily be identified, then mapping it yields ∂N from which the whole template can be constructed.

The *central* result is that $f^{-1}(\partial N)$ is included in, but not in general equal to, the set of *critical points* of the Nyquist mapping. These are the points where the rank of the Jacobian of the Nyquist map drops. The locus of critical points is a network of *critical curves*. Some of these curves could be interconnected, in a pattern specified by rigid rules of differential topology, around the so-called *degenerate* critical points. Mapping those curves yields a complicated network of *critical values curves* in the template. Some of the critical value curves constitute the boundary of the template, that consists of many differentiable curves pieced together in some nondifferentiable pattern around the degenerate critical values. In addition to the curves mapped to ∂N, *some other critical value curves are running inside the template.*

While this was not part of our initial objective, but as we shall see a couple of chapters down the road, it turns out that those curves running inside the template are responsible for the failure of the real μ-function to be continuous relative to problem data (see [Barmish et al. 1990]).

To clarify the degenerate, nondifferentiable phenomena at the interconnections between many differentiable curves, we have to resort to a somewhat clumsy approach. We have to look at the template from a projection angle θ and define the map $f_\theta : D \times \Omega \to \mathbb{R}$ that takes $p \in D \times \Omega$ to the orthogonal projection of $f(p)$ onto the line with argument θ. Since now we are dealing with an $\mathbb{R}$-valued, rather than $\mathbb{R}^2$-valued, map, we can exploit some advanced concepts from the *Morse theory* to clarify the critical phenomena. Roughly speaking, the Morse theory deals with C^∞ functions defined on a C^∞ manifold and relates the critical points of the function to the homology of the manifold.

Finally, with the "Morse data"—that is, the critical points together with their critical values—we construct a "cellular" decomposition of both the uncertainty manifold and the Nyquist template. The Nyquist map preserves this decomposition and, restricted to these building blocks, the Nyquist map commutes with the boundary. In a sense, this natural cell decomposition provides an answer to our query as to *what* is the minimum amount of decomposition required to construct a map that commutes with the boundary.

21.1 Compact Differentiable Uncertainty Manifold

Consider the phase margin problem with two uncertain phase angles, in which case $D = \mathbb{T}^2$, the 2-torus. Consider the intuitive picture of the 2-torus embedded in the ambient Euclidean space. The 2-torus inherits from $\mathbb{R}^3$ a *relative* topology—that is, the topology in which the open sets are the intersections of the usual open sets of $\mathbb{R}^3$ with the 2-torus surface. If at each point $p \in \mathbb{T}^2$ we choose a neighborhood of $\mathbb{T}^2$, small enough, so that its orthogonal projection on the tangent plane to $\mathbb{T}^2$ at p is a homeomorphism, we reach the conclusion that every point has a neighborhood homeomorphic to an open set of $\mathbb{R}^2$ or, equivalently, the whole $\mathbb{R}^2$ space. The latter is essentially the concept of a *manifold.*

Definition 21.1. *An n-dimensional* **manifold** *is a Hausdorff topological space with a countable basis so that every point has a neighborhood homeomorphic to an open subset of* $\mathbb{R}^n$*, or the whole* $\mathbb{R}^n$.

If ∂ denotes the boundary for the *neighborhood topology* of a manifold M, it follows that $\partial M = \overline{M} \setminus \mathrm{int}(M) = \emptyset$. If the manifold, say $\mathbb{D}$, is embedded in a Euclidean space, say $\mathbb{R}^2 \cong \mathbb{C}$, and if we choose the usual topology of $\mathbb{R}^2$ and the associated point set topological boundary, it follows that $\partial\mathbb{D} = \mathbb{T} \neq \emptyset$. It appears that we are back to the point already made in Section 3.2 that we have to be *explicit* as to what topology is invoked when we use the boundary.

Let p^1, p^2 be two points on the manifold, and let O_{p^1}, O_{p^2} be two open, intersecting neighborhoods. Let h_{p^1}, h_{p^2} be the homeomorphic mappings of O_{p^1}, O_{p^2}, respectively, onto $\mathbb{R}^n$. From these data, we can define the so-called **gluing map**

$$h_{p^2} \circ h_{p^1}^{-1} : h_{p^1}(O_{p^1} \cap O_{p^2}) \to h_{p^2}(O_{p^1} \cap O_{p^2})$$

Observe that the gluing map is homeomorphic. The Euclidean space $\mathbb{R}^n$ homeomorphic to O_{p^1} is conveniently charted with orthonormal coordinates $p_1, ..., p_n$ which, traced back to O_{p^1}, become the so-called **local Riemann coordinates**.

The gluing maps provide much of the additional structure a manifold can be endowed with.

Definition 21.2. *A manifold is said to be* **differentiable** *if its gluing maps are at least* C^1. *The manifold is said to be* **smooth** *if the gluing maps are* C^∞.

Observe that the Jacobian of a gluing map of a differentiable manifold has nonvanishing determinant.

Definition 21.3. *A differentiable manifold is said to be* **orientable** *if the determinants of the Jacobians of its gluing maps all have the same sign.*

To link the concept of a manifold with the simplicial techniques of the previous chapters, observe that from an intuitive point of view the 2-torus is *triangulable*—that is, there exists a simplicial complex K and a homeomorphism $|K| \to \mathbb{T}^2$. One such simplicial complex K is the "simplicial torus" of Figure 4.3. Actually, a much more general result holds:

Theorem 21.4. *A differentiable manifold, typically* $D \times \Omega$, *is* **triangulable**; *that is, there exists a (finite) simplicial complex* K *and a homeomorphism* $|K| \to D \times \Omega$.

Proof. See [Bott and Tu 1982]. ∎

It can be shown that, for a differentiable manifold triangulable by a pseudomanifold, the above differential definition of orientability is consistent with the combinatorial definition of Section 18.1.

Coming back to the 2-torus, it appears to be *compact*, since any covering of the 2-torus by open sets has a finite subcover. More formally, we have the following definition:

Definition 21.5. *A manifold is said to be* **compact** *if every open covering has a finite subcover.*

Such a compact differentiable manifold as $\mathbb{T}^2$ embedded in Euclidean space $\mathbb{R}^3$ appears to be *closed* from the point set topological point of view, as a subset of $\mathbb{R}^3$. On the other hand, if its triangulation K is the simplicial torus of Figure 4.3 and if z_2 is the correct assembly of all of its 2-simplexes, we have $\partial_2 z_2 = 0$, in other words, the torus is *homologically closed*. Therefore, for a compact differentiable manifold, the word "closed" has a double meaning; for an interesting discussion regarding this subtle issue, see [Stocker 1989, page 217] and [Munkres 1984, page 199].

Among compact differentiable manifolds, we will mention the unit circle $\mathbb{T}$, the n-torus $\mathbb{T}^n$, the n-sphere S^n, and so on. It follows that $\Omega = \mathbb{T}$ and, in the case of the phase margin problem, $D = \mathbb{T}^n$ are compact differentiable manifolds. A useful result is the following:

Theorem 21.6. *The topological product of two compact differentiable manifolds, typically $D \times \Omega$, is a compact differentiable manifold.*

21.2 Singularity Analysis of Nyquist Map

21.2.1 Variational Interpretation of Template Boundary

This chapter hinges on a variational interpretation of $f^{-1}\partial N$. First of all, since $P = D \times \Omega$ is compact and since the map f is continuous, it follows from a classical result of point set topology (see [Dugundji 1970]) that $N = f(P)$ is also compact. To grasp the variational problem in a simple context, we consider a boundary point $z \in \partial N$ around which N is locally convex and ∂N is locally differentiable. (We will show later that modulo an arbitrarily small deformation of the map, ∂N is differentiable except at *finitely* many points.) Draw the line normal to N at z, and let θ be the argument of the direction pointing outward the Horowitz template. Let $\pi_\theta : \mathbb{R}^2 \to \mathbb{R}$ be the orthogonal projection on the line with argument θ, namely, $\pi_\theta(z) = (\Re z)\cos\theta + (\Im z)\sin\theta$. Define (see Figure 21.2)

$$\begin{aligned} \pi_\theta \circ f =: f_\theta : D \times \Omega &\to \mathbb{R} \\ p &\mapsto f_\theta(p) = \Re(f(p))\cos\theta + \Im(f(p))\sin\theta \end{aligned}$$

It is not hard to see that any $p \in f^{-1}(z)$ is a local maximum of the performance index function f_θ. Since the problem is assumed to be differentiable, this implies that $\frac{\partial f_\theta}{\partial p} = 0$. To deal with the second-order conditions, we assume that f_θ is a function that has *nonsingular* Hessian at the points where

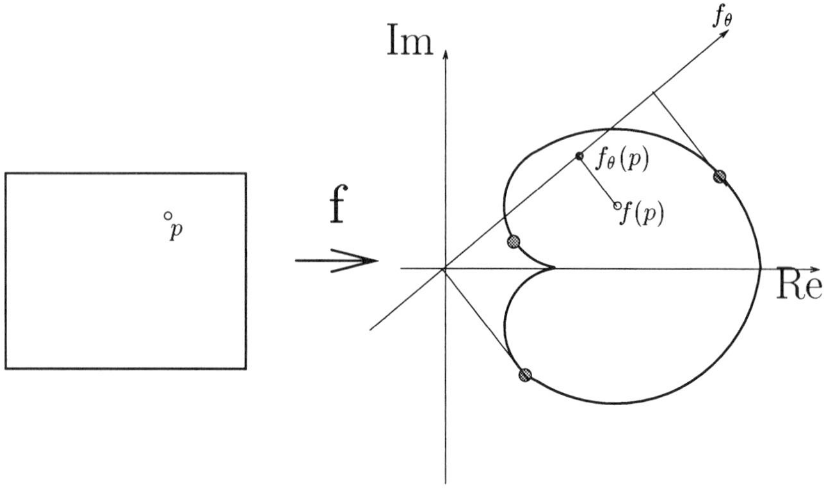

Fig. 21.2. Projection of Nyquist template.

its gradient vanishes; such functions are called **Morse functions**; to justify the assumption that f_θ is Morse, we invoke the fact that any smooth function can be arbitrarily closely approximated by a Morse function. Assuming f_θ is Morse, the second-order (necessary and sufficient) condition for local minimum reads $\frac{\partial^2 f_\theta}{\partial p_i \partial p_j} > 0$. By the same token, p is also a local *minimum* of $f_{\theta+\pi}$, and this implies that $\frac{\partial f_\theta}{\partial p} = 0$ together with $\frac{\partial^2 f_\theta}{\partial p_i \partial p_j} < 0$. We already perceive at this stage a very close connection between ∂N and the critical points of f_θ.

If the boundary is locally *concave* (and differentiable), the argument is essentially the same, except that nothing can be said about the sign of the Hessian. This situation typically happens when the template has a hole (open-loop unstable system) and when we take a point on the boundary of the hole.

Now we rephrase the results carefully.

Theorem 21.7. *Let $f : D \times \Omega \to N$ be a differentiable map defined over a differentiable manifold. Then, $\forall p^* \in f^{-1}(\partial N)$, there exists a θ such that $\frac{\partial f_\theta}{\partial p}(p^*) = 0$.*

Proof. Take $z^* \in \partial N$ and $p^* \in f^{-1}(z)$. Consider the bundle of all differentiable curves on $D \times \Omega$ passing through p^* and all tangent to each other at p^*. The collection of all such bundles is, by definition, the **tangent space** at p^* to $D \times \Omega$, written $T_{p^*} D \times \Omega$. For the conceptualization of this idea, see [Golubitsky and Guillemin 1973]. Likewise, consider the collection of all bundles of pairwise tangent differentiable curves at $z^* = f(p^*)$ in $\mathbb{R}^2$.

This is $T_{z^*}\mathbb{R}^2$, the tangent space at $z^* = f(p^*)$ to $\mathbb{R}^2$. A bundle of curves at p^* is mapped into a bundle of curves at $z^* = f(p^*)$ via the differential,

$$d_{p^*}f : T_{p^*}D \times \Omega \to T_{z^*}\mathbb{R}^2$$

defined, in matrix representation, by the **Jacobian matrix**,

$$J_f(p^*) = \begin{pmatrix} \frac{\partial \Re f}{\partial p_1} & \frac{\partial \Re f}{\partial p_2} & \cdots \\ \frac{\partial \Im f}{\partial p_1} & \frac{\partial \Im f}{\partial p_2} & \cdots \end{pmatrix}$$

If $d_{p^*}f$ had rank 2, by appropriate choice of a bundle through p^*, the image bundle through $z^* = f(p^*)$ could be rotated in all possible directions and z^* could not be a boundary point of N. If rank $d_{p^*}f = 1$, then, choose θ such that $(\cos\theta \sin\theta)d_{p^*}f = 0$ and the theorem is proved. If rank $d_{p^*}f = 0$, then any θ satisfies $(\cos\theta \sin\theta)d_{p^*}f = 0$, and the theorem is proved. ■

At this stage, there is enough motivation for introducing the concept of *critical points and values.*

Definition 21.8. *A* **critical point** *of a differentiable map $f : D \times \Omega \to \mathbb{R}^2$ defined over a differentiable manifold $D \times \Omega$ is a point p^* where the rank of the differential or Jacobian*

$$d_{p^*}f = \begin{pmatrix} \frac{\partial \Re f}{\partial p_1} & \frac{\partial \Re f}{\partial p_2} & \frac{\partial \Re f}{\partial p_3} & \cdots \\ \frac{\partial \Im f}{\partial p_1} & \frac{\partial \Im f}{\partial p_2} & \frac{\partial \Im f}{\partial p_3} & \cdots \end{pmatrix}$$

drops (see [Bochnak, Coste, and Roy 1987, Section 9.5] and [Goresky and MacPherson 1988, I.12.1]). A **critical value** *is the value of the map $f(p^*)$ at a critical point p^*.*

Definition 21.9. *A* **critical point** *of a differentiable function (typically, f_θ) defined on a differentiable manifold (typically, $D \times \Omega$) is a point where the differential vanishes, typically, $df_\theta = 0$; that is, in terms of Riemann coordinates, we have*

$$\frac{\partial f_\theta}{\partial p} = \begin{pmatrix} \frac{\partial f_\theta}{\partial p_1} & \cdots & \frac{\partial f_\theta}{\partial p_{n_p}} \end{pmatrix} = 0$$

A **critical value** *is the value taken by a differentiable function at a critical point.*

A *critical point* of the Nyquist map can be defined either as a critical point of $f : D \times \Omega \to \mathbb{R}^2$ or as a critical point of $f_\theta : D \times \Omega \to \mathbb{R}$ for some θ, because indeed,

$$\begin{pmatrix} \cos\theta & \sin\theta \end{pmatrix} d_{p^*}f = \frac{\partial f_\theta}{\partial p}$$

With these concepts, we can reformulate a key result:

Corollary 21.1. *If $f : D \times \Omega \to \mathbb{C}$ is a smooth function defined on a smooth manifold, then*

$$f^{-1}(\partial N) \subseteq \cup_\theta\{\textit{critical points of } f_\theta\} = \{\textit{critical points of } f\}$$

$$\partial N \subseteq \cup_\theta\{\textit{critical values of } f_\theta\} = \{\textit{critical values of } f\}$$

Now, we somewhat specialize the concept of critical points:

Definition 21.10. (Morse or Nondegenerate Critical Points) *A critical point of a smooth map $f_\theta : D \times \Omega \to \mathbb{R}$ defined over a compact smooth manifold is* **nondegenerate** *if the Hessian, typically,*

$$\frac{\partial^2 f_\theta}{\partial p_i \partial p_j}$$

is nonsingular. A **Morse function** *is a smooth real-valued function (typically f_θ) defined on a compact smooth manifold (typically $D \times \Omega$) that has only nondegenerate critical points. The* **Morse index** λ *of a nondegenerate critical point is the number of negative eigenvalues of the Hessian.*

Now, we can rephrase some variational properties of ∂N in the language of the Morse critical points.

Theorem 21.11. *Assume $D \times \Omega$ is a compact smooth manifold and that $f : D \times \Omega \to \mathbb{R}^2$ is smooth. Then*

$$f^{-1}((\partial N)_{c,d}) \setminus \{\textit{deg pts}\} \subseteq \cup_{\theta\in[0,2\pi]}\{\textit{Morse critical points of index } 0, n_p \textit{ of } f_\theta\} \setminus \{\textit{deg pts}\}$$

$$f^{-1}((\partial N)_d) \setminus \{\textit{deg pts}\} \subseteq \cup_\theta\{\textit{Morse critical points of } f_\theta\} \setminus \{\textit{deg pts}\}$$

where c, d denotes the convex, differentiable part, respectively, of ∂N and $\{deg\ pts\}$ denotes the set of degenerate points.

Proof. The first claim has already been proved. In the second claim, N need not be convex and nothing can be said about the Hessian and hence the Morse index. ■

Here, we are at a critical juncture. $f^{-1}(\partial N)$ is clearly embedded in the critical points of f_θ. However, it turns out that to get the "big picture" we need to consider *all* critical points, whether they be of Morse type or degenerate, and in the former case regardless of their indexes. Indeed, a Morse critical point in $f^{-1}(\partial N)$ does not in general have index 0 or n_p, unless the boundary is locally convex. The preimage of a point on the inner boundary of an annulus-shaped template is neither a maximum nor a minimum. It is a saddle kind of point. These elementary considerations force us to chase *all* Morse critical points of *all* indexes.

Besides these elementary considerations, there is a *more compelling* reason for chasing *all* critical points. Arguing intuitively, assume that, for a given θ, f_θ has finitely many critical points of the Morse type. (The Morse type is the generic situation.) As we perturb θ, we draw finitely many critical curves on the manifold $D \times \Omega$. It is easily seen that the Morse index is locally constant, so that as long as the Hessian does not vanish, a Morse index can be associated with each curve. The difficulties start when two curves "hit" each other, which typically happens at a degenerate critical point. Such a phenomenon, when mapped to the complex plane, yields such a pathology as a "kink" in the boundary of the template. Our experience has shown that it is extremely difficult to compute numerically the critical points around degeneracy; the best one can hope for is to reliably compute some pieces of nondegenerate critical curves and then attempt to assemble the "jigsaw puzzle." There are plenty of differential topology rules telling us how to "piece together" the nondegenerate critical curves around a degenerate critical point; however, these rules involve *all* curves, regardless of their Morse indexes. Therefore, we will have to consider *all* critical points, even those *not* in the preimage of ∂N. Mapping the latter curves to the complex plane yields a network of critical value curves inside the template, with some of these curves connecting to the boundary at a singular point. Probably the most important motivation for looking at those curves not part of the boundary is that they are crucially related to structural stability of the crossover $f^{-1}(0+j0)$, as we will see in Section 23.15. The latter is the key to the explanation of the lack of continuity of the real μ function to problem data.

21.2.2 Basic Facts of Morse Theory

It already transpires that we will often use, instead of the *map* f, the one-parameter family of *functions* $\{f_\theta : \theta \in [0, 2\pi)\}$. Intuitively, the family $\{f_\theta\}$ might be thought of as a "tomographic representation" of f. Clearly, from $\{f_\theta\}$ the map f can be reconstructed. The reason why we have elected to use the somewhat cumbersome tomographic representation is that for such a *function* as f_θ defined over a compact differentiable manifold and taking values in $\mathbb{R}$, we have at our disposal the powerful *Morse Theory*. Essentially the Morse theory deals with such a Morse function as f_θ defined over such a topological space as $D \times \Omega$ and relates the critical points/values of f_θ to the topology of $D \times \Omega$.

The connection between the critical points of a smooth function and the topology of its domain is probably most intuitively illustrated by the "cell" decomposition due to [Thom 1949]. To be more precise, Thom uses the Morse critical points of a function, defined on a smooth manifold and taking value in $\mathbb{R}$, to define the "cells" of a decomposition of the smooth manifold. The specific feature of Thom's decomposition is that each cell contains *exactly one* critical point.

Another illustration of the deep connection between the critical points of a smooth function and the topology of its domain is the celebrated "handle decomposition:"

Lemma 21.12. (Morse) *Let $f_\theta : D\times\Omega \to \mathbb{R}$ be a Morse function defined on a compact smooth manifold $D\times\Omega$. Let p^* be a Morse critical point with index λ. Then there exists a diffeomorphic change of coordinates $p(\overline{p})$ such that $\overline{p}(p^*) = 0$ and the Morse function takes the canonical form:*

$$f_\theta(p(\overline{p})) = \sum_{i=1}^{\lambda} \overline{p}_i^2 - \sum_{i=\lambda+1}^{n_p} \overline{p}_i^2$$

Proof. See [Hirsch 1976, Chapter 6, Section 1.1, page 145]. ∎

Theorem 21.13. (Morse–Smale) *Let $f_\theta : D \times \Omega \to \mathbb{R}$ be a smooth map defined on a compact smooth manifold $D \times \Omega$. As v varies within the open interval between two adjacent critical values, the homotopy type of the sublevel set $P_{\leq v} = \{p \in P : f_\theta(p) \leq v\}$ remains unchanged. Furthermore, if f_θ is a Morse function, the change of homotopy type of $P_{\leq v}$ as v crosses the critical value $v^* = f_\theta(p^*)$ with Morse index λ is equivalent to attaching the "handle" $\mathbf{D}^\lambda \times \mathbf{D}^{n_p-\lambda}$ along $(\partial\mathbf{D}^\lambda) \times \mathbf{D}^{n_p-\lambda}$, where $\mathbf{D}^i$ denotes the closed i-dimensional disk on the manifold. The pair*

$$(\mathbf{D}^\lambda \times \mathbf{D}^{n_p-\lambda}, (\partial\mathbf{D}^\lambda) \times \mathbf{D}^{n_p-\lambda})$$

is called **(tangential) Morse data.**

Proof. That the sublevel sets $P_{\leq v_1}, P_{\leq v_2}$ have the same homotopy type between adjacent critical values, $v_1^* < v_1 \leq v_2 < v_2^*$, is easy to prove. Indeed, consider the vector field ∇f_θ between two adjacent critical levels on the manifold. Since $f_\theta^{-1}((v_1^*, v_2^*))$ has no critical points, it follows that the vector field ∇f_θ is smooth, without equilibrium points. Therefore, the trajectories of the differential equation $\dot{p} = -\nabla f_\theta(p)$ defines a deformation retract of $P_{\leq v_2}$ onto $P_{\leq v_1}$. See [Bott and Tu 1982]. The modern proof of the handle decomposition under the stronger Morse function condition goes as follows: For the critical value v^*, choose ϵ such that $[v^* - \epsilon, v^* + \epsilon]$ contains no other critical values than v^*. Define the Morse data

$$(f_\theta^{-1}([v^* - \epsilon, v^* + \epsilon]), f_\theta^{-1}(v^* - \epsilon))$$

Use the lemma of Morse to put the function in canonical coordinates and consider the map

$$p \mapsto (\sum_{i=1}^{\lambda} p_i^2, \sum_{i=\lambda+1}^{n_p} p_i^2)$$

Manipulating the Morse data expressed in terms of preimages of the above map yields the result. For details on this modern proof, see [Goresky and MacPherson 1988, Section 4.5]. See also [Bott and Tu 1982, Theorem 17.16] for a different proof, but still in the same spirit. ∎

The above result is useful to understand the local behavior of the manifold around the inverse image of a ∂N point. Besides this handle decomposition, there are plenty of other results of the Morse theory that we will use and introduce in due course.

Corollary 21.2. *Let $f_\theta : D \times \Omega \to \mathbb{R}$ be a smooth map defined on a compact smooth manifold $D \times \Omega$. As v varies in the open interval between two adjacent critical values, $v_1^* < v_2^*$, the level set $P_v = \{p \in D \times \Omega : f_\theta(p) = v\}$ changes homeomorphically.*

Proof. Consider the level sets P_{v_1}, P_{v_2} between two adjacent critical values, $v_1^* < v_1 < v_2 < v_2^*$. The (nonintersecting!) trajectories of the differential equation $\dot{p} = -\nabla f_\theta(p)$ connects every point of P_{v_2} to a unique point of P_{v_1}, and vice versa. This, together with continuity of the solution to the initial condition, establishes the homeomorphism between P_{v_1} and P_{v_2}. ■

The big problem in applying the Morse theory is that we cannot apriori rule out the Hessian of f_θ becoming singular for some θ at some critical point.

21.2.3 Degeneracy Phenomena

We now proceed to the analysis of the critical curves and related degeneracy phenomena. Consider the basic critical points equation, namely,

$$\frac{\partial f_\theta}{\partial p} = 0$$

If the Hessian is nonsingular around θ, then, by the Implicit Functions Theorem, the critical points can be written as functions of θ, namely,

$$p^{*k}(\theta), \ \ k = 1, 2, ...$$

Therefore, we could think of plotting the critical point curves on the manifold $D \times \Omega$. This is of course hard to visualize. Furthermore, we have to watch the Hessian for possible degeneracy, and this could be done by plotting the eigenvalues of the Hessian

$$\lambda_l(\frac{\partial^2 f_\theta}{\partial p_i \partial p_j}(p^{*k}(\theta)))$$

versus θ. At this stage, we cannot exclude the possibility that some eigenvalues might vanish for some θ's, in which case the explicit function $p^{*k}(\theta)$ ceases to exist. We therefore face the problem of "piecing together" on the manifold pieces of curves that have to interconnect some way around the degenerate points. To develop a convenient graphical display of related critical phenomena, which can get very complicated, we follow the procedure suggested by [Cerf 1970] in a seminal article dealing with critical point

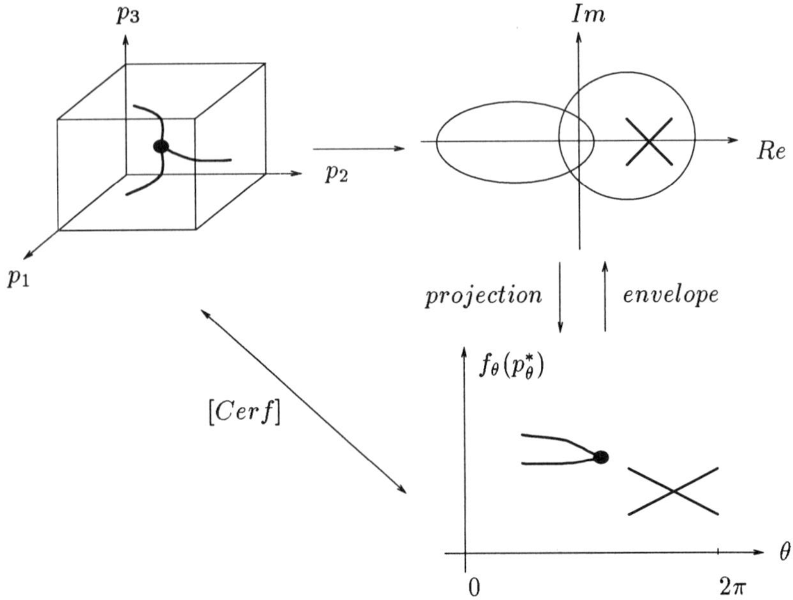

Fig. 21.3. "Tomographic" projection of supertemplate, its reconstruction as an "envelope," and the Cerf critical values plots.

locus of a one-parameter family of functions. It turns out that degeneracy and related phenomena can be visualized by plotting the critical *values*,

$$f_\theta(p^{*k}(\theta))$$

in the two-dimensional plane.

As we shall see later, taking rounding errors into consideration, it can be assumed without loss of generality that the Hessian becomes singular, only at *finitely* many θ's, θ_i. Therefore, the Morse theory will be applicable between two adjacent degenerate θ's. Furthermore, in the same seminal article, [Cerf 1970] showed that there are generically two cases of convergence of the plot to the degenerate points. From there on, we will know how to proceed by continuity from Morse to degenerate critical points.

Once the plots of critical values are drawn, we will have to single out those critical points that are in $f^{-1}(\partial N)$. Finally, we will be able to reconstruct ∂N as the envelope of the lines perpendicular to the direction θ at a distance from the origin equal to the selected critical value. It order to "piece together" the curves in the complex plane, it is convenient—and sometimes necessary—to also reconstruct by the same envelope procedure *all* critical curves of *all* indexes of the Nyquist template. This is yet another motivation for plotting *all* critical *values* of f_θ.

21.2.4 Three Approaches to Singularity Analysis

The reader will already have realized that we are jumping between the one-parameter family of functions into $\mathbb{R}$ and the original Nyquist map into $\mathbb{R}^2$. Actually, to completely elucidate the singularity structure of the Nyquist map, we will develop *three* different approaches, by order of increasing sophistication:

- The *"projection"* approach (Section 21.6)—that is, the singularity analysis of the one-parameter family of maps $f_\theta : D \times \Omega \to \mathbb{R}$.
- The *singularity subsets* approach (Section 21.11), which applies to one single, nominal Nyquist map $f : D \times \Omega \to \mathbb{R}^2$.
- The *jet space* approach (section 23.1), which applies to the nominal Nyquist map $f : D \times \Omega \to \mathbb{R}^2$ and its perturbation due to variation of "certain" parameters.

21.3 Nyquist Curve as Critical Value Plot

For purpose of illustration, we consider the extreme case where $p = \omega$ is the only "uncertain" parameter, in which case the Nyquist "template" reduces to the Nyquist curve. The critical point equation yields

$$\frac{\partial f_\theta}{\partial \omega} = 0 \Leftrightarrow \cos\theta \frac{\partial \Re f}{\partial \omega} + \sin\theta \frac{\partial \Im f}{\partial \omega} = 0$$

Therefore, $\forall\omega$, there exists at least one critical point at an angle θ given by

$$\tan\theta = -\frac{\frac{\partial \Re f}{\partial \omega}}{\frac{\partial \Im f}{\partial \omega}}$$

In other words, *the Nyquist curve is a critical values plot.* Generically, these are nondegenerate critical values. As we shall see later, degeneracy typically occurs at infinite frequency when the Nyquist curve has a "kink" across the real axis.

It is easily seen by elementary geometry that θ is the argument of the direction perpendicular to the plot. Hence if we consider the line perpendicular to θ at a distance of $f_\theta(\omega^*(\theta))$ from the origin, and if we compute the envelope of all such lines as θ varies, we recover the Nyquist curve.

21.4 Nash Functions

In this section we introduce a class a functions that are very useful in singularity and root-locus analyses. Polynomial/rational functions occur most naturally in control engineering. The crossover X_ω is given by a polynomial equation whenever $\det(I+L\Delta)$ is rational. Likewise, the critical points of the Nyquist map are given by polynomial equations whenever the Nyquist map is rational. The problem is that polynomial functions do not have good closure properties under some of the most fundamental operations of control engineering. This is most clearly exemplified by the root-locus procedure: The roots of a polynomial equation with coefficients depending in a polynomial fashion on, say, a gain parameter k are not polynomial, not even rational, in k.

The above motivates the following definitions: A **Nash function** $\varphi : \mathbb{R}^n \to \mathbb{R}$ is a function that is an analytic solution to a polynomial equation of the form

$$a_0(p)\varphi^k + a_1(p)\varphi^{k-1} + \ldots + a_{k-1}(p)\varphi + a_k(p) = 0$$

where the $a_i(p)$'s are polynomials, not all of them vanishing; these functions are also called **analytic-algebraic** (see [Bochnak, Coste, and Roy 1987, page 143]). Clearly, polynomials and rational fractions are Nash functions. Nash functions form a *ring* (see [Bochnak, Coste, and Roy 1987, page 147, Definition 8.1.8]); the partial derivatives of a Nash function are Nash functions (see [Bochnak, Coste, and Roy 1987, page 205]).

It is easy to devise a system-theoretic proof of the ring property of Nash functions. Let $\varphi(p)$, $\gamma(p)$ be Nash functions, analytic solutions to the polynomial equations:

$$\begin{aligned} P(\varphi, p) \equiv \sum_i a_i(p)\varphi^{n-i} &= 0 \\ Q(\gamma, p) \equiv \sum_i b_i(p)\gamma^{m-i} &= 0 \end{aligned}$$

Define $\sigma = \varphi + \gamma$ and let us prove that σ is Nash. View $P(\sigma - \gamma, p) = 0$ as a polynomial equation $P'(\gamma, \sigma, p) = 0$ in γ, σ, p. Therefore, we get

$$\begin{aligned} P'(\gamma, \sigma, p) &= 0 \\ Q(\gamma, p) &= 0 \end{aligned}$$

Now, do a Sylvester elimination of γ between $P' = 0$ and $Q = 0$. More specifically, write the Sylvester matrix $S(P', Q)$ of P', Q viewed as polynomials in γ. For example, if

$$\begin{aligned} P' &\equiv P'_0\gamma^2 + P'_1\gamma + P'_2 \\ Q &\equiv Q_0\gamma^3 + Q_1\gamma^2 + Q_2\gamma + Q_3 \end{aligned}$$

the Sylvester matrix is

$$S(P', Q) = \begin{pmatrix} P'_0 & P'_1 & P'_2 & 0 & 0 \\ 0 & P'_0 & P'_1 & P'_2 & 0 \\ 0 & 0 & P'_0 & P'_1 & P'_2 \\ 0 & Q_0 & Q_1 & Q_2 & Q_3 \\ Q_0 & Q_1 & Q_2 & Q_3 & 0 \end{pmatrix}$$

The polynomials P', Q have a common root γ iff $\det S(P', Q) = 0$. The latter is a polynomial equation in σ, p. Write it as a polynomial equation in σ with coefficients polynomials in p. It follows that σ is solution of a polynomial equation. Since by definition φ, γ are analytic, so is $\sigma = \varphi + \gamma$. Therefore, $\sigma = \varphi + \gamma$ is a Nash function.

To prove that $\pi = \varphi\gamma$ is Nash, consider $P(\frac{\pi}{\gamma}, p) = 0$. Multiplying it by an appropriate power of γ yields the polynomial equation $P'(\pi, \gamma, p) = 0$. Again, doing a Sylvester elimination of γ between $P'(\pi, \gamma, p) = 0$ and $Q(\gamma, p) = 0$ yields the required property of π.

To prove that the partial derivative of a Nash function is Nash, observe that

$$\frac{\varphi(p + \epsilon e^i) - \varphi(p)}{\epsilon}$$

is a Nash function. Using the Sylvester procedure, it is easy to find the polynomial equation that has the above Nash function as solution. From there it is easy to see that the limit as $\epsilon \to 0$ of the coefficients exists. Hence $\frac{\partial \varphi}{\partial p_i}$ is a Nash function.

The *local* properties of Nash functions are captured by the concept of *germs* of Nash functions (see Appendix D) and are characterized by formal algebraic series. A formal series $g \in \mathbb{R}[[p]]$ is said to be **algebraic** if it satisfies a polynomial equation like $a_0(p)g^k + a_1(p)g^{k-1} + \ldots + a_k(p) = 0$, where the $a_i(p)$'s are polynomials. The ring of germs of Nash functions around $p = 0$ is isomorphic to the ring of formal algebraic series (see [Bochnak, Coste, and Roy 1987, Section 8.2]).

For the specific purposes of this chapter, we will define a **Nash manifold** $M \subseteq \mathbb{R}^{n'}$ to be a smooth manifold that is also a semialgebraic set—that is, a set characterized by finitely many polynomial equations and inequalities (see [Bochnak, Coste, and Roy 1987, page 148, Example 8.1.11]). Clearly, the spheres are Nash manifolds. Less obvious is the fact that the 2-torus $\mathbb{T}^2$ is a Nash manifold.

As we already know, $\mathbb{T}^2$ is a smooth manifold, so that it remains to show that it is also a semialgebraic set. Considered the 2-torus, embedded in the 3-D Euclidean space charted with coordinates (x_1, x_2, x_3). For this specific purpose, the torus is defined as the product of two circles. The "small" circle has its center at $a = (a_1, a_2, 0)$, has a radius of 1, and is contained in the (a, x_3)-plane. This "small" circle is algebraically described by the set of polynomial equations:

$$(x_1 - a_1)^2 + (x_2 - a_2)^2 + x_3^2 = 1 \tag{21.1}$$

$$x_1 a_2 = x_2 a_1 \tag{21.2}$$

The torus can be thought of as generated by the above "small" circle as its center moves around the "big" circle

$$a_1^2 + a_2^2 = 4 \tag{21.3}$$

More formally, the equation of the 2-torus is obtained by elimination of the *real* variables a_1, a_2 among the above three equations. To put it more formally, the algebraic structure of the torus is the condition for Equations 21.1, 21.2, and 21.3, viewed as polynomials equations in a_1 and a_2, to have *real* common roots. It should be stressed that the variables a_1 and a_2 that are eliminated must be *real*, because otherwise the algebra has another geometric interpretation. In such classical elimination procedures as Bezout, Sylvester, and resultant matrices, there is no guarantee that the common root among a set of polynomials is *real*. Eliminating *real* variables among polynomials—that is, deriving conditions for a set of polynomials to have some *real* common root, if any—is the so-called *Tarski–Seidenberg decision problem* that is computationally very intensive (see Chapter 25). Here, however, it is possible to eliminate (a_1, a_2) while at the same time ensuring their real property using elementary analysis. Consider the elimination of a_1 between Equations 21.2 and 21.3. If these two equations have a common root, it could only be real because Equation 21.2 is linear in a_1. Elimination of a_1 between Equations 21.2 and 21.3 yields

$$a_2^2 = \frac{4x_2^2}{x_1^2 + x_2^2} \tag{21.4}$$

The same argument holds for the elimination of a_1 between Equations 21.1 and 21.2, and this yields

$$(x_2 - a_2)^2(x_1^2 + x_2^2) = x_2^2(1 - x_3^2) \tag{21.5}$$

Now the problem is to eliminate a_2 between Equations 21.4 and 21.5. Again, if these two equations have a common root at all, this common root could only be real as seen from Equation 21.4. Therefore, elementary elimination yields

$$16(x_1^2 + x_2^2) = (3 + x_1^2 + x_2^2 + x_3^2)^2$$

Clearly, the torus is an *algebraic variety*. This, in addition to its smooth manifold structure, makes it a *Nash manifold*.

If M is a Nash manifold, a map $f : M \to \mathbb{R}^n$ is said to be a **Nash mapping** if its graph is a semialgebraic set and the components of the mapping between underlying real spaces are Nash functions (see [Bochnak, Coste, and Roy 1987, page 148, Definition 8.1.10]). The critical points of a Nash function are defined as those points where the rank of the Jacobian drops.

21.5 Sard's Theorem

In this section, which can be skipped at a first reading, we address a technicality that we have deliberately ignored in Subsection 21.2.3—namely, in what kind of set does the index k of $f_\theta(p^{*k}(\theta))$ run? If k is meant to be the index of the set of critical *values*, then we have the most traditional theorem:

Theorem 21.14. (Sard) *If for a fixed* θ, $f_\theta \in C^\infty(D\times\Omega, \mathbb{R})$, *where* $D\times\Omega$ *is a smooth manifold, then the set of critical values* $\{f_\theta(p^{*k}(\theta)) : k\}$ *has zero measure in* $\mathbb{R}$.

Proof. See [Golubitsky and Guillemin 1973, Chapter II, Theorem 1.12, page 34] or [Milnor 1963, I, Theorem 6.1]. Observe that f_θ need not be a Morse function for the result to hold. ∎

The fact that the set of critical values is of zero measure does not mean much. It is even harder to see what is happening as θ sweeps $[0, 2\pi]$. In particular, it is tricky to say anything about the dimension of the resulting critical values set. To make progress along those lines, we need additional hypotheses and a version of the theorem of Sard more powerful than the traditional one. A more powerful version of Sard's theorem is available in [Bochnak, Coste, and Roy 1987, Theorem 9.5.2]. The relevant theorem is the following:

Theorem 21.15. (Sard—Real Algebraic Geometry Version) *Let* $f : X \to Y$ *be a Nash function between two Nash manifolds. Then the set of critical values is a semialgebraic set, of dimension strictly smaller than the dimension of* Y.

Proof. See [Bochnak, Coste, and Roy 1987, page 205, Theorem 9.5.2.]. ∎

Corollary 21.3. *Let* $f : D \times \Omega \to \mathbb{C}$ *be a Nash function defined on the Nash manifold* $D \times \Omega$. *Then the dimension of the set of critical values of* f *is exactly one.*

Proof. By the real algebraic version of Sard's theorem ([Bochnak, Coste, and Roy 1987, Theorem 9.5.2]), it follows that the dimension of the set of critical values of f is strictly less than the dimension of N, which is 2. On the other hand, the dimension of the set of critical values could not be zero, because it contains ∂N of dimension one. Therefore the dimension of the set of critical values is exactly one. ∎

In general not much can be said about the dimension of the inverse image of the critical values, f^{-1}(critical values of f). It strongly depends on the index and on whether some transversality conditions are satisfied.

21.6 Critical Values Plots

21.6.1 The Problem

In this section, we address the technicality that if we want to do a Morse analysis on f_θ it is allowed to have only *nondegenerate* critical points. The elementary variational interpretation of ∂N requires the Hessian to be non-singular at the critical points (because otherwise it is necessary to go up to fourth order). The essential point of this section is that, up to a small deformation of the map, f_θ can be assumed to lose its Morse property only at *finitely* many θ's.

The technicality that a Morse function is restricted to have only *nondegenerate* critical points can be circumvented by invoking the *Morse approximation lemma* [Morse 1925]: Any smooth function can always be approximated in the C^∞ sense with a function that has only nondegenerate critical points:

Lemma 21.16. (Morse Approximation Lemma) *Let $f_\theta : D \times \Omega \to \mathbb{R}$ be a smooth—that is, C^∞—function. Then for almost every $(\epsilon_1, ..., \epsilon_{n_p}) \in \mathbb{R}^{n_p}$ the function $f_\theta + \epsilon_1 p_1 + ... + \epsilon_{n_p} p_{n_p}$ is a Morse function.*

Proof. See [Bott and Tu 1982, Proposition 17.18]. ■

The Morse Approximation Lemma allows us to "fix" the function f_θ, for a precise θ. Here, we instead have a one-parameter family of functions that we want to be Morse for *all* θ's. If we use the Morse approximation lemma to fix the function around, say θ_1, the problem is that the degeneracy phenomenon is going to reappear at a nearby θ. Therefore, unfortunately, it is not, in general, possible to fix f_θ with a *small* perturbation so that the resulting function be Morse for all θ.

To analyze the above problem more formally, we consider f_θ to be a *path of functions* in $C^\infty(D \times \Omega, \mathbb{R})$:

$$\begin{array}{rcl} [0, 2\pi] & \to & C^\infty(D \times \Omega, \mathbb{R}) \\ \theta & \mapsto & f_\theta \end{array}$$

Actually, f_θ is a little more than a path: By periodicity in θ, it is a *loop* in $C^\infty(D \times \Omega, \mathbb{R})$.

Even though C^∞ is a space of "nice" functions, not all functions in C^∞ are equally "nice." A function that has degenerate critical points is not as "nice" as a function that has only nondegenerate critical points. Besides degenerate critical points, other singularities—like multiple critical values—could occur. Therefore, the path f_θ in C^∞ might contain functions that have varying degrees of singularity. The remainder of this section is devoted to the following program: First, we measure the "degree of singularity" of a function by its codimension. Next, we consider the "strata" of $C^\infty(D \times \Omega, \mathbb{R})$ consisting of all functions of a given codimension and investigate how these "strata" are "assembled." Finally, we investigate how the loop f_θ moves across strata.

21.6.2 Classification of Critical Points by Their Codimensions

A critical point p^* is a solution to $\frac{\partial f_\theta}{\partial p} = 0$, and the nature of the critical point depends strongly on the "multiplicity" of the solution p^* to the gradient equation. The concept of multiplicity is well-understood in the rational case. In the case where $f_\theta(p)$ is not rational in p, there are of course some questions as to what the "multiplicity" of a critical point is. Formally, the **multiplicity** of a solution p^* to $\frac{\partial f_\theta}{\partial p} = 0$ is the codimension (defect number) of the ideal generated by the germs of $\frac{\partial f_\theta}{\partial p_i}$ in the ring C_x of germs of C^∞ functions $D \times \Omega \to \mathbb{R}$. (See Appendix D for the precise definition of C_x.) To be more specific, define the local variables $x_i = p_i - p_i^*$, evaluate the partial derivatives of f_θ around $p^* + x$, and form the ideal generated by those partial derivatives in the ring C_x; then the multiplicity is defined as

$$\text{multiplicity } (p^*) = \dim_{\mathbb{R}}(C_x/\text{Id}(\frac{\partial f_\theta}{\partial p_1}(p^* + x), \frac{\partial f_\theta}{\partial p_2}(p^* + x), ...))$$

The dimension is to be understood as the dimension of the quotient ring as a vector space over $\mathbb{R}$. See [Cerf 1970, page 23], [Arnold, Gusein-Zade, and Varchenko 1985, page 121], [Golubitsky and Guillemin 1973, page 168], and [Castrigiano and Hayes 1993, page 127]. See also Appendix D for the bigger algebraic picture and, in particular, for the formal power series $\mathbb{R}[[x]]$ representation of the ring of germs of smooth functions modulo the "flat" functions. In case $f_\theta(p)$ is polynomial, this algebraic definition of multiplicity reduces to the usual concept of multiplicity.

More formally, critical points are classified by their **codimension**, defined as

$$\begin{aligned}\text{codim}(p^*) &= \dim_{\mathbb{R}} \mathcal{M}_x/\text{Id}(\{\frac{\partial f_\theta}{\partial p_i}\}) \\ &= \dim_{\mathbb{R}} C_x/\text{Id}(\{\frac{\partial f_\theta}{\partial p_i}\}) - 1\end{aligned}$$

where $\mathcal{M}_x \subseteq C_x$ denotes the maximal ideal of germs of smooth functions vanishing at $x = 0$ (see Appendix D). From the above we derive the intuitive definition

$$\text{codimension} = \text{multiplicity} - 1$$

The motivation for the terminology of *codimension* of a singularity is as follows: Consider in C^∞ the subset of all functions that have their respective singularities of codimensions adding up to, say, i; then this subset has (geometric) codimension, or defect number, i in C^∞.

A critical point p^* that makes the Hessian nonsingular is said to be a **critical point of the Morse type**. Because the Hessian is nonsingular, it is a *simple* point, of multiplicity 1; that is, of codimension 0. This can be seen from the implicit function theorem: Let p^{*1} be a Morse critical point corresponding to θ_1. The implicit function theorem applied to $\frac{\partial f_\theta}{\partial p} = 0$ yields a *unique* continuously differentiable function $p^*(\theta)$ such that $p^*(\theta_1) = p^{*1}$. In other words, a θ perturbation of the critical point problem will result in a *unique* curve departing from p^{*1}.

Following the *Lemma of Morse*, there exists a canonical form of a singularity of codimension 0 in terms of some Riemann coordinates $\bar{p}$ such that $\bar{p}$(critical point) $= 0$ and in which the function takes the form

$$f_\theta(p(\bar{p})) = c - \sum_{k=1}^{\lambda} \bar{p}_k^2 + \sum_{k=\lambda+1}^{n} \bar{p}_k^2$$

If a critical point makes the Hessian singular, it is said to be **degenerate**. Because the Hessian vanishes, it is of a multiplicity greater than one; that is, of a codimension greater than 0. This can be seen from the ring-theoretic argument explained above. A somewhat more concrete approach to the same problem is provided by the so-called *Malgrange preparation theorem*. Since the Hessian vanishes, the implicit function theorem does not apply. Assume, however, that by continuing to differentiate, a nonvanishing partial derivative is found. For illustrative purposes, assume first that there is only one variable,

$$\frac{\partial f_\theta}{\partial p} = 0, \quad \frac{\partial^2 f_\theta}{\partial p^2} = 0, \ldots, \quad \frac{\partial^k f_\theta}{\partial p^k} = 0, \quad \frac{\partial^{k+1} f_\theta}{\partial p^{k+1}} \neq 0$$

Therefore, the *Malgrange preparation theorem* asserts that the most "explicit" form of the solution p to the gradient equation is in the form of the root of a polynomial equation of degree k with coefficients continuously depending on θ, namely,

$$p^k + a_1(\theta)p^{k-1} + \ldots + a_k(\theta) = 0$$

In other words, the gradient equation is locally equivalent to the polynomial equation. At $\theta = \theta_1$, the above polynomial has a k-fold root, and as we perturb θ near θ_1, *several* curves will emanate from the degenerate critical point.

We hasten to say that the above situation is unlikely to occur in a Nyquist problem where the *frequency* is the only variable, $p = \omega$, in which case the Nyquist "template" becomes the Nyquist "curve." To see this, consider the SISO problem $L(s) = \frac{1}{(s+1)^k}$. It is known that for $k \geq 2$ and even, the Nyquist plot has a "kink" across the real axis occurring at $\omega = \infty$. It is easily seen that

$$\frac{\partial f_\theta}{\partial \omega}\Big|_{\omega=\infty} = 0$$

so that this "kink" is indeed a critical value with $\omega = \infty$ the critical point. If we continue to differentiate relative to ω, every new differentiation increases the degree of the denominator relative to that of the numerator, so that all partial derivatives vanish at $\omega = \infty$, and the Malgrange analysis fails.

However, if we compactify the imaginary axis via the bilinear transformation $s = \frac{1-z}{1+z}$, we get

$$L(s(z)) = \frac{(1+z)^k}{2^k}$$

and it is trivial that

$$\frac{\partial f}{\partial z} = 0, \ldots, \frac{\partial^{k-1} f}{\partial z^{k-1}} = 0, \frac{\partial^k f}{\partial z^k} \neq 0$$

Next, since $dz = jzd\zeta$, we get

$$\frac{\partial f_\theta}{\partial \zeta} = 0, \ldots, \frac{\partial^{k-1} f_\theta}{\partial \zeta^{k-1}} = 0, \frac{\partial^k f_\theta}{\partial \zeta^k} \neq 0$$

Therefore, for $k > 2$, $z = -1, \zeta = 0$ is a degenerate critical point, amenable to the Malgrange analysis. Observe that this point is critical degenerate for all θ. Doing a Malgrange analysis for $\theta = \pi/2$, it follows that there is at least one critical value curve $f_\theta(\zeta^*(\theta))$ that has an inflection point around $\theta = \pi/2$ with 0 critical value; this is a typical degeneracy phenomenon.

Conversely, constructing the envelope of all lines perpendicular to the line of constant argument θ at a distance of $f_\theta(p^*(\theta))$, we recover the Nyquist curve, and the Morse degeneracies of $f_{\pi/2}$ are related to the "kink." The singularity of all the lines perpendicular to θ clustering around the kink is called **caustics** (see [Arnold, Gusein-Zade, and Varchenko 1985]).

Next, consider a degenerate case with two variables. Assume the Hessian has rank 1 and that after a linear change of coordinates, one gets

$$\frac{\partial^2 f_\theta}{\partial p_i \partial p_j} = \begin{pmatrix} 1 & 0 \\ 0 & 0 \end{pmatrix}$$

Since $\frac{\partial^2 f_\theta}{\partial p_1^2} \neq 0$, we use the implicit function theorem on $\frac{\partial f_\theta}{\partial p_1} = 0$ and derive the explicit form $p_1 = p_1(p_2, \theta)$. Define

$$g(p_2, \theta) = \frac{\partial f_\theta}{\partial p_2}(p_1(p_2, \theta), p_2)$$

Now, we perform a Malgrange analysis on g viewed as a function of p_2. We differentiate relative to p_2 until a nonvanishing partial derivative is found. Then we express p_2 as a root of a polynomial equation with coefficients smoothly depending on θ. This polynomial reveals the degeneracy occurring on the variable p_2 and finally we substitute for p_2 in $p_1(p_2, \theta)$. The reader is referred to Section D.8 for more details on this kind of manipulations.

If p^* has multiplicity 2, or codimension 1, the gradient equation is equivalent to

$$p^2 + a_1(\theta)p + a_2(\theta) = 0$$

As we perturb θ, it follows from elementary quadrature that the critical point of codimension 1 splits into two curves. Such points are therefore called **points of birth or death**. To put it more formally, as we perturb the *one* parameter θ, the degenerate critical point of *codimension one "un-*

folds" into two nondegenerate critical points. This is called **unfolding** and θ is called **control parameter**. In case of a higher-codimensional critical point, the number of parameters for the unfolding equals the codimension (see Appendix D). Around such a singularity as birth or death, there exists a system of Riemann coordinates $\bar{p}$ such that $\bar{p} = 0$ at the singularity of codimension 1 and the function has the canonical representation [Cerf 1970, page 23]

$$f_{\theta_1}(p(\bar{p})) = c - \sum_{k=1}^{\lambda} \bar{p}_k^2 + \sum_{k=\lambda+1}^{n-1} \bar{p}_k^2 + \bar{p}_n^3$$

where c is the critical value.

Next, if p^* has multiplicity 3, or codimension 2, it has a local canonical representation [Cerf 1970, page 23]

$$f_{\theta_1}(p(\bar{p})) = c - \sum_{k=1}^{\lambda} \bar{p}_k^2 + \sum_{k=\lambda+1}^{n-1} \bar{p}_k^2 - \bar{p}_n^4$$

This kind of a singularity is also called "swallow tail." To visualize the swallow tail phenomenon, we first need an initial perturbation that would split the triple critical point of f_{θ_1} into three distinct critical points. Then plotting the critical points/values versus θ around $\theta = \theta_1$ results in a network of curves having the deeply forked shape of the tail of a swallow bird. Again, this is an **unfolding** process—that is, the process of perturbing a function around a multiple critical point so as to reveal the *real nature* of the singularity (see [Bruce and Giblin 1992] and Appendix D). Observe that in this case, in addition to θ, we need one additional control parameter, the initial perturbation, because the singularity has codimension 2. We will come back to the swallow tail phenomenon later.

As an example will prove, critical points of codimension 1,2 are *very likely* to appear in robust stability. However, the odds of encountering critical points of codimension 3,4,... appear very slim and we will no longer pursue this classification any further.

21.6.3 Isotopy

Given topological spaces X, Y, the question arises as to whether the space $C^0(X, Y)$ of continuous maps $X \to Y$ is connected. This problem is tackled using the concept of *homotopy*. The space $C^0(X, Y)$ is connected iff, given two arbitrary maps f, g, they can be connected by a homotopy—that is, a *continuous* path of maps.

In the realm of spaces of smooth maps between smooth manifolds, we require the path to be *smooth*. However, the mere concept of "smooth homotopy" is not very useful without some restrictions on the kind of maps f, g it applies to. One such restriction that makes the concept useful is the concept of *embedding*: A smooth map $f : X \to Y$ between smooth manifolds X, Y is said to be an **embedding** iff f maps X homeomorphically onto its image $f(X)$ and the differential $d_x f : T_x X \to T_{f(x)} Y$ is injective $\forall x \in X$. This leads to the concept of *isotopy*: Given two embeddings $f, g \in C^\infty(X, Y)$ an **isotopy** between f and g is a smooth map

$$\begin{array}{rcl} H : X \times [0,1] & \to & Y \times [0,1] \\ (x,t) & \mapsto & (H_t(x), t) \end{array}$$

such that

$$H_t \text{ is an embedding } \forall t \in [0,1]$$

and

$$\begin{array}{rcl} H_0 & = & f \\ H_1 & = & g \end{array}$$

Under those circumstances, the embeddings f, g are said to be **isotopic**. It requires a little bit of work to prove that "being isotopic" is an equivalence relation.

As an illustration of the usefulness of this concept, we will mention the classical result that two orientation-preserving embeddings $\mathbb{D} \to M$, where M is an orientable 2-manifold, are isotopic (see [Hirsch 1976, Chapter 8, Theorem 3.1]). In other words, up to orientation, there is just *one* way to embed a disk into a 2-manifold.

As asserted by [Hirsch 1976, page 177], it turns out that an isotopy is, in most cases, realizable by some very special transformations, called *diffeotopies*, of the domain and the target. We now proceed to define these transformations.

We first need some background definitions and concepts. Let Diff(M) be the group of diffeomorphisms of the compact, smooth, orientable manifold M. A diffeomorphism has the property that $\det J(m) \neq 0$, $\forall m \in M$, where $J(m)$ is the Jacobian of the transformation. Clearly, Diff(M) could not possibly be connected. Let $\text{Diff}_+(M)$ be the subgroup of *orientation-preserving* diffeomorphisms of M. These can be identified with the diffeomorphisms with the property that $\det J(m) > 0$, $\forall m \in M$. Observe that $\text{Diff}_+(\mathbb{R}^n)$ is connected. To see this, consider first a map such that $f(0) = 0$; observe that the diagonal elements of $J(x)$ are positive; then wipe out the off-diagonal elements by a smooth homotopy, and this yields the (diagonal) Jacobian of an orientation-preserving diffeomorphism; finally, by another smooth homotopy, scale the diagonal elements to 1, and this yields the Jacobian of the identity transformation; therefore, all orientation-preserving maps $f : \mathbb{R}^n, 0 \to \mathbb{R}^n, 0$ are in the connected component of the identity so that the space of orientation-preserving smooth maps $f : \mathbb{R}^n, 0 \to \mathbb{R}^n, 0$ is connected; to deal with the case $f(0) \neq 0$, just wipe out the $f(0)$ term by a smooth homotopy. Observe, however, that $\text{Diff}_+(M)$ is usually not connected when M is a manifold. For example, $\mathrm{Diff}_+(S^6)$ and $\mathrm{Diff}_+(S^{10})$ are not connected, and this fact has far-reaching consequences in physics (see [Nash 1991]). Therefore, we define $\mathrm{Diff}_e(M)$ to be the connected component of $\mathrm{Diff}_+(M)$ that contains 1_M.

Let M be a compact, smooth, orientable manifold. A **diffeotopy** $d : M \to M$ of the manifold M is defined via a level-preserving diffeomorphism of the cylinder $M \times [0,1]$,

$$\begin{aligned} H : M \times [0,1] &\to M \times [0,1] & (21.6)\\ (m,t) &\mapsto (H_t(m),t) & (21.7)\end{aligned}$$

such that

$$H_t \in \mathrm{Diff}_+(M), \ \forall t \in [0,1]$$

Furthermore, the diffeomorphism of the cylinder is the identity mapping at the base of the cylinder,

$$H_0 = 1_M$$

and the diffeotopy d is the mapping at the top of the cylinder,

$$d = H_1$$

In other words, a diffeotopy of M is a diffeomorphism of M that is isotopic to the identity on M. It is easily seen that the diffeotopies of M form a group, $\mathrm{Difftop}(M)$.

The group $\mathrm{Difftop}(M)$ of diffeotopies acts on $\mathrm{Diff}_+(M)$ as follows:

$$\begin{aligned} \mathrm{Difftop}(M) \times \mathrm{Diff}_+(M) &\to \mathrm{Diff}_+(M)\\ (H_1, g) &\mapsto H_1 \circ g\end{aligned}$$

The relevant fact is that the orbits of $\mathrm{Diff}_+(M)$ under the action of the group $\mathrm{Difftop}(M)$ are the connected components of $\mathrm{Diff}_+(M)$. Applying the preceding to the orbit of 1_M, it follows that $\mathrm{Difftop}(M) = \mathrm{Diff}_e(M)$.

Two diffeomorphisms $f, g \in \mathrm{Diff}_+(M)$ in the same connected component or the same orbit in $\mathrm{Diff}_+(M)$ are said to be **diffeotopic**.

It is a rather formidable problem [Cerf 1970] to determine whether the "level preserving" property in Equation 21.7 can be dispensed of. Dropping the level preserving property yields the group of *pseudodiffeotopies* of M, and the big question is whether the classification of orientation-preserving diffeomorphisms $M \to M$ using diffeotopies and pseudodiffeotopies are the same. The only clearcut fact is that whenever $\pi_1(M) = 0$ and $\dim(M) \geq 5$, the two classifications are the same. In our case, $M = D \times \Omega$, and since Ω has been compactified to the circle, we have $\pi_1(D \times \Omega) \neq 0$ so that the two classifications cannot be guaranteed to coincide. We therefore cannot in this control context drop the level preserving property.

With these concepts, we can define isotopy between two functions in $C^\infty(M,\mathbb{R})$. First, observe that two functions $f, g \in C^\infty(M,\mathbb{R})$ can be connected by a smooth homotopy—for example, $(1-t)f(m) + tg(m)$— so that the space $C^\infty(M,\mathbb{R})$ in general and the space $C^\infty(D \times \Omega, \mathbb{R})$ in particular are connected. Two functions $f, f' \in C^\infty(M,\mathbb{R})$ are said to be **isotopic** iff there exist diffeotopies $d_1 \in \mathrm{Diff}_e(M)$, $d_2 \in \mathrm{Diff}_e(\mathbb{R})$ such that the following diagram commutes:

$$\begin{array}{ccc} M & \xrightarrow{f} & \mathbb{R} \\ d_1 \downarrow & & \downarrow d_2 \\ M & \xrightarrow{f'} & \mathbb{R} \end{array}$$

An alternate definition is as follows: Let G_e be the connected component of $\mathrm{Diff}_+(M) \times \mathrm{Diff}_+(\mathbb{R})$ that contains the neutral element e of the group. Define the **action** of G_e on $C^\infty(M,\mathbb{R})$ as

$$\begin{array}{rcl} G_e \times C^\infty(M,\mathbb{R}) & \to & C^\infty(M,\mathbb{R}) \\ ((d_1, d_2), f) & \mapsto & d_2 \circ f \circ d_1^{-1} \end{array}$$

Two functions $f, f' \in C^\infty(M,\mathbb{R})$ are **isotopic** if they lie in the same orbit under the action of the group G_e.

21.6.4 Stratification of Space of Differentiable Functions

The nature of the singularity associated with a critical *point*, as defined above, does not reveal the whole singularity that the *function* might exhibit. As an example, let us consider a function that has two distinct, simple critical points that are mapped to the same critical value. The two distinct critical points are of the mildest possible singularity (codimension 0), yet the function exhibits more singularity than a function that has distinct critical values.

These considerations lead us to the definition of the multiplicity of a critical value. The multiplicity of a critical value c is the number, multiplicity counted, of critical points in $f_\theta^{-1}(c)$. The codimension of a critical value is its multiplicity minus 1. The codimension of a function can be defined, at least for small values, as the sum of the codimensions of its critical values.

In a seminal paper, [Cerf 1970] showed that the space $C^\infty(D \times \Omega, \mathbb{R})$ can be decomposed as the disjoint union

$$C^\infty(D \times \Omega, \mathbb{R}) = F^0 \cup F^1 \cup F^2 \cup \ldots$$

where each F^i is the subset of C^∞ functions of codimension i. Each such subset is open; and furthermore,

$$F^{i+1} \subseteq \overline{F^i}$$

Furthermore, F^i is a subvariety of codimension (defect number) i in either

$F^0 \cup ... \cup F^i$ or $C^\infty(D \times \Omega, \mathbb{R})$. Such a decomposition is called **stratification**. The geometry of this stratification is heavily $(D \times \Omega)$-dependent.

To be more specific, F^0 is the space of Morse functions that have distinct critical values. These functions have been called **excellent** by Thom. With this concept, we can reformulate the Morse approximation lemma, in a slightly stronger version, as follows:

Lemma 21.17. (Morse Approximation Lemma) *F^0 is everywhere dense in $C^\infty(D \times \Omega, \mathbb{R})$.*

Proof. See [Cerf 1970]. ■

The second space F^1 is the space of functions with codimension 1. In codimension 1, the singularity emanates from either a double critical point or distinct critical points mapped to the same critical value. Therefore, we break F^1 in yet another disjoint union

$$F^1 = F^1_\alpha \cup F^1_\beta$$

F^1_α is the set of functions that have a *point of birth or death*, all of the other critical points being of the Morse type with distinct critical values. A **point of birth (death)** is a θ at which a pair of critical values with Morse indices $\lambda, \lambda+1$ appear (or annihilate). It is not hard to see from the implicit function theorem that, at a point of birth (death), two eigenvalues of the Hessian collide and disappear on (emerge from) the $\lambda_i = 0$ axis. The converse—that is, that a singular Hessian implies a point of birth (death)—is, however, much harder to prove (see [Cerf 1970] and [Boardman 1967]).

F^1_β is the subset of Morse functions that have distinct critical values, except exactly two of them that are equal. These are called **points of crossing**, because in the diagram of the critical values versus θ, the plots of two critical values *cross*.

If we proceed toward more complex singularities, the next set F^2 is the set of functions with singularities of codimension 2. This means that the plot of critical values has a threefold multiplicity for some $\theta = \theta_1$.

In higher codimension, it is usually not possible to characterize the functions in F^i, $i \geq 3$ by a simple rule expressible in terms of "multiplicities" of critical points and critical values. To define the codimension of a function in C^∞, we have to proceed more formally. Given a function $f \in C^\infty(D \times \Omega, \mathbb{R})$, the **action** of the group $G = \text{Diff}_+(D \times \Omega) \times \text{Diff}_+(\mathbb{R})$ on f is formally defined by the mapping:

$$\begin{aligned} a_f : G &\rightarrow C^\infty(D \times \Omega, \mathbb{R}) \\ (g_1, g_2) &\mapsto g_2 \circ f \circ g_1^{-1} \end{aligned}$$

The action of (g_1, g_2) on f is depicted by the following commutative diagram:

$$\begin{array}{ccc} D \times \Omega & \xrightarrow{f} & \mathbb{R} \\ g_1 \downarrow & & \downarrow g_2 \\ D \times \Omega & \xrightarrow{g_2 \circ f \circ g_1^{-1}} & \mathbb{R} \end{array}$$

Let (TG, π_g, G) be the tangent bundle to the group G and let $(TC^\infty(D \times \Omega, \mathbb{R}), \pi_c, C^\infty(D \times \Omega, \mathbb{R}))$ be the tangent bundle to $C^\infty(D \times \Omega, \mathbb{R})$. The action mapping a_f induces a tangent bundle morphism

$$\begin{array}{ccc} TG & \xrightarrow{da_f} & TC^\infty(D \times \Omega, \mathbb{R}) \\ \pi_g \downarrow & & \downarrow \pi_c \\ G & \xrightarrow{a_f} & C^\infty(D \times \Omega, \mathbb{R}) \end{array}$$

The map da_f is linear from the fiber over (g_1, g_2) to the fiber over $g_2 \circ f \circ g_1^{-1}$. The **codimension** of the function f is, by definition, the codimension of the image $da_f(TG)$ in $TC^\infty(D \times \Omega, \mathbb{R})$. This definition is due to Mather.

The lower-codimensional strata enjoy some stability properties under the action of the group $\text{Diff}_+(D \times \Omega) \times \text{Diff}_+(\mathbb{R})$. Define the **action** of $\text{Diff}_+(D \times \Omega) \times \text{Diff}_+(\mathbb{R})$ on $C^\infty(D \times \Omega, \mathbb{R})$ by

$$\begin{array}{rcl} (\text{Diff}_+(D \times \Omega) \times \text{Diff}_+(\mathbb{R})) \times C^\infty(D \times \Omega, \mathbb{R}) & \to & C^\infty(D \times \Omega, \mathbb{R}) \\ ((g, g'), f) & \mapsto & g' \circ f \circ g^{-1} \end{array}$$

It can be shown [Cerf 1970, page 204] that the strata $F^0, F^1_\alpha, F^1_\beta$ are stable under the action of $\text{Diff}_+(D \times \Omega) \times \text{Diff}_+(\mathbb{R})$. Furthermore, for any $f \in F^0 \cup F^1$, the stratum of f coincides with the orbit of f under the action of $\text{Diff}_+(D \times \Omega) \times \text{Diff}_+(\mathbb{R})$. However, for higher-codimensional functions, it might happen that the orbit is strictly contained in the connected component.

It can be shown that two elements of $F^0 \cup F^1 \cup F^2$ are isotopic iff they lie in the same connected component of a stratum of $C^\infty(D \times \Omega, \mathbb{R})$.

The motivation for the concept of isotopic functions is the classification of functions by their *differential type*. Two functions are of the same differential type whenever they are isotopic. This is similar to the related concept that two functions are of the same topological type whenever they are homotopic.

We will tacitly assume that the odds of hitting a singularity of a codimension greater than 2 are very slim and we will therefore not go into the details of the strata F^3, F^4, and so on.

21.6.5 Stratification of the Space of Morse Functions

We could apply the same technique as in the preceding subsection to stratify the space of all *Morse* functions, M,

$$M = M^0 \cup M^1 \cup M^2 \cup \ldots$$

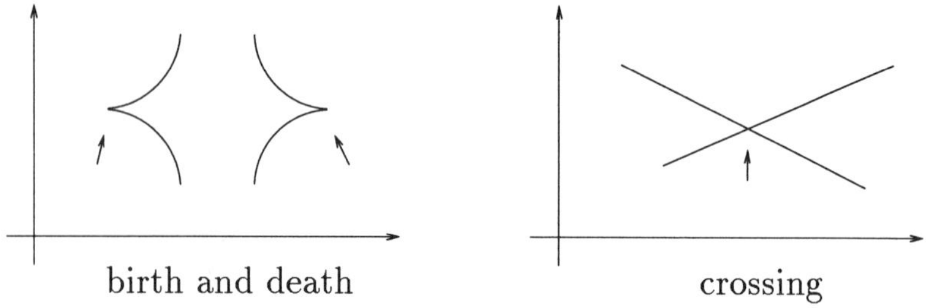

Fig. 21.4. Codimension 1 singularities.

Equivalently, this stratification could be obtained by intersecting the stratification of $C^\infty(D \times \Omega, \mathbb{R})$ with M, namely,

$$M^i = M \cap F^i$$

M^i is the space of Morse functions of codimension i—that is, the space of Morse functions that have critical values of codimension i. Under the action of $\text{Diff}_+(D \times \Omega) \times \text{Diff}_+(\mathbb{R})$, the space M and its strata remain stable. Furthermore, for any $f \in M$ the stratum that contains f coincides with the orbit of f under the action of $\text{Diff}_+(D \times \Omega) \times \text{Diff}_+(\mathbb{R})$. Finally, two Morse functions f, f' are isotopic iff they belong to the same connected component of the stratification of M.

Observe that the salient difference between the stratification of M as opposed to the stratification of $C^\infty(D \times \Omega, \mathbb{R})$ is that in the former case stability and related properties are not restricted to codimension ≤ 2.

21.6.6 Local Properties of Family

In the same seminal article, [Cerf 1970] introduces the concept of *good path of functions.* This is a one-parameter family of functions that loses the "Morse with distinct critical values" property only at a finite set of points.

Definition 21.18. (Cerf 1970, page 24) *A* **good path**, *typically f_θ, is a one-parameter family of functions such that f_θ is in F^0 for all but a finite number of θ's, and for those exceptional θ_i's, f_{θ_i} is in F^1 and f_θ crosses F^1 transversally.*

In other words, if we follow a good path, when it crosses F^1, either of the following occurs

- Either f_θ has a point of birth (death); that is, at θ a pair of critical points with indices $\lambda, \lambda + 1$ appears (disappears),
- or f_θ is a Morse function with exactly two critical values that are equal; in other words, at θ two critical values cross, all of the other critical values being distinct.

The motivation for introducing the *good paths* is the following:

Lemma 21.19. (Cerf 1970) *The set of good paths is dense in the space of all paths taking value in $C^\infty(D \times \Omega, \mathbb{R})$ and equipped with the C^0 topology; that is, the topology induced by the "distance" $d(f_\theta, g_\theta) = \sup_{p,\theta} |f_\theta(p) - g_\theta(p)|$.*

Proof. See [Cerf 1970, page 24]. ■

In view of this lemma, we can assume, without loss of generality, that f_θ is a *good path.* Under those circumstances, we remove from $[0, 2\pi)$ a finite set of points θ_i's

$$[0, 2\pi)^* = [0, 2\pi) - \cup_i \theta_i$$

Therefore, the Morse theory is applicable, provided that we restrict ourselves to

$$\theta \in [0, 2\pi)^*$$

and proceed by continuity to the degenerate critical points.

The above might be called the "local" approach. Indeed, by an arbitrarily small deformation—which can be thought of as modeling rounding errors—the loop f_θ can be made to remain entirely within F^0 and F^1, thereby avoiding functions with more complex singularities. Another justification for the terminology of "local" is that we have only considered θ-local singularities without bothering how θ-local singularities should be "pieced together" to provide a θ-global pattern.

21.6.7 Global Properties of Family

As we have seen in the preceding subsection, an ϵ-perturbation can remove all singularities of codimension ≥ 2. This leaves plots of critical values versus θ, devoid of the worst singularities, but that still have "mild" singularities of codimension 1—that is, points of birth or death and points of crossing. Although, locally, around a θ_i, the singularities of codimension 1 appear "mild," the *global* pattern as θ runs from 0 to 2π of all curves with many births, deaths, and crossings can still be quite involved. The best example of this situation is the *swallow tail.*

The question is whether via some transformation of f_θ, yet to be determined, a complicated global pattern of crossings, births, and deaths could be simplified, even totally eradicated. This may require large perturbations. Indeed, splitting two curves that share two points of crossing requires a large perturbation that would bring the two points of crossing together, annihilate them, and then pull the two curves apart.

[Cerf 1970] precisely investigates whether there exists a *homotopy* that takes the initial path f_θ and deforms it to g_θ devoid of points of crossing, birth, or death. It turns out that it is not, in general, possible to totally eradicate all points of birth, death and crossing, because the connected components of F^0, F^1 are not in general acyclic. To be more specific, since f_θ is a *loop* over $C^\infty(D \times \Omega, \mathbb{R})$, it naturally embeds into an element $\{f_\theta\} \in \pi_1(F^0 \cup F^1, f_0)$. $\{f_\theta\}$ can be thought of as the "obstruction" to deforming f_θ into a function with uniform critical values plot. If $\{f_\theta\} = 0$, the loop f_θ can be deformed into a uniform critical value plot while remaining within $F^0 \cup F^1$ during the entire homotopy. If $\{f_\theta\} \neq 0$, it is not possible to simplify f_θ while remaining within $F^0 \cup F^1$. We have to go to higher-dimensional strata—that is, cross higher-codimensional singularities—before the path f_θ can be simplified. If adjunction of the stratum F^2 "kills" the homotopy—that is, $\pi_1(F^0 \cup F^1 \cup F^2, f_0) = 0$—then the loop f_θ can be simplified by allowing to cross singularities of codimension 2.

Beyond the homotopy groups $\pi_1(F^0 \cup F^1, f_0)$ and $\pi_1(F^0 \cup F^1 \cup F^2, f_0)$ the most *comprehensive* way to cope with these *global* issues is to consider the *nerve of the stratification.* The **nerve of the stratification** $F^0 \cup F^1 \cup F^2 \cup ...$ is the simplicial complex that has as "simplexes" the connected components of $F^0, F^1, ...$ and is endowed with the facing relation

$$A \preceq B \Leftrightarrow A \subseteq \overline{B}$$

Clearly, the presence of a nonvanishing obstruction that prevents the singularities of codimension 1 to be removed endows the robustness analysis with better structural stability properties than in the case where there are no obstruction.

As the homotopy begins to move around the curves, we are likely to cross new singularities before the pattern is simplified. There are two cases to be considered: Either during the homotopy, all singularities encountered remain in F^1, or during the homotopy a singularity of codimension 2 need to be crossed.

We first formulate the global patterns the removal of which necessitates crossing a singularity of codimension 1. We refer to [Cerf 1970] for the details as to whether the singularity can be crossed.

- *Crossing Removal:* This is the situation where there are two critical values, c_1, c_2, such that $c_1 < c_2$ for $0 \leq \theta < \theta_1$, $c_1 > c_2$ for $\theta_1 < \theta < \theta_2$, and $c_1 < c_2$ for $\theta_2 < \theta \leq 2\pi$. The problem is whether it is possible to find a homotopy that "pushes c_1 up" and "pulls c_2 down" so that the two graphs become disjoint and defined all over $[0, 2\pi]$. The answer depends heavily on the indices of the critical points. See [Cerf 1970, II.4.1, Propositions 3,4, pages 54–55].
- *Death–Birth Fusion:* This is the situation where a pair of critical values c_1, c_2 with indices $\lambda, \lambda + 1$ exist for $0 \leq \theta < \theta_1$, die at $\theta = \theta_1$, is nonexistent for $\theta_1 < \theta < \theta_2$, reappear at $\theta = \theta_2$, and exist for $\theta_2 < \theta \leq 2\pi$. The question is whether it is possible to "fuse" the point of death θ_1 and the point of birth θ_2 so as to have two nonintersecting graphs for c_1, c_2 with indices $\lambda, \lambda + 1$ defined all over $[0, 2\pi]$. The criterion is

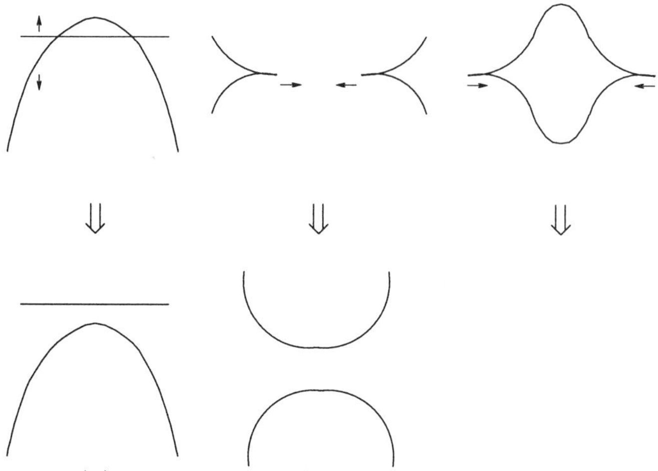

Fig. 21.5. Global approach I.

in [Cerf 1970, III.1.3, Corollary 2, page 67].

- *Birth–Death Annihilation:* This is the situation where at $\theta = \theta_1$ there is the birth of a pair of critical values. The pair of critical values is present for $\theta_1 < \theta < \theta_2$, and for $\theta = \theta_2$ the critical values die. The problem is whether we can bring together θ_1 and θ_2 and annihilate both critical values. The answer is given in [Cerf 1970, III.2.4, Proposition 4, page 73].

Remember that in our case, f_θ is periodic. It follows that any birth is associated with a death and vice versa. The "death–birth fusion" and the "birth–death annihilation" cases are therefore the same, modulo a shift transformation of the parameter θ. To remove these singularities we have the choice between "fusion" or "annihilation," whichever has no obstruction.

The situation is somewhat more complicated with some *global* patterns, which *locally* consist of singularities of codimension 1 but which require crossing a singularity of codimension 2 for their removal. In other words, in the process of deforming the critical value curves, we need to cross a threefold critical value. While the following situations might appear highly pathological, they are, however, very likely to occur in robustness problems, as an example will indicate.

- *The "Beak" Lemma:* This is the situation where a decreasing critical value curve crosses a pair of newly born critical values with indexes $\lambda, \lambda + 1$. The question is whether the descending critical value curve can be moved backward while the birth is delayed to greater θ's. As

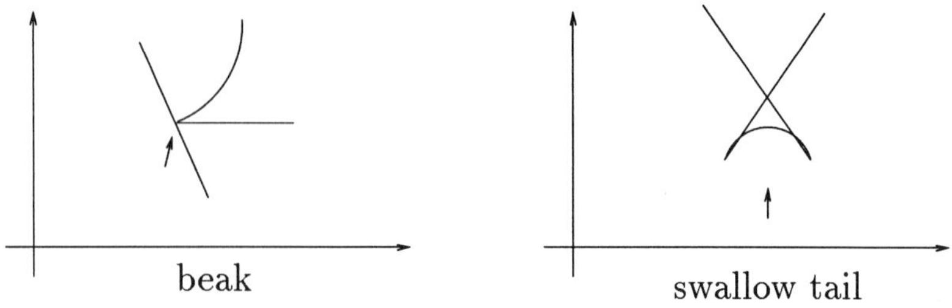

Fig. 21.6. Codimension 2 singularities.

we attempt to separate the descending curve from the birth, it is clear that we have to go through the situation where the descending critical curve passes *exactly* through the birth. This latter is a threefold critical value, hence a singularity of codimension 2. The reader is referred to [Cerf 1970, page 80] for sufficient conditions under which the singularity of codimension 2 can be crossed.

- *The "Swallow Tail":* In this pattern, a λ, $\lambda+1$ birth occurs underneath a descending $\lambda+1$ critical curve; then the descending $\lambda+1$ curve crosses the newly born, ascending $\lambda+1$ curve; and finally, the descending $\lambda+1$ curve and the newly born λ curve annihilate underneath the ascending $\lambda + 1$ curve. The question is whether the tail can be removed, leaving one single $\lambda+1$ curve. Clearly, as the birth, crossing, and death become closer and closer, the tail becomes smaller and smaller, and we need to cross a threefold critical value before the tail can be completely removed. See [Cerf 1970, page 86] for sufficient conditions under which this can be done.

21.6.8 Effect of Variation of "Certain" Parameters

Assume that the nominal Nyquist map, disregarding rounding errors, has its f_θ loop precisely crossing a singularity of codimension 1 or 2. Clearly, an arbitrarily small rounding error could cause the path f_θ to go on either side of the codimension 1 or 2 singularity, resulting in two kinds of critical value plots *that are no longer topologically equivalent.* This is an example of an *unstable Nyquist map* (see Chapter 23). One already perceives at this stage that the case of critical value plots not topologically stable under rounding errors does not bode well on the sensitivity property of the robustness analysis, but we defer the precise analysis of this to Chapter 23.

21.7 Loops of Critical Points

In this section, we focus on the critical *points* rather than the critical values.

From [Cerf 1970], we have to make a distinction between two cases:

- The path f_θ never crosses F^1_α. In this case, there are no births, no deaths, and because of the periodicity in θ, the locis of the critical points are *loops* parameterized by $[0, 2\pi)$.
- If the path f_θ penetrates in F^1_α, there are births and deaths, and eventually the births and deaths combine to give loops of locis of critical points, parameterized by a *subset* of $[0, 2\pi)$.

We begin by deriving some results for the number of critical *points* of the Morse function f_θ for a fixed θ. Then we investigate more carefully how these points draw loops as θ varies.

It is a general feature of Morse theory that the number of critical points is related to the homology of the manifold of definition, regardless of what the Morse function is. To illustrate this, define

$$\{f_\theta(p) : p \in D \times \Omega\} = [f_{\theta,min}, f_{\theta,max}]$$

We already know that $f_{\theta,min}, f_{\theta,max}$ are Morse critical values with indexes $n_p, 0$, respectively. However, there are, in general, *other* Morse critical values within this interval, and some of these critical values may have index 0 or n_p. To see this, one defines, as in the general Morse theory, the "sublevel sets,"

$$M_{\leq v} = \{p \in D \times \Omega : f_\theta(p) \leq v\}$$

The Morse–Smale theory states that the topological type of $M_{\leq v}$ remains unchanged as long as v remains between adjacent critical values. Furthermore, as v crosses over a critical value, the topological type varies in the same manner as attaching the "handle" $D^\lambda \times D^{n_p-\lambda}$ along $(\partial D^\lambda) \times D^{n_p-\lambda}$, where D^i denotes the closed i-disk. For $v < f_{\theta,min}$, $M_{\leq v}$ is the empty set and for $v > f_{\theta,max}$, $M_{\leq v}$ is the full manifold. Therefore, in the interval $[f_{\theta,min}, f_{\theta,max}]$ we must have enough critical values, with the appropriate indexes, in order to have constructed the full manifold by the time we have crossed over $f_{\theta,max}$. As a corollary to this argument, it follows that the number of Morse critical points must be, one way or the other, bounded from below by the topology of the manifold $D \times \Omega$. This result is made precise by the following:

Theorem 21.20. (Morse Fundamental Inequality) *Let $D \times \Omega$ be a compact, differentiable manifold and let f_θ be a good path of functions $: D \times \Omega \to \mathbb{R}$. Then $\forall \theta \in [0, 2\pi)^*$ we have*

$$\#\{\textit{Morse critical points with index } \lambda \textit{ of } f_\theta\} \geq b_\lambda(D \times \Omega)$$

where b_λ denotes the λth Betti number of the manifold $D \times \Omega$.

Proof. See [Milnor 1963, page 29, Theorem 5.2]. ■

Following in the footsteps of this theorem, assume p is a critical point of index λ of f_θ. Then use the lemma of Morse to write the local expansion of f_θ around p. Changing the sign of both sides of the equation and observing that $-f_\theta = f_{\theta+\pi}$, it follows that p is also a critical point of index $n_p - \lambda$ of $f_{\theta+\pi}$. Therefore, any critical point of index λ of f_θ is also a critical point of index $n_p - \lambda$ of $f_{\theta+\pi}$. With this latter argument and the preceding theorem, we reach the conclusion that the number of Morse critical points of index λ of f_θ is greater than b_λ and $b_{n_p-\lambda}$. To avoid inconsistency, we have to invoke the *Poincaré duality theorem*:

Theorem 21.21. (Poincaré Duality) *Let $D \times \Omega$ be a compact, differentiable, orientable manifold of dimension n_p. We have*

$$b_\lambda(D \times \Omega) = b_{n_p-\lambda}(D \times \Omega)$$

Proof. This is one of the oldest, and most celebrated, result of algebraic topology. The early formulation of this result is purely combinatorial and relies on the notion of *triangulated homology manifold*—that is, a simplicial complex enjoying some specific properties one would expect from the triangulation of a *topological manifold*; see [Munkres 1984, Section 63] for the precise definition and the equivalence between a triangulated homology manifold and a pseudomanifold without boundary in the sense of Definition 18.1. For a triangulated homology manifold, the result is proved using two mutually dual dissections of $D \times \Omega$; this duality is essentially the same as the Voronoi diagram/Delaunay triangulation duality (see [Hilton and Wylie 1965, Theorem 4.4.13] and [Munkres 1984, Section 65]). However, more recently, a (complicated!) counterexample (in dimension 5) revealed that a triangulated topological manifold is not always a triangulated homology manifold (see [Munkres 1984, Section 63, Example 2]), so that the combinatorial proof falls short of being general. The proof for a genuine topological manifold is apparently due to Milnor and is less intuitive; it relies on singular homology (see [Massey 1980, Chapter IX, Section 4, Theorem 4.1] and [Dubrovin, Fomenko, and Novikov III 1990b, Section 18]). ■

For an enlightening discussion regarding the role of Poincaré duality in Morse theory, the reader is referred to [Goresky and MacPherson 1988, Introduction, Section 2.7].

At this juncture, the number and the indexes of the critical points of f_θ, for a fixed θ, are well-understood. It remains to allow θ to sweep the interval $[0, 2\pi]$ in order to plot the locus of the critical points versus θ on the uncertainty manifold. Let us assume, to begin with, that there are no births, no deaths. Take a critical point p^* with index λ of f_θ. We call this a (θ, λ) point. The relevant fact is that it is also a critical point with index $n_p - \lambda$ of $f_{\theta+\pi}$. As θ increases, this point traces a curve that goes back to the original point p^*. This generates a "loop." This *single* loop is, however, the common locus of *two* critical points—one critical point with index λ and another critical point with index $n_p - \lambda$. Each locus can be thought of as being charted by θ, and the two charts are 180 degrees off phase. There are, in general, many such loops on the uncertainty manifold. Doing the book-keeping yields the following result:

Theorem 21.22. *Assume f_θ is a good path of functions $D \times \Omega \to \mathbb{R}$ that never crosses F^1_α. Assume each critical point is critical for exactly two θ's differing by π. The number of loops of Morse critical points in $D \times \Omega$ as θ goes from 0 to 2π is greater than or equal to*

$$\frac{1}{2}\sum_{k=0}^{n_p} b_k(D \times \Omega)$$

Proof. Consider first the case where n_p is odd. Observe that by the Poincaré duality the above division is exact. Consider a (θ, λ) critical point where $0 \leq \lambda \leq \frac{n_p-1}{2}$. By the Morse weak inequality, there are at least b_λ such points. As θ increases, the critical point traces a curve and eventually returns to the departure point. Remember that the (θ, λ) starting point is also a $(\theta + \pi, n_p - \lambda)$ point. It follows that starting at those critical points with indexes λ, $0 \leq \lambda \leq \frac{n_p-1}{2}$, we cover all critical points of all indexes. Therefore, the number of loops of critical points is at least $b_0 + \ldots + b_{\frac{n_p-1}{2}}$. Invoking the Poincaré duality, the result, for n_p odd, follows.

We now consider the case where n_p is even. Consider the (θ, λ) critical points, for $0 \leq \lambda \leq \frac{n_p}{2} - 1$. These starting points are also $(\theta + \pi, n_p - \lambda)$ points, so that we cover all critical points, except those with index $\frac{n_p}{2} = n_p - \frac{n_p}{2}$, and this reveals at least $b_0 + \ldots + b_{\frac{n_p}{2}-1}$ loops. It remains to evaluate the number of loops with index $\frac{n_p}{2} = n_p - \frac{n_p}{2}$. The salient result [Massey 1980, IX, Proposition 5.4] is that the Betti number $b_{\frac{n_p}{2}}$ is *even*. Again, starting at a $(\theta, \frac{n_p}{2})$ point, which is also a $(\theta + \pi, n_p - \frac{n_p}{2} = \frac{n_p}{2})$ point, we generate a loop with unique index $\frac{n_p}{2}$. Therefore, the number of all such loops with index $\frac{n_p}{2}$ is at least $\frac{b_{\frac{n_p}{2}}}{2}$. Therefore, the number of loops of critical points of all indexes is at least

$$b_0 + \ldots + b_{\frac{n_p}{2}-1} + \frac{b_{\frac{n_p}{2}}}{2}$$

Appealing one more time to the Poincaré duality, the result follows. ■

Broadly speaking, the preceding results have related the *sum* over all λ's of the number of critical points with index λ to the *sum* of the Betti numbers of the manifold $D \times \Omega$. There are similar results relating the *alternate sum* of the number of critical points to the *alternate sum* of the Betti numbers. The alternate sum of the Betti numbers of a manifold is the celebrated *Euler characteristic*:

Definition 21.23. *The* **Euler characteristic** χ *of a manifold, typically $D \times \Omega$, is defined as*

$$\chi(D \times \Omega) = \sum_k (-1)^k b_k(D \times \Omega)$$

At a much more practical level, the Euler characteristic can be related to the triangulation:

Theorem 21.24. *If the manifold $D \times \Omega$ is triangulable, then*

$$\chi(D \times \Omega) = \sum_n (-1)^n \#(\sigma_n)$$

where $\#(\sigma_n)$ is the number of n-simplexes of the triangulation of the manifold. Furthermore, the right-hand side of the above equation does not depend on the particular triangulation chosen.

Proof. See Section B.2. ■

The result we are aiming at relates the alternate sum of the number of critical points to the Euler characteristic:

Theorem 21.25. *Let f_θ be a good path of functions $D \times \Omega \to \mathbb{R}$. For an arbitrary $\theta \in [0, 2\pi)^*$ let $\#_\lambda(f_\theta)$ be the number of critical points with index λ. Then, for all $\theta \in [0, 2\pi)^*$, we have*

$$\chi(D \times \Omega) = \sum_\lambda (-1)^\lambda \#_\lambda(f_\theta)$$

Proof. This is a direct corollary of a result of [Milnor 1963]. This is, actually, a corollary of the very general Hopf theorem for the index of vector fields (see Section 21.9). ■

In general, computing the critical points of f_θ is very hard and is plagued with sensitivity problems, and even worse the possibility that some critical points might have been missed cannot be ruled out. The above equality relation comes handy as a test which, although not totally foolproof, nevertheless gives some reassurance in the sense that the computed critical set is consistent with the homology of the manifold.

21.8 Degree Approach to Critical Points

The differential topology of the critical points can be approached from a more intuitive point of view than the previous sections by resorting to the analytical degree of Section 18.3.

Consider, locally, the gradient map of f_θ. The local gradient map is viewed as defined on the Euclidean image of a neighborhood of p in $D \times \Omega$ and taking value in $\mathbb{R}^{n_p}$,

$$\nabla f_\theta : O_p \to \mathbb{R}^{n_p}$$

This local version is enough to reinterpret, within the degree context, such phenomena as birth and death of pairs of critical points. Should we need *all* critical points, we need several local versions as the above; the details are left to the reader.

Consider a critical point p^{*i} of f_θ; that is, $p^{*i} \in (\nabla f_\theta)^{-1}(0)$. The linkage between the Morse theory approach and the degree approach is the fact that the Hessian of f_θ is the Jacobian of ∇f_θ. Assume that p^{*i} is a critical point of the *Morse* type; that is, 0 is a *regular* value of the gradient map. Let λ^i be the Morse index of p^{*i} viewed as a critical point of f_θ. Let $J(p)$ be the Jacobian of the gradient map evaluated at p. Clearly,

$$\operatorname{sign}\det J(p^{*i}) = (-1)^{\lambda^i} \tag{21.8}$$

It follows that

$$\operatorname{degree}(\nabla f_\theta) = \sum_i (-1)^{\lambda^i}$$

where λ^i is the Morse index of p^{*i} and the sum is extended over all critical points.

Clearly, variation of θ in ∇f_θ is a homotopy; furthermore, any "certain" parameter continuously deforming ∇f_θ has the same effect as a homotopy. By the fundamental homotopy invariance of the degree, the latter should remain unchanged under variation of θ or any "certain" continuous deformation parameter.

Consider two critical points p^{*i}, p^{*j} of f_θ on a collision course under variation of θ or a "certain" parameter. If the two critical points annihilate, for the degree of the gradient map to remain unchanged, we must have

$$\operatorname{sign}\det J(p^{*i}) + \operatorname{sign}\det J(p^{*j}) = 0$$

From Equation 21.8, it follows that

$$(-1)^{\lambda^i} + (-1)^{\lambda^j} = 0$$

In other words,

$$\lambda^i = (\lambda^j + 1) + 2m, \quad m \in \mathbb{N}$$

Hence, from an elementary degree argument, we recover the result of [Cerf 1970], except that the latter analysis is somewhat more refined and tells us that, under generic circumstances, $m = 0$.

21.9 Vector Field Approach to Critical Points

To proceed more formally than in the previous section, ∇f_θ should be viewed as a *vector field* in the sense of Subsection 13.1.7. It is actually a very specific vector field—it is a *gradient* vector field. This vector field has an index in the sense of Section 18.6. The linkage between the Morse index and the gradient vector field index is the following:

Lemma 21.26. *Let p^* be a zero point of the vector field ∇f_θ with Morse index λ as a nondegenerate critical point of f_θ. Then*

$$ind(p^*, \nabla f_\theta) = (-1)^\lambda$$

Proof. See [Dubrovin, Fomenko, and Novikov 1990a, Corollary 15.2.5]. ■

Now, we quote a very general result about vector fields:

Theorem 21.27. (Hopf) *Let X be a compact, orientable manifold, let V be a vector field, and let $\{x^{*i}\}$ be the set of zero points of the vector field, assumed to be finite. Then*

$$\chi(M) = \sum_i ind(x^{*i}, V)$$

Proof. See [Bott and Tu 1982, Theorem 11.25] or [Dubrovin, Fomenko, and Novikov 1990a, Theorem 15.2.7]. ■

Clearly, combining the above lemma and theorem yields Theorem 21.25. The above theorem also has the following corollary:

Corollary 21.4. *The compact, orientable manifold M has an everywhere nonvanishing vector field iff $\chi(M) = 0$.*

21.10 Quadratic Differential of Nyquist Map

Here we provide a deeper interpretation of the intuitively motivated Hessian of $f_\theta : D \times \Omega \to \mathbb{R}$ in terms of the so-called *quadratic differential* of the *full* map $f : (D \times \Omega) \to \mathbb{R}^2$.

The **quadratic differential** of a map is a family of matrices constructed from second order partial derivatives of the components of the maps, restricted to $\ker(J)$ and mapping into $\text{coker}(J)$. More specifically, consider the Nyquist map $f : D \times \Omega \to \mathbb{R}^2$ with Jacobian J. The quadratic differential is the family of quadratic forms constructed from the family of matrices

$$\alpha \frac{\partial^2 \Re f}{\partial p_i \partial p_j} + \beta \frac{\partial^2 \Im f}{\partial p_i \partial p_j} : \ker(J) \to \text{coker}(J)$$

where the allowable α's and β's are chosen so as to ensure that the map is into $\text{coker}(J)$. Following [Arnold, Gussein-Zadeh, and Varchenko 1985], the signature of the family of quadratic forms is an invariant that provides for a finer classification than the Thom–Boardman classification. Since the relevant invariant is the signature of the quadratic form, there is no harm to normalize the coefficients α, β as

$$(\alpha \; \beta) = (\cos\theta \; \sin\theta)$$

(If $(\alpha, \beta) = 0$, define the quadratic differential to be null.) Therefore, the signature of the Hessian of f_θ

$$\frac{\partial^2 f_\theta}{\partial p_i \partial p_j} : \ker(J) \to \text{coker}(J)$$

restricted to ker(J) with θ adjusted so that the map is into coker(J) is actually the quadratic differential of the full map!

The reader can easily verify for himself that the quadratic differential of a smooth map is essentially the same concept as the Levi form of a biholomorphic map (see Section 3.16).

21.11 Thom–Boardman Singularity Sets

The above procedure of plotting the critical values of f_θ versus θ is merely an intuitive approach to the deeper problem of determining the *singularity sets* of the mapping

$$f : D \times \Omega \to \mathbb{R}^2$$

To define the singularities of such a map, let $T_p(D \times \Omega)$ be the tangent space to $D \times \Omega$ at the point $p \in D \times \Omega$. The tangent space $T_{f(p)}N$ to the Nyquist template N at the point $f(p)$ is defined similarly. We define the differential of the map f at the point p, namely,

$$d_p f : T_p D \times \Omega \to T_{f(p)} N$$

by its matrix representation in terms of partial derivatives,

$$d_p f = \begin{pmatrix} \frac{\partial \Re f}{\partial p_1} & \cdots \\ \frac{\partial \Im f}{\partial p_1} & \cdots \end{pmatrix}$$

Following the Thom–Boardman definition, p is a critical point if the rank of the differential drops at p. We define the **singularity subset** S_r to be the set of points where the rank of the differential drops of r,

$$S_r = \{p \in D \times \Omega : \text{rank } d_p f = 2 - r\}$$

It is important to observe that the singularity set is *not always a manifold*; nevertheless, it is *generically* a smooth manifold (see Section 23.5 or [Golubitsky and Guillemin 1973] and [Boardman 1967] for the concept of "generic"). In this particular Nyquist problem, where the target space is two-dimensional, we only have S_1, S_2 singularity sets.

From

$$\frac{\partial f_\theta}{\partial p} = \begin{pmatrix} \cos\theta & \sin\theta \end{pmatrix} \begin{pmatrix} \frac{\partial \Re f}{\partial p_1} & \cdots \\ \frac{\partial \Im f}{\partial p_1} & \cdots \end{pmatrix}$$

it follows that the rank of the differential $d_p f$ drops at p iff p is a critical point of f_θ for some θ. Therefore,

$$\cup_{k,\theta} p^{*k}(\theta) = S_1 \cup S_2$$

A natural question is whether the intuitive analysis of the critical points of f_θ versus θ could allow for the S_1 versus S_2 discrimination. We have the following:

Theorem 21.28. *$p^\bullet \in S_2$ iff $p^\bullet$ does not depend on θ.*

Proof. If $p^\bullet \in S_2$, it follows that $d_{p^\bullet} f = 0$, so that $(\cos\theta \sin\theta) d_{p^\bullet} f = \frac{\partial f_\theta}{\partial p}|_{p^\bullet} = 0, \forall\theta$, and $p^\bullet$ as a critical point of f_θ does not depend on θ. Conversely, if $p^\bullet$ does not depend on θ as a critical point of f_θ, it follows that $\frac{\partial f_\theta}{\partial p}|_{p^\bullet} = (\cos\theta \sin\theta) d_{p^\bullet} f = 0, \forall\theta$ and hence $d_{p^\bullet} f = 0$; that is, $p^\bullet \in S_2$. ■

If the critical point $p^\bullet \in S_2$ then its critical value becomes

$$f_\theta(p^\bullet) = \cos\theta \Re f(p^\bullet) + \sin\theta \Im f(p^\bullet)$$

and since $p^\bullet$ does not depend on θ, its critical value plot is a pure sine wave.

Corollary 21.5. *Let $p^*(\theta_1) \in S_1$. Then in a neighborhood of θ_1, $p^*(\theta)$ depends on θ.*

Some caution needs to be exercised if we want to use θ as parameter of the critical curve.

Despite the light they shed on the problem, the singularity sets S_r do not reveal the full singularity structure of the map. The Thom–Boardman theory probes deeper by considering, in a nested fashion, the singularities of the map *restricted to its singularity sets.* To define the differential of, say $f|S_r$, we have to assume that S_r is a manifold. This leads to a new singularity set, $S_{r,s}$, defined as the set of points where the rank of the differential of $f|S_r$ drops by s. To be more precise,

Definition 21.29. *Assuming S_r is a manifold, the $S_{r,s}$ singularity set is*

$$\begin{aligned} S_{r,s} &= \{p \in D \times \Omega : \mathit{rank}\ d_p(f|S_r) = 2 - r - s\} \\ &= \{p \in D \times \Omega : \dim(\ker d_p f \cap T_p S_r(f)) = s\} \end{aligned}$$

(For the equivalence between the two definitions, see [Golubitsky and Guillemin 1973, page 152].)

In this particular Nyquist problem, assuming S_1 is one-dimensional, the only nontrivial case is the situation where the rank of the differential of the map

$$f|S_1 : S_1 \to \mathbb{R}^2$$

drops by 1, leading to an $S_{1,1}$ set, which is called **cusp**. Observe that since S_1 is, generically, a smooth manifold, $S_{1,1}$ has to embed into S_1 in such as way as to make it, generically, a smooth manifold. However, mapping this singularity set into $\mathbb{C}$ does not result in a smooth object; typically, $f(S_{1,1})$ is a singular point of $f(S_1)$, as we shall see in the next section.

Clearly, analyzing the singularity of $f|S_1$ is, besides the f_θ analysis, yet another way to capture the singularities (angular points, "kinks,"...) of ∂N.

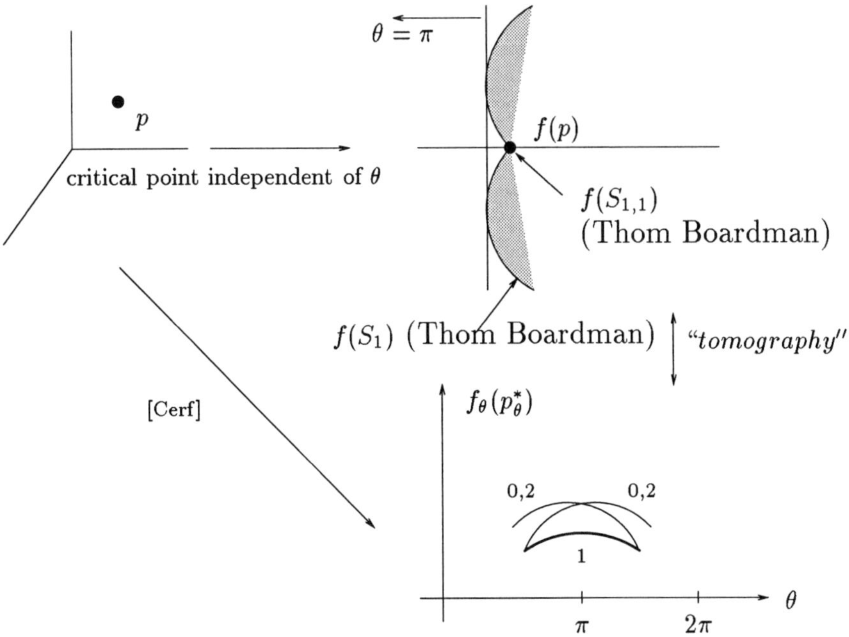

Fig. 21.7. Connection between the "swallow tail" in the Cerf critical value plot, the "kink" in the supertemplate, and the Thom–Boardman singularity sets.

21.12 The Case of Two Uncertain Parameters

If we examine the case of *two* uncertain parameters in the light of the Thom–Boardman singularity theory of the previous section, we immediately run into the celebrated Whitney classification of singularities between 2-manifolds:

Theorem 21.30. (Whitney) *Let $D \times \Omega$ be a smooth 2-manifold and let the Nyquist map $f : (D \times \Omega), p^0 \to \mathbb{R}^2, 0$ be smooth. Then, up to diffeomorphic changes of variables,*

$$\begin{aligned} (p - p^0), 0 &\leftrightarrow x, 0 \\ (\Re, \Im), 0 &\leftrightarrow y, 0 \end{aligned}$$

the only generic cases of local behavior of the map around p^0 are

- *the* **regular point**, *with local canonical form,*

$$y = \begin{pmatrix} x_1 \\ x_2 \end{pmatrix}$$

- *the* **fold**—*that is, p^0 is a S_1 singularity of f and a regular point of $f|S_1$, and the local canonical form of the map is*

$$y = \begin{pmatrix} x_1^2 \\ x_2 \end{pmatrix}$$

- *the* **(Whitney) pleat** *or* **cusp**—*that is,* p^0 *is a* $S_{1,1}$ *singularity, and the local canonical form of the map is*

$$y = \begin{pmatrix} x_1^3 + x_1 x_2 \\ x_2 \end{pmatrix}$$

Proof. See [Whitney 1955]. ■

As we shall see in Section 23.9, the set of all "generic" cases, to which Whitney's theorem is relevant, is everywhere dense in $C^\infty(D \times \Omega, \mathbb{C})$.

The reader can verify for himself that the S_1 singularity set of the fold, called *fold line*, is smooth around p^0 and that its image $f(S_1)$ is also smooth around 0, because p^0 is a regular point of $f|S_1$. The geometric model of the fold (in canonical coordinates) is a sheet of paper folded along a line parallel to the horizontal plane and orthogonally projected to the horizontal plane.

The situation is different with the pleat. While S_1 is a smooth parabola (in the canonical coordinates), its image $f(S_1)$ is not smooth, because $f|S_1$ has a singularity at p^0. The geometric model of the cusp is a sheet of paper with two fold lines annihilating each other at the pleat point and orthogonally projected onto the horizontal plane.

From the above insight, it follows that, when $D \times \Omega$ is compact, smooth, and 2-dimensional, the generic case of a *smooth* boundary point of the template is the critical value associated with a Whitney fold. On the other hand, the generic case of a *nonsmooth* boundary point of the template is the crossing of (the images of) two fold lines.

The Whitney theorem immediately leads to the classification of the structurally stable imaginary axis crossings by root-locis of the form $d(s) + kn(s) = 0$. Indeed, imaginary axis crossing is equivalent to existence of a zero-point of the following Nyquist map, subject to the Whitney theorem:

$$\begin{aligned} f : \mathbb{R}^2 &\rightarrow \mathbb{R}^2 \\ (k, \omega) &\mapsto \begin{pmatrix} \Re d(j\omega) + kn(j\omega) \\ \Im d(j\omega) + kn(j\omega) \end{pmatrix} \end{aligned}$$

From the Whitney theorem, we learn that there are only three structurally stable ways for a branch of the locus to cross the imaginary axis at, say, $j\omega^0$ for, say, k^0:

- Nonsingular crossing—that is, the plot crosses the imaginary axis transversally; (k^0, ω^0) is a regular point of f and a single root of $f(k, \omega) = 0$. From a template point of view, $(0,0) \in \text{int}(f(\mathbb{R}^2))$. Under perturbation of the data $n(s), d(s)$, $f^{-1}(0,0)$ remains a single point, continuously depending on the perturbation, so that the stability margin, measured as the smallest k of imaginary axis crossing, remains continuous.

- Fold crossing—that is, the plot is tangent to the imaginary axis, entirely contained on either side of the imaginary axis; in other words, (k^0, ω^0) is a critical point of f and a regular point of $f|S_1$; furthermore, (k^0, ω^0) is a double root of $f(k, \omega) = 0$. From a template point of view, $(0,0) \in \partial f(\mathbb{R}^2)$. It easy to see that such a plot creates a lack of continuity of the stability margin relative to perturbation. Indeed, under small perturbation of the data, $f^{-1}(0,0)$ either consists of two points or is empty, in which case the imaginary axis crossing disappears and the stability margin changes abruptly from k^0 to a different k corresponding to another crossing.
- Cusp crossing—that is, the plot has a threefold intersection with the imaginary axis, three different branches converging to the threefold point; (k^0, ω^0) is a critical point of both f and $f|S_1$. From the template point of view, $(0,0) \in \text{int}(f(\mathbb{R}^2))$ and $(0,0)$ is the point where two fold lines annihilate. Under perturbation of the data, $f^{-1}(0,0)$ could have three, two, or one point around (k^0, ω^0); in other words, no lack of continuity of the stability margin is associated with a cusp.

As an illustration of a cusp crossing, consider

$$s^3 + k(s+1)$$

From elementary root-locus analysis, three branches of the locus diverge from the triple open-loop pole at $s = 0$ for $k = 0$. Chasing imaginary axis crossing yields the Nyquist map

$$f : (k, \omega) \mapsto \begin{pmatrix} k \\ -\omega^3 + \omega k \end{pmatrix}$$

Clearly, $(0,0)$ is a cusp singularity of the map.

21.13 Template Boundary Revisited

With the machinery developed in the previous two sections, we are in a position to formulate, more completely, the differential topology of ∂N.

Theorem 21.31. *Assume $D \times \Omega$ is a compact, smooth manifold and that f_θ is a good path of functions $D \times \Omega \to \mathbb{R}$. Then ∂N is a piecewise differentiable curve; the differentiable points of ∂N are critical values of the Morse type of f_θ. Furthermore, the differentiable curves interconnect at angular points that are either Morse degenerate critical values or Morse critical values in case two differentiable boundary curves cross.*

21.14 Example I

Consider the following 2×2 loop matrix:

$$L(s,q) = \begin{pmatrix} \frac{1}{s+1}\cos q & -\sin q \\ \sin q & \frac{1}{s+a}\cos q \end{pmatrix}$$

q is the nominal, uncertain parameter. The open-loop pole $-a$ is a parameter that is considered "certain"—that is, a parameter that will be kept constant in the course of the robustness against q analysis. However, we will repeat the robustness analysis for several values of a running in the interval $[1, 2]$ in an attempt to assess the sensitivity of a structured stability margin to parameters such as a that have been, quite subjectively, declared "certain."

For the sake of the mathematical analysis, the loop transmission is factored as follows:

$$L(s)\Delta(q) = \underbrace{\begin{pmatrix} \frac{1}{s+1} & 1 & 0 & 0 \\ 0 & 0 & 1 & \frac{1}{s+a} \end{pmatrix}}_{L(s)} \underbrace{\begin{pmatrix} \Re\exp(jq) & 0 \\ 0 & -\Im\exp(jq) \\ \Im\exp(jq) & 0 \\ 0 & \Re\exp(jq) \end{pmatrix}}_{\Delta(q)}$$

From the above reformulation of the uncertainty Δ in terms of $\exp(jq)$, it follows that, regardless of what $L(s)$ is, we have $D = \mathbb{T}$. Next, the classical compactification of the imaginary axis argument yields $\Omega = \mathbb{T}$. It follows that $D \times \Omega = \mathbb{T}^2$, the 2-torus. The latter is clearly a *compact differentiable* manifold. Remember, the Betti numbers of the 2-torus are

$$b_0 = 1,\ b_1 = 2,\ b_2 = 1$$

Therefore, the Euler characteristic of the 2-torus is

$$\chi(D \times \Omega) = 1 - 2 + 1 = 0$$

It follows from the above homological properties of the domain of uncertainty that the number of critical points of f_θ for an arbitrary $\theta \in [0, 2\pi)^*$ is bounded from below by the inequalities

$$\begin{aligned} \#_0(f_\theta) &\geq 1 \\ \#_1(f_\theta) &\geq 2 \\ \#_2(f_\theta) &\geq 1 \end{aligned}$$

These inequalities remain satisfied regardless of what $L(s)$ is, regardless of the value of the "certain" parameter a. Along the same lines of ideas, assuming that f_θ never crosses F^1_α, the *lower bound* on the number of loops of critical points of all indexes is

$$\frac{1}{2}\sum_k b_k(D \times \Omega) = 2$$

Again, the above result is $L(s)$-independent.

Finding the critical points of indexes 0 and 2 of f_θ is a relatively easy

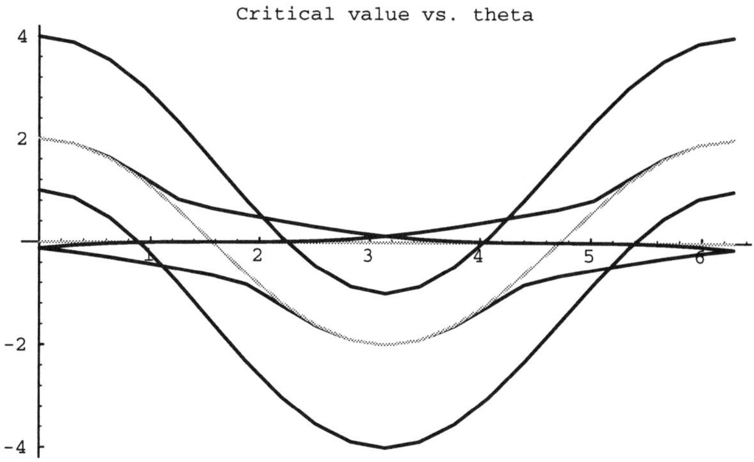

Fig. 21.8. Plot of critical values of f_θ versus θ for $a = 1$ (Example I).

task since these critical points are local minima and maxima, respectively, of f_θ that can be computed using min–max techniques. However, computing the critical points of index 1 is quite another matter since these are *saddle* points that are neither minima nor maxima. At this point, the Euler characteristic argument comes to the rescue. It yields

$$\#_1(f_\theta) = \#_2(f_\theta) + \#_0(f_\theta)$$

The interesting thing with the above is that it provides the *exact* number of critical points with index 1 to be expected, so that we could make sure that no such critical points have been missed. Of course, the above provides only the number of critical *points* and does not preclude several critical *points* being mapped to the same critical *value*.

Observe that the above *homological* results depend *only on the uncertainty*. They do not depend on the loop function $L(s)$; a fortiori, they do not depend on the "certain" parameter a. It is a general pattern of this algebraic/differential topological approach that many structural properties of robustness depend more on the topological type of the uncertainty than on the loop $L(s)$. In the obstruction approach of Chapter 17, Corollary 17.1 of the celebrated *Hopf classification theorem* appeared as an early manifestation of this trend.

The plots of the critical values of f_θ versus θ are shown in Figure 21.8 for $a = 1$ and in Figure 21.9 for $a = 2$. The solid lines are the critical values of indexes 0, 2 while the dotted curves are the critical values of index 1. A first

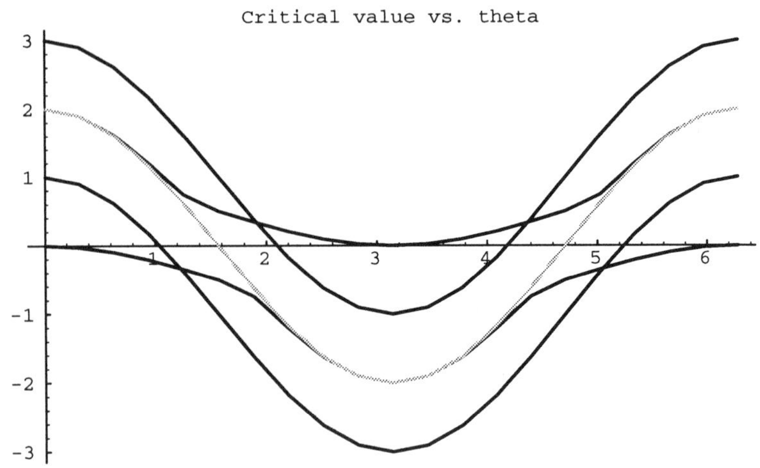

Fig. 21.9. Plot of critical values of f_θ versus θ for $a = 2$ (Example I).

observation is that these plots of critical values—which are nothing other than a graphical representation of the Horowitz template—can get very complicated even for problem with an "easy" external appearance. Observe that the plot for $a = 1$ is the most complicated one, with singularities of codimension greater 2, so that f_θ fails to be a good path.

The plots of the critical *points* of f_θ (or f) for $a = 1$ drawn on a planar model of the 2-torus are shown in Figure 21.10. Again, the solid lines are the critical points of Morse indexes $\lambda = 0, 2$, while the dotted lines are the critical points of Morse index $\lambda = 1$. Observe that the critical points form loops on the torus. Also observe that one of the $\lambda = 1$ loops has tangential contact with a $\lambda = 0, 2$ loop. Because of this tangency situation, the singularity set fails to be manifold. It follows that this $a = 1$ case is *nongeneric* in the sense of Section 23.5.

Mapping the critical point curves to the complex plane using the map f yields the boundary of the template, as shown in Figure 21.11.

We now follow the approach of [Cerf 1970] and use the plots of the critical *values* of f_θ versus θ as the *central* object of focus of our analysis.

The solid $\lambda = 0, 2$ curves are all of multiplicity 1, or codimension 0, except when they cross other curves. It is necessary to take a sharper look at the dotted $\lambda = 1$ curves since they hide some extra pathologies. The "flat" $\lambda = 1$ curve that coincides with the θ-axis in Figure 21.8 is the image under f_θ of one *single* critical point, $p^\bullet = (q = \pi, \omega = 0)$, that does not depend on θ and such that $f(p^\bullet) = 0$. Therefore, the flat $\lambda = 1$

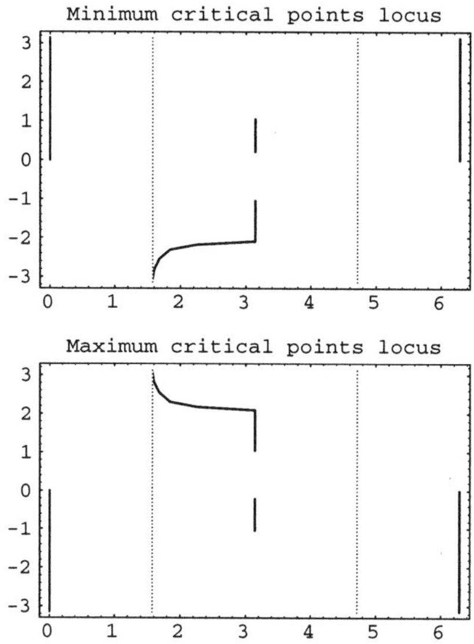

Fig. 21.10. Plots of critical points of f_θ versus θ on a planar model of the 2-torus of uncertainty for $a = 1$; the horizontal axis is q; the vertical axis is ω (Example I).

curve has codimension 0, except at the points where it crosses other curves. The situation is more involved with the cosine-wave-like $\lambda = 1$ curve of Figure 21.8. It turns out that there are two more critical points $p^\bullet = (q = \pm\pi/2, \omega = \infty)$ that do not depend on θ, and such that $f(p^\bullet) = 2$, so that $f_\theta(p^\bullet) = 2\cos\theta$, so that these two constant critical points are mapped to the same cosine wave curve. (The infinite frequency point $\omega = \infty$ has the same status as a finite frequency point since we have compactified the imaginary axis.) Besides the constant critical points, there are two critical points $p^* = (q = \pm\pi/2, \omega^*(\theta))$ which, although they do depend on θ, yield $f(p^*) = 2$, so that $f_\theta(p^*) = 2\cos\theta$, and these two points are also mapped to the same cosine wave curve. It follows that the cosine-wave $\lambda = 1$ curve is a *fourfold* critical value plot.

In the language of singularity sets, the solid $\lambda = 0, 2$ curves are images of critical points in the S_1 singularity set. The flat $\lambda = 0$ curve reveals a critical point in the S_2 singularity set. The cosine wave curve is the image of critical points in S_2 that happens in this example to coincide with the image of critical points in the S_1 set.

For $a = 1$, singularities of f_θ occur at $\theta = 0, \pi/2, \pi, 3\pi/2$. For $\theta = 0$, a singularity is the tangency between the index 2 and the index 1 curves. Likewise, for $\theta = \pi$, a singularity is the tangency between the index 1 and the index 0 curves. A mild kind of singularity (i.e, a singularity of

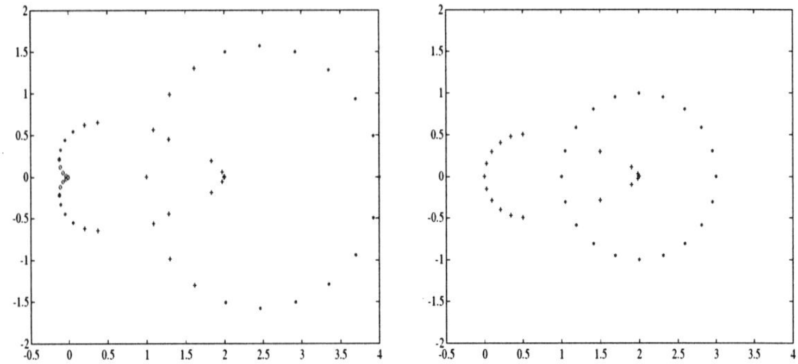

Fig. 21.11. Nyquist template reconstructed by mapping the critical point curves to the complex plane. The left graph is the $a = 1$ case; the right graph is the $a = 2$ case. Observe the "kink" across the real axis in the left graph; observe that the "kink" is not present in the right graph.

codimension 1) that occurs at $\theta = 0, \pi$ is the crossing of two curves with the same index and with distinct critical points. Similar crossings also occur roughly at $\theta = 1, 2, 2\pi - 1, 2\pi - 2$ rad. However, a much harder singularity occurs at $\theta = \pi/2$. The singularity is indeed a combination of a death, a birth, and a crossing. A similar situation occurs at $\theta = 3\pi/2$. From a more global point of view, observe that this plot involves a double "swallow tail." One is centered around $\theta = \pi$, while the other, that is "inverted," occurs around $\theta = 0$. The extra difficulty of this $a = 1$ case is the fact that the two tails are connected to each other at $\theta = \pi/2, 3\pi/2$. All of these pathological phenomena occurring at $a = 1$ and probably due to the double open-loop pole at -1 makes this case very intricate, even beyond the repertoire of singularities of [Cerf 1970].

For $a = 1$, there are $4 + 1 = 5$ critical points of index 1. Therefore, $\#_1(f_\theta) = 5, \forall\theta$. Furthermore, we count a total of five critical values with indexes $0, 2$, so that $\#_0(f_\theta) + \#_2(f_\theta) = 5, \forall\theta$. Clearly, the equality $\#_1 = \#_0 + \#_2$ is satisfied.

To remove the singularities of codimension ≥ 2, we use the perturbation afforded by the freedom in the a-parameter. As the parameter a goes from 1 to 2, observe the gradual simplification of the plots of critical values of f_θ versus θ. In particular, observe the disappearance of the flat $\lambda = 1$ curve.

As we increase a from 1 to 2, we observe that the "hard" singularities at $\theta = \pi/2, 3\pi/2$ disappear because the two "swallow tails" clearly split and move away from $\theta = \pi/2, 3\pi/2$.

After the splitting of the connected swallow tails when a increases from 1, there are two more singularities of codimension 2 that need to be crossed before we reach the $a = 2$ situation. The first singularity of codimension 2 appearing as a goes from 1 to 2 is the so-called singularity of the "beak" lemma. For $a > 1$, there is indeed a descending curve of index 2 crossing the newly born pair of critical values of the swallow tail. As a increases, the curve of index 2 moves backward relative to the birth, and there will be a value of the parameter a at which the curve of index 2 crosses *exactly* the birth. For that particular value of a, f_θ has a singularity of codimension 2; to be more precise, the path f_θ crosses the "stratum" F^2 of the space of differentiable functions. In this particular example, it is clear that there is no "obstruction" to crossing that singularity, although, as shown by [Cerf 1970], crossing a singularity of codimension 2 is not always possible.

For $1 < a < 2$ the critical value count changes around the death and the birth. Before the first death, the count is $\#_1 = \#_0 + \#_2 = 5$, and immediately after this death, the count is $\#_1 = \#_0 + \#_2 = 4$.

Finally, as a approaches 2, the $\lambda = 1$ component of the swallow tail becomes smaller and smaller and the question is whether the $\lambda = 1$ component could completely disappear, leaving one single $\lambda = 2$ curve. Again, this final hurdle is equivalent to the path f_θ crossing over F^2 in the stratified space of differentiable functions. In this particular example, this can be done, although, as proved by [Cerf 1970], this is not a general feature.

For $a = 2$ the Euler alternate sum argument is $\#_1(f_\theta) = \#_2(f_\theta) + \#_0(f_\theta) = 4, \forall\theta$.

Observe that even for $a = 2$ we still have tangency situations at $\theta = 0, \pi$, at which points the Hessian is singular and the plots of corresponding critical *points* are also tangent.

The above complicated situation is compounded by the fact that we still have the cosine wave curve of multiplicity 4 that prevents f_θ to be a good path. From the article of [Cerf 1970], we learn that there is an ϵ-perturbation that splits the fourfold cosine curve into four distinct curves with at most double crossings. After this perturbation we have a path that never crosses F^1_α. Therefore, the argument of the number of loops would yield a lower bound of 2. Counting the number of loops, we see that there are actually $4 + 4 = 8$ of them. Therefore, the lower bound is satisfied, but appears somewhat conservative.

The lesson to be learned from the present exercise is that for relatively simple robust stability problems, the critical points/values plots can become very complicated, thereby making the robustness problem very difficult.

21.15 Example II

This second example is characterized by

$$L(s)\Delta(q) = \begin{pmatrix} \frac{1}{s+1} & 1 \\ 1 & \frac{1}{s+a} \end{pmatrix} \begin{pmatrix} \cos q & -\sin q \\ \sin q & \cos q \end{pmatrix}$$

As in the previous example, after compactification of the imaginary axis to the unit circle, the domain of uncertainty becomes the 2-torus $\mathbb{T}^2$. The open-loop pole $-a$ is a "certain" parameter that is allowed to vary in $[1, 2]$.

The critical points of f_θ for $a = 1$ plotted on a planar model of $\mathbb{T}^2$ are shown in Figure 21.12.

The plots of the critical values of f_θ versus θ are shown in Figures 21.13, 21.14, 21.15, and 21.16, for $a = 1$, $a = 1.05$, $a = 1.3$, and $a = 2$, respectively. The major phenomenon to be observed is the appearance of a swallow tail for $a = 1.05$; the swallow tail progressively shrinks as a increases, and eventually disappears for $a = 2$.

It is important to observe that the disappearance of the swallow tail in the critical value plots is associated with the disappearance of the kink across the real axis in the supertemplate, as shown by Figures 21.17 and 21.18.

It should be stressed that computing the critical points and the critical values is very tedious, because indeed these computations are plagued with sensitivity problems; for details, see [Cheng 1994].

Besides the boundary of the Horowitz supertemplate, there are other motivations for looking at the singularity structure of the Nyquist map. In Chapter 23, we will develop the structural stability of the crossover as a byproduct of the singularity structure. In the mean-time, there is an easy interpretation of the critical value curves drawn on the supertemplate. Indeed, when the manifold of uncertainty is uniformly gridded, the images of the sample points are clustering around the critical value curves of the supertemplate, as shown in Figure 21.20.

21.16 Cell Decomposition

To provide an intuitive motivation for this section, consider the problem of constructing a simplicial approximation that requires a minimum amount of decomposition of $D \times \Omega$ and N. Clearly, such a decomposition for N would hinge on ∂N. ∂N would be approximated with a string of 1-simplexes. Since the simplicial map would map some simplexes of $D \times \Omega$ to those approximating ∂N, it would be a wise idea to triangulate $f^{-1}(\partial N)$, get a simplicial map $f^{-1}(\partial N) \to \partial N$, and then "hinge" the decomposition of $D \times \Omega$ on the decomposition of $f^{-1}(\partial N)$.

The only problem with this intuitive idea is Zeeman's counterexample to the relative simplicial approximation: There are problems at extending the simplicial approximation $\partial f^{-1}(\partial N) \to \partial N$ to a simplicial approximation $D \times \Omega \to N$. Nevertheless, the intuitive idea can be made to work, *provided that we renounce to rectilinear subdivisions* and use $f^{-1}(\partial N)$ as the "wall" of a decomposition of $D \times \Omega$ into curvilinear "building blocks." The Nyquist map would become "simplicial" in the sense that it would map a "building block" of $D \times \Omega$ to a "building block" of N.

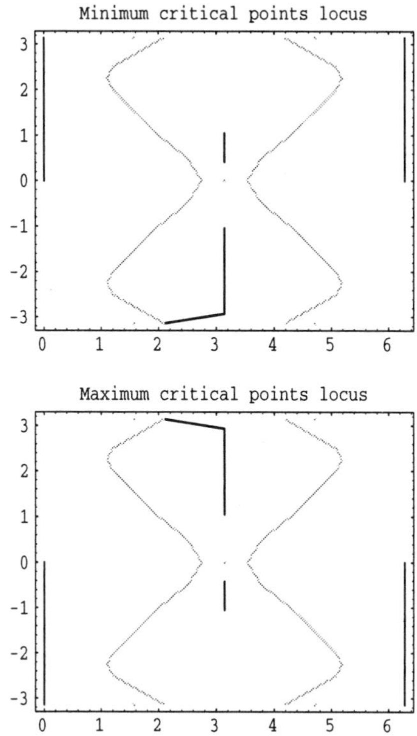

Fig. 21.12. Plots of critical points of f_θ versus θ on a planar model of the torus of uncertainty for $a = 1$; the horizontal axis is q; the vertical axis is ω (Example II).

In this section, we make the above intuitive ideas precise and aim at a cellular decomposition of the Nyquist map in the spirit of Section 4.8. In contrast to *rectilinear* subdivisions, here we allow for decompositions into curvilinear objects, or "cells," so as to make the Nyquist map decomposition preserving. This new decomposition hinges on the critical points/values. Restricted to each cell, the inverse image and the boundary commute.

As argued in Section 3.2, it is necessary to be specific as to *what* topologies are used for $D \times \Omega$ and N. Here we use the intersection topology of N together with the topology of $D \times \Omega$ that comes from the topological space part of the manifold structure. Combining these two topologies make any reasonably well-behaved Nyquist map continuous. A related issue is the extent to which the neighborhood topology of $D \times \Omega$ reveals the boundaries. The ∂ operator of the neighborhood topology of $D \times \Omega$ yields the correct result that $\partial D \times \Omega = \emptyset$. Next, if we break $D \times \Omega$ into pieces, each piece is

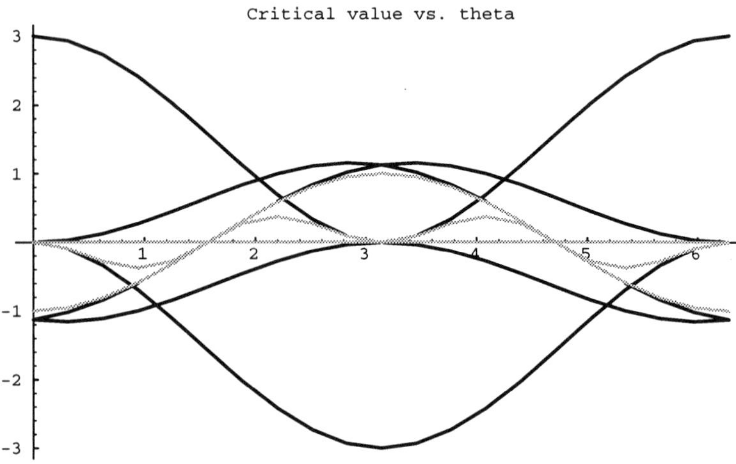

Fig. 21.13. Plot of critical values of f_θ versus θ for $a = 1$ (Example II).

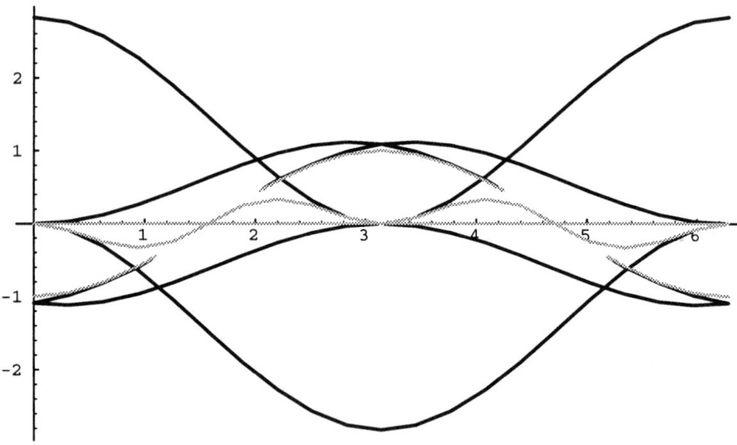

Fig. 21.14. Plot of critical values of f_θ versus θ for $a = 1.05$ (Example II). Observe the swallow tail around $\theta = \pi$.

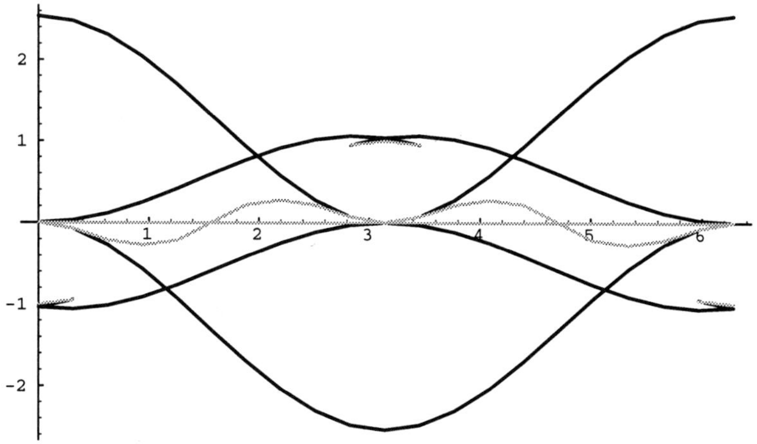

Fig. 21.15. Plot of critical values of f_θ versus θ for $a = 1.3$ (Example II). Observe that the swallow tail is shrinking.

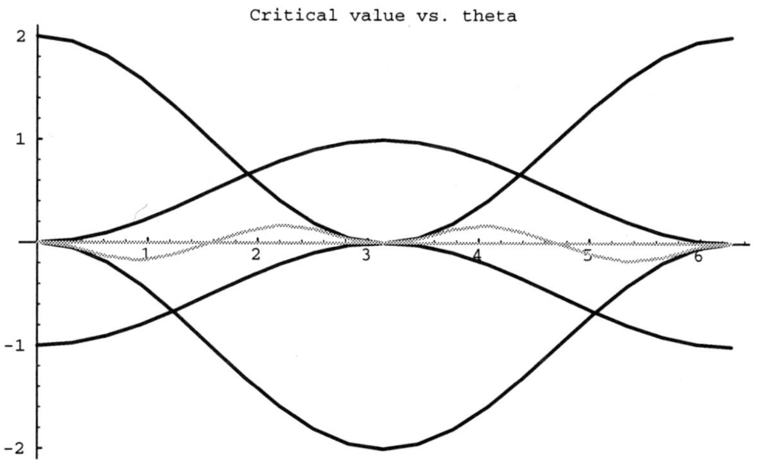

Fig. 21.16. Plot of critical values of f_θ versus θ for $a = 2$ (Example II). Observe that the swallow tail has disappeared.

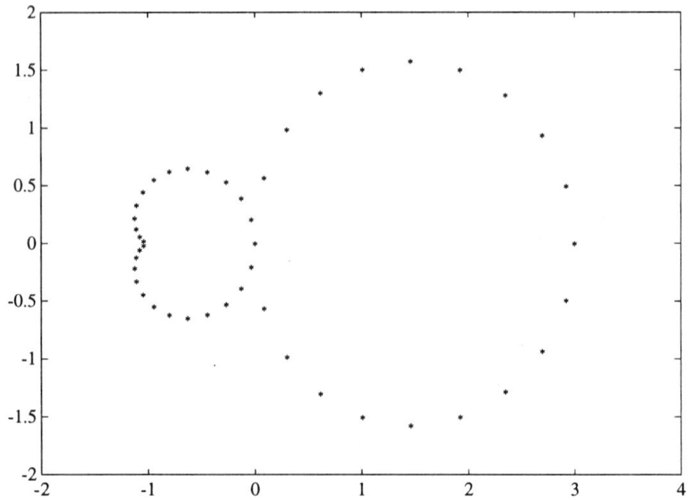

Fig. 21.17. The part of the boundary of the supertemplate for $a = 1$ (Example II), obtained as image of critical point curves of Morse index $\lambda = 0, 2$. Observe the "kink" across the real axis.

embedded in the bigger topological space $D \times \Omega$, so that the intersection topology of any piece does reveal its boundary. The reader unfamiliar with these concepts is urged to go back to Section 3.2.

Remember that the critical values set is one-dimensional; it is thus a network of curves within the Nyquist template N. Because of the periodicity of the critical values of f_θ in θ and the general restrictions on the critical values of a good path of function, it turns out that the critical value curves all close on themselves. Therefore, by the Jordan curve theorem, these curves partition N into a certain number of regions, N_2^i, where the subscript 2 refers to the dimension. Let N_1^i be the curves after removal of the points of crossing, birth, and death; again, the subscript 1 in N_1^i stands for the dimension. Finally, let all the points of crossing, birth, and death be lumped in N_0^i, where the superscript 0 stands as before for the dimension of the set. Therefore, we have a *partitioning* of N as

$$N = \cup_{n,i_n} N_n^{i_n}$$

Define

$$P_n^{i_n} = f^{-1}(N_n^{i_n})$$

For convenience of the notation, we also define

$$N_n = \cup_{i_n} N_n^{i_n}, P_n = \cup_{i_n} P_n^{i_n}$$

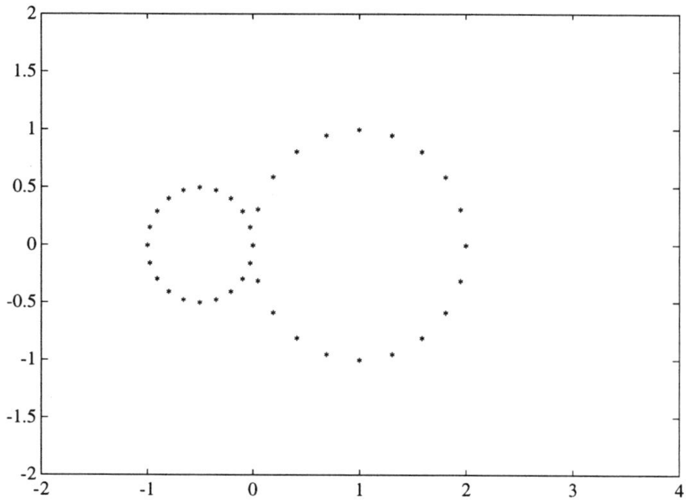

Fig. 21.18. The part of the boundary of the supertemplate for $a = 2$ (Example II), obtained as image of critical point curves of Morse index $\lambda = 0, 2$. Observe that the "kink" has disappeared.

Observe that $N_0 \cup N_1$ is the set of *all* critical values; to be more specific, N_1 is made up with all critical values of the *Morse* type, whereas N_0 consists of all degenerate critical values. Therefore,

$$P_2 \subseteq \{\text{regular points of } f\}$$

$$P_0 \cup P_1 = f^{-1}(\{\text{critical values}\}) \supseteq \{\text{critical points}\}$$

Clearly, $N_2^{i_2}$ is open, and by definition of continuity of the function f, it follows that $P_2^{i_2}$ and hence P^2 are open. Furthermore, since $N_0 \cup N_1$ is closed for the usual topology of $\mathbb{R}^2$, continuity of f implies that $P_0 \cup P_1$ is also closed. ($P_0 \cup P_1$ is also closed as the complement of the open set P_2 in $D \times \Omega$.) It is a general feature of continuous maps that the inverse image of a partitioning of the image yields a partitioning of the domain of definition,

$$D \times \Omega = \cup_{n,i_n} P_n^{i_n}$$

Therefore, we have a partitioning $\cup_{n,i_n} P_n^{i_n}$ of the uncertainty space, as well as a partitioning of the Nyquist template $\cup_{n,i_n} N_n^{i_n}$, and the partitioning is preserved under the map f, because indeed

$$f(P_n^{i_n}) = N_n^{i_n}$$

Such a map is said to be *cellular*.

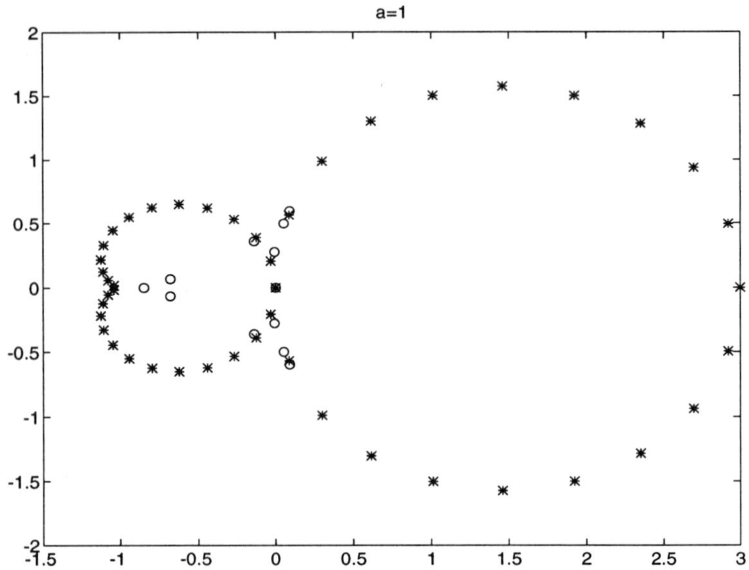

Fig. 21.19. Critical value curves drawn in the supertemplate for $a = 1$ (Example II). Observe the critical value curves around the "kink" matching the area of clustering of sample image points $\{b^k\}$ in the Voronoi diagram of Figure 6.3.

We now examine the commutativity properties to be expected from such a decomposition. We first need the following lemma:

Lemma 21.32. $\overline{f^{-1}(N_2^i)} \subseteq f^{-1}(\overline{N_2^i})$

Proof. This is a general feature of continuous maps (see [Dugundji 1970, page 80]). ■

Observe that, in general, and even in this context, it is not possible to strengthen the result to equality. To get the feeling for this, consider first the following simple example of the map $g : \mathbb{R} \to \mathbb{R}$ defined by

$$g(x) = x^2, \qquad x \leq 0$$
$$g(x) = 0, \qquad 0 \leq x \leq 1$$
$$g(x) = (x-1)^2, \qquad x \geq 1$$

This map is clearly C^∞. Take $A = (0, 1)$. Clearly,

$$g^{-1}(\overline{A}) = [-1, 2]$$

while

$$\overline{g^{-1}(A)} = [-1, 0] \cup [1, 2]$$

We can also construct another example, conceptually much closer to the Nyquist problem. Construct in 3-D Euclidean space with coordinates

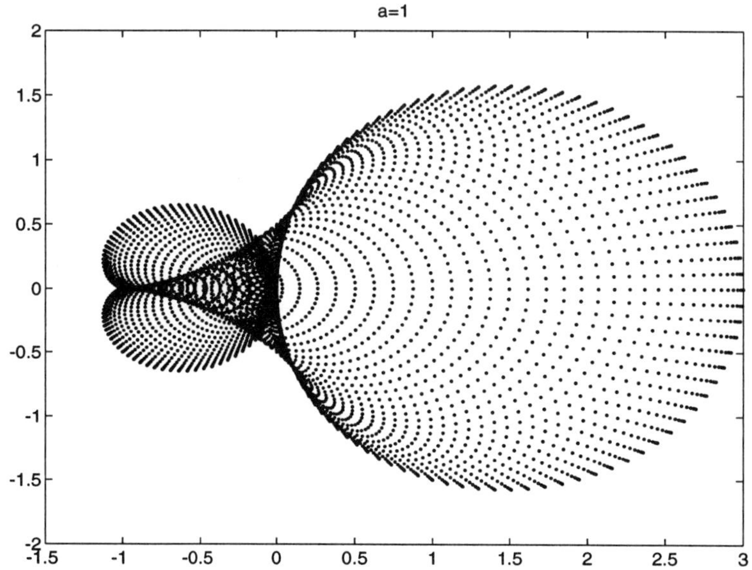

Fig. 21.20. Images of the sample points of a uniform gridding of the domain of uncertainty $[0, 2\pi]^2$ of Example II. Observe that the image points are clustering around the critical value curves shown in Figure 21.19. See also Figure 6.3.

(x, y, z) the following compact differentiable 2-manifold: Define the cylinder:

$$C = \{(x, y, z) : x^2 + y^2 = 1,\ 2 \leq z \leq 3\}$$

The top and the bottom of the cylinder are closed with the (open) half-spheres

$$\begin{aligned} H^+ &= \{(x, y, z) : x^2 + y^2 + (z-3)^2 = 1, z > 3\} \\ H^- &= \{(x, y, z) : x^2 + y^2 + (z-2)^2 = 1, z < 2\} \end{aligned}$$

respectively. Define the 2-manifold P to be the disjoint union

$$P = C \cup H^+ \cup H^-$$

Define the projection map

$$\begin{aligned} f : P &\rightarrow \mathbb{R}^2 \\ (x, y, z) &\mapsto (x, y) \end{aligned}$$

Clearly, the "template" is

$$f(P) = \{(x, y) : x^2 + y^2 \leq 1\}$$

The critical point set is clearly C and the critical values set is

$$\{(x, y) : x^2 + y^2 = 1\} = N_1$$

Clearly,

$$N_2 = \{(x,y) : x^2 + y^2 < 1\}$$

Clearly, we have

$$f^{-1}(N_2) = H^+ \cup H^-$$

so that

$$\overline{f^{-1}(N_2)} = \overline{H^+} \cup \overline{H^-}$$

On the other hand,

$$f^{-1}(\overline{N_2}) = P$$

The discrepancy is obvious. This example is, however, very pathological. It is an example of an *unstable* map (see Chapter 23). A slight tilting of the cylinder will correct the situation.

With the preceding lemma, we can establish a general fact:

Theorem 21.33. $f^{-1}(\partial N_i^2) \supseteq \partial f^{-1}(N_i^2)$.

Proof. The proof relies on the general fact that all Boolean operations on sets are preserved under the inverse image of a continuous map:

$$\begin{aligned}
f^{-1}(\partial N_i^2) &= f^{-1}(\overline{N_i^2} \setminus N_i^2) \\
&= f^{-1}(\overline{N_i^2}) \setminus f^{-1}(N_i^2) \\
&\supseteq \overline{f^{-1}(N_i^2)} \setminus f^{-1}(N_i^2) \\
&= \overline{P_i^2} \setminus P_i^2 \\
&= \partial P_i^2 \\
&= \partial f^{-1}(N_i^2)
\end{aligned}$$

∎

This inclusion is *opposite* to that which would indicate existence of a robust stability test. Nevertheless, this theorem singles out a *general* fact about decomposed maps, *general* in the sense that nowhere in the proof of this result have we made use the variational properties of the decomposition. Decomposing the map starting from any network of closed curves in the image would yield the same result.

Subject to some extra condition, we can get an *equality* relationship, relevant to robust control:

Theorem 21.34. *If the Nyquist map has the additional property that*

$$\overline{f^{-1}(N_2^i)} = f^{-1}(\overline{N_2^i})$$

then

$$f^{-1}(\partial N_2^i) = \partial f^{-1}(N_2^i)$$

Proof. Trivial modification of the above proof. ∎

Again, the above result remains valid for *any* cell decomposition of the template, provided the technical requirement $\overline{f^{-1}(N_2^i)} = f^{-1}(\overline{N_2^i})$ be satisfied. However, to get the correct *forward* image property of the map, we do need the cell decomposition that hinges on the critical values.

Lemma 21.35. *If $f : P \to \mathbb{R}^2$ is continuous and if $f(A)$ is not open for some open subset A of P, then A contains a critical point.*

Proof. If $f(A)$ is not open, there there exists a point $z \in \partial f(A), \in f(A)$. Let θ be the direction normal to $\partial f(A)$ at z. Take $p \in f^{-1}(z)$. A classical variational argument shows that $p \in A$ is critical for f_θ and hence for f. ■

Theorem 21.36. *With the cell decomposition hinged on the critical values, we have*

$$f(\partial P_2^i) \supseteq \partial f P_2^i$$

Proof. Take

$$z \in \partial f P_2^i = \partial N_2^i \subseteq N_1$$

We must show that $\exists p \in \partial P_2^i$ such that $z = f(p)$. Assume $\nexists p \in \partial P_2^i$ such that $z = f(p)$. Hence $f^{-1}(z) \subseteq P_2^i$. Therefore, any preimage p of z is a regular value. Let O_p be a neighborhood of p contained in P_2^i. Since p is regular, $f(O_p) = O_z$ is an (open) neighborhood of z (see Lemma). Since $z \in \partial N_2^i$, it follows that a subset of P_2^i is not mapped to $f(P_2^i) = N_2^i$. A contradiction. ■

NOTES

The crucial fact that the boundary of the template is related to a rank deficient Jacobian has been around for a while; see, for example, [Ackerman 1993], [Anderson, Jury, and Mansour 1987], and [Kraus, Anderson, and Mansour 1989]. However, the deeper singularity analysis of the boundary of the template and the other critical value curves running inside the template is believed to be novel.

The idea of looking at the template at an angle θ was introduced in [Jonckheere and Cheng 1993].

22

SINGULARITY OVER STRATIFIED UNCERTAINTY SPACE

To get the gist of things, the reader should be able to see stratified spaces almost everywhere.

S. Weinberger,
The Topological Classification of Stratified Spaces,
Chicago Lecture in Mathematics, 1994,
page 7.

SUMMARY

Originally the Morse theory was developed for real-valued functions defined on *compact, differentiable* manifolds [Morse 1925], [Milnor 1963]. Unfortunately, this does not lead us further than the multivariable phase margin problem where, typically, $f_\theta : \mathbb{T}^{n_q} \times \mathbb{T} \to \mathbb{R}$. In many other structured stability problems, the set of uncertainties $D \times \Omega$ is more complicated than a mere compact differentiable manifold. It suffices that the uncertainty manifold has a boundary for the "global" results of the previous chapter to fail, although the "local" results remain valid modulo some revisions. Early generalizations of the Morse theory to include the case of manifolds-with-boundary are due to Morse himself. This effort and other early attempts to extend the Morse theory have, more recently, crystallized around the so-called *the Stratified Morse Theory* [Goresky and MacPherson 1988]. The stratified Morse theory extends the classical Morse theory to the so-called *stratified* spaces, that encompass practically all sets of uncertainties encountered in robust stability. The idea of the stratified Morse theory, which comprises the case of an uncertainty manifold with boundary, is to decompose, or *"stratify,"* the space in a disjoint union of submanifolds, or *"strata,"* of varying dimensions. Typically, the boundary would be a low-dimensional stratum while the interior of the uncertainty space is the highest-dimensional stratum. Since each stratum is open, the usual local variational principles of the classical Morse theory apply to the *restriction* of the Nyquist map to each stratum. Therefore, applying the local concepts of the previous chapter, we plot $f^{-1}\partial N$ on the uncertainty manifold. The *specific* feature of the *stratified* Morse theory is the *global* behavior of the critical points that could hit the boundary and, more generally, move across the strata. Should the critical points remain within the low-dimensional strata, it would suffice to check stability on those low-dimensional strata. At the other extreme of the spectrum, the higher dimensional the strata in which the critical points penetrate, the higher dimensional the subsets of uncertainties on which a stability test has to be conducted.

22.1 (Whitney) Stratified Uncertainty Space

Consider the usual Kharitonov cube. Remember, the Kharitonov cube is *closed*. In this case, D is *not* a manifold, because none of its boundary points has a neighborhood homeomorphic to (an open subset of) the Euclidean space. The next attempt would be the concept of *manifold-with-boundary*:

Definition 22.1. (Spivak 1965, page 113) *An n-dimensional* **manifold-with-boundary** *is a Hausdorff topological space where every point has a neighborhood homeomorphic to an open set of the Euclidean half-space defined as*

$$\mathbb{HR}^n = \{(x_1, ..., x_n) \in \mathbb{R}^n, x_n \geq 0\}$$

Observe that the Kharitonov cube not only has boundaries (its faces) but also has "corners" (its edges and vertices). This leads to the more general concept of *manifold-with-corner*:

Definition 22.2. (Spivak 1965, pages 131 and 137) *An n-dimensional* **manifold-with-corners** *is a Hausdorff topological space where every point has a neighborhood homeomorphic to an open set (for the relative topology) of a subset of the Euclidean space $\mathbb{R}^n$ bounded by $(n-1)$-D planes.*

Observe that, from a purely topological point of view, there is no difference between a manifold-with-boundary and a manifold-with-corners. Indeed, the Euclidean half-space is homeomorphic to that part of the Euclidean space bounded by $(n-1)$-D planes. For example,

$$\mathbb{HR}^2 = \{(x_1, x_2) \in \mathbb{R}^2 : x_2 \geq 0\} \cong \{(x_1, x_2) \in \mathbb{R}^2 : x_1 \geq 0, x_2 \geq 0\}$$

To show homeomorphism, use, for example, a Schwarz–Christoffel conformal transformation that extends to a homeomorphism of the boundaries (see Section 3.11). Despite this homeomorphism, differences between manifolds-with-boundaries and manifolds-with-corners occur at differential analysis.

A difficulty in dealing with the concept of manifold-with-boundary is that it does not appear to be known whether a general manifold-with-boundary is triangulable (see [Munkres 1984, page 200]). For that reason, and also because of the clumsiness of the concept of manifold-with-corner and the apparent lack of theory of functions defined on such manifolds (apparently, only [Spivak 1965] and [Mather 1969a] introduce the latter

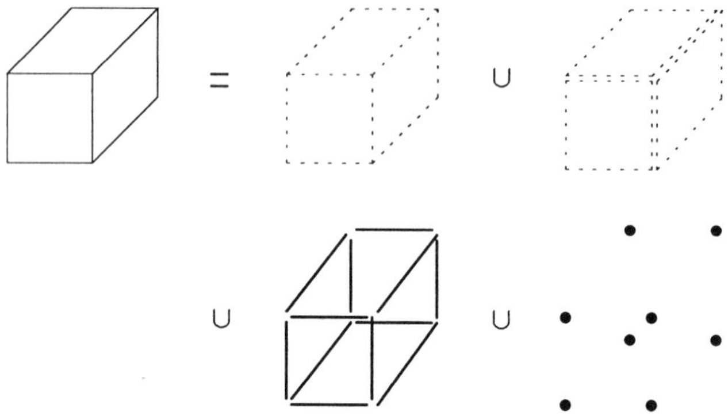

Fig. 22.1. Stratification of Kharitonov cube.

notion), we prefer to resort to a more modern concept of singular spaces—the concept of *stratified* spaces.

To grasp this concept, we come back to the Kharitonov cube. Although the closed Kharitonov cube is *not* a manifold, it can nevertheless be decomposed into manifolds,

$$\text{closed cube} = \{\text{vertices}\} \cup \{\text{open edges}\} \cup \{\text{open faces}\} \cup \text{open cube}$$

This decomposition is illustrated in Figure 22.1. That each piece of the above disjoint union is a manifold is not hard to see. Less trivial is the fact that these are *smooth* manifolds.

Theorem 22.3. *The vertices, open edges, open faces, and open cube of the decomposition of the closed Kharitonov cube are smooth manifolds.*

Proof. This is done by proving that an open face of the cube, for example, is diffeomorphic to the Euclidean space. This is achieved by "smoothing" the corners. ■

The above decomposition of the Kharitonov cube is just a manifestation of the modern concept of *Whitney stratification*, which we will use to deal with the Kharitonov cube and other singular uncertainty spaces.

Definition 22.4. (Goresky and MacPherson 1988, Part I, 1.1,1.2) *A closed subset D of a smooth manifold $\mathcal{X}$ is said to be a* **(Whitney) stratified space** *if it is the disjoint union*

$$D = \cup_\alpha S_\alpha$$

of locally closed, smooth submanifolds of $\mathcal{X}$, subject to the condition

$$S_\alpha \cap \overline{S_\beta} \neq \emptyset \Leftrightarrow S_\alpha \subseteq \overline{S_\beta}$$

Furthermore, the above defines a partial ordering, facing relation on the collection $\{S_\alpha\}$ of subspaces,

$$S_\alpha \preceq S_\beta \Leftrightarrow S_\alpha \subseteq \overline{S_\beta}$$

The submanifolds S_α *are called* **strata**.

The **tangent space** *at a point* p *of a (Whitney) stratified space is defined as the usual tangent space to the unique stratum that contains* p. *A* **generalized tangent space** *to the point* $p \in S_\alpha \subset \overline{S_\beta}$ *is any space of the form*

$$Q = \lim_{p^k \to p} T_{p^k} S_\beta$$

where

$$S_\beta \ni p_k \to p \in S_\alpha$$

A stratification subject to the **Whitney condition**

$$T_p S_\alpha \subseteq Q$$

is called **Whitney stratification**.

In the above definition, we omitted one extra Whitney condition which we will not need in the sequel; for details, see [Bochnak, Coste, and Roy 1987, Section 9.6] and [Goresky and MacPherson 1988, I, 1.2].

Observe the need to define the Whitney stratified space as embedded in a big manifold $\mathcal{X}$, together with its topology $\mathcal{T}(\mathcal{X})$, from which the strata and the whole Whitney stratified space inherit their topology through relativization.

The Kharitonov cube is a stratified space. The open cube is the *largest stratum*; the vertices constitute the *smallest strata*, and the open faces and the open edges constitute the *middle strata*. This stratification is shown in Figure 22.1.

The boundary of a Whitney stratified space can be defined as follows: Let S_ω be the largest stratum. We have

$$\partial D = \cup_{S \prec S_\omega} S$$

Observe that the strata of the decomposition need not be *compact* manifolds as is the case in the classical Morse theory. However, since the strata are smooth manifolds, any function defined on the stratification behaves locally around a point p, and provided that the function is restricted to the unique stratum that contains p, like a classical smooth function defined on a smooth manifold. The difficulty of course is to ensure enough consistency of the differentiable structure at the interconnections of the strata. There, the Whitney condition becomes instrumental.

The Kharitonov cube topologically multiplied by the unit circle to include the frequency parameter is also a stratified space. Indeed, it is easily verified that

$$(\cup_\alpha S_\alpha) \times \mathbb{T} = \cup_\alpha (S_\alpha \times \mathbb{T})$$

is a stratification of Cube $\times\ \mathbb{T}$. The above is, actually, a particular case of the formula of [Cerf 1970, page 16], expressing the stratification of the topological product of two stratified spaces as the "convolution" between the strata of the factors.

The concept of *Whitney* stratified space provides some sort of a "bridge" between the simplicial techniques of Part I and the present "manifold" approach. This is revealed by the following:

Theorem 22.5. *A Whitney stratified space (typically $D \times \Omega$) is triangulable; that is, there exists a simplicial complex K and a homeomorphism $h : |K| \to D \times \Omega$.*

Proof. See [Goresky and MacPherson 1988, Part I, Section 1.2] and [Munkres 1984]. ■

Corollary 22.1.

$$\partial D \times \Omega = h(\partial |K|)$$

Proof. Follows from the triangulability of Whitney stratified spaces together with the Brouwer domain invariance. ■

From the above, it is not hard to see that the Poincaré duality does not in general hold for a Whitney stratified space. Since the Poincaré duality played a crucial role in the *global* behavior of the critical points on a compact, differentiable uncertainty manifold, we already perceive that the global behavior in the stratified case will be quite different than that in the manifold case.

After having emphasized the Whitney stratified property of the domain of uncertainty, we observe that the *template* provides another, quite natural example of a Whitney stratified space. The interior of the template is the highest stratum, the smooth part of the boundary is the middle stratum, and the singular part of the boundary ("kink" points) constitute the lowest stratum.

This leads to the concept of *stratifying map.* If $D \times \Omega$ and N are stratified spaces, the Nyquist map f is said to be **stratifying** iff for any stratum S'_α of N, $f^{-1}(S'_\alpha)$ is the union of connected components of strata, each component mapping submersively to S'_α (see [Goresky and MacPherson 1988, 1.6]). Clearly, a Nyquist map is not always stratifying. However, it can be shown that any algebraic map between algebraic varieties can be made stratifying for some stratifications of both the domain and the target (see [Goresky and MacPherson 1988, 1.7]). The decomposition of $D \times \Omega$ and N of Section 21.16 can be formalized along those lines.

22.2 Stratified Morse Theory

To extend such notions as Morse function, Morse index, Morse data, and so on, to stratified spaces, observe that *any* $p \in D \times \Omega$ belongs to a *unique* stratum of $D \times \Omega$. This leads to the following:

Definition 22.6. (Tangential Morse Data) *$p^* \in D \times \Omega$ is a* **critical point** *of the smooth function $f_\theta : D \times \Omega \to \mathbb{R}$ defined over a Whitney stratified space if p^* is a critical point (in the usual sense) of the restriction $f_\theta|S$ of f_θ to the unique stratum S that contains p^* or, more formally, if $df_\theta|T_{p^*}S = 0$. A* **critical value** *is the value taken by f_θ at a critical point. f_θ is a* **Morse function** *iff*

- *for any critical point $p^* \in S$, the Hessian of $f_\theta|S$ at p^* is nonsingular,*
- *for any generalized tangent space Q at $p^* \in S$, $d_{p^*}f_\theta(Q) \neq 0$, except for $Q = T_{p^*}S$, where $T_{p^*}S$ is the tangent space to S at $p^* \in S$ which can be defined by $d_{p^*}f_\theta(T_{p^*}S) = 0$.*

The **Morse index** *at a critical point p^* of a stratified space is the usual Morse index of $f_\theta|S$. The corresponding Morse data are referred to as* **tangential**.

A difficulty with the above definition is how to define a critical point of a zero-dimensional stratum—that is, a vertex. If we trace back to the most fundamental definition of a critical point as a point where the differential vanishes, it follows that *all zero-dimensional strata (i.e., vertices) are critical* (see [Goresky and MacPherson 1988, page 4]).

The motivation for the terminology of *tangential* Morse data stems from the fact that the function f_θ is analyzed in the stratum that contains the critical point p^*. However, in a stratified space, the stratum $S \ni p^*$ fits in between other strata, and it is imperative to know how f_θ behaves in the neighboring strata. This is accomplished by taking a slice perpendicular to S and checking the behavior there.

Definition 22.7. (Normal Morse Data) *Assume that the stratified space is embedded in a Euclidean space. Take a small ball $B_p(\epsilon)$ transverse to $S \ni p$. (The dimension of $B_p(\epsilon)$ is the dimension of the Euclidean space minus the dimension of the stratum $S \ni p$.) The* **normal slice** *$\mathcal{N}$ is defined to be the intersection of the small ball with the stratified space. The boundary $\partial\mathcal{N}$ is called the* **link** *$\mathcal{L}$. $\mathcal{N}$ is the cone over the link $\mathcal{L}$ with vertex p. The* **normal Morse data** *are the pair $(\mathcal{N}, \mathcal{L})$.*

The major result of [Goresky and MacPherson 1988] is that the topological change of the sublevel set as a critical value is crossed is the product of the tangential and normal Morse data.

As already argued, the issue in robust stability problems is whether the inverse image of the boundary of the Nyquist template N is within the boundary of $D \times \Omega$. In the context of the stratified Morse theory, we reformulate this as follows: *In what stratum do the Morse critical points of f_θ lay?* In the case of Kharitonov's theorem, the situation is straightforward:

Theorem 22.8. *The Morse critical points of the fixed-frequency ($\omega \neq 0$) mapping $f_\theta : D \to \mathbb{C}$, $\theta \neq 0, \pi$, where D is the usual Kharitonov cube, are all in the smallest, zero-dimensional strata (i.e., the vertices) of D.*

Proof. Because of the Cartesian geometry of the cube, any stratum of dimension greater than zero has for Riemann coordinates a subset of $\{a_i\}$, the set of coefficients of the polynomial. Therefore, to chase critical points, it suffices to look at the partial derivatives $\frac{\partial f_\theta}{\partial a_i}$. Because of the linear dependency of f_θ on a_i, none of these partial derivatives vanishes. It follows that the strata of dimension greater than zero have no critical points. Therefore, the critical points could only be in the zero-dimensional strata (i.e., the vertices), and by definition all of the vertices are critical points. ■

Theorem 22.9. *The mapping $f_\theta : D \times \Omega \to \mathbb{C}$, where D is the usual Kharitonov cube and Ω the unit circle, provided that $\omega \neq 0$ and $\theta \neq 0, \pi$, has its Morse critical points in the one-dimensional strata of $D \times \Omega$; that is, the strata generated by the vertices of D topologically multiplied by Ω.*

Proof. Remember, a stratum of $D \times \Omega$ is a stratum of D topologically multiplied by Ω. Therefore, the Riemann coordinates of a stratum of $D \times \Omega$ is a (possibly nonproper) subset of $\{a_i\}$ together with ω. Therefore, a Morse critical point in such a stratum is characterized by $\frac{\partial f_\theta}{\partial a_i} = 0$, $\frac{\partial f_\theta}{\partial \omega} = 0$ where i runs into a (possibly nonproper) subset of indexes. By the linearity of the map (see previous theorem), any relation of the form $\frac{\partial f_\theta}{\partial a_i} = 0$ is impossible. Therefore, the only possible Morse points are those characterized by $\frac{\partial f_\theta}{\partial \omega} = 0$; that is, Morse points in a one-dimensional stratum of $D \times \Omega$ generated by a vertex of the cube D topologically multiplied by Ω. ■

The Kharitonov situation is exceptional. In the general situation, the Morse critical points can be distributed over *all* strata. Should some Morse critical points be in the *largest* stratum, the problem becomes difficult. Indeed, the Morse critical points, which encompass the inverse image of the boundary of the Nyquist template N, are locked in the *highest-dimensional* stratum. They cannot be narrowed down to a lower-dimensional space. Consequently a test over a lower-dimensional space than $D \times \Omega$ is bound to fail.

22.3 Boundary Singularity

A phenomenon typical of the case where the uncertainty is a manifold-with-boundary is a branch of the plot of critical points versus θ "crashing" on the boundary. More generally, we have a similar phenomenon of "crashing" every time a branch of the plot hits a stratum of dimension lower than the current stratum.

To be more precise, assume we have a plot of critical points $p^{*k}(\theta)$ that remains within the n-dimensional stratum S_n for $\theta \in [\theta_1, \theta_2)$. As long as the plot remains within S_n, it is given in terms of the Riemann coordinates $(p_1, ..., p_n)$ of S_n as

$$\begin{aligned} \frac{\partial f_\theta}{\partial p_1} &= 0 \\ &\vdots \\ \frac{\partial f_\theta}{\partial p_{n-1}} &= 0 \\ \frac{\partial f_\theta}{\partial p_n} &= 0 \end{aligned}$$

In other words, for fixed θ, we have n equations in n variables, resulting, generically, in a finite set of critical points that includes $p^{*k}(\theta)$. Let the branch $p^{*k}(\theta), \theta \in [\theta_1, \theta_2)$ of the plot converge to $\partial S_n \subseteq S_{n-1} \subset \overline{S_n}$ as θ goes to θ_1. Let $S_{n-1} \subset \overline{S_n}$ be the stratum to which $p^{*k}(\theta)$ converges. Therefore, $\lim_{\theta \to \theta_2} p^{*k}(\theta) \in S_{n-1}$.

Once we are in the lower-dimensional stratum S_{n-1}, charted by Riemann coordinates $(q_1, ..., q_{n-1})$, the critical points are given by

$$\begin{aligned} \frac{\partial f_\theta}{\partial q_1} &= 0 \\ &\vdots \\ \frac{\partial f_\theta}{\partial q_{n-1}} &= 0 \end{aligned}$$

This amounts to $(n-1)$ equations with $(n-1)$ variables—that is, generically, a finite set of critical points, the locus of which remains in S_{n-1} for $\theta \in [\theta_2, \theta_2 + \epsilon)$ for some sufficiently small ϵ.

To prove that the limit of the plot on S_n "connects" to a plot on S_{n-1}, we have to proceed a little more formally and invoke the Whitney conditions on the interconnection between the strata S_n and S_{n-1}. Let T_pS be the plane tangent to the stratum S at the point p. Let $d_pf_\theta : T_pS \to \mathbb{R}$ be the differential of the real-valued smooth function f_θ evaluated at a point p on the manifold of definition (see [Golubitsky and Guillemin 1973, Chapter 1]). In terms of these concepts, the definition of the critical points $q^*(\theta_2)$ on the stratum S_{n-1} is

$$d_{q^*(\theta_2)}f_{\theta_2}(T_{q^*(\theta_2)}S_{n-1}) = 0 \tag{22.1}$$

Similarly, the definition of the critical points $p^*(\theta)$ on the stratum S_n is

$$d_{p^*(\theta)}f_\theta(T_{p^*(\theta)}S_n) = 0$$

Now, we take the limit of the above as $\theta \to \theta_2$,

$$\lim_{\theta \to \theta_2} d_{p^*(\theta)} f_\theta (T_{p^*(\theta)} S_n) = 0$$

The limit

$$\lim_{\theta \to \theta_2} T_{p^*(\theta)} S_n = Q$$

is a *generalized tangent space* (see [Goresky and MacPherson 1988, page 44]). It follows that

$$(\lim_{\theta \to \theta_2} d_{p^*(\theta)} f_\theta)(Q) = 0$$

Next, the *Whitney condition* (see [Goresky and MacPherson 1988, page 37]) on the stratification of the domain of uncertainty stipulates that

$$T_{q^*(\theta_2)} S_{n-1} \subseteq \lim_{\theta \to \theta_2} T_{p^*(\theta)} S_n$$

Therefore,

$$(\lim_{\theta \to \theta_2} d_{p^*(\theta)} f_\theta)(T_{q^*(\theta_2)} S_{n-1}) = 0 \tag{22.2}$$

Comparing Equation 22.1 and Equation 22.2, it follows that

$$\lim_{\theta \to \theta_2} p^*(\theta) = q^*(\theta_2)$$

In other words, the limit of the plot on S_n is within the plot on S_{n-1}, as claimed.

The conclusion is that, when the locus of critical points hits a lower-dimensional stratum, *it does not disappear*; it only confines itself to the lower-dimensional stratum.

Observe that, as a corollary of the above transition across strata analysis, we have

$$d_{q^*(\theta_2)} f_{\theta_2}(Q) = 0$$

where, as we remember, Q is a generalized tangent space at $q^*(\theta_2)$. The above reveals that at the point of transition of the critical plot from S_n to S_{n-1} the function f_{θ_2} has some "degeneracy." Indeed, following Definition 22.6 (see also [Goresky and MacPherson 1988, page 52]), for a function to qualify as a "Morse function defined over a Whitney stratified space," it has to satisfy, in addition to the usual C^∞ and nonsingularity of the Hessian requirements, the extra requirement that the differential at a point be nonvanishing over any generalized tangent space at that point. Clearly, this requirement fails, so that every time the critical plot transits across strata the function f_θ loses its "Morse function defined over a Whitney stratified space" property.

This is a situation reminiscent to the situation of the classical Morse theory where, in general, f_θ loses its Morse property for finitely many θ's, an issue that was dealt with in [Cerf 1970]. Whether an analysis, similar to that of [Cerf 1970], can be conducted for a one-parameter family of functions defined over a stratified space that could potentially loses its stratified Morse property for some θ's is an issue we will not explore, at least in this book.

22.4 Application to Mapping Theorems

In most of the control applications, the stratum S_n is, typically, an open cube and the stratum S_{n-1} is an open face of the cube, the only purpose of which is to impose a "hard" bound on the values of the parameters. In this problem setup, by the Whitney extension theorem (see, e.g., [Mather 1969a]), f_θ can be extended from S_n to a bigger manifold $\mathcal{S}_n$ that includes S_{n-1}. Now, let a critical curve transit from S_n to S_{n-1} at $p^* \in S_{n-1} \subset \overline{S_n} \subseteq \mathcal{S}_n$. It can be shown that the restriction of the Hessian of $f_\theta : \mathcal{S}_n \to \mathbb{R}$ to S_{n-1} is the Hessian of $f_\theta : S_{n-1} \to \mathbb{R}$ (see [Goresky and MacPherson 1988, Part II, Section 4.A]). Therefore, if $\mathcal{S}_n$ and S_{n-1} are charted by n and $n-1$ Riemann coordinates, respectively, the change of Morse index associated with the transition from S_n to S_{n-1} is ± 1.

In the compact differentiable case, there are already nontrivial singularity phenomena occurring on the critical value curves on the boundary of the template. In the stratified case, on top of this, there are the singularity phenomena associated with the transitions across strata. All of these phenomena explain the complicated singularity structure (e.g., "sharp" points) of the template associated with a cube of uncertainty (see [Jonckheere and Ke 1995]).

HISTORICAL NOTES

The difficulty of dealing with *singular spaces* was already observed by Poincaré, who showed that singularities (typically, boundaries, corners, etc.) destroy the duality named after him. A milestone was Smale's h-cobordism theorem asserting that a high-dimensional, simply connected manifold-with-boundary homotopically equivalent to a cylinder is homeomorphic to a cylinder. An accurate theory of manifolds-with-corners specially devised for singularity theory was developed in [Mather 1969a]. The concept of Whitney stratified spaces arose from Whitney's investigation of complex-analytic varieties. The Morse theory of Whitney stratified spaces was codified in [Goresky and MacPherson 1988], which has become a standard reference. There are, however, signs that the concept of Whitney stratified spaces is too restrictive, because indeed, as argued in [Weinberger 1994], it relies on too much differential structure that is not topologically invariant. In [Weinberger 1994], another theory of stratified spaces is in the making, a theory where the strata need not even be manifolds.

23

STRUCTURAL STABILITY OF CROSSOVER

The openness of transversality often can be used to show that corresponding models have the important property of robustness. A model is robust if the properties under study remain after perturbation.

S. Smale,
The Mathematics of Time,
Springer-Verlag, New York, 1980,
page 135.

SUMMARY

In this chapter we address some potentially "catastrophic" sensitivity issues in robustness analysis in the best of the tradition initiated by Whitney, Thom, Zeeman, and Arnold. While some results overlap with those of Chapter 21, the latter are nevertheless worth rephrasing in the very specific context of the *theory of stability of maps* and the *structural stability of solutions to equations.*

All of these issues have to do with the topological consequences of slight perturbation of the map f. In the control context, the slight perturbation of the map f is due to variation of parameters that have been, quite subjectively, declared "certain." In the realm of numerical computation, the perturbation of the map f is due to rounding errors.

The *stability* of such maps as $f : D \times \Omega \to \mathbb{R}^2$ is, roughly speaking, the issue as to whether there could be qualitative changes in the set of critical points and critical values of the maps f under perturbation. The issue of *structural stability* of the solution set $\{p : f(p) = 0 + j0\}$, somewhat interrelated with the previous one, is whether the crossover $X_\omega = f^{-1}(0 + j0)$ could undergo a qualitative change under variation of "certain" parameters.

The ultimate objective of such an analysis is whether the structured stability margin is continuous relative to "certain" parameters. An early counterexample by [Barmish, Khargonekar, Shi, and Tempo 1990] shows that robustness margin *need not* be a continuous function of the problem data. Our contention is that the theory of structural stability provides the mathematical foundation upon which the lack of continuity phenomenon can be explained. To be more specific, we reformulate the continuity problem as the problem of structural stability of the crossover—that is, whether X_ω remains topologically unchanged under rounding errors—and show that the lack of continuity of the stability margin can be traced back to lack of structural stability of the crossover.

This chapter is organized as a spinoff of the Morse theory and the singularity analysis the Nyquist map. Indeed, in the previous chapters, we have already been "nibbling around" the issue of stability of the Nyquist map. For example, the path f_θ crossing a higher-order singularity as a result of the drifting of the "certain" pole parameter $-a$ was an example of lack of stability of the f_θ map. Roughly speaking, crossing a higher-order singularity is accompanied by a *qualitative* change in the critical value plots and reflects a change in the *differential type* of the mapping.

23.1 Jet Space

Since the purpose of this exercise is to perturb the Nyquist map, we cannot focus our analysis on *one single*, "nominal" Nyquist map. We instead have to consider *all* maps from $D \times \Omega$ to Y. In this control context, the target Y could be either $\mathbb{R}$ or $\mathbb{R}^2$, depending on whether we work with f_θ or f. Since singularity is a *local* concept that relies on partial derivatives, it is natural to use the *jet* local representation of functions—roughly speaking, the representation of functions by (possibly nonconvergent) Taylor series around the singularity. (The reader is invited to spend some time on Section D.3 to get all of the subtlety involved in the concept of jet.) Since the Nyquist map is to be perturbed, the critical points could potentially move all across $D \times \Omega$, so that we have to consider the space of *all* jets of functions around *all* points of $D \times \Omega$. This intuitive idea is formalized in the concepts of *jet space*, $J^\infty(D \times \Omega, Y)$. A basic attribute of the jet space is that it depends *only* on the domain and target spaces, $D \times \Omega$ and Y. It does not depend on some "nominal" Nyquist map. All of our perturbation analysis is going to refer to this (very big!) space.

Now, we introduce the formal definitions. A **k-jet** $j^k f(p)$ is an equivalence class of C^k functions, of which f is a representative, all functions agreeing together with their derivatives up to and including order k at p. The **infinite-jet** $j^\infty f(p)$ is the equivalence class of C^∞ functions that agree, together with all derivatives of all order, with f at p. The materialization of this idea requires choosing charts for the manifolds $D \times \Omega$ and Y around p and $f(p)$, respectively, and this is left to the reader. The reader can also verify that the property "agreeing ... up to and including order k" is independent of the chart.

The **jet space** is defined as the collection of all equivalence classes of all functions and all $p \in D \times \Omega$,

$$J^k(D \times \Omega, Y) = \{j^k f(p) : f \in C^k(D \times \Omega, Y), p \in D \times \Omega\}$$

A "point" on the jet space is an equivalence class of functions mapping some point p to some point y such that all derivatives of order $\leq k$ agree at p. An element of $J^k(D \times \Omega, Y)$ is called a **k-jet**.

Observe that

$$J^0(D \times \Omega, Y) = (D \times \Omega) \times Y$$

On the other hand, identifying $J^1(D \times \Omega, Y)$ requires some more work. Let $T(D \times \Omega)$, TY denote the tangent bundle to $D \times \Omega$, Y, respectively. Any map $f : D \times \Omega \to Y$ induces a bundle morphism df that makes the following diagram commute:

$$\begin{array}{ccc} T(D \times \Omega) & \xrightarrow{df} & TY \\ \downarrow & & \downarrow \\ D \times \Omega & \xrightarrow{f} & Y \end{array}$$

df is linear on the fibers and when it is restricted to the fiber over p it becomes the usual differential, $d_p f : T_p(D \times \Omega) \to T_{f(p)}Y$. It is not hard to show that

$$J^1(D \times \Omega, Y) = \hom(T(D \times \Omega), TY)$$

where hom denote the collection of all tangent bundle morphisms.

The **infinite jet space** $J^\infty(D \times \Omega, Y)$ is defined as the collection of all infinite jets $j^\infty f(p)$ for all $f \in C^\infty(D \times \Omega, Y)$ and all $p \in D \times \Omega$.

The collection of finite jet spaces comes naturally equipped with projections maps:

$$\ldots \to J^{k+1}(D \times \Omega, Y) \overset{\pi_{k+1,k}}{\longrightarrow} J^k(D \times \Omega, Y) \to \ldots$$

It can be shown (see [Boardman 1967]) that the infinite jet space is isomorphic to the inverse limit of the above system:

$$J^\infty(D \times \Omega, Y) \cong \lim_{\leftarrow} J^k(D \times \Omega, Y)$$

The concept of *jet* was introduced by [Ehresmann 1951] and further developed by [Boardman 1967] into a more elegant approach to the somewhat clumsy concept of singularity sets $S_{r,s}$.

A useful feature is that the jet space is a *manifold.* To prove this, we have to define a neighborhood of an arbitrary k-jet, say $j^k f(p)$, in $J^k(D \times \Omega, Y)$ and show that this neighborhood maps bijectively into some Euclidean space. Since $p \in D \times \Omega$ and since $D \times \Omega$ is a manifold, there exists an open set $O_p \subseteq D \times \Omega$ containing p. We define the canonical neighborhood of $j^k f(p)$ as $J^k(O_p, Y)$, the set of k-jets of functions $O_p \to Y$ at $p' \in O_p$. By the manifold property of $D \times \Omega$, there exists a homeomorphism $h : O_p \to \mathbb{R}^{n_p}$. This in turn induces a homeomorphism that maps any k-jet of functions $O_p \to Y$ at p' into a k-jet of functions $\mathbb{R}^{n_p} \to Y$ at $h(p')$,

$$\begin{array}{rcl} jh : J^k(O_p, Y) & \to & J^k(\mathbb{R}^{n_p}, Y) \\ j^k f(p') & \mapsto & j^k f \circ h^{-1}(x), x = h(p') \end{array}$$

A jet of $J^k(\mathbb{R}^{n_p}, Y)$ can clearly be identified with finitely many truncated Taylor series. Therefore, $J^k(\mathbb{R}^{n_p}, Y)$ obviously has a finite-dimensional Euclidean structure. Therefore, the neighborhood $J^k(O_p, Y)$ of $j^k f(p)$ has been mapped bijectively into the Euclidean space $J^k(\mathbb{R}^{n_p}, Y)$. This endows $J^k(D \times \Omega, Y)$ with the C^0 manifold structure. The last step is to show that the C^∞ structure of $D \times \Omega$ induces a C^k structure on $J^k(D \times \Omega, Y)$. The reader can easily verify this by checking the "gluing maps."

To insert a *given* smooth Nyquist map $f : D \times \Omega \to Y$ into $J^k(D \times \Omega, Y)$ we introduce the mapping

$$\begin{aligned} j^k f : D \times \Omega &\to J^k(D \times \Omega, Y) \\ p &\mapsto j^k f(p) \end{aligned}$$

where $j^k f(p)$ is defined as the equivalence class of $f(\cdot)$ in $J^k(D \times \Omega, Y)$. $j^k f$ is called **k-jet** of f. The map $p \mapsto j^k f(p)$ has originally been called *k-prolongation* by [Ehresmann 1951]. It has also been called **jet section**. Observe that $j^0 f$ is the graph of f.

We also need a couple of projection maps: Define

$$\begin{aligned} \pi_{J,Y} : J^\infty(D \times \Omega, Y) &\to Y \\ jf(p) &\mapsto f(p) \end{aligned}$$

and

$$\begin{aligned} \pi_{J,X} : J^\infty(X, Y) &\to X \\ jf(p) &\mapsto p \end{aligned}$$

23.2 Whitney Topology

To make the intuitive concept of "small perturbation" precise, it is necessary to define a topology on $C^\infty(D \times \Omega, \mathbb{R}^2)$. Topologizing such a space as $C^k(D \times \Omega, Y)$, where Y could be either $\mathbb{R}^2$ or $\mathbb{R}$ depending on whether we work with f or f_θ, respectively, is a standard problem. There are essentially two ways to go:

- The weak or compact-open topology
- The strong or Whitney topology

23.2.1 C^0 Case

To get the feeling for these topologies, we illustrate them in the C^0 case. We exploit the fact that Y, being either $\mathbb{R}$ or $\mathbb{R}^2$, is a metric space with distance function $\rho(\cdot)$.

In the case $D \times \Omega$ is compact, the **weak** topology is generated by open neighborhoods of the form

$$O_\epsilon(f) = \{f' \in C^0(D \times \Omega, Y) : \rho(f(p) - f'(p)) < \epsilon, \forall p \in D \times \Omega\}$$

The basis for the topology is generated as ϵ runs in $(0, \infty)$ and f runs in $C^0(D \times \Omega, Y)$. This topology is *metrizable*; actually, a metric $d(\cdot, \cdot)$ for this topology is given by $d(f, f') = \sup\{\rho(f(p), f'(p)) : p \in D \times \Omega\}$.

The above topology is also called **compact-open**, because it has a sub-basis comprising all sets of the form $\{f \in C^0(D \times \Omega) : f(K) \subseteq O\}$ where K is a *compact* subset of $D \times \Omega$ and O an *open* subset of Y.

Assuming the manifold $D \times \Omega$ is paracompact, the **strong** topology can be generated by such basis sets as

$$O_{\epsilon(\cdot)}(f) = \{f' \in C^0(D \times \Omega, Y) : \rho(f(p) - f'(p)) < \epsilon(p), \forall p \in D \times \Omega\}$$

As before, f runs through $C^0(D \times \Omega, Y)$, but the big difference is that $\epsilon(\cdot)$ runs through all *continuous functions*: $D \times \Omega \to (0, \infty)$. Clearly, with $\epsilon(p)$, we have arbitrary control at infinity, so that this topology happens to be very fine; actually, it is even not metrizable without extra assumptions.

The weak topology is nominally devised for a *compact*, or at least a *locally compact*, domain of definition. In the latter case, the compact-open topology deals only with uniform convergence over compact subsets. To be specific, $f_n \to f$ in the weak C^0 topology iff for any compact subset $K \subseteq (D \times \Omega)$ and $\epsilon > 0$ there exists an integer $N(\epsilon, K)$ such that $\rho(f_n(p) - f(p)) < \epsilon$, $\forall n > N(\epsilon, K)$, $\forall p \in K$ (see [Dugundji 1970, Definition 7.1, Theorem 7.2, pages 267–268] and [Hirsch 1976, Section 2.4, pages 58–59]).

The Whitney topology is a much stronger topology that nominally applies to a *noncompact* domain of definition. It coincides with the weak topology over a compact domain of definition.

23.2.2 C^k and C^∞ Cases

The next step is to define the weak and the strong C^k topologies; that is, we introduce differentiability into the picture. We are heading towards the conceptualization of the intuitive idea that two maps f, f' are "close" in the C^k topology whenever $f(p), f'(p)$, along with their derivatives up to and including order k, are "close." The problem is that partial derivatives require charts for the domain and target manifolds. To remove the dependency on the charts, we prefer to use the k-jet $j^k f(p)$ as an invariant description of the derivatives of f from order 0 to k at the point p. Remember, the collection of all such jets is the *jet space*, $J^k(D \times \Omega, Y)$, introduced in Section 23.1. Since $J^k(D \times \Omega, Y)$ is a manifold, it is metrizable and let ρ be a metric compatible with its topology.

The **weak** C^k topology is defined as the topology generated by neighborhoods of the form,

$$O_\epsilon(f) = \{f' \in C^\infty(D \times \Omega, Y) : \rho(j^k f(p) - j^k f'(p)) < \epsilon, \forall p \in D \times \Omega\}$$

for all $f \in C^\infty(D \times \Omega, Y)$ and all $\epsilon > 0$.

The **Whitney** C^k topology is by definition the topology generated by all open sets

$$O_{\epsilon(\cdot)}(f) = \{f' \in C^\infty(D \times \Omega, Y) : \rho(j^k f(p) - j^k f'(p)) < \epsilon(p), \forall p \in D \times \Omega\}$$

for all $f \in C^\infty(D \times \Omega, Y)$ and all continuous mappings $\epsilon : D \times \Omega \to (0, \infty)$.

It can be shown (see [Mather 1969a, Section2] or [Golubitsky and Guillemin 1973, pages 42–43]), that the Whitney C^k topology is generated by sets of the form $\{f \in C^\infty(D \times \Omega, Y) : j^k f(D \times \Omega) \subseteq O\}$, where O is an arbitrary open set of $J^k(D \times \Omega, Y)$.

Let $\mathcal{O}^k$ denote the set of open subsets of $C^\infty(D \times \Omega, Y)$ for the Whitney C^k topology. The Whitney C^∞ topology is defined as the topology with basis

$$\cup_{k=0}^{\infty} \mathcal{O}^k$$

If the domain of definition $D \times \Omega$ is compact, $f_n \to f$ in the Whitney C^k topology iff $j^k f_n \to j^k f$, uniformly over $D \times \Omega$. Thus, this is a pretty standard topology. However, whenever the domain is not compact, the Whitney topology becomes quite another thing, because we have arbitrary control at ∞. Actually, we have the following:

Theorem 23.1. *$f_n \to f$ in the Whitney C^k topology iff there exists a compact subset $K \subseteq (D \times \Omega)$ such that $j^k f_n \to j^k f$, uniformly over K, and $f_n(p) = f(p)$, $\forall p \in (D \times \Omega) \setminus K$, for all but a finite number of n's.*

Proof. See [Golubitsky and Guillemin 1973, pages 43–44]. ■

This theorem has some important consequences in case of a continuous-time Nyquist map where the frequency makes the domain noncompact. To be illustrative, let us ask the question as to whether we have the following convergence in the Whitney C^k topology:

$$\frac{1}{j\omega + q + \frac{1}{n}} \overset{?}{\to} \frac{1}{j\omega + q}$$

In the above $-q$ is an uncertain pole running in a compact subset of $\mathbb{R}$, while ω runs in the noncompact space $\mathbb{R}$. $\frac{1}{n}$ denotes a rounding error. Clearly, $f_n(p) - f(p)$, together with its partial derivatives up to and including order k, converges to 0, uniformly over $D \times \Omega$. However, there is no compact subset $K \subset D \times \Omega$ such that $f_n(p) = f(p), \forall p \in (D \times \Omega) \setminus K$, and for all but a finite number of n's. Therefore, the above is *not* a convergence in the Whitney C^k topology. To make it a convergence in the Whitney C^k topology, we have to exploit the roll-off at infinite frequency; more specifically, we require the deformation of the map to be "killed" beyond a finite frequency ω_B. For example, the following is a Whitney C^k convergence,

$$\frac{1}{j\omega + q + \frac{1}{n} I_{[-\omega_B, +\omega_B]}(\omega)} \to \frac{1}{j\omega + q}$$

where $I_{[a,b]}$ is the indicator of $[a, b]$.

23.3 (Elementary) Transversality

Thom's concept of *transversality* is relevant to the following theorem, the relevance of which to control should be obvious:

Theorem 23.2. *Let $f : X \to Y$ be a smooth map between smooth manifolds. Let $y \in Y$ be a regular value. Then $f^{-1}(y)$ is a smooth submanifold of X.*

Proof. See [Hirsch 1976, page 22, Theorem 3.2]. ■

This theorem tells us that sufficient for the crossover X_ω to be a smooth manifold is that no critical value curves cross over $0 + j0$. This condition is, however, not necessary; see [Bruce and Giblin 1992, page 207, Sections 8.2 and 8.10(1)] for a counterexample.

Transversality turns out to be the generalized condition that guarantees that $f^{-1}(Z)$ is a smooth submanifold of X whenever Z is a smooth submanifold of Y.

Definition 23.3. (Transversality) *The smooth map $f : X \to Y$ between smooth manifolds intersects* **transversally** *the smooth submanifold $Z \subseteq Y$ at $x \in X$ iff*

$$T_{f(x)}Y = d_x f(T_x X) + T_{f(x)}Z$$

Theorem 23.4. *Under the same conditions as in the above definition, $f^{-1}(Z)$ is a smooth submanifold of X whenever f intersects Z transversally $\forall x \in X$.*

Proof. See [Golubitsky and Guillemin 1973, page 52, Theorem 4.4]. ■

It is easily seem that the map $f : D \times \Omega \to \mathbb{R}^2$ intersects $y \in \mathbb{R}^2$ *transversally* iff y is a *regular* value. Therefore, a fundamental crossover problem can be rephrased in the context of transversality.

Corollary 23.1. *The crossover $X_\omega = f^{-1}(0 + j0)$ is a smooth manifold if the Nyquist map $f : D \times \Omega \to \mathbb{R}^2$ intersects $0 + j0$ transversally.*

The celebrated transversality theorem of Thom asserts that smooth maps transverse to a submanifold of Y are plentiful.

Theorem 23.5. (Elementary Transversality) *Let X, Y be smooth manifolds and let $Z \subseteq Y$ be a smooth submanifold. Then the set*

$$\{f \in C^\infty(X, Y) : f \text{ is transverse to } Z\}$$

is dense. Furthermore, if in addition Z is a closed submanifold, then this set is open for the Whitney C^∞ topology.

Proof. See [Golubitsky and Guillemin 1973, page 56]. ■

The control corollary of the above is that, when $0 + j0$ happens to be a critical value of the Nyquist map, an arbitrarily small perturbation of the map should be able to remove this unfortunate situation.

There are generalizations of these transversality results to the case of manifolds-with-boundaries (see Definition 22.1). As one might expect, if the domain is a manifold-with-boundary, the inverse image of a point—the crossover—could "crash" on the boundary of the domain of uncertainty. We need a concept that indicates that the boundary of the inverse image is nicely positioned within the boundary of the domain.

Definition 23.6. *Let X be a manifold-with-boundary and let $W \subseteq X$ be a submanifold-with-boundary. W is said to be a* **neat submanifold** *if $\partial W = W \cap \partial X$; moreover, the charts of X can be used as charts for W; for details, see [Hirsch 1976, page 30].*

Theorem 23.7. *Let X, Y be C^r manifolds-with-boundary, and let $f : X \to Y$ be a C^r map. Let $y \in Y \setminus \partial Y$ be a regular value for both f and $f|\partial X$. Then $f^{-1}(y)$ is a neat submanifold of X.*

Proof. See [Hirsch 1976, page 31, Theorem 4.1]. ■

We shall see in Section 23.15 that X_ω being a neat submanifold in $D \times \Omega$ will play a crucial role.

One of the advantages of the concept of Whitney stratified space is that it allows for the formulation of transversality results quite similar to those of the manifold case.

Definition 23.8. *Let $f : \mathcal{X} \to \mathcal{Y}$ be a smooth map between smooth manifolds and let X, Y be Whitney stratified submanifolds of $\mathcal{X}, \mathcal{Y}$, respectively. The map $f|X$ is said to be* **transverse** *to Y iff for any stratum W of X and any stratum Z of Y the map $f|W$ intersects Z transversally.*

Theorem 23.9. *Let $\mathcal{X}, \mathcal{Y}$ be smooth manifolds and let X, Y be closed Whitney stratified submanifolds of $\mathcal{X}, \mathcal{Y}$, respectively. Then the set*

$$\{f \in C^\infty(\mathcal{X}, \mathcal{Y}) : f|X \text{ is transverse to } Y\}$$

is dense and open for the Whitney C^∞ topology.

Proof. See [Goresky and MacPherson 1988, page 38]. ■

The above theorem is relevant to the situation where $0 + j0$ is on the boundary, or is one of the singular points of the boundary, of the template. The details are left to the reader.

23.4 Singularity Sets Revisited

Here, we redefine the Thom–Boardman singularity sets in a more algebraic fashion which will lead in the next section to yet another definition of the singularity set independent on perturbation of the Nyquist map.

23.4.1 Iterated Jacobi Extensions

Consider an ideal I of C_x. (See Appendix D for the definition of C_x.) Let $\varphi_i : \mathbb{R}^n \to \mathbb{R}$ denote an arbitrary function in the ideal. The kth **Jacobi extension** of I, $\Delta_k(I)$, is the ideal of C_x generated by I and the determinants of all $k \times k$ Jacobian made up with partial derivatives of functions in I,

$$\Delta_k(I) = I + \sum_{i_1<i_2<\ldots<i_k;j_1<j_2<\ldots<j_k} \det \begin{pmatrix} \frac{\partial \varphi_{i_1}}{\partial x_{j_1}} & \cdots & \frac{\partial \varphi_{i_1}}{\partial x_{j_k}} \\ \vdots & & \vdots \\ \frac{\partial \varphi_{i_k}}{\partial \varphi_{j_1}} & \cdots & \frac{\partial \varphi_{i_k}}{\partial x_{j_k}} \end{pmatrix} C_x$$

By definition $\Delta_0(I) = C_x$. The sequence is easily seen to be decreasing, i.e., $\Delta_{k+1}(I) \subseteq \Delta_k(I)$. The extension $\Delta_k(I)$ is said to be **critical** iff $\Delta_{k-1} = C_x$, $\Delta_k \overset{\subset}{\neq} C_x$.

The motivation for this definition is the following: Consider a map $f : \mathbb{R}^n \to \mathbb{R}^m$ and let $I = \text{Id}(f_1, ..., f_m)$ be the ideal generated by the components of the map. $\Delta_1(I)$ critical means that all partial derivatives of the components of the map vanish at $x = 0$; that is, the rank of the Jacobian drops to zero at $x = 0$, or, in other words, the dimension of the kernel of $d_0 f : \mathbb{R}^n \to \mathbb{R}^m$ is n.

More generally, $\Delta_k(I)$ critical means that $d_0 f$ has rank $k - 1$; that is, the dimension of the kernel of the Jacobian is $n - (k - 1)$.

This suggests that we should re-index the Jacobi extension so as to exhibit the dimension of the kernel *:

$$\Delta^{n-k+1} = \Delta_k$$

Next, we iterate the Jacobi extension: We compute the Jacobi extension of the critical $\Delta_k(I)$; that is, we compute $\Delta_l(\Delta_k(I)) = \Delta_l \Delta_k(I)$. Again, $\Delta_l(\Delta_k(I))$ is critical iff $\Delta_{l-1}\Delta_k(I) = C_x$ and $\Delta_l \Delta_k(I) \neq C_x$. The process is iterated until we find a sequence of critical extensions:

$$\ldots \Delta_{k_3} \Delta_{k_2} \Delta_{k_1}(I)$$

23.4.2 Jacobi Extension Definition of Singularity Sets

Now, it is easy to redefine the singularity set $S_{r,s}$ of the Nyquist map using the concept of critical Jacobi extension. Consider a point $p^* \in D \times \Omega$ and define, locally, $x = p - p^*$ and $F(x) = f(p^* + x) - f(p^*)$. This yields, at least locally, a map $F : \mathbb{R}^n, 0 \to \mathbb{R}^2, 0$. Let $I = \text{Id}(F_1, F_2)$ be the ideal generated by the components of the map. It is easily seen that $p^* \in S_{r,s}$ iff the sequence of critical Jacobi extensions is

*Indexing by the dimension of the kernel of the Jacobian is [Boardman 1967]'s notation.

$$\Delta_{2-s+1}\Delta_{2-r+1}(I)$$

It is possible to redefine the Jacobi extension in a coordinate independent, *invariant* matter. To provide an idea as to what the coordinate independent approach is, assume we do not want to rely on the "components of the map," since that latter require charting the target manifold. Consider the smooth map $f : X, x^0 \to Y, y^0$. Let $\mathcal{M}_{y-y^0}$ be the maximal ideal of functions $Y \to \mathbb{R}$ vanishing at y^0. (See Appendix D for the underlying algebraic concepts.) f induces a homomorphisn f^* as shown by the following diagram:

$$\begin{array}{ccc} X & \xrightarrow{f} & Y \\ \downarrow & & \downarrow \\ \mathcal{M}_{x-x^0} & \xleftarrow{f^*} & \mathcal{M}_{y-y^0} \end{array}$$

Now, it is easily seen that $\mathrm{Id}(f_1, f_2) = f^*\mathcal{M}_{y-y^0}$, and the *crucial point* is to observe that the latter definition does not rely on the "components of the map."

Unfortunately, removal of the dependency on coordinates when taking partial derivatives is technically very awkward and we shall not pursue this investigation any further and rather refer the interested reader to [Boardman 1967].

23.4.3 Jacobian of a Set of Functions

Given a set of functions $A = \{\phi_i : X \to \mathbb{R}\}$ defined on the compact, smooth n-manifold X, we define the Jacobian of the set of functions as the matrix obtained by "stacking together" the row of partial derivatives of the functions relative to Riemann coordinates of X,

$$J(A) = \begin{pmatrix} \frac{\partial \phi_1}{\partial x_1} & \cdots & \frac{\partial \phi_1}{\partial x_n} \\ \frac{\partial \phi_2}{\partial x_1} & \cdots & \frac{\partial \phi_2}{\partial x_n} \\ \vdots & & \vdots \end{pmatrix}$$

The rank of the Jacobian of the set of functions is simply defined as the rank of the above matrix.

A more formal definition of the rank of $J(A)$ proceeds via Jacobi extensions. Let I be the ideal generated by the functions in A. Then $J(A)$ has rank $k-1$ iff $\Delta_k(I)$ is critical.

23.5 Universal Singularity Sets

The concept of singularity subsets $S_{r,s}$ introduced in Section 21.11 is somewhat clumsy, the main reason, as argued by [Boardman 1967], being that the singularity subsets are not in general manifolds. In this control context, there is an even stronger sign of the inadequacy of the singularity subsets: It is unclear how the singularity subsets are affected under deformation of the Nyquist map which typically happens under variation of "certain" parameters.

This motivates our program to define the singularity sets "universally," without reference to some nominal Nyquist map. For each sequence of integers, $(i_1, i_2, ...)$, we define the subset $\Sigma^{i_1,i_2,\cdots}$ of $J^\infty(D \times \Omega, Y)$ consisting of all jets that have iterated Jacobians with $i_1, i_2, ...$ as dimensions of their kernels. The inverse images of the universal singularity sets $\Sigma^{i_1 i_2 \cdots}$ under the jet section jf of a nominal Nyquist map yield, under generic transversality conditions, the Thom–Boardman singularity subsets $S_{r,s,\ldots}(f)$.

To define the singularity sets "universally," we have to work with the jet space $J^\infty(D \times \Omega, Y)$. We define the "dimension of the kernel" of a set A of functions defined on the manifold $J^\infty(D \times \Omega, Y)$ and define the universal singularity sets from a sequence of critical Jacobi extension of the ideal generated by functions in A.

23.5.1 Total Jacobi Extension

Consider the set of functions A be defined in a neighborhood of $j^\infty f(p)$. A induces a set of functions $(j^\infty f)^*(A)$ defined in a neighborhood O_p of $p \in D \times \Omega$, as shown by the following commutative diagram:

$$\begin{array}{ccc} O_p & \xrightarrow{j^\infty f} & J^\infty(O_p, Y) \\ \downarrow & & \downarrow \\ C(O_p, \mathbb{R}) & \xleftarrow{(j^\infty f)^*} & C(O_{j^\infty f(p)}, \mathbb{R}) \end{array}$$

We define the Jacobian $J((j^\infty f)^* A)$ of the set of functions defined on $O_p \subseteq D \times \Omega$, by "stacking" on top of each other the rows of partial derivatives of every single function in $(j^\infty f)^* A$. (For a more formal exposition of this process, see [Boardman 1967].) Hence we define the **total** dimension of the kernel of A as

$$\operatorname{tdim}\ker A = \dim\ker J((j^\infty f)^*(A))$$

(In general, the adjective "total" refers to the jet space.)

23.5.2 Universal Singularity Sets

Now, for each sequence of integers $(i_1, i_2, ...)$, we define a subset $\Sigma^{i_1,i_2,\cdots} \subseteq J^\infty(D \times \Omega, Y)$. This subset is called "universal singularity subset," in the sense that it does not refer to one single map, but instead to the family of differentiable maps $D \times \Omega \to Y$. This subset is constructed from iterated Jacobi extensions of some ideals. Take a jet,

$$j \in J^\infty(D \times \Omega, Y)$$

and let us spell out the conditions for j to be in $\Sigma^{i_1,i_2,\cdots}$. This jet $j = j^\infty f(p)$ induces a point $y = f(p) \in Y$. Consider the ideal $\mathcal{M}_y$ of functions $Y \to \mathbb{R}$ vanishing at y. The pull-back of the projection $\pi_{J,Y}$ yields a subset of functions defined on the manifold $J^\infty(D \times \Omega, Y)$. This is shown by the following diagram:

$$\begin{array}{ccc} J^{\infty}(D\times\Omega,Y) & \stackrel{\pi_{J,Y}}{\longrightarrow} & Y \\ \downarrow & & \downarrow \\ C(J^{\infty}(D\times\Omega,Y),\mathbb{R}) & \stackrel{\pi^*_{J,Y}}{\longleftarrow} & C(Y,\mathbb{R}) \end{array}$$

The universal singularity set $\Sigma^{i_1,i_2,i_3,\ldots}$ of $J^{\infty}(D\times\Omega,Y)$ is defined as follows:

$$j\in\Sigma^{i_1,i_2,\ldots}$$

iff

$$\begin{array}{rcl} \operatorname{tdim}\ker\pi^*_{J,Y}\mathcal{M}_y & = & i_1 \\ \operatorname{tdim}\ker\Delta^{i_1}\pi^*_{J,Y}\mathcal{M}_y & = & i_2 \\ \operatorname{tdim}\ker\Delta^{i_2}\Delta^{i_1}\pi^*_{J,Y}\mathcal{M}_y & = & i_3 \\ \ldots & = & \ldots \end{array}$$

equivalently, iff the sequence of Jacobi extensions

$$\ldots\Delta^{i_2}\Delta^{i_1}\pi^*_{J,Y}\mathcal{M}_y$$

is critical. To be formal, we also define

$$\Sigma^{\emptyset}=J^{\infty}(D\times\Omega,Y)$$

If we want indexes in terms of the drop r_j in the rank of the Jacobian *, we re-index as follows:

$$\Sigma_{r_1,r_2,\ldots}=\Sigma^{i_1,i_2,\ldots}$$

where

$$r_j=m-n+i_j$$

Indeed, if $m\leq n$, as usually happens in control, the maximal rank of the Jacobian is $m=\dim Y$; the rank of the Jacobian is $n-i_j$, where i_j is the dimension of the kernel; therefore, the rank deficiency (or rank drop) of the Jacobian is $m-(n-i_j)=r_j$.

23.5.3 (Strong) Transversality

Next, given any map $f:D\times\Omega\rightarrow Y$, we define the subset

$$\Sigma_{r,s,\ldots}(f)=(j^{\infty}f)^{-1}(\Sigma_{r,s,\ldots})\subseteq D\times\Omega$$

where $j^{\infty}f$ is the jet section of the nominal Nyquist map $f:D\times\Omega\rightarrow Y$.

Theorem 23.10. (Boardman 1967) *For any sequence of integers $r,s,\ldots$, $\Sigma_{r,s,\ldots}$ is a (not necessarily closed) submanifold of the jet space $J^{\infty}(D\times$*

*Indexing by the drop of rank of the Jacobian, when $n\geq m=\dim Y$, which is the case in the control problem, is [Thom 1955–1956]'s notation.

Ω, Y). If the nominal Nyquist map $f : D \times \Omega \to Y$ has its jet section $j^\infty f$ transverse to $\Sigma_{r,s,\ldots}$, then $\Sigma_{r,s,\ldots}(f) = (j^\infty f)^{-1}(\Sigma_{r,s,\ldots})$ is a submanifold of $D \times \Omega$. In this case,

$$\Sigma_{r,s,\ldots,t} = \Sigma_t(f|\Sigma_{r,s,\ldots}(f))$$

In particular, since $\Sigma_\emptyset(f) = D \times \Omega$ is a manifold, we have

$$\Sigma_r(f) = \{p \in D \times \Omega : \textit{rank}\ d_p f = \dim Y - r\}$$

Finally, any Nyquist map $f : D \times \Omega \to Y$ can be approximated in the fine C^∞ sense by a map f' whose jet section $j^\infty f'$ is transverse to all manifolds $\Sigma_{r,s\ldots}$.

Proof. See [Boardman 1967, page 408, Theorems 6.1, 6.2, and 6.3]. ∎

There are several things we learn from Boardman's theorem: First, $\Sigma_{r,s,\ldots}(f)$ coincides with the singularity subsets introduced in the previous section. Next, and more importantly, we learn that, generically, the singularity subset S_1 is a *submanifold* of $D \times \Omega$. Remember, however, that mapping the manifold S_1 back into the template results in an assembly of curves that is, in general, far from being a manifold. Therefore, the critical value curves are in general more complicated than the critical curves.

23.6 Stability of Nyquist Map

The previous sections of this chapter have been dominated by the concept of transversality and its genericity. Here, we introduce the related concept of *stability of maps* which—for the specific dimensions of the target spaces $\mathbb{R}$, $\mathbb{R}^2$ of the Nyquist related maps f_θ, f—turns out to provide a more tractable concept of "genericity."

Definition 23.11. *A smooth map $f : X \to Y$ between smooth manifolds X, Y is said to be* **stable** *whenever there exists a neighborhood O_f of f in $C^\infty(X, Y)$ equipped with the Whitney topology such that $\forall f' \in O_f$ there exist diffeomorphisms d_1, d_2 such that the following diagram commutes:*

$$\begin{array}{ccc} X & \xrightarrow{f} & Y \\ d_1 \downarrow & & \downarrow d_2 \\ X & \xrightarrow{f'} & Y \end{array}$$

Rephrased in the language of Section 21.6 and Subsection 21.6.4, the map f is stable whenever a Whitney small perturbation is equivalent to the action of the group Diff(X) $\times$ Diff(Y).

The importance of this concept in robust control is the following: Let $X = D \times \Omega$, $f(\Delta, \omega) = \det(I + L(j\omega)\Delta)$, and $Y = \mathbb{R}^2$. Anticipating variation of "certain" parameters, we will have to consider the perturbed map f' and ensure that both f and f' enjoy some common properties ensuring that the robustness analysis is not too dependent on rounding errors. One such property shared by f, f' in the stable case refers to the singularity structure.

Theorem 23.12. *Let $f : D \times \Omega \to \mathbb{R}^2$ be stable. Let S be its singularity set and $f(S)$ be its critical values set. Then for all f' in a sufficiently small open set O_f of the Whitney topology of $C^\infty(D \times \Omega, \mathbb{R}^2)$, the singularity sets S and S' of the maps f and f', respectively, are diffeomorphic, namely, $S' = d_1(S)$. Furthermore, the critical values sets are also diffeomorphic, namely, $f'(S') = d_2(f(S))$.*

Proof. The chain rule yields

$$J_{d_2 \circ f \circ d_1^{-1}}(x') = J_{d_2}(f(x)) J_f(x) J_{d_1}^{-1}(x'), \quad (x = d_1^{-1}(x'))$$

from which the result is obvious. ∎

It is desirable that the set of stable maps be dense so that a small perturbation of a stable map will keep the map stable. The issue as to whether stable maps are *dense* in the space of all smooth maps is a highly generic issue that ultimately depends on the dimensions of the manifolds X, Y (see [Golubitsky and Guillemin 1973, page 163]). In this control context, we have *two* maps: the original Nyquist map into $\mathbb{R}^2$ and a one-parameter family of maps into $\mathbb{R}$. Since the two maps are into spaces of different dimensions, we have to break the stability issue into two cases.

Theorem 23.13. *The set of stable maps $D \times \Omega \to \mathbb{R}^2$ is dense in $C^\infty(D \times \Omega, \mathbb{R}^2)$ equipped with the Whitney topology.*

Proof. See [Golubitsky and Guillemin 1973, page 163]. ∎

The good news is that stable Nyquist maps are plentiful. The bad news, however, is that stability of such a map as f is a *local* concept that relates to some neighborhood of f. To illustrate this, let $f : D \times \Omega \to \mathbb{R}^2$ be a stable Nyquist map and let O_f be the neighborhood of f containing perturbed maps related to f by $f' = d_2 f d_1^{-1}$. If this open set is so small or the map is so sensitive that, say, rounding errors throw the map outside O_f and cause the loop f_θ to cross a singularity of codimension 2—for example, a $\lambda, \lambda + 1$ swallow tail—the perturbed map f'' will miss a λ critical point curve. The initial map f and the perturbed map f'' will no longer have diffeomorphic singularity sets. It follows that the two maps will fail to be related by $f'' = d_2 f d_1^{-1}$.

Regarding the one-parameter family of maps, the following result is a classic:

Theorem 23.14. *A smooth map $X \to \mathbb{R}$, defined over a smooth compact manifold, typically $f_\theta : D \times \Omega \to \mathbb{R}$, is stable whenever it is a Morse function with distinct critical values.*

Proof. See [Golubitsky and Guillemin 1973, page 79]. ■

Corollary 23.2. *The set of stable maps* $f_\theta : D \times \Omega \to \mathbb{R}$ *is dense in* $C^\infty(D \times \Omega, \mathbb{R})$.

Proof. This is a consequence of the strong version of the Morse approximation lemma (see Lemma 21.17). ■

Since stable maps into $\mathbb{R}$ and $\mathbb{R}^2$ are dense, we can safely identify "stable" and "generic" for such maps as f_θ and f. As we shall see in the next few sections, deciding whether a map is "stable" is a computationally tractable problem.

As we have seen, generically, the function f_θ fails to be a Morse function with distinct critical values (i.e., goes outside F^0) at finitely many θ's. It follows that, generically, the family f_θ contains finitely many unstable maps. This is in apparent contradiction with the fact that the full map $f : D \times \Omega \to \mathbb{R}^2$ is generically stable. What happens is that, even though the full map f is stable, the freedom in θ could make the map f_θ look at f through the worst possible angle. Assume that the Horowitz template of a stable Nyquist map has a "kink." (By the example of Section 21.15 such maps exist.) Draw the line tangent to the boundary of the template at the kink and choose this as the direction θ_0. In the critical value plot, θ_0 corresponds to the death (or the birth) in a swallow tail. It is easily seen that a small perturbation of θ around θ_0 results in a small perturbation of f_{θ_0} in the Whitney topology. This small perturbation, however, adds or deletes a pair of $\lambda, \lambda + 1$ critical points, and maps differing in their number of critical points are not related by diffeomorphisms of $D \times \Omega$ and $\mathbb{R}$. Therefore, f_{θ_0} is unstable. On the other hand, a small perturbation in the Whitney topology of f does not alter the basic features of the critical values plots; all it does is delay (or advance) the birth (or the death). This explains how the full map into $\mathbb{R}^2$ remains stable.

23.7 Infinitesimal Stability

Despite the conceptual appeal of the notion of stability of maps and its relevance to the problem of variation of "certain" parameters, stability of a map is, unfortunately, extremely difficult, if not impossible, to check. In this section, we introduce the related concept of *infinitesimal stability*. The latter will lead in the next section to *local infinitesimal stability*, which can be checked via computationally tractable rings and modules manipulation.

Consider the tangent bundle morphism df induced by $f : D \times \Omega \to Y$ as displayed in the following diagram:

$$\begin{array}{ccc} T(D\times\Omega) & \xrightarrow{df} & TY \\ h\uparrow\downarrow\pi_P & \overset{a}{\nearrow} & \pi_Y\downarrow\uparrow g \\ D\times\Omega & \xrightarrow{f} & Y \end{array}$$

In the above, π denotes the bundle projection. h is a cross section through the tangent bundle of $D \times \Omega$; in other words, it is a *vector field* on $D \times \Omega$ (see Subsection 13.1.7). Likewise, g is a vector field on Y. The morphism a is said to be a **vector field along** f iff the following diagram commutes:

$$\begin{array}{ccc} & & TY \\ & \overset{a}{\nearrow} & \downarrow\pi_Y \\ D\times\Omega & \xrightarrow{f} & Y \end{array}$$

Definition 23.15. *The smooth map $f : D \times \Omega \to Y$ between smooth manifolds is said to be* **infinitesimally stable** *iff for any vector field a along f there exist vector fields g, h such that*

$$a = df(h) + g \circ f \tag{23.1}$$

Theorem 23.16. (Mather) *If $D\times\Omega$ is compact smooth and Y is smooth, the smooth map $f : D \times \Omega \to Y$ is stable iff it is infinitesimally stable.*

Proof. See [Mather 1969a] or [Golubitsky and Guillemin 1973, Chapter V]. ■

23.8 Local Infinitesimal Stability

In this section, we formulate a *strictly local* version of infinitesimal stability that applies to *map germs*; see Appendix D for the concept of *germ* and the formal power series representation of smooth functions, soon to come. Consider a point p^0 in the domain of uncertainty, and let $z^0 = f(p^0)$ be its image. The point p^0 has a neighborhood homeomorphic to $\mathbb{R}^n$ and let x be the Euclidean coordinates in $\mathbb{R}^n$. (If $D \times \Omega$ has Euclidean structure, then $x = p - p^0$.) Clearly, the Nyquist map induces a map germ

$$\begin{array}{rcl} F : (\mathbb{R}^n, 0) & \to & (\mathbb{R}^2, 0) \\ x & \mapsto & F(x) = f(p^0 + x) - f(p^0) \end{array}$$

As the terminology suggests, local infinitesimal stability of the map germ $F(x)$ around $x = 0$ is a local version of infinitesimal stability of the map f.

Definition 23.17. *The map germ $(x, 0) \mapsto (F(x), 0)$ is* **locally infinitesimally stable** *(around 0) iff for any $a \in (\mathbb{R}[[x]])^2$, there exist formal series $g_i \in \mathbb{R}[[y]], h_i \in \mathbb{R}[[x]]$ such that*

$$a(x) = \sum_i \frac{\partial F(x)}{\partial x_i} h_i(x) + \begin{pmatrix} g_1(F(x)) \\ g_2(F(x)) \end{pmatrix} \tag{23.2}$$

We leave it to the reader to verify that Equation 23.2 is just a local version of Equation 23.1.

The definition of local infinitesimal stability is somewhat restrictive in the sense that it involves variations in the ring $\mathbb{R}[[x]]$ of formal power series (see Appendix D). Actually, depending of the level of smoothness that we want to achieve, we could consider variations in any of the following rings:

- The ring $\mathbb{R}[x]$ of (finite) polynomials in x
- The ring $\mathbb{R}[[x]]$ of formal power series in x
- The ring H_x of power series in x converging in some neighborhood of 0
- The ring C_x of smooth map germs at $x = 0$

Observe that

$$\mathbb{R}[x] \subseteq H_x \subseteq \mathbb{R}[[x]] \subseteq C_x$$

As shown in Appendix D, the so-called Borel map $C_x \to \mathbb{R}[[x]], \phi \mapsto \sum_i (\phi^{(i)}(0)/(i!))x^i$ is *surjective*, but the map has a kernel, the ideal $\mathcal{M}_x^\infty$ of all functions all derivatives of which vanish at $x = 0$. Therefore, $\mathbb{R}[[x]] \cong C_x/\mathcal{M}_x^\infty$, which explains the last inclusion. Therefore, $\mathbb{R}[[x]]$ can be thought of as a "representation" of C_x that does not distinguish two functions differing by an $\mathcal{M}_x^\infty$ function. This representation is useful to get down to the algebraic essence of local infinitesimal stability. This explains our preference for $\mathbb{R}[[x]]$. We could as well use H_x, in which case we reach a higher level of smoothness.

Again, the local infinitesimal stability criterion is still hard to check. To justify this, we need to introduce some module concepts. Consider the free module $(\mathbb{R}[[x]])^2$ over the ring $\mathbb{R}[[x]]$. Clearly, $\frac{\partial F(x)}{\partial x_i} h_i(x)$ is the multiplication of the module element $\frac{\partial F(x)}{\partial x_i}$ by the ring element $h_i(x)$ and is hence linear in the unknown $h_i(x)$. The difficulty, however, is the term $g_i(F(x))$ that is not linear, in the sense of module theory, in the unknown $g_i(x)$. It is, however, possible to rewrite the above equation into a linear polynomial equation.

Theorem 23.18. *The map* $(x, 0) \mapsto (F(x), 0)$ *is locally infinitesimally stable iff for every* $a \in (\mathbb{R}[[x]])^2$ *there exists solution* $c_i \in \mathbb{R}, h_i, g_{i,j} \in \mathbb{R}[[x]]$ *to the linear equation*

$$a(x) = \sum_i e^i c_i + \sum_i \frac{\partial F(x)}{\partial x_i} h_i(x) + \sum_{i,j} e^i F_j(x) g_{i,j}(x) \tag{23.3}$$

where $e^1 = \begin{pmatrix} 1 \\ 0 \end{pmatrix}, e^2 = \begin{pmatrix} 0 \\ 1 \end{pmatrix}$ *is the usual Euclidean basis.*

Proof. Equation 23.2 can be rewritten

$$a(x) = \sum_i e^i g_{i0} + \sum_i \frac{\partial F(x)}{\partial x_i} h_i(x) + \begin{pmatrix} F_{j_1}(x)\gamma_1(F(x)) \\ F_{j_2}(x)\gamma_2(F(x)) \end{pmatrix}$$

In the above, $g_{i0} = g_i(0)$, and $\gamma_i(F(x)) = \frac{g_i(F(x))-g_{i0}}{F_{j_i}(x)} \in (\mathbb{R}[[x]])^2$. Clearly, *any* $a(x) \in (\mathbb{R}[[x]])^2$ is expressible this way. In particular, $\gamma(x)$ should also be expressible this way. Upon substitution and repetition of the procedure, it is easily seen that the right-hand side converges to the result, except for arbitrarily high powers of the components of $F(x)$. Intuitively, since $F(0) = 0$, the "tail" terms should not matter. Formally, we have to make use of the *generalized Malgrange preparation theorem* (see Section D.8) to justify that we can safely "throw away the tail." Consider the $\mathbb{R}[[x]]$-module $A = (\mathbb{R}[[x]])^2/\{\frac{\partial F}{\partial x_i}\}$, where $\{\frac{\partial F}{\partial x_i}\}$ denotes the submodule generated by $\frac{\partial F}{\partial x_i}$. Equation 23.2 is clearly equivalent to saying that the module A is finitely generated over $\mathbb{R}[[y]]$, where $y = F(x)$. By the generalized Malgrange preparation theorem (see Section D.8), this means that $A/\mathcal{M}[[y]]A$, where $\mathcal{M}[[y]]$ is the maximal ideal of $\mathbb{R}[[y]]$, is a vector space over $\mathbb{R}$. Observe that

$$A/\mathcal{M}[[y]]A = (\mathbb{R}[[x]])^2/\{e^i F_j, \frac{\partial F}{\partial x_i}\}$$

From the above, it is clear that $A/\mathcal{M}[[y]]A$ being a vector space is equivalent to Equation 23.3 having a solution. This completes the proof. (See [Arnold, Gusein-Zade, and Varchenko 1985, Section 6.6, pages 128 and 129] for the original proof.) ■

This theorem can be reformulated in terms of quotient modules.

Theorem 23.19. *The map germ* $(\mathbb{R}^n, 0) \to (\mathbb{R}^2, 0), (x, 0) \mapsto (F(x), 0)$ *is locally infinitesimally stable iff*

$$\mathbb{R}^2 \supseteq (\mathbb{R}[[x]])^2/\{\frac{\partial F}{\partial x_i}, F_j e^k\}$$

where $\{\frac{\partial F}{\partial x_i}, F_j e^k\}$ *denotes the submodule generated by* $\frac{\partial F}{\partial x_i}, F_j e^k$ *for all* i, j, k.

Proof. This is just an algebraic reformulation of the preceding theorem. ■

Observe that the above stability criterion is *purely algebraic*. Essentially, we give ourselves a *ground ring* R_x (that could be either $\mathbb{R}[x]$, $\mathbb{R}[[x]]$, H_x, or C_x), construct the module $(R_x)^2$, compute the submodule $\{.\} \subseteq (R_x)^2$, and then compute the quotient module $(R_x)^2/\{.\}$. The above criterion involves explicitly the ring $R_x = \mathbb{R}[[x]]$. If the map germ $(x, 0) \mapsto (F(x), 0)$ is *polynomial*, then we could run the above test using the ground ring $R_x = \mathbb{R}[x]$ which is much simpler. In principle, any C^∞ map germ can be reduced to the so-called *normal form* which is polynomial (see [Arnold, Gusein-Zade, and Varchenko 1985, Chapter 14]). This reduction procedure is essentially a sequence of diffeomorphic changes of variables that do not alter the stability issue. The problem is that the reduction to normal form is computationally very involved.

If we have a rational function, as is the case in most robust control problems, we could easily get a formal or converging power series representation of the function and apply the algebraic criterion over the appropriate module. Unfortunately, the criterion is still difficult to handle using infinite power series.

When we have to check local infinitesimal stability of a rational Nyquist map germ, we propose to first get rid of the denominator using the following lemma:

Lemma 23.20. *Let $w(x)$ be a map such that both $w(x)$ and $1/w(x)$ have formal power series representation. Then $(x,0) \mapsto (F(x),0)$ is locally infinitesimally stable iff $(x,0) \mapsto (F(x)w(x),0)$ is locally infinitesimally stable.*

Proof. If $F(x)w(x)$ is locally infinitesimally stable, then $\forall a \in (\mathbb{R}[[x]])^2$, there exists $h_i, g_{i,j} \in \mathbb{R}[[x]]$ such that

$$\begin{aligned} a &= \sum_i e^i c_i + \sum_i \frac{\partial(Fw)}{\partial x_i} h_i + \sum_{i,j} e^i (F_j w) g_{i,j} \\ &= \sum_i e^i c_i + \sum_i \frac{\partial F}{\partial x_i} w h_i + \sum_k F \frac{\partial w}{\partial x_k} h_k + \sum_{i,j} e^i F_j w g_{i,j} \\ &= \sum_i e^i c_i + \sum_i \frac{\partial F}{\partial x_i} (w h_i) + \sum_{i,j} e^i F_j \left(\delta_{i,j} \sum_k \frac{\partial w}{\partial x_k} h_k + w g_{i,j} \right) \end{aligned}$$

The last equation means that $F(x)$ is locally infinitesimally stable. Conversely, if $F(x)$ is locally infinitesimally stable, we have

$$\begin{aligned} a &= \sum_i e^i c_i + \sum_i \frac{\partial F}{\partial x_i} h_i + \sum_{i,j} e^i F_j g_{i,j} \\ &= \sum_i e^i c_i + \sum_i \frac{\partial F}{\partial x_i} w \frac{h_i}{w} + \sum_{i,j} e^i (F_j w) \frac{g_{i,j}}{w} \\ &= \sum_i e^i c_i + \sum_i \frac{\partial(Fw)}{\partial x_i} \frac{h_i}{w} - \sum_k F \frac{\partial w}{\partial x_k} \frac{h_k}{w} + \sum_{i,j} e^i (F_j w) \frac{g_{i,j}}{w} \\ &= \sum_i e^i c_i + \sum_i \frac{\partial(Fw)}{\partial x_i} \left(\frac{h_i}{w} \right) + \sum_{i,j} e^i (F_j w) \left(-\delta_{i,j} \sum_k \frac{\partial w}{\partial x_k} \frac{h_k}{w^2} + \frac{g_{i,j}}{w} \right) \end{aligned}$$

Since $\frac{1}{w} \in \mathbb{R}[[x]]$, the last equation is the criterion for local infinitesimal stability of Fw over the ground ring $\mathbb{R}[[x]]$. ∎

Therefore, when we have a rational representative of a map germ, as is the case with most Nyquist problems, we remove the denominators using the above recipe and run the quotient module test on $\mathbb{R}[x]$.

In the case of such a map as $f_\theta : D \times \Omega \to \mathbb{R}$—that is, a map into $\mathbb{R}$—computing the quotient *module* reduces to computing the quotient *ring* $\mathbb{R}[x]/\mathrm{Id}(\{\frac{\partial f_\theta}{\partial x_i}\}, f_\theta)$ which can be efficiently done by computing a Gröbner basis (see [Becker and Weispfenning 1993]) for the denominator ideal.

Having reviewed local infinitesimal stability and its computational implementation, we still have to go from *local* infinitesimal stability to (global) infinitesimal stability (and, via Mather's theorem, to stability). Clearly, local infinitesimal stability could only break down for a germ from a critical point to the corresponding critical value. Should the germ of a map fail to be locally infinitesimally stable around a critical point, the map could not be stable. The converse is somewhat more tricky: Essentially, for any critical value y^* and any finite set of critical points p^{*k} in $f^{-1}(y^*)$, we have to check *simultaneous* local infinitesimal stability of the germs: $p^{*k} \mapsto y^*$. *Simultaneous* means that all solutions to Equation 23.2 for the multiple map germs must share the same vector field on Y; for details, see [Golubitsky and Guillemin 1973, Chapter V]. Since the relevant control examples are cases of *lack* of stability of the map germ, and hence lack of stability of the Nyquist map, we will not pursue any further the issue of local to global infinitesimal stability; we refer the interested reader to [Golubitsky and Guillemin 1973, Chapter V].

23.9 Stability of Whitney Fold and Cusp

Remember the Whitney fold and cusp that are likely to occur in the case of two uncertain parameters.

Theorem 23.21. *The Whitney fold and cusp are locally infinitesimally stable. Furthermore, the Whitney fold and cusp are, up to diffeomorphisms, the only stable maps : $\mathbb{R}^2, 0 \to \mathbb{R}^2, 0$ with a singularity at 0. Finally, stable maps $\mathbb{R}^2 \to \mathbb{R}^2$ are dense in $C^\infty(\mathbb{R}^2, \mathbb{R}^2)$ and therefore any map with a singularity at 0 can be reduced by an arbitrarily small perturbation and diffeomorphisms to either a fold or a cusp.*

Proof. We will prove local infinitesimal stability of the fold in the next section, as an illustration of the polynomial criterion. For the other claims, see [Whitney 1955]. ■

23.10 Example (Simple)

Here, a very simple example serves to illustrate the polynomial test for local infinitesimal stability; it also reveals an algebraic problem typical of this kind of manipulations.

We want to test local infinitesimal stability of the germ

$$(x, 0) \mapsto (\frac{x^2}{2} - \frac{x^3}{3}, 0)$$

using the ground ring of formal power series. We have to test whether

$$\mathbb{R}[[x]]/\mathrm{Id}((\frac{x^2}{2} - \frac{x^3}{3}), (x - x^2)) \subseteq \mathbb{R}$$

Clearly, it suffices to show that

$$\mathbb{R}[[x]]/\mathrm{Id}(x - x^2) \subseteq \mathbb{R} \tag{23.4}$$

The direct approach would be to take any formal power series, substitute x^2 for x and hope that all terms of degree ≥ 1 would be deleted. Unfortunately, this seems to be a never ending process,

$$\begin{aligned} & a_0 + a_1 x + a_2 x^2 + \dots \\ = & \quad a_0 + a_1 x^2 + a_2 x^4 + a_3 x^6 + \dots \\ = & \quad a_0 + a_1 x^4 + a_2 x^8 + a_3 x^{12} + \dots \\ = & \quad \dots \end{aligned}$$

Clearly, it does not seem to be possible to remove the tail.

To formalize the above intuitive argument, we need some algebra. Observe that Equation 23.4 can be rewritten

$$\mathbb{R}[[x]] \subseteq \mathbb{R} + \mathrm{Id}(x - x^2)$$

Equivalently,

$$\mathcal{M}[[x]] \subseteq \mathrm{Id}(x - x^2)$$

where $\mathcal{M}[[x]] = x\mathbb{R}[[x]]$ is the unique maximal ideal of $\mathbb{R}[[x]]$ (see Appendix D). The *crucial* point is to invoke Nakayama's lemma (see Appendix D) that says that it suffices to show that

$$\mathcal{M}[[x]] \subseteq \mathrm{Id}(x - x^2) + \mathcal{M}[[x]]\mathcal{M}[[x]]$$

Less formally, the above can be rewritten

$$x\mathbb{R}[[x]] \subseteq (x - x^2)\mathbb{R}[[x]] + x^2\mathbb{R}[[x]]$$

and the latter is trivial.

Therefore, the map germ is locally infinitesimally stable.

23.11 Example (Whitney Fold)

Remember, the Whitney fold is

$$(x_1, x_2), 0 \mapsto \begin{pmatrix} x_1^2 \\ x_2 \end{pmatrix}, 0$$

Following the polynomial criterion, we have to show that

$$(\mathbb{R}[[x_1, x_2]])^2/\{\underbrace{\begin{pmatrix} x_1^2 \\ 0 \end{pmatrix}, \begin{pmatrix} 0 \\ x_1^2 \end{pmatrix}, \begin{pmatrix} x_2 \\ 0 \end{pmatrix}, \begin{pmatrix} 0 \\ x_2 \end{pmatrix}}_{F_i e^j}, \underbrace{\begin{pmatrix} 2x_1 \\ 0 \end{pmatrix}, \begin{pmatrix} 0 \\ 1 \end{pmatrix}}_{\frac{\partial F}{\partial x_i}}\} \subseteq \mathbb{R}^2$$

Clearly, because of the "denominator" term

$$\begin{pmatrix} 0 \\ 1 \end{pmatrix}$$

all terms, polynomial and constant, in the bottom position of $(\mathbb{R}[[x]])^2$ are canceled. Because of the terms

$$\begin{pmatrix} 2x_1 \\ 0 \end{pmatrix}, \begin{pmatrix} x_2 \\ 0 \end{pmatrix}$$

all polynomial (i.e., degree ≥ 1) terms are canceled in the top position. It follows that

$$(\mathbb{R}[[x]])^2)/\{\frac{\partial F}{\partial x_i}, F_i e^j\} = \begin{pmatrix} \mathbb{R} \\ 0 \end{pmatrix} \subseteq \mathbb{R}^2$$

and the germ of the fold is locally infinitesimally stable.

23.12 Example (Phase Margin)

To illustrate the concept of *stable map germ* in the robust stability context, we come back to the example of Section 21.15:

$$L(s)\Delta(q) = \begin{pmatrix} \frac{1}{s+1} & 1 \\ 1 & \frac{1}{s+1} \end{pmatrix} \begin{pmatrix} \cos q & -\sin q \\ \sin q & \cos q \end{pmatrix}$$

Viewing f as a map into $\mathbb{R}^2$, we find

$$f(\omega, q) = \begin{pmatrix} \frac{2\cos q}{1+\omega^2} + \frac{1-\omega^2}{(1+\omega^2)^2} \\ -\frac{2\omega\cos q}{1+\omega^2} - \frac{2\omega}{(1+\omega^2)^2} \end{pmatrix}$$

It is easily seen that the following are critical point/critical value pairs:

$$\begin{array}{rcl} (q=\pi, \omega=0) & \mapsto & -1 \\ (q=\pm\pi/2, \omega=\infty) & \mapsto & 0 \end{array}$$

We analyze local infinitesimal stability of the germ of f around $(\pi, 0) \mapsto -1$. First, we introduce the shift $q = \pi + q^0$ and consider the map germ $((q^0, \omega), 0) \mapsto (f(\pi + q^0, \omega) + 1, 0)$. By the lemma, stability is preserved if we multiply $(f(\pi+q^0, \omega)+1)$ by $(1+\omega^2)^2$, because indeed this function has together with its inverse a formal or convergent power series in ω. Define

$$F(q^0, \omega) = (f(\pi + q^0, \omega) + 1)(1 + \omega^2)^2$$

and the problem is reduced to stability of the map germ $((q^0, \omega), 0) \rightarrow (F(q^0, \omega), 0)$, where

$$F(q^0, \omega) = \begin{pmatrix} -2\cos q^0(1+\omega^2) + (1-\omega^2) + (1+\omega^2)^2 \\ 2\omega\cos q^0(1+\omega^2) - 2\omega \end{pmatrix}$$

To get rid of the transcendental function, observe that the substitution $\cos q^0 = 1 - x^2$ is in fact a diffeomorphic change of variable $q^0 \leftrightarrow x$.

Therefore the problem is reduced to stability of the map germ

$$\begin{pmatrix} -\omega^2 + 2x^2 + 2x^2\omega^2 + \omega^4 \\ 2\omega^3 - 2\omega x^2 - 2x^2\omega^3 \end{pmatrix}$$

It is easily computed that

$$\begin{pmatrix} \frac{\partial F}{\partial x} \frac{\partial F}{\partial \omega} \end{pmatrix} = \begin{pmatrix} 4x + 4x\omega^2 & -2\omega + 4x^2\omega + 4\omega^3 \\ -4\omega x - 4x\omega^3 & 6\omega^2 - 2x^2 - 6x^2\omega^2 \end{pmatrix}$$

Next consider the submodule generated by $\frac{\partial F}{\partial x}$, $\frac{\partial F}{\partial \omega}$, $F_i e^j$; that is, the submodule generated by the columns,

$$\begin{pmatrix} 4x + 4x\omega^2 \\ -4\omega x - 4\omega^3 x \end{pmatrix}, \begin{pmatrix} -2\omega + 4\omega x^2 + 4\omega^3 \\ 6\omega^2 - 2x^2 - 6\omega^2 x^2 \end{pmatrix}$$

$$\begin{pmatrix} -\omega^2 + 2x^2 + 2x^2\omega^2 + \omega^4 \\ 0 \end{pmatrix}, \begin{pmatrix} 0 \\ -\omega^2 + 2x^2 + 2x^2\omega^2 + \omega^4 \end{pmatrix}$$

$$\begin{pmatrix} 0 \\ 2\omega^3 - 2\omega x^2 - 2\omega^3 x^2 \end{pmatrix}, \begin{pmatrix} 2\omega^3 - 2\omega x^2 - 2\omega^3 x^2 \\ 0 \end{pmatrix}$$

Clearly, the submodule $\{\frac{\partial F}{\partial x}, \frac{\partial F}{\partial \omega}, e^i F_j\}$ does not contain

$$\begin{pmatrix} 0 \\ \omega \end{pmatrix}, \begin{pmatrix} 0 \\ x \end{pmatrix}$$

and therefore the quotient module $(\mathbb{R}[x, \omega])^2/\{\}$ has some polynomial (i.e., degree ≥ 1) terms and the germ is not stable.

23.13 Example (Uncertain Degree)

Intuitively, the map

$$\begin{aligned} \mathbb{R}^2 &\to \mathbb{C} \\ (\epsilon, \omega) &\mapsto \frac{1}{j\omega\epsilon + 1} \end{aligned}$$

should involve some odd features, because indeed the degree of the denominator drops to zero for $\epsilon = 0$. The question is how these "odd" features can be formalized in the language of singularity and stability of maps. Here, rather than using Lemma 23.20, we exploit the fact that $1 + j\omega\epsilon \neq 0$ so that

$$\frac{1}{j\omega\epsilon + 1} \mapsto j\omega\epsilon + 1$$

is a diffeomorphism of $\mathbb{C}$ that should not affect the stability issue. Hence the problem is reduced to stability of the map

$$(\epsilon, \omega) \mapsto \begin{pmatrix} 1 \\ \omega\epsilon \end{pmatrix}$$

The Jacobian is

$$J = \begin{pmatrix} 0 & 0 \\ \omega & \epsilon \end{pmatrix}$$

The odd feature is that the whole plane is critical. To be more specific, in terms of Thom–Boardman singularity sets, $S_1 = \mathbb{R}^2 \backslash (0,0)$ and $S_2 = (0,0)$. Observe that the critical value set, $\begin{pmatrix} 1 \\ \mathbb{R} \end{pmatrix}$, has zero measure in $\mathbb{R}^2$, as predicted by Sard's theorem. We analyze local infinitesimal stability of the map germ

$$((\epsilon, \omega), (0,0)) \mapsto (\begin{pmatrix} 0 \\ \omega\epsilon \end{pmatrix} = F(\epsilon, \omega), \begin{pmatrix} 0 \\ 0 \end{pmatrix})$$

The crucial issue is the submodule of $(\mathbb{R}[\epsilon, \omega])^2$ generated over $\mathbb{R}[\epsilon, \omega]$ by the columns

$$\underbrace{\begin{pmatrix} 0 \\ \omega\epsilon \end{pmatrix}, \begin{pmatrix} \omega\epsilon \\ 0 \end{pmatrix}}_{F_i e^j}, \underbrace{\begin{pmatrix} 0 \\ \epsilon \end{pmatrix}, \begin{pmatrix} 0 \\ \omega \end{pmatrix}}_{\frac{\partial F}{\partial x_j}}$$

Clearly, this submodule is

$$\begin{pmatrix} \epsilon\omega\mathbb{R}[\epsilon, \omega] \\ 0 \end{pmatrix} \oplus \begin{pmatrix} 0 \\ \epsilon\mathbb{R}[\epsilon, \omega] \end{pmatrix} \oplus \begin{pmatrix} 0 \\ \omega\mathbb{R}[\epsilon, \omega] \end{pmatrix}$$

In the first component of the quotient module, the terms of degree 1 are clearly not canceled. Therefore, the quotient module could not be included in $\mathbb{R}^2$ and the map germ is unstable.

23.14 Example (Pole/Zero Cancellation)

Consider the map

$$\begin{array}{rcl} \mathbb{R}^2 & \to & \mathbb{C} \\ (\epsilon, \omega) & \mapsto & \dfrac{s+1+\epsilon^3}{s+1} = \dfrac{1+\epsilon^3+\omega^2 - j\omega\epsilon^3}{1+\omega^2} \end{array}$$

The potential for pole/zero cancellation is obvious. The Jacobian is

$$J = \begin{pmatrix} \frac{3\epsilon^2}{1+\omega^2} & \frac{-2\omega\epsilon^3}{(1+\omega^2)^2} \\ \frac{-3\omega\epsilon^2}{1+\omega^2} & \frac{-\epsilon^3+\omega^2\epsilon^3}{1+\omega^2} \end{pmatrix}$$

Clearly, $(\epsilon, \omega) = (0,0)$ is critical. Therefore, we analyze local infinitesimal stability of the map germ,

$$((\epsilon, \omega), (0,0)) \mapsto (\begin{pmatrix} \frac{1+\epsilon^3+\omega^2}{1+\omega^2} - 1 \\ -\frac{\omega\epsilon^3}{1+\omega^2} \end{pmatrix}, \begin{pmatrix} 0 \\ 0 \end{pmatrix})$$

By invoking Lemma 23.20 the problem is reduced to stability of the polynomial map germ

$$(\epsilon, \omega) \mapsto \begin{pmatrix} \epsilon^3 \\ -\omega\epsilon^3 \end{pmatrix} = F(\epsilon, \omega)$$

The crucial issue is to compute the submodule of $(\mathbb{R}[\epsilon, \omega])^2$ induced over $\mathbb{R}[\epsilon, \omega]$ by the columns,

$$\underbrace{\begin{pmatrix} \epsilon^3 \\ 0 \end{pmatrix}, \begin{pmatrix} 0 \\ \epsilon^3 \end{pmatrix}, \begin{pmatrix} -\omega\epsilon^3 \\ 0 \end{pmatrix}, \begin{pmatrix} 0 \\ -\omega\epsilon^3 \end{pmatrix}}_{F_i e^j}, \underbrace{\begin{pmatrix} 3\epsilon^2 \\ -3\omega\epsilon^2 \end{pmatrix}, \begin{pmatrix} 0 \\ -\epsilon^3 \end{pmatrix}}_{\frac{\partial F}{\partial x_j}}$$

Clearly, all polynomials in that submodule are made up of terms of degree 3 at least. Therefore, in the quotient module the terms of degree 1 and 2 are certainly not canceled, so that the stability test fails.

23.15 Structural Stability of Crossover

Recall that the *crossover* X_ω is defined as $f^{-1}(0 + j0)$. Anticipating variation of "certain" parameters, we would have to consider the crossover computed from the perturbed map $X'_\omega = f'^{-1}(0 + j0)$. It is important to understand under what conditions the crossovers X_ω and X'_ω are "close" and in what sense they are "close."

Essentially, we introduce a so-called *tubular neighborhood* of X_ω in $D \times \Omega$ and show that the variation of X_ω under variation of "certain" parameters remains confined within this tubular neighborhood and is an isotopy.

Definition 23.22. *A* **tubular neighborhood** *of the (neat) submanifold X_ω in the bigger manifold (with boundary) $D \times \Omega$ consists of*

- *A vector bundle (E, π, X_ω) over the crossover;*
- *An embedding $g : E \to D \times \Omega$ such that $g(E)$ is a neighborhood—the tubular neighborhood—of X_ω in $D \times \Omega$ and g maps the zero section of E homeomorphically to X_ω. (The* **zero section** *is the disjoint union over all $x \in X_\omega$ of the origins of all vector spaces E_x.)*

For the details, see [Hirsch 1976, page 109].

The situation can be depicted by the following diagram:

$$\begin{array}{ccc} & E & \\ \overset{\pi}{\swarrow} & & \overset{g}{\searrow} \\ X_\omega & \overset{i}{\longrightarrow} & D \times \Omega \end{array}$$

which does not commute, except on the zero section of E. Figure 23.1 attempts to depict the concept of tubular neighborhood. We will make use of the following lemma:

Lemma 23.23. *A neat submanifold X_ω of $D \times \Omega$ has a tubular neighborhood.*

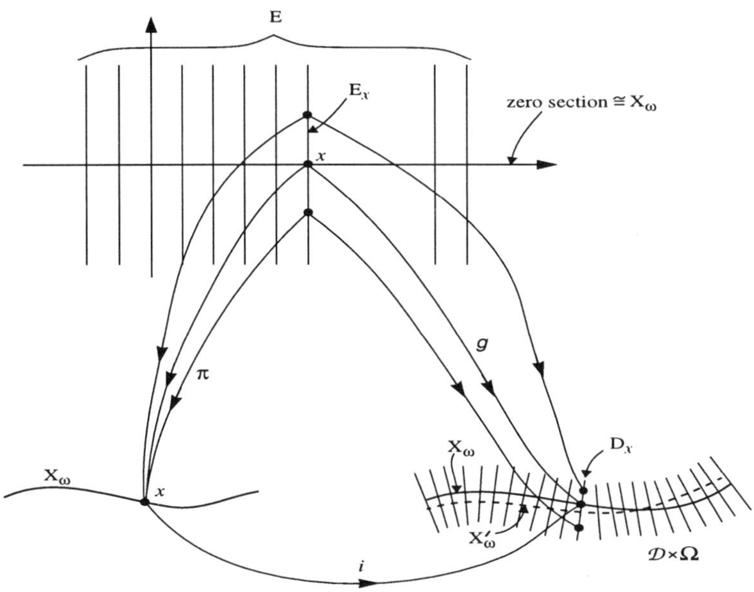

Fig. 23.1. A tubular neighborhood.

Proof. To construct this tubular neighborhood, assume first that X_ω is a submanifold of dimension $(n_p - 2)$ in the manifold $D \times \Omega$ of dimension n_p. $\forall x \in X_\omega$, choose E_x to be a 2-D plane *transverse* to X_ω. (For example, we could take the normal plane.) Define the disjoint union $E = \sqcup_{x \in X_\omega} E_x$ in $X_\omega \times \mathbb{R}^2$ and let $\pi : E \to X_\omega$ be the natural projection. Clearly, (E, π, X_ω) is a vector bundle over the crossover. (If E_x is chosen to be the normal plane, the bundle (E, π, X_ω) is called **normal bundle**.) Clearly, π maps the zero section of the bundle homeomorphically to X_ω. It follows from [Hirsch 1976, page 110] that π extends to an embedding $g : E \to D \times \Omega$ such that $W = g(E)$ is the required tubular neighborhood. This takes care of the case where X_ω is a manifold. For the case where X_ω is a neat submanifold-with-boundary in the manifold-with-boundary $D \times \Omega$, see [Hirsch 1976, page 114, Theorem 6.3]. ■

If we define $\mathbf{D}_x = g(E_x)$, all $\mathbf{D}_x$'s are open, are disjoint, and their (disjoint) union is the tubular neighborhood W.

Next, we need to specialize the concept of *isotopy* developed in Subsection 21.6.3 so that it can be used to formalize the deformation of the crossover.

Definition 23.24. *Let X be a submanifold of W. An* **isotopy** *of X in W is a map*

$$H : X \times [0, 1] \quad \to \quad W$$

$$(x,t) \quad \mapsto \quad H_t(x)$$

such that

$$H_t \textit{ is an embedding } \forall t \in [0,1]$$

and

$$H_0 = 1_X$$

Furthermore, if $X' \subseteq W$ is another submanifold, X and X' are said to be **isotopic** *iff there exists an isotopy H of X in W such that*

$$H_1(X) = X'$$

For details, see [Bochnak, Coste, and Roy 1987, pages 237, 238, 317] or [Hirsch 1976, pages 111 and 178].

Now we are in a position to state the result.

Theorem 23.25. *Let the Nyquist map $f : D \times \Omega \to \mathbb{R}^2$ be stable, and let the crossover X_ω contain no critical points, or, equivalently, let $0 + j0$ be a regular value. Then the crossover X'_ω computed from a perturbed Nyquist map f' in some sufficiently small open set O_f of the Whitney topology of $C^\infty(D \times \Omega, \mathbb{R}^2)$ is an isotopy of X_ω in some tubular neighborhood of X_ω; furthermore X_ω and X'_ω are isotopic.*

Proof. Since $0 + j0$ is not a critical value, it follows that X_ω is a neat submanifold of $D \times \Omega$ (see [Hirsch 1976, page 31]) and hence that X_ω has a *tubular neighborhood*—that is, a transverse bundle (E, π, X_ω) together with an embedding $g : E \to D \times \Omega$ such that $g(E) = W$ is the required neighborhood of X_ω. Since $g : E \to D \times \Omega$ is an embedding, we can do the deformation analysis in E instead of $D \times \Omega$. More specifically, consider the map $f \circ g : E \to \mathbb{R}^2$. Clearly, $(f \circ g)^{-1}(0)$ is the zero section of the bundle E and the problem is to assess the deformation of $(f \circ g)^{-1}(0)$ under deformation of f. Let E be charted by (x, z), where x is charting X_ω and $z = (z_1, z_2)$ is charting $E_x \cong \mathbb{R}^2$. Clearly, the Jacobian of $f \circ g(x, z)$ relative to z is nonsingular because z is the transverse component. Now, choose O_f, small enough, such that $\forall f' \in O_f$ and if we connect f and f' by the homotopy

$$f_t = (1-t)f + tf'$$

we have

$$\text{rank} \begin{pmatrix} \frac{\partial \Re f_t \circ g}{\partial z_1} & \frac{\partial \Re f_t \circ g}{\partial z_2} \\ \frac{\partial \Im f_t \circ g}{\partial z_1} & \frac{\partial \Im f_t \circ g}{\partial z_2} \end{pmatrix} = 2, \quad \forall t, \forall x$$

From the implicit function theorem (see, e.g., [Bruce and Giblin 1992, page 68, Theorem 4.14]), it follows that the solution to $f_t \circ g(x, z) = 0$ can be written as $(x, z^\bullet(x, t))$, where $z^\bullet$ is smooth in x, t. It follows that the deformation in $D \times \Omega$ is given by $p^\bullet(x, t) = g(x, z^\bullet(x, t))$. For fixed $x \in X_\omega$, $p^\bullet(x, t)$ defines a *track* from $x \in X_\omega$ to the corresponding point in

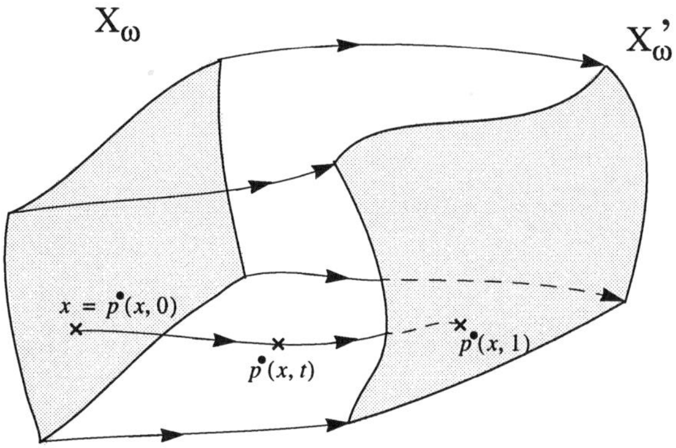

Fig. 23.2. The isotopy from the nominal to the perturbed crossover.

X'_ω. Clearly, $p^\bullet(\cdot,1) : X_\omega \to X'_\omega$ is a differentiable map. Furthermore, none of the tracks cross; indeed, any two different tracks, $p^\bullet(x',t)$ and $p^\bullet(x,t)$, are contained within different disks, $\mathbf{D}_x$ and $\mathbf{D}_{x'}$, that are disjoint by the definition of the tubular neighborhood. Hence the tracks are reversible and the map $X_\omega \to X'_\omega$ is differentiably invertible. Observe that $p^\bullet(x,0)$ is the identity map $1_{X_\omega} : X_\omega \to X_\omega$; furthermore, each map $p^\bullet(x,t)$ is easily seen to be an embedding; therefore, $p^\bullet(x,t)$ is an *isotopy* from X_ω to X'_ω. ■

The isotopy from X_ω to X'_ω is illustrated in Figure 23.2.

In its broader interpretation, the above result turns out to be *Thom's isotopy theorem* (see [Bochnak, Coste, and Roy 1987, Theorem 14.1.1] for a statement of Thom's isotopy theorem in the same spirit as the above). In case $D \times \Omega$ is a Whitney stratified space, a similar result can be proved by making crucial use of Thom's first isotopy lemma (see [Goresky and MacPherson 1988, Chapter 4] for some indications as to how this program can be achieved).

As we will soon see, in the counterexample of [Barmish, Khargonekar, Shi, and Tempo 1990]—a counterexample to the continuity of structured stability margin relative to certain parameters—a critical value curve in the Nyquist template precisely crosses over $0 + j0$.

23.16 The Counter-Example

The Barmish–Khargonekar–Shi–Tempo counterexample reveals that the stability margin relative to a collection of uncertain parameters need not be a continuous function of the problem data. The counterexample can be rewritten in our notation as follows: The loop transmission is

$$L\Delta(s,q) = \frac{N(s,q)}{D(s,q)}$$

where

$$\begin{aligned} N(s,q) &= 4a^2 + 10a^2 q_1 \\ D(s,q) &= s^4 + (20 - 20q_2)s^3 \\ &\quad +(44 + 2a + 10q_1 - 40q_2)s^2 \\ &\quad +(20 + 8a + 20aq_1 - 20q_2)s + a^2 \end{aligned}$$

In the above, $q = \begin{pmatrix} q_1 \\ q_2 \end{pmatrix}$ is a vector of real, uncertain parameters relative to which we perform a robustness analysis. On the other hand, a denotes the "problem data"—that is, a "certain" parameter that is nevertheless allowed to drift to model, for example, rounding errors. The robustness margin $r_{\max}$ is defined as the largest bound such that, for all q, $||q||_\infty < r_{\max}$, the system remains stable. The robustness margin of course depends on the problem data, which we write as $r_{\max}(a)$. The issue is whether $r_{\max}(a)$ is a continuous function. It turns out that $r_{\max}(a)$ lacks continuity at $a^* = 3 + 2\sqrt{2}$. To be more specific,

$$\lim_{a \to a^*} r_{\max}(a) = 0.417$$

while

$$r_{\max}(a^*) = 0.234$$

To make things worse, the robustness margin *drops* at $a = a^*$ so that if we proceed by continuity to a^* we are misled by a false sense of security.

To rephrase these issues in our context, consider the crossover $f_a^{-1}(0 + j0)$, where $f_a(q,s) = 1 + L\Delta(q,s)$. This crossover depends on the parameter a. This crossover, projected along the ω-axis to the q_1, q_2 plane, yields a curve that decomposes this plane into a region of stability and a region of instability (Jordan curve theorem), as shown in Figure 23.3. To define $r_{\max}$ in our context, consider the square $[-r, r] \times [-r, r]$ in the q_1, q_2 plane. Define $r_{\max}$ to be the supremum of all r's such that the square $[-r, r] \times [-r, r]$ does not intersect the curve. For $a < a^*$, the supremum occurs at $r_{\max} = 0.417$, as shown in Figure 23.4. What's happening at $a = a^*$ is that the crossover contains an extra point $(q_1, q_2, \omega) = (0.234, 0.234, \sqrt{a^*})$ that is reducing $r_{\max}$ down to 0.234, as shown in Figure 23.5.

The sudden appearance of an extra point in the crossover at $a = a^*$ is of course a change that is not a diffeomorphism. Therefore, in view of the analysis of the previous section, we should look for a critical point mapped into $0 + j0$ under the Nyquist map for $a = a^*$. It is easily verified that $(q_1 = q_2 = 0.234, \omega = \sqrt{a^*})$ is in the inverse image of $0 + j0$. Computing the matrix of partial derivatives around this point yields

$$\begin{pmatrix} \frac{\partial \Re f}{\partial q_1} & \frac{\partial \Re f}{\partial q_2} & \frac{\partial \Re f}{\partial \omega} \\ \frac{\partial \Im f}{\partial q_1} & \frac{\partial \Im f}{\partial q_2} & \frac{\partial \Im f}{\partial \omega} \end{pmatrix} = \begin{pmatrix} -1.30654 & -1.08139 & 0.828896 \\ -1.30694 & -1.08172 & 0.829083 \end{pmatrix}$$

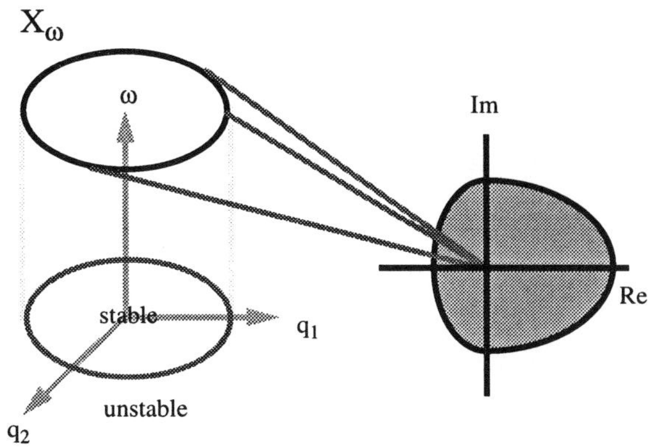

Fig. 23.3. Crossover X_ω region and its projection on the space of uncertain parameters.

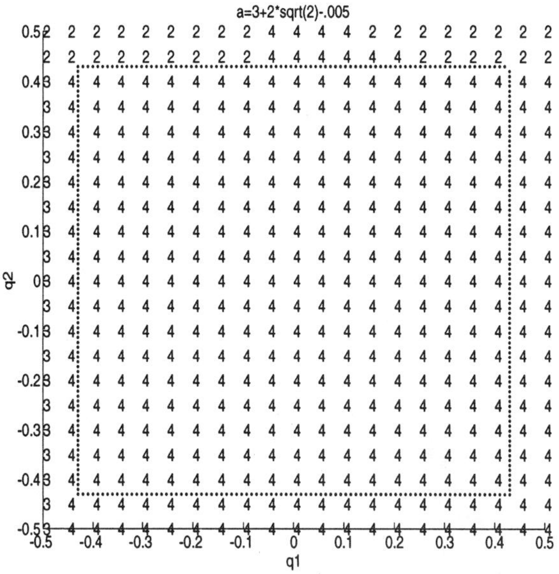

Fig. 23.4. The largest square that can be inserted within the stable region for $a < a^*$; the numbers indicate the number of LHP closed-loop poles.

Clearly, $(q_1 = q_2 = 0.2, \omega = \sqrt{a^*})$ is a *critical point* of f. To go deeper into the nature of this critical point, we perform the f_θ analysis. $(q_1 = q_2 = 0.234, \omega = \sqrt{a^*})$ is a critical point of f_θ for $\theta = -45$ degrees, because indeed

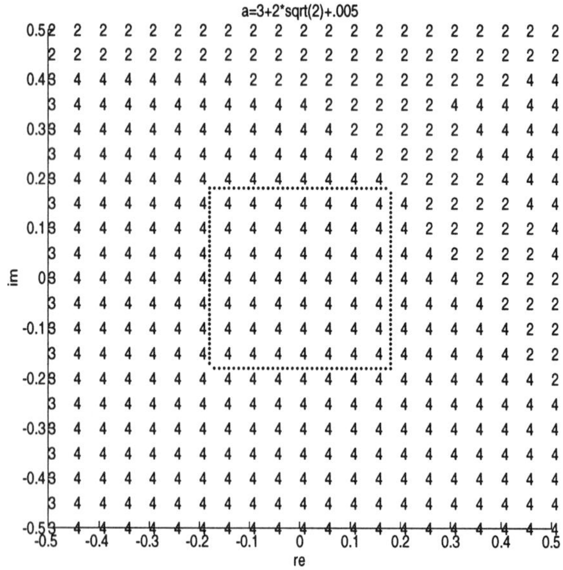

Fig. 23.5. Largest square that can be inserted in stable region for $a = a^*$.

$$\frac{\partial f_\theta}{\partial p} = \begin{pmatrix} \cos\theta & \sin\theta \end{pmatrix} \begin{pmatrix} -1.30654 & -1.08139 & 0.828896 \\ -1.30694 & -1.08172 & 0.829083 \end{pmatrix}$$

Next we compute the eigenvalues of the Hessian,

$$\begin{pmatrix} 0.00000966989539 & -0.00757797136297 & 6.68266330146758 \end{pmatrix}$$

Clearly, this is a *degenerate* critical point—that is, a point of death (or birth). It follows that two critical value curves of index 2, 1 are annihilating at $0+j0$, following a path of approach at an angle of 45 degrees with the real axis (remember, the θ direction is orthogonal to the critical value curve). Because their Morse indexes are not $0, 3 = \dim(D \times \Omega)$, these curves could *not* be critical boundary curves of the convex side of the Horowitz template. Caution should be exercised here, since average resolution graphical display of the Horowitz template seems to indicate that $0 + j0$ is on the boundary of the Horowitz template. Actually, $0+j0$ is very close to, but not on, ∂N. Deeper numerical exploration reveals that the preimage of that part of ∂N around $0 + j0$ is within $\partial D \times \Omega$, where $D = [0, 0.5] \times [0, 0.5]$. Therefore, the relevant tool to analyze the singularity structure of ∂N is the stratified Morse theory, but we won't discuss this here. It goes without saying that there is a very complicated situation, from the point of view of differential topology, that develops around $0 + j0$ in the Horowitz template.

HISTORICAL AND BIBLIOGRAPHICAL NOTES

Singularity theory grew up from men's fascination for curves and surfaces and has roots tracing back to Newton (see [Bruce and Giblin 1992]).

Some of the deeper analytical issues in singularity theory were foreseen by Weierstrass in his now famous "preparation theorem."

A forerunner to the modern theory of stability of maps, the now famous *Morse Theory*, evolved from about 1925 until about 1940 (see [Morse 1925]), but was "popularized" only later in [Milnor 1963].

It is fair to say that modern singularity theory started in the 1950s with the work of Whitney—more specifically, the classification of the generic singularities of maps between 2-manifolds. The rather formidable program of Thom on stability and singularity of differentiable maps and transversality was initiated at about the same time. The concept of *jet space*, due to [Ehresmann 1951], also appeared at about the same time.

The *Malgrange preparation theorem*, another cornerstone of the analytic theory of singularity, was developed within the context of differential geometry in 1962–1964.

The concept of the universal singularity sets as submanifolds of the jet space is due to [Boardman 1967].

The algebraic codification of singularity, stability, and determinacy of map germs was developed by [Mather 1968b, 1969b] in a famous series of papers.

Out of all of this work grew the controversial Thom–Zeeman catastrophe theory. A *catastrophe* may be defined as a bifurcation in a gradient dynamical system.

The book by [Golubitsky and Guillemin 1973] put under the same cover much of the material that had remained scattered across a voluminous amount of literature. An easily readable, fairly complete account of singularity theory, in the same spirit as [Golubitsky and Guillemin 1973], is [Arnold, Gusein-Zade, and Varchenko 1985], with special emphasis on canonical forms, spectral sequence for reduction to canonical forms, and applications.

For an excellent account of catastrophe theory in the spirit of Mather's algebraic theory, see [Castrigiano and Hayes 1993].

Probably the first intrusion of structural stability in control was [Bucy 1975], where it was shown that a continuum of mixed solutions to the Riccati equation is not structurally stable. Probably the first accounts of structural stability in continuity of robustness margin were [Cheng 1994] and [Jonckheere and Cheng 1994]. The same approach was developed independently by [Vicino and Tesi 1995].

Another application of differential topology to engineering is the inverse kinematic problem in robotics *; see, for example, [Tchon 1995]. The relevant mapping is from the space of joint angles and translation lengths to the position of the end effector. Specifying the end effector position and computing its preimage is the *inverse kinematic problem.* Differential topology analysis of the kinematic mapping reveals that the inverse kinematic problem is not that easy. See [Kwoh, Hu, Jonckheere, and Hayati 1988] for a neurosurgical application of the inverse kinematic problem.

*This was brought to the author's attention by an anonymous referee.

Part V

ALGEBRAIC GEOMETRY OF CROSSOVER

24

GEOMETRY OF CROSSOVER

La géométrie algébrique réelle se développe en relation avec de nombreux autres domaines des mathématiques: topologie différentielle y compris la théorie des singularités, algèbre commutative, géométrie analytique mais aussi topologie algébrique, formes quadratiques, théorie des modèles, analyse.

J. Bochnak, M. Coste, and M.-F. Roy,
Géométrie Algébrique Réelle,
Springer-Verlag, New York, 1987,
page 4.

SUMMARY

In Parts I and II, the computation of the homology of the crossover was purely combinatorial-numeric. To be more specific, such computation presupposes the existence of a triangulation of $D \times \Omega$ as well as a test to chase those simplexes that contain some piece of crossover. From the way these simplexes are assembled, it is possible to compute the homology of the crossover without the need to "visualize" the situation; see [Donald and Chang 1991].

The question now is whether it is possible to get some a priori estimate of the homology of the crossover *before* gearing up for triangulation and labeling test. As we shall see in this chapter, this is workable if, possibly through some change of variables, the problem can be formulated as one in which L and Δ are rational in ω and q. In this case, the combinatorics of Parts I, II receives an algebraic geometry interpretation from which bounds on the complexity of the crossover, as measured by its homology, follow.

In this chapter, we take the point of view that uncertainty is more accurately modeled with *real* variables. Keeping all variables, all coefficients, and all solutions *real* leads to *real*, as opposed to complex, algebraic geometry.

24.1 Crossover as a Real Algebraic Set

In the case when $L(j\omega)\Delta(q)$ is rational in the *real* variables q and ω, we write

$$\det(I + L\Delta) = \frac{r(q,\omega) + ji(q,\omega)}{d(j\omega)}$$

where $r(\cdot)$ and $i(\cdot)$ are the real and imaginary parts, respectively, of the numerator polynomial of $\det(I + L\Delta)$, while $d(\cdot)$ is the denominator polynomial. In this polynomial case, assuming there are no purely imaginary pole/zero cancellations, the crossover in $D \times \Omega \subseteq \mathbb{R}^{n_p}$ can be characterized by the system of equations

$$r(q,\omega) = 0, \quad i(q,\omega) = 0$$

From this polynomial equation representation, it follows that X_ω is a *real algebraic set* (see [Bochnak, Coste, and Roy 1987] and [Akbulut and King 1992]). By definition, an **(affine) real algebraic set** is the locus in $\mathbb{R}^{n_p}$ of all real solutions to a system of polynomial equations with real coefficients in finitely many real variables.

A useful feature of *real* algebraic sets is that they can be characterized by *one single* polynomial, namely, $r^2 + i^2$.

The deeper algebra stems from the observation that the real algebraic set X_ω depends only on the *ideal* $\mathrm{Id}(r, i)$ generated by the polynomials r, i, and *not* on the specific generators. To formalize this observation, given an ideal $I \subseteq \mathbb{R}[x_1, x_2, ..., x_n]$, we define $\mathcal{S}(I)$ to be the set of all real simultaneous solutions to all polynomials in the ideal, namely, $\{x \in \mathbb{R}^n : \phi(x) = 0, \forall \phi \in I\}$. With this concept, we rewrite, more formally, $X_\omega = \mathcal{S}(\mathrm{Id}(r, i))$. Conversely, given a subset $S \subseteq \mathbb{R}^n$, we define $\mathcal{I}(S)$ to be the ideal of polynomials $\phi \in \mathbb{R}[x_1, ..., x_n]$ such that $\phi(x) = 0, \forall x \in S$. Therefore X_ω is a real algebraic set iff $X_\omega = \mathcal{S}(\mathcal{I}(X_\omega))$. The real algebraic set X_ω is said to be **irreducible** iff $\mathcal{I}(X_\omega)$ is a prime ideal. It can be shown that any real algebraic set can be decomposed as the union of irreducible real algebraic sets, none of which is contained in another; quite contrary to the complex case, the irreducible real algebraic sets need not be connected (see [Akbulut and King 1992, page 20]).

Given a polynomial ideal I, it is easily seen that the ideal $\mathcal{I}(\mathcal{S}(I))$ is, in general, bigger than the original ideal I. The issue as to what the ideal $\mathcal{I}(\mathcal{S}(I))$ is is addressed by the famous *Hilbert Nullstellensatz theorem*: The *real* ($I \subseteq \mathbb{R}[x_1, ..., x_n]$) Nullstellensatz theorem states that $\mathcal{I}(\mathcal{S}(I)) = \sqrt{I}$, where $\sqrt{I}$ denotes the **real radical** of I (not to be confused with the Jacobson radical); that is, $\{\phi \in \mathbb{R}[x_1, ..., x_n] : \exists m \in \mathbb{N}, \exists \gamma_1, \gamma_2, ... \in \mathbb{R}[x_1, ..., x_n] : \phi^{2m} + \gamma_1^2 + \gamma_2^2 + ... \in I\}$ (see [Bochnak, Coste, and Roy 1987, Chapter 4]). The usefulness of the Nullstellensatz theorem is that it provides a bijective mapping between the set of (isomorphisms classes of) real algebraic varieties and the set of real radical ideals (those ideals that equal their own real radicals). The complex ($I \subseteq \mathbb{C}[p_1, ..., p_n]$) Nullstellensatz theorem states that $\mathcal{I}(\mathcal{S}(I)) = \{\phi \in \mathbb{C}[x_1, ..., x_n] : \exists m \in \mathbb{N} : \phi^m \in I\}$ (see [Munford 1991, page 3]).

A typical feature of such a real algebraic set as $X_\omega = \phi^{-1}(0)$, $\phi = \begin{pmatrix} r \\ i \end{pmatrix}$ is that it has a *stratification*,

$$X_\omega = S_\alpha \sqcup S_\beta \sqcup S_\gamma$$

where

$$\begin{aligned} S_\alpha &= \{p \in X_\omega : \text{ rank } (J_\phi(p)) = 0\} \\ S_\beta &= \{p \in X_\omega : \text{ rank } (J_\phi(p)) = 1\} \\ S_\gamma &= \{p \in X_\omega : \text{ rank } (J_\phi(p)) = 2\} \end{aligned}$$

Clearly, $S_\alpha \sqcup S_\beta$ is characterized by $\phi = 0$ together with the extra condition that all 2×2 submatrices of J have vanishing determinant, so that the lowest two strata together form a real algebraic set. The lowest stratum S_α is characterized by $\phi = 0$ and $J = 0$ so that this stratum is a real algebraic set. The reader can easily understand why S_β, S_γ are not real algebraic sets for indeed, their characterization requires a Boolean combination of inequalities; they are *semi*algebraic sets in the sense of Chapter 25.

The problem with the above stratification is that it depends on the particular polynomial representation we have singled out in $\mathcal{I}(X_\omega)$. More intrinsically, S_γ is redefined as the set of points $p \in X_\omega$ that have a neighborhood O_p such that $X_\omega \cup O_p = \phi^{-1}(0)$, where $\phi = \begin{pmatrix} \phi_1 \\ \phi_2 \end{pmatrix}$, $\phi_i \in \mathcal{I}(X_\omega)$, $\text{rank} J(\phi) = 2$, and furthermore p has no neighborhood that can be represented with a number of polynomials < 2. The reader can supply for himself the redefinition of S_α, S_β. This new stratification is called **singular stratification** (see [Akbulut and King 1986, page 23]).

24.2 Triangulation of Real Algebraic Sets

Our objective is to do some homology analysis of the crossover X_ω viewed as an algebraic set. Despite the potential complexity of algebraic sets, their homology analysis can be reduced to most conventional simplicial theory.

Theorem 24.1. *Let $S \subseteq \mathbb{R}^n$ be a bounded, closed algebraic set. Then it has a* **semialgebraic triangulation***; that is, there exists a simplicial complex $K \subseteq \mathbb{R}^n$ and a homeomorphism $K \to S$.*

Proof. See [Bochnak, Coste, and Roy 1987, Theorem 9.2.1]. ∎

Therefore, we define the homology of the crossover $H_*(X_\omega)$ as the homology, in the usual combinatorial sense, of its homeomorphic simplicial complex, $H_*(K)$.

24.3 Local Euler Characteristic of Real Algebraic Crossover Set

Let K be a simplicial complex homeomorphic to X_ω and let a^0 be one of its vertices. The *global* structure of K is characterized by its homology groups. In addition to this global structure, the complex K might have some interesting *local* structure that needs to be axiomatized.

Definition 24.2. *The* **local homology group** *of K at its vertex a^0 is the relative homology group $H_n(K, K \setminus star(a^0))$.*

Definition 24.3. *The* **local Euler characteristic** *of K at a^0 is the alternate sum*

$$\chi(K, K \setminus star(a^0)) = \sum_n (-1)^n \; rank \; (H_n(K, K \setminus star(a^0)))$$

Theorem 24.4.

$$\chi(K, K \setminus star(a^0)) = \sum_n (-1)^n \#(a^0\sigma'_{n-1})$$

where $\#(a^0\sigma'_{n-1})$ is the number of n-simplexes that have a^0 as a vertex.

Proof. This is a corollary of $\chi(K, K \setminus \mathrm{star}(a^0)) = \chi(K) - \chi(K \setminus \mathrm{star}(a^0))$ (see [Bochnak, Coste, and Roy 1987, Proposition 11.2.1, page 232]). ■

Theorem 24.5. *The local Euler characteristic of the algebraic set X_ω is odd.*

Proof. See [Bochnak, Coste, and Roy 1987, Theorem 11.2.2, page 232]. ■

In dimension not exceeding 2, a stronger statement can be made:

Theorem 24.6. *A necessary and sufficient condition for a simplicial complex K of dimension ≤ 2 to be homeomorphic to an algebraic set of dimension ≤ 2 is that the complex has odd local Euler characteristic.*

Proof. See [Bochnak, Coste, and Roy 1987, Remark 11.2.4, page 234]. ■

24.4 Betti Numbers of Real Algebraic Crossover Set

With this real algebraic geometry formulation, it is possible to derive *some bounds on the complexity of the crossover* in terms of its homology. However, here we have to deal with chain groups $C_*(X_\omega, \mathbb{Z} \bmod 2)$ with coefficients in $\mathbb{Z} \bmod 2$. By definition, $C_n(X_\omega, \mathbb{Z} \bmod 2)$ is the formal Abelian group of linear combinations of n-simplexes of X_ω subject to the *relation* $c_n + c_n = 0$. A boundary homomorphism $\partial_n : C_n(X_\omega, \mathbb{Z} \bmod 2) \to C_{n-1}(X_\omega, \mathbb{Z} \bmod 2)$ is defined in the usual way and we still have $\partial_{n-1}\partial_n = 0$. From there, homology groups, $H_*(X_\omega, \mathbb{Z} \bmod 2)$, are defined.

Finally, the Betti number $b_n(X_\omega, \mathbb{Z} \bmod 2)$ is defined as the dimension of $H_n(X_\omega, \mathbb{Z} \bmod 2)$ as a vector space over the Galois field $\mathbb{Z} \bmod 2$. Of course the homology with coefficients in $\mathbb{Z} \bmod 2$ provides a coarser picture of the topological space than does the usual homology with coefficients in $\mathbb{Z}$. The homology modulo 2 is used when the intricacies of *orientation* are to be avoided.

Theorem 24.7. *Let the crossover X_ω be a bounded algebraic set of dimension n. Let $(X_\omega)_n$ be the sum, with coefficients in $\mathbb{Z} \bmod 2$, of the n-simplexes of X_ω for the semialgebraic triangulation of X_ω. Then $\partial_n(X_\omega)_n = 0$, where ∂_n is the boundary operator for the homology modulo 2. Furthermore, the cycle $(X_\omega)_n$ determines a nontrivial element of the homology group $H_n(X_\omega, \mathbb{Z} \bmod 2)$.*

Proof. This is a basic result on the topology of algebraic sets (see [Bochnak, Coste, and Roy 1987, Proposition 11.3.1]). ■

Theorem 24.8. *Let n_q be the number of variables in q. Let d be the degree of r and i. Then*

$$\sum_k b_k(X_\omega, \mathbb{Z} \bmod 2) \leq d(2d-1)^{n_q}$$

Proof. This is a trivial consequence of the result of [Milnor 1964] bounding the Betti numbers of an algebraic set in terms of the number of polynomial variables and the degree. For a more modern exposition, see [Bochnak, Coste, and Roy 1987, Section 11.5] and [Risler 1988, Theorem 1.2]. ■

We observe that the complexity of the crossover, as measured by the upper bounds on the Betti numbers, is *exponential* in the dimension and *polynomial* in the degree.

The above result is in terms of the homology modulo 2, not the usual homology over $\mathbb{Z}$. Nevertheless, some relations between the two homologies can be derived. These are the so-called *universal coefficient theorems*; see, for example, [Hilton and Wylie 1965, Section 5.2] or Appendix A. In a few words, these theorems assert that the homology for an arbitrary coefficient group G (e.g., $\mathbb{Z} \bmod 2$) can be derived from the usual homology over $\mathbb{Z}$. The relevant result is the following:

Theorem 24.9. *Let K be a simplicial complex and let G be an arbitrary (Abelian) coefficient group. We have*

$$H_n(K, G) = (H_n(K) \otimes G) \oplus (H_{n-1}(K) * G)$$

where $*$ denotes the torsion product of Abelian groups. The **torsion product** *can be defined by the following rules:*

- $\mathbb{Z} * G \cong G * \mathbb{Z} \cong 0$, *for any Abelian group G*
- $(\mathbb{Z} \bmod k) * (\mathbb{Z} \bmod l) \cong (\mathbb{Z} \bmod \gcd(k, l))$

Furthermore, the torsion product is commutative, up to an isomorphism, and is distributive, up to an isomorphism, relative to the direct sum $\oplus$.

Proof. See [Munkres 1984, page 331], [Hilton and Wylie 1965, page 171], [Hocking and Young 1961, page 248], or Appendix A, which provides the deeper interpretation of the torsion product. ■

In many cases, the usual Betti number (over $\mathbb{Z}$) equals the Betti number over $\mathbb{Z}$ mod 2:

Theorem 24.10. *If* $H_*(X_\omega)$ *has no torsion, then* $b_n(X_\omega) = b_n(X_\omega, \mathbb{Z} \bmod 2)$.

Proof. Since $H_{n-1}(X_\omega)$ has no torsion, it is isomorphic to the direct product of several copies of $\mathbb{Z}$. Therefore, it follows from the rules of the torsion product that $H_{n-1}(X_\omega) * (\mathbb{Z} \bmod 2) = 0$, so that

$$H_n(X_\omega, \mathbb{Z} \bmod 2) = H_n(X_\omega) \otimes (\mathbb{Z} \bmod 2)$$

Since $H_n(X_\omega)$ is free and has rank b_n, the usual Betti number, it follows from [Hilton and Wylie 1965, Theorem 5.2.7] that the right-hand side tensor product is isomorphic to the direct sum of b_n copies of $\mathbb{Z}$ mod 2, namely,

$$H_n(X_\omega, \mathbb{Z} \bmod 2) = \bigoplus_{1}^{b_n} (\mathbb{Z} \bmod 2)$$

Since $b_n(X_\omega, \mathbb{Z} \bmod 2)$ is by definition the dimension of $H_n(X_\omega, \mathbb{Z} \bmod 2)$ as a vector space over the Galois field $\mathbb{Z}$ mod 2, the result follows. ■

In the general situation where $H_*(X_\omega)$ could have torsion, we can only get an inequality:

Theorem 24.11.

$$b_n(X_\omega) \leq b_n(X_\omega, \mathbb{Z} \bmod 2)$$

Proof. In the general case, we have to take a closer look to the torsion product. We have

$$H_{n-1}(X_\omega) = \mathbb{Z} \oplus \ldots \oplus (\mathbb{Z} \bmod k_1) \oplus \ldots$$

It follows that

$$\begin{aligned} & H_{n-1}(X_\omega) * (\mathbb{Z} \bmod 2) \\ = \; & (\mathbb{Z} \oplus \ldots \oplus (\mathbb{Z} \bmod k_1) \oplus \ldots) * (\mathbb{Z} \bmod 2) \\ = \; & \bigoplus \mathbb{Z} \bmod \gcd(k_i, 2) \end{aligned}$$

Therefore,

$$H_{n-1}(X_\omega) * (\mathbb{Z} \bmod 2) = \bigoplus (\mathbb{Z} \bmod 2)$$

where the number of terms in direct sum equals the number of even torsion coefficients. We now examine the tensor product

$$H_n(X_\omega) \otimes (\mathbb{Z} \bmod 2) = (\mathbb{Z} \oplus \ldots \oplus (\mathbb{Z} \bmod k_1) \oplus \ldots) \otimes (\mathbb{Z} \bmod 2)$$

From [Hilton and Wylie 1965, Theorem 5.2.6] the tensor product is distributive relative to the direct sum; hence,

$$\begin{aligned} & H_n(X_\omega) \otimes (\mathbb{Z} \bmod 2) \\ &= (\mathbb{Z} \otimes (\mathbb{Z} \bmod 2)) \oplus \ldots \oplus ((\mathbb{Z} \bmod k_1) \otimes (\mathbb{Z} \bmod 2)) \oplus \ldots \\ &= (\mathbb{Z} \bmod 2) \oplus \ldots \oplus (\mathbb{Z} \bmod \gcd(k_1, 2))\ldots \end{aligned}$$

Finally, we get

$$\begin{aligned} & b_n(X_\omega, \mathbb{Z} \bmod 2) \\ &= \dim_{\mathbb{Z} \bmod 2} H_n(X_\omega, \mathbb{Z} \bmod 2) \\ &\geq \dim_{\mathbb{Z} \bmod 2} H_n(X_\omega) \otimes (\mathbb{Z} \bmod 2) \\ &= \dim_{\mathbb{Z} \bmod 2}(\underbrace{\mathbb{Z} \oplus \ldots}_{b_n(X_\omega, \mathbb{Z}) \text{ terms}}) \otimes (\mathbb{Z} \bmod 2) \\ &\geq b_n(X_\omega, \mathbb{Z}) \end{aligned}$$

■

Corollary 24.1.

$$\sum_k b_k(X_\omega) \leq d(2d-1)^{n_q}$$

24.5 Algebraic Crossover Curve

In the case of three uncertainties, namely, (ω, q_1, q_2), the full power of real algebraic geometry might not be absolutely necessary. Indeed, three uncertainties together with two equations, $r = 0$ and $i = 0$, yields an *algebraic (space) curve* crossover. Some good insight into the algebraic geometry of the crossover can be gained by resorting to traditional algebraic curve theory; see, for example, [Walker 1978]. See Subsection 9.2.3 and Subsection 9.2.4 for such analysis of a selected robust stability problem.

24.6 Example

Now we illustrate these features by coming back to the multivariable phase margin ("2-torus") problem of Subsection 9.2.2, Equations 9.1 and 9.2. Observe that $\det(I + L\Delta) = 0$ is equivalent to $\det(\Delta^{-1} + L) = 0$. Now, remember that the determinant of the sum of two matrices equals the sum of all determinants made up with some columns of the first matrix together with the complementary columns of the second matrix. In other words,

$$\det(\Delta^{-1} + L) = \det \Delta^{-1} + \det L + \det(\Delta_1^{-1}|L_2) + \det(L_1|\Delta_2^{-1})$$

where Δ_i^{-1} and L_i denote the ith column of Δ^{-1} and L, respectively. Furthermore, as a consequence of the Binet–Cauchy theorem and Jacobi's theorem on adjoint determinants, it is easily seen that

$$\det L(s) = \det(\frac{N(s)}{d(s)}) = \frac{-0.25}{d(s)}$$

Therefore,

$$\begin{aligned}
\det(\Delta^{-1} + L) = & \\
& \frac{1}{d(s)}(s^4 + 1.75s^3 + 7.5s^2 + 4s + 8 - 0.25 \\
+ & \cos\theta \exp(-j\phi)(s+4) \\
- & \sin\theta(0.25s^2 + 0.187s + 0.75) \\
+ & (0.0625s + 0.25)\cos\theta\exp(j\phi) + \sin\theta(s^2 + s + 4))
\end{aligned}$$

The above is, in part, a trigonometric polynomial. To convert it into a polynomial we introduce the ad hoc change of variable

$$\cos\theta = q_1, \ \sin\theta = q_2, \ q_1^2 + q_2^2 = 1$$

$$\cos\phi = q_3, \ \sin\phi = q_4, \ q_3^2 + q_4^2 = 1$$

Therefore the *polynomial equations representing* X_ω *as a real algebraic set* are

$$r(q,\omega) \equiv \omega^4 - 0.75\omega^2 q_2 - 7.5\omega^2 + 0.9375\omega q_1 q_4 + 4.25 q_1 q_3 + 3.25 q_2 + 7.75 = 0$$

$$i(q,\omega) \equiv -1.75\omega^3 + 1.0625\omega q_1 q_3 - 3.75 q_1 q_4 + 0.8125 q_2 \omega + 4\omega = 0$$

$$q_1^2 + q_2^2 = 1$$

$$q_3^2 + q_4^2 = 1$$

Now, it is clear that the degree $d = 4$. Unfortunately, the number of real variables in q has increased by two in the process of converting the trigonometric part into a usual polynomial. Thus $n_q = 4$. Therefore, the bound becomes

$$\sum_k b_k(X_\omega) \leq d(2d-1)^{n_q} = 4(8-1)^4 = 9604$$

The conservativeness of this result is not as much due to the conservativeness of the bound itself as it is due to the procedure of deliberately increasing the number of variables to get a polynomial formulation in a problem that is exponential in the number of variables. The same spring-mass-dashpot plant together with a real structured singular value perturbation, namely, $\Delta = \text{diag}\{k_1, k_2\}$, would yield $n_q = 2$ and a bound of $4(8-1)^2 = 196$, which is much more reasonable.

We note, as an aside, that Hovansky (see [Risler 1988]) derived some results related to the complexity of varieties described by polynomial equations in real variables *and* sine and cosine of the real variables. However, a Tarski–Seidenberg-like principle for varieties described by transcendental equations has not yet been proved.

NOTES

Real algebraic geometry has roots tracing back to the theorem of Sturm, predicting the number of real roots of a real polynomial within an interval of the real line. Following [Bochnak, Coste, and Roy 1987], it is probably the extraordinary success of complex-analytic function theory that led algebraic geometry on the path of algebraically closed ground fields like $\mathbb{C}$. Removing this assumption—that is, working with such a ground field as $\mathbb{R}$ that is not algebraically closed—requires nontrivial modification of the foundational Hilbert Nullstellensatz, resulting in an algebraic geometry quite different from the traditional one. It was Thom who, in the 1950s, re-emphasized the importance of the real nature of things, against the then prevailing popular acceptance of complex solutions. Real algebraic geometry, and in particular the topology of real algebraic sets, was taken up by Milnor, Nash, and Whitney. Probably the first modern comprehensive survey of real algebraic geometry was [Bochnak, Costes, and Roy 1987]. Real algebraic geometry has now becomes an established independent mathematical discipline—as exemplified by [Akbulut and King 1992] and [Akbulut 1995].

The specific computational problems associated with real (semi)algebraic sets have crystallized in an independent discipline of computer sciences, sometimes referred to as computational algebraic geometry (see [Risler 1988]).

25

GEOMETRY OF STABILITY BOUNDARY

SUMMARY

In the previous chapter, we have derived bounds on the homology of the neutral stability region X_ω in the $D \times \Omega$ space. What still needs to be done is the elimination of ω in order to get bounds on the homology of the stability boundary $X \subseteq D$ in the physical parameters space. In Chapter 11, we developed a geometric–combinatorial approach based on the idea of projecting X_ω on the first factor of $D \times \Omega$. Here, we approach the same problem from the real algebraic geometry point of view: We view the projection on the first factor of $D \times \Omega$ as the so-called *Tarski–Seidenberg* elimination of the *real* variable ω among the polynomial equations describing X_ω.

25.1 Tarski–Seidenberg Elimination

The Tarski–Seidenberg elimination of the *real* variable ω between $r(q, \omega)$ and $i(q, \omega)$ results in a Boolean combination $\mathcal{B}(q)$ of polynomial equations and inequalities such that $r(q, \omega) = 0, i(q, \omega) = 0$ has a *real* solution q for some *real* ω iff $\mathcal{B}(q)$ is true (see [Bochnak, Coste, and Roy 1987, Chapters 2 and 4]). It follows that the set X of q's leading to a $j\omega$ instability can be characterized by $\mathcal{B}(q)$. This elimination procedure is also called *projection*, because X is indeed the projection—in the usual geometric sense—of X_ω along the ω axis on the q plane. Since $\mathcal{B}(q)$ incorporates polynomial equations *and* inequalities, it follows that X is a set of a more complicated nature than X_ω. Such a set as X characterized by real polynomial equations and/or polynomial *inequalities* is called a **semialgebraic set** (see [Bochnak, Coste, and Roy 1987]).

A well-known and fundamental fact of real algebraic geometry is that the projection of a real algebraic or semialgebraic set is *semialgebraic*. Next, it is shown that a semialgebraic set can be decomposed as the disjoint union of finitely many semialgebraic sets, each of which is homeomorphic to an open subset of $\mathbb{R}^n$, for some $n \in \mathbb{N}$. These elementary properties are proved in [Bochnak, Coste, and Roy 1987, Chapter 2] and their proofs rely on the Tarski–Seidenberg elimination. Less trivial to prove is the fact that a semialgebraic set has a semialgebraic triangulation (see [Bochnak, Coste, and Roy 1987, Chapter 9]).

25.2 Complexity

Bounds on the Betti numbers of semialgebraic sets are not as clearcut as in the algebraic case (see [Risler 1988]). Nevertheless, the most relevant result appears to be the following theorem:

Theorem 25.1. *Let $\mathcal{B}(q)$ be*

$$f_1(q) = \dots = f_m(q) = 0$$

$$f_{m+1}(q) > 0, \dots, f_{m+u}(q) > 0$$

$$f_{m+u+1}(q) \geq 0, \dots, f_{m+u+v}(q) \geq 0$$

then

$$b_0(X) \leq (2dv + 1)^{n_q + 2}$$

Proof. See [Risler 1988, Theorem 1.2 (d)]. ∎

From this theorem, we learn that the complexity of the crossover depends critically on the number of polynomial inequalities ($\geq$), v, resulting from the Tarski–Seidenberg elimination.

25.3 Example

We now illustrate the Tarski–Seidenberg elimination of the real variable ω on the "2-torus" example of Subsection 9.2.2, Equations 9.1 and 9.2. We proceed from the equations, derived in Section 24.6, describing X_ω as a real algebraic set. To simplify the notation we write

$$\begin{aligned} r &\equiv \omega^4 + A_2\omega^2 + A_1\omega + A_0 \\ i' \equiv \frac{i}{-1.75} &\equiv \omega^3 + B_1\omega + B_0 \end{aligned}$$

where

$$\begin{aligned} A_2 &\equiv -0.75q_2 - 7.5 \\ A_1 &\equiv 0.9375q_1q_4 \\ A_0 &\equiv 4.25q_1q_3 + 3.25q_2 + 7.75 \\ B_1 &\equiv -0.6071q_1q_3 - 0.464q_2 - 2.285 \\ B_0 &\equiv 2.14q_1q_4 \end{aligned}$$

To eliminate the real variable ω on this particular example, we compute the greatest common divisor of r and i', viewed as polynomials in ω, using a parameterized version of the Euclidean algorithm (see [Jacobson 1974, Chapter 5]). If we can show that this greatest common divisor has degree one in ω, then r and i' share one and only one common root that could only be real!

The Euclidean division of r by i' yields

$$\gcd(r, i') = \gcd(i', (A_2 - B_1)\omega^2 + (A_1 - B_0)\omega + B_0)$$

The *significant* fact that simplifies the Tarski–Seidenberg elimination in this particular example is

$$A_2 - B_1 \equiv -0.28q_2 - 5.22 + 0.607q_1q_3 \neq 0, \forall q$$

Therefore, we can proceed with the Euclidean division

$$\frac{i'}{(A_2 - B_1)\omega^2 + (A_1 - B_0)\omega + A_0}$$

and this yields

$$\begin{aligned} &\gcd(i', (A_2 - B_1)\omega^2 + (A_1 - B_0)\omega + A_0)) \\ &\quad = \gcd((A_2 - B_1)\omega^2 + (A_1 - B_0)\omega + A_0, C\omega + D) \end{aligned}$$

where

$$\begin{aligned} C &\equiv (A_2 - B_1)^2 B_1 - A_0(A_2 - B_1) + (A_1 - B_0)^2 \\ D &\equiv (A_2 - B_1)^2 B_0 + A_0(A_1 - B_0) \end{aligned}$$

The symbolic computation of C and D, using μmath of Microsoft, yields

$C =$

$$\begin{aligned} &-59641/2744 - 1499/5488q_2 + 799/686q_2q_1q_3 + 867/21952q_2q_1^2q_3^2 \\ &+42313/2744q_1q_3 + 153/1372q_2^2q_1q_3 + 18225/12544q_1^2q_4^2 + 289/686q_1^2q_3^2 \\ &-4913/21952q_1^3q_3^3 - 220/343q_2^2 - 13/343q_2^3 \end{aligned}$$

$D =$

$$\begin{aligned} &54165/21952q_2q_1q_4 - 255/343q_2q_1^2q_4q_3 + 1073895/21952q_1q_4 \\ &+60/343q_2^2q_1q_4 - 410295/21952q_1^2q_4q_3 + 4335/5488q_1^3q_4q_3^2 \end{aligned}$$

From the expression for C, bounding the polynomial terms by their absolute values for $q_i = 1$, we get

$$C \leq -1.94871765670554, \forall q$$

In other words, C *never vanishes.* Therefore, r and i' have a real root in common iff the remainder of the Euclidean division

$$\frac{(A_2 - B_1)\omega^2 + (A_1 - B_0)\omega + A_0}{C\omega + D}$$

vanishes; that is,

$$F \equiv (A_2 - B_1)D^2 - (A_1 - B_0)DC + A_0C^2 = 0$$

Therefore, the crossover X in the q parameter space is the algebraic set $F = 0$.

The μmath symbolic computation of F yields

$F =$

$$
\begin{aligned}
&110268515311/30118144\\
&+24506546541/15059072q_2 - 19469431271/7529536q_2q_1q_3\\
&-4307574986325/1927561216q_2q_1^2q_4^2 + 87413410083/120472576q_2q_1^2q_3^2\\
&+3542717138175/1927561216q_2q_1^3q_4^2q_3 + 56917237651/240945152q_2q_1^3q_3^3\\
&-3120897048525/7710244864q_2q_1^4q_4^2q_3^2 + 2694110625/240945152q_2q_1^4q_4^4\\
&-3799954937/240945152q_2q_1^4q_3^4 + 17265633075/550731776q_2q_1^5q_4^2q_3^3\\
&-627395625/481890304q_2q_1^5q_4^4q_3 - 2721865869/963780608q_2q_1^5q_3^5\\
&-1259079075/1927561216q_2q_1^6q_4^2q_3^4 + 24137569/275365888q_2q_1^6q_3^6\\
&-95992726269/30118144q_1q_3 - 32396933235/120472576q_2^2q_1q_3\\
&-482281338375/1927561216q_2^2q_1^2q_4^2 + 8637859443/240945152q_2^2q_1^2q_3^2\\
&+285326253675/1927561216q_2^2q_1^3q_4^2q_3 + 1759212649/60236288q_2^2q_1^3q_3^3\\
&-167965232175/7710244864q_2^2q_1^4q_4^2q_3^2 + 36905625/120472576q_2^2q_1^4q_4^4\\
&+95464503/1927561216q_2^2q_1^4q_3^4 + 51954975/60236288q_2^2q_1^5q_4^2q_3^3\\
&-72412707/275365888q_2^2q_1^5q_3^5 - 1426543865/15059072q_2^3q_1q_3\\
&-5358971025/240945152q_2^3q_1^2q_4^2 + 856869683/240945152q_2^3q_1^2q_3^2\\
&+3560091975/481890304q_2^3q_1^3q_4^2q_3 + 149998803/60236288q_2^3q_1^3q_3^3\\
&-7868025/15059072q_2^3q_1^4q_4^2q_3^2 - 13112797/275365888q_2^3q_1^4q_3^4\\
&-132657715/15059072q_2^4q_1q_3 - 121928175/120472576q_2^4q_1^2q_4^2\\
&-5032935/15059072q_2^4q_1^2q_3^2 + 141525/941192q_2^4q_1^3q_4^2q_3\\
&+1523303/1229312q_2^4q_1^3q_3^3 - 574821/941192q_2^5q_1q_3\\
&-1125/67228q_2^5q_1^2q_4^2 - 11271/2151296q_2^5q_1^2q_3^2\\
&-2873/134456q_2^6q_1q_3 - 858356817975/60236288q_1^2q_4^2\\
&-34574548235/30118144q_1^2q_3^2 + 24029142686775/1927561216q_1^3q_4^2q_3\\
&+133584131003/120472576q_1^3q_3^3 - 478879538625/137682944q_1^4q_4^2q_3^2\\
&+196670075625/1927561216q_1^4q_4^4 + 2677432697/60236288q_1^4q_3^4\\
&+3123739811475/7710244864q_1^5q_4^2q_3^3 - 45799880625/1927561216q_1^5q_4^4q_3\\
&-3619215493/120472576q_1^5q_3^5 - 142651779975/7710244864q_1^6q_4^2q_3^4\\
&+10665725625/7710244864q_1^6q_4^4q_3^2 - 796539777/1927561216q_1^6q_3^6\\
&+319467825/1927561216q_1^7q_4^2q_3^5 + 410338673/1927561216q_1^7q_3^7\\
&+30750657379/120472576q_2^2 + 12811302085/120472576q_2^3\\
&+37053095/3764768q_2^4 + 6705491/3764768q_2^5\\
&+79599/470596q_2^6 + 2197/470596q_2^7
\end{aligned}
$$

The lesson we learn from the above exercise is that real algebraic geometry is computationally extremely intensive!

Part VI

EPILOGUE

26

EPILOGUE

To conclude let me just say that the analysis and topology are now inextricably mixed.

Michael F. Atiyah,
"Algebraic topology and elliptic operators,"
Communications on Pure and Applied Mathematics,
vol. XX, pp. 237–249, 1967.

Having reached the end of our journey through the topology of robust stability, we would like to devote the last chapter of this book to a few remarks pertaining to those topics that the lack of space and the lack of time have prevented us from developing in this book.

Contrary to the popular belief (especially among electrical engineering students) that topology is relevant to pure mathematics only, topology actually emerged out of practical needs—topology à la Euler emerged out of the Königsberg bridge problem, topology à la Poincaré emerged out of the qualitative theory of dynamical systems, and so on. All along this century, Poncaré's vision of the fundamental role topology should play in all science consolidated. Modern physics has turned out to be a fertile ground for applications of topology (see [Nash and Sen 1983] and [Nash 1991]). It did not take much time for operator theory investigators to realize that the index of operators is a fundamentally topological problem. Closer to our line of investigations, a long chain of events culminated in the topology of the space of transfer matrices of fixed McMillan degree. This book has followed this general trend. One can only hope that the range of applications of topology will keep on increasing.

Among the topics of topological robust stability which we feel should have been investigated in more detail, we will mention

- the interface between the complex μ and the real μ problems from the point of view of CR geometry, where CR stands for either Complex-Real or Cauchy–Riemann (see [D'Angelo 1993]),
- the systematic utilization of such modern computational algebra tools as Gröbner bases and Collins cylindrical algebraic decomposition for the ring and module manipulations pertaining to stability and singularity of the Nyquist map and related computational issues in algebraic

crossover sets,

- the homotopy of the return difference matrix map,
- the structural stability issue in the realm of stratified spaces, or what has been called the "boundary singularity" problem by [Arnold, Gusein-Zade, and Varchenko 1985, 1988],
- the Jordan–Brouwer separation problem is the parameter space D.

Finally, a few remarks about design, which this book has totally ignored. They are reasons to believe that the $\mathcal{H}^\infty$ design, viewed as a mapping from the space of plants of McMillan degree d to the space of compensators of McMillan degree $(d-1)$, is some sort of a "cellular map." The "cellular" property would mean that the map respects the decomposition into cells, cells in which some signature related to the Cauchy index remains constant. These features become clearer when the $\mathcal{H}^\infty$ design is written out in the LQG balanced coordinates. (See [Furhmann and Ober 1993] for these insights.)

Part VII

APPENDICES

APPENDIX A

HOMOLOGICAL ALGEBRA OF GROUPS

A.1 Abelian Groups and Homomorphisms

Definition A.1. *An* **Abelian group** $(G, +)$ *is a set* G *together with an internal law* $+ : G \times G \to G$ *such that*

- $(a + b) + c = a + (b + c)$
- $a + 0 = 0 + a = a$
- $a + (-a) = (-a) + a = 0$
- $a + b = b + a$

Definition A.2. *The* **direct sum** $A \oplus B$ *of two Abelian groups* A, B *is the group of all ordered pairs* (a, b), $a \in A$, $b \in B$ *together with the additive law* $(a, b) + (c, d) = (a + c, b + d)$.

Definition A.3. *A group is said to be* **finitely generated** *iff there are finitely many elements* $g_1, ..., g_n$, *called* **generators**, *such that any element* $a \in G$ *can be written (not necessarily in a unique fashion) as* $\sum_{i=1}^{n} \alpha_i g_i$, *where the* α_i*'s are integers. A finitely generated group is said to be* **free** *iff*

$$\sum_{i=1}^{n} \alpha_i g_i = 0; \quad g_i \neq g_j, \text{ for } i \neq j; \quad \alpha_i \in \mathbb{Z}$$

implies

$$\alpha_i = 0$$

The opposite case of a free group is a group that is specified in terms of generators and relations: We are given a finite set of generators $g_1, ..., g_n$; any linear combination with integral coefficients of the g_i's is an element of G; however, the number of different elements that can be created is limited by the relations

$$\sum_{j=1}^{n} a_{i,j} g_j = 0, \quad i = 1, 2, ..., r$$

Clearly, the relations can be written in matrix form as

$$A \begin{pmatrix} g_1 \\ g_2 \\ \vdots \\ g_n \end{pmatrix} = 0$$

The matrix A is defined over the ring of integers. To clarify the fine structure of an Abelian group defined in terms of generators and relations, we compute the Smith form of A over the ring of integers. The Smith form is obtained by premultiplying/postmultiplying the A matrix by unimodular matrices. The latter are matrices over the ring of integers with determinant 1, so that unimodular matrices are invertible over the ring of integers. Postmultiplication by unimodular matrices is equivalent to changing the set of generators to, say, $\{h_i\}$; premultiplication amounts to redefining a new, equivalent set of relations. This process yields

$$\begin{pmatrix} k_1 & 0 & \dots & 0 & 0 & \dots & 0 \\ 0 & k_2 & \dots & 0 & 0 & \dots & 0 \\ \vdots & \vdots & \ddots & \vdots & \vdots & & \vdots \\ 0 & 0 & \dots & k_m & 0 & \dots & 0 \\ 0 & 0 & \dots & 0 & 0 & \dots & 0 \\ \vdots & \vdots & & \vdots & \vdots & \ddots & \vdots \\ 0 & 0 & \dots & 0 & 0 & \dots & 0 \end{pmatrix} \begin{pmatrix} h_1 \\ h_2 \\ \vdots \\ h_n \end{pmatrix} = 0$$

In the above, k_1 divides k_2, k_2 divides k_3, and so on. It follows from the Smith form equation that $k_1 h_1 = 0$. In other words, the subgroup generated by h_1 is isomorphic to $\mathbb{Z} \bmod k_1$. Furthermore, there are $b = n - m$ generators, $h_{m+1}, \dots, h_n$, that are free (i.e., not subject to relations), so that the subgroup they generate is isomorphic to the direct sum of b copies of $\mathbb{Z}$. Therefore, we have the following:

Theorem A.4. *A finitely generated Abelian group G can be decomposed as*

$$G \cong (\oplus_{i=1}^{b} \mathbb{Z}) \oplus (\mathbb{Z} \bmod k_1) \oplus (\mathbb{Z} \bmod k_2) \oplus \dots \oplus (\mathbb{Z} \bmod k_m)$$

where k_i divides k_{i+1}. The subgroup $\oplus_{i=1}^{b} \mathbb{Z}$ is the **free part** *of the group; the rank of the free part, b, is called* **Betti number**; *$\oplus_{i=1}^{m} \mathbb{Z} \bmod k_i$ is called* **torsion subgroup**; *and the k_i's are called* **invariant factors** *or* **torsion coefficients**.

There is another way to decompose an Abelian group. We need the following definition:

Definition A.5. *A* **p-primary component** *in a torsion group is a sub (torsion) group of the form $\mathbb{Z} \bmod p^n$, where p is a prime and n is a natural number.*

Theorem A.6. *A torsion group can be decomposed as the direct sum of p-primary components.*

A.2 Chain Complexes

Algebraically, a **chain complex** (C_*, ∂_*) is a collection of groups $\{C_n\}$ and group homomorphisms $\partial_n : C_n \to C_{n-1}$ such that $\partial_n \partial_{n-1} = 0$. Whenever we have a chain complex structure, we can define the **nth homology group of the chain complex** as

$$H_n(\{C_*, \partial_*\}) = \ker \partial_n / \text{im } \partial_{n+1}$$

A **cochain complex** $\{C^*, \delta^*\}$ is a collection of groups and group homomorphisms $\delta^n : C^n \to C^{n+1}$ such that $\delta^n \delta^{n+1} = 0$. With these data we define the **nth cohomology group**

$$H^n(\{C^*, \delta^*\}) = \ker \delta^n / \text{im } \delta^{n-1}$$

A.3 Tensor Product

Definition A.7. *The* **tensor product** $A \otimes B$ *of two Abelian groups is the Abelian group generated by elements of the form* $a \otimes b$*, where* $a \in A, b \in B$*, subject to the relations*

- $(a_1 + a_2) \otimes b = a_1 \otimes b + a_2 \otimes b$
- $a \otimes (b_1 + b_2) = a \otimes b_1 + a \otimes b_2$

Given two chain complexes $\{C_*, \partial_*\}$ and $\{C'_*, \partial'_*\}$, we define their tensor product $\{C_*^{\otimes}, \partial_*^{\otimes}\}$ as the chain complex with chain groups

$$C_n^{\otimes} = \oplus_i C_i \otimes C'_{n-i}$$

To define the boundary of this chain complex, consider the chain

$$c_{k+l}^{\otimes} = c_k \otimes c'_l$$

Therefore, the boundary is defined as

$$\partial^{\otimes} c_{k+l}^{\otimes} = (\partial c_k) \otimes c'_l + (-1)^k c_k \otimes (\partial' c'_l)$$

The boundary of the sum of many tensor products is defined by linearity from the above. It is easily checked that

$$\partial^{\otimes} \partial^{\otimes} = 0$$

so that $\{C^{\otimes}, \partial^{\otimes}\}$ is indeed a chain complex.

A.4 Categories and Functors

Definition A.8. *A* **category** *consists of the following data:*

- *A collection of* **objects***,* $A, B, C, \ldots$
- *A collection* $\hom(A, B)$ *of* **arrows** *from* A *to* B*, namely,* $A \to B$*, for arbitrary objects* A, B*.*

- *A mapping, or composition of arrows,* $\hom(A,B) \times \hom(B,C) \to \hom(A,C)$, *for arbitrary objects* A, B, C*; the image of* $(f : A \to B, g : B \to C)$ *in* $\hom(A,C)$ *is written* $g \circ f$.

The data are subject to the following axioms:

- *The composition of arrows is associative,* $(h \circ g) \circ f = h \circ (g \circ f)$.
- *For every object* A *there exists an arrow* $1_A \in \hom(A,A)$ *such that* $f \circ 1_A = f, \forall f \in \hom(A,B), \forall B$ *and* $1_A \circ g = g, \forall g \in \hom(Z,A), \forall Z$.

Among the classical examples of categories, we will mention

- The category of topological spaces and continuous maps
- the category of polyhedra and simplicial maps
- the category of Abelian groups and group homomorphisms

Definition A.9. *A* **(covariant) functor** $\mathcal{F}$ *from the category* $\mathcal{A}$ *to the category* $\mathcal{B}$ *is a rule that assigns to any object* A *of* $\mathcal{A}$ *an object* $\mathcal{F}(A)$ *of* $\mathcal{B}$ *and to any arrow* $f : A_1 \to A_2$ *of* $\mathcal{A}$ *an arrow* $\mathcal{F}(f) : \mathcal{F}(A_1) \to \mathcal{F}(A_2)$ *such that*

$$\begin{aligned} \mathcal{F}(1_A) &= 1_{\mathcal{F}(A)} \\ \mathcal{F}(g \circ f) &= \mathcal{F}(g) \circ \mathcal{F}(f) \end{aligned}$$

A functor is better visualized by the diagram

$$\begin{array}{ccc} A_1 & \xrightarrow{f} & A_2 \\ \mathcal{F}\downarrow & & \downarrow\mathcal{F} \\ \mathcal{F}(A_1) & \xrightarrow{\mathcal{F}(f)} & \mathcal{F}(A_2) \end{array}$$

Intuitively, it can be said that the functor $\mathcal{F}$ provides a "picture" of the category $\mathcal{A}$ in the category $\mathcal{B}$.

The most traditional example is the homology functor that maps the category of topological spaces and continuous maps to the category of Abelian groups and homomorphisms.

A **contravariant functor** is essentially the same concept as a (covariant) functor, except that the arrows in the target category are *reversed.* The most traditional example of a contravariant functor is cohomology.

A.5 Exact Sequences

Definition A.10. *A sequence of groups and group homomorphisms*

$$\ldots \xrightarrow{h_{i-2}} A_{i-1} \xrightarrow{h_{i-1}} A_i \xrightarrow{h_i} A_{i+1} \xrightarrow{h_{i+1}} \ldots$$

is said to be **exact** *iff*

$$im(h_{i-1}) = \ker(h_i), \ \ \forall i$$

Some exact sequences are infinite in both directions; some other exact sequences have a specified starting point and/or end point. A typical example of an exact sequence of Abelian groups with specified starting and ending points is one like

$$0 \to A \xrightarrow{\alpha} B \xrightarrow{\beta} C \to 0$$

Such a sequence is called **short exact sequence**. Clearly, α is a monomorphism and β is an epimorphism. Furthermore, it is easily seen that β induces an isomorphism $B/\alpha(A) \to C$. Conversely, if $A \subseteq B$ as a subgroup and if $C = B/A$, the relevant group, subgroup, and quotient group can be linked by the short exact sequence

$$0 \to A \to B \to B/A \to 0$$

An important class of short exact sequences is given by the following theorem-definition:

Theorem A.11. *For a short exact sequence of Abelian groups,*

$$0 \to A \xrightarrow{\alpha} B \xrightarrow{\beta} C \to 0$$

the following statements are equivalent:

- $B = \alpha(A) \oplus X$ *for some group* X.
- $\alpha^\dagger\alpha = 1_A$, *for some morphism* $\alpha^\dagger : B \to A$.
- $\beta\beta^\dagger = 1_C$, *for some morphism* $\beta^\dagger : C \to B$.

If either statement holds, the sequence is said to **split**. *In this case, there exists an isomorphism* $B \to A \oplus C$. *Furthermore, if* C *is free, then the sequence splits.*

Proof. See [Munkres 1984, Section 23]. ■

Finally, a useful recipe is the following: Consider an exact sequence,

$$A \xrightarrow{\alpha} B \to C \xrightarrow{\delta} E$$

It is easily seen that the above yields the exact sequence,

$$0 \to \mathrm{coker}(\alpha) \to C \to \ker(\delta) \to 0$$

A.6 Free Resolution

As already said, every Abelian group A has a presentation in terms of generators and relations. Let $\{g_i\}$ be a set of generators of A, and let F denote the *free* group generated by the g_i's. We define the homomorphism $F \xrightarrow{\epsilon} A$, which is the identity on the generators and that is clearly epimorphic. Therefore, the sequence $F \to A \to 0$ is exact. Next, let R be the kernel of the arrow $F \xrightarrow{\epsilon} A$, observe that R is free, and define the injection map $R \to F$. Everything fits within the short exact sequence

$$0 \to R \to F \xrightarrow{\epsilon} A \to 0$$

Therefore, the presentation of a group in terms of generators and relations can be expressed in terms of a short exact sequence of free groups. The above is also referred to as **canonical resolution of the group**. Since the arrow $R \to F$ is just the injection, it follows that $A = F/R$. In other words, *any group can be presented as the quotient of two free groups.*

Besides the *canonical* resolution, there are other free resolutions of the group A. Take a free group F and an epimorphism $\epsilon : F \to A$. Define $R = \ker(\epsilon)$. Clearly, $0 \to R \to F \to A \to 0$ is another free resolution of the group.

A.7 Connecting Morphism

In this section, we are getting to the heart of homological algebra. Modern methods in homological algebra rely on "diagram chasing." The following lemma is probably the most celebrated example:

Lemma A.12. (Snake Lemma) *Consider the following commutative diagram where the two middle rows (and the columns) are exact:*

$$\begin{array}{ccccccccc}
 & & \ker f & \to & \ker g & \to & \ker h & & \\
 & & \downarrow & & \downarrow & & \downarrow & & \\
0 & \to & A & \to & B & \to & C & \to & 0 \\
 & & \downarrow & & \downarrow & & \downarrow & & \\
0 & \to & A' & \to & B' & \to & C' & \to & 0 \\
 & & \downarrow & & \downarrow & & \downarrow & & \\
 & & coker f & \to & coker\ g & \to & coker\ h & &
\end{array}$$

Then there exists a connecting morphism $\partial : \ker h \to coker\ f$ that makes the following sequence exact:

$$0 \to \ker f \to \ker g \to \ker h \xrightarrow{\partial} coker\ f \to coker\ g \to coker\ h \to 0$$

Proof. For an elementary proof, see [Hilton and Wu 1974, page 207] or [Mac Lane 1971, pages 203–204]. ■

The following result is considered to be the beginning of all homological algebra:

Theorem A.13. *Given a short exact sequence of chain complexes and chain maps,*

$$0 \to C_*(A) \to C_*(B) \to C_*(C) \to 0$$

there exists a long exact sequence of homology groups:

$$\ldots \to H_{n+1}(C) \xrightarrow{\partial_*} H_n(A) \to H_n(B) \to H_n(C) \xrightarrow{\partial_*} H_{n-1}(A) \to \ldots$$

where ∂_* is the so-called **connecting homomorphism**.

Proof. Consider the commutative diagram *:

$$\begin{array}{ccccccccc}
 & & 0 & & 0 & & 0 & & \\
 & & \downarrow & & \downarrow & & \downarrow & & \\
0 & \to & Z_n(A) & \to & Z_n(B) & \to & Z_n(C) & & \\
 & & \downarrow & & \downarrow & & \downarrow & & \\
0 & \to & C_n(A) & \to & C_n(B) & \to & C_n(C) & \to & 0 \\
 & & \downarrow & & \downarrow & & \downarrow & & \\
0 & \to & C_{n-1}(A) & \to & C_{n-1}(B) & \to & C_{n-1}(C) & \to & 0 \\
 & & \downarrow & & \downarrow & & \downarrow & & \\
 & & \frac{C_{n-1}(A)}{B_{n-1}(A)} & \to & \frac{C_{n-1}(B)}{B_{n-1}(B)} & \to & \frac{C_{n-1}(C)}{B_{n-1}(C)} & \to & 0 \\
 & & \downarrow & & \downarrow & & \downarrow & & \\
 & & 0 & & 0 & & 0 & &
\end{array}$$

By hypothesis, the two middle rows are exact. Furthermore, it is clear that the columns are exact. Therefore, by the snake lemma, the middle two rows of the following diagram are exact:

$$\begin{array}{ccccccccc}
 & & \ker \partial_A & & \ker \partial_B & & \ker \partial_C & & \\
 & & \downarrow & & \downarrow & & \downarrow & & \\
 & & \frac{C_{n+1}(A)}{B_{n+1}(A)} & \to & \frac{C_{n+1}(B)}{B_{n+1}(B)} & \to & \frac{C_{n+1}(C)}{B_{n+1}(C)} & \to & 0 \\
 & & \downarrow & & \downarrow & & \downarrow & & \\
0 & \to & Z_n(A) & \to & Z_n(B) & \to & Z_n(C) & & \\
 & & \downarrow & & \downarrow & & \downarrow & & \\
 & & \text{coker } \partial_A & & \text{coker } \partial_B & & \text{coker } \partial_C & &
\end{array}$$

The downarrows between the middle two rows are those induced by the boundaries $\partial_A, \partial_B, \partial_C$ of the chain complexes of A, B, C, respectively. By another application of the snake lemma, it follows that the following sequence is exact

$$\ker \partial_A \to \ker \partial_B \to \ker \partial_C \xrightarrow{\partial_*} \text{coker } \partial_A \to \text{coker } \partial_B \to \text{coker } \partial_C$$

Looking more carefully at the kernels and cokernels, it is easily seen that the above is, actually,

$$H_{n+1}(A) \to H_{n+1}(B) \to H_{n+1}(C) \xrightarrow{\partial_*} H_n(A) \to H_n(B) \to H_n(C)$$

■

The above "diagram chasing" proof is apparently due to [Mac Lane 1975]; see also [Weibel 1994].

The connecting morphism is probably best illustrated by rearranging the relevant groups and morphisms as follows:

* $\frac{A}{B}$ is a more compact notation for the quotient group A/B.

$$\begin{array}{ccccc}
H_n(A) & \rightarrow & H_n(B) & \rightarrow & H_n(C) \\
\uparrow & & & & \\
& \leftarrow & \overset{\partial_*}{\longleftarrow} & \leftarrow & \\
& & & & \uparrow \\
H_{n+1}(A) & \rightarrow & H_{n+1}(B) & \rightarrow & H_{n+1}(C)
\end{array}$$

A.8 Torsion Product

From the previous section, much information can be drawn from exact sequences. When manipulating sequences—typically when tensoring an exact sequence with another group—it is important to know how to keep the sequence exact, possibly at the expense of making it longer.

As a first illustration, consider a short exact sequence *that splits*,

$$0 \rightarrow A \xrightarrow{\alpha} A' \xrightarrow{\alpha'} A'' \rightarrow 0$$

Tensoring this sequence with another group, say B, results in a new sequence *that remains exact*:

$$0 \rightarrow A \otimes B \xrightarrow{\alpha \otimes 1_B} A' \otimes B \xrightarrow{\alpha' \otimes 1_B} A'' \otimes B \rightarrow 0$$

Next, consider the presentation of the group A:

$$0 \rightarrow R \xrightarrow{i} F \rightarrow A \rightarrow 0$$

If we tensor this sequence with B, in other words if we apply the covariant functor $\otimes B$, the resulting sequence,

$$0 \rightarrow R \otimes B \xrightarrow{i \otimes 1_B} F \otimes B \rightarrow A \otimes B \rightarrow 0$$

is exact, except possibly at $R \otimes B$. To overcome this lack of exactness, we have to insert between 0 and $R \otimes B$ the kernel of $i \otimes 1_B$. We define the **torsion product** of two Abelian groups, $A * B$, precisely as

$$A * B = \ker(i \otimes 1_B)$$

With this concept, we get the exact sequence:

$$0 \rightarrow A * B \rightarrow R \otimes B \rightarrow F \otimes B \rightarrow A \otimes B \rightarrow 0$$

Finally, consider an exact sequence of Abelian groups

$$0 \rightarrow A \rightarrow A' \rightarrow A'' \rightarrow 0$$

Tensoring the above with another Abelian group B, in other words applying the functor $\otimes B$, does not always result in an exact sequence. However, inserting some extra terms in the tensored sequence makes it exact.

Theorem A.14. *The sequence*

$$0 \rightarrow A * B \rightarrow A' * B \rightarrow A'' * B \rightarrow A \otimes B \rightarrow A' \otimes B \rightarrow A'' \otimes B \rightarrow 0$$

is exact.

A.9 Universal Coefficient Theorem

Theorem A.15. (Universal Coefficient Theorem) *If C is a free chain complex and G an Abelian group, there exists a split short exact sequence:*

$$0 \to H_n(C) \otimes G \to H_n(C \otimes G) \to H_{n-1}(C) * G \to 0$$

Proof. Consider the split short exact sequence:

$$0 \to Z_n \xrightarrow{i} C_n \xrightarrow{\partial} B_{n-1} \to 0$$

In the above, i is the inclusion map and ∂ is the boundary of the chain complex C. We endow Z, B with a chain complex structure by defining the "boundary" to be the trivial, vanishing homomorphism. The above therefore can be viewed as a short exact sequence of *chain complexes and chain maps.* Tensoring it with G, we obtain another short exact sequence:

$$0 \to Z \otimes G \to C \otimes G \to B \otimes G \to 0$$

Going to the homology yields the long exact sequence:

$$\begin{aligned} \ldots \to B_n(C) \otimes G &\xrightarrow{j_n \otimes 1_G} Z_n(C) \otimes G \to H_n(C \otimes G) \to B_{n-1}(C) \otimes G \\ &\xrightarrow{j_{n-1} \otimes 1_G} Z_{n-1}(C) \otimes G \to \ldots \end{aligned}$$

The connecting homomorphism is $j \otimes 1_G$, where $j_n : B_n \to Z_n$ is the inclusion map. From the above, we obtain the short exact sequence:

$$0 \to \text{coker}(j_n \otimes 1_G) \to H_n(C \otimes G) \to \ker(j_{n-1} \otimes 1_G) \to 0$$

Finally, observe that $\text{coker}(j_n \otimes 1_G) = H_n(C) \otimes G$ and $\ker(j_{n-1} \otimes 1_G) = H_{n-1}(C) * G$. Substitution hence yields the required short exact sequence. It remains to prove the split property; this is left to the reader; for details, see [Spanier 1989, page 222, Theorem 8]. ■

A.10 Künneth Formula

The universal coefficient theorem deals with the homology of the tensor product of a chain complex by an Abelian group. The next generalization—Künneth's theorem—deals with the homology of the tensor product of two chain complexes. Künneth's theorem asserts that the homology of the tensor product of two complexes can be obtained as the direct sum of the tensor and torsion products of the homology of the factors.

Theorem A.16. (Künneth) *Let $\{C, \partial\}$ and $\{C', \partial'\}$ be two chain complexes where C' is free. Let $H_*(C \otimes C')$ denote the homology of the tensor product of the two complexes. Then there exists a split short exact sequence,*

$$0 \to \bigoplus_i H_{n-i}(C) \otimes H_i(C') \to H_n(C \otimes C') \to \bigoplus_i H_{n-i-1}(C) * H_i(C') \to 0$$

Proof. The proof follows the same lines as the universal coefficient theorem. Remember, the latter is essentially the issue of the homology of the tensor product of a complex and an *Abelian group*. Here, we are dealing with the tensor product of two complexes. So the only thing to watch while extending the universal coefficient theorem to the Künneth theorem is that the second factor is a *chain complex* that no longer has trivial boundary. The details are left to the reader. ∎

BIBLIOGRAPHICAL AND HISTORICAL NOTES

Homological algebra can be loosely defined as the impact of the formalism of algebraic topology on algebra. The "classics" are [Cartan and Eilenberg 1970] and [Mac Lane 1975]; see also [Hilton and Stammach 1970]. For a more modern exposition, that goes beyond the scope of this book, see [Weibel 1994].

For an exposition of the fundamental algebraic concepts that underline homology and algebraic topology, the reader is referred to [Hilton and Wu 1974]. For shorter expositions following the same line of thoughts, see [Hilton and Wylie 1965, Chapter 5] or [Munkres 1984, Chapters 6 and 7]. For a treatment of presentation of groups in terms of generators and relations, see [Magnus, Karras, and Solitar 1976].

Categories can be defined as the formalization of proving theorems by "diagram chasing." It was formulated by [Eilenberg and Mac Lane 1945], as a spinoff of algebraic topology. For an account of classical categories, see [Mac Lane 1971]. As an aside, we note that, more recently, categories have been proposed as a language to be used in the Foundations of Mathematics.

APPENDIX B

MATRIX ANALYSIS OF INTEGRAL HOMOLOGY GROUPS AND HOPF TRACE THEOREM

SUMMARY

This Appendix deals with matrix representations of the boundary homomorphisms and the chain maps occuring in a diagram like

$$\begin{array}{ccccc} \to & C_n & \xrightarrow{\partial_n} & C_{n-1} & \to \\ & \downarrow \varphi_n & & \downarrow \varphi_{n-1} & \\ \to & C_n & \xrightarrow{\partial_n} & C_{n-1} & \to \end{array}$$

The essential idea is to choose bases for all of the groups involved, and write the matrix representations of all homomorphisms in the chosen bases. Matrix analysis of the top row—to be more specific, a matrix representation of ∂_n in a carefully chosen basis—yields a matrix computation of the homology groups. Next the matrix representation for ∂_n combined with the matrix representation for the chain map φ_n yields the so-called *Hopf Trace Theorem.*

B.1 Matrix Computation of Homology Groups

Here we summarize the algorithm of [Munkres 1984, Section 11], itself based on [Hilton and Wylie 1965, Section 5.1], from which it follows that the homology groups of finite complexes are computable via finitely many arithmetic operations.

Let each group C_n be free and finitely generated. Consider the boundary homomorphism

$$\partial_{n+1} : C_{n+1} \to C_n$$

Since C_{n+1} and C_n are finitely generated, we can find bases, and relative to the chosen bases, the boundary homomorphism has a representation in terms of a matrix with integral coefficients. Apply the Smith algorithm over the ring of integers to put this matrix in Smith form:

$$\partial_{n+1} = \begin{pmatrix} 1 & \dots & 0 & 0 & \dots & 0 & 0 \\ \vdots & \ddots & \vdots & \vdots & & \vdots & 0 \\ 0 & \dots & 1 & 0 & \dots & 0 & 0 \\ 0 & \dots & 0 & k_1 & \dots & 0 & 0 \\ \vdots & & \vdots & \vdots & \ddots & \vdots & 0 \\ 0 & \dots & 0 & 0 & \dots & k_l & 0 \\ 0 & \dots & 0 & 0 & \dots & 0 & 0 \end{pmatrix}$$

In the above, the k_i's are integers, $k_i > 1$; furthermore, k_1 divides k_2, which itself divides k_3, and so on. The k_i's are called **invariant factors**.

The Smith form is obtained by pre/postmultiplication of the original matrix representation of ∂_{n+1} by unimodular matrices—that is, matrices with determinant 1 that are invertible over the ring of integers. Postmultiplication by unimodular matrices amounts to changing the basis of C_{n+1}. Premultiplication amounts to changing the basis for C_n. Combine the new generators conformably with the partitioning of ∂_{n+1} and we get

$$C_{n+1} = D_{n+1} \bigoplus E_{n+1} \bigoplus Z_{n+1}$$

Likewise, combining the new generators of C_n conformably with the column partitioning of ∂_{n+1} yields

$$C_n = V_n \bigoplus W_n \bigoplus 0$$

More concretely, all of these things can be written as

$$\begin{pmatrix} v_n \\ w_n \\ 0 \end{pmatrix} = \begin{pmatrix} I_{n+1} & 0 & 0 \\ 0 & K_{n+1} & 0 \\ 0 & 0 & 0 \end{pmatrix} \begin{pmatrix} d_{n+1} \\ e_{n+1} \\ z_{n+1} \end{pmatrix}$$

In the above I_{n+1} denotes an identity matrix, but the subscript $n+1$ *does not* indicate its size; it indicates that it is an identity block appearing in the matrix representation of ∂_{n+1}. Same remark for K_{n+1}; the subscript $n+1$ *does* not denote its size, but indicates that it is the diagonal matrix of invariant factors of ∂_{n+1}.

Looking at the Smith form representation of ∂_{n+1}, it is clear that

$$V_n \bigoplus W_n \subseteq B_n \subseteq Z_n$$

Therefore, there exists a subgroup A_n such that

$$Z_n = A_n \bigoplus V_n \bigoplus W_n$$

Let $\mathbf{d}, \mathbf{e}, \mathbf{v}, \mathbf{w}$ be basis elements of $D_{n+1}, E_{n+1}, V_n, W_n$, respectively. From the matrix representation of the boundary operator, it follows that

$$\partial_{n+1}\mathbf{d} = \mathbf{v}$$

$$\partial_{n+1}\mathbf{e} = k\mathbf{w}$$

Therefore,

$$\partial_{n+1} D_{n+1} = V_n$$

$$\partial_{n+1} E_{n+1} \subseteq W_n$$

Clearly, V_n is a group of *boundaries*. The **w**'s are called *weak boundaries* because, although they do not bound, a multiple of them does bound. In other words,

$$B_n = V_n \bigoplus \{k\mathbf{w}\}$$

Remember, the cycle group is decomposed as

$$Z_n = A_n \bigoplus V_n \bigoplus W_n$$

Combining the above two group decompositions, we get

$$H_n \cong A_n \bigoplus (\mathbb{Z} \bmod k_1 \bigoplus \ldots \bigoplus \mathbb{Z} \bmod k_l)$$

The first element is the free subgroup while the second is the torsion subgroup; clearly, the torsion coefficients at n are the invariant factors of the Smith form of ∂_{n+1}.

For the complexity aspects of these computations, the reader is referred to [Donald and Chang 1991].

B.2 Hopf Trace Theorem

The *Hopf trace theorem* reveals a homological invariant that is associated with a chain map,

$$\varphi : C \to C$$

from a free, finitely generated, finite complex into itself. The complex C is said to be **finite** if only finitely many C_n's are nontrivial. The Hopf theorem relies on matrix representations for ∂ and φ and their interplay.

In the matrix representation of $\partial_{n+1} : C_{n+1} \to C_n$ developed in the previous section, the decomposition of C_n and C_{n+1} were inconsistent. We now derive matrix representations for consistent decompositions of C_n, C_{n+1}. Let the subgroups D_n, E_n, V_n, W_n, A_n be defined as in the previous section. Clearly,

$$C_n = D_n \bigoplus E_n \bigoplus V_n \bigoplus W_n \bigoplus A_n$$

Let d_n, e_n, v_n, w_n, a_n be coordinate vectors relative to chosen basis of the subgroups. Rearranging the matrix representation of the previous section, we get

$$\begin{pmatrix} d_n \\ e_n \\ v_n \\ w_n \\ a_n \end{pmatrix} = \begin{pmatrix} 0 & 0 & 0 & 0 & 0 \\ 0 & 0 & 0 & 0 & 0 \\ I_{n+1} & 0 & 0 & 0 & 0 \\ 0 & K_{n+1} & 0 & 0 & 0 \\ 0 & 0 & 0 & 0 & 0 \end{pmatrix} \begin{pmatrix} d_{n+1} \\ e_{n+1} \\ v_{n+1} \\ w_{n+1} \\ a_{n+1} \end{pmatrix}$$

Relative to the same basis, the matrix representation Φ_n of $\varphi_n : C_n \to C_n$ reads

$$\Phi_n = \begin{pmatrix} \Phi_{n,DD} & \Phi_{n,DE} & \Phi_{n,DV} & \Phi_{n,DW} & \Phi_{n,DA} \\ \Phi_{n,ED} & \Phi_{n,EE} & \Phi_{n,EV} & \Phi_{n,EW} & \Phi_{n,EA} \\ \Phi_{n,VD} & \Phi_{n,VE} & \Phi_{n,VV} & \Phi_{n,VW} & \Phi_{n,VA} \\ \Phi_{n,WD} & \Phi_{n,WE} & \Phi_{n,WV} & \Phi_{n,WW} & \Phi_{n,WA} \\ \Phi_{n,AD} & \Phi_{n,AE} & \Phi_{n,AV} & \Phi_{n,AW} & \Phi_{n,AA} \end{pmatrix}$$

Next we need to exploit the chain map relation $\varphi_n \partial_{n+1} = \partial_{n+1} \varphi_{n+1}$ to get more accurate information about the structure of the matrix representation in the chosen basis. Writing the chain map relation in the matrix representation yields

$$\begin{pmatrix} \Phi_{n,DV} & \Phi_{n,DW}K_{n+1} & 0 & 0 & 0 \\ \Phi_{n,EV} & \Phi_{n,EW}K_{n+1} & 0 & 0 & 0 \\ \Phi_{n,VV} & \Phi_{n,VW}K_{n+1} & 0 & 0 & 0 \\ \Phi_{n,WV} & \Phi_{n,WW}K_{n+1} & 0 & 0 & 0 \\ \Phi_{n,AV} & \Phi_{n,AW}K_{n+1} & 0 & 0 & 0 \end{pmatrix} =$$

$$\begin{pmatrix} 0 & 0 & 0 & 0 & 0 \\ 0 & 0 & 0 & 0 & 0 \\ \Phi_{n+1,DD} & \Phi_{n+1,DE} & \Phi_{n+1,DV} & \Phi_{n+1,DW} & \Phi_{n+1,DA} \\ K_{n+1}\Phi_{n+1,ED} & K_{n+1}\Phi_{n+1,EE} & K_{n+1}\Phi_{n+1,EV} & K_{n+1}\Phi_{n+1,EW} & K_{n+1}\Phi_{n+1,EA} \\ 0 & 0 & 0 & 0 & 0 \end{pmatrix}$$

It is easily seen from the above that

$$\begin{aligned} \Phi_{n,VV} &= \Phi_{n+1,DD} \\ \Phi_{n,WW}K_{n+1} &= K_{n+1}\Phi_{n+1,EE} \end{aligned}$$

Remember that K_{n+1} is a diagonal matrix of invariant factors $k_i > 1$; therefore the second equality implies that $\Phi_{n,WW}$ and $\Phi_{n+1,EE}$ are equal and diagonal. Hence we get

$$\begin{aligned} \text{Trace}(\Phi_{n,VV}) &= \text{Trace}(\Phi_{n+1,DD}) \\ \text{Trace}(\Phi_{n,WW}) &= \text{Trace}(\Phi_{n+1,EE}) \end{aligned}$$

Clearly, for *any* matrix representation Φ_n of φ_n we have

$$\begin{aligned} &\text{Trace}(\Phi_n) = \\ &\quad \text{Trace}(\Phi_{n,DD}) + \text{Trace}(\Phi_{n,EE}) + \text{Trace}(\Phi_{n,VV}) \\ &\quad +\text{Trace}(\Phi_{n,WW}) + \text{Trace}(\Phi_{n,AA}) \end{aligned}$$

Combining the above three equations yields

$$\sum_n (-1)^n \ \text{Trace}(\Phi_n) = \sum_n (-1)^n \ \text{Trace}(\Phi_{n,AA})$$

Remember, A_n is the free part of the homology group $H_n(C)$. Therefore, $\Phi_{n,AA}$ is the endomorphism induced by Φ_n on the free part of $H_n(C)$. This yields the following "traditional" result:

Theorem B.1. (Hopf Trace Theorem) *Consider a chain map* $\varphi : C \to$

C of the free, finitely generated, finite chain complex C. Let Φ_n be a matrix representation of φ_n relative to an arbitrary basis of C_n. Let $\Phi_{\star,n}$: free $(H_n) \to$ free (H_n) be the induced endomorphism on the free part of $H_n(C)$. Then

$$\sum_n (-1)^n \, Trace(\Phi_n) = \sum_n (-1)^n \, Trace(\Phi_{\star,n})$$

The above is called **Lefschetz number**, *$\Lambda(\varphi)$, of the chain map φ.*

Taking $\varphi = 1$, the identity chain mapping on C, yields the following:

Corollary B.1.

$$\sum_n (-1)^n rank(C_n) = \sum_n (-1)^n b_n(C)$$

where $b_n(C)$ is the nth Betti number—that is, the rank of the free part of $H_n(C)$. The above is called **Euler characteristic**, *$\chi(C)$, of the complex C.*

APPENDIX C

HOMOLOGICAL ALGEBRA OF MODULES

SUMMARY

In this Appendix, we reformulate homological algebra in terms of modules, rather than Abelian groups. This is motivated by the Eilenberg–Moore spectral sequence. All along this book, except for the Eilenberg–Moore spectral sequence, it is possible to work solely with the concept of Abelian group, so that the more general concept of *module* can be dispensed of. However, a first sign of the inadequacy of Abelian groups already shows when using other coefficient groups than $\mathbb{Z}$. An additive group structure together with coefficients in an *arbitrary ring* R is formalized in the concept of *module over* R. From that point of view, an Abelian group is a module over $\mathbb{Z}$. A typical feature of a (co)homology module is that it is a *differential graded module.* A cohomology module with the cup product structure is a *differential graded algebra.* When the coefficients of a cohomology module are themselves in a cohomology algebra, as typically happens in spectral sequences, one obtains a *differential graded module over a differential graded algebra.* This Appendix focuses more specifically on the higher torsion products defined on relevant modules. Higher torsion products of differential graded modules over a differential graded algebra are instrumental in identifying the E_2 term of the Eilenberg–Moore spectral sequence.

C.1 Modules

C.1.1 Modules and Projective Modules

All along this Appendix, R denotes a commutative ground ring with multiplicative unit.

The formal definition of an R-module is provided in Definition D.25.

In Abelian group theory, the concept of *free* Abelian group is central. In module theory, we prefer to work with a concept slightly more general than that of free module:

Definition C.1. *The R-module M is said to be* **projective** *iff any of the following equivalent statements is satisfied:*

- *M is a direct summand in a free R-module; that is, there exists a R-module A such that $M \oplus A$ is a free R-module.*

- *For any surjective module homomorphism $\phi : A \to B$, for any morphism $\gamma : M \to B$, there exists a morphism $\tilde{\gamma} : M \to A$ such that the following diagram commutes:*

$$\begin{array}{ccc} & & A \\ & \overset{\tilde{\gamma}}{\nearrow} & \downarrow \phi \\ M & \overset{\gamma}{\to} & B \end{array}$$

A free module is always projective. The converse is not always true; only for "nice" rings R (i.e., $\mathbb{Z}$, $\mathbb{Q}$, division rings, and so on) is a projective R-module free.

This generalization of the concept of free module still enjoys the crucial property that any short exact sequence of *projective* modules, namely,

$$0 \to A \to B \to C \to 0$$

splits; that is,

$$B \cong A \oplus C$$

C.1.2 Projective Resolution

Remember that an Abelian group can be specified by its *resolution*—that is, a short exact sequence of *free* groups. The corresponding concept in the realm of modules is an exact sequence of projective modules that is, in general, longer than that in the case of a group.

Definition C.2. *A* **projective resolution** *of the module M over R is an exact sequence*

$$\ldots \overset{d^{-3}}{\longrightarrow} P^{-2} \overset{d^{-2}}{\longrightarrow} P^{-1} \overset{d^{-1}}{\longrightarrow} P^{0} \overset{\epsilon}{\longrightarrow} M \longrightarrow 0$$

where the P's are projective R-modules.

Observe that $\{P, d\}$ is a (co)chain complex with its (co)homology concentrated in dimension 0: $H^n(\{P,d\}) = 0$, $n \neq 0$, and $H^0(\{P,d\}) = P^0/\text{im}\, d^{-1} = P^0/\ker\epsilon = M$.

Theorem C.3. *Every R-module M has a projective resolution.*

Proof. Take a projective module P^0 and an epimorphism $\epsilon : P^0 \to M$. Define $M^{-1} = \ker\epsilon$. Construct the short exact sequence:

$$0 \to M^{-1} \overset{i}{\to} P^0 \overset{\epsilon}{\to} M \to 0$$

In general, M^{-1} is not a projective module. Therefore, choose a projective module P^{-1} and an epimorphism $\epsilon^{-1} : P^{-1} \to M^{-1}$, define $M^{-2} = \ker(\epsilon^{-1})$, and construct the exact sequence:

$$0 \to M^{-2} \to P^{-1} \to M^{-1} \to 0$$

Combine the short exact sequences in a longer one and construct the morphism d as shown below:

$$\begin{array}{ccccccccccccc}
0 & \to & M^{-2} & \to & P^{-1} & \to & M^{-1} & \to & P^{0} & \stackrel{\epsilon}{\to} & M & \to & 0 \\
 & & & & \downarrow & & & & \uparrow & & & & \\
 & & & & & \to & \stackrel{d}{\to} & \to & & & & &
\end{array}$$

The above yields the sequence

$$0 \to M^{-2} \to P^{-1} \to P^{0} \stackrel{\epsilon}{\to} M \to 0$$

Iterating this process yields the required long sequence that is easily seen to be exact. For more details, see [Weibel 1994, Lemma 2.2.5, page 34], [Hilton and Wu 1974, page 201], and [Lluis-Puebla 1992]. ■

C.1.3 Tensor Product

Definition C.4. *Let M be a right R-module and let N be a left R-module. The* **tensor product** *$M \otimes_R N$ is defined by the generators $m \otimes n$, $\forall m \in M, \forall n \in N$ subject to the relations*

- $(m_1 + m_2) \otimes n = m_1 \otimes n + m_2 \otimes n$
- $m \otimes (n_1 + n_2) = m \otimes n_1 + m \otimes n_2$
- $(m \cdot r) \otimes n = m \otimes (r \cdot n)$

and the action of R on $M \otimes_R N$ is given by

$$r(m \otimes n) = (m \cdot r) \otimes n = m \otimes (r \cdot n)$$

It can be shown (see [Hilton and Wu 1974, page 141]) that the module action defined on $M \otimes_R N$ is compatible with the defining relations.

(The notation $\otimes_R$ means that the tensor product is bilinear relative to multiplication by elements in the ring R. Using this convention, the usual tensor product of Abelian groups is written $\otimes_{\mathbb{Z}}$. We shall sometimes drop the subscript when there is no danger of confusion or when the subscript should be clear from the context.)

C.1.4 Higher Torsion Products

In Abelian groups, the concept of torsion product is motivated by the lack of exactness of the sequence resulting from tensoring a short exact sequence with another Abelian group. Torsion product terms have to be inserted in the tensor sequence in order to keep it exact. To compute the torsion product of two Abelian groups, we need the short exact sequence resolution of any of the groups. In module theory, the resolution sequence is longer and therefore there is a hierarchy of torsion products needed to keep the sequence exact.

Consider two R-modules, M and N, together with their projective resolutions:

$$\ldots \xrightarrow{d_P} P^{-1} \xrightarrow{d_P} P^{0} \xrightarrow{\epsilon} M \to 0$$

$$\ldots \xrightarrow{d_Q} Q^{-1} \xrightarrow{d_Q} Q^{0} \xrightarrow{\epsilon} N \to 0$$

Observe that

$$(M \otimes_R Q^{-n}, 1 \otimes d_Q^{-n})$$

is a complex graduated by n and the differential has degree $+1$. This leads to the following definition:

Definition C.5. *The* **higher torsion products** *of the modules M, N are defined by*

$$Tor\,_R^{-n}(M, N) = H^n(M \otimes Q^*, 1 \otimes d_Q^*)$$

This definition apparently lacks symmetry; however, we could equally take the resolution of M and tensor it on the right by N; there is also a most symmetric form of the definition involving *both* projective resolutions:

Theorem C.6.

$$\begin{aligned} Tor\,_R^{-n}(M, N) &= H^n(P^* \otimes N, d_P^* \otimes 1) \\ &= H^n(\oplus_{i+j=n} P^{-i} \otimes Q^{-j}, d_P^{-i} \otimes 1 + (-1)^i 1 \otimes d_Q^{-j}) \end{aligned}$$

Proof. See [McCleary 1985, Proposition 3.12]. ■

From Definition C.5, it is easy to see that

$$\mathrm{Tor}\,_R^0(M, N) = M \otimes_R N$$

With the above definition of the higher torsion products, we get the required property:

Theorem C.7. *Let*

$$0 \to A \to B \to C \to 0$$

be a short exact sequence of right R-modules. Let M be a left R-module. Then there exists a long exact sequence

$$\begin{array}{ccccccc} \ldots & \to & Tor\,_R^{-n}(A, M) & \to & Tor\,_R^{-n}(B, M) & \to & Tor\,_R^{-n}(C, M) \\ & \to & Tor\,_R^{-n+1}(A, M) & \to & Tor\,_R^{-n+1}(B, M) & \to & Tor\,_R^{-n+1}(C, M) \\ \ldots & & \ldots & \ldots & \ldots & \ldots & \ldots \\ & \to & Tor\,_R^{-1}(A, M) & \to & Tor\,_R^{-1}(B, M) & \to & Tor\,_R^{-1}(C, M) \\ & \to & A \otimes_R M & \to & B \otimes_R M & \to & C \otimes_R M \\ & \to & 0 & & & & \end{array}$$

Proof. This is fairly easy to prove from the definition of the torsion product; see, for example, [McCleary 1985, page 66] and [Lluis-Puebla 1992, Theorem 3.3, page 11]. ■

Applying the above to Abelian groups M, N, we recover the definition of the torsion product $*$ of Abelian groups, developed in Appendix A; to be specific,

$$M * N = \mathrm{Tor}\,_{\mathbb{Z}}^{-1}(M, N)$$

C.2 Algebra

Definition C.8. *An R-***algebra** *Λ consists of the following structure:*

- *Λ is a R-module.*
- *There is an associative multiplication defined on Λ; that is, an R-module homomorphism*

$$m : \Lambda \otimes_R \Lambda \to \Lambda$$

such that the following diagram commutes:

$$\begin{array}{ccc} \Lambda \otimes_R \Lambda \otimes_R \Lambda & \xrightarrow{m\otimes 1} & \Lambda \otimes_R \Lambda \\ 1 \otimes m \downarrow & & \downarrow m \\ \Lambda \otimes_R \Lambda & \xrightarrow{m} & \Lambda \end{array}$$

C.3 Differential Graded Modules (Over Differential Graded Algebra)

(Co)chain modules are naturally *graded* and are equipped with a *differential*; such modules are said to be *differential graded.* The minimum ground structure that is needed for a differential graded module is a mere ground *ring* R; however, in some applications, we require the ground structure to be an *algebra* Λ. Tensor and torsion products are developed for the specific case of modules over an *algebra*, and we leave it to the reader to scale down the relevant concepts to the case of a ground *ring.*

(Co)homology modules are also naturally *graded.* It is convenient to consider them as *differential graded* modules by endowing them with the trivial differential.

C.3.1 dg Modules

Definition C.9. *A* **differential graded (dg)** R**-module**, (M, d), *consists of the following structure:*

- *M is an R-module.*
- *The module is graded as $M = \oplus_n M^n$ and there exists a module homomorphism $d : M^n \to M^{n+1}$ such that $dd = 0$.*

C.3.2 dg Algebra

Definition C.10. *A* **differential graded algebra** *over R, (Λ, d), consists of the following structure:*

- *(Λ, d) is a differential graded R-module; that is, $\Lambda = \oplus_n \Lambda^n$, $d^2 = 0$, and d has degree $+1$.*
- *Λ is an R-algebra and the algebra multiplication maps as an R-module homomorphism $m : \Lambda^n \otimes \Lambda^m \to \Lambda^{n+m}$.*
- *The differential satisfies the Leibniz rule: $d(a^n \cdot b^m) = (da^n) \cdot b^m + (-1)^n a^n \cdot db^m$, where $a \cdot b = m(a \otimes_R b)$.*
- *Λ has a unit; that is, there exists a ring injection morphism $R \to \Lambda$.*

C.3.3 dg Module Over dg Algebra

Definition C.11. *A* **right differential graded module** (M, d) **over the differential graded algebra** (Λ, d) *has the following structure:*

- (M, d) *is a differential graded R-module.*
- (Λ, d) *is a differential graded R-algebra.*
- *There is a right action $a_r : M \otimes_R \Lambda \to M$, which is a differential graded R-module homomorphism with the associativity specified by the following commutative diagram:*

$$\begin{array}{ccc} M \otimes_R \Lambda \otimes_R \Lambda & \xrightarrow{1 \otimes m} & M \otimes_R \Lambda \\ a_r \otimes 1 \downarrow & & \downarrow a_r \\ M \otimes_R \Lambda & \xrightarrow{a_r} & M \end{array}$$

C.3.4 Tensor Product

We define the tensor product of two differential graded modules, M and N, both defined over the same differential graded algebra Λ. To be more specific, M is a *right* Λ-module while N is a *left* Λ-module. The right Λ action on M is viewed as a homomorphism:

$$a_r : M \otimes_R \Lambda \to M$$

and the left action on N is viewed as

$$a_l : \Lambda \otimes_R N \to N$$

Define the morphism:

$$a_r \otimes 1 - 1 \otimes a_l : M \otimes_R \Lambda \otimes_R N \to M \otimes_R N$$

Therefore, the **tensor product** of the differential graded modules M and N over the differential graded algebra Λ is defined by

$$M \otimes_\Lambda N = M \otimes_R N / \operatorname{im} (a_r \otimes 1 - 1 \otimes a_l)$$

C.3.5 Torsion Product

Consider the differential graded module N. This module also comes with a projective resolution, assumed to be proper (see [McCleary 1985, page 224]). To exhibit the interplay between the graduation and the resolution, we introduce the following diagram:

$$\begin{array}{ccccccccc} & & \uparrow & & \uparrow & & \uparrow & & \\ \longrightarrow & & (Q^{-1})^1 & \xrightarrow{d_Q} & (Q^0)^1 & \xrightarrow{\epsilon} & N^1 & \longrightarrow & 0 \\ & & \uparrow \delta_Q & & \uparrow \delta_Q & & \uparrow \delta_N & & \\ \longrightarrow & & (Q^{-1})^0 & \xrightarrow{d_Q} & (Q^0)^0 & \xrightarrow{\epsilon} & N^0 & \longrightarrow & 0 \end{array}$$

The horizontal arrows refer to the resolution while the vertical arrows are the differentials of the graduation. With this diagram, we construct a new cochain complex $\mathcal{Q}$,

$$\mathcal{Q} = \{\oplus_{i+j=n}(Q^{-i})^j, \delta_{\mathcal{Q}} = \oplus_{i+j=n}(d_Q^{-i,j} + \delta_Q^{-i,j})\}$$

This complex is graduated by n, and the differential increases the graduation by 1. Clearly, this complex has a cohomology, and this leads to the following definition:

Definition C.12. *The* **higher torsion products** *of the dg modules M, N over the dg algebra Λ are defined as*

$$Tor_{\Lambda}^{*}(M,N) = H^*(M \otimes_\Lambda \mathcal{Q}, \delta_M \otimes 1 \pm 1 \otimes \delta_{\mathcal{Q}})$$

To be more specific,

$$\begin{aligned}&\mathrm{Tor}_{\Lambda}^{u}(M,N) = \\ &\quad \ker(\delta_M \otimes 1 + (-1)^i 1 \otimes \delta_{\mathcal{Q}} : \oplus_{i+j=u}(M^i \otimes \mathcal{Q}^j) \to \oplus_{i+j=u+1}(M^i \otimes \mathcal{Q}^j))/ \\ &\quad \mathrm{im}(\delta_M \otimes 1 + (-1)^i 1 \otimes \delta_{\mathcal{Q}} : \oplus_{i+j=u-1}(M^i \otimes \mathcal{Q}^j) \to \oplus_{i+j=u}(M^i \otimes \mathcal{Q}^j))\end{aligned}$$

The above defines the graduation of the torsion product. The module is also bigraduated as follows: Remember,

$$\mathcal{Q}^j = \oplus_{j_1+j_2=j}(Q^{-j_1})^{j_2}$$

Define $\mathrm{Tor}_{\Lambda}^{u,v}(M,N)$ to be the subset of $\mathrm{Tor}_{\Lambda}^{u}(M,N)$ consisting of elements such that $i + j_2 = v$.

NOTES

The material of this Appendix dealing with differential algebras is distilled from [Mac Lane 1995, Chapter 6], [Weibel 1994, Chapter 2], and [Mc Leary 1985, Section 7.1].

For an example of a modern application of differential algebras to a problem at the cutting edge of homotopy theory, see [Anick 1993].

APPENDIX D

ALGEBRAIC SINGULARITY THEORY

SUMMARY

In this Appendix, we develop tools for analyzing the local structure of a map. Relevant to this local structure are the implicit function theorem, singularity, local infinitesimal stability, and so on. Since our concern is the *local* structure, the maps are assumed to be between Euclidean spaces.

We first develop the Weierstrass/Malgrange preparation theorem as an analytical result and we subsequently reformulate it in terms of rings and modules. Then we develop purely algebraic tools that are not only convenient algebraic reformulations of facts of analysis but that are sometimes leading us beyond what the analysis could do.

D.1 Weierstrass Preparation Theorem

The Weierstrass preparation theorem deals with implicit equations of the form $F(q,s) = 0$, where $F : \mathbb{C}^n \times \mathbb{C}, (0,0) \to \mathbb{C}, 0$ is holomorphic in a neighborhood of $(0,0)$. The problem is the holomorphic dependency of the solution s relative to the complex vector q.

Assume that, as we keep on differentiating relative to s, we eventually find a nonvanishing partial derivative

$$\frac{\partial F}{\partial s}|_{(0,0)} = 0, ..., \frac{\partial^{k-1} F}{\partial s^{k-1}}|_{(0,0)} = 0, \frac{\partial^k F}{\partial s^k}|_{(0,0)} \neq 0$$

Then the Weierstrass preparation theorem says that the most explicit form of the solution s as a function of q is given by the polynomial equation

$$s^k + a_{k-1}(q)s^{k-1} + ... + a_0(q) = 0$$

The above is a manifestation of the following theorem:

Theorem D.1. (Weierstrass' Division Theorem) *Let*

$$F : \mathbb{C}^n \times \mathbb{C}, (0,0) \to \mathbb{C}, 0$$

be a holomorphic function subject to

$$F(0,s) = s^k F_0(s), \quad F_0(0) \neq 0 \tag{D.1}$$

or equivalently,

$$\frac{\partial F}{\partial s}|_{(0,0)} = 0, ..., \frac{\partial^{k-1} F}{\partial s^{k-1}}|_{(0,0)} = 0, \frac{\partial^k F}{\partial s^k}|_{(0,0)} \neq 0 \tag{D.2}$$

Then given any holomorphic function $G(q,s)$ there exist holomorphic functions $Q(q,s)$ and $r(q,s)=\sum_{i=0}^{k-1} s^i r_i(q)$ such that

$$QF = G + r$$

Setting $G = s^k$, we get the most famous theorem:

Theorem D.2. (Weierstrass Preparation Theorem) *Under the same conditions as the above, the implicit equation $F(q,s)=0$ is equivalent to*

$$s^k + \sum_{i=0}^{k-1} r_i(q)s^i = 0$$

where the r_i's are holomorphic functions.

In the case $k=1$—that is, $\frac{\partial F}{\partial s}|_{(0,0)} \neq 0$—we find that the equation $F(q,s)=0$ is locally equivalent to $s=-r_0(q)$. This is the so-called *Implicit Function Theorem for holomorphic functions.*

D.1.1 Example (Root-Locus Breakaway Point)

To make the Weierstrass preparation theorem more palatable to the control-oriented reader, consider the following root-locus example:

$$F(k,s) \equiv (s+2)^2(s-1)+(k+4)=0$$

Clearly, this root-locus has a breakaway point at 0 for $k=0$. The Weierstrass preparation theorem precisely provides for a local analysis of the breakaway. If we define $q=k+j\Im q$, we clearly have

$$F(q,s) \equiv s^3+3s^2+q=0$$

The key feature in the above is that the analytic expansion of $F(q,s)$ starts at s^2 which is written, more formally, as

$$\frac{\partial F}{\partial s}|_{(0,0)}=0, \quad \frac{\partial^2 F}{\partial s^2}|_{(0,0)} \neq 0$$

Because the analytic expansion starts at s^2, we can express s^2 in terms of higher-order terms:

$$s^2 = -\frac{q}{3}-\frac{s^3}{3}$$

Substituting the right-hand side of the above for the s^2 part of s^3 yields

$$\begin{aligned} s^2 &= -\frac{q}{3}-\frac{s}{3}\left(-\frac{q}{3}-\frac{s^3}{3}\right) \\ &= \left(-\frac{q}{3}\right)+\left(\frac{q}{9}\right)s+\frac{1}{9}s^4 \end{aligned}$$

Another round of substitutions yields

$$s^2 = \underbrace{(-\frac{q}{3} + \frac{q^2}{81})}_{\approx -r_0} + \underbrace{(\frac{q}{9} - \frac{2q^2}{243})}_{\approx -r_1} s + \underbrace{(\frac{s^6}{81} - \frac{2qs^4}{243})}_{\text{"tail"}}$$

Clearly, we begin to see the build up of the analytic expansion of the r_i's. The only problem with the above intuitive substitution argument is that, while the order of the tail terms can be made arbitrarily high, we can never completely get rid of them. In a certain sense, the Weierstrass preparation theorem tells us that we can safely "throw away the tail." It follows that the breakaway is the bifurcation of the solution to $s^2 + r_1(q)s + r_0(q) = 0$ around $q = 0$.

D.1.2 Proofs

All we really have to prove is the Weierstrass division theorem.

D.1.2.1 *Equivalence Between Key Conditions*

The formulation of the division theorem involves the equivalence D.1 $\Leftrightarrow$ D.2. Since F is holomorphic in s in a neighborhood of 0, it has a power series converging in a neighborhood of $s = 0$. From this, the equivalence D.1 $\Leftrightarrow$ D.2 is obvious. ■

D.1.2.2 *Polynomial Division Theorem*

We first prove the division theorem when the divisor is a so-called *Weierstrass polynomial,*

$$P_k(s, \lambda) = s^k + \sum_{i=0}^{k-1} \lambda_i s^i$$

where

$$\lambda = (\lambda_0, \ldots, \lambda_{k-1})$$

is a vector of auxiliary holomorphic variables.

Theorem D.3. (Polynomial Division Theorem) *Given a holomorphic function $G(q, s)$ defined in a neighborhood of $(0, 0)$, there exist holomorphic functions $Q(q, s, \lambda)$ and $r(s, q, \lambda)$ such that*

$$Q(q, s, \lambda)P_k(s, \lambda) = G(q, s) + r(q, s, \lambda)$$

$$r(q, s, \lambda) = \sum_{i=0}^{k-1} r_i(q, \lambda)s^i$$

Proof. By Cauchy's theorem, we have

$$Q(q, s) = \frac{1}{2\pi j} \oint \frac{G(q, z)}{P_k(q, z, \lambda)(z - s)} dz$$

where the contour of integration is a circle around 0, small enough not to contain any root of the Weierstrass polynomial. Consider

$$\begin{aligned}
&P_k(q,s,\lambda)Q(q,s) - G(q,s)\\
&= \frac{1}{2\pi j}\oint \frac{P_k(q,s,\lambda)G(q,z) - P_k(q,z,\lambda)G(q,z)}{P_k(q,z,\lambda)(z-s)}dz\\
&= \frac{1}{2\pi j}\oint \frac{(P_k(q,s,\lambda) - P_k(q,z,\lambda))G(q,z)}{P_k(q,z,\lambda)(z-s)}dz
\end{aligned}$$

The crucial point is to observe that $(P_k(q,s,\lambda) - P_k(q,z,\lambda))$, as a polynomial of degree k in the variable s, is divisible by the monomial $(s-z)$. Therefore, the quotient

$$\frac{P_k(q,s,\lambda) - P_k(q,z,\lambda)}{(z-s)} = b_{k-1}(q,z,\lambda)s^{k-1} + \ldots + b_0(q,z,\lambda)$$

is obviously a polynomial of degree $(k-1)$ in the variable s. Therefore,

$$P_k(q,s,\lambda)Q(q,s) - G(q,s) = \frac{1}{2\pi j}\oint \frac{(b_{k-1}(q,z,\lambda)s^{k-1} + \ldots + b_0(q,z,\lambda))G(q,z)}{P_k(q,z,\lambda)}dz$$

is obviously a polynomial of degree $k-1$ in the variable s. Write it $r(q,s,\lambda)$, and the result is proved. ■

D.1.2.3 *Polynomial Division Implies Division Theorem*

From the Division Theorem when the divisor is a Weierstrass polynomial, it is easy to prove the Division Theorem for an arbitrary holomorphic function. The proof is exactly the same as that of the real case; the transcription from real to complex is left to the reader. ■

D.2 Malgrange Preparation Theorem

The Implicit Function Theorem asserts that whenever $\frac{\partial F(x,y)}{\partial y}|_{(0,0)} \neq 0$, then the implicit equation $F(x,y) = 0$ is locally equivalent to an explicit relationship of the form $y(x)$, with $y(0) = 0$, around $x = 0$. One of the issues addressed by the Malgrange preparation theorem is what kind of explicit relation, if any, can be expected from $F(x,y) = 0$ when $\frac{\partial F}{\partial y}|_{(0,0)} = 0$. A corollary of the Malgrange preparation theorem reads as follows: Assume

$$\frac{\partial F}{\partial y}|_{(0,0)} = 0, \ldots, \frac{\partial^{k-1} F}{\partial y^{k-1}}|_{(0,0)} = 0, \; \frac{\partial^k F}{\partial y^k}|_{(0,0)} \neq 0$$

Then the most explicit relation is y given as any of the roots of a polynomial of degree k with coefficients smoothly depending on x,

$$y^k + a_{k-1}(x)y^{k-1} + \ldots + a_1(x)y + a_0(x) = 0$$

where the a_i's are smooth functions.

The above is just a manifestation of the following monumental result:

Theorem D.4. (Mather's Division Theorem) *Let*

$$F : \mathbb{R}^n \times \mathbb{R}, (0,0) \to \mathbb{R}, 0$$

be a smooth function subject to

$$F(0, y) = y^k F_0(y), \quad F_0(0) \neq 0 \tag{D.3}$$

or equivalently

$$\frac{\partial F}{\partial y}|_{(0,0)} = 0, \ldots, \quad \frac{\partial^{k-1} F}{\partial y^{k-1}}|_{(0,0)} = 0, \quad \frac{\partial^k F}{\partial y^k}|_{(0,0)} \neq 0 \tag{D.4}$$

Then given any smooth function $G(x, y)$ there exist smooth functions $q(x, y)$ and $r(x, y) = \sum_{i=0}^{k-1} y^i r_i(x)$ such that

$$qF = G + r$$

The Malgrange preparation theorem is a corollary of the above. Take $G = y^k$. We first want to show that $q(0,0) \neq 0$. Indeed, the Mather division theorem written for $x = 0$ yields $q(0, y) y^k F_0(y) = y^k + \sum_{i=0}^{k-1} r_i(0) y^i$. Since both sides of the equality are smooth, we take the partial derivative of order k relative to y evaluated at $y = 0$, and this yields $q(0,0) F_0(0) \neq 0$. Since F_0 is smooth in a neighborhood of 0, we get $q(0,0) \neq 0$. With the latter, the implicit equation $F(x, y) = 0$ is clearly equivalent to the more explicit equation $G + r = 0$. Therefore, we have the following:

Theorem D.5. (Malgrange Preparation Theorem) *Under the same conditions as the above, the implicit equation $F(x, y) = 0$ is equivalent to*

$$y^k + \sum_{i=0}^{k-1} r_i(x) y^i = 0$$

where the r_i's are smooth functions.

The above version of the Malgrange preparation theorem attempts to write an explicit formulation of *just one* among many variables linked by *one* implicit relation. It is as yet unclear how to derive an "explicit" form of some selected variables among a set of variables linked by *many* implicit relations. The generalized version of the Malgrange preparation theorem addresses this among other issues, but its formulation requires the algebraic machinery of rings and modules.

D.2.1 Proofs

Our proofs follow [Mather 1968a], [Golubitsky and Guillemin 1973], and [Castrigiano and Hayes 1993].

D.2.1.1 *Hadamard's Lemma*

The statement of the Malgrange preparation theorem involves the apparently innocent equivalence D.3 ⇔ D.4. That D.3 ⇒ D.4 is trivial. (Just differentiate equation D.3). The proof of the converse, however, needs some caution. *If $F(x,y)$ is analytic—that is, if $F(x,y)$ equals a converging power series*—then D.4 $\Rightarrow$ D.3 is trivially proved using a power series argument. However, when $F(x,y)$ fails to be analytic, the proof cannot rely upon power series and we have to argue as follows: Define

$$F^{(i)}(y) = \frac{\partial^i F(0,y)}{\partial y^i}$$

We have

$$\begin{aligned} F^{(k-1)}(y) &= \int_0^y F^{(k)}(y)dy \\ &= y\phi(y) \end{aligned}$$

where

$$\phi(y) = \frac{1}{y}\int_0^y F^{(k)}(y)dy$$

Furthermore, from L'Hôpital's rule it follows that $\phi(y)$ is smooth. It is also easily shown that $\phi(0) \neq 0$. Indeed, differentiating the expression for $F^{(k-1)}$ yields

$$F^{(k)}(y) = \phi(y) + y\phi'(y)$$

It follows that $F^{(k)}(0) = \phi(0) \neq 0$. From here on, we iterate. The next step is

$$\begin{aligned} F^{(k-2)}(y) &= \int_0^y y\phi(y)dy \\ &= \frac{1}{2}(y^2\phi(y) - \int_0^y y^2\phi'(y)dy) \\ &= y^2\phi_1(y) \end{aligned}$$

$\phi_1(y)$ is defined as before; furthermore, some tricks with L'Hôpital's rule show that ϕ_1 is smooth; and also, differentiating $F^{(k-2)}(y) = y^2\phi_1(y)$ twice and evaluating the result at $y = 0$ yields $\phi_1'(0) \neq 0$. The recursion should now be obvious: We continue until we reach $F^0(y) = y^k F_0(y)$. ■

This result is sometimes referred to as *Hadamard's lemma* (see [Bruce and Giblin 1992, page 50, 3.4]).

D.2.1.2 *Polynomial Division Theorem*

All we have to prove is the Mather division theorem. The first and *crucial* step in the proof of the Mather division theorem is to demonstrate the result in the particular case where the "divisor" $F(x,y)$ is a polynomial of the form

$$P_k(y,\lambda) = y^k + \sum_{i=0}^{k-1} \lambda_i y^i$$

where

$$\lambda = (\lambda_0, ..., \lambda_{k-1})$$

The fact that any smooth function $G(x, y)$ can be "divided" by P_k in a way similar to the Euclidean algorithm is formulated in the following:

Theorem D.6. (Polynomial Division Theorem) *Given a smooth (real-valued) function $G(x, y)$ defined in a neighborhood of $(0, 0)$, there exist (real-valued) smooth functions $q(y, x, \lambda)$ and $r(y, x, \lambda)$ such that*

$$q(x, y, \lambda)P_k(y, \lambda) = G(x, y) + r(x, y, \lambda)$$

$$r(x, y, \lambda) = \sum_{i=0}^{k-1} r_i(x, \lambda)y^i$$

Proof. The proof of this result relies on a highly technical, not quite elegant transcription of the proof of the Weierstrass theorem into the real, smooth language. The crucial point is to find a real, smooth substitute for Cauchy's integral formula. See, for example, [Golubitsky and Guillemin 1973, Chapter 4, Section 2] or [Castrigiano and Hayes 1993, Chapter 9]. ■

D.2.1.3 *Proof That Polynomial Division Implies Mather's Division Theorem*

The Malgrange preparation theorem is a corollary of the polynomial division theorem: Write the polynomial division theorem for both both G and F:

$$G(x, y) = q^G(x, y, \lambda)P_k(y, \lambda) + r^G(x, y, \lambda)$$
$$F(x, y) = q^F(x, y, \lambda)P_k(y, \lambda) + r^F(x, y, \lambda)$$

Write $r^F(x, y, \lambda) = \sum_{i=0}^{k-1} r_i^F(x, \lambda)y^i$. It is easily seen that because of the conditions on F, the matrix of partial derivatives,

$$\frac{\partial r_i^F(0, \lambda)}{\partial \lambda_j}(0)$$

is lower triangular and nonsingular. Therefore, by the implicit function theorem, there exists a function $\phi(x)$ such that

$$r_i^F(x, \phi(x)) = 0$$

in a neighborhood of 0. Therefore, the F-equation yields

$$\begin{aligned} F(x, y) &= q^F(x, y, \lambda)P_k(y, \lambda) + r^F(x, y, \lambda) \\ &= q^F(x, y, \phi(x))P_k(y, \phi(x)) \end{aligned}$$

Therefore,

$$P_k(y, \phi(x)) = \frac{F(x, y)}{q^F(x, y, \phi(x))}$$

and the equation for $G(x, y)$ yields

$$G(x,y) = \frac{q^G(x,y,\phi(x))}{q^F(x,y,\phi(x))}F(x,y) + r^G(x,y,\phi(x))$$

If we define

$$q(x,y) = \frac{q^G(x,y,\phi(x))}{q^F(x,y,\phi(x))}$$

$$r(x,y) = r^G(x,y,\phi(x))$$

the result is proved. ■

D.3 Jets and Germs

In this section, we develop two different formulations of the "local structure" of a map.

The *local structure* of a function around $x = 0$ can be defined as what is shared in common by the collection of all functions agreeing in a neighborhood of $x = 0$.

Definition D.7. *A* **germ** *of C^∞ functions around $x = 0$ is an equivalent class of C^∞ functions under the following equivalence relation: Two functions f, g are equivalent iff there exists a neighborhood O_0 of 0 such that $f(x) = g(x), \forall x \in O_0$.*

We denote by C_x the set of germs of (real-valued) functions around $x = 0$.

Another way of defining the local structure of a function is by means of partial derivatives.

Definition D.8. *A* **k-jet** *of functions is an equivalence class of C^∞ functions agreeing, together with their derivatives up to an including order k, at $x = 0$. The k-jet of f at $x = 0$, $j^k f(0)$, is the equivalence class of functions g such that $f^i(0) = g^i(0), 0 \leq i \leq k$.*

A k-jet can be represented by the vector $(a_0, ..., a_k)$ where a_i is meant to be the ith derivative of the function at $x = 0$. Another representation of the k-jet is the polynomial $a_0 + a_1 x + \frac{1}{2!}a_2 x^2 + ... + \frac{1}{k!}a_k x^k$. This representation forges an obvious connection between k-jets and truncated Taylor expansions. We could also define an infinite jet of function j^∞ to be a one-sided infinite sequence $(a_0, a_1, ...)$ such that a_i is the ith derivative of a certain function. But then an existence problem immediately crops up.

Theorem D.9. *Given an infinite sequence of numbers, a_i, there exists a smooth function such that $f^{(i)}(0) = a_i$.*

Proof. Define a smooth, "bell-shaped" function

$$\beta(x) \quad = \quad 1, \ |x| \leq \frac{1}{2}$$

$$\beta(x) = 0, \ |x| \geq 1$$

Define the function

$$f(x) = \sum_{i=0}^{\infty} \frac{x^i}{i!} \beta(M_i x) a_i$$

It is a relatively easily exercise of real analysis to prove that there exists a sequence, $\lim_{i \to \infty} M_i = \infty$, such that the function $f(x)$ is smooth. Then showing that $\frac{\partial^i f}{\partial x^i}(0) = a_i$ is trivial. ∎

In fact, we can strengthen this result to the following:

Corollary D.1. (Borel) *The* **Borel map**

$$\begin{array}{rcl} C_x & \rightarrow & \mathbb{R}[[x]] \\ \phi & \mapsto & \sum_k \frac{1}{k!} \sum_{k_1+k_2+\ldots+k_n=k} \frac{\partial^{k_1+k_2+\ldots+k_n}\phi(0)}{\partial x_1^{k_1} \partial x_2^{k_2} \ldots \partial x_n^{k_n}} x_1^{k_1} x_2^{k_2} \ldots x_n^{k_n} \end{array}$$

is a surjective homomorphism. Furthermore, its kernel is the set $\mathcal{M}_x^\infty$ of all functions, all partial derivatives of which vanish at $x = 0$.

Proof. The bulk of the proof is surjectivity which is addressed in the previous theorem. The last claim regarding the kernel is trivial. For the details, see [Castrigiano and Hayes 1993, page 76, Lemma 3]. ∎

This result has some surprising consequences. Even if we take a fast-growing sequence like $a_i = (i!)^2$, there still exists a function $f(x)$ such that $f^{(i)}(0) = a_i$. However, because the sequence of derivatives grows too fast, the Taylor expansion $\sum_i (i!) x^i$ converges nowhere except at $x = 0$. Because of the discrepancy between the function and its Taylor expansion, the function could not be analytic.

By definition, a C^∞ function is **analytic** iff it equals its Taylor expansion. Not all C^∞ functions equal their Taylor expansions, the most classical counterexample being $x \mapsto \exp(-\frac{1}{x^2})$.

There is a difference between an infinite jet and a germ. A function in a germ is C^∞, but it need not be analytic. The infinite jet of $x \mapsto \exp(-\frac{1}{x^2})$ vanishes, whereas the germ of the same function contains other functions than those identically vanishing. Therefore, an infinite jet is not a faithful representation of a germ.

We could, however, ask the question as to *what* germs are uniquely characterized by a k-jet. This is *Mather's k-determinacy problem*, to which we shall return later.

D.4 Rings and Ideals of Functions

The motivation for defining rings and ideals of functions defined on a manifold goes as follows: To each manifold, we can associate the ring of real-valued functions defined on it; to each submanifold, we can associate the ideal of functions vanishing on the submanifold.

Definition D.10. *A* **ring** *is a set R endowed with an additive internal law $+$ and a multiplicative internal law $\cdot$ subject to the following conditions:*

- *$(R, +)$ is a commutative group.*
- *$(R, \cdot)$ is a semigroup.*
- *The multiplication is distributive relative to the addition, namely,*
  - $(a + b) \cdot c = a \cdot c + b \cdot c$
  - $c \cdot (a + b) = c \cdot a + c \cdot b$

Among the rings of relevance to singularity theory, we mention

- $\mathbb{R}[x]$, the ring of polynomials in x;
- $\mathbb{R}[[x]]$, the ring of formal power series in x;
- $\mathbb{R}[[x]]_{alg}$, the ring of algebraic formal power series in x;
- $\mathcal{N}_x$, the ring of germs of Nash functions in x;
- H_x, the ring of power series in x converging in some neighborhood of $x = 0$;
- C_x, the ring of germs of C^∞ functions around $x = 0$.

Observe that

$$\mathbb{R}[x] \subseteq \mathbb{R}[[x]]_{alg} = \mathcal{N}_x \subseteq H_x \subseteq \mathbb{R}[[x]] \subseteq C_x$$

The equality involving formal algebraic series and Nash functions is proved in [Bochnak, Coste, and Roy 1987, page 150]. The second inclusion relates to the fact that Nash functions are, by definition, analytic; see [Bochnák, Coste, and Roy 1987, page 143]. The last inclusion is from Corollary D.1.

Next, we introduce the notion of ideal.

Definition D.11. *An* **ideal** *I in a ring $(R, +, \cdot)$ is a subset $I \subseteq R$ such that*

- *$\forall i, j \in I \Rightarrow i + j \in I$;*
- *$\forall i \in I, \forall r \in R \Rightarrow r \cdot i \in I$.*

A typical example is the ideal of germs of C^∞ functions vanishing at $x = 0$.

Definition D.12. *A* **principal ideal** *in R is an ideal of the form aR, for some $a \in R$, and is written $Id(a)$. More generally, given ring elements $a_1, ..., a_n$, the ideal formed by all linear combinations of the form $\sum_i a_i r_i : r_i \in R$ is called the* **ideal generated by** *a_i and is written $Id(a_1, ..., a_n)$.*

Definition D.13. *A* **maximal ideal** *$\mathcal{M}$ in a ring $(R, +, \cdot)$ is an ideal such that any other ideal that contains $\mathcal{M}$ as a proper subset is the full ring R. A ring that has a unique maximal ideal is called* **local ring**.

In a local ring, the unique maximal ideal is precisely the set of noninvertible elements.

Theorem D.14. *Let $\mathcal{M}$ be an ideal in the ring $(R, +, \cdot)$. Then $\mathcal{M}$ is maximal iff $R/\mathcal{M}$ is a field.*

Proof. See [Hilton and Wu 1974, Theorem 1.8, page 129]. ■

To illustrate this theorem, we invite the reader to prove as an exercise that in the commutative ring $\mathbb{R}[x]$ of polynomials in one variable, the (principal) ideal $d(x)\mathbb{R}[x]$ is maximal iff $\mathbb{R}[x]/d(x)\mathbb{R}[x]$ is a field iff the polynomial $d(x)$ is irreducible over $\mathbb{R}$. In particular, $\mathbb{R}[x]/(x^2+1)\mathbb{R}[x] \cong \mathbb{C}$. It should be clear from these observations that $\mathbb{R}[x]$ is *not* a local ring because it has plenty of maximal ideals.

On the other hand, in the ring $\mathbb{R}[[x]]$ of *formal* power series, an ideal of the form $(x-a)\mathbb{R}[[x]]$ is not proper unless $a = 0$. From this observation, it is not hard to see that $\mathcal{M}[[x]] = x\mathbb{R}[[x]]$ is the *unique* maximal ideal in $\mathbb{R}[[x]]$, so that $\mathbb{R}[[x]]$ is a *local* ring (see [Bhattacharya, Jain, and Nagpaul 1986, page 189]). *

The local ring most relevant to singularity is revealed by the following:

Theorem D.15. *The ideal of germs of C^∞ functions vanishing at $x = 0$ is maximal in C_x. (This maximal ideal is written $\mathcal{M}_x$.) Furthermore, $\mathcal{M}_x$ is the unique maximal ideal of C_x. In other words, C_x is a local ring.*

Proof. See [Bruce and Giblin 1992, pages 269–270], [Golubitsky and Guillemin 1973, Lemma 3.2, page 103], or [Castrigiano and Hayes 1993, page 82]. ■

In addition to $\mathcal{M}_x$, there are other ideals in C_x. Define

$$\mathcal{M}_x^k = \{f \in C_x : \frac{\partial^{k_1+k_2+\ldots} f(0)}{\partial x_1^{k_1} \partial x_2^{k_2} \ldots} = 0, \quad k_1 + k_2 + \ldots < k\}$$

An equivalent definition is

$$\mathcal{M}_x^k = \underbrace{\mathcal{M}_x \cdot \mathcal{M}_x \cdot \ldots \cdot \mathcal{M}_x}_{k}$$

where $\cdot$ denotes the ring multiplication. This last definition justifies the notation. Clearly, $\mathcal{M}_x^k$ is an ideal.

Theorem D.16. *The quotient ring $C_x/\mathcal{M}_x^k$ is isomorphic to the ring of (real coefficient) polynomials of degree $k-1$; equivalently, $C_x/\mathcal{M}_x^k \cong \mathbb{R}^k$.*

Proof. First, take a function $\phi : \mathbb{R} \to \mathbb{R}$ defined in a neighborhood of $x = 0$. By elementary analysis, we have

$$\phi(x) = \phi(0) + \int_0^1 \frac{d\phi(tx)}{dt} dt = \phi(0) + \int_0^1 \phi'(tx) x dt$$

Clearly, the function $x \mapsto \int_0^1 \phi'(tx)xdt$ is smooth and vanishes at $x = 0$, so that it is in $\mathcal{M}_x$. Therefore, the above equation means that any function

*I thank Prof. I. S. Reed for helping me clarify this point.

can be decomposed as a real constant plus a function in the ideal; in other words, $C_x/\mathcal{M}_x \cong \mathbb{R}$. The case $k > 1$ is by induction on k and is omitted. The case of more than one variable is basically the same and the details are omitted. ■

Now, we come back to Corollary D.1. With the above concepts, we redefine $\mathcal{M}_x^\infty = \cap_{k=1}^\infty \mathcal{M}_x^k$, and the latter is obviously another ideal of C_x. Therefore, the Borel map induces an isomorphism

$$\mathbb{R}[[x]] \cong C_x/\mathcal{M}_x^\infty$$

In a ring that has plenty of maximal ideals, there is still an ideal that can be singled out.

Theorem D.17. *Let R be a commutative ring with unit. We have*

$$\begin{aligned} J(R) &= \cup \text{ all maximal ideals of } R \\ &= \{r \in R : (\forall s \in R)\,(1 + rs) \text{ is invertible}\} \end{aligned}$$

The ideal $J(R)$ is called **Jacobson ideal** *or* **Jacobson radical** *of R.*

Proof. See [Weibel 1994] or [Hilton and Wu 1974]. ■

A concept related to maximal ideal is the following:

Definition D.18. *A* **prime ideal** *I in a commutative ring $(R, +, \cdot)$ is an ideal such that*

$$a \cdot b \in I \Rightarrow a \in I \text{ or } b \in I$$

It is easily seen that any maximal ideal is a prime ideal.

The following lemma might appear to be just one of those obscure algebraic results. However, as we will see in the next section, it allows us to get out of some dead-end of analysis.

Lemma D.19. (Nakayama) *Let R be a commutative ring with unit and let $\mathcal{I}$ be an ideal subject to either of the following conditions:*

1. *$\mathcal{I} \subseteq J(R)$.*
2. *All elements of $1 + \mathcal{I}$ are invertible.*
3. *If R is a local ring, $\mathcal{I}$ is the unique maximal ideal.*

Let A be a finitely generated R-module. Then

$$A \subseteq \mathcal{I}A \Rightarrow A = \{0\}$$

Furthermore, if B is another finitely generated R-module, then

$$A \subseteq B + \mathcal{I}A \Rightarrow A \subseteq B$$

Proof. For the proof of the first claim, under the most general conditions, see [Hilton and Wu 1974], [Rosenberg 1994], or [Weibel 1994]. For proof of the first claim under more restrictive conditions, see [Golubitsky and Guillemin 1973, Lemma 3.4, page 104], [Castrigiano and Hayes 1993, Lemma 15, page 86], or [Bruce and Giblin 1992, 11.16, page 271]. The second claim is just a corollary of the first; see [Castrigiano and Hayes 1993, Corollary 16, page 87] or [Bruce and Giblin 1992, 11.17, page 272]. ■

D.5 Formal Inverse Function Theorem

As an illustration of Nakayama's lemma, let us prove the formal inverse function theorem. To give a taste of the problem, consider a function, for example,

$$y = x + x^2$$

specified by a formal power series (remember Borel's lemma!), with non-vanishing first degree term, and let us solve the above for x. Clearly, we have

$$x = y - x^2$$

To obtain a satisfactory inverse function, we have to "get rid" of x^2 in the right-hand side. This is attempted by doing yet another round of substitutions,

$$\begin{aligned} x &= y - (y - x^2)^2 \\ &= y - y^2 + 2x^3 + x^4 \end{aligned}$$

Clearly, we could continue this process *ad infinitum*; at each step, the degree of the x-terms in the right-hand side would increase, but it seems that we will never be able to "get rid of the tail." Here algebra comes to the rescue.

Consider a formal power series:

$$y = \sum_{k=1}^{\infty} a_k x^k, \ a_1 \neq 0$$

To prove that the above can be solved for x, we have to show that introducing the relation $y - \sum_{k=1}^{\infty} a_k x^k = 0$ in any formal power series in x, y yields a formal power series in y; algebraically,

$$\mathbb{R}[[x, y]]/\text{Id}(y - \sum_{k=1}^{\infty} a_k x^k) \subseteq \mathbb{R}[[y]]$$

equivalently,

$$\mathbb{R}[[x, y]] \subseteq \text{Id}(y - \sum_{k=1}^{\infty} a_k x^k) + \mathbb{R}[[y]]$$

or,

$$\mathcal{M}[[x, y]] \subseteq \text{Id}(y - \sum_{k=1}^{\infty} a_k x^k) + \mathcal{M}[[y]]$$

Here we are at the *crucial point*. Invoking Nakayama's lemma, to prove the above, it suffices to show that

$$\mathcal{M}[[x,y]] \subseteq \mathrm{Id}(y - \sum_{k=1}^{\infty} a_k x^k) + \mathcal{M}[[y]] + \mathcal{M}[[x,y]]^2$$

To prove this, the only nontrivial thing to show is that the right-hand side contains $x\mathbb{R}[[x,y]]$. Clearly, any term of degree at least one in x is generated in the right-hand side as follows:

$$x = -\frac{1}{a_1}(\underbrace{(y - \sum_{k=1}^{\infty} a_k x^k)}_{\in \mathrm{Id}(y - \sum a_k x^k)} - \underbrace{y}_{\in \mathcal{M}[[y]]} + \underbrace{\sum_{k=2}^{\infty} a_k x^k}_{\in \mathcal{M}^2[[x,y]]})$$

And we are done.

Theorem D.20. (Formal Inverse Function Theorem) *Given a function specified by a formal power series,*

$$y = Ax + \begin{pmatrix} \vdots \\ \sum_{d=2}^{\infty} \sum_{k_1+\ldots+k_n=d} a^i_{k_1\ldots k_n} x^{k_1} \ldots x^{k_n} \\ \vdots \end{pmatrix}$$

where the matrix A is nonsingular, the above equation can be solved for x as a formal power series in the components of y.

Proof. The proof involves just a trivial extension of the above argument to the case of many x and y variables and this is left to the reader. ■

The formal inverse function theorem is the key step in proving the formal implicit function theorem; the formulation of the formal implicit function theorem is left to the reader.

D.6 Local Ring of a Map

The philosophy behind the *local ring* of a map goes as follows: Like in cohomology, to study a manifold around a point, we could study the ring of real-valued functions defined in a neighborhood of that point. Going one step further, to study a map between manifolds around a singularity, it would suffice to study the induced ring homomorphism.

Consider a smooth map $F : X, 0 \to Y, 0$ and let C_x, C_y be the rings of germs of smooth functions defined in a neighborhood of $x = 0, y = 0$, respectively. Then the smooth map F induces a ring homomorphism:

$$\begin{aligned} F^* : C_y &\to C_y \\ g(y) &\mapsto F^*(g)(x) = g(F(x)) \end{aligned}$$

The situation can be depicted by the following diagram:

$$\begin{array}{ccc} X & \xrightarrow{F} & Y \\ \downarrow & & \downarrow \\ C_x & \xleftarrow{F^*} & C_y \end{array}$$

Definition D.21. *Consider a smooth map $F : X, 0 \to Y, 0$. Let C_x, C_y be the rings of germs of smooth functions around $x = 0, y = 0$, respectively. Let $\mathcal{M}_y$ be the maximal ideal in C_y. The* **local ring** *of F is defined, invariantly, by*

$$Q_F = C_x / C_x F^* \mathcal{M}_y \tag{D.5}$$

Equivalently, the local ring can be defined, in terms of the coordinates F_i of F, as

$$Q_F = C_x / Id(\{F_i\}) \tag{D.6}$$

The reader should understand the subtle difference between the two definitions: Definition D.5 is the so-called *invariant* definition, while Definition D.6 relies on the choice of coordinates. See [Mather 1969b] or [Golubitsky and Guillemin 1973] for the invariant definition and see [Arnold, Gusein-Zade, and Varchenko 1985, pages 74–75] or [Castrigiano and Hayes 1993] for the coordinatewise definition.

The importance of the concept of local ring of a map $F : X, 0 \to Y, 0$ is that it provides the concept of multiplicity of the solution $x = 0$ to $F(x) = 0$ in case the function is not polynomial.

Definition D.22. *Let $F : X, 0 \to Y, 0$ be a smooth map. The* **multiplicity** *of the solution $x = 0$ to $F(x) = 0$ is the dimension of the local ring as a vector space over $\mathbb{R}$.*

Theorem D.23. *Let $F : X, 0 \to Y, 0$ be a smooth map and let μ be the dimension of the local ring as a vector space over $\mathbb{R}$. Then there exist neighborhoods O_x, O_y of 0 in X, Y, respectively, such that $\forall y \in O_y$, $F^{-1}(y) \cap O_x$ has at most μ elements.*

Proof. See [Golubitsky and Guillemin 1973, page 168, Proposition 2.4]. ■

A deeper motivation for the local ring is the following theorem:

Theorem D.24. *Let F, F' be two stable germs of smooth functions such that*

$$Q_F / \mathcal{M}_x^3 = Q_{F'} / \mathcal{M}_x^3$$

Then the germs are **equivalent**—*that is, there exist diffeomorphisms g, h such that $F = h \circ F' \circ g$.*

Proof. See [Mather 1969b, Theorem A]. ■

D.7 Modules Over Rings of Functions

Intuitively speaking, a *module* has all of the features of a vector space, except that the "scalars" form a *ring*, not a field.

Definition D.25. *A* **(right) module over** R, $(M, +, R, \cdot)$, *consists of the following structure:*

- *A commutative group* $(M, +)$;
- *A ring* $(R, +, \cdot)$;
- *An external law or* **action** $\cdot : M \times R \to M$ *that satisfies the compatibility conditions*
  - $(m_1 + m_2) \cdot r = m_1 \cdot r + m_2 \cdot r$,
  - $m \cdot (r_1 + r_2) = m \cdot r_1 + m \cdot r_2$,
  - $m \cdot (r_1 \cdot r_2) = (m \cdot r_1) \cdot r_2$,
  - $m \cdot 1 = m$.

A module over the ring R *is also called* R-**module**.

A typical module that is often encountered in singularity theory and stability of maps is the module of n-tuples of polynomials in x over the ring of polynomials in x, namely, $((\mathbb{R}[x])^n, +, \mathbb{R}[x], \cdot)$. Also important and equally typical are the modules $((\mathbb{R}[[x]])^n, +, \mathbb{R}[[x]], \cdot)$, $((H_x)^n, +, H_x, \cdot)$, and $((C_x)^n, +, C_x, \cdot)$.

The following theorem will play a crucial role in the module-theoretic generalization of the Malgrange preparation theorem:

Theorem D.26. *Let* A *be a module over the local ring* R. *Let* $\mathcal{M}$ *be the unique maximal ideal in* R. *Then* $A/\mathcal{M}A$ *is a finite-dimensional vector space over the field* $R/\mathcal{M}$. *Furthermore, let* $\pi : A \to A/\mathcal{M}A$ *be the natural projection; choose a basis* $\{e^i\}$ *of the vector space* $A/\mathcal{M}A$; *and choose* g^i *such that* $\pi(g^i) = e^i$; *then* $\{g^i\}$ *generates* A *as a module over* R.

Proof. See [Golubitsky and Guillemin 1973, page 105, Corollary 3.5]. ■

D.8 Generalized Malgrange Preparation Theorem

Let A be a submodule of $C_{x,y}$, finitely generated over $C_{x,y}$. Take an arbitrary $a(x, y) \in A$. Very often the question arises as to whether $a(x, y)$ can be written as $\sum_{i=0}^{k-1} r_i(x) y^i$. This last issue is equivalent to whether the module A is finitely generated over C_x. The latter is an algebraic reformulation of the fact that, in the module A, all variables $y^i, i \geq k$ can be written explicitly in terms of x.

Now, we generalize the above. First, observe the following technical detail: Given a module A over C_{x_1} and a smooth mapping $\phi : X_1 \to X_2$, the module A can be considered as a module over C_{x_2}; indeed, it suffices to define the external law $a \cdot \alpha(x_2) = a \cdot \alpha(\phi(x_1))$, where $a \in A$ and $\alpha \in C_{x_2}$. It is instructive to look at the categorical interpretation of this:

$$\begin{array}{ccc} X_1 & \xrightarrow{\phi} & X_2 \\ \downarrow & & \downarrow \\ C_{x_1} & \xleftarrow{\phi^*} & C_{x_2} \\ \downarrow & & \downarrow \\ (A, +, C_{x_1}, \cdot) & \xrightarrow{\phi^{**}} & (A, +, C_{x_2}, \cdot) \end{array}$$

The top downarrows denote the contravariant functor $\hom(\cdot, \mathbb{R})$ while the bottom downarrows denote the contravariant functor $\hom(\cdot \times A, A)$, which is not to be confused with the algebraic K-theoretic functor K_0 of Section 20.15. The composition of these two functors is covariant, and ϕ^{**} makes the C_{x_1}-module A a C_{x_2}-module. With this concept, a generalization of the issue raised in the preceding paragraph is whether a module finitely generated over C_{x_1} is finitely generated over C_{x_2}. (Set $x_1 = (x, y)$ and $\phi(x, y) = x$ to recover the preceding paragraph.)

Theorem D.27. (Generalized Malgrange Preparation Theorem) *Let*

$$\begin{array}{rcl} \phi : \mathbb{R}^{n_1}, 0 & \to & \mathbb{R}^{n_2}, 0 \\ x_1 & \mapsto & x_2 \end{array}$$

be a smooth mapping. Let A be a finitely generated module over C_{x_1}. Then A is finitely generated over C_{x_2} iff $A/\mathcal{M}_{x_2}A$ is a finite-dimensional vector space. Furthermore, let $\pi : A \to A/\mathcal{M}_{x_2}A$ be the natural projection; choose a basis $\{e^i\}$ of the vector space $A/\mathcal{M}_{x_2}A$; and choose $g^i \in A$ such that $\pi(g^i) = e^i$; then $\{g^i\}$ generates A as a module over C_{x_2}.

If A is a finitely generated module over $\mathbb{R}[[x_1]]$, the same statements hold.

Proof. See [Mather 1968b, page 132], [Golubitsky and Guillemin 1973], or [Castrigiano and Hayes 1993] for modules over the ground ring C_x. See [Arnold, Gusein-Zade, and Varchenko 1985, page 130] for modules over other ground rings. ■

To grasp the feeling for this theorem, let us come back to the problem we started from, the "generalized implicit function theorem." We are given a function $F(x, y)$ that satisfies the condition D.3 of the Malgrange preparation theorem. The smooth mapping is $\phi(x, y) = x$. We form the local ring of the map F—that is, $C_{x,y}/F$. The latter can be regarded as a finitely generated module over $C_{x,y}$. The issue addressed by the generalized Malgrange preparation theorem is whether the module A is finitely generated *over* C_x. Before proving that A is indeed finitely generated over C_x, let us see what kind of information can be derived from the latter.

If $A = C_{x,y}/F$ is finitely generated over C_x, this means that any element of the quotient module can be written as

$$-1r_0(x) - yr_1(x) - \dots - y^{k-1}r_{k-1}(x)$$

Therefore, any smooth function, in particular y^k, can be decomposed as an element of the quotient module $C_{x,y}/F$ and an element of the ideal $\mathrm{Id}(F)$; that is,

$$y^k = -1r_0(x) - yr_1(x) - \dots - y^{k-1}r_{k-1}(x) + q(x,y)F(x,y)$$

Furthermore, repeatedly differentiating the above relative to y and using Condition D.3 yields

$$q(0,0) \neq 0$$

Therefore, we recover the result that the equation $F(x,y) = 0$ is locally equivalent to a polynomial equation in y with coefficients smoothly depending on x.

Now, we have to prove that $A = C_{x,y}/F$ is finitely generated. The *essential point* of the generalized Malgrange preparation theorem is that it is *easier* to check that $A/\mathcal{M}_x A$ is a finite-dimensional vector space. The following string is pure algebra:

$$A/\mathcal{M}_x A = (C_{x,y}/F)/\mathcal{M}_x(C_{x,y}/F) = C_{x,y}/\mathrm{Id}(F,x)$$

Next, we prove that $\mathrm{Id}(F,x) = \mathrm{Id}(y^k, x)$. Indeed, consider

$$F(x,y) - F(0,y) = \sum_i \int_0^{x_i} \frac{\partial F}{\partial x_i} dx_i$$

Using Rolle's theorem, we write

$$\int_0^{x_i} \frac{\partial F}{\partial x_i} dx_i = x_i \phi_i(x_i)$$

By L'Hôpital's rule, the ϕ_i's are smooth. Putting the pieces together, we get

$$F(x,y) = y^k F_0(y) + \sum_i x_i \phi_i(x_i)$$

From the latter it easily follows that $\mathrm{Id}(F,x) = \mathrm{Id}(y^k, x)$. Finally, $A/\mathcal{M}_x A = C_{x,y}/\mathrm{Id}(y^k, x)$ is clearly a finite-dimensional vector space.

D.9 Jacobi Ideal, Codimension, and Determinacy

A smooth function germ $f : \mathbb{R}^n, 0 \to \mathbb{R}, 0$ is said to be **finitely determined** or, more specifically, to be ***k*-determined** ($k < \infty$) iff for any other smooth function germ g that has the same k-jet as f there exists a diffeomorphism h such that $f = g \circ h$. (The germs f and g are said to be **right-equivalent**.) It is easily seen that only functions with isolated critical points can have finite determinacy. Observe that, by Lemma 21.12

of Morse, Morse functions have determinacy 2.

Determinacy of a germ of smooth functions f relies crucially on its **Jacobi ideal** defined as the ideal generated by the partial derivatives of the function, namely,

$$\mathrm{Id}(\{\frac{\partial f}{\partial x_i}\})$$

An *invariant* definition of the Jacobi ideal—that is, a definition that does not rely on coordinates—does exist but is hard to come by. With the above definition we define the **codimension** of a germ of functions as

$$\begin{aligned}\mathrm{codim}(f:\mathbb{R}^n,0\to\mathbb{R},0) &= \dim_{\mathbb{R}}(\mathcal{M}_x/\mathrm{Id}(\{\frac{\partial f}{\partial x_i}\}))\\ &= \dim_{\mathbb{R}}(C_x/\mathrm{Id}(\{\frac{\partial f}{\partial x_i}\}))-1\end{aligned}$$

Theorem D.28. *A germ in $\mathcal{M}_x^2$ has finite determinacy iff it has finite codimension.*

Proof. See [Mather 1968b, Theorem 3.5 (right-equivalence case), page 141] or [Castrigiano and Hayes 1993, Corollary 8, page 130]. ■

Theorem D.29. *The germ f has finite determinacy iff $\mathcal{M}_x^k \subseteq Id(\{\frac{\partial f}{\partial x_i}\})$ for some finite integer k. Furthermore, in this case, the determinacy is either k or $k+1$.*

Proof. See [Castrigiano and Hayes 1993, Corollary 40, page 110]. ■

D.10 Universal Unfolding

The *unfolding* problem is the following: Assume we have a (germ of a) function that has a degenerate critical point. This situation is unstable in the sense that a small perturbation will in general remove, "unfold," the degeneracy and reveal to what stable singularities the degeneracy can reduce.

Definition D.30. *Let $f:\mathbb{R}^n\to\mathbb{R}$ be a map germ. A r-parameter* **unfolding** *is a map germ $F:\mathbb{R}^n\times\mathbb{R}^r\to\mathbb{R},(x,u)\mapsto F(x,u)$ such that $f(x)=F(x,0)$ in a neighborhood of $x=0$. The vector u is referred to as* **control parameter**. *The unfolding is said to be* **universal** *if it induces all possible unfoldings with a minimum number of control parameters.*

Theorem D.31. (Universal Unfolding) *Let $(x,0)\mapsto(f(x),0)$ be a map germ of codimension r. Then its universal unfolding has exactly r parameters. Furthermore, if we take $g_1,g_2,\ldots,g_r$ as a basis of*

$$\mathcal{M}_x/Id(\{\frac{\partial f}{\partial x_i}\})$$

then the universal unfolding is given by

$$F(x, u) = f(x) + \sum_{i=1}^{r} u_i g_i(x)$$

Proof. See [Castrigiano and Hayes 1993, page 231]. ■

NOTES

Historically, the Malgrange preparation theorem, that appeared chronologically first, and the Mather division theorem were derived using independent proofs. However, Mather [1968a, page 92] correctly observed that the Malgrange preparation theorem could be derived from the Mather division theorem. It is this latter approach that we have adopted, along with [Golubitsky and Guillemin 1973] and [Castrigiano and Hayes 1993].

Following [Mather 1968a, page 92], it seems unlikely that the Mather division theorem could be easily derived from the Malgrange preparation theorem, so that the division theorem should be considered as a result more fundamental than the preparation theorem.

The generalized Malgrange preparation theorem, also referred to as *Mather–Malgrange preparation theorem*, was apparently formulated for the first time in [Mather 1968b, page 132].

BIBLIOGRAPHY

1. Ackerman, J. [1993]. *Robust Control: Systems with Uncertain Physical Parameters*, Springer-Verlag, New York.
2. Adams, J.F. [1972]. *Algebraic Topology—A Student's Guide*, London Mathematical Society Lecture Note Series 4, Cambridge University Press, Cambridge, England.
3. Ahlfors, L.V. [1966]. *Complex Analysis*, McGraw-Hill, New York.
4. Ahlfors, L.V. [1973]. *Conformal Invariants—Topics in Geometric Function Theory*, McGraw-Hill, New York.
5. Akbulut, S., and King, H. [1992]. *Topology of Real Algebraic Sets*, Mathematical Sciences Research Institute Publications, Vol. 25, Springer-Verlag, New York.
6. Akbulut, S. (Ed.) [1995]. *Real Algebraic Geometry and Topology*, Contemporary Mathematics 182, American Mathematical Society, Providence, RI.
7. Alexander, J.C. [1978]. "The topological theory of an embedding method," in *Continuation Methods* (Wacker, H., Ed.), Academic Press, pp. 37–68.
8. Alexandroff, P. [1961]. *Elementary Concepts of Topology*, Dover, New York.
9. Alexandroff, P.S. [1972]. "Poincaré and topology," *Russian Math. Surveys*, Vol. 27, pp. 157–168.
10. Anderson, B.D.O., Jury, E.I., and Mansour, M. [1987]. "On robust Hurwitz polynomials, *IEEE Trans. Autom. Contr.*, Vol. AC-32, pp. 909–912.
11. Anick, D. [1993]. *Differential Algebras in Topology*, AK Peters, Wellesley, MA.
12. Arnold, V.I., Gusein-Zade, S.M., and Varchenko, A.N. [1985]. *Singularities of Differentiable Maps*, Vol. I, *The Classification of Critical Points, Caustics and Wave Fronts*, Birkhäuser, Boston.
13. Arnold, V.I., Gusein-Zade, S.M., and Varchenko, A.N. [1988]. *Singularities of Differentiable Maps*, Vol. II, *Monodromy and Asymptotics of Integrals*, Birkhäuser, Boston.
14. Arnon, D.S., and Buchberger, B. (Eds.) [1988]. *Algorithms in Real Algebraic Geometry*, Academic Press, New York.
15. Artin, E. [1988]. *Geometric Algebra*, Wiley Classics Library, Wiley, New York.

16. Atiyah, M.F., and Hirzebruch, F. [1961]. "Vector bundles and homogeneous spaces," *Proceedings, Symposium in Pure Mathematics*, Vol. 3, *American Mathematical Society*, Providence, R.I., pp. 7–38.
17. Atiyah, M.F., and Bott, R. [1964]. "On the periodicity theorem for complex vector bundles," *Acta Math.*, Vol. 112, pp. 299–347.
18. Atiyah, M.F. [1966a]. "K-theory and reality," *Q. J. Math., Oxford (2)*, Vol. 17, pp. 367–386.
19. Atiyah, M.F. [1966b]. "Global aspects of the theory of elliptic differential operators," *Proceedings, International Symposium of Mathematicians*, Moscow, pp. 57–64.
20. Atiyah, M.F. [1966c]. "The role of algebraic topology in mathematics," *Journal London Math. Society*, Vol. 41, pp. 63–69.
21. Atiyah, M.F. [1967a]. *K-Theory*, W.A. Benjamin, New York.
22. Atiyah, M.F. [1967b]. "Algebraic topology and elliptic operators," *Commun. Pure Appl. Math.*, Vol. 20, pp. 237–249.
23. Atiyah, M.F. [1968]. "Bott periodicity and the index of elliptic operators," *Q. J. Math.*, Vol. 19 (2), pp. 113–140.
24. Atiyah, M.F., and Singer, I.M. [1968]. "The index of elliptic operators: I," *Annals of Mathematics*, Vol. 87, pp. 484–530.
25. Atiyah, M.F. [1969]. *Algebraic Topology and Operators in Hilbert Spaces*, Lectures in Modern Analysis and Applications, I, Springer, Berlin, pp. 101–121.
26. Atiyah, M.F. [1973]. "The index of elliptic operators," *Colloquium Lectures, American Mathematical Society*, Dallas, pp. 477–495.
27. Atiyah, M.F. [1977]. "A survey of K-theory," *Proceedings, Conference on K-Theory and Operator Algebras*, Athens, Georgia, Lecture Notes in Mathematics, 575, Springer-Verlag, Berlin, pp. 821–829.
28. Barratt, M.G., Gugenheim, V.K.A.M., and Moore, J.C. [1959]. "On semisimplicial fiber-bundles," *Am. J. Math.*, Vol. 81, pp. 639–657.
29. Barmish, R.R., Khargonekar, P.P., Shi, Z.C., and Tempo, R. [1990]. "Robustness margin need not be a continuous function of the problem data," *Syst. Contr. Lett.*, Vol. 15, pp. 91–98.
30. Barmish, B. R. [1994]. *New Tools for Robustness of Linear Systems*, MacMillan, New York.
31. Bar-on, J. [1990]. "Phase and gain margins for multivariable systems," Ph. D. dissertation, Department of Electrical Engineering – Systems, University of Southern California, Los Angeles, CA.
32. Bar-on, J., and Jonckheere, E. A. [1990]. "Phase margins for multivariable systems," *Int. J. Contr.*, Vol. 52, pp. 485–498.
33. Bar-on, J., and Jonckheere, E.A. [1991]. "Multivariable gain margin," *Int. J. Contr.*, Vol. 54 (2), pp. 337–365; see also Bar-on, J., and Jonckheere, E.A. [1990]. *Proceedings, IEEE Conference on Decision and Control*, Honolulu, Hawaii, pp. 340–347.

34. Bar-on, J., and Jonckheere, E.A. [1992]. "The geometry of the multivariable phase margin," *IEEE Trans. Autom. Contr.*, Vol. AC-37, pp. 798–800.
35. Bartlett, A.C., Hollot, C.V., and Lin, L. [1988]. "Root locations of an entire polytope of polynomials: It suffices to check the edges," *Math. Contr., Signals, Syst.*, Vol 1, pp. 61–71.
36. Barut, A.O., and Raczka, R. [1986]. *Theory of Group Representations and Applications*, World Scientific, Singapore.
37. Beals, M., Fefferman, C., and Grossman, R. [1983]. "Strictly pseudoconvex domains in $\mathbb{C}^n$," *Bulletin Am. Math. Soc.*, Vol. 8, pp. 125–322.
38. Becker, T., and Weispfenning, V. (in cooperation with Kredel, H.) [1993]. *Gröbner Bases—A Computational Approach to Commutative Algebra*, Graduate Texts in Mathematics 141, Springer-Verlag, New York.
39. Berlekamp, E.R., Conway, J.H., and Guy, R.K. [1982]. *Winning Ways for your Mathematical Plays*, in two volumes, Academic Press, New York.
40. Bhattacharya, P.B., Jain, S.K., and Nagpaul, S.R. [1986]. *Basic Abstract Algebra*, Cambridge University Press, Cambridge, England.
41. Bhattacharyya, S.P., Chapellat, H., and Keel, L.H. [1995]. *Robust Control—The Parametric Approach*, Prentice-Hall, Englewood Cliffs, NJ.
42. Bitmead, R.R., and Anderson, B.D.O. [1977]. "The matrix Cauchy index: Properties and Applications," *SIAM J. Applied Mathematics*, Vol. 33, pp. 655–672.
43. Blackadar, B. [1986]. *K-Theory for Operator Algebras*, Springer-Verlag, New York.
44. Boardman, J.M. [1967]. "Singularities of differentiable maps," *Publications Mathématiques, Institut des Hautes Etudes Scientifiques (I.H.E.S.)*, No. 33, pp. 21–57.
45. Bochnak, J., Coste, M., and Roy, M.-F. [1987]. *Géométrie Algébrique Réelle*, Springer-Verlag, New York.
46. Borgwardt, K. H. [1988]. *The Simplex Method—A Probabilistic Analysis*, Algorithms and Combinatorics 1, Springer-Verlag, New York.
47. Bott, R. [1957]. "The stable homotopy of the classical groups," *Proc. Natl. Acad. Sci. USA*, Vol. 43, pp. 933–935.
48. Bott, R. [1958]. "The space of loops on a Lie group," *Michigan Math. J.*, Vol. 5, pp. 35–61.
49. Bott, R. [1959]. "The stable homotopy of the classical groups," *Ann. Math.*, Vol. 70, pp. 313–337.
50. Bott, R. [1970]. "The periodicity theorem for the classical groups and some of its applications," *Adv. Math.*, Vol. 4, pp. 353–411.
51. Bott, R., and Tu, L.W. [1982]. *Differential Forms in Algebraic Topology*, Graduate Texts in Mathematics, Springer-Verlag, New York.

52. Bredon, G.E. [1993]. *Topology and Geometry*, Springer-Verlag, New York.
53. Brockett, R. W. [1976]. "Some geometric questions in the theory of linear systems," *IEEE Trans. Autom. Contr.*, Vol. AC-21, pp. 449–455.
54. Brown, L.G., Douglas, R.G., and Fillmore, P.A. [1977]. "Extensions of C^*-algebras and K-homology," *Ann. Math.*, Vol. 105, pp. 265–324.
55. Bruce, J.W., and Giblin, P.J. [1992]. *Curves and Singularities*, second edition, Cambridge University Press, Cambridge, England.
56. Bucy, R.S. [1975]. "Structural stability for the Riccati equation," *SIAM J. Contr.*, Vol. 13, pp. 749–753.
57. Bucy, R.S. [1991]. "Geometry and multiple direction estimation," *Inf. Sci.*, Vol. 57–58, pp. 145–158.
58. Callier, F., and Desoer, C.A. [1982]. *Multivariable Feedback Systems*, Springer-Verlag, New York.
59. Carathéodory, C. [1913]. "Über die Begrenzung einfach zusammenhängender Gebiete," *Math. Ann.*, Vol. 73, pp. 323–370.
60. Cartan H., and Eilenberg, S. [1970]. *Homological Algebra*, Princeton University Press, Princeton, NJ.
61. Castrigiano, D. P.L., and Hayes, S. A. [1993]. *Catastrophe Theory*, Addison-Wesley, Reading, MA.
62. Cerf, J. [1970]. "La stratification naturelle des espaces de fonctions différentiables réelles et le théorème de la pseudo-isotopie," *Publications Mathématiques, Institut des Hautes Etudes Scientifiques (I.H.E.S.)*, No. 39, pp. 5–173.
63. Cheng, C.-Y. [1994]. "Robust stability—A topological approach," Ph. D. dissertation, Department of Electrical Engineering – Systems, University of Southern California, Los Angeles, CA.
64. Chu, C.-K. [1996]. "Algebraic Structure and Stochastic Complexity of Hex Game Model of Robust Stability Algorithms," Ph. D. dissertation, Department of Electrical Engineering—Systems, University of Southern California, Los Angeles, CA.
65. Clark, J.M. [1976]. "The consistent selection of local coordinates in linear system identification," *Proceedings, Joint Automatic Control Conference*, pp. 576–580.
66. Coffin, S. T. [1990]. *The Puzzling World of Polyhedral Dissections*, Oxford University Press, New York.
67. Cohen, F.R., Cohen, R.L., Mann, B.M., and Milgram, R.J. [1991]. "The topology of rational functions and divisors of surfaces," *Acta Math.*, Vol. 166, pp. 163–221.
68. Cohn, H. [1967]. *Conformal Mapping on Riemann Surfaces*, Dover, New York.
69. Colgren, R. D. [1993]. "H^∞ Control of Nonlinear Systems using Describing Functions and Simplicial Algorithms," Ph.D. dissertation,

Department of Electrical Engineering—Systems, University of Southern California, Los Angeles, CA.

70. Conway, J.H., and Sloane, N.J.A. [1982]. "Voronoi regions of lattices, second moments of polytopes, and quantization," *IEEE Trans. Inf. Theory*, Vol. IT-28, pp. 211–226.
71. Conway, J.H., and Sloane, N.J.A. [1991]. *Sphere Packings, Lattices and Groups*, Springer-Verlag, New York.
72. Coutinho, M.G. [1996]. "Computational Geometry Implementation of Topological Robust Control," Ph.D. dissertation, Department of Electrical Engineering—Systems, University of Southern California, Los Angeles, CA.
73. Coutinho, M.G., and Jonckheere, E.A. [1995]. "SimplicialVIEW—A package for robust stability analysis," *Proceedings, IEEE Conference on Decision and Control*, New Orleans, LA.
74. Coxeter, H.M.S. [1973]. *Regular Polytopes*, Dover, New York.
75. Coxson, G. E., and DeMarco, C. L. [1991]. " Testing robust stability of general matrix polytopes in an NP-hard computation," *Proceedings, Twenty-Ninth Annual Allerton Conference on Communication, Control, and Computing*, Coordinated Science Laboratory, University of Illinois, Urbana-Champaign, Illinois, October 2-4, pp. 105–106.
76. D'Angelo, J. P. [1993]. *Several Complex Variables and the Geometry of Real Hypersurfaces*, Studies in Advanced Mathematics, CRC Press, Ann Arbor.
77. Dantzig, G.B. [1963]. *Linear Programming and Extensions,* Princeton University Press, Princeton, New Jersey.
78. DeCarlo, R., and Saeks, R. [1977]. "The encirclement condition—An approach using algebraic topology," *Int. J. Contr.*, Vol. 26, pp. 279–287.
79. de Gaston, R., and Safonov, M.G. [1988]. "Exact calculation of the multiloop stability margin," *IEEE Trans. Autom. Contr.*, Vol. AC-33, pp. 156–171.
80. Delchamps, D.F., and Byrnes, C.I. [1982]. "Critical point behavior of objective functions defined on spaces of multivariable systems," *Proceedings, IEEE Conference on Decision and Control*, Orlando, FL, pp. 937–943.
81. de Rham, G. [1984]. *Differentiable Manifolds*, Springer-Verlag, New York.
82. Dieudonné, J. [1989]. *A History of Algebraic and Differential Topology 1900–1960*, Birkhäuser, Boston.
83. do Carmo, M. P. [1994]. *Differential Forms and Applications*, Springer-Verlag Universitext, New York.
84. Doležal, V. [1964]. "The existence of a continuous basis of a certain linear subspace of E_r which depends on a parameter," *Cas. Pro. Pest. Mat.*, Vol. 89, pp. 466–468.

85. Donald, B. R., and Chang, D. R. [1991]. "On the complexity of computing the homology type of a triangulation," *Proceedings, IEEE 32nd Annual Symposium on Foundations of Computer Science*, San Juan, Puerto Rico, October 1–4, 1991, pp. 650–661; see also Cornell Computer Science Department, Technical Report, 1991.

86. Douglas, R. G. [1972]. "Banach algebra techniques in the theory of Toeplitz operators," Regional Conference Series, Vol. 15, American Mathematical Society, Providence, RI.

87. Douglas, R. G. [1980]. *C^*-Algebra Extensions and K-Homology*, Princeton University Press, Princeton, NJ.

88. Doup, T. [1988]. *Simplicial Algorithms on the Simplotope*, Springer-Verlag, New York.

89. Doyle, J. [1982]. "Analysis of feedback systems with structured uncertainties," *Proc. of IEE*, Vol. 129, pp. 242–250.

90. Dubrovin, B.A., Fomenko, A.T., and Novikov, S.P. [1990a]. *Modern Geometry—Methods and Applications. Part II: The Geometry and Topology of Manifolds*, Springer-Verlag, New York.

91. Dubrovin, B.A., Fomenko, A.T., and Novikov, S.P. [1990b]. *Modern Geometry—Methods and Applications. Part III: Introduction to Homology Theory*, Springer-Verlag, New York.

92. Dugundji, J. [1970]. *Topology*, Allyn and Bacon, Boston.

93. Eaves, B.C. [1971]. "Computing Kakutani fixed points," *SIAM J. Appl. Math.*, Vol. 21, pp. 236–244.

94. Eaves, B.C. [1972]. "Homotopies for computation of fixed points," *Mathematical Programming*, Vol. 3, pp. 1–22.

95. Eaves, B.C., and Saigal, R. [1972]. "Homotopies for computation of fixed points on unbounded regions," *Mathematical programming*, Vol. 3, pp. 225–237.

96. Eaves, B. C. [1984]. *A Course in Triangulations for Solving Equations with Deformation*, Lecture Notes in Economics and Mathematical Systems, 234, Springer-Verlag, New York.

97. Edelsbrunner, H. [1987]. *Algorithms in Combinatorial Geometry*, Monographs on Theoretical Computer Sciences (Brauer, Rozenberg, Salomaa, Eds.), Springer-Verlag, New York.

98. Ehresmann, C. [1951]. "Les prolongements d'une variété différentiable," *C. R. Acad. Sci. Paris*, Vol. 233, pp. 598–600.

99. Eilenberg, S., and Mac Lane, S. [1945]. "General theory of natural equivalences," *Trans. Am. Math. Soc.*, Vol. 58, pp. 231–294.

100. Eilenberg, S., and Zilber, J.A. [1950]. "Semi-simplicial complexes and singular homology," *Ann. Math.*, Vol. 51, pp. 499–513.

101. Eilenberg, S., and Steenrod, N. [1952]. *Foundations of Algebraic Topology*, Princeton University Press, Princeton, NJ.

102. Erdos, P. , Gruber P.M., and Hammer, J. [1989]. *Lattice Points*, Pitman Monographs and Surveys in Pure and Applied Mathematics, Vol. 39, Longman Scientific and Technical, New York.

103. Even, S., and Tarjan, R.E. [1976]. "A combinatorial problem which is complete in polynomial space," *J. Assoc. Comput. Mach.*, Vol. 23, pp. 710–719.

104. Fortune, S. [1987]. "A sweepline algorithm for Voronoi diagrams," *Algorithmica*, Vol. 2, pp. 153–174.

105. Fraenkel, A.S. [1991]. "Selected bibliography on combinatorial games and some related material," in *Combinatorial Games* (Guy, R.K., Ed.), American Mathematical Society, Providence, RI.

106. Freudenthal, H. [1937a]. "Über die Klasse der Sphärenabbildungen," *Comp. Math.*, Vol 5, pp. 299–314.

107. Freudenthal, H. [1937b]. "Eine Simplizialzerlegung des Cartesischen Produktes zweier Simplexe," *Fund. Math.*, Vol. 29, pp. 138–144.

108. Freudenthal, H. [1942]. "Simplizialzerlegungen von beschränkter Flachheit," *Ann. Math.*, Vol. 43 (3), pp. 580–582.

109. Freund, R. M. [1986]. "Combinatorial theorems on the simplotope that generalize results on the simplex and cube," *Math. Oper. Res.*, Vol. 11 (1), pp. 169–179.

110. Fuks, D.B. [1973]. "Homotopic topology," *Itogi Nauki Tekh.*, Ser. Algebra Topologiya Geom., pp. 71–122, 1969; English translation: *J. Soviet Math.*, Vol. 1 (3), pp. 333–362.

111. Fuks, D.B., and Rokhlin, V.A. [1984]. *Beginner's Course in Topology*, Universitext, Springer-Verlag, New York.

112. Furhmann, P.A., and Ober, R. [1993]. "A functional approach to LQG balancing," *Int. J. Contr.*, Vol. 57(3), pp. 627–714.

113. Gale, D. [1979]. "The game of hex and the Brouwer fixed point theorem," *Am. Math. Monthly*, pp. 818–827.

114. Georgiou, T., and Smith, M.C. [1990]. "Optimal robustness in the gap metric," *IEEE Trans. Autom. Contr.*, Vol. AC-35, pp. 673–686.

115. Gillet, H. [1992]. "Comparing algebraic and topological K-theory," in *Higher Algebraic K-Theory: An Overview* (Lluis-Puebla, E., et al., Eds.), Springer-Verlag, pp. 55–99.

116. Glaser, L.C. [1970]. *Geometrical Combinatorial Topology*, Vol. 1, Van Nostrand Reinhold Mathematical Studies No. 27; Van Nostrand Reinhold, New York.

117. Goldberg, S.I. [1982]. *Curvature and Homology*, Dover, New York.

118. Goldberg, S. [1966]. *Unbounded Linear Operators—Theory and Applications*, Dover, New York.

119. Golubitsky, M., and Guillemin, V. [1973]. *Stable Mappings and Their Singularities*, Graduate Texts in Mathematics 14, Springer-Verlag, New York.

120. Goresky, M., and MacPherson, R. [1988]. *Stratified Morse Theory,* Springer-Verlag, New York.

121. Grauert, H., and Remmert, R. [1984]. *Coherent Analytic Sheaves,* Springer-Verlag, New York.

122. Green, M.L. [1993]. "Infinitesimal methods in Hodge theory," in *Algebraic Cycles and Hodge Theory* (Albano, A., and Bardelli, F., Eds.), Springer-Verlag, New York.

123. Griffiths, P., and and Harris, J. [1994]. *Principles of Algebraic Geometry,* Wiley Classic Library, Wiley, New York.

124. Gugenheim, V.K.A.M. [1957]. "On supercomplexes," *Trans. Am. Math. Soc.*, Vol. 85, pp. 35–51.

125. Gugenheim, V.K.A.M., and Moore, J.C. [1957]. "Acyclic models and fiber spaces," *Trans. Am. Math. Soc.*, Vol. 85, pp. 265–306.

126. Hahn, A.J., and O'Meara, O.T. [1989]. *The Classical Groups and K-Theory,* A Series of Comprehensive Studies in Mathematics, 291, Springer-Verlag, New York.

127. Helmke, U. [1991]. "A compactification of the space of rational transfer functions by singular systems," *J. Math. Syst. Estimation Contr.*, Vol. 3(4), pp. 459–472.

128. Henle, M. [1979]. *A Combinatorial Introduction to Topology,* Dover, New York.

129. Hilton P.J., and Wylie S. [1965]. *Homology Theory—An Introduction to Algebraic Topology,* Cambridge University Press, Cambridge, England.

130. Hilton, P.J., and Stammach, U. [1970]. *A Course in Homological Algebra,* Springer-Verlag, New York.

131. Hilton, P. [1971]. *General Cohomology Theory and K-Theory,* London Mathematical Society Lecture Note Series, 1, Cambridge University Press, Cambridge, England.

132. Hilton, P.J., and Wu, Y.-C. [1974]. *A Course in Modern Algebra,* Wiley, New York.

133. Hirsch, M.W. [1976]. *Differential Topology,* Graduate Texts in Mathematics, 33, Springer-Verlag, New York.

134. Hocking, J.G., and Young, G.S. [1961]. *Topology,* Dover, New York.

135. Horn, R.A., and Johnson, C.R. [1991]. *Topics in Matrix Analysis,* Cambridge University Press, Cambridge, England.

136. Horowitz, I. [1963]. *Synthesis of Feedback Systems,* Academic Press, New York.

137. Horowitz, I. [1982]. "Quantitative feedback theory," *Proc. of IEE,* Vol. 29, pp. 215–226.

138. Hovansky, A.G. [1980]. "On a class of systems of transcendental equations," *Soviet Math. Doklady,* Vol. 22, pp. 762–765.

139. Hsiang, W. Y. [1975]. *Cohomology Theory of Topological Transformation Groups*, Ergebnisse der Mathematik und iherer Grenzgebiete, Band 85, Springer-Verlag, New York.

140. Hu, S.-T. [1959]. *Homotopy Theory*, Pure and Applied Mathematics, A series of monographs and books, No. 8, Academic Press, New York.

141. Husemoller, D. [1994]. *Fibre Bundles (Third Edition)*, Springer-Verlag, New York.

142. Jacobson, N. [1974]. *Basic Algebra, I*, W.H. Freeman, New York.

143. James, I.M. [1984]. *General Topology and Homotopy Theory*, Springer-Verlag, New York.

144. Jänich, K. [1965]. "Vectorraumbündel und der Raum der Fredholm-Operatoren," *Math. Ann.*, Vol. 161, pp. 129–142.

145. Jänich, K. [1984]. *Topology*, Undergraduate Texts in Mathematics, Springer-Verlag, New York.

146. Jonckheere, E.A., and Ke, N.-P. [1995]. "Stability margin relative to stratified uncertainty space," *Proceedings, IEEE Conference on Decision and Control*, New Orleans, LA.

147. Jonckheere, E.A., and Silverman, L.M. [1981]. "Spectral theory of the linear-quadratic optimal control problem: Analytic factorization of rational matrix-valued functions," *SIAM J. Contr. Optim.*, Vol. 19 (2), pp. 262–281.

148. Jonckheere, E.A., and Silverman, L.M. [1983]. "A new set of invariants for linear systems—Application to reduced order compensator design," *IEEE Trans. Autom. Contr.*, Vol. AC-28, pp. 953–964.

149. Jonckheere, E.A. [1984]. "Principal component analysis of flexible systems—Open-loop case," *IEEE Trans. Autom. Contr.*, Vol. AC-29, pp. 1095–1097.

150. Jonckheere, E.A., and Bar-on, J. [1991]. "Algebraic topology in robust control," *Workshop on Robust Control*, San Antonio, Texas, March 13–15, 1991; see also *Control of Uncertain Dynamic Systems* (Bhattacharryya, S.P., and Keel, L.H., Eds.), CRC Press, Boca Raton, FL, pp. 339–358.

151. Jonckheere, E.A., and Cheng, C.-Y. [1993]. "Robust stability, Morse theory and singularity" in Proceedings of the *Proceedings, IEEE Conference on Decision and Control*, San Antonio, Texas, FP-2, pp. 3453-3458.

152. Jonckheere, E.A., and Cheng, C.-Y. [1994]. "Structural stability of robustness margin," *Proceedings, IFAC Symposium on Robust Control Design*, Rio de Janeiro, Brazil, pp. 423–428.

153. Kan, D.M. [1955a]. "Abstract homotopy I," *Proc. Natl. Acad. Sci. USA*, Vol. 41, pp. 1092–1096.

154. Kan, D.M. [1955b]. "Abstract homotopy II," *Proc. Natl. Acad. Sci. USA*, Vol. 42, p. 255.

155. Kan, D.M. [1958]. "A combinatorial definition of homotopy groups," *Ann. Math.*, Vol. 67, pp. 282–312.
156. Karoubi, M. [1978]. *K-Theory, An Introduction*, A Series of Comprehensive Studies in Mathematics, 226, Springer-Verlag, New York.
157. Kharitonov, V. [1979]. "Asymptotic stability of an equilibrium position of a family of systems of linear differential equations," *Differ. Equations*, Vol. 14, pp. 1483–1485.
158. Kitapci, A., Jonckheere, E., and Silverman, L. [1986]. "Singular optimal filtering," in *Frequency Domain and State Space Methods for Linear Systems* (Byrnes, C.I., and Lindquist, A., Eds.), North-Holland, New York, pp. 271–284.
159. Klein, R. [1987]. *Concrete and Abstract Voronoi Diagrams*, Lecture Notes in Computer Science 400, Springer-Verlag, New York.
160. Kraus, F.J., Anderson, B.D.O., and Mansour, M. [1989]. "Robust stability of polynomials with multilinear parameter dependence," *Int. J. Contr.*, Vol. 50 (5), pp. 1745–1762.
161. Kuhn, H.W. [1968]. "Simplicial approximation of fixed points," *Proc. Natl. Acad. Sci. USA*, Vol. 61, 1238–1242.
162. Kuhn, H.W., and MacKinnon, J.G. [1975]. "Sandwich method for finding fixed points," *J. Optim. Th. Appl.*, Vol. 17, pp. 189–204.
163. Kuiper, N.H. [1965]. "The homotopy type of the unitary group of Hilbert space," *Topology*, Vol. 3, pp. 19–30.
164. Kwoh, Y.S., Hu, J., Jonckheere, E.A., and Hayati, S. [1988]. "A robot with improved absolute positional accuracy for CT-guided stereotactic brain surgery," *IEEE Trans. Biomed. Eng.*, Vol. BME-35, pp. 153–160.
165. Lefschetz, S. [1927]. "The residual set of a complex on a manifold and related questions," *Proc. Natl. Acad. Sci. USA*, Vol. 13, pp. 614–622; Vol. 13, pp. 805–807.
166. Lefschetz, S. [1949]. *Introduction to Topology*, Princeton University Press, Princeton, NJ.
167. Lloyd, N.G. [1978]. *Degree Theory*, Cambridge University Press, Cambridge, England.
168. Lluis-Puebla, E. [1992]. "Introduction to algebraic K-theory," in *Higher Algebraic K-Theory: An Overview* (Lluis-Puebla, E., et al., Eds.), Springer-Verlag, New York, pp. 1–30.
169. Lluis-Puebla, E., Loday, J.L., Gillet, H., Soule, C., and Snaith, V. [1992]. *Higher Algebraic K-Theory: An Overview*, Lecture Notes in Mathematics, 1491, Springer-Verlag, New York.
170. Lomadze, V.G. [1990]. "Finite-dimensional time-invariant linear dynamical systems: Algebraic theory," *Acta Appl. Math.*, Vol. 19, pp. 149–201.
171. Mac Lane, S. [1971]. *Categories for the Working Mathematician*, Springer-Verlag, New-York.

172. Mac Lane, S. [1975]. *Homology*, Springer-Verlag, New York.
173. Magnus, W., Karras, A., and Solitar, D. [1976]. *Combinatorial Group Theory—Presentations of Groups in Terms of Generators and Relations*, Dover, New York.
174. Mann, B.M., and Milgram, R.J. [1991]. "Some spaces of holomorphic maps to complex Grassmann manifolds," *J. Differ. Geom.*, Vol. 33, pp. 301–324.
175. Marcus, M., and Minc, H. [1964]. *A Survey of Matrix Theory and Matrix Inequalities*, Dover, New York.
176. Martin, C., and Hermann, R. [1978]. "Applications of algebraic geometry to systems theory: The McMillan degree and the Kronecker indices of transfer functions as topological and holomorphic system invariants," *SIAM J. Contr. Optim.*, Vol. 16 (5), pp. 743–755.
177. Martinetz, T., and Schulten, K. [1994]. "Topology representing networks," *Neural Networks*, Vol. 7(3), pp. 507–522.
178. Massey, W. [1980]. *Singular Homology Theory*, Springer-Verlag, New York.
179. Mather, J.N. [1968a]. "Stability of C^∞ mappings: I. The division theorem," *Ann. Math.*, Vol. 87, pp. 89–104.
180. Mather, J.N. [1969a]. "Stability of C^∞ mappings: II. Infinitesimal stability implies stability," *Ann. Math.*, Vol. 89, pp. 254–291.
181. Mather, J.N. [1968b]. "Stability of C^∞ mappings, III: Finitely determined map-germs," *Publications Mathématiques, Institut des Hautes Etudes Scientifiques (I.H.E.S.)*, No. 35, pp. 127–156.
182. Mather, J.N. [1969b]. "Stability of C^∞ mappings, IV: Classification of stable germs by R-algebras," *Publications Mathématiques, Institut des Hautes Etudes Scientifiques (I.H.E.S.)*, No. 37, pp. 223–248.
183. May, J.P. [1967]. *Simplicial Objects in Algebraic Topology*, Van Nostrand Mathematical Studies No. 11, Princeton, NJ.
184. May, J.P. [1992]. *Simplicial Objects in Algebraic Topology*, Chicago Lectures in Mathematics.
185. McCleary, J. [1985]. *User's Guide to Spectral Sequences*, Mathematical Lecture Series 12, Publish or Perish, Inc., Wilmington, DE.
186. Milnor, J. [1963]. *Morse Theory*, Princeton University Press, Princeton, NJ.
187. Milnor, J. [1964]. "On the Betti numbers of real algebraic varieties," *Proc. Am. Math. Soc.*, Vol. 15, pp. 275–280.
188. Mimura, M., and Toda, H. [1991]. *Topology of Lie Groups, I and II*, Translations of Mathematical Monographs, Vol. 91, American Mathematical Society, Providence, RI.
189. Morse, M. [1925]. "Relations between the critical points of a real function of n independent variables," *Trans. Am. Math. Soc.*, Vol. 27, pp. 345–396.

190. Mumford, D. [1991]. *Algebraic Geometry I: Complex Projective Varieties*, Springer-Verlag, New York.

191. Munkres, J. [1984]. *Elements of Algebraic Topology*, Addison-Wesley, Reading, MA.

192. Murnaghan, F.D. [1938]. *The Theory of Group Representations*, Dover, New York.

193. Nash, C., and Sen, S. [1983]. *Topology and Geometry for Physicists*, Academic Press, New York.

194. Nash, C. [1991]. *Differential Topology and Quantum Field Theory*, Academic Press, New York.

195. Nwokah, O.D.I., and Thompson, D.F. [1989]. "Algebraic and topological aspects of quantitative feedback theory," *Proceedings, 1989 American Control Conference*, pp. 953–959.

196. Opdenacker, Ph., and Jonckheere, E.A. [1985]. "LQG balancing and reduced LQG compensation of symmetric passive systems," *Int. J. Contr.*, Vol. 41, pp. 73–109.

197. Packard, A., and Pandey, P. [1993]. "Continuity properties of the real/complex structured singular value," *IEEE Trans. Autom. Contr.*, Vol. AC-38, No. 3, pp. 415–428.

198. Pommerenke, C. [1992]. *Boundary Behavior of Conformal Maps*, A Series of Comprehensive Studies in Mathematics 299, Springer-Verlag, New York.

199. Preparata, F., and Shamos, P. [1985]. *Computational Geometry—An Introduction*, Texts and Monographs in Computer Sciences, Springer-Verlag, New York.

200. Rabindranathan, M. [1969]. "On the inversion of Toeplitz operators," *J. Math. Mech.*, Vol. 19, pp. 195–206.

201. Ravi, M.S., and Rosenthal, J. [1994]. "A smooth compactification of the space of transfer functions with fixed McMillan degree," *Acta Appl. Math.*, Vol. 34, pp. 329–352.

202. Reisch, S. [1981]. "Hex ist PSPACE vollständig," *Acta Inf.*, Vol. 15, pp. 167–191.

203. Richter, S., and DeCarlo, R. [1984]. "A homotopy method for eigenvalue assignment using decentralized state feedback," *IEEE Trans. Autom. Contr.*, Vol. AC-29, pp. 148–158.

204. Risler, J.-J. [1988]. "Some aspects of complexity in real algebraic geometry," *J. Symb. Comput.*, Vol. 5, pp. 109–119.

205. Rosenberg, J. [1994]. *Algebraic K-Theory and Its Applications*, Springer-Verlag, New York.

206. Rudin, W. [1969]. *Function Theory in Polydisks*, W.A. Benjamin, New York.

207. Rudin, W. [1987]. *Real and Complex Analysis*, McGraw-Hill, New York.

208. Rushing, T. B. [1973]. *Topological Embeddings*, Academic Press, New York.

209. Scarf, H. [1967]. "The approximation of fixed points of a continuous mapping," *SIAM J. Appl. Math.*, Vol. 15, pp. 1328–1343.

210. Sefton, J.A., and Ober, R.J. [1994]. "Hankel-norm approximation and control systems," *Linear Algebra and Its Applications*, Vol. 205–206, pp. 1081–1120.

211. Segal, G. [1979]. "The topology of spaces of rational functions," *Acta Math.*, Vol. 143, pp. 39–72.

212. Siegel, C.L. [1969]. *Topics in Complex Functions Theory—Volume I: Elliptic Functions and Uniformization Theory*, Wiley Classics Library, Wiley.

213. Smale, S. [1976]. "A convergent process of price adjustment and global Newton methods," *Journal of Mathematical Economics*, Vol. 3, pp. 107–120.

214. Solian, A. [1977]. *Theory of Modules-An Introduction to the Theory of Module Categories*, Wiley, New York.

215. Spanier, E.H. [1989]. *Algebraic Topology*, Springer-Verlag, New York.

216. Sperner, E. [1928]. "Neuer Beweis für die Invarianz der Dimensionszahl und des Gebietes," *Abh. Math. Sem. Univ. Hamburg*, Vol. 6, pp. 265–272.

217. Spivak, M. [1965]. *Calculus on Manifolds*, Addison-Wesley, Reading, MA.

218. Steenrod, N. [1951]. *The Topology of Fibre Bundles*, Princeton University Press, Princeton, NJ.

219. Stocker, J.J. [1989]. *Differential Geometry*, Wiley Classics Library, Wiley, New York.

220. Tchon, [1995]. *Differential topology of redundant robot manipulator*, Engineering Cybernetics, Technical University of Wroclaw, Poland.

221. Thom, R. [1949]. "Sur une partition en cellules associée à une fonction sur une variété," *C. R. Acad. Sci. Paris*, Vol. 228A, pp. 973–975.

222. Thom, R. [1955–1956]. "Les singularités des applications différentiables," *Ann. Inst. Fourier*, Vol. 6, pp. 43–87.

223. Todd, M. [1976]. *The Computation of Fixed Points and Applications*, Springer-Verlag, New York.

224. Toussaint, G.T. [1988]. *Computational Morphology*, Machine Intelligence and Pattern Recognition 6, North-Holland.

225. Verma, M., Helton, J.W., and Jonckheere, E.A. [1986]. "Robust stabilization of a family of plants with varying number of right half plane poles," *Proceedings, American Control Conference*, Seattle, Washington, June 18–20, FP3, pp. 1827–1832.

226. Vicino, A., and Tesi, A. [1995]. "Regularity conditions for the stability margin problem with linear dependent perturbations," *SIAM J. Contr. Optim.*, Vol. 33, pp. 1000-10016.
227. Wacker, H. (Ed.) [1978]. *Continuation Methods*, Proceedings of a Symposium at the University of Linz, Austria, October 3-4, 1977, Academic Press Rapid Manuscript Reproduction, Academic Press, New York.
228. Walker, R.J. [1978]. *Algebraic Curves*, Springer-Verlag, New York.
229. Wegge-Olsen, N.E. [1993]. *K-Theory and C^*-Algebras*, Oxford Science Publications, Oxford University Press, New York.
230. Weibel, C.A. [1994]. *An Introduction to Homological Algebra*, Cambridge Studies in Advanced Mathematics 38, Cambridge University Press, Cambridge, England.
231. Weinberger, S. [1994]. *The Topological Classification of Stratified Spaces*, Chicago Lectures in Mathematics, Chicago and London.
232. Whitehead, G.W. [1966]. *Homotopy Theory*, The MIT Press, Cambridge, MA.
233. Whitehead, G.W. [1978]. *Elements of Homotopy Theory*, Graduate Text in Mathematics 61, Springer-Verlag, New York.
234. Whitney, H. [1955]. "On singularities of mappings of Euclidean spaces. I. Mappings of the plane into the plane," *Annals of Mathematics*, Vol. 62, No. 3, pp. 374–410.
235. Whitney, H. [1972]. *Complex Analytic Varieties*, Addison-Wesley Publishing Company, Menlo Park, CA.
236. Zeeman, E.C. [1962a]. "Dihomology I. Relations between homology theories," *Proc. London Math. Soc.*, Vol. 12 (3), pp. 609–638.
237. Zeeman, E.C. [1962b]. "Dihomology II. The spectral theories of a map," *Proc. London Math. Soc.*, Vol. 12 (3), pp. 639–689.
238. Zeeman, E.C. [1964]. "Relative simplicial approximation," *Proc. Camb. Philos. Soc.*, Vol. 60, pp. 39–43.
239. Zornig, P. [1991]. *Degeneracy Graphs and Simplex Cycling*, Lecture Notes in Economics and Mathematical Systems, 357, Springer-Verlag, New York.

INDEX